HVACR

Level 4

SIXTH EDITION

Pearson

NCCER
National Center for Construction Education and Research

NCCER
President and Chief Executive Officer: Boyd Worsham
Vice President of Innovation and Advancement: Jennifer Wilkerson
Chief Learning Officer: Lisa Strite
Senior Curriculum Leader: Chris Wilson
HVACR Project Manager: Patrick Bruce
Technical Writing Manager: Gary Ferguson
Technical Writer: Troy Staton
Art Manager: Bree St. Germain
Technical Illustrators: Judd Ivines, Lisa Wailes
Production Artist: Chris Kersten
Permissions Specialists: Sherry Davis
Project Coordinator: Emma Morse

Pearson
Commercial Product Manager-Associations: Andy Dunaway
Associate Project Manager: Monica Perez-Kim
Senior Digital Content Producer: Shannon Stanton
Content Producer: Alexandrina Wolf, Mary Roth
Designer: Mary Siener
Rights and Permissions: Kaitlyn Coddington
Composition: Integra Software Services
Printer/Binder: Lakeside Book Company
Cover Printer: Lakeside Book Company
Text Typefaces: Palatino LT Pro and Helvetica Neue

Cover Image
Berg Chilling Systems was selected by Canlan Ice Sports—the largest private recreational ice operator in North America—to design and build an ammonia-based ice rink refrigeration system. The state-of-the-art system includes energy-efficient reciprocating compressors, upgraded plate-and-frame chillers, and advanced variable frequency drives to match the refrigeration load to compressor capacity. This customized solution reduced peak energy demand by 45%, achieving over $300,000 in annual savings. Berg's system design is representative of the company's expertise in developing sustainable, efficient refrigeration systems for large-scale applications.

10 9 8 7 6 5 4 3 2 1

3 2025

Pearson NCCER National Center for Construction Education and Research

ISBN-10: 0-13-821060-8
ISBN-13: 978-0-13-821060-1

PREFACE

To the Trainee

The heating, ventilating, air-conditioning, and refrigeration (HVACR) industry has a broader focus than just building comfort. HVACR technicians also maintain and install hydronic systems, sophisticated controls, and various energy conservation devices. Consistent technological advances require today's technicians to access updated information and embrace changes to enhance their skillset and value.

Technological advances also mean employers must recognize the need for standardized training and continuing education to keep up with the latest equipment, techniques, and regulations. Technical training and apprenticeship programs provide employers with skilled craftworkers who are well-prepared to lead the way and confront new challenges.

NCCER's HVACR program has been designed by highly qualified subject matter experts. The four levels support an in-depth apprenticeship approach to the HVACR craft, including both theoretical and practical skills essential to success as an HVACR technician.

We wish you the best as you continue your training and begin an exciting and promising career. This newly revised curriculum presents advanced HVACR concepts that will help you enter the workforce with the knowledge and skills needed to perform successfully in any facet of the HVACR craft.

New with *HVACR Level 4*

NCCER is proud to present the 6th edition of *HVACR Level 4*. This revision includes updates to the curriculum that will help engage you and provide the best training possible. All modules include modernized art and updated tables. The content has been updated to include the most recent information regarding hydronics, psychrometrics, HVACR support systems, and installation planning.

We wish you success as you progress through this training program. NCCER appreciates and welcomes its customers' feeback. If you have any comments on how NCCER might improve upon this textbook, please complete the Online Curriculum Feedback Form using the QR code on this page. You may also submit your feedback by emailing **support@nccer.org**. When doing so, please identify feedback on this title by listing *#HVACRL4* in the subject line.

Our website, **www.nccer.org**, has information on the latest product releases and training.

NCCER Standardized Curricula

NCCER is a not-for-profit 501(c)(3) education foundation established in 1996 by the world's largest and most progressive construction companies and national construction associations. It was founded to address the severe workforce shortage facing the industry and to develop a standardized training process and curricula. Today, NCCER is supported by hundreds of leading construction and maintenance companies, manufacturers, and national associations. The NCCER Standardized Curricula was developed by NCCER in partnership with Pearson, the world's largest educational publisher.

Some features of the NCCER Standardized Curricula are as follows:

- An industry-proven record of success
- Curricula developed by the industry, for the industry
- National standardization providing portability of learned job skills and educational credits
- Compliance with the Office of Apprenticeship requirements for related classroom training (*OSHA Standard CFR 29:29*)
- Well-illustrated, up-to-date, and practical information

NCCER maintains a secure online database that provides certificates, digital credentials, transcripts, and wallet cards to individuals who successfully complete programs under an NCCER-accredited organization or through one of NCCER's self-paced, online programs. This system also allows individuals and employers to track and verify industry-recognized credentials and certifications in real time.

For information on NCCER's credentials, contact NCCER Customer Service at 1.888.622.3720 or visit **www.nccer.org**.

Digital Credentials

Show off your industry-recognized credentials online with NCCER's digital credentials!

NCCER is now providing online credentials. Transform your knowledge, skills, and achievements into digital credentials that you can share across social media platforms, send to your network, and add to your resume. For more information, visit **www.nccer.org**.

Cover Image Provider

BERG CHILLING SYSTEMS
Expert thermal process energy management

For over 50 years, Berg Chilling Systems custom designs, manufactures, installs, and services dependable and durable industrial chilling solutions for demanding applications around the world. Working closely with clients, Berg sales engineers apply unmatched energy management experience to fully understand and satisfy clients' cooling challenges. Visit **www.berg-group.com**

DESIGN FEATURES

Content is organized and presented in a functional structure that allows trainees to access the information where they need it.

Source: Baloncici/iStock/Getty Images

Objectives

Successful completion of this module prepares you to do the following:

1. Explain how to store, start up, and shut down boiler systems.
   a. Explain how to shut down and prepare boilers for storage.
   b. Explain how to prepare and start up steam and hot-water boilers.
2. Explain how to start up and shut down chillers and cooling towers.
   a. Explain how to start up and shut down reciprocating chillers.
   b. Explain how to start up and shut down centrifugal chillers.
   c. Explain how to start up and shut down cooling towers and clean water-cooled condensers.
3. Explain how to start up air handling units and packaged rooftop units.
   a. Explain how to start up an air handling unit.
   b. Explain how to start up packaged rooftop units.

Performance Tasks

Under supervision, you should be able to do the following:

1. Prepare and complete the startup or shutdown procedures for at least one of the following:
   - Steam or hot-water boiler
   - Chiller
   - Cooling tower or evaporative condenser
2. Prepare and complete the startup procedures for an air handling unit or packaged HVACR unit.

Overview

When boilers, chillers, and other commercial HVAC systems are shut down for extended periods and then returned to service, specific routines must be followed to prepare the system for operation. This helps to ensure things go smoothly when very large, expensive, and critical equipment is at stake. This module provides an overview of the startup and shutdown requirements for various types of equipment.

Digital Resources for HVACR

Scan this code using the camera on your phone or mobile device to view the digital resources related to this craft.

SCAN ME

Trainees can navigate *HVACR Level 4* using color-coded tabs on the upper right-hand corner.

The Objectives list the skills and knowledge trainees need in order to complete the module successfully.

The Performance Tasks give you an opportunity to apply your knowledge to real-world tasks.

Section Openers provide a visual organizational structure for the information. Objectives and Performance Tasks are broken out for each section.

1.0.0 Construction Drawings

Performance Tasks	Objective
1. Identify and interpret information found on architectural drawings such as elevations and floor plans. 2. Identify and interpret information found on mechanical drawings such as air distribution systems and piping drawings. 3. Identify and interpret information found on HVACR-related shop drawings.	Describe the types of construction drawings HVACR technicians encounter and use. a. Explain how to approach a set of construction drawings for an initial review. b. Describe site plans. c. Describe plan views, elevations, detail drawings, and section drawings. d. Describe mechanical, electrical, and plumbing drawings. e. Describe shop drawings. f. Describe as-built drawings. g. Identify and explain the importance of building codes to the design and construction process.

When a sectional cut is made along the long dimension of a building, it is called a **longitudinal section**. When it is made through the short dimension of a building, it is called a **transverse section**. The point on the original drawing that the section was taken from, showing where the imaginary cut has been made, is indicated by a *section line*. A section line is a special line with arrows on each end to indicate the direction of view. Letters and numbers identifying the section drawing are placed on the arrows or adjacent to the line. *Figure 10* shows a transverse section designated as Section 2 that is found on drawing A303.

Longitudinal section: A section drawing on which the cut is made parallel to the longest dimension of a structure.

Transverse section: A section drawing on which the cut is made along the short dimension of the structure.

Trade Terms appear on the page adjacent to the text where they are first presented.

Step-by-step presentations help make the concepts clear and easy to grasp.

The following procedure outlines the preparation for wet storage of a typical steam boiler:

Step 1 Shut down, drain, and clean the boiler as outlined in the dry storage procedure.

Step 2 With the boiler still valved off from the rest of the system, refill it to overflowing from the top with treated water. If de-aerated water is not available, fire the boiler for a short time until the water reaches a temperature of about 200°F (about 90°C) to drive dissolved gases and oxygen from the water.

Step 3 Let the boiler cool back down to room temperature, and then add enough water until it again overflows from the top. If the amount of water added is significant, add additional chemical treatment to compensate. Circulate or stir the boiler water to ensure that the chemicals are properly dispersed.

Step 4 With the boiler flooded (filled to the top, beyond the normal operating water line), close all the boiler openings.

Step 5 Maintain a slight pressure (1–3 psig) on the waterside of the boiler. Nitrogen is typically used for this purpose because it is inert, does not support corrosion, and is readily available.

Step 6 Protect the exterior surfaces and components from rust by coating them with a corrosion inhibitor or paint. Cover all tube and firebox surfaces with a coating of mineral oil or similar corrosion inhibitor.

Step 7 Leave the flue and fireside/combustion chamber doors open during the shutdown. This helps to keep the doors dry. Keep the boiler room dry and well ventilated.

Step 8 Prominently display a tag on the boiler declaring that it must not be operated until the combustion chamber is closed and the boiler is drained down to the normal waterline.

SCAN ME

QR codes link trainees directly to digital resources that highlight current content.

Important information is highlighted, illustrated, and presented to facilitate learning.

Placement of images near the text description and details such as callouts and labels help trainees absorb information.

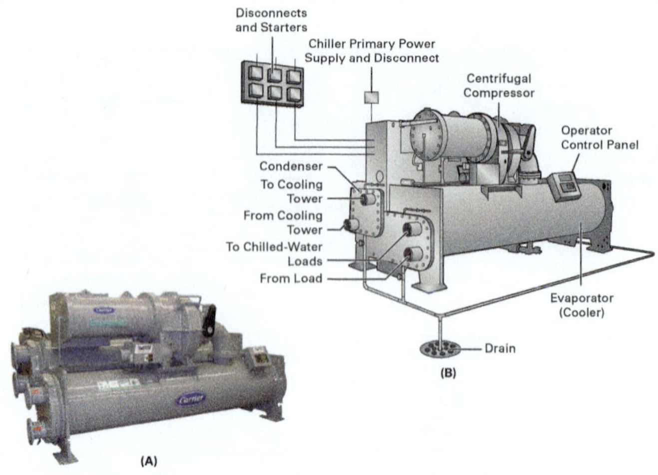

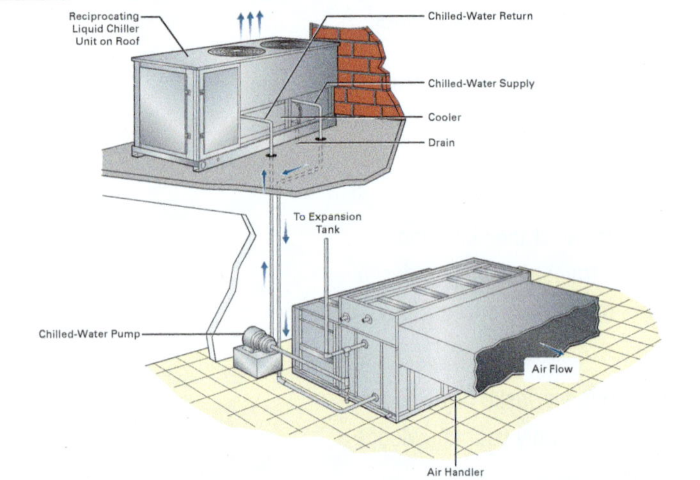

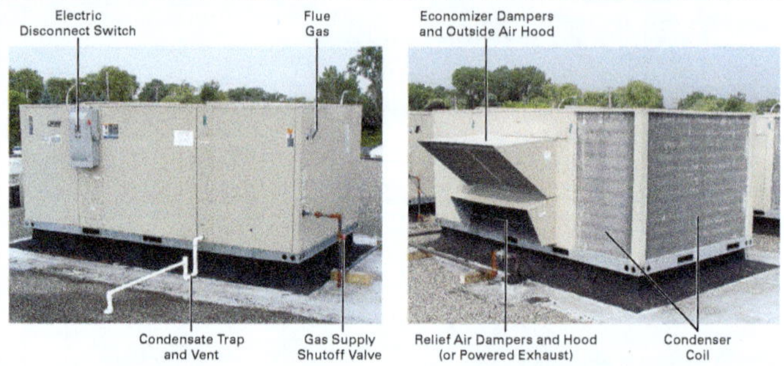

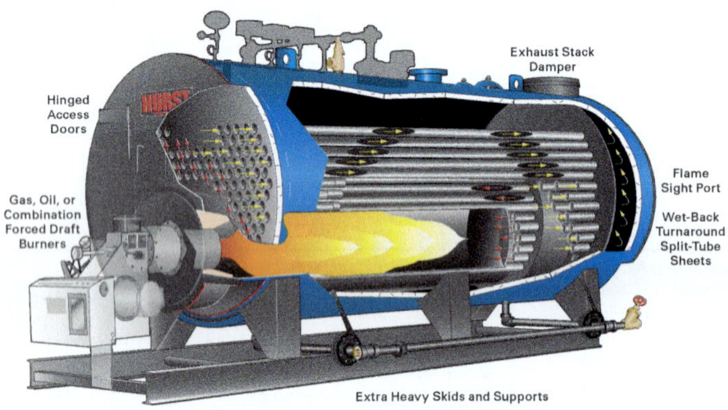

New boxes highlight safety and other important information for trainees. Warning boxes stress potentially dangerous situations, while Caution boxes alert trainees to dangers that may cause damage to equipment. Notes boxes provide additional information on a topic.

WARNING!

Preparing a boiler for storage involves working with energized equipment containing pressurized and/or hot components. Follow all manufacturer's instructions, OSHA regulations, and jobsite safety requirements.

CAUTION

A boiler should be started and shut down as directed in the boiler manufacturer's instructions. Do not operate, repair, or dismantle a boiler before reviewing the manufacturer's service literature.

NOTE

Note that the times and other values provided in the sequence are for example only. Any events in the sequence that fail to take place typically result in an aborted startup sequence and a fault message on the control panel display.

Going Green

Impact of LEED on System Commissioning

The Leadership in Energy and Environmental Design (LEED) green-building rating system was introduced earlier in the HVACR craft training program. LEED requirements have an impact on every phase of building construction, from design through commissioning. The HVACR commissioning process is now more extensive than ever before.

In many cases, a third-party commissioning agent is employed by the building owner for the task, in addition to the traditional system testing and balancing contractor. Filling these new positions requires both experience and higher levels of training. Today's commissioning agents are often on the go and travel to every corner of the country.

Going Green looks at ways to preserve the environment, save energy, and make good choices regarding the health of the planet.

On Site

Silica Gel

Silica gel is one of the safest and most popular materials to use for boilers in dry storage. About 5 lb of silica gel is generally recommended per 30 ft^3 of boiler volume.

Silica gel does not visibly change as it absorbs moisture. Some products, however, are modified to change color when the gel needs to be replaced or dehydrated. The product shown here changes to green when it is saturated.

Silica gel can absorb up to 40% of its weight in water vapor and has an almost indefinite shelf life when stored airtight. Gentle heating drives the absorbed moisture out, making it reusable.

Source: Dry & Dry

These boxed features provide additional information that enhances the text.

Did You Know?

Taking Good Care of the Wine

The proper storage of wine as it ages is an interesting field of science. A certain amount of wine stored in barrels evaporates over time—as much as 12% of the contents may be lost to evaporation.

To limit the loss, wine in barrels needs to be stored in a very humid but cool environment. An area maintained at 55°F (13°C) and 80% relative humidity is considered desirable. However, mechanical cooling processes generally result in dehumidification, which works against the target conditions.

A famous California winery had literally hundreds of storage units used for aging. Rooftop-mounted cooling units were being used to maintain the proper conditions, but with limited success. Over $500,000 of product per year was being lost to evaporation losses alone.

Johnson Air-Rotation® proposed that air-turnover units be used and the organization guaranteed the results. How did that go, you ask? Wine evaporation was reduced by 75%, saving over $375,000 per year in evaporation losses at each of the company's storage units. Because the ductwork from the rooftop units was also eliminated, the horsepower requirement to move the needed volume of air was also cut in half.

Source: vanbeets/iStock

Review Questions at the end of each section and module allow trainees to measure their progress.

2.0.0 Section Review

1. For what purpose must power be provided to reciprocating and scroll chillers well in advance of startup?
   a. To energize the crankcase heaters
   b. To help humid air escape from the control panel
   c. To ensure the electronic components work reliably
   d. To increase the pressure of the charged refrigerant
2. When a recycle shutdown mode is initiated, which of the following best describes the status of the chiller?
   a. It is under a light cooling load.
   b. It is under a heavy cooling load.
   c. The chilled-water temperature is excessively low.
   d. The chilled-water temperature has risen well above the setpoint.
3. What will cause the condenser water pump to draw air into the water from a cooling tower sump?
   a. Scale on the fill
   b. Blocked spray nozzles
   c. Incorrect float valve setting
   d. Misaligned drift eliminators

Module 03406 Review Questions

1. Thermal shock tends to have a more significant effect on areas where dissimilar metals meet and on which of these materials?
   a. Cast iron
   b. All gaskets
   c. Carbon steel
   d. Flue piping
2. When a boiler is prepared for wet storage, a slight pressure should be maintained on the waterside using _____.
   a. oxygen
   b. nitrogen
   c. compressed air
   d. carbon dioxide (CO_2)
3. When preparing a steam boiler for startup, the main steam stop valve should be _____.
   a. fully open
   b. 10% open
   c. 50% open
   d. fully closed

NCCERconnect

This interactive online course is a unique web-based supplement in the form of an electronic book that provides a range of visual, auditory, and interactive elements to enhance training. Visit **www.nccerconnect.com** for more information!

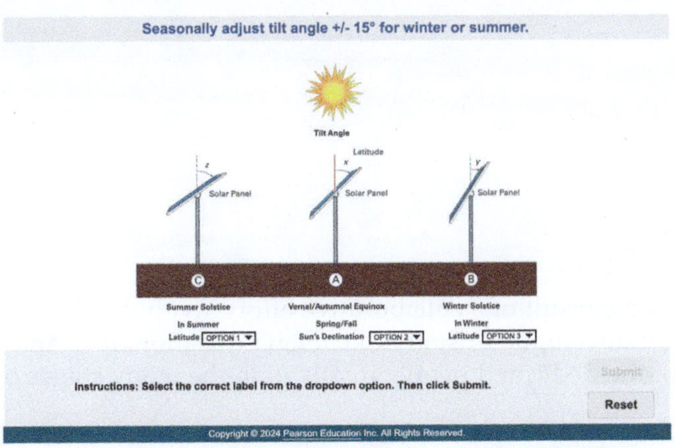

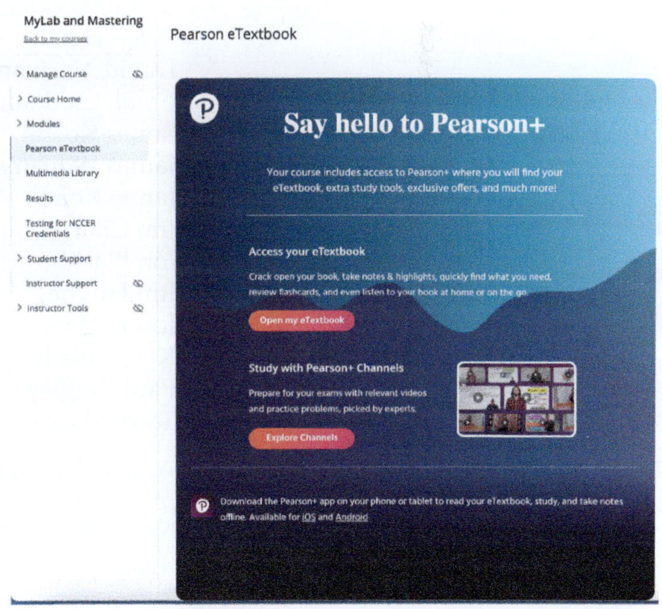

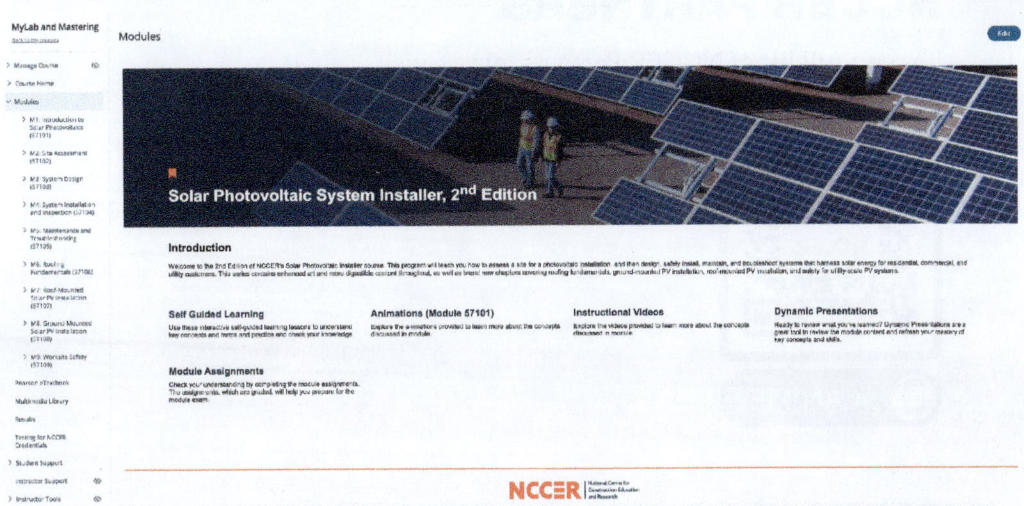

ACKNOWLEDGMENTS

This curriculum was revised as a result of the farsightedness and leadership of the following sponsors:
AGC of Oklahoma Education Foundation
Brown and Root Industrial Services
Canadian Valley Technology Center
College of Southern Maryland
Construction Industry Training Council of Washington
Epling Distributing
HVAC Technical Institute
International Code Council
LaGuardia Community College
Lennox Industries
Mielke Holdings
Pacific Northwest National Laboratory
Service Experts
Standard Heating and Air Conditioning
Southern Careers Institute
Vista College

This curriculum would not exist were it not for the dedication and unselfish energy of those volunteers who served on the Authoring Team. A sincere thanks is extended to the following:
Allen Heath
David Norman
Frank Calderone
James Koehn
James Williams
Jamie Kono
Jim Cika
Jon Blackburn
Jim Hughes
Joseph Moravek
Kasey Wade
Paul Ringley
Terry Mielke

A final note: This book is the result of a collaborative effort involving the production, editorial, and development staff at Pearson Education, Inc., and NCCER. Thanks to all of the dedicated people involved in the many stages of this project.

NCCER PARTNERS

To see a full list of NCCER Partners, please visit
www.nccer.org/about-us/partners.

SCAN ME

CONTENTS

Module 03308 Water Treatment

Module 03404 Energy Conservation Equipment

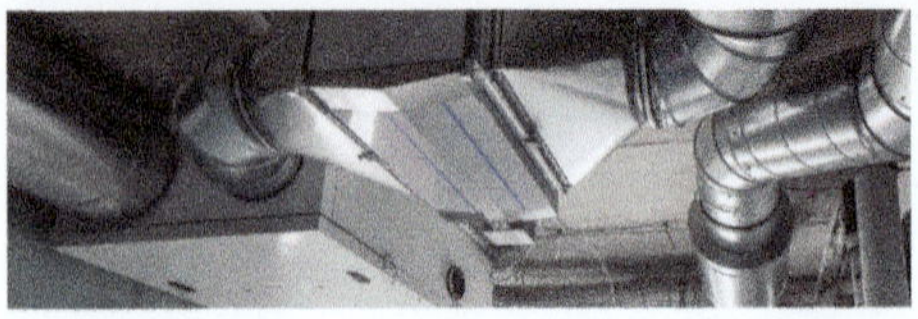

Module 03405 Building Management Systems

Module 03402 System Air Balancing and Psychrometrics

Module 03406 System Startup and Shutdown

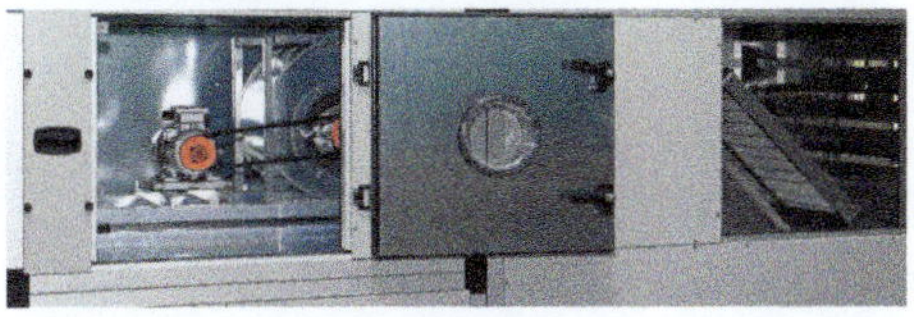

Module 03401 Construction Drawings and Specifications

Module 03407 Comfort System Design and Planning

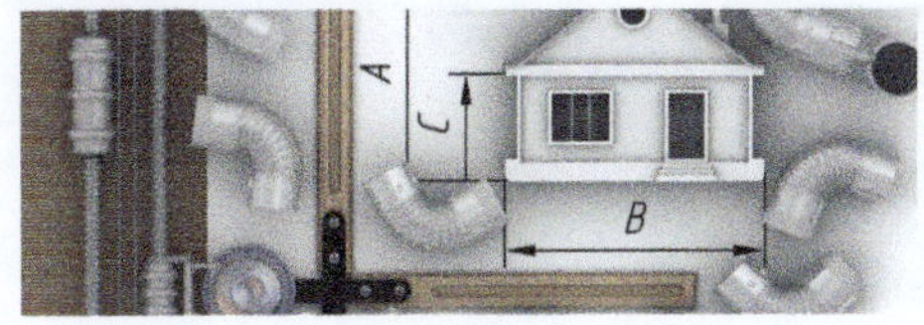

Module 03408 Commercial/Industrial Refrigeration

Module 03409 Alternative and Specialized Heating and Cooling Systems

Module 03305 Commercial Hydronic Systems

Module 03201 Commercial Airside Systems

Module 46100 Introduction to Leadership

MODULE 03308

Water Treatment

Source: vorclub/iStock/Getty Images

Objectives

Successful completion of this module prepares you to do the following:

1. Describe the challenge of water treatment and filtration and explain how water is tested.
   a. Describe the properties of water that relate to water treatment in HVACR systems.
   b. Identify problems with water quality that affect HVACR system performance.
   c. Describe how water test kits are used to collect samples for analysis.
   d. State common safety precautions related to working with water treatment chemicals.
2. Identify various types of mechanical water treatment equipment and devices.
   a. Identify types of filtration devices.
   b. Identify other types of mechanical water treatment equipment.
3. Identify and describe how to address water-related problems in specific types of hydronic and steam systems.
   a. Identify and describe how to treat water-related problems in open water systems.
   b. Identify and describe how to treat water-related problems in closed water systems.
   c. Identify and describe how to treat water-related problems in steam systems.

Performance Tasks

Under supervision, you should be able to do the following:

1. Use a water analysis test kit to test water.
2. Inspect a cooling tower or steam boiler and the related water system and identify signs of water treatment issues.

Overview

This module focuses on the methods and devices used to cleanse and treat water in HVACR systems. It introduces the characteristics of water and describes how untreated water can be harmful to both people and HVACR systems. The module also presents common water treatment chemicals, devices, and methods used in HVACR and other industrial applications.

Digital Resources for HVACR

Scan this code using the camera on your phone or mobile device to view the digital resources related to this craft.

SCAN ME

NCCER Industry-Recognized Credentials

If you are training through an NCCER-accredited sponsor, you may be eligible for credentials from NCCER. The ID number for this module is 03308. Note that this module may have been used in other NCCER curricula and may apply to other level completions. Contact NCCER at 1.888.622.3720 or go to **www.nccer.org** for more information.

You can also show off your industry-recognized credentials online with NCCER's digital credentials. Transform your knowledge, skills, and achievements into credentials that you can share across social media platforms, send to your network, and add to your resume. For more information, visit **www.nccer.org**.

1.0.0 Introduction to Water Treatment

Performance Tasks

1. Use a water analysis test kit to test water.
2. Inspect a cooling tower or steam boiler and the related water system and identify signs of water treatment issues.

Objective

Describe the challenge of water treatment and filtration and explain how water is tested.

a. Describe the properties of water that relate to water treatment in HVACR systems.
b. Identify problems with water quality that affect HVACR system performance.
c. Describe how water test kits are used to collect samples for analysis.
d. State common safety precautions related to working with water treatment chemicals.

HVACR system water treatment is essential to maintain efficient system operation and prevent premature equipment failure. Some form of water treatment is needed for all steam, hot water, chilled water, and cooling water systems, regardless of the application.

Water treatment is often required to alter the water characteristics and destroy harmful organisms. Approaches to water treatment include chemical treatment, filtration, and water **bleed-off**. These three approaches are often combined to form a complete treatment program.

Bleed-off: Consistently draining (bleeding off) a controlled amount of water from a water circuit to carry away minerals and other impurities that remain as water evaporates from a cooling tower or evaporative condenser, reducing the concentration of the minerals in the remaining water.

Applying the proper water treatment involves making a detailed analysis of the water, considering its intended use, then implementing the program consistently. A qualified water treatment specialist normally analyzes a water source and develops an appropriate treatment program. However, HVACR technicians need to be familiar with the topic and the equipment involved. In some cases, commercial HVACR technicians maintain the water treatment equipment and monitor water conditions as part of a comprehensive maintenance program. Water treatment specialists are only called in when their expertise is needed.

1.1.0 Water Characteristics

Water naturally contains a variety of dissolved substances. Pure water is obtained only through extensive and expensive treatment. Water is an effective solvent that dissolves gases from the air, minerals from the soil, and even materials from the piping and other components through which it flows. Water easily dissolves various minerals and organic chemicals. These impurities, and not the water itself, are the main cause of problems in HVACR systems. Even though the quality of a potable water supply makes it safe to drink, the water can still contain minerals and gases that can be harmful to HVACR systems if left untreated.

The relative volumes of dissolved inorganic materials, including gases, are quantified by testing water samples. The strength or relative amount of an element present in a water solution is commonly referred to as its **concentration**.

Concentration: An indication of the strength or relative amount of a substance present in a volume of water.

The unit *parts per million* (ppm) is used to express the concentration of a specific substance. For example, one ounce of a given contaminant mixed with 7,500 gallons of water equals a concentration of about one ppm. One drop of a given material or element mixed in 60 quarts of water also represents about one ppm.

An alternate unit of measure called *grains per gallon* (gpg) is sometimes used instead of parts per million to express concentration. Grains per gallon can be converted to parts per million by multiplying grains per gallon by 17. Concentrations in water are also frequently expressed using the metric unit of measure *milligrams per liter* (mg/L). One mg/L is equal to one ppm. This is based on there being 1,000,000 mg in one liter of fluid. When the concentration of a substance is very small, the common unit of measure is *micrograms per liter* (µg/L).

Dissolved substances include both **dissolved solids** and **dissolved gases**. Dissolved solids regularly contribute to problems in water systems. Common dissolved solids include calcium, iron, manganese, magnesium, and zinc, but these are just a few of the many possibilities. It is important to remember that the characteristics of water in a system can vary widely by location and over time.

Some dissolved gases, such as nitrogen, have virtually no effect on water or its suitability for HVACR use. Other gases such as oxygen, carbon dioxide (CO_2), and hydrogen sulfide (H_2S) can and do affect water systems. Levels of oxygen and H_2S are often measured when the water sample is collected because of the effect of these gases on water systems. Oxygen from the atmosphere easily dissolves into and saturates water, and it is an essential component in most corrosive activities.

The amount of CO_2 in water can either be measured directly or estimated from the water's **pH** and **alkalinity**. The pH is a measure of the acidity of water. The pH scale ranges from 0 (extremely acidic) to 14 (extremely alkaline), with a pH of 7 being neutral (*Figure 1*). Specifically, pH defines the relative concentration of hydrogen ions and hydroxide ions. The concentration of hydroxide ions increases as pH increases.

The alkalinity of a water sample is a measure of water's **buffering capacity**, which allows it to counteract a reduction in pH from the addition of acids. Water with low alkalinity will experience a greater reduction in pH from the addition of acid than water of higher alkalinity.

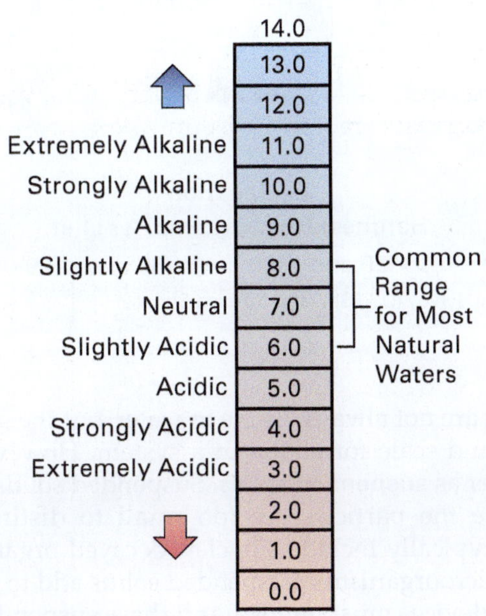

Figure 1 The pH scale.

Dissolved solids: Solid substances such as calcium, magnesium, chloride, and sulfates dissolved as tiny particles in a water sample. Dissolved solids contribute to corrosion and scale formation in a water system.

Dissolved gases: Gases such as oxygen and carbon dioxide that are dissolved in a water sample.

pH: Abbreviation for *potential of hydrogen*, which is a measure of the alkalinity or acidity of water. The pH scale ranges from 0 (extremely acidic) to 14 (extremely alkaline), with a pH of 7 considered neutral. The pH defines the relative concentration of hydrogen and hydroxide ions.

Alkalinity: A measure of water's ability to neutralize or resist a change to increased acidity.

Buffering capacity: A measure of water's ability to resist reductions in its pH, which lead to acidity.

Hardness: A measure of the amount of calcium and magnesium contained in water, which affects the formation of scale in a water system. As hardness increases, the potential for scale development increases.

Scale: A layered coating of mineral matter that separates from water, settling on and adhering to internal water-system surfaces. Scale development is often accelerated by rising or falling water temperatures.

The **hardness** of a water sample indicates the amount of dissolved calcium (limestone) salts and magnesium in the water. Both calcium and magnesium ions may result from several dissolved chemical compounds. For example, **scale** is typically comprised of a calcium compound such as calcium bicarbonate, calcium chloride, or calcium sulfate.

The volume of calcium and magnesium in a sample is often reported as if both of these solids are in the form of calcium carbonate ($CaCO_3$). This allows the total hardness to be reported as a single value. Many water hardness classification systems are available. *Table 1* shows the classification of water hardness according to the American Society of Agricultural and Biological Engineers (ASABE). As hardness increases, the amount of scale that develops in a system increases. Conversely, a decrease in hardness results in less scaling.

TABLE 1 Classification of Water Hardness

Hardness Classification	Total Hardness (as $CaCO_3$)
Soft	0.0 to 60 ppm or mg/L
Moderately hard	61 to 120 ppm or mg/L
Hard	121 to 180 ppm or mg/L
Very hard	Over 180 ppm or mg/L

Did You Know?

Water Hardness in the United States

A study from the National Water-Quality Assessment (NAWQA) Project assessed water-quality conditions for about 2,100 domestic wells across the United States. Water hardness was one of the many parameters studied, with the results shown on the map.

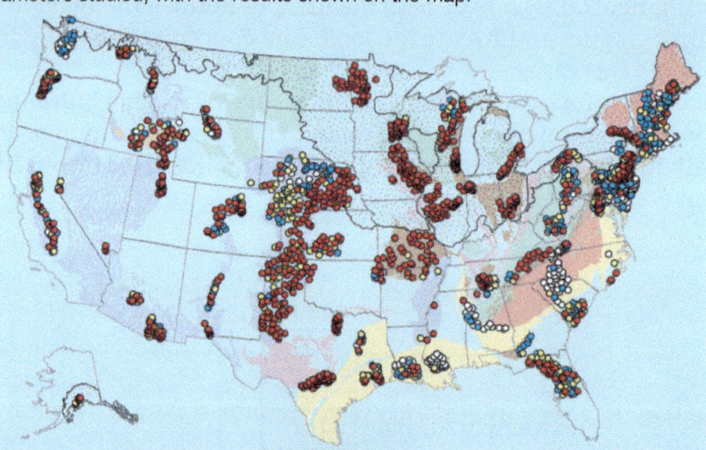

Explanation

Hardness, in Miligrams Per Liter

>180   >120 and ≤180   >60 and ≤120   ≤60

Source: US Geological Survey (USGS)/Water Science School

Suspended solids: Individual particles suspended in water that give the water a cloudy appearance.

Total solids: A measure of the volume of both dissolved and suspended solids contained in a water sample.

Dissolved solids are not always visible in water, but the solids still contribute to both corrosion and scale formation in a system. However, dissolved solids may appear in water as **suspended solids**. Suspended solids are usually seen as cloudy water, since the particles are too small to distinguish individually. Suspended solids typically include silt, clay, decayed organisms, iron, manganese, sulfur, and microorganisms. Suspended solids add to the volume of material that treatment devices must remove, and these suspended solids can shield microorganisms from disinfection. The combination of dissolved solids and suspended solids is referred to as the **total solids**.

The term **cycle of concentration** is frequently used when determining the total dissolved solids (TDS) in system water. This cycle expresses a ratio that indicates when the concentration of a given substance in an HVACR water system has risen above the concentration contained in the makeup water source.

Cycle of concentration: A measurement of the ratio of dissolved solids contained in water to the volume of dissolved solids contained in the makeup water source. The term *cycle* is used to express the ratio.

The cycles of concentration are determined by the ratio of dissolved solids in the existing system water to the quantity of dissolved solids contained in the makeup water source. For instance, if the hardness in the system water is determined to be two times as great as the hardness in the makeup water, then the system water is said to have two cycles of hardness. If you know the volume of TDS in the makeup water, it can be multiplied by the ratio to determine how much of the TDS is being contributed by the makeup water source. Note that this does not take into account the TDS contributed by water treatment additives.

Living organisms such as bacteria, algae, fungi, and protozoa can also be present in water systems. Problems resulting from these organisms are usually limited to cooling water systems, since these organisms tend to die in systems operating at high temperatures.

1.2.0 Problems Related to Water Quality

Water problems in HVACR systems reduce system efficiency and can lead to premature system failure. Some symptoms that generally indicate water quality problems include the following:

- Reduced heat transfer due to scale on heat exchanger surfaces
- Reduced waterflow resulting from partial or complete blockage of pipelines, condenser tubes, and other water system passages
- Rapid wear and failure of moving parts such as pump shafts and shaft seals

The most common problems that untreated water causes in HVACR systems include the following:

- Corrosion
- Scale formation
- Biological growth
- Suspended solid matter

1.2.1 Corrosion

Corrosion is a natural process that results in the decomposition of metal surfaces. Corrosion results from chemical or electrochemical reactions with other substances. **Electrolysis** is the process of changing the chemical composition of a metal when an electrical current develops. Electrolysis is caused by the existence of potential current between two different metals, with an **electrolyte** acting as a conductor. *Galvanic corrosion* is a type of corrosion that is associated with electrolysis.

Electrolysis: The process of changing the chemical composition of a metal by passing an electrical current through it that initiates an exchange of molecules with a second metal.

Electrolyte: A substance that conducts an electric current between positive and negative ions when dissolved in water.

Corrosion can occur both internally and externally in a system. Internal corrosion is most often caused by acids, oxygen, and other gases present in the water. Many forms of corrosion are related directly to an interaction with oxygen. Rust, for example, is technically referred to as iron oxide. Similar examples of oxygen-related corrosion are aluminum oxide and copper oxide.

Corrosion in HVACR-related water systems generally occurs in the following forms:

- *Air corrosion* — The carbon dioxide and oxygen compounds in air released from water while inside piping systems corrode pump impellers and other components.
- *Uniform corrosion* — Common corrosion caused by acids.
- *Pitting corrosion* — A localized type of corrosion normally caused by the breakdown of the passive film at the metal surface, resulting in a blister. The blister concentrates ions, which accelerates metal loss and typically leads to a pinhole leak.

- *Galvanic corrosion* — An accelerated electrochemical corrosive process produced when one metal is in contact with a dissimilar metal, and both are in contact with the same electrolyte. The corrosion is caused by *electrolytic action* that transfers the molecules from one metal to the other metal.
- *Grooving* — A uniform type of corrosion indicated by narrow grooves. Groove corrosion commonly appears along the edges of riveted joints.
- *Embrittlement (caustic cracking)* — A crystallizing or hardening process that causes fine cracks in metal, especially around riveted joints and holes. This is a dangerous type of corrosion because it weakens the metal walls and joints in equipment, increasing vulnerability to stress when subjected to high operating pressures.

On Site

Cooling Tower Construction

Cooling towers are consistently exposed to water and all the substances the water contains. In some applications, the best approach to cooling tower longevity is to construct the tower of materials that either resist corrosion or do not corrode at all.

Stainless steel and fiberglass are often used in cooling tower construction because both materials are strong and are not susceptible to corrosion. Although fiberglass cooling towers do not have the overall durability of stainless steel, they are valued in applications where weight and cost are significant factors. Fiberglass cooling towers that require some metal components to improve structural integrity usually contain stainless steel instead of galvanized steel.

Source: Berg Chilling Systems Inc.

1.2.2 Scaling

Scaling is a significant problem in HVACR water systems. Water is primarily used in the HVACR industry to transfer heat, and scale development accelerates as water temperatures rise and fall. The mineral content and structure of scale makes it a very poor heat transfer medium. Thus, scale in piping systems and heat exchangers significantly reduces heat transfer. In addition, even a small buildup of scale reduces waterflow due to increased friction and narrowing of the inside diameter of piping and passages.

As shown in *Table 2*, energy consumption increases as the scale thickness increases significantly. When many heat exchangers are present in an industrial setting, the energy losses associated with fouled piping and heat exchangers can be staggering.

TABLE 2 Example Increases in Energy Consumption as a Function of Scale

Scale Thickness	Increase in Energy Consumption
1/32"	8.5%
1/16"	12.4%
1/8"	25.0%
1/4"	40.0%

1.2.3 Biological Growth

Temperatures in water systems ranging from 40°F to 120°F (~5°C to 50°C) favor bacterial growth. Most waters contain organisms capable of producing a slimy film when the conditions are favorable. Those conditions also include sufficient nutrients to support growth. Where the water is exposed, algae use sunlight to convert bicarbonate or CO_2 into the energy needed for growth.

Biological growth interferes with the operation of a cooling system when it clogs distribution passages and restricts waterflow. A biological film, or *biofilm*, acts as an insulator on heat transfer surfaces. In addition, accumulations of organic matter can cause localized corrosion.

1.2.4 Suspended Solids

Suspended solids may include silt, sand, grease, bacteria, algae, and corrosion byproducts. The matter can accumulate in a system and clog nozzles, pipes, and filters, restricting waterflow and reducing heat transfer. Blockages and obstructions that result from suspended solids are referred to as *fouling*.

1.3.0 Water Testing

Water systems are commonly tested using commercially available test kits (*Figure 2*). Field kits allow quick on-site analysis of the water, although lab testing typically provides more data and greater accuracy. These kits typically test pH, alkalinity, hardness, chlorides, and sulfites.

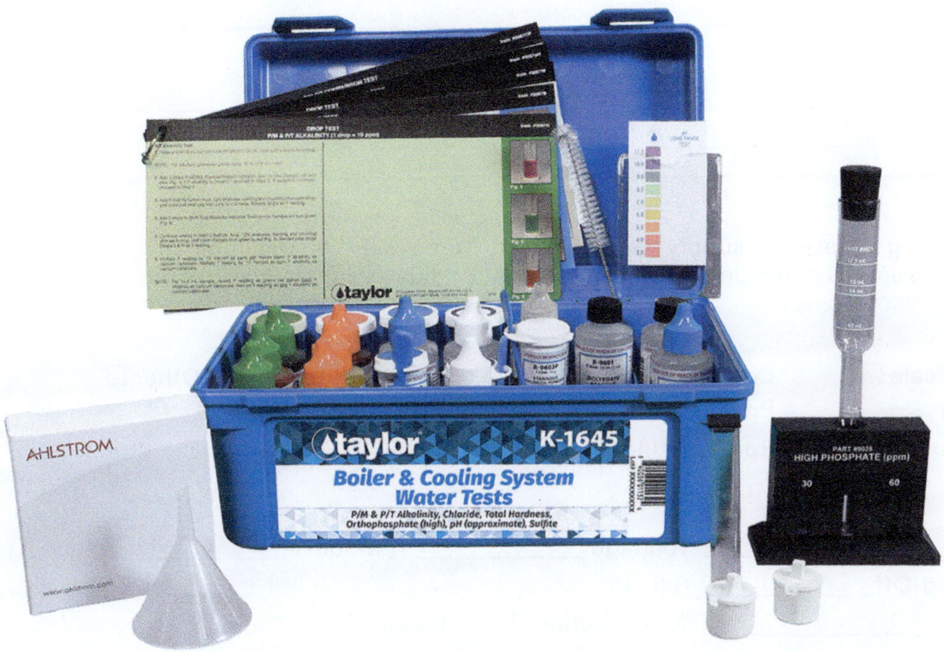

Figure 2 Water test kit for boilers and cooling towers.
Source: Taylor Water Technologies, LLC

More comprehensive testing requires samples to be sent to a lab for analysis. Field sampling kits contain directions for gathering the water sample and include the containers and packaging necessary to ship the samples back to the lab. Most also contain a data sheet that provides the testing facility with important information about the system being tested (*Figure 3*). The lab returns a report showing the measured or calculated amount of each contaminant.

HVACR technicians may conduct simple tests using commercially available test kits. When more extensive or specialized testing is required, a water chemist should perform the tests. Water is often sampled both before and after it passes through treatment equipment to ensure that a treatment device or program is working properly.

When using test kits or preparing samples, remember the following:

- Properly collecting, handling, and preserving water samples is crucial for an accurate analysis.
- Use only those containers provided or recommended by a testing laboratory.
- Carefully follow the laboratory and/or the test kit manufacturer's instructions. Any deviation can lead to inaccurate results.

Sampling containers used to collect water for analysis may contain a *fixing compound* that prevents the breakdown of the contents. These containers should never be emptied or rinsed before use, or filled to overflowing with sample fluid. Laboratories prepare containers specifically for each category of contaminant. The type of container provided depends on the tests to be performed. Microbiological testing, for example, requires a sterile container, and it may contain a chemical that deactivates any residual chlorine.

Date of Sampling: ______________

Water Analysis Data Sheet

STOP! This page must be filled out completely. Failure to do so will prevent an analysis and sample will be discarded.

	Name	Street	City	State
Job				
Contractor / Dealer				
Wholesaler				

Report Should Be Mailed To

Company: ______________ Fax No: ______________

Attention: ______________ E-Mail: ______________

Water Sample:

STOP! Make sure that you are sending a make-up (supply) water sample. Do not send recirculating or "system" sample by itself. Failure to send a make-up sample will prevent or delay your analysis report.

Water:

Problem Involved: Lime Scale ☐ Corrosion ☐ Algae ☐ Slime ☐ Other ☐

Source of Water: City ☐ Private Well ☐

Source of Sample: Make-up ☐ Recirculating ☐

Equipment Involved:

1. Evap. Condenser: Make ______________ Tonnage ______________ Model # ______________
   Existing Bleed-Off ______________ Gal / Hr.
2. Ice Machine: Make ______________ Type: Cuber ☐ Flaker ☐ Drum ☐
   Model # ______________ Capacity ______________ lbs. Ice / Day
3. Closed System: Chilled ☐ Hot Water Boiler ☐ Both ☐
   Size ______ Tons ______ Btu/h (output): Capacity ______ gals.
4. Steam Heating Boiler Make ______________ Btu/h ______________
5. Evaporative (Swamp) Cooler: Make ______________ Size ______________ cfm

TREATMENT:

1. Chemical: Brand ______________ Type or Name ______________
   Part/Code No. ______________ Results: Good ☐ Bad ☐
2. Feed Equip: Chemical Feed Pump (size ______ GPD) Drip Feed ______
   TDS Monitor/Controller ______ Lockout Timer ______
3. Date System Last Cleaned: ______________

Test Results (Field or Office Use)

pH: ______________
Alk: ______________ ppm
Hard: ______________ ppm
Cl: ______________ ppm
SiO2: ______________ pprr
Cond: ______________ mmhos
______________ ppm
______________ ppm

Nu-Calgon Wholesaler, Inc.

Calgon is a licensed trade name.

Figure 3 Example of a water sample data sheet.
Source: Courtesy of Nu-Calgon

Common procedures for three basic tests are discussed further in this section:

- pH testing
- Alkalinity testing
- Hardness testing

1.3.1 pH Testing

Water pH can be tested very simply using one of the following three methods:

- *pH test paper* — A small strip of paper is briefly immersed in the water, then removed and compared to a color chart to determine the pH (*Figure 4*). Most test papers are designed to test across the entire pH range. However, papers designed for a smaller range, such as a pH of 5.5 to 8.0, can provide more accurate results when it is important.
- *Dye* — Add dye to the water sample per the manufacturer's instructions and compare the color with a color chart to determine the pH (*Figure 5*).
- *Electronic pH meter* — Electronic meters are compact and easy to use (*Figure 6*). The tip is simply placed in a water sample and the pH value is read once the display stabilizes. However, the meters must be properly maintained and stored. Rinse and clean the instrument after each use. Some have a special solution that is placed on the tip before the protective cap is installed over it. This keeps the tiny electrode at the tip hydrated. Use and care for a pH meter as directed by the manufacturer.

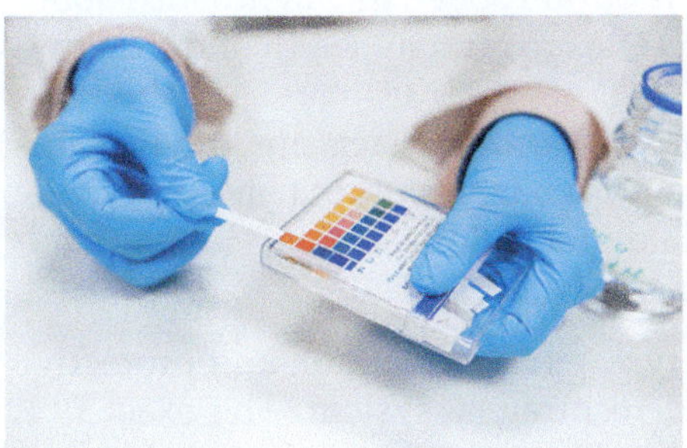

Figure 4 Using pH test papers.
Source: Volodymyr Kalyniuk/iStock/Getty Images

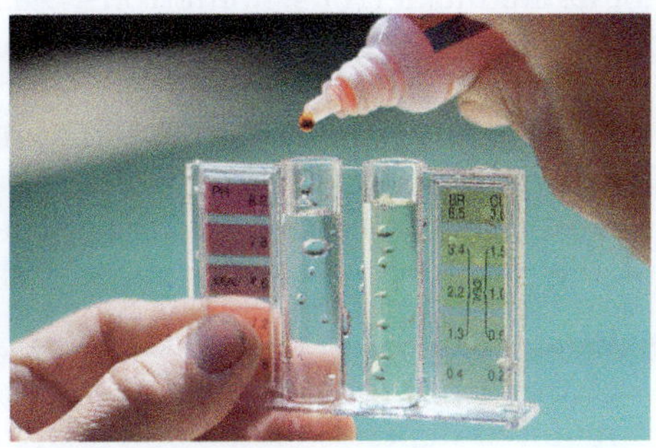

Figure 5 pH testing dye.
Source: StefanieDegner/iStock/Getty Images

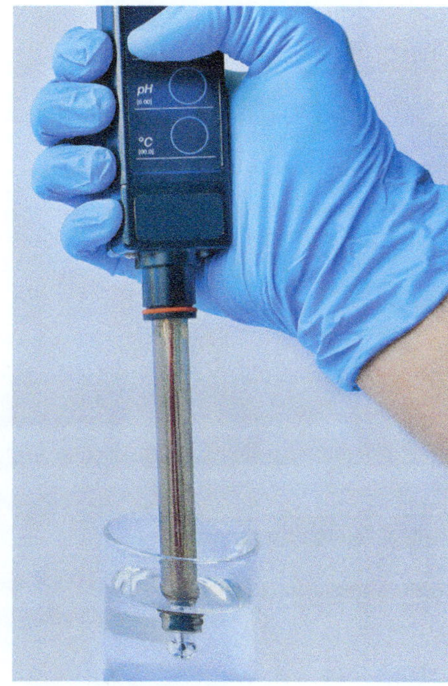

Figure 6 pH meter.
Source: Eakkalak Khiwsuwannasri/iStock/Getty Images

1.3.2 Alkalinity Testing

Remember that the alkalinity of a water sample is a measure of the water's buffering capacity, which allows it to counteract a reduction in pH from the addition of acids. Water with low alkalinity will experience a proportionately greater reduction in pH from the addition of acid than water of higher alkalinity.

For boilers, water treatment specialists must generally maintain alkalinity below a maximum level prescribed by the boiler's manufacturer. Alkalinity can be measured three different ways, and testing using more than one of these ways is common. The phenolphthalein alkalinity (P alkalinity) measures the strongest alkali present. The methyl orange alkalinity (M alkalinity) measures all the alkalinity in the water. The total alkalinity (T alkalinity) represents the alkalinity from all three sources of alkalinity: carbonate (CO_3), bicarbonate (HCO_3), and hydroxide (OH).

Measuring alkalinity in any one of the three ways is not difficult. Alkalinity is measured by mixing various **reagents** with the water sample. To complete a test, specific reagents are added to a measured water sample in sequence. The user observes and records the reactions along with the count of reagent drops added to reach a reaction. Following the test kit manufacturer's directions precisely is essential.

Reagents: Substances that facilitate or initiate a predictable chemical reaction, used to test for the presence of other substances.

WARNING!

Carefully read and follow all the precautions found on reagent bottles and test kits. Some reagents can be hazardous. Wear proper gloves, safety goggles, and any other PPE that may be recommended by the test kit manufacturer. Refer to the safety data sheet (SDS) that accompanies the chemical.

1.3.3 Hardness Testing

Water hardness is also measured using reagents added to a water sample, observing the results, and counting the number of drops required to achieve a specific water color. The following procedure is an example of the instructions you might find in a test kit:

Step 1 Rinse and fill a 25 mL test tube with the source water.

Step 2 Add the prescribed number of drops of a hardness buffer. Gently swirl the tube to mix.

Step 3 Add the prescribed amount of hardness indicator powder. Again, swirl the tube gently to mix. Assuming some level of hardness is present, the mixture will turn red.

Step 4 Begin adding a hardness reagent, one drop at a time, while watching for the color to change to blue. Count the number of drops required to reach the proper shade of blue.

Step 5 Multiply the required number of drops by 10 and record it as the total hardness, in the form of calcium carbonate, in parts per million.

1.4.0 Chemical Safety Precautions

HVACR technicians using chemicals in water treatment programs must be certain that the chemicals will not harm system components. This includes, but is not limited to, ferrous, nonferrous, plastic, natural rubber, and synthetic rubber components. The safety of a chemical as it relates to human health and the environment is even more important.

All chemicals should be treated as hazardous until the facts are known. This knowledge won't make a given chemical any less hazardous, but you will know the extent of the hazard and the appropriate precautions to take. Take the following precautions when working with chemicals:

- Wear specified PPE such as gloves, safety goggles, respirator, and protective clothing to protect your skin, eyes, and respiratory system. The exact PPE required depends on the potential hazards.
- Acquire and carefully review the product SDS, where you will find information about the personal and environmental hazards associated with a specific water treatment chemical.
- Do not mix or add chemicals other than as directed by the manufacturer or supplier. Chemicals must usually be added to a solution tank that is already holding water. Adding the chemicals first and then adding the water on top can produce a violent reaction and tremendous heat.
- When using a mixture of chemicals, make sure that any interaction between the components will not be harmful to yourself, the equipment, or the environment.
- Make sure to follow all federal, state, and local regulations that govern chemical disposal or the discharge of chemicals into the environment.

1.0.0 Section Review

1. What substances often found in water does water hardness represent?
   a. Calcium and zinc
   b. Zinc and potassium
   c. Iron and magnesium
   d. Calcium and magnesium

2. The development of scale in a piping system accelerates due to ______.
   a. consistent bleed-off
   b. electrochemical corrosion
   c. rising and falling water temperatures
   d. the presence of biological contaminants

3. Alkalinity in a water sample is measured by using ______.
   a. reagents
   b. test paper
   c. infrared light
   d. a spectrometer

4. Information about the personal and environmental hazards associated with a specific water treatment chemical is provided ______.
   a. by OSHA
   b. in the relevant SDS
   c. in water sampling kits
   d. by the cooling tower manufacturer

2.0.0 Mechanical Water Treatment Equipment

Objective

Identify various types of mechanical water treatment equipment and devices.

a. Identify types of filtration devices.

b. Identify other types of mechanical water treatment equipment.

Performance Tasks

There are no Performance Tasks in this section.

In addition to chemical water treatment methods, many mechanical devices are used to filter and treat water. These devices may be used alone or in conjunction with a chemical treatment program. Common mechanical water treatment equipment includes the following:

- Filtration equipment
- Evaporators
- Water softeners
- Deaerators
- Automatic chemical feeder systems
- Blowdown controllers and separators

2.1.0 Water Filtration Equipment

Mechanical filtration is used to trap suspended solids circulating in a system. Strainers, filters, and separators are commonly used to reduce suspended solids to an acceptable level.

Water filters are carefully selected to ensure that certain size particles are captured. In most cases, the smallest particles that could represent a problem for the system must be captured, but particles smaller than that can be dismissed.

This factor in filter selection is important because filtration efficiency directly affects pressure drop across the filter and the energy required to move fluid through it. Note that this is true for air filters as well as those that filter liquid.

For example, the common particle size in a cooling tower system ranges from 1 to 50 micron (μm), or micrometers, in size. However, 90% of those particles are smaller than 10 microns. These smaller particles are best addressed through water treatment to further break down any that are harmful. Only the larger, and fewer, particles need to be addressed by physical filtration.

Thus, when filters are selected, it is important to avoid over-filtering the water—attempting to remove particles that are not concerning—to minimize filter pressure drop and wasted pump energy. In the case of a cooling tower then, it would be a waste of energy and material cost to try to remove particles as small as 10 microns.

NOTE

One micron is equal to $^{1}/_{25,400}$" or one-millionth of a meter (0.000001 meter).

Did You Know?

Steam Filtration

Steam is used in most industrial and manufacturing processes for a variety of purposes. Steam used in the chemical, pharmaceutical, food and beverage, and power generation industries, as well as others, must be filtered to remove impurities and suspended solids, maintaining product integrity and protecting the related equipment.

Source: Steam Filter Elements. ©Donaldson Company, Inc., www.donaldson.com/enus/compressed-airprocess/products/steam/filterelements/

2.1.1 Strainers

The term *strainer* refers to a cleanable screening element designed to remove foreign particles as small as 0.01" in diameter. However, in most cases, strainers are used to remove relatively large particles from a water stream.

(A) Duplex Strainer Assembly **(B) Strainer**

Figure 7 Duplex strainer assembly.
Source: Hayward Flow Control

Strainers for HVACR water systems can be made from cast iron, bronze, stainless steel, alloys, or plastic. Some strainers have magnetic inserts installed to capture tiny iron or steel particles that may be present in the water. Strainer assemblies are available as single or duplex units, containing one or two strainers, respectively (*Figure 7*). Strainers usually must be removed and cleaned manually, but some are self-cleaning.

Strainers are simple to disassemble and clean. For single-basket strainers, the system usually must be shut down before isolating and removing the strainer for cleaning. It is very important to ensure that the internal pressure has been relieved before opening the basket housing. A gauge may be installed on the housing to check the pressure. A vent valve is also common, which is opened as a positive means of ensuring that the vessel is not under pressure. Follow the manufacturer's instructions carefully.

Duplex strainers, which contain two independent strainers, can be serviced without taking the system offline. Cleaning generally consists of opening the vessel containing the basket, removing the basket, and rinsing it with water to remove accumulated debris. An indicator, such as a lever, typically indicates which of the two baskets on a duplex strainer is active. The active strainer is unaffected by the removal and cleaning of the other strainer. Once one strainer is cleaned, the flow can be diverted through it while the second strainer is isolated and removed for cleaning.

Self-cleaning strainers do not require disassembly for cleaning. They are automatically cleaned by periodically **backwashing** the strainer using a separate water source and sending the debris to a suitable drain. Although more piping is required, self-cleaning strainers pay for themselves quickly in applications that require cleaning multiple times per day.

Backwashing: A procedure used periodically to reverse the normal direction of waterflow through a strainer or filter to flush accumulated debris.

WARNING!

Serious personal injury can be caused by liquid and chemical solutions escaping from a strainer housing opened under pressure. Wear the proper PPE, including chemical-resistant gloves, safety goggles, and a face shield when opening a strainer housing. Ensure that the vessel is not under pressure; a pressure gauge mounted on the strainer housing is recommended.

2.1.2 Filters

Various types of filters are used to remove smaller particles than strainers can capture. Both strainers and filters may be employed in the same system. The strainer captures larger particles that would clog the filter too quickly, extending the life of the filter that is capturing the smaller particles.

Cartridge Filters

Cartridge filters are commonly used to remove suspended particles ranging between 1 μm and 100 μm in size. These filters are typically used where the contamination levels are less than 100 ppm (0.01% by mass). Higher concentrations usually clog these filters too quickly. Where particle concentrations are high, strainers are usually placed upstream of the filter to capture the bulk of the contaminants.

Filter cartridges vary widely in construction and are made from many different materials. *Figure 8* shows some common types of filter housings and cartridges.

(A) Multi-Cartridge Filter Housings **(B) Pleated Cartridge Filters** **(C) Absorbent Filter Cartridges**

Figure 8 Examples of filter housings and cartridges.
Source: Courtesy of Sporlan Division - Parker Hannifin Corporation

Filters either capture debris on the surface or throughout the depth of a thicker media. The latter type is typically made from wound yarns or resin-bonded fibers that increase in density as the center of the cartridge is approached. Large particles enter and are captured on the surface, while smaller particles are captured in the denser media (*Figure 9*).

Filters that capture particles only on the surface are made from pleated paper or a similar material (*Figure 10*). These filters can generally handle higher flow rates because of a lower resistance to waterflow than dense-media filters. However, these filters are less effective in capturing debris and have far less holding capacity.

Once a filter cartridge becomes plugged, as indicated by a reduction in waterflow or an increase in pressure drop through it, the cartridge is replaced. Some manufacturers provide guidelines for filter replacement by rating the cartridges according to the number of gallons that can be treated. In practice, the frequency of replacement is determined by the size and concentration of the suspended solids in the water and the holding capacity of the filter element.

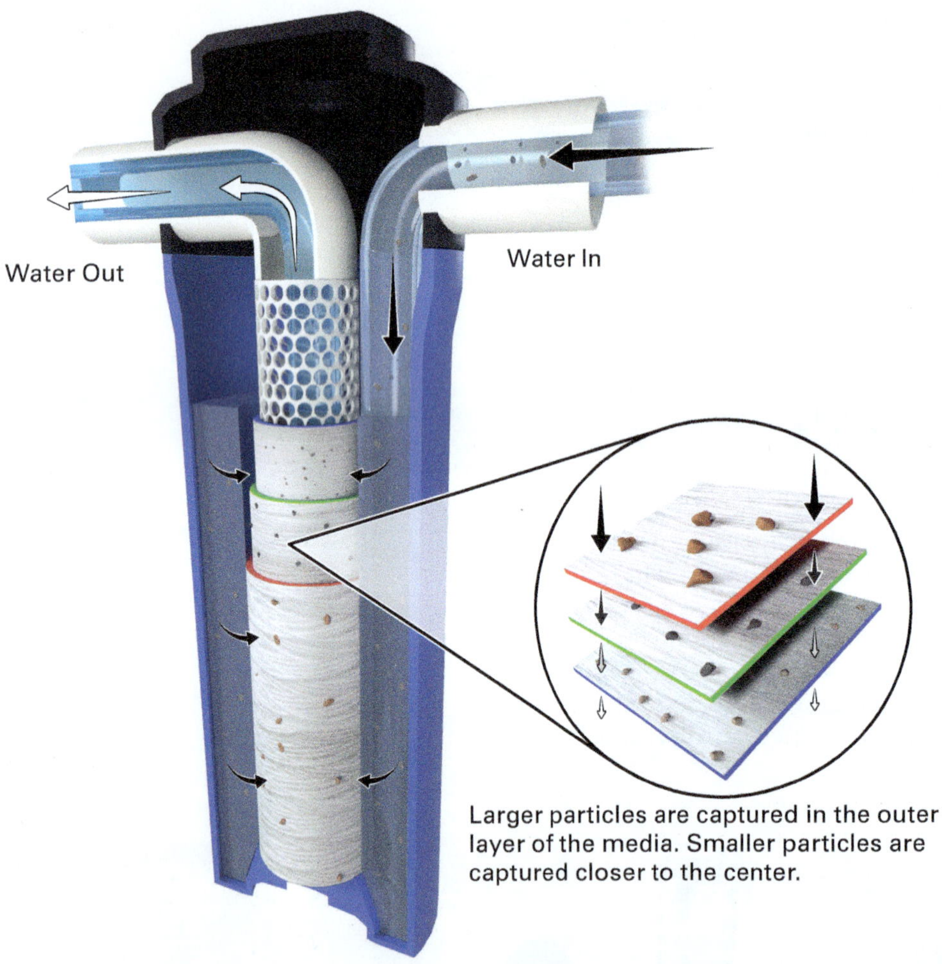

Figure 9 Waterflow through a dense filter cartridge.

Figure 10 Pleated cartridge filter.
Source: Ganna Zelinska/iStock/Getty Images

Multimedia Filters

Multimedia filters consist of a filter tank containing multiple layers of different media. The different grain sizes and types of filtering media provide progressive filtration. In the multimedia filter shown in *Figure 11(A)*, the coarsest filtering material is at the top with each successive layer being a finer-grained material. The media used in this filter are bituminous coal, anthracite coal, sand, and garnet particles. The final layer is a coarse quartz gravel.

Untreated water enters the top of the filter housing under pressure and flows through each of the media layers on its way to the pipe in the center, where clean water exits. At each layer, smaller and smaller particles are trapped.

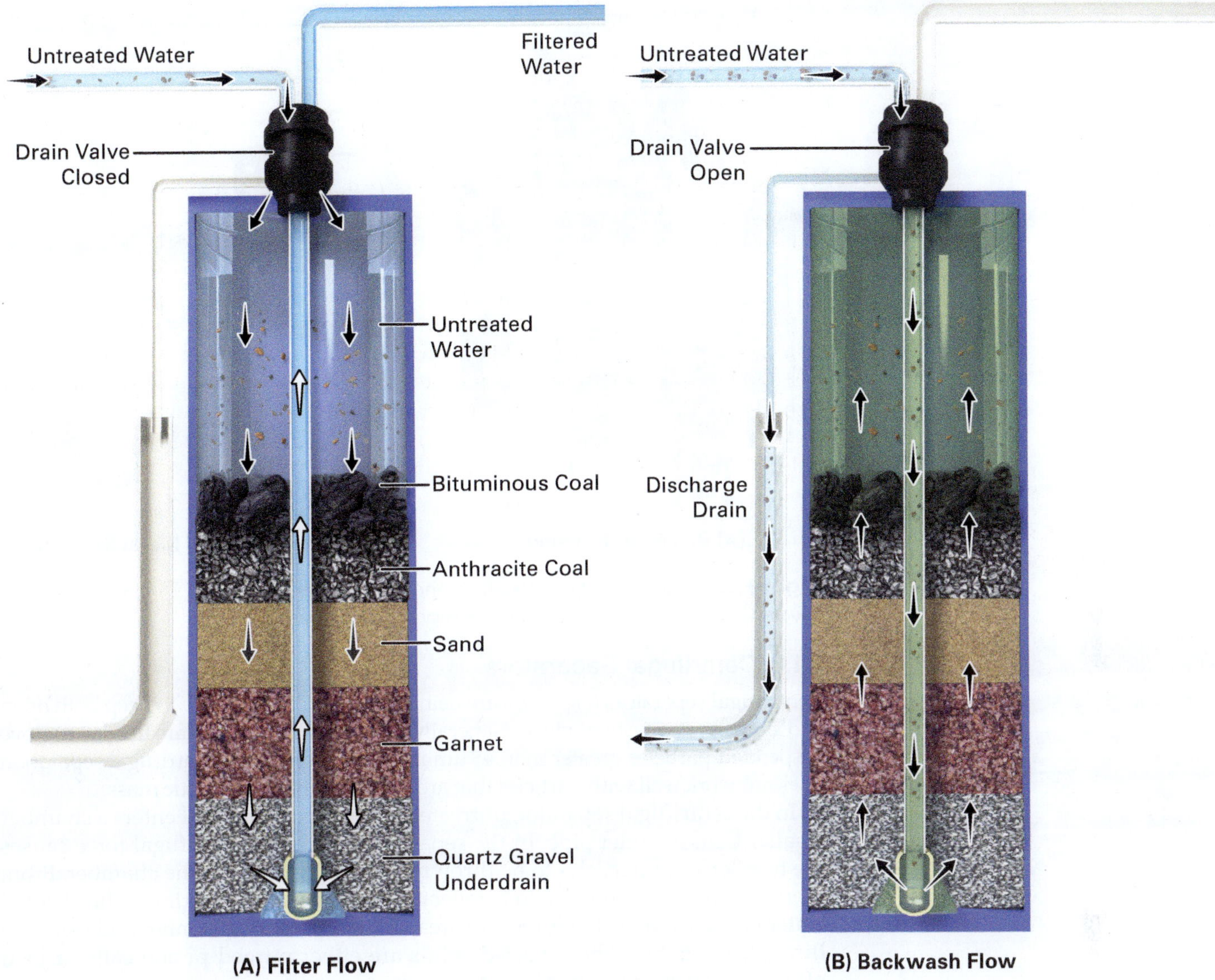

Figure 11 Multimedia filter operation.

Multimedia filters are almost exclusively cleaned by backwashing, as shown in *Figure 11(B)*. The waterflow through the media is reversed to flush accumulated debris from the media. The debris-filled water is routed to a drain line by the control valve at the top.

Backwashing must be performed on a regular basis to prevent accumulated particles from clogging the media. The media is replaced only as needed. When multimedia filters are properly maintained, the media can remain in service for 10 to 30 years.

> **On Site**
>
> **Duplex Multimedia Filter**
>
> Duplex multimedia filters provide two or more complete filter assemblies. The assemblies can be configured using valves so that all assemblies filter the system water simultaneously, or so that one assembly is in the backwash mode while another filters the system water.

Bag Filters

Bag filters rely on a mesh or felt bag supported by a metal basket enclosed in a tubular housing (*Figure 12*). The bag support basket is usually made from perforated stainless steel. The metal basket alone can often be used without a filter bag. For example, this type of filter may be used with a bag during the startup of a new closed piping system to capture debris. Once the system is clean, the bag may be removed to reduce flow resistance for normal operation.

Some manufacturers make a two-stage bag filter consisting of an inner basket and bag that fits inside a larger outer basket and bag. This provides for a coarse filtering stage (inner bag) followed by a finer one (outer bag).

Filter bags are made from materials such as cotton, nylon, polypropylene, and polyester. Bags made from felt material are common because of the fine pores. When they are full of debris, increasing the pressure drop and reducing the flow rate, the bags must be removed and replaced with new ones. Some bags can be washed and reused, depending on the application.

(A) Bag Filter Housings **(B) Filter Bags and Metal Strainers**

Figure 12 Typical bag filter housings and filter inserts.
Source: Courtesy of Sporlan Division - Parker Hannifin Corporation

2.1.3 Centrifugal Separators

Centrifugal separators (*Figure 13*) are being used more often today to keep both open and closed water systems clean. Generally, centrifugal separators are used to remove suspended particles greater than 45 µm in size from liquids. Centrifugal separation does not work well with particles that are small and have very little mass.

In the centrifugal separator, untreated water accelerates as it enters a chamber smaller than the inlet pipe. In the separation chamber, centrifugal force causes the heavier solid particles to be thrown toward the edges of the chamber. From there, the particles migrate to a collection chamber at the bottom. The cleaned water is drawn to the low-pressure area of the separation chamber and flows up through the outlet. The collected solids are either purged periodically or bled off continuously from the separator sump.

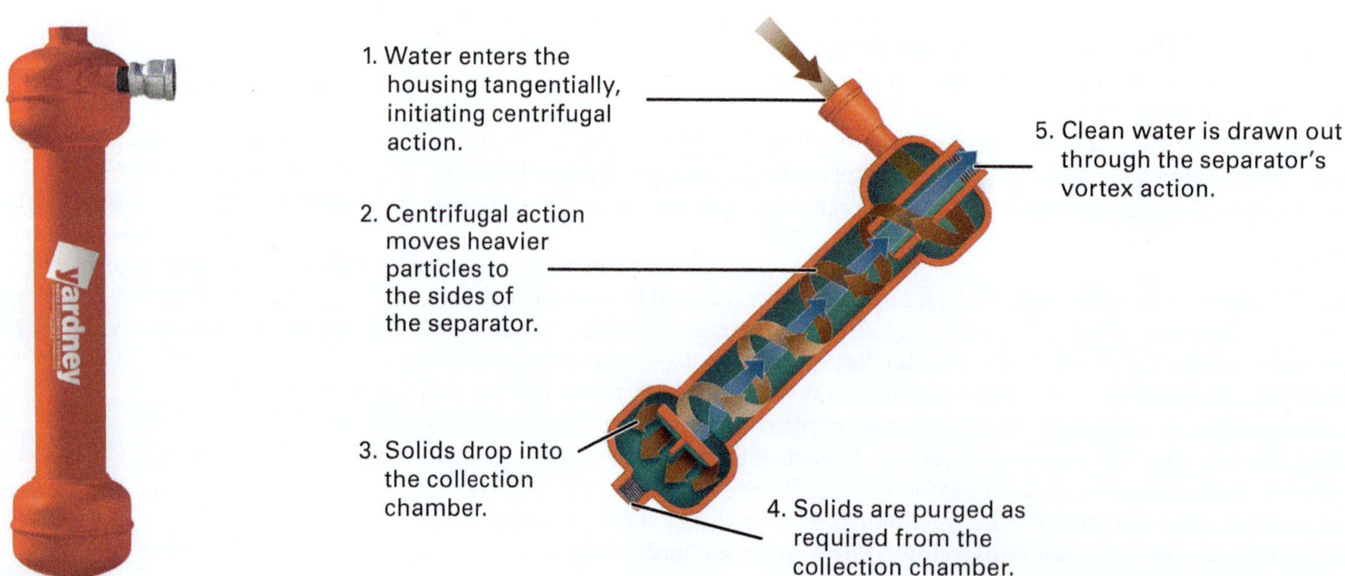

(A) Centrifugal Separator **(B) Centrifugal Separator Diagram**

Figure 13 Typical centrifugal separator operation.
Source: Yardney Water Management Systems Inc.

2.2.0 Other Water Treatment Equipment

Water treatment equipment employs various processes beyond simple filtration to maintain water quality. Equipment commonly used for water treatment includes the following:

- Evaporators
- Water softeners
- Automated chemical feed systems
- Deaerators
- Steam blowdown controllers and separators

2.2.1 Water Treatment Evaporators

Evaporators use **distillation** to remove solids. Distillation can also remove some organic contaminants and disinfect the water. Evaporators are installed in source water lines that supply makeup or feedwater to various pieces of equipment to remove solids before they can enter the system.

Distillation: The process of purifying a liquid by successive operations to heat and vaporize the liquid, then cool it to allow the vapor to condense back to its liquid form. Solid contaminants remain behind as the liquid is initially vaporized.

Evaporators are available in several types. *Figure 14* shows one common type. It consists of a distillation chamber and an air-cooled condenser section. Untreated water enters the distillation chamber and is heated to boiling. This vaporizes the water, causing all the solids to remain and settle to the bottom of the distillation chamber. The collected solids are flushed from the chamber periodically. The water vapor, now free of particulate matter, flows through an air-cooled condenser where it condenses back to water. The treated water exits and is used downstream.

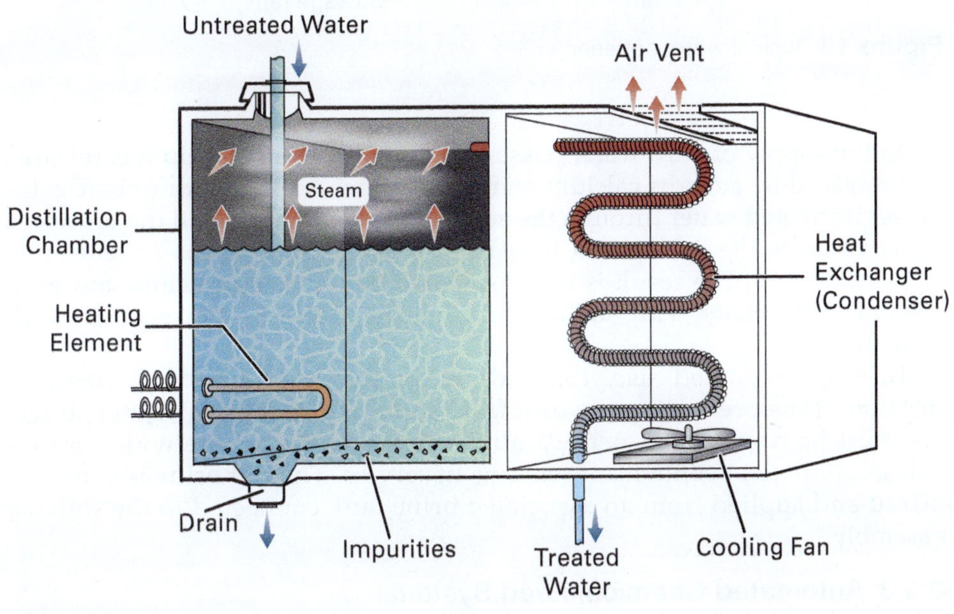

Figure 14 Typical water treatment evaporator.

The evaporator unit shown uses an electrical element to heat the water, but other heat sources can be used. A water-cooled condenser can also be used in place of an air-cooled condenser.

2.2.2 Water Softeners

Water softeners are used to remove the minerals from hard water that form scale and similar deposits. Water softeners are inline, point-of-entry devices installed in the water lines that supply makeup or feedwater to various kinds of equipment. Many homes are also served by water softeners.

Zeolite: A natural volcanic mineral that possesses a rare natural negative charge. Its honeycomb structure and negative charge allows zeolite to adsorbs liquids and adsorbs elements.

Many water softeners use a sand-like substance called **zeolite** that is saturated with sodium (salt) and arranged in a filter bed (*Figure 15*). Zeolite adsorbs calcium and magnesium, the main components of hardness, more readily than sodium.

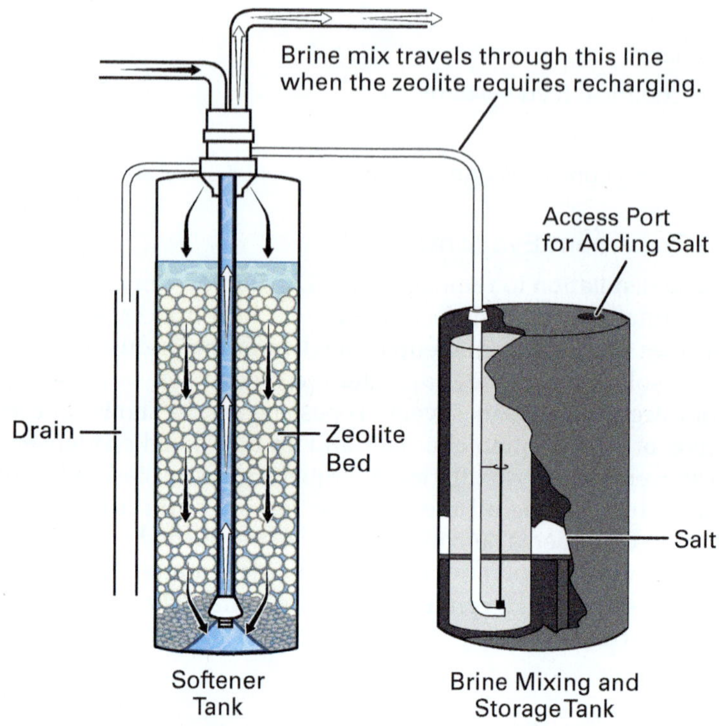

Figure 15 Typical water softener.

As the supply of hard water passes through the zeolite, sodium is released and replaced by adsorbs calcium and magnesium. A distributor or baffle disperses untreated water through the zeolite. This ensures that all the untreated water contacts the zeolite bed instead of passing through the softener unit without contact. The result is water softened by replacing calcium and magnesium bicarbonate with sodium bicarbonate that does not support scale formation.

Through extended use, the sodium in the zeolite bed is gradually depleted, thus preventing the unit from softening effectively. Water softeners must be regenerated periodically by treating the zeolite with a strong salt solution; the zeolite itself remains in service. The salt brine is generally mixed and applied from an associated brine tank connected to the softener assembly.

2.2.3 Automated Chemical Feed Systems

Adding chemical treatment to water systems manually is impractical because inconsistent conditions and spikes in chemical levels result, which could be detrimental to system components. Automatic chemical feed controllers use sensors to monitor water conditions and control injection pumps that add various chemicals in precise doses.

An example of an automated chemical treatment system is shown in *Figure 16*. Similar systems are found in many commercial and industrial HVACR equipment rooms. Automated chemical feeders are primarily used for open water systems, such as cooling towers. Closed water systems, such as chilled and hot water systems, do not typically require the same level of diligence in chemical treatment. Closed systems are not constantly exposed to the atmosphere or to fresh water.

Figure 16 Automated chemical feed system.
Source: Berg Chilling Systems Inc.

Simple controllers can provide pump control to add chemicals based solely on time and programmed dosage. Others have multiple inputs to monitor conductivity, pH, and other related water characteristics. In some cases, the controller also monitors the volume of makeup water added to the system, which may affect both the volume of chemicals to be added and the volume of bleed-off required. Most chemical feed controllers are capable of also controlling a cooling tower bleed-off solenoid valve to maintain proper conductivity and dissolved solid levels, further automating the water treatment and maintenance process.

Pumps used for this application are capable of very precise metering (*Figure 17*). Coupled with extremely sensitive and accurate controllers, the pumps help save on chemical costs and maintain consistent water conditions.

Chemical feed systems can also be connected to building management systems that allow the operator to monitor water conditions and chemical usage remotely. Consistent monitoring, especially on larger systems, can result in dramatic savings in chemical costs and minimize the environmental impact of biocide use.

Figure 17 Diaphragm metering pump.
Source: Nikita Rublev/Shutterstock

Designing and programming a chemical feed system requires a water system specialist or other responsible party to be aware of the types of water problems to expect in the local area and how to correct them. The chemicals used and the dosage levels will vary based on a variety of conditions, such as the following:

- Chemistry of the local water source
- Overall system design, including anticipated water temperatures
- Materials of construction used in system components such as the piping, cooling towers, boilers, heat exchangers, and pumps
- Potential for interaction between chosen chemicals

Because of the diversity in water chemistry and the depth of knowledge required, detailed coverage and installation of chemical feed systems is beyond the scope of this module. More detailed information on automatic chemical feed systems can be found in resources regarding water treatment referenced at the end of this module. HVACR technicians are not typically called upon to design these systems, select the equipment, or determine the chemicals and dosage required.

On Site

Typical Automatic Control and Feed System

In many cases, pre-engineered automatic systems for boilers, chillers, cooling towers, and process applications are available from the same equipment manufacturers. These systems include sensors, controllers, and the appropriate chemical metering pumps. Completely assembled systems make installation and startup relatively easy.

2.2.4 Deaerators

Deaerators (*Figure 18*) are typically used in steam systems to remove air and other non-condensable gases from the system feedwater. Oxygen is an essential component in many corrosive processes; its removal slows or stops corrosion.

Oxygen becomes less soluble as the water temperature increases; therefore, oxygen can be removed by heating the incoming feedwater to its boiling point.

(A)

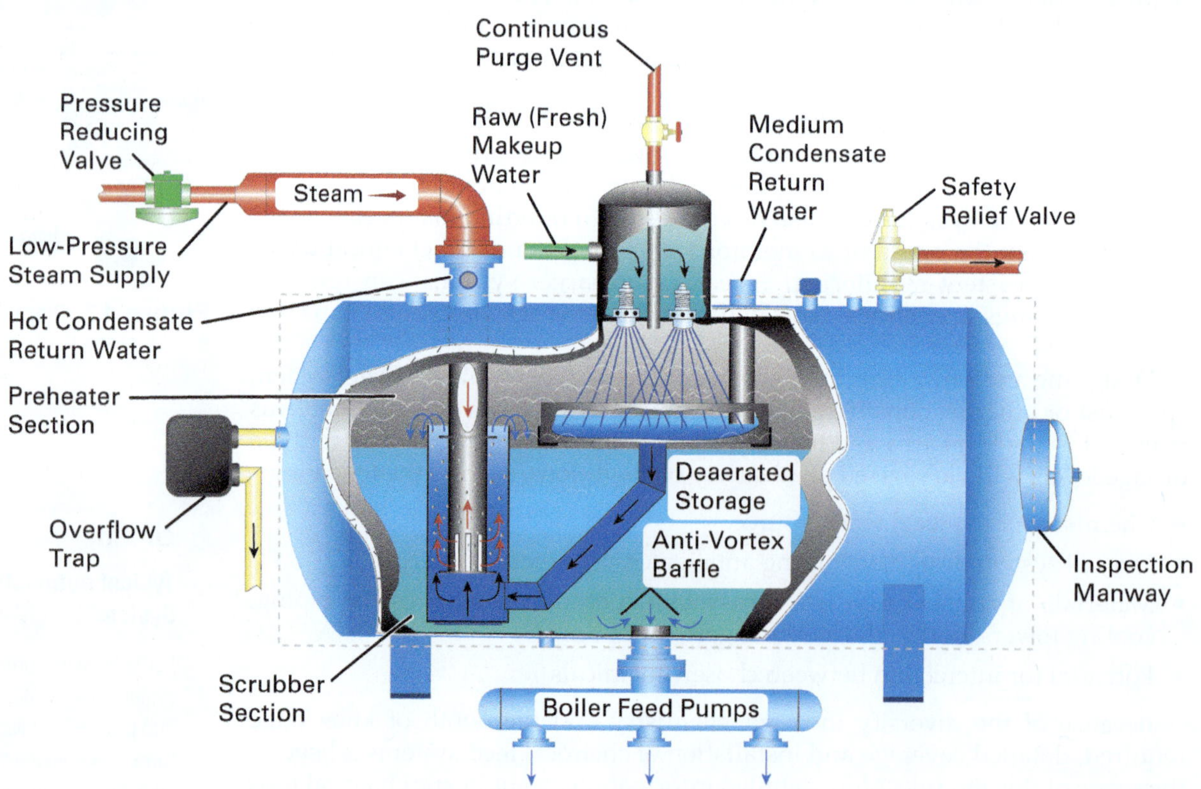

(B)

Figure 18 Spray deaerator.

Source: Images courtesy of Hurst Boiler Welding & Co., Inc. of Coolidge, GA

The oxygen and other gases are driven out and vented into the atmosphere. Other impurities that increase hardness settle to the bottom of the unit. A filtration system captures the solidified impurities.

Several types of deaerators are available. One type uses system steam as its heat source. In this type, the incoming feedwater is discharged as a fine spray into the upper portion of a closed chamber. The steam enters and mixes with the sprayed water, subjecting the water to an intense scrubbing (cleaning) action. While the cleaning action continues, another traversing stream of steam heats the water to free the oxygen and other non-condensable gases.

After contacting the feedwater, most of the steam cools and condenses. Any remaining steam vapor, along with separated oxygen and other gases, enters a vented condenser where the steam condenses as it contacts tubes that contain cool, incoming feedwater. The liberated oxygen and other gases are vented into the atmosphere.

2.2.5 Steam Blowdown Controllers and Separators

Steam boilers require blowdown from both the surface and the bottom of the boiler to remove solids and other contaminants. *Blowdown* refers to the controlled draining of boiler water to reduce the concentration of suspended and dissolved solids in the boiler. Although a blowdown can be done manually, substantial cost savings can result from the use of an automated process.

Since the liquid and other material discharged during blowdown is at the same temperature as the boiler water, heat energy lost during the blowdown process can be significant. Steam boilers that have little or no condensate return (steam is consumed and lost) typically need more blowdowns due to the increased volume of contaminants entering. In addition, boiler operation and loads that vary from one day to the next affect the accumulation of solids and impurities.

Automated systems with sensors provide another level of control. When a blowdown is accomplished manually, the operator cannot be certain how long the blowdown should continue, as they have no way to determine how much of the solids have been removed or how much remains. Since a significant amount of energy can be lost during each blowdown cycle, the number and length of cycles must be minimized. Automated systems can address this issue.

An example of a packaged blowdown system is shown in *Figure 19*. With automated blowdown, probes from the controller sense the conductivity of the water, which represents the volume of dissolved solids and other contaminants. The controller compares this information to preset parameters, and a modulating valve is opened and closed to provide the necessary blowdown. Blowdown can be done from the surface of the water, the bottom of the boiler, or both.

Steam boilers may also be equipped with a blowdown separator (*Figure 20*). These vessels are centrifugal separators that help reduce corrosion and scaling

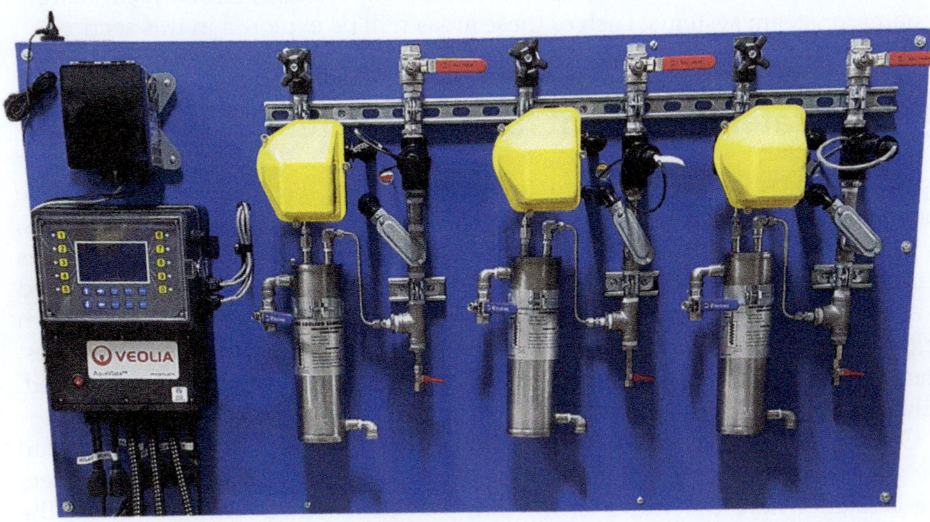

Figure 19 Boiler blowdown control system.
Source: Veolia Water Technologies

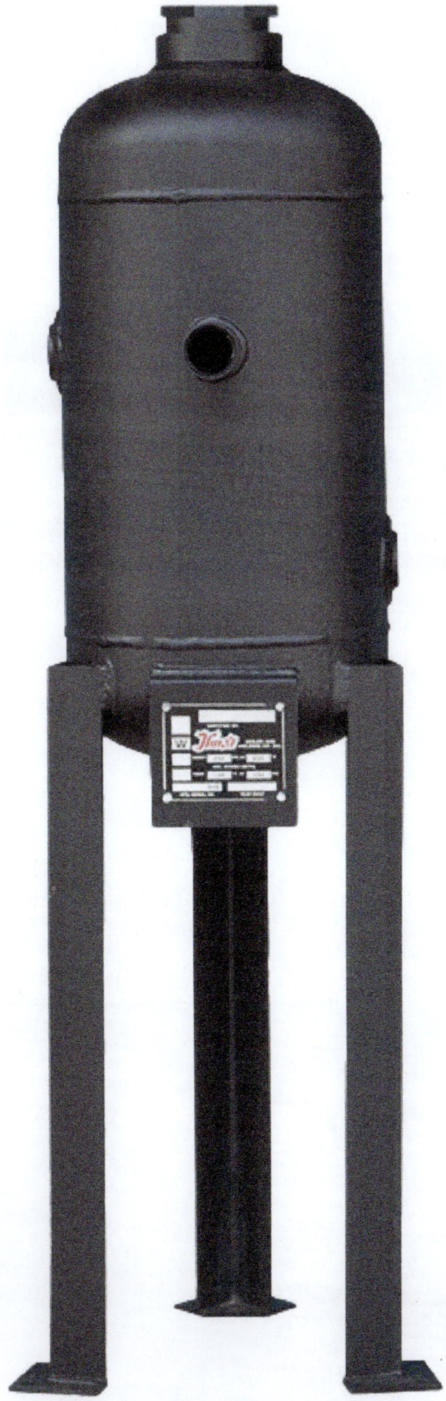

Figure 20 Blowdown separator.
Source: Image courtesy of Hurst Boiler Welding & Co., Inc of Coolidge, GA

downstream by separating steam from water and collecting the dissolved solids. Blowdown separators also represent an opportunity to reduce the temperature of the water being discharged, which is often necessary to comply with environmental regulations and municipal codes.

2.0.0 Section Review

1. What is the best way to make a cartridge filter last longer in systems with high concentrations of suspended particles?
   a. Install a strainer upstream.
   b. Backwash cartridge filters more often.
   c. Install multiple cartridge filters in series.
   d. Install multiple cartridge filters in parallel.

2. Some automatic chemical feed controllers also monitor the ______.
   a. age of scale deposits
   b. depth of zeolite in the filter bed
   c. type of bacteria in the recirculated water
   d. volume of makeup water added to the system

3.0.0 System-Specific Water Treatment Problems

Performance Task

2. Inspect a cooling tower or steam boiler and the related water system and identify signs of water treatment issues.

Objective

Identify and describe how to address water-related problems in specific types of hydronic and steam systems.

a. Identify and describe how to treat water-related problems in open water systems.
b. Identify and describe how to treat water-related problems in closed water systems.
c. Identify and describe how to treat water-related problems in steam systems.

Water-related problems, such as scale and corrosion, adversely affect HVACR system performance. The problems that are most likely to occur depend on the type of system. Likewise, the specific steps for addressing water problems depend on whether the problems occur in open recirculating systems, closed recirculating systems, or steam systems. Each of these areas will be explored in this section.

3.1.0 Water Treatment for Open Water Systems

Open systems are usually related to cooling towers (*Figure 21*) or evaporative condensers that provide cooling for industrial process equipment and products, as well as to condense refrigerants. Heat in the water is transferred to the air as water is sprayed inside the cooling tower, where individual droplets are exposed to air. As a result, the composition of the water is changed by evaporation, which helps cool the water but leaves the minerals and solids behind as it evaporates. The water is also aerated in the process, with fresh oxygen readily absorbed. Other contaminants and gases in the air can also be absorbed into the water.

Open cooling towers and evaporative condensers are highly susceptible to the growth of algae, bacteria, and similar organisms, especially if the towers are located where the water in the sump is exposed to sunlight. Thus, the main goals of water treatment in an open recirculating system are to minimize corrosion and scale development and eliminate algae, bacteria, and similar organisms.

(A)

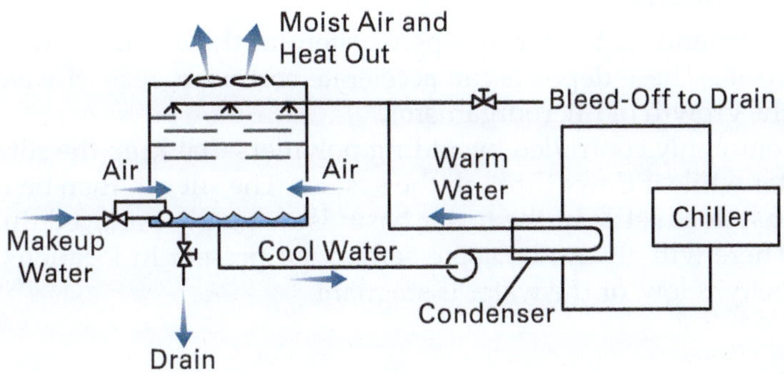

(B)

Figure 21 Modular cooling tower.
Source: vorclub/iStock/Getty Images (A)

3.1.1 Corrosion and Scale Deposits

Cooling tower systems are exposed to many types of corrosion, from simple electrochemical corrosion to pitting caused by microorganisms. Microbiologically induced corrosion is one of the main types of corrosion found in cooling towers. It results when *biocides* are not used or properly managed, which can lead to a high algae and/or bacterial count in the water.

The buildup of solids contributes to both corrosion and scaling in an open recirculating water system. These solids and other impurities are left behind as water evaporates from the tower. Typically, these solids include calcium and magnesium that were in the makeup water naturally.

Remember that calcium carbonate and calcium sulfate are the main ingredients in scale deposits. Additional calcium and other minerals are consistently introduced into the system as fresh makeup water is added to compensate for evaporation and the controlled bleed-off of water.

Chemical treatment of open recirculating systems to control corrosion and scale involves blends of phosphates, phosphonate, molybdate, zinc, silicate, and various polymers. Totyltriazole or benzotriazole is added to these blends to protect copper and copper alloys from corrosion. These two chemicals bond to the metals to prevent destabilization through corrosion.

While these chemicals may not be familiar to you, they would be familiar to water-treatment craft professionals with the significant training required to prescribe chemical treatments for water systems. A complete study of water chemistry is beyond the scope of this module. Before prescribing any chemical for water treatment, the water chemist or other responsible person must be aware of all local, state, and federal environmental guidelines and regulations.

Going Green

Using Chromates for Corrosion Control

In the past, the use of chromates as a corrosion inhibitor in water systems was common. However, the EPA eventually identified chromates—salts of chromic acid—as a suspected carcinogen. As a result, the EPA banned chromate use in cooling towers related to comfort cooling many years ago. Chromates can still be used in industrial cooling tower applications.

Chromate coatings are still used in many other ways. Items such as household door hinges, screws, and similar hardware items are often protected from corrosion with a chromate coating.

The range of pH in cooling tower water is typically maintained between 7.5 and 8.5. Sulfuric acid is generally used when the pH needs to be reduced because of high hardness or alkalinity in the makeup water. Sulfuric acid increases the solubility of calcium, thereby reducing the potential for scaling.

If enough acidic gases are absorbed, the water can become very acidic with a very low pH. The pH of the recirculating water must be carefully increased by adding caustic or alkali. Because acid is introduced into the recirculating water from the surrounding air rather than from the makeup water, the exact volume of alkali to be added cannot be calculated; it is usually determined through conservative trial and error as well as experience.

Beyond chemical additives, corrosion and scaling are also controlled by bleeding off a consistent volume of water. A metered amount of water is discharged from the system to carry away excess minerals and other impurities that remain as water evaporates from the cooling tower. The bleed rate for most systems should be about 0.5 to 1 gallon per hour per ton of capacity.

Although bleeding water from a system is somewhat wasteful, and more minerals and impurities enter with the makeup water added, bleeding from the right location discharges water that has a significantly higher mineral content than the fresh water.

3.1.2 Silt Deposits

Airborne dust and dirt enter an open system at the cooling tower and settle as silt deposits. These deposits can accelerate corrosion, restrict waterflow, and promote the growth of microorganisms.

Silt is commonly controlled by adding polymers that keep the silt suspended in the water while it flows through the system. The silt can then be filtered out, and any that does settle in the tower basin is manually removed during maintenance. There will always be some settled silt present in locations where the water velocity is low, or the water is stagnant.

Going Green

Bleed Water Control

People are aware of the need to conserve the precious resource of clean water. Older cooling tower systems were often fitted with a simple manual valve to control bleed-off. In most cases, the valve was set by HVACR technicians and system operators to a flow rate far greater than was required. In some cases, calculations were never made to even determine the correct volume, nor was the bleed-off water volume measured. As long as the objective of reducing impurities was reached, this method seemed sufficient.

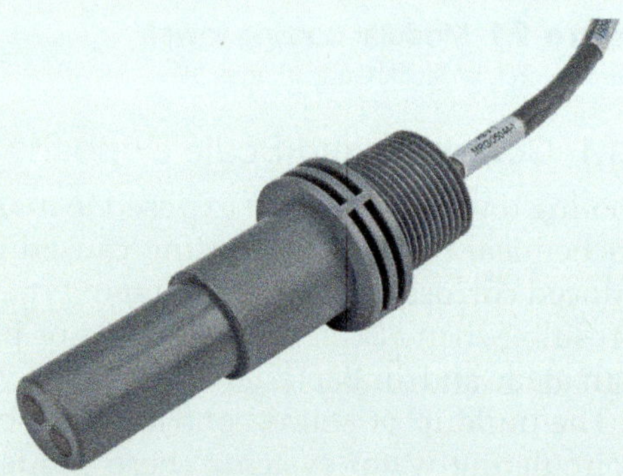

Source: WALCHEM, IWAKI America Inc.

With water conservation in mind, proper methods of monitoring the level of total dissolved solids through conductivity testing are readily available today, allowing automated control of the bleed water volume. These monitors use a probe like the one shown to measure conductivity. A solenoid valve in the bleed-off line is opened only when impurity levels indicate the need for dilution. As a result, literally millions of gallons of water across the country are saved each year.

3.1.3 Biological Growth

As water is broken into droplets and sprayed into a cooling tower, airborne organic matter such as bacteria, algae, and similar organisms are introduced in the system water. Algae are primitive plants that require sunlight for growth. Algae tend to grow in the perforated distribution pans of cooling towers and in stagnant areas of the tower sump. Algae is a problem for open systems where the water is exposed to light—an essential ingredient for its development.

Slime is a jelly-like material produced by bacterial growth. Slime clings to many parts of the system. If various slimes are allowed to flourish, waterflow through the nozzles can be restricted. Microbiological activity can also support corrosion.

The control of algae and slime can be difficult because they react differently to various chemicals. Both types of growth can also build up resistance to chemicals used repeatedly. Treatment with chemical algaecides or bactericides prevents or slows the growth. Regular **shock treatments** with chlorine or other chemicals are commonly used for control. Typically, two different chemicals are alternated to ensure that the algae or microorganisms do not build up a resistance to any one chemical.

Shock treatments: Water system treatments with a single substantial dose of chemical(s), as opposed to consistent dosing over time.

Larger doses of chemicals are needed if biological growth has been allowed to accumulate. The dosage volume depends on the amount of water held in the system. Therefore, the system water volume must be calculated and documented for future use. The treatment frequency is usually determined by trial and error, and by monitoring the rate of growth.

Note that chemical treatment alone does not usually control algae or slime. Biological growth is nearly impossible to prevent without using chemicals in hazardous doses. For this reason, the equipment must be thoroughly cleaned at regular intervals to control growth.

3.1.4 Cooling Tower and Open Recirculating System Treatment

Regular maintenance is necessary in any water system, but it cannot be stressed enough in open systems because of health and safety issues related to **Legionnaires' disease** and similar illnesses. Poorly maintained cooling towers are known to support the growth of *Legionella* bacteria. The aerosol produced by the bacteria can potentially infect not only the people at the equipment site, but also those in the surrounding area.

Legionnaires' disease: A serious type of pneumonia developing 2 to 14 days after exposure to the airborne bacteria, although many healthy people exposed do not get sick. The *Legionella* bacteria was identified and named following a 1976 outbreak at an American Legion convention in Philadelphia.

Legionnaires' disease can be fatal, but our knowledge and understanding of the disease has largely eliminated fatalities today. However, the illness remains a concern, and outbreaks do still occur. If cooling towers are not properly maintained and treated, a building owner and/or contractor may be held liable for outbreaks.

Recommended maintenance actions and tests related to cooling towers and other open recirculating water systems are summarized in *Table 3*. Note that this is only an example, as the maintenance needs for specific systems and environments can differ significantly.

TABLE 3 Cooling Tower/Open Recirculating Water System Maintenance

Maintenance Action	Frequency
Check that dosing chemical containers are full, pumps are operating properly, and supply lines are not blocked or kinked.	Weekly
Check the bleed-off control equipment to make sure it is operating properly and remains calibrated. Manually operate the bleed solenoid valve to observe that the drain is clear.	Weekly
Check the system for algae and slime. There should be no algae growth in the towers or slimy feel to the fill pack, tower sump, or side walls.	Weekly
Test and record bacteriological quality.	Biweekly
Test and record biocide and inhibitor reserves, pH, and conductivity.	Biweekly
Check the condition and operation of sump immersion heaters. Ensure the temperature control is functional.	Monthly
Check the operation and condition of spray nozzles, fans, and drift eliminators. There should be no physical damage, and the components should be free of deposits. The distribution system should provide a balanced flow of water across areas of the tower.	Monthly
Drain, clean, and disinfect cooling towers and associated pipe work in accordance with the method approved for the site.	Semiannually
Review maintenance and water treatment program performance, including the quality of the results obtained and the frequency of tasks.	Annually

After the installation or major repair of an open system, it is necessary to clean the internal water system of protective oils, grease, welding flux, dirt, and other debris. Cleaning is typically done according to the following guidelines:

Step 1 Fill, vent, and leak test the system.

Step 2 Add the chemical cleaning agent prescribed for the system in the amount and manner directed by the manufacturer.

Step 3 Operate all system pumps with the strainers installed for the time interval prescribed by the responsible water treatment specialist.

Step 4 Drain the system; remove, clean, and replace all strainers, and then refill the system.

Step 5 Operate all system pumps for at least 30 minutes. During this time, temporarily open all dead ends, drain valves, and strainer flush valves.

Step 6 Drain and refill the system. Repeat *Step 5*. On a newly installed system, it may take as many as four flushes to adequately clean the system.

Step 7 Test the system water and certify its cleanliness using a procedure that measures the residual cleaning agent in the water. The final system water pH should be within ±0.3 pH of the makeup water pH.

Step 8 After the system has been tested and certified to be free of cleaning agent, take samples of the system water for testing and future use.

Step 9 Initiate the approved water treatment program specified for use during system operation. This may include the removal of bag filters used during the initial cleanup.

Scale

Scale is a very common problem for systems that have been in operation for a while. Scale doesn't develop overnight. If the amount of scale inhibitor prescribed for a water treatment program fails to control the scale buildup as expected, the following factors should be checked:

Step 1 Check that the prescribed rate of bleed-off is being maintained.

Step 2 Ensure that the prescribed dosage of scale inhibitor is being applied. Take samples of both the recirculated water and the makeup water, and have them analyzed to check the concentration of scale-forming salts. Look for changes in the concentration levels by comparing them with the concentration previously used as the basis for the original water treatment prescription. Determine if something has changed that would indicate the need for higher doses of scale inhibitor.

Step 3 Because scale forms more rapidly at higher water temperatures, compare the temperature of the water in the sump with the ambient wet-bulb temperature. If the sump water temperature is more than 10°F (5.5°C) above the ambient wet-bulb temperature, the following are possible reasons why the water is not cooling as expected:

- The cooling tower capacity is insufficient for the flow rate.
- A fan is not working properly due to loose belts, worn bearings, or damaged blades.
- The air inlet to the tower is restricted.
- The distribution of the water is inadequate due to insufficient waterflow, plugged nozzles, slime-covered water distribution orifices, or worn-out tower fill.
- Moist air from the discharge of the cooling tower is being recirculated back to the air intake, often due to nearby structures coupled with the prevailing wind.

Step 4 Because scale may be forming if the temperature rise of the water flowing through the system condenser is too high (more than 10°F, or 5.5°C), check for the following:

- Slime or debris is covering the pump inlet screen or strainer.
- The water lines and/or condenser are undersized or currently restricted by corrosion or scale.
- The pump impeller is worn, or the pump does not have enough capacity for the application.
- The condenser flow control valve is not admitting enough water to the condenser or other heat exchanger.

3.2.0 Water Treatment for Closed Water Systems

Hot-water, chilled-water, and dual-temperature systems are examples of closed recirculating systems. Unlike open systems, closed recirculating systems do not experience constant water loss through evaporation or spillage, and the water is not in constant contact with fresh air. These characteristics significantly reduce the problems that occur, but corrosion and scale formation must still be managed.

Because the water does not evaporate, the concentration of minerals does not consistently increase. However, no matter how tight a system is, every closed system will eventually need makeup water to compensate for leakage or water lost during service operations.

An undiscovered leak is a significant problem when the addition of makeup water is automated (*Figure 22*). Although the concentration of minerals will not likely change, dissolved oxygen and other impurities flow into a closed system in the makeup water. Oxygen is one of the main agents of corrosion. If the leak is not discovered for weeks or months, the damage from corrosion can be significant.

Figure 22 Even a small leak in a closed water system can lead to extensive corrosion.
Source: bancha singchai/iStock/Getty Images

3.2.1 Corrosion

Several chemicals can help prevent corrosion in water systems. In the past, the use of chromates and hydrazine was common. However, the use of chromates in cooling towers related to comfort systems, such as the cooling tower for a water-cooled chiller serving an office building, is prohibited by the EPA. Inhaling air containing chromates increases the risk of lung cancer. Chromates can be used, however, in industrial cooling towers that support manufacturing processes or power generation, and in closed systems where they cannot become airborne. Hydrazine is a highly reactive and flammable substance that also presents health concerns, but its use in closed systems is permitted.

Sodium nitrite is commonly used instead of chromates as an inhibitor in closed loop systems. Sodium nitrite has been proven nearly as effective as chromate for protecting steel and other ferrous metal components, but it provides little or no protection for components made of nonferrous metals. Other inhibitors must be included in the mixture to protect these components. One such nitrite-based mixture includes borax as a pH buffer and sodium tolyltriazole as an inhibitor to protect nonferrous metals. These metals are generally found in condensers and heat exchangers.

3.2.2 Scale Deposits

In chilled-water systems, scale formation is normally not a problem even if the makeup water is hard, due to the low water temperature. Hot-water systems are more likely to suffer from scale formation. Because a high pH also promotes the buildup of scale, avoid using water with a high pH in hot-water systems whenever possible.

3.3.0 Water Treatment for Steam Systems

Steam systems are also closed recirculating systems, except in the rare cases where the steam is expelled and isn't returned to the boiler as condensate. The water used in a boiler must be kept clean for efficient operation. Add only clean water to a steam boiler.

Scale, corrosion, fouling, and foaming can all interfere with boiler operation and heat transfer. Scale deposits on heating surfaces increase boiler temperatures and lower the operating efficiency. Corrosion can damage metal surfaces, resulting in metal fatigue and failure. Fouling clogs nozzles and pipes with solid materials, thereby restricting circulation and reducing heat transfer efficiency.

Foaming results in overheating and can cause water impurities to be carried along with the steam into the system. Foaming is a condition caused by excessive water surface tension due to the accumulation of scum comprised of oil, grease, and/or sediment floating on the surface of the boiler water. Foaming prevents steam bubbles from breaking through the water surface and hinders steam production. The trapped bubbles eventually rise through the foam, resulting in impurities being entrained in the steam and then carried over into the steam system.

Hard water may eventually interfere with the efficient operation of a boiler due to scale development. For this reason, hard makeup water should be chemically pretreated with water softeners.

3.3.1 Corrosion

Corrosion erodes away the boiler metal and appears in different forms, including pitting, *grooving*, and *embrittlement*. Grooving refers to corrosion that occurs along the seam of electric resistance welded (ERW) steel pipe and appears as a round-bottomed groove. Embrittlement is typically caused by the diffusion of hydrogen in metal and is more likely to be found in high-strength steels.

Corrosion can be caused by acidic (low pH) boiler feedwater. Another cause is excessive dissolved oxygen in the feedwater. Oxygen as well as carbon dioxide and other gases present in the feedwater enable and support corrosion. These gases must be eliminated using a mechanical deaerator (deaerating heater) and/or scavenger chemicals such as sodium sulfite and hydrazine to remove some of the oxygen from the water.

Most steam systems do experience some losses and makeup water is added, although it is added at a much lower rate compared to open systems. Periodic blowdowns are therefore necessary to control chloride and hardness levels. To minimize corrosion, the boiler water pH should be maintained between 11 and 12 by an alkaline water treatment using soda ash, caustic soda, sodium silicate, or sodium phosphates.

Maintaining high pH levels helps prevent corrosion, but it unfortunately promotes the formation of scale. Therefore, scale inhibitors must be used in conjunction with elevated pH levels in a complete program.

3.3.2 Scale Deposits

When water vaporizes into steam in a boiler, the steam moves into the system toward the point of usage. Any calcium, magnesium, and other salts dissolved in the boiler water remain behind and form scale on the boiler's tubes and other surfaces.

This condition is aggravated, particularly in hard water areas, when steam losses are significant, requiring the frequent addition of makeup water. The greater the hardness of the makeup water, the faster scale forms. If scaling becomes heavy enough in firetube boilers, the tubes can bulge due to overheating, because the tube cannot effectively transfer the heat to the water on the opposite side. The tubes can burn out or burst. Scale clogs the inside of the tubes on watertube boilers, interfering with flow (*Figure 23*).

Figure 23 Scale deposits on boiler tubes and tube sheets.
Source: RYosha/iStock/Getty Images

Scale formation is best avoided by removing all the calcium, magnesium, and other mineral salts from the feedwater before it enters the boiler. Chemicals commonly added to prevent scale formation are sodium carbonate and phosphate.

Another water treatment for scale prevention involves the use of a **colloidal substance** such as tannin or lignin. A colloidal substance is a jelly-like material made up of very small, insoluble, and non-diffusible particles that are larger than molecules but small enough to remain suspended in a fluid without settling to the bottom. When added to boiler water, these substances react with and absorb the calcium, magnesium, and other salts. The entire mass then forms a sludge that is easily removed from the boiler during a blowdown. Another common approach is to add polymers that keep the sludge in suspension while it flows through the system. Moving debris can be captured through filtration.

Colloidal substance: A jelly-like material comprised of very small, insoluble, non-diffusible particles that are larger than molecules but small enough to remain suspended in a fluid without settling.

3.3.3 Sludge Deposits

Sludge is a deposit formed by salts and other solids present in boiler water. It appears as lumps or thick masses of material that settle in the low points of a boiler.

Some sludge in a system occurs naturally. Other sludge is created with intent by water treatment, as mentioned in the previous section. Minerals that exist in water cannot be destroyed, but with the use of proper chemicals, the properties of these minerals can be modified to make them more manageable. Turning the minerals into a non-adhering sludge that is less likely to settle allows them to be removed by blowdown.

Regardless of its source or method of creation, sludge can insulate tubes and interfere with steam and water circulation. Sludge remaining in an empty boiler may bake out and harden.

Sludge is removed by periodic boiler blowdown. In extreme cases, a thorough cleaning of the boiler with a chemical product may be necessary to ensure its complete removal.

3.3.4 Foaming

Foaming can result in impurities from the boiler water becoming entrained in the steam where it is carried into the system. Once in the system, the impurities can form deposits that damage heat exchangers and other system components. Foaming is commonly controlled by filtration to remove solid impurities and chemical antifoaming agents. Periodic surface blowdown also helps remove and minimize foaming.

3.3.5 Steam Boiler Water Treatment Guidelines

A steam boiler water treatment program may include pretreatment of the raw makeup water using filters, water softeners, and de-alkalizers before it enters the feedwater system. Chemicals can also be used to treat both the boiler feedwater and boiler water, and to condition any sludge to ensure it remains in circulation rather than settling out in an inaccessible area.

Boiler water treatment and maintenance also relies on periodic blowdown. The water that is removed by blowdown is then replaced with an equal amount of fresh, clean feedwater.

The volume of water removed during blowdown is normally determined by the concentration of total dissolved solids in the boiler water. Enough water must be removed to keep the TDS below the level at which foaming and other problems are likely to occur. Refer to *Table 4* for the limits of total and suspended solids recommended by the American Boiler Manufacturers Association (ABMA) for different boiler operating pressures. Most steam systems operate at a low pressure, well within the 0 to 300 psig range shown in the first row of *Table 4*.

TABLE 4 Maximum Allowable Total and Suspended Solids in Boiler Water

Boiler Pressure	TDS	Suspended Solids
0 to 300 psig	3,500 ppm	300 ppm
301 to 450 psig	3,000 ppm	250 ppm
451 to 600 psig	2,500 ppm	150 ppm
601 to 750 psig	2,000 ppm	60 ppm
751 to 900 psig	1,500 ppm	40 ppm

Boilers are normally equipped with two kinds of blowdown valves: a bottom blowdown valve and a surface blowdown valve. Surface blowdown is also referred to as *skimming*. The bottom blowdown valve is located at the lowest part of the boiler. It is used to drain water carrying sediment, scale, and other debris that accumulates at the bottom of the boiler.

The skimming valve is connected to a pipe protruding into the boiler just below the normal water level. Skimming often requires a controlled and continuous draining of surface water to remove foam and the impurities it contains. The skimming water flow rate is controlled manually using the valve. Periodic adjustments of the valve must be made to increase or decrease the flowrate, as determined by monitoring the TDS. Note that larger boilers may be equipped with more than one blowdown valve at the bottom and at the surface. Smaller boilers may not be equipped with a skimming valve.

Always perform skimming and blowdown tasks as directed in the boiler manufacturer's service instructions. To review blowdown and skimming procedures in greater detail, refer to NCCER Module 03306, *Steam Systems*.

CAUTION

Typical plumbing codes require that the temperature of water entering a drainage system be 140°F (60°C) or less.

3.0.0 Section Review

1. Silt deposits in open recirculating water systems are commonly controlled by adding ______.
   a. algae
   b. polymers
   c. phosphates
   d. makeup water

2. True or False? Sodium nitrite is commonly used as an inhibitor in closed loop systems.
   a. True
   b. False

3. Where does a form of corrosion called grooving typically occur?
   a. Around the circumference of copper tubing
   b. Where dissimilar metals contact each other
   c. Wherever water is moving at an extreme velocity
   d. Along the seam of electric resistance welded (ERW) steel pipe

Module 03308 Review Questions

1. Water with a pH of 9 is considered ______.
   a. acidic
   b. neutral
   c. alkaline
   d. very acidic

2. As the hardness of water decreases, the potential for scaling ______.
   a. increases
   b. decreases
   c. fluctuates
   d. remains the same

3. What type of corrosion results from electrolysis?
   a. Grooving
   b. Embrittlement
   c. Pitting corrosion
   d. Galvanic corrosion

4. An excessive amount of suspended solids in water is likely to result in ______.
   a. fouling
   b. scaling
   c. electrolysis
   d. caustic cracking

5. Which of the following is *true* regarding the use of water sampling containers that contain a fixing compound?
   a. Fill the container with the sample until it overflows.
   b. Always preheat the fixing compound before adding the water sample.
   c. Never empty or rinse the container before adding the water sample.
   d. A wax seal must be used to seal the container after adding the water sample.

6. True or False? Unless documentation indicates otherwise, all water treatment chemicals should be handled as if they are hazardous.
   a. True
   b. False

7. Compare a dense-media cartridge filter to a cartridge filter that captures debris only on the surface. Which of the following is *true*?
   a. Surface-capture filters are more efficient at removal.
   b. Surface-capture filters generally handle higher flow rates.
   c. Dense-media filters are never filled and can be used indefinitely.
   d. Dense-media filters offer less resistance to flow than surface-capture filters.

8. Centrifugal separators are used to remove _____.
   a. the tiniest particles from water
   b. water and impurities from steam
   c. larger suspended particles from water
   d. entrained gases during boiler blowdown

9. The zeolite bed in a typical water softener _____.
   a. distills minerals from hard water
   b. is regenerated by backwashing
   c. must be replaced when its sodium charge is depleted
   d. absorbs calcium and magnesium while releasing sodium

10. Deaerators are used to remove air and other gases from _____.
    a. cooling towers
    b. hot-water systems
    c. chilled-water systems
    d. steam system feedwater

11. Algae is more likely to be found in _____.
    a. steam piping
    b. cooling towers
    c. hot-water heating systems
    d. chilled-water cooling systems

12. The use of chromates for corrosion control in cooling tower systems is _____.
    a. the preferred and safest approach for all towers
    b. prohibited in cooling towers serving comfort systems
    c. prohibited in cooling towers serving industrial processes
    d. acceptable in all towers when prescribed by a certified chemist

13. True or False? Scale formation is more likely to be a problem in a closed hot-water system than in a closed chilled-water system.
    a. True
    b. False

14. True or False? The water in a steam boiler should be kept slightly acidic to help prevent corrosion.
    a. True
    b. False

15. The volume of water removed from a steam boiler during blowdown is normally determined by the ______.
    a. pH of the water
    b. water hardness
    c. volume of fresh makeup water added
    d. concentration of total dissolved solids

Answers to odd-numbered questions are found in the Module Review Answer Key at the back of this book.

Answers to Section Review Questions

Answer	Section	Objective
Section One		
1. d	1.1.0	1a
2. c	1.2.2	1b
3. a	1.3.2	1c
4. b	1.4.0	1d
Section Two		
1. a	2.1.2	2a
2. d	2.2.3	2b
Section Three		
1. b	3.1.2	3a
2. a	3.2.1	3b
3. d	3.3.1	3c

MODULE 03404

Energy Conservation Equipment

Source: ronstik/iStock/Getty Images

Objectives

Successful completion of this module prepares you to do the following:

1. Identify and describe the operation of various HVACR-related energy recovery and reclamation systems.
    a. Identify and describe the operation of energy- and heat-recovery ventilators.
    b. Identify and describe the operation of condenser heat-recovery systems.
    c. Identify and describe the operation of coil energy-recovery loops.
    d. Identify and describe the operation of heat-pipe heat exchangers.
    e. Identify and describe the operation of waterside economizers.
    f. Identify and describe the operation of flue-gas heat-recovery systems.
    g. Identify and describe the operation of steam heat-recovery systems.
    h. Identify and describe the operation of ice storage systems.

Performance Tasks

This is a knowledge-based module. There are no Performance Tasks.

Overview

Because HVACR systems consume a great deal of energy, these systems are primary targets in the battle for energy conservation. Beyond the research and development that has resulted in HVACR systems that consume far less energy than the predecessors, manufacturers have developed innovative devices that recover and use energy that would otherwise be wasted. HVACR installers and service technicians need to be aware of these devices, how the devices affect the related HVACR systems, and how they contribute to energy conservation.

NCCER Industry-Recognized Credentials

If you are training through an NCCER-accredited sponsor, you may be eligible for credentials from NCCER. The ID number for this module is 03404. Note that this module may have been used in other NCCER curricula and may apply to other level completions. Contact NCCER at 1.888.622.3720 or go to **www.nccer.org** for more information.

You can also show off your industry-recognized credentials online with NCCER's digital credentials. Transform your knowledge, skills, and achievements into credentials that you can share across social media platforms, send to your network, and add to your resume. For more information, visit **www.nccer.org**.

Digital Resources for HVACR

Scan this code using the camera on your phone or mobile device to view the digital resources related to this craft.

SCAN ME

1.0.0 Energy Recovery and Reclamation

Performance Tasks

There are no Performance Tasks in this section.

Objective

Identify and describe the operation of various HVACR-related energy recovery and reclamation systems.

a. Identify and describe the operation of energy- and heat-recovery ventilators.
b. Identify and describe the operation of condenser heat-recovery systems.
c. Identify and describe the operation of coil energy-recovery loops.
d. Identify and describe the operation of heat-pipe heat exchangers.
e. Identify and describe the operation of waterside economizers.
f. Identify and describe the operation of flue-gas heat-recovery systems.
g. Identify and describe the operation of steam heat-recovery systems.
h. Identify and describe the operation of ice storage systems.

The high cost of energy, the need to conserve energy, and government-mandated efficiency standards are all factors that have increased the use of heat recovery and other energy-saving devices in HVACR systems. Heat-recovery devices typically save energy through the capture and reuse of heat that would otherwise be wasted. Other energy-saving devices affect the operation of an HVACR system in a way that increases the efficiency of the connected system(s).

The focus of this module is to introduce some common components and systems used in HVACR systems to conserve or recover energy. Put simply, heat-recovery equipment captures heat that would otherwise be wasted and uses it for a secondary purpose. Many kinds of heat-recovery devices and processes are in use, including these presented in this module:

- Energy and heat-recovery ventilators and wheels
- Condenser heat-recovery systems
- HVAC coil energy-recovery loops
- Heat-pipe thermosiphon heat exchangers
- Waterside economizers
- Flue-gas heat-recovery systems
- Steam heat-recovery systems
- Thermal (ice) storage systems

1.1.0 Energy- and Heat-Recovery Ventilators

To maintain good indoor air quality, the American Society of Heating, Refrigerating and Air Conditioning Engineers (ASHRAE) standards recommend that a home's indoor air be exchanged for fresh outdoor air at a rate of 0.35 air changes per hour (ACH). An alternative ASHRAE recommendation calls for an exchange rate of 15 cubic feet per minute (cfm) per person, 20 cfm per bathroom, and 25 cfm per kitchen. Recovery ventilators are used to bring a controlled amount of outside air into a space, exhaust a comparable volume of inside air, and facilitate an exchange of heat energy between the two airstreams.

Two types of recovery ventilators are energy-recovery ventilators (ERVs) and heat-recovery ventilators (HRVs). ERVs supply fresh air, exhaust stale indoor air, and exchange both sensible heat and latent heat (heat contained in water vapor) between the two airstreams. HRVs are similar in construction and operation to ERVs, but HRVs are designed to transfer only sensible heat between the airstreams. This makes ERVs popular in warm, humid climates, such as in the southern United States. HRVs are typically applied in colder climates only.

Air from the occupied space passes through the ERV or HRV unit core and is exhausted outside (*Figure 1*). At the same time, fresh air is brought in and passes through the core in the opposite direction.

(A)

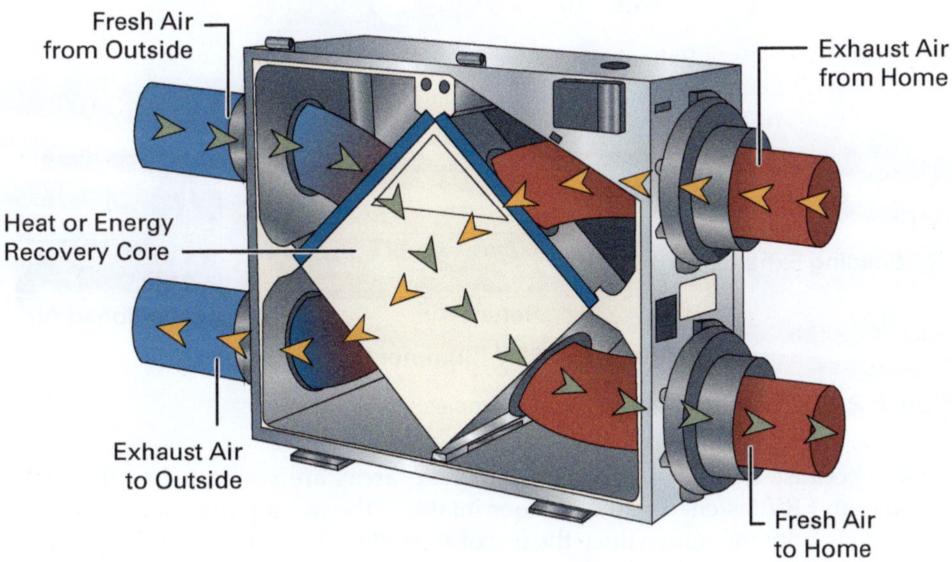

(B)

Figure 1 Energy-recovery ventilator (ERV).
Source: Image courtesy of RenewAire LLC (A).

When the two airstreams pass through the heat exchanger core, heat is transferred from the warmer airstream to the cooler airstream. The core design allows this heat transfer while maintaining separation between the airstreams; the two airstreams do not mix.

The result is a constant stream of fresh air delivered to the occupied space that has been tempered, reducing the load on the building's HVAC systems. According to the US Department of Energy, most recovery ventilators can transfer about 60% to 80% of the heat energy between the airstreams, which is used to precondition the incoming fresh air.

Some commercial/industrial air conditioning systems use an ERV unit in conjunction with a packaged rooftop air conditioning unit (*Figure* 2). The ERV may be a stand-alone unit, as shown in *Figure 2A*. Alternatively, the ERV is fastened directly over the outdoor intake of the rooftop unit (*Figure 2B*).

One of the benefits of a stand-alone ERV is that it allows a traditional outside-air economizer to remain attached to the rooftop unit, as shown in *Figure 2A*. When the economizer is active, the rooftop unit typically keeps the ERV exhaust fan on, but the supply fan and recovery wheel are shut down. All outside air, which is already cool enough to cool the building, enters through the economizer.

Another benefit of a stand-alone unit is that it can eliminate the need for separate exhaust fans to serve areas such as the building restrooms and small kitchens.

On Site

Fume Hood Exhaust Heat Recovery

Laboratories often require 100% makeup air for workstations where fume hood exhaust must be prevented from becoming re-entrained into the building's ventilation system. The makeup air generally needs to be preheated or precooled.

At its 20,000-square-foot facility for chemical research in Branford, CT, Neurogen Research Foundation had Tri-Stack® heat-recovery systems from Strobic Air Technologies installed to help cut energy costs. In these systems, heat-recovery coils filled with a solution of glycol and water extract heat from a workstation fume-hood exhaust before it is discharged. These systems increased the intake air temperature by 10°F (5.6°C), reducing the need to use natural gas to preheat the incoming makeup air to a comfortable temperature. As a result, a 30% reduction in energy costs was achieved.

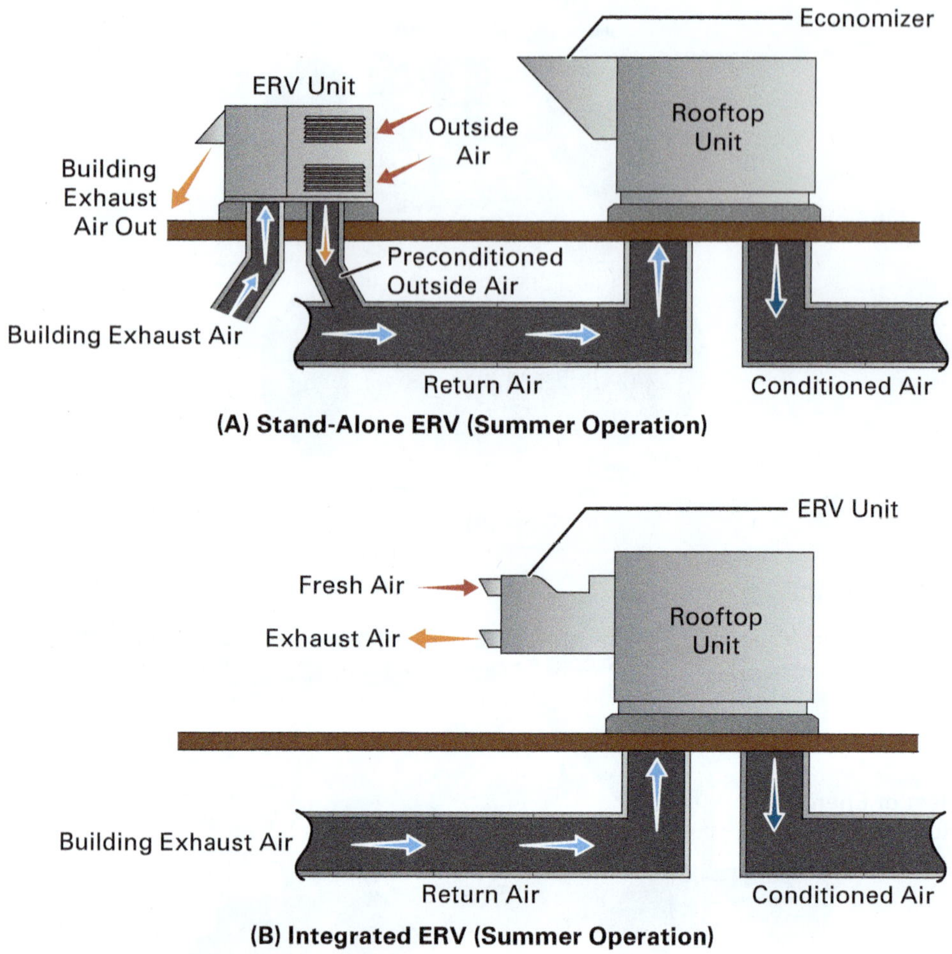

Figure 2 ERV commercial application examples.

This is because the exhaust ducts from these areas are connected to the ERV. Because an ERV fastened to the outdoor intake of the rooftop unit can only draw exhaust air from the return duct, the use of separate exhaust fans is still required.

When performing periodic maintenance on ERV and HRV equipment, always refer to the manufacturer's manual and instructions. The equipment typically requires periodic inspection and maintenance as follows:

- Inspect and clean or replace the intake and exhaust filters or insect/bird screens.
- Inspect and service the motors, belts (if any), and drive mechanisms.
- Inspect and clean the exchanger cores periodically. Annual cleaning is often recommended, but the frequency depends largely on the quality of filtration and the operating hours. Some cores may require removal for thorough cleaning.
- Restart the equipment and confirm proper operation after completing any maintenance.

1.1.1 Static-Plate Heat Exchangers

Static-plate heat exchangers (*Figure 3*) are commonly used in smaller ERV units. The heat exchanger core has no moving parts. The heat exchanger consists of alternately layered plates, separated and sealed, that form exhaust and supply airstream passages. Sensible heat is transferred through the plates from the warmer airstream to the cooler airstream, while latent heat is transferred by an exchange of water vapor through the unique plates.

The direction of the supply and exhaust airflow through the heat exchanger can be parallel, counterflow, or crossflow. The heat exchanger shown in *Figure 3* has a crossflow configuration. In a parallel configuration, both airstreams move in the same direction. While this does work, it is usually far more practical to design equipment with a crossflow or counterflow configuration.

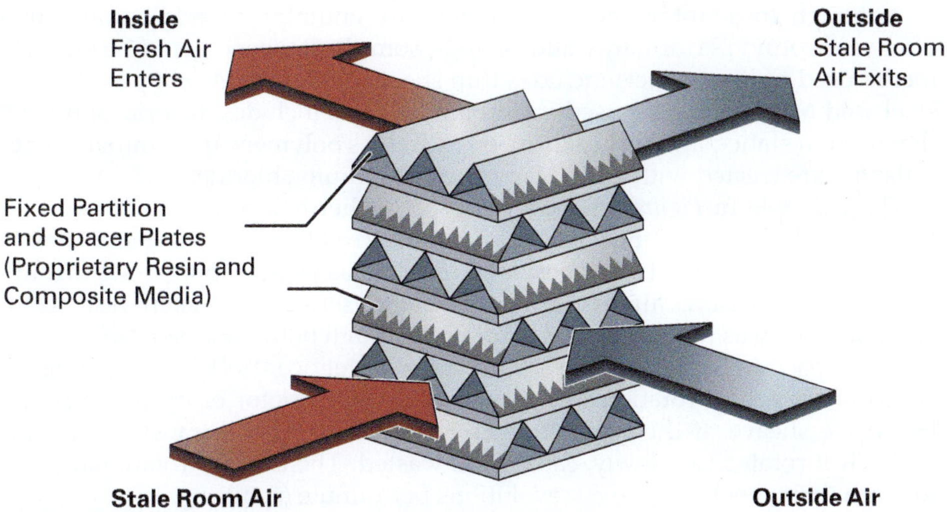

Figure 3 Static-plate heat exchanger core construction.
Source: Image courtesy of RenewAire LLC

The layered plates that separate the airstreams are often made from an engineered resin material that transfers sensible heat by conduction and latent heat by passing water vapor from one airstream to the other. Another approach is to use a water-permeable paper to transfer water vapor and the latent heat it contents.

1.1.2 Rotary Heat Exchangers

A rotary air-to-air heat exchanger is more often referred to as an *enthalpy wheel* or a *heat wheel*. **Enthalpy**, you may recall, refers to the total heat in a substance, including both latent heat and sensible heat. An enthalpy wheel can transfer both sensible and latent heat, while a heat wheel only transfers sensible heat.

Enthalpy: The total heat content (sensible and latent) of a refrigerant or other substance.

An enthalpy wheel or heat wheel consists of a motor-driven (usually via a drive belt) revolving disk or drum containing heat transfer media through which the airstreams pass (*Figure 4*). The supply and exhaust airstreams each flow through half of the wheel, usually in a counterflow arrangement with the two airstreams parallel to each other but flowing in opposite directions (*Figure 5*). A gasketed partition separates the two airstreams.

Figure 4 Rotary air-to-air heat exchanger (enthalpy wheel).
Source: Airxchange, Inc.

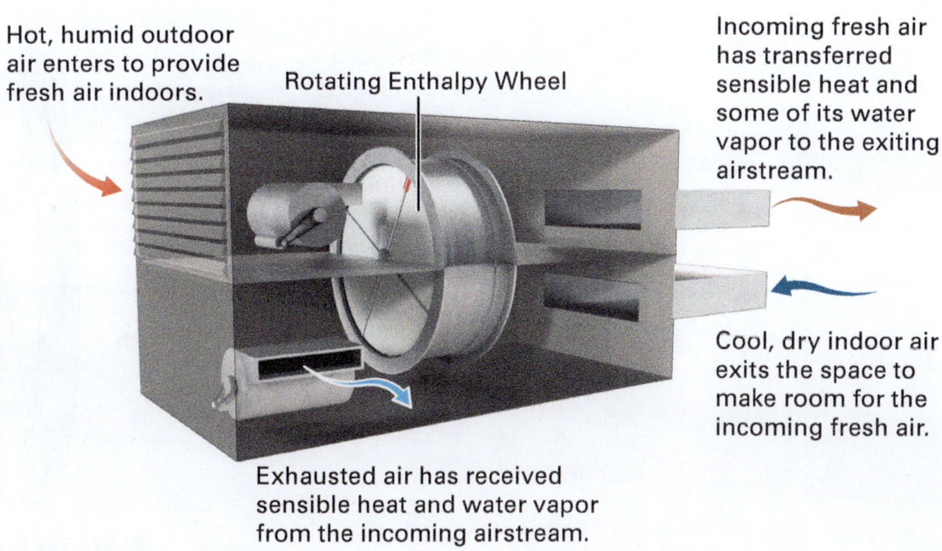

Figure 5 Counterflow air movement through an enthalpy wheel.

Monel®: An alloy that is very resistant to corrosion, made from nickel, copper, iron, manganese, silicon, and carbon.

Although most rotary heat exchangers are enthalpy wheels because they offer year-round performance and savings, some heat wheels are in service. The media used in heat wheels includes thin sheets of aluminum, copper, stainless steel, and **Monel®**. Media used in enthalpy wheels includes material similar to that used in static-plate heat exchangers as well as polymers and synthetic materials that are treated with a desiccant, such as lithium chloride.

The example in *Figure 6* shows the type of thermal performance expected from an enthalpy wheel. Sensible heat is transferred from the warmer airstream to the cooler airstream. Latent heat is transferred as water vapor passes through the media to the dryer airstream. The transfer of sensible and latent heat occurs simultaneously as the wheel slowly rotates through both airstreams.

The capacity of rotary heat exchangers is largely controlled by varying the speed of the wheel rotation using a variable-speed motor or by changing the belt drive sheave. If the wheel rotates too quickly, energy transfer is diminished. If it rotates too slowly, capacity is wasted. The average rotational speed for enthalpy wheels is 20 to 50 revolutions per minute (rpm).

At low outdoor temperature, frost can develop on a wheel. However, the temperature at which frost will usually begin to accumulate on the surface is generally 20°F to 30°F (11°C to 17°C) lower than the temperature at which frost develops on a typical static-plate heat exchanger. Therefore, enthalpy wheels can be used without frost protection in many areas where static-plate heat exchangers require protection.

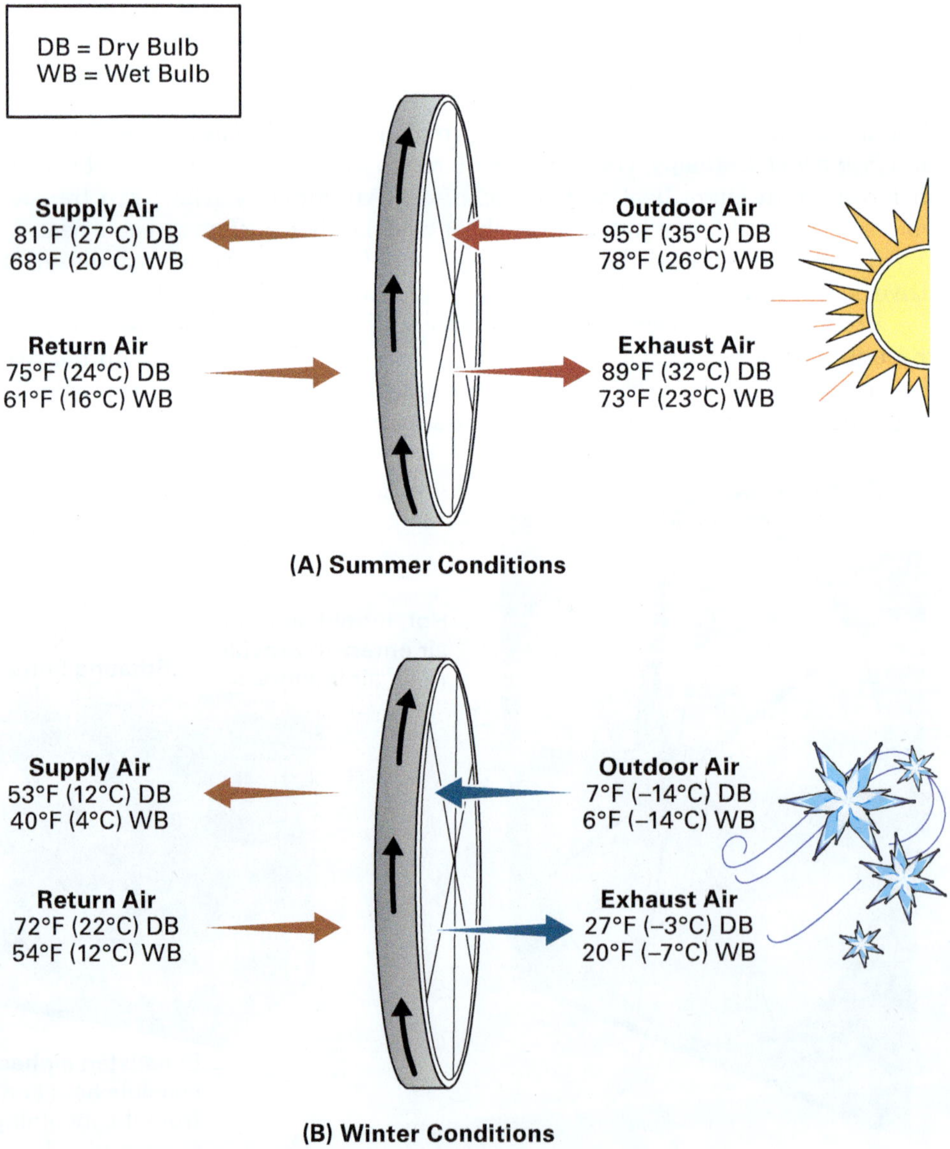

Figure 6 Enthalpy wheel thermal performance.
Source: Airxchange, Inc.

Rotary heat exchangers should be maintained as directed by the manufacturer. The wheel media must be periodically cleaned. Maintaining good air filtration is essential to avoid clogging the wheel with dust and debris. Media treated with desiccants to aid in the transfer of water vapor must not be wetted with cleaning agents.

The media in most rotary heat exchangers is assembled in sections. Each section can be removed and cleaned, as opposed to removing the entire wheel at once.

1.2.0 Condenser Heat-Recovery Systems

A great deal of heat is rejected through HVACR condensers. Rejected condenser heat can be a valuable resource in large commercial and industrial systems. The recovered heat can be used in a variety of ways.

Essentially, a second air- or water-cooled condenser is used to extract heat from the hot refrigerant vapor before it enters the primary condenser where heat is rejected to the outside air or a water source. The recovered heat is then transferred to air or water that is used for another productive purpose.

1.2.1 Air Conditioning/Refrigeration System Condenser Heat Recovery

A dual-condenser refrigeration system that might be used in larger supermarkets is shown in *Figure 7*. To maintain the many refrigerated fixtures, the refrigeration

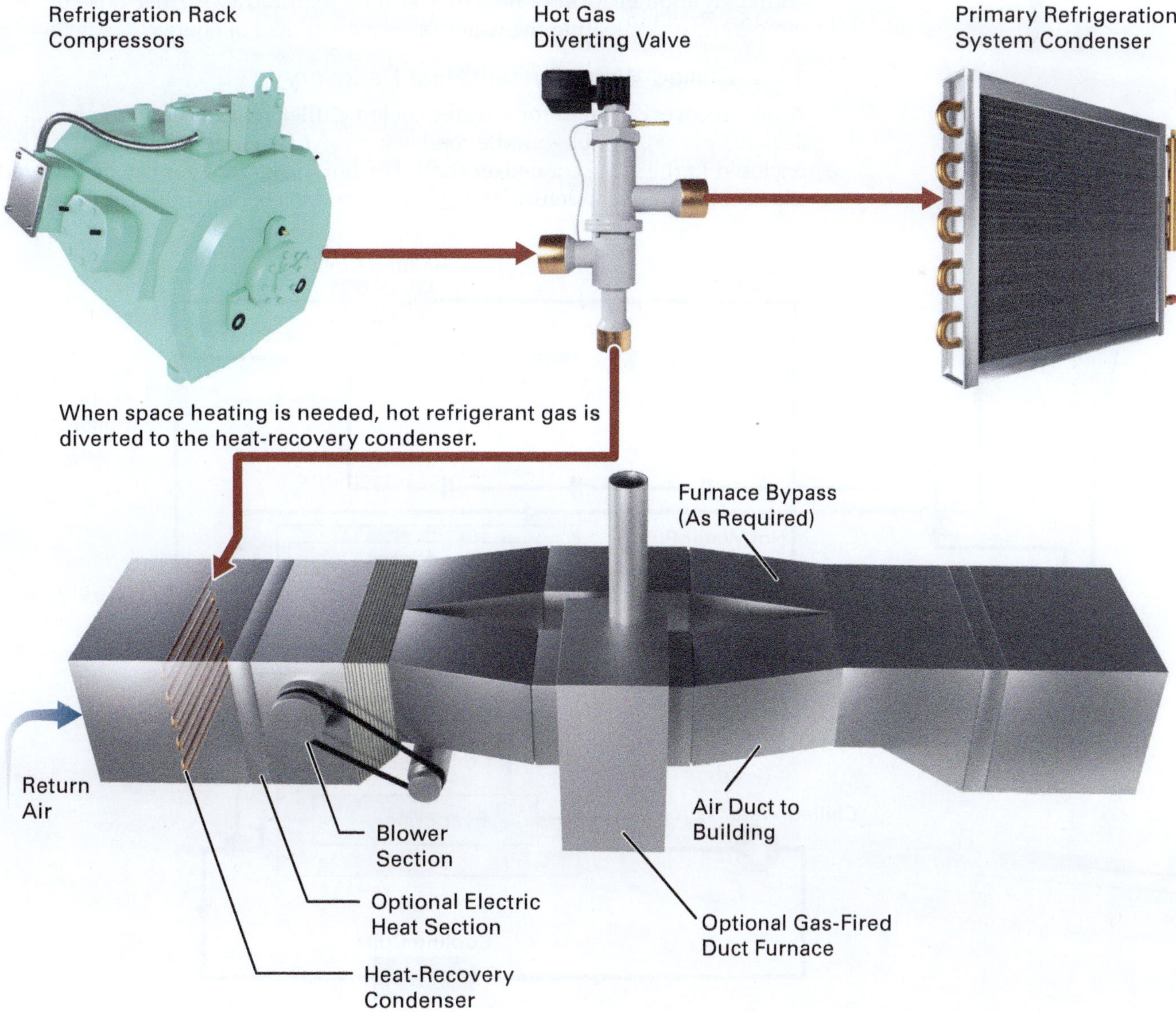

Figure 7 Dual-condenser system for refrigeration-system heat recovery.

system must run constantly year-round. The open display cases also tend to cool the store, and in many cases, the store requires some heat to maintain reasonable comfort. Rather than use more energy (electric or gas) to heat the store, rejected heat from the refrigeration systems can be used to provide space heating.

When the building thermostat calls for heat, a hot-gas diverting valve changes position and routes the refrigeration-system hot gas through the recovery condenser inside the air handling unit. Building air is now serving as condenser air for the refrigeration system. As building air flows through the recovery condenser, the heat rejected from the condenser warms the air. The heated air is then distributed throughout the building or to needy zones.

When the heating demand is greater than that supplied by the recovery condenser, the difference can be made up by optional gas or electric heaters. When the building heating requirement is satisfied, the hot-gas diverting valve directs the hot gas back to the primary air-cooled condenser.

In *Figure 7*, note the bypass duct around the gas-fired duct furnace. When the system is in the cooling mode, with cold, saturated air leaving the evaporator, it can lead to premature corrosion in the duct furnace. For that reason, air is typically bypassed around a duct furnace using motorized dampers (not shown) in the cooling mode.

The heat rejected by refrigeration systems can also be used to heat water. In this case, the recovery condenser is water-cooled. This arrangement can provide preheated water needed for cleaning, dishwashing, and other common tasks. Although another form of heat may still be required to further raise the water temperature, preheating the water saves a great deal of energy.

1.2.2 Chilled-Water System Heat Recovery

A heat-recovery system for a water-cooled chiller relies on a *double-bundle condenser* (*Figure 8*). Double-bundle condensers have two independent water circuits enclosed in the same condenser shell. The hot refrigerant gas can be directed to the heat-recovery condenser, the primary condenser, or to both condensers.

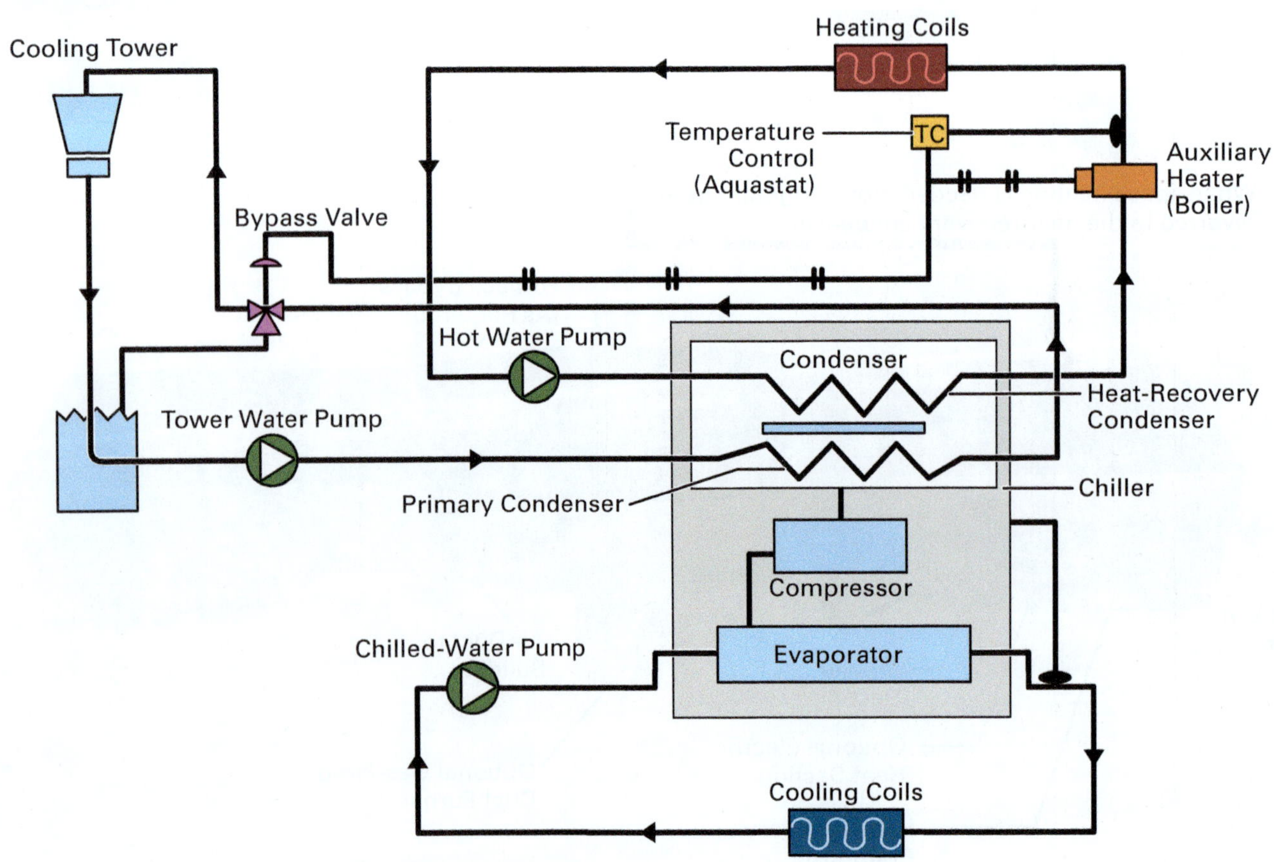

Figure 8 Double-bundle heat-recovery system.

The heat-recovery condenser is part of the building hydronic heating circuit. When heat generated from the chiller exceeds the building needs, and when no heat is required, the bypass valve modulates as needed to route water through the cooling tower and maintain the proper chiller head pressure.

When some portion of the building requires heat and the chiller is also running, the building heating system extracts heat from the heat-recovery condenser, condensing the refrigerant in the process. The bypass valve reduces water flow through the primary condenser as it diverts some or all of the heat to the building heating system. Note that it is not unusual for both heating and cooling to be required at the same time in large buildings, especially multistory buildings.

When the building heating load exceeds the heat recovered from the chiller, or when the chiller is not running, the auxiliary heater (boiler) cycles on and heats the building water as necessary. In the summer, when little or no comfort heating is needed, the recovered heat may be used for another purpose. Where practical, such as in hotels and medical facilities with significant hot water demands, a third heat exchanger (condenser) can be added to preheat the domestic water supply or even a process-related liquid.

To summarize, the chiller is providing chilled water to cool some part of a building or industrial process, while also recovering heat to warm another part of the building or heat water and other fluids.

The system is controlled by a water temperature controller that interfaces with the tower bypass valve and auxiliary heat source. Although such a system can be controlled using traditional electromechanical controls, programmable controllers are far more accurate, flexible, and reliable.

Did You Know?

Evaporative Precoolers for Air-Cooled Condensers

Given rising outdoor temperatures and seemingly endless heat waves, summer can challenge air-cooled condensers to deliver maximum performance. This is especially true in buildings that have been expanded without increasing cooling capacity.

Especially in hot, dry regions, evaporative precoolers can be used to cool the outside air before it enters the condenser coil. Cooler air entering the condenser causes the head pressure to drop and improves subcooling, increasing the system cooling capacity and reducing power consumption.

Evaporative precoolers consist of one or more panels that contain absorbent pads. Tubing connects a water source to a header at the top of each panel. The precoolers are positioned on the entering side of the condenser coil. Water, controlled by a solenoid valve, is applied from the top to wet the pads. The air being drawn across the condenser coil first passes through the precooler, providing significantly cooler condenser air.

Like all evaporative coolers, precoolers are less effective as the relative humidity increases, and precoolers can consume a significant amount of water in hot, dry weather.

Source: Courtesy of Peak+

1.2.3 Swimming Pool Heat-Recovery Systems

Indoor swimming pools provide an excellent opportunity for heat recovery. Specialized high-capacity dehumidification systems, such as the one shown in *Figure 9*, are specifically designed for these wet environments.

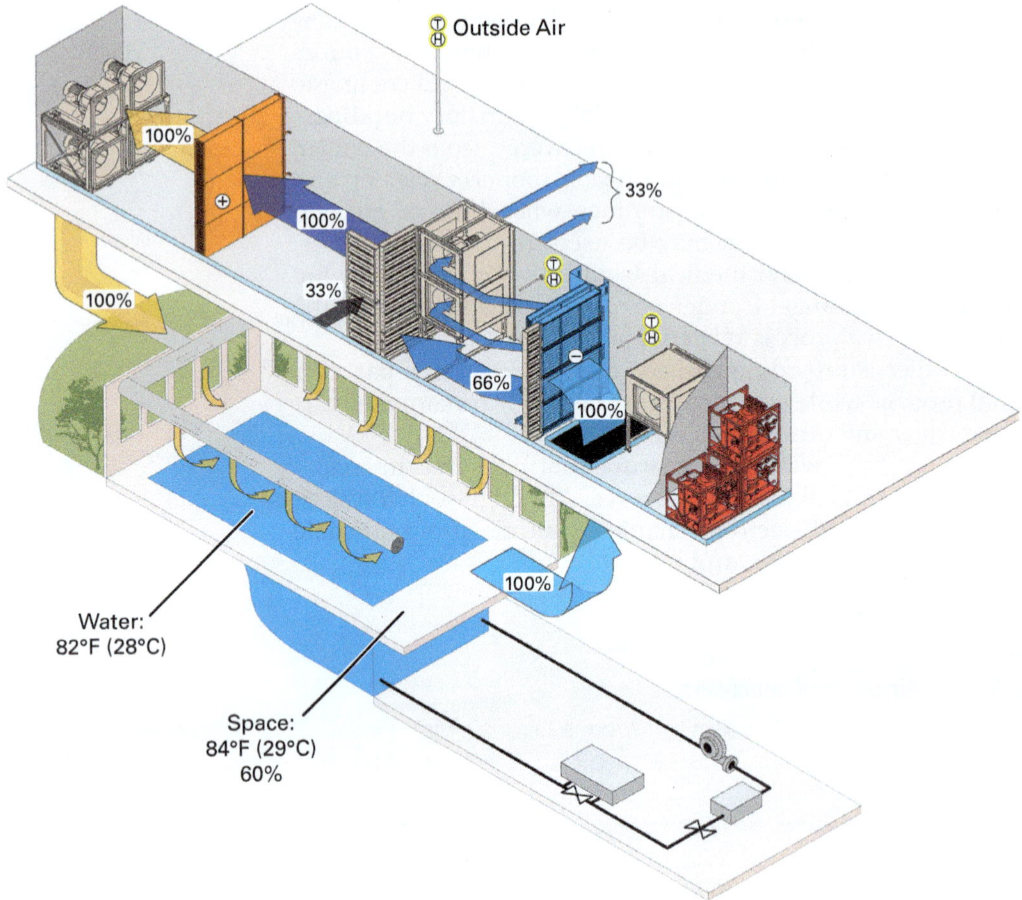

Figure 9 Swimming pool heat-recovery system installation.
Source: PoolPak LLC

The system primarily cools and dehumidifies air in the pool enclosure while the rejected heat is used to heat the pool water. However, these systems have an array of operating modes depending on the conditions and the season.

As shown in *Figure 10*, the hot discharge vapor leaving the compressor can be routed to the air- or water-cooled primary condenser, the condenser reheat coil, the pool water condenser, or even an optional auxiliary condenser that uses the heat for another purpose. The pool water temperature is normally maintained by the recovered heat. Microprocessor-based controls activate the solenoid valves based on where recovered heat can be best used. If there is no demand for space or pool heating, the hot refrigerant gas is routed to the primary condenser. A separate heater keeps the pool water warm when the system compressor is not running.

Subcooled liquid refrigerant leaving all the condensers is stored in the receiver. The refrigerant is fed through the expansion valve to the evaporator coil. The evaporator coil tends to be thick, as dehumidification is an essential function. Thicker coils allow the water vapor more time to condense as the air passes through. *Figure 11* shows the physical arrangement of the unit.

Other environments in which this dehumidification and heat-recovery strategy can be used include museums, printing facilities, specialized warehouses, plywood manufacturing facilities, and pharmaceutical plants. All of these represent applications where humidity control is often critical.

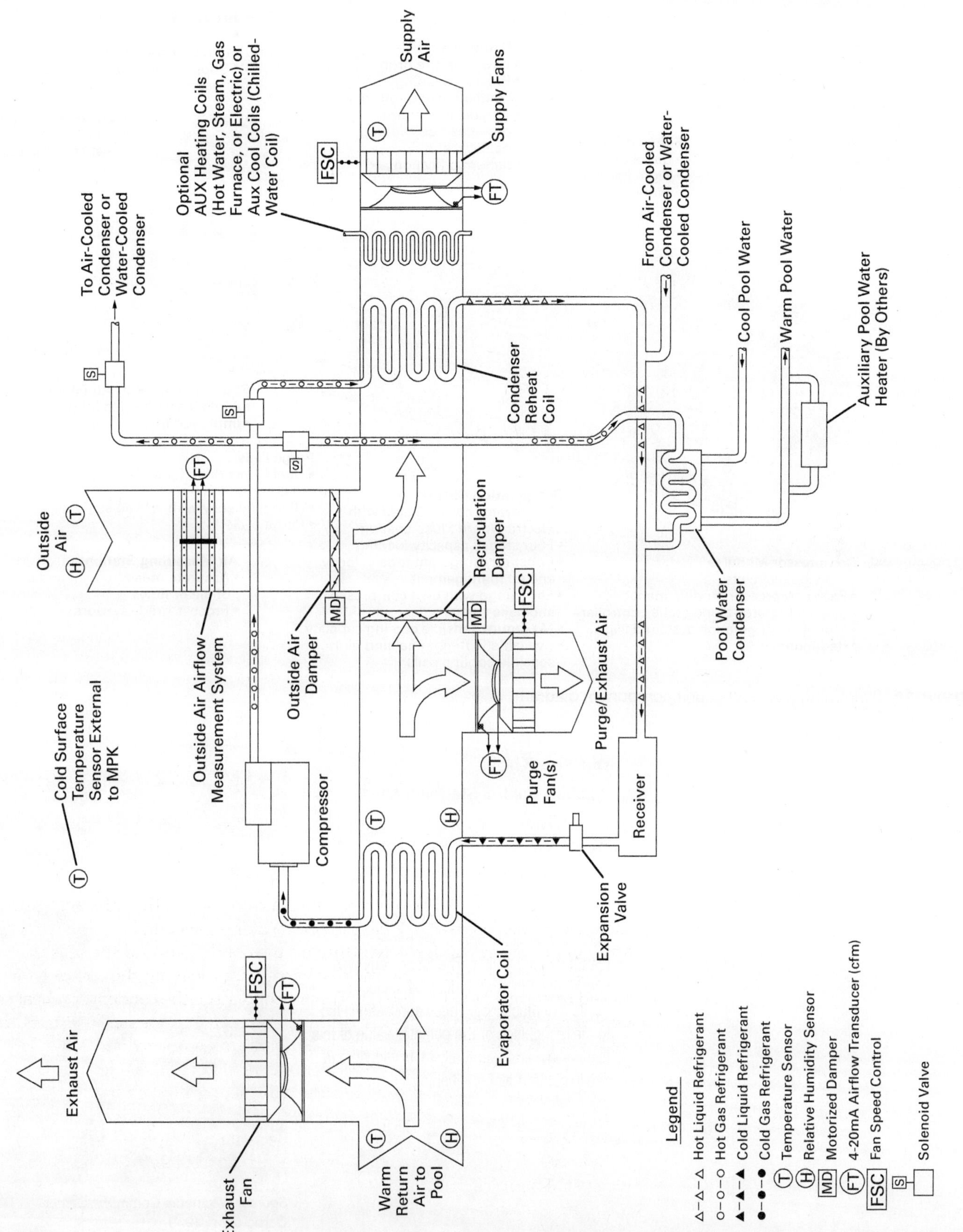

Figure 10 Swimming pool HVAC unit schematic.
Source: PoolPak LLC

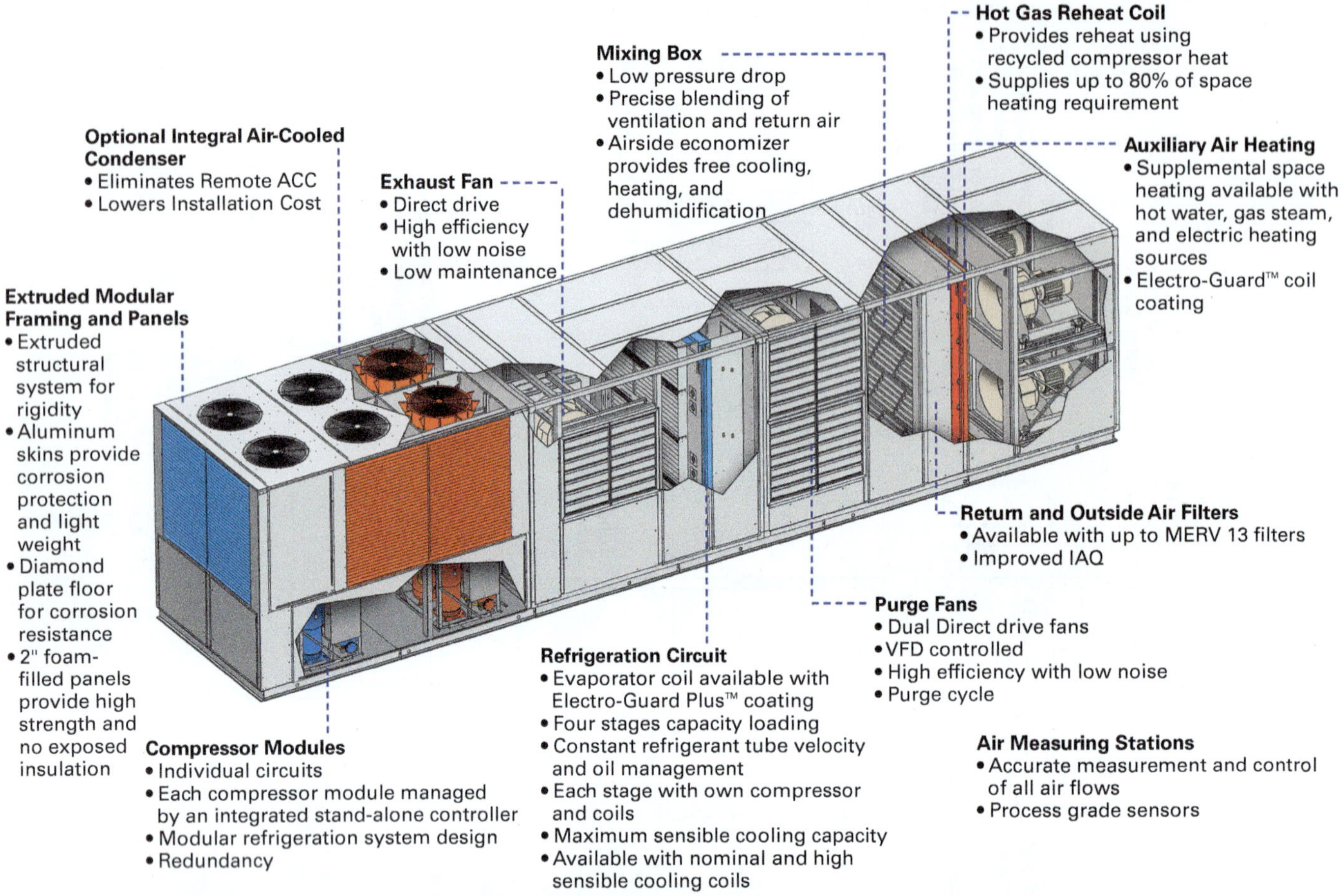

Figure 11 Swimming pool HVAC unit component diagram.
Source: PoolPak LLC

Did You Know?

Humidity Control Started It All

You may have noticed that some paper tends to curl in hot, humid weather. This fact led to the invention of modern air conditioning. The air conditioning system designed by Dr. Willis Carrier in 1902 was developed specifically to control humidity and cool a paper manufacturing facility. The technology was later applied to other industries, including textile manufacturing. These applications led to the recognition of the potential value of this technology, and eventually to its widespread use in comfort air conditioning.

Source: Wikimedia Commons/Carrier Corporation, 1915

1.3.0 Coil Energy-Recovery Loops

In many manufacturing and industrial environments, stale or contaminated air must be constantly exhausted and fresh air brought in to replace it. The volume of air that must be exhausted and replaced in these applications can be massive—many times more than the small amount of air brought in to satisfy fresh air requirements in commercial buildings.

The exhausted air contains a lot of energy, whether it has been heated or cooled. That energy can be used to temper the fresh air as it enters. This simple arrangement is referred to as an *energy-recovery loop,* and more casually as a **runaround loop**.

Runaround loop: A closed-loop energy-recovery system in which finned-tube water coils in supply and exhaust airstreams, connected by piping, exchange heat.

The typical runaround loop consists of two finned-tube water coils—one in the exhaust duct and one in the makeup air duct (*Figure 12*). The coils are connected by a piping loop with fluid flow provided by a pump. A thermostatically controlled three-way mixing valve controls the fluid flow path.

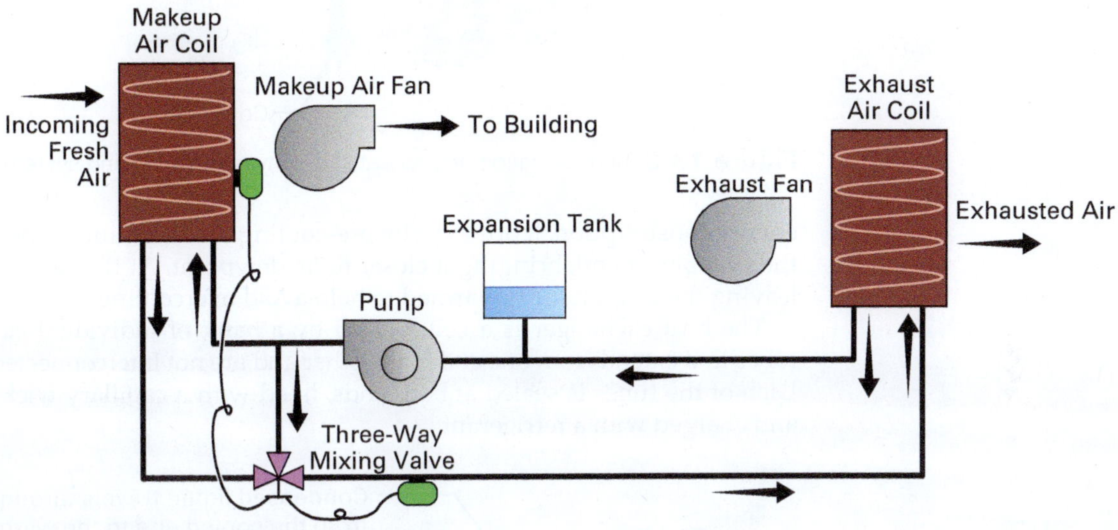

Figure 12 Runaround loop.

Sensible heat is transferred between the exhaust and incoming supply airstreams without any cross contamination. The system heats or cools the incoming air, depending on the conditions. The recovery efficiency in the winter for runaround loops averages between 40% and 65%, but it largely depends on the outdoor temperature in the winter. The efficiency is lower in the summer because the temperature difference between the two airstreams is not as high.

A glycol solution is often used in the loop to prevent freezing. However, moisture must not be allowed to freeze on the coil face either, as it would block airflow. The three-way mixing valve prevents the exhaust coil from developing frost due to extremely cold conditions outdoors. This valve is controlled to maintain the water entering the exhaust coil at a temperature no less than 32°F (0°C). Bypassing some of the warmer solution around the supply air coil ensures the exhaust coil remains frost-free.

1.4.0 Heat-Pipe Heat Exchangers

In a conventional air conditioning system, dehumidification of the air occurs at the system's evaporator coil. The cold coil, below the dewpoint of the indoor air, removes both sensible and latent heat as water condenses on the coil face.

When the dehumidification load is very high, it is difficult to remove the moisture without overcooling the space. An air conditioner needs to operate longer to remove large amounts of moisture in a hot, humid environment; therefore, it consumes more energy. One way to avoid this is to reheat the air with an electric heater or a hot-water coil before it enters the space (*Figure 13*). However, this also results in a significant amount of wasted energy, because energy is being used to both cool and heat the air.

A heat-pipe heat exchanger is a thermosiphon device that can address this problem. It increases the dehumidification capacity of a system while reducing

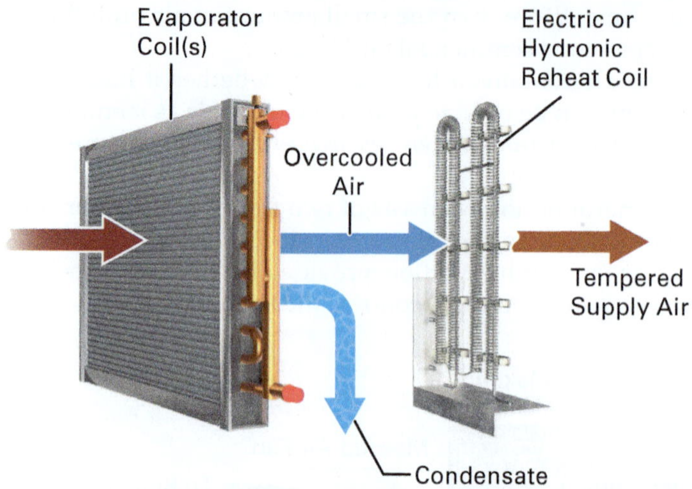

Figure 13 Dehumidification and reheat in a conventional cooling system.

energy consumption. It does this by pre-cooling the incoming air before it enters the evaporator coil, bringing it closer to its dewpoint. At the same time, the air leaving the evaporator is warmed to help avoid overcooling the space.

The heat exchanger is a coil formed by a bank of individual closed copper tubes that extend from one end to the other and are not interconnected (*Figure 14*). Each of the tubes is sealed at both ends, lined with a capillary wick, evacuated, and charged with a refrigerant.

Figure 14 Heat-pipe heat exchanger tube construction.

When the coil assembly, comprised of a number of individual heat pipes, is mounted in the system ductwork, half of the coil is exposed to the return airstream while the other half is exposed to the supply airstream. The end of each heat pipe acts as an evaporator in the return air duct, while the opposite end acts as a condenser in the supply air duct (*Figure 15*).

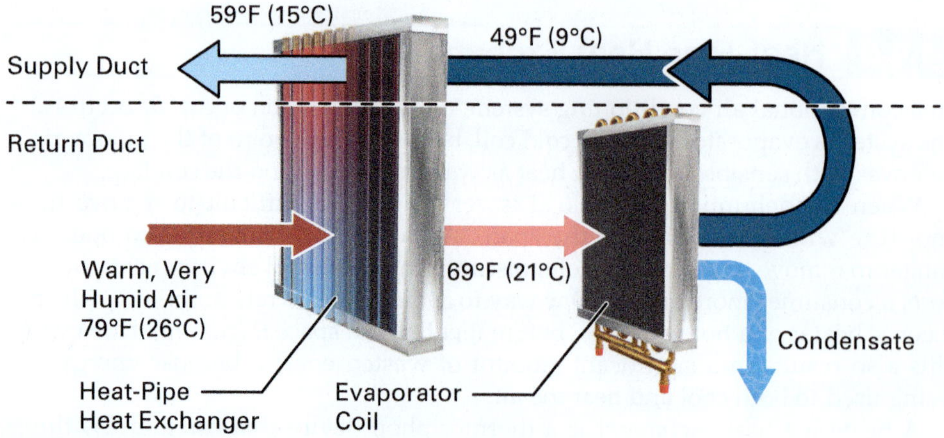

Figure 15 Heat-pipe heat exchanger in operation.

During operation, the evaporator side of the heat pipes are exposed to the warm return airstream. The refrigerant vaporizes, absorbing heat from the air, and the resulting vapor naturally migrates through the center of the tube to the cooler end of the pipe. As the refrigerant vapor inside reaches the opposite end and is exposed to the cool airstream, the refrigerant condenses, transferring heat to the air. The condensed refrigerant travels through the wick back to the warmer end of the coil. Note that the ends of the tubes in the return airstream need to be lower than the opposite end, allowing gravity to assist in the movement of the liquid.

This evaporating and condensing process continues as long as there is a sufficient temperature difference between the two airstreams. There is nothing to turn on or off. As a result, heat-pipe exchangers require little maintenance beyond ensuring the coil remains clean. The fins are typically set farther apart than on a traditional evaporator or condenser coil. The tubes are sealed and will likely remain leak-free. Since the tubes are not connected to each other, the rare loss of refrigerant from one tube does not affect the performance of the others.

Manufacturers recommend testing the performance of the heat exchanger tubes annually to ensure their thermal performance. Tubes that have lost their refrigerant charge can generally be replaced.

Another configuration of heat pipe coils is the wrap-around style (*Figure 16*). This approach is often used in large modular air handling units. Remember that heat-pipe heat exchangers have no real value for comfort cooling. Heat pipes are applied when the dehumidification load is substantial and humidity is an important factor in the application.

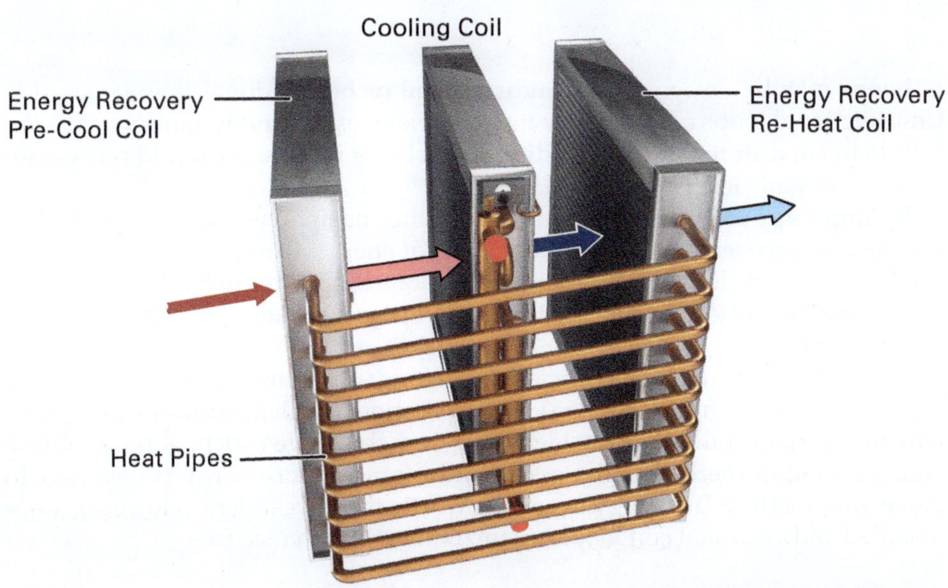

Figure 16 Wrap-around heat-pipe heat exchanger for dehumidification applications.

1.4.1 Coil-Loop Thermosiphons

Coil-loop thermosiphons are used when the two ducts are not adjacent to each other. Coil-loop thermosiphons are similar to runaround loops, but there is no pump to move the fluid. The fluid moves naturally, through thermodynamics, as it does in a heat pipe. The temperature difference between the two airstreams creates the flow of vapor in one direction and liquid in the opposing direction.

The loop consists of two coils interconnected by vapor and condensate return piping. The loop is charged with refrigerant in saturation, meaning that both liquid and vapor are present in normal conditions.

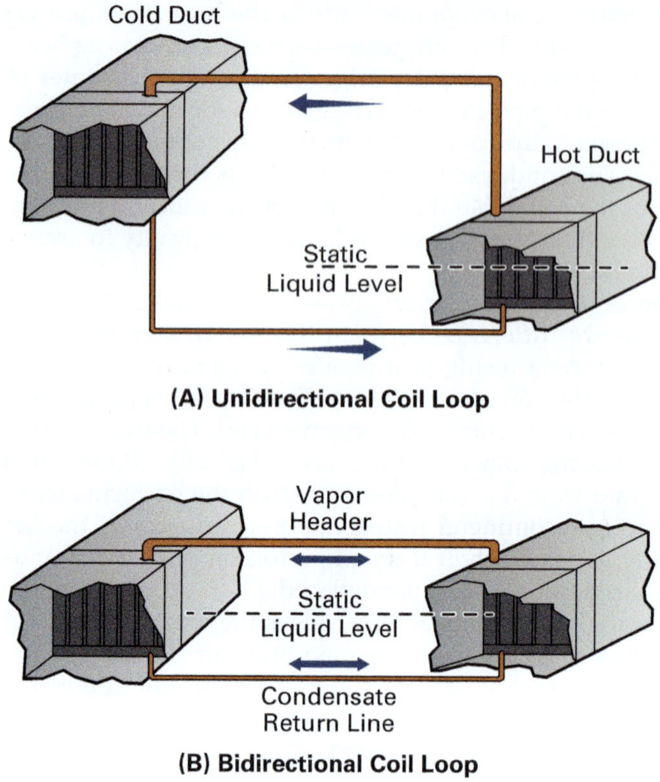

Figure 17 Coil-loop thermosiphon systems.

Loops may be designed for unidirectional or bidirectional flow (*Figure 17*). Unidirectional loops are normally more efficient, as thermodynamics favors the coils being at different heights. Bidirectional loops require the liquid refrigerant to be at the same level in each coil.

In larger applications, several coil-loop thermosiphons can be mounted in the ducts to recover the maximum amount of energy. Coil-loop thermosiphons can be applied in dehumidification applications but are more likely to be found transferring energy from an exhaust airstream to a makeup airstream, like the runaround loop.

The routing of the interconnected tubing must be considered because ambient conditions can interfere with system operation. Insulation may be necessary. Due to all the installation considerations and challenges such as pipe routing and coil height, there must be a significant amount of energy transferred to make this method of energy recovery worth the investment. However, once installed, bidirectional coil loops are virtually maintenance free.

1.5.0 Waterside Economizers

Waterside economizers are often used in buildings that are cooled with chilled water, because many chillers are water-cooled. Waterside economizers can save enormous amounts of energy because they can provide cooled water without using the chiller. Data centers, for example, have tremendous heat loads year-round. During the cooler months, a great deal of the cooling can be done by a waterside economizer and powering only the pumps while the chiller remains idle or provides any additional capacity needed.

When it is cool and dry outside, the condenser water from the cooling towers can be used to cool the load directly (*Figure 18*). The condenser water can be redirected to a coil in an air duct, or to a heat exchanger in the chilled-water piping. This bypasses the chiller entirely. Waterside economizers also offer some cooling redundancy by providing some cooling if the chiller fails.

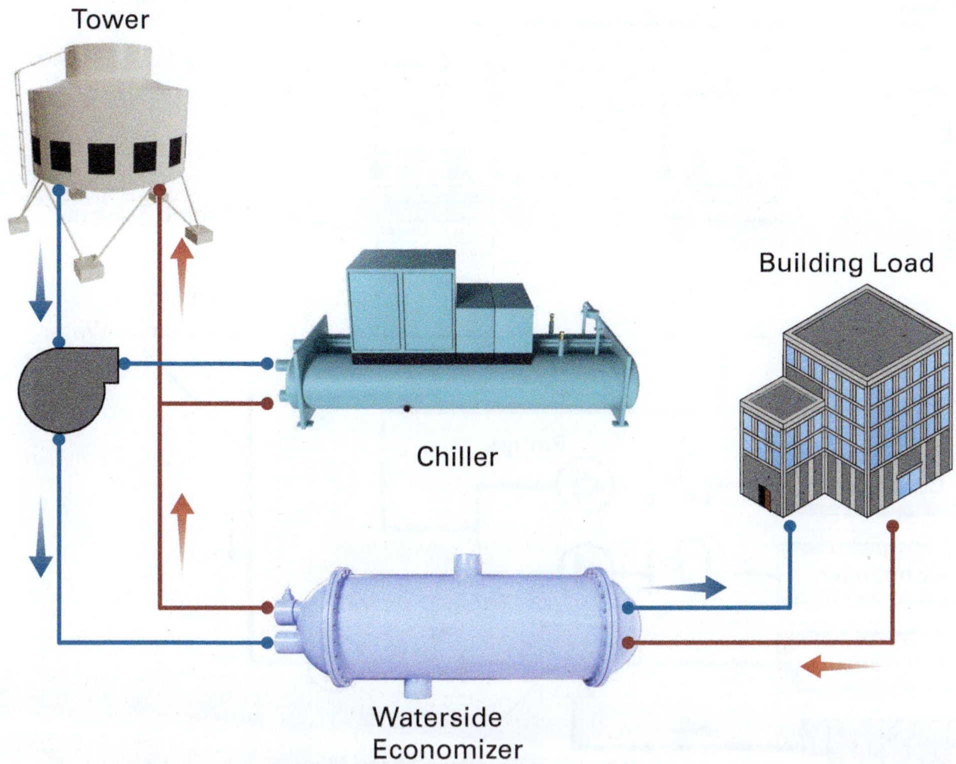

Figure 18 Basic waterside economizer arrangement.

Waterside economizers work best where the outdoor wet-bulb temperature is lower than 55°F (13°C) for a significant portion of the year. The wet-bulb temperature affects cooling tower operation, as dryer air increases the cooling capacity of the tower. The needed conditions apply to a significant portion of the United States. The cost of operating a typical chilled-water plant can be reduced up to 70% when it is equipped with a waterside economizer.

The system shown in *Figure 18* is a very basic representation. In practice, applying a waterside economizer is more complicated and requires a significant amount of control. The application guidelines found in ASHRAE Standard 90.1, *Energy Standard for Sites and Building Except Low-Rise Residential Buildings* currently require a properly designed system to be able to cool supply air by indirect evaporation and be capable of providing up to 100% of the expected system cooling load at outdoor air temperatures of 50°F dry bulb / 45°F wet bulb and below.

In addition, ASHRAE now requires waterside economizers to be integrated with the mechanical system in a manner that allows cooling to be derived from both sources (tower water and chillers) at the same time. The simple arrangement shown in *Figure 18* does not meet this standard.

In *Figure 19,* notice that there is one cooling tower dedicated to one chiller and the waterside economizer. The other two chillers are connected to the two remaining cooling towers. A three-way diverting valve directs cooling tower water through the active load—the chiller or the plate heat exchanger.

In this arrangement, the condenser water from the dedicated cooling tower bypasses the chiller and flows through the heat exchanger when the system is economizing. That cooling tower is typically set to supply the coldest water possible. The other two towers supply somewhat warmer water to the remaining chillers. When the water from the cooling tower is not cool enough to provide significant cooling to the system, the third chiller is enabled and runs as needed.

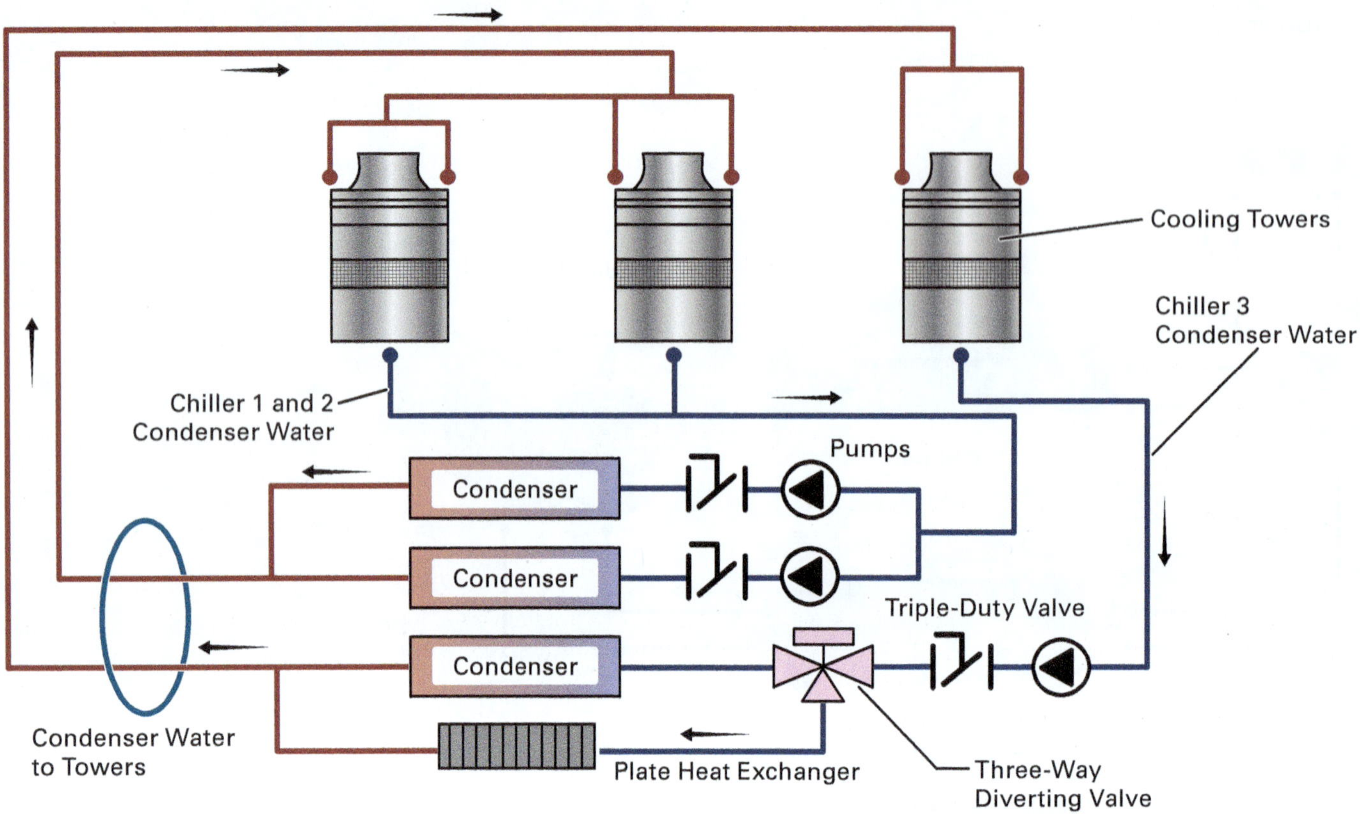

Figure 19 Waterside economizer with three chillers—condenser water piping arrangement.

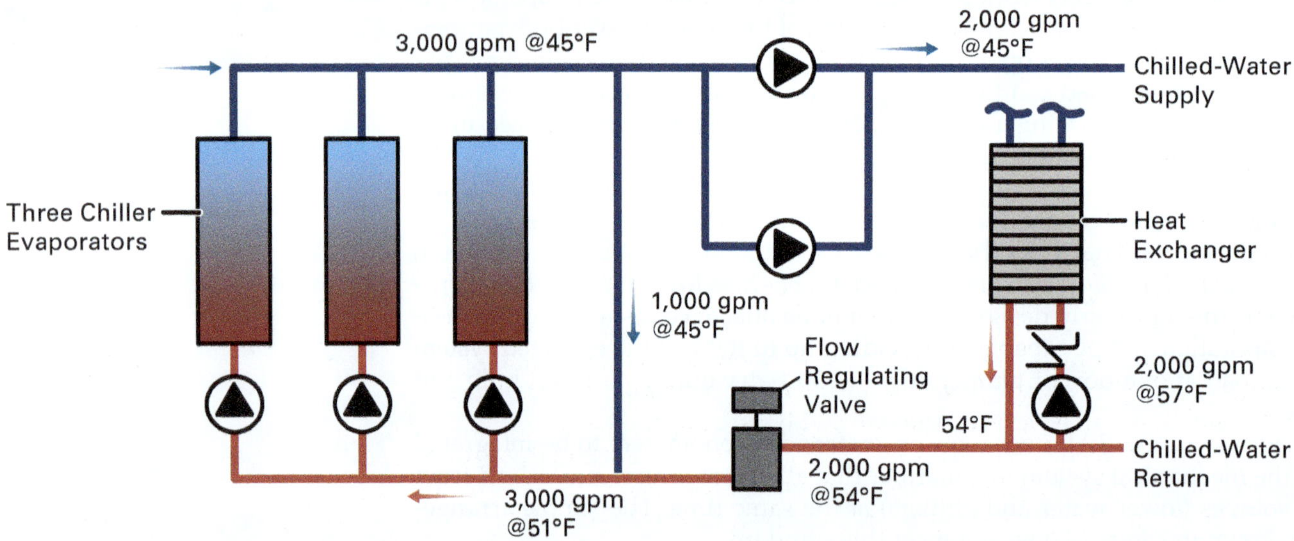

Figure 20 Waterside economizer with three chillers—evaporator water piping arrangement.

Figure 20 shows the load side of the chiller arrangement shown in *Figure 19*. The diagram shows a total chilled-water flow of 3,000 gallons per minute (gpm). From the chillers, 2,000 gpm flow out to the system load. The chilled water returning from the load at 57°F (14°C) flows through the heat exchanger where it is cooled to 54°F (12°C). The 54°F return water then mixes with 1,000 gpm of the 45°F (7°C) water from the chilled-water supply.

In a system like this, the economizer can assist if the water temperature from the dedicated cooling tower is lower than the chilled water returning from the load. When it is not, the third chiller is enabled. Of course, this example is just one of many possible scenarios.

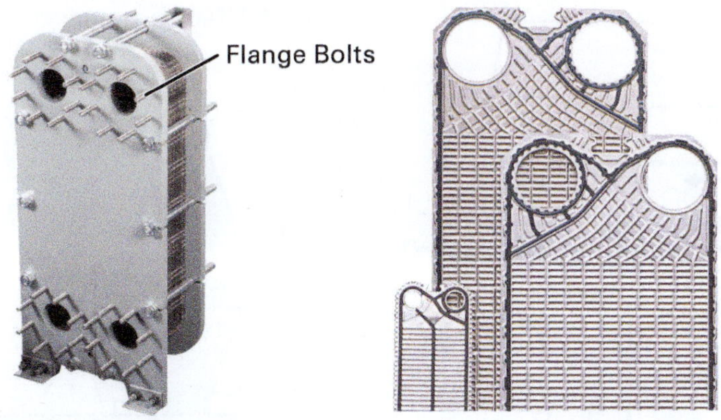

(A) Plate-and-Frame Heat Exchanger

(B) Plateflow® Plates

Figure 21 Plate-and-frame heat exchanger and plates.
Source: Courtesy of Xylem Inc.

Figure 22 Assembled and installed plate-and-frame heat exchanger.
Source: Aliaksandr Yarmashchuk/iStock/Getty Images

Plate-and-frame heat exchangers (*Figure 21*) are typically used in applications like these to transfer heat from the chilled-water system to the cooling tower. These heat exchangers consist of multiple thin, corrugated metal plates to transfer heat between two separate fluid streams. Plate-and-frame heat exchangers expose the two fluid streams to a great deal of shared surface area, making them surprisingly efficient.

Gaskets are placed between each of the plates. The plates are then stacked and compressed together using the through-bolts that pull the ends of the frame together (*Figure 22*). The flow paths through the plates are intricate, contributing to their heat transfer efficiency.

Plate-and-frame heat exchangers are also designed for easy capacity expansion. Engineers can accurately predict the heat transfer each plate provides, based on flow rates and the temperature difference of the fluids. One of the few disadvantages is that cleaning a fouled heat exchanger requires complete disassembly and a lot of effort. Thus, it is essential to ensure the water or other fluid is well filtered. Remember that water flowing through an open cooling tower can capture a great deal of dirt and debris in the process.

1.6.0 Flue-Gas Heat-Recovery Systems

The high-temperature flue gases from steam and hot-water boilers offer another opportunity to recover energy. Sensible heat can be reclaimed by inserting a heat exchanger in the flue between the boiler exhaust and the stack.

Figure 23 shows a steam boiler with a heat exchanger in the flue. In this example, the recovered heat is used to preheat the boiler's feedwater supply. This is the most common use for the heat recovered by the heat exchanger.

In this system, water from the boiler feed pump first enters the reheat tank. A portion of this water is pumped to the stack reclaimer heat exchanger, where it absorbs heat from the exhaust flue. A thermostat is normally used to control the reheat pump. The heated water flows back to the reheat tank, heating the incoming feedwater from the feedwater pump. The warmed feedwater then enters the boiler.

By preheating the feedwater using waste heat, less fuel is needed to produce steam. The recovered heat can also be used to heat domestic water or preheat makeup air for the building.

A major consideration when using a flue-mounted heat exchanger is that it must offer negligible resistance to the flow of flue gases. A heat exchanger that offers too much resistance to flow can adversely affect the boiler.

Consistent cleaning and maintenance of the heat exchanger are required to ensure that it does not become blocked with soot and debris. Thus, it is very important that the heat exchanger be easily accessible to maintenance personnel.

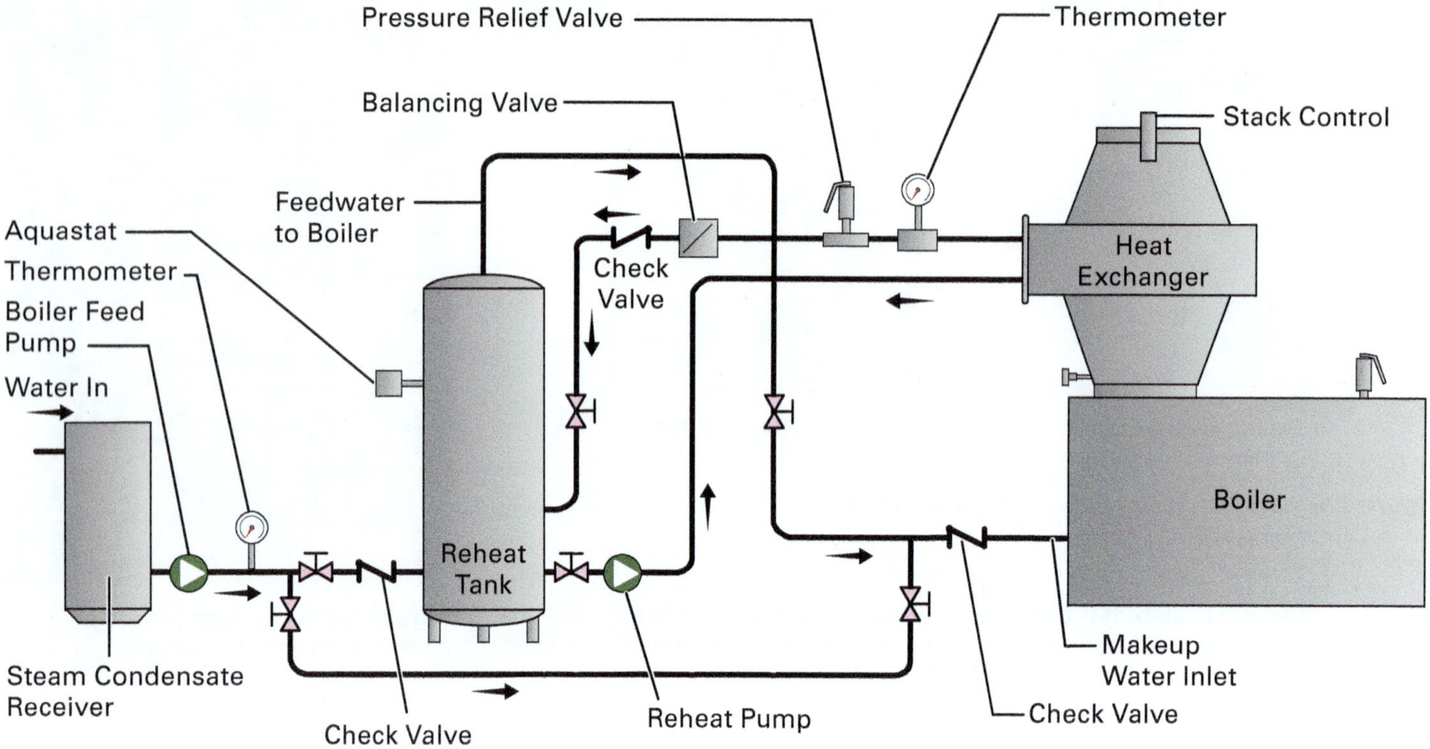

Figure 23 Flue-gas heat-recovery system.

1.7.0 Steam Heat-Recovery Systems

Heat recovery in steam systems can take several forms, as there are several points in larger systems where energy can be recovered and repurposed. Recovering heat from the flue gases, introduced in the previous section, is just one of the possibilities. In medium-pressure and high-pressure process steam systems, heat can also be recovered from a system's condensate and from the blowdown water.

1.7.1 Flash Steam Heat Recovery

Hot condensate in high-pressure steam systems is usually at or just below the saturation temperature of the low-pressure condensate return system. This means that the water is very close to its boiling point—a small change in pressure causes it to flash to steam very quickly. Steam in the condensate system creates problems, causing pressure surges and preventing condensate from reaching traps so it can move out of the steam lines.

Engineers avoid these problems by directing hot condensate and steam trap discharges to *flash tanks* (*Figure 24*). In a flash tank, flash steam can vent directly to the atmosphere or flow into a low-pressure steam main to serve a purpose.

The heat content of flash steam is no different than that of live steam from the boiler at the same pressure. Routing the flash steam into a low-pressure steam supply allows this heat to be used rather than wasted. Low-pressure steam is commonly used for space heating, heating water or other liquids, and even to drive some machinery.

High-pressure condensate enters the flash tank via the high-pressure condensate line. Vertical flash tanks, as shown in *Figure 24*, provide a better separation of steam and liquid condensate. As the condensate encounters the reduced pressure, it flashes to steam. The low-pressure flash steam then enters the low-pressure steam header to serve the connected loads downstream. Any remaining condensate drains into the low-pressure condensate return line.

When more than one high-pressure condensate line feeds into the flash tank, a check valve must be installed in each line to prevent backflow. The top of the flash tank has a thermostatic air vent to release any air that might accumulate.

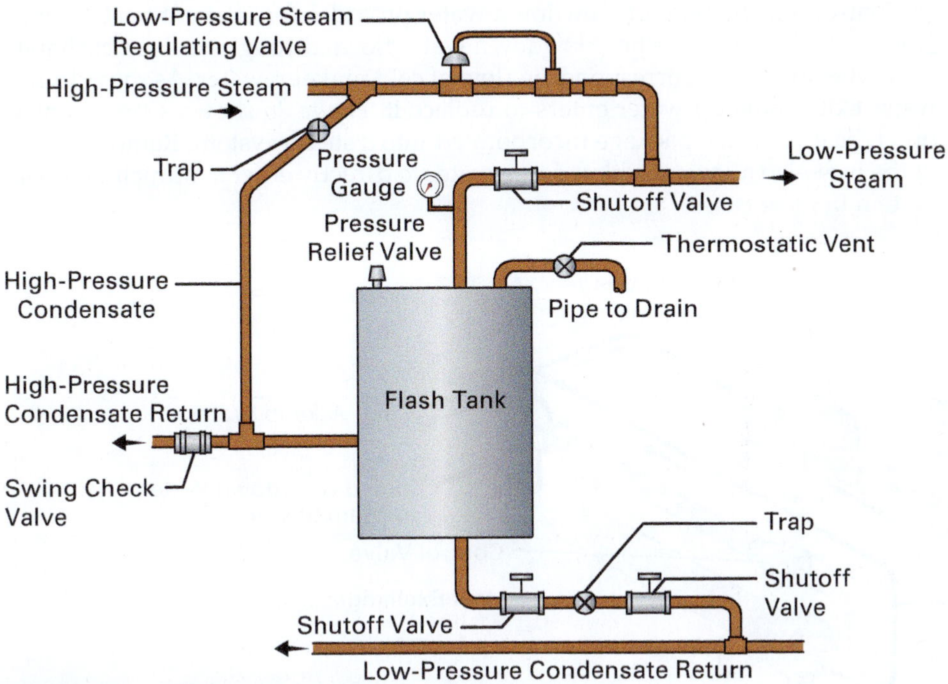

Figure 24 Steam system flash-tank schematic.

The flash steam produced is usually less than the amount of low-pressure steam that is needed. The supply can't be depended upon to serve important loads. To ensure there is always enough low-pressure steam to support the load, a low-pressure steam regulating valve admits steam from the high-pressure system as needed to maintain a consistent pressure. The regulating valve will close completely when a sufficient supply of flash steam is available to maintain the pressure.

In a properly designed system, it is very rare that the volume of flash steam generated is greater than the low-pressure steam load. However, it is possible, especially if someone inadvertently closes a low-pressure system valve. For this reason, a pressure relief valve is installed at the top of the flash tank to vent the steam to the atmosphere when the pressure exceeds a safe level.

1.7.2 Blowdown Water Heat Recovery

Blowdown heat-recovery systems recover heat from the boiler blowdown water and use it to preheat the makeup water. Surface blowdown, better known as *skimming*, is needed to purge the floating solids from the water's surface.

Not only is this energy often wasted, but the blowdown water also cannot be routed to a sewer or municipal drain unless it is below 140°F (60°C). Therefore, the blowdown water needs to be cooled, and it contains energy that can be recovered and used to preheat the boiler's makeup water.

Automatic control of the skimming process is a part of a blowdown heat-recovery package (*Figure 25*). A control valve in the unit controls the flow of skimmed blowdown water and positions itself to maintain the desired ratio of makeup water. As a result, the concentration of dissolved solids within the boiler is maintained automatically.

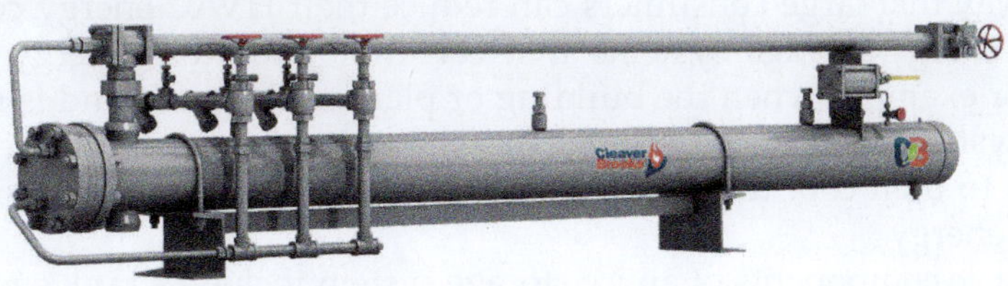

Figure 25 Blowdown heat-recovery package.
Source: The Cleaver-Brooks Company, Inc.

Controlling the flow of blowdown water provides for efficient heat recovery (about 90%) because the hot blowdown water flows through the heat exchanger only when there is a corresponding flow of cold makeup water. As the skimmed water exits, makeup water enters to replace it. *Figure 26* shows a typical blowdown heat-recovery package incorporated into a steam system. Removing most of the heat also allows the blowdown water to directly enter a municipal sewage system in most cases.

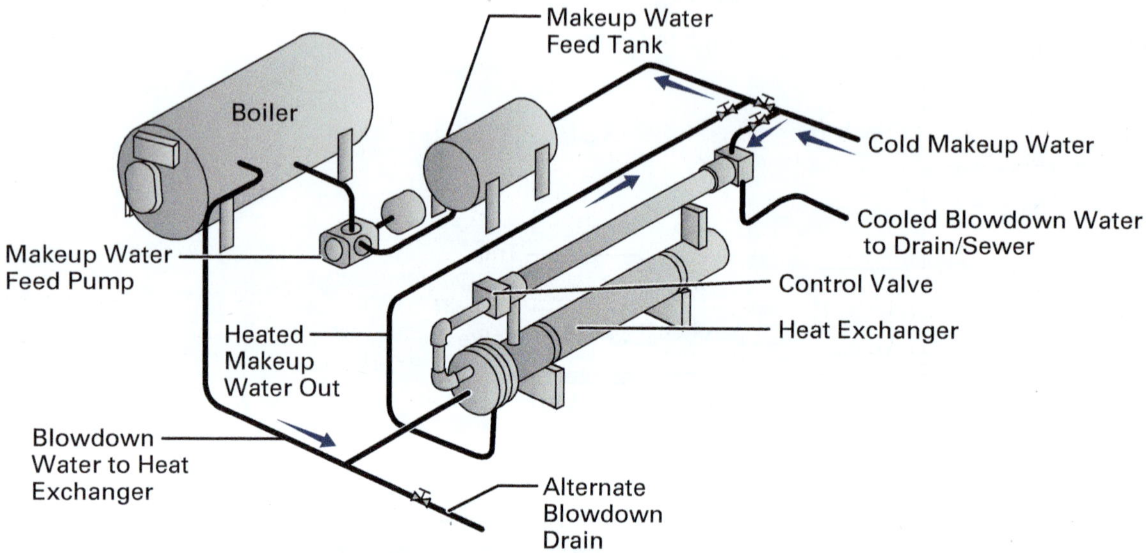

Figure 26 Blowdown water heat-recovery system for a steam boiler.

1.8.0 Ice Storage Systems

Utilities typically charge industrial and manufacturing electricity users differently than they charge a residential customer. For your home, you are usually invoiced through *volumetric billing*. This simply means that you pay for the total amount of electricity consumed, measured in kilowatt-hours (kWh). To lower your bill, you must reduce the total volume of electrical power used during each billing period.

However, industrial and manufacturing clients are generally billed a different way, which leads to different opportunities to lower their power bill. These customers are billed based on *peak demand*—the maximum amount of power consumed during a single hour (or other short period) across the entire billing period. Although they still pay for the total amount of power consumed, the rate per kWh is largely based on the peak demand and at what time of day it occurs.

Thus, industrial consumers are incentivized to reduce the amount of power demanded at any one moment. They can generally do this, assuming it is practical, by spreading the electrical demand across more hours of the day. In other words, they are encouraged to shift their demand to *off-peak hours*. Off-peak hours are at night, when there is less activity and the demand for cooling is reduced.

One way that large consumers can reduce their HVAC energy cost is to use thermal energy storage systems that consume and store energy during the night. For example, when the building or plant cooling demand is reduced, ice storage systems chill water to build ice. The stored energy in the ice is used the next day to then cool the facility. This reduces energy costs by using cheaper off-peak energy.

The basic components of an ice storage system include a tank containing the water, a chiller capable of chilling a glycol solution well below freezing, and

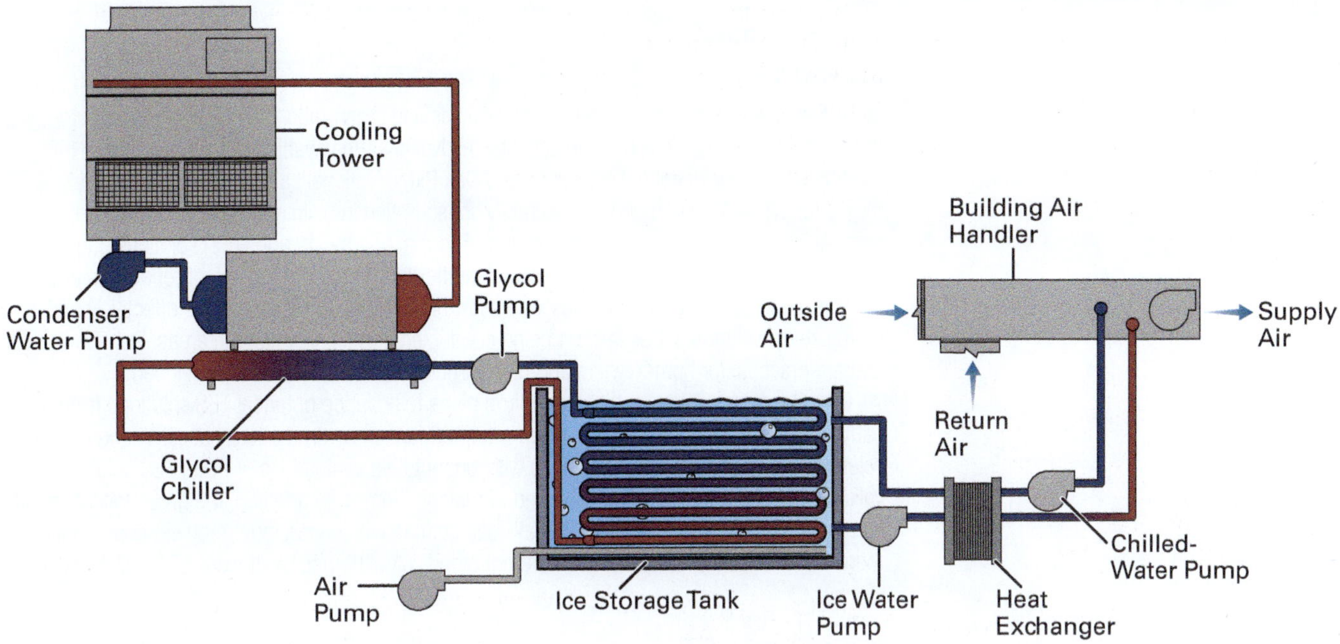

Figure 27 Basic HVAC ice storage system.
Source: Courtesy of EVAPCO, Inc.

the interconnecting piping, valves, pumps, and controls. *Figure 27* shows a basic arrangement for an ice storage system. Note the air pump connected to a manifold at the bottom of the ice storage tank. The air pump helps create some water movement in the tank when the system is in the ice-building mode, enhancing heat transfer.

Ice storage is often the system of choice for HVAC thermal energy storage because the apparatus can fit into relatively small spaces. The ice storage tanks can even be buried in the ground. The chiller must be able to chill the fluid to a temperature of 15°F to 26°F (–9°C to –3°C), which is below the normal operating range of conventional stock chillers. Therefore, special ice-making equipment or chillers suitable for low-temperature applications must be used. The heat transfer fluid is typically a mixture of glycol and water.

Several technologies are available for ice storage. The most common in commercial HVAC applications is *internal melt ice-on-coil.* In this approach, ice forms on the surface of submerged pipes or tubes through which the sub-freezing glycol is circulated. The chiller operates at a normal temperature during normal daytime hours, producing a 45°F (7°C) chilled-water supply.

At night, the chilled-water temperature is reduced to the sub-freezing range and the chilled-water supply is diverted away from the building load to the ice-building tanks. The next day, water in the tanks is circulated through a plate-and-frame heat exchanger to cool the normal chilled-water supply, while the chiller remains off until all the ice is depleted.

Operating strategies are generally classified as either *full storage* or *partial storage.* These terms refer to the amount of cooling load that is transferred from on-peak to off-peak. In a full-storage design, the entire on-peak cooling load is shifted to off-peak hours. In other words, enough thermal energy must be stored at night to satisfy the total demand for the next day. A full-storage system requires a relatively large chiller, plenty of storage capacity, and a larger financial investment.

In the more widely used partial-storage design, the ice storage system need only meet some portion of the total cooling load the next day. This approach reduces the size of the chiller and the size/number of ice storage tanks, and it also reduces the size of the investment. Remember that energy management and recovery, while always prudent and seemingly sensible, must also provide a financial return on the investment to be widely embraced.

Did You Know?

Stor4Build

Stor4Build is a coalition of like-minded companies and organizations focused on energy storage solutions for buildings. The mission is led by the National Renewable Energy Laboratory (NREL), Lawrence Berkeley National Laboratory, and Oak Ridge National Laboratory.

Currently, as much as 50% of electricity consumption in buildings is dedicated to meeting thermal load demands (heating and cooling). Thermal energy storage (TES) solutions offer a cost-effective alternative that can increase load flexibility, promote the use of renewable energy sources, decrease carbon emissions, and allow heat pumps to function more effectively and in more extreme climates. Research is being conducted in multiple research areas. But TES solutions that are practical for homeowners and/or communities have never been developed.

Over the next few years, the consortium plans to advance potential TES solutions from early- to medium-stage development to market adoption by dramatically accelerating the science-to-system timeline. The group will conduct field evaluations of novel TES solutions to validate the TES benefits when deployed in multiple homes. In addition, the group will complete a community-scale demonstration of peak-load shifting while reducing the power consumption for thermal loads (air conditioning and heating) of individual homes by at least 50% for 4 to 6 hours.

1.8.1 Ice Storage System Operation

The operation of one manufacturer's ice storage system, referred to as IceBank® energy storage, is illustrated in *Figure 28*. This example is a partial-storage system that uses the internal melt ice-on-coil technique. However, IceBank® energy storage systems can also be configured for full-storage performance.

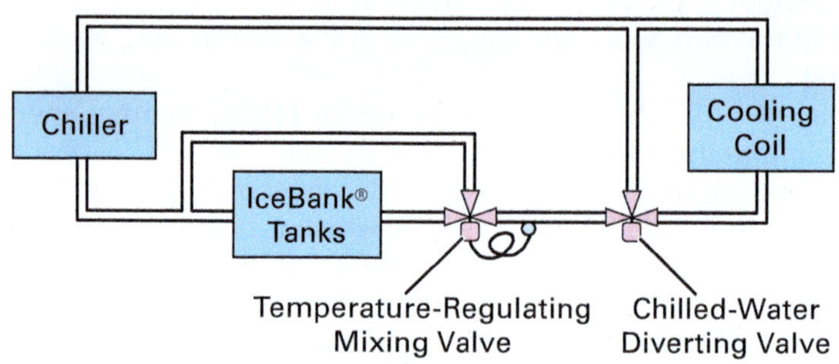

Figure 28 IceBank® partial-storage system simplified schematic.

During off-peak hours, the system chiller is used to build ice in large tanks (*Figure 29*). The tanks are insulated, water-filled polyethylene tanks, each containing many loops of submerged plastic tubing surrounded by water. The tanks may be installed indoors, outdoors, or even underground.

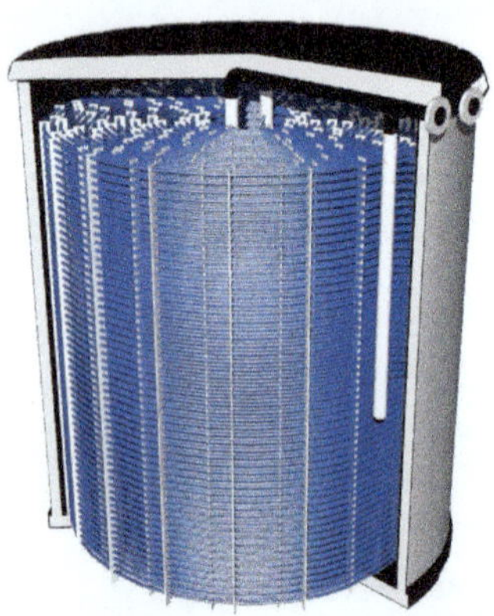

(A) IceBank® Ice Storage Tank

(B) IceBank® Tank Farm

Figure 29 IceBank® energy storage tanks.

Source: CALMAC Corp.

NOTE

Today's ice storage tanks often have more innovative plastic heat exchanger designs to enhance the ice-building process.

During the various modes of system operation, the chilled-water diverting valve controls the direction of the chilled-water supply. The chilled fluid is a glycol solution, with a 25% glycol concentration generally preferred for ice storage systems. After the chilled solution passes through the ice tank, the diverting valve can send the solution directly back to the chiller in the ice-building mode, or through the cooling coil when the chiller is in the cooling mode. The temperature-regulated mixing valve controls the downstream temperature of the glycol solution by mixing the flows through the chiller and ice tank.

In the ice-building mode, often referred to as the *charge cycle*, the chiller produces fluid at around 25°F (–4°C). The fluid circulates through the ice tank tubing (*Figure 30*). The mixing valve allows full flow through the ice tank, and the

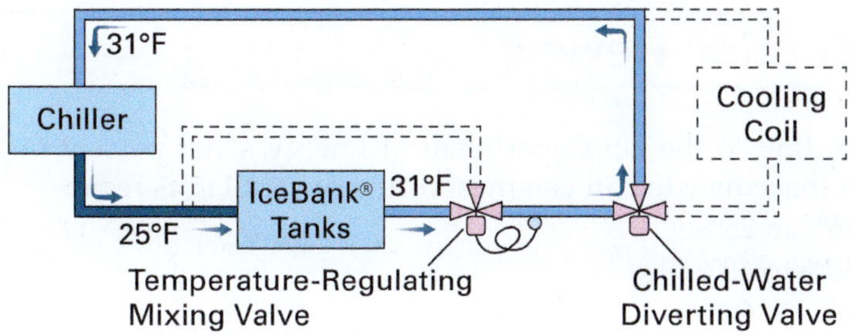

Figure 30 Flow path and temperatures for the charge cycle.

diverting valve directs all flow directly back to the chiller. The 25°F solution causes the water in the tanks to begin freezing. The charge cycle ends when the ice cannot be subcooled further, or when the off-peak hours have elapsed.

As the building is occupied and the cooling load increases, the chiller may or may not be operating. For some time, the ice alone may handle the cooling load. This is referred to as the *discharge cycle*. In *Figure 31*, the chiller is running to reduce the chilled-water temperature to 52°F (11°C). Note that this requires far less energy than reducing the temperature to 45°F (7°C).

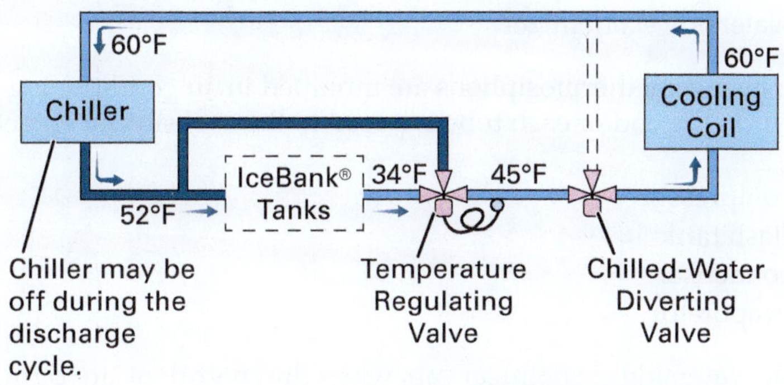

Figure 31 Flow path and temperatures for the discharge cycle.

As the warm solution circulates through the ice tank tubing, the ice in the tanks cools it to a temperature of about 34°F (1°C). The cold solution output from the tanks is then mixed with some of the warmer 52°F (11°C) solution to produce a solution of about 45°F (7°C). The diverting valve routes all flow through the cooling coil. This is the normal design chilled-water temperature range for the cooling coil.

Once the ice is depleted and it can no longer contribute to cooling the facility, the mixing valve stops accepting flow from the tanks and forces the solution to flow around them (*Figure 32*). This mode of operation is called the *bypass cycle*. The diverting valve routes all flow through the cooling coil. In a partial-storage system, the system will normally be found operating in the bypass cycle late in the day.

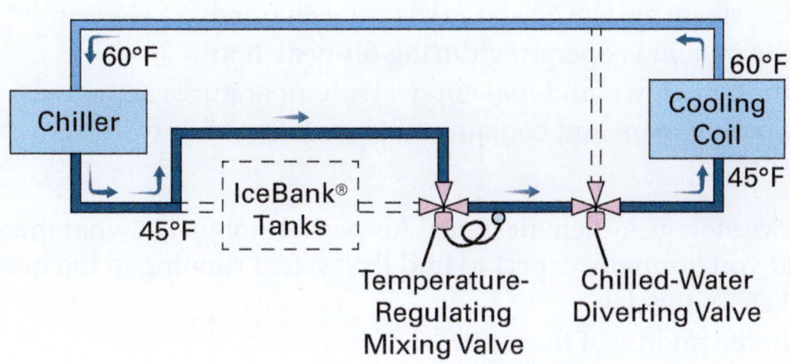

Figure 32 Flow path and temperatures for the bypass cycle.

1.0.0 Section Review

1. According to the US Department of Energy, what percent of the energy from the exhausted air can most recovery ventilators recover?
   a. 15% to 25%
   b. 30% to 45%
   c. 45% to 60%
   d. 60% to 80%

2. With which piece of equipment is a double-bundle condenser associated?
   a. Steam flash tanks
   b. HRVs and ERVs
   c. Water-cooled chillers
   d. Conventional condensing units

3. Manufacturing and industrial exhaust and makeup air applications can benefit from heat recovery using coils in the ducts and a ______.
   a. flash tank
   b. runaround loop
   c. cooling tower
   d. waterside economizer

4. When heat-pipe thermosiphons are mounted in air conditioning system ductwork, the end of each tube exposed to the cool supply air acts as a(n) ______.
   a. compressor
   b. flash tank
   c. condenser
   d. evaporator

5. A basic waterside economizer can, when the conditions are right, provide cooling for a building using ______.
   a. outside air
   b. a thermosiphon
   c. a special chiller
   d. cooling tower water

6. What is the most common use for the heat recovered by a flue-gas heat-recovery system installed on a steam boiler?
   a. Deaerate the feedwater to reduce system corrosion.
   b. Produce flash steam to serve low-pressure steam loads.
   c. Preheat the boiler feedwater to reduce fuel consumption.
   d. Reduce the flue gas temperature before it enters the environment.

7. In a blowdown water heat-recovery system, makeup water flows into the boiler when ______.
   a. the steam system flash tank is fully charged
   b. the system is operating during off-peak hours
   c. the blowdown and makeup water temperatures are equal
   d. floating foam and contaminants are being skimmed from the water surface

8. In an ice storage system designed for partial storage, at what time of day would you normally expect to find the system running in the bypass cycle?
   a. Late in the day
   b. In the middle of the night
   c. As soon as the building is occupied in the morning
   d. When the building becomes unoccupied in the evening

Module 03404 Review Questions

1. One advantage of using a stand-alone ERV, as opposed to an integrated ERV, with a packaged rooftop unit is that it _____.
   a. eliminates the need for exhaust fans
   b. maximizes recovery wheel efficiency
   c. increases airflow through the supply duct
   d. allows the use of an outside-air economizer

2. A static-plate heat exchanger in an ERV can recover _____.
   a. latent heat only
   b. sensible heat only
   c. both sensible and latent heat
   d. neither sensible nor latent heat

3. A type of heat-recovery device in which the hot discharge gas from a compressor is used to heat building air or water is called a _____.
   a. thermosiphon
   b. recovery condenser
   c. dual-evaporator system
   d. water-cooled condenser

4. HVAC systems designed specifically for the pool environment are typically equipped with _____.
   a. a steam humidifier
   b. desiccant dehumidifiers
   c. integrated heat-recovery ventilators
   d. thicker evaporator coils to enhance dehumidification

5. How many coils are found in a typical runaround loop system?
   a. One coil
   b. Two coils
   c. Three coils
   d. Six coils

6. A heat-pipe heat exchanger _____.
   a. transfers only latent heat
   b. is equipped with an electric reheat coil
   c. can increase a system's dehumidification capacity
   d. equalizes the supply and exhaust airstream temperatures

7. When are waterside economizers most effective?
   a. During the hottest part of the summer
   b. When the chiller is operating at 100% capacity
   c. Where the outdoor wet-bulb temperature is often lower than 55°F (13°C)
   d. Where the building dehumidification load is greater than the sensible load

8. Which of the following statements about waterside economizers is *correct*?
   a. Waterside economizers can provide both heating and cooling to a building.
   b. ASHRAE standards require today's waterside economizers to be able to support 50% of the cooling load throughout the summer.
   c. New waterside economizers must be integrated to contribute cooling capacity while the chillers are also running.
   d. Thermosiphons are typically used with a waterside economizer to transfer heat from the chilled-water system to the cooling tower.

9. Flash steam derived from hot condensate in high-pressure steam systems can often be used to ______.
   a. test the steam traps
   b. reduce the need for skimming
   c. support low-pressure steam loads
   d. support high-pressure steam loads

10. Which of the following is a method used to reduce the cost of the electrical power required to produce cooling capacity?
    a. Off-peak ice storage
    b. Flue-gas heat recovery
    c. Off-peak heat pump defrost
    d. Flash tank steam production

Answers to odd-numbered questions are found in the Module Review Answer Key at the back of this book.

Answers to Section Review Questions

Answer	Section	Objective
Section One		
1. d	1.1.0	1a
2. c	1.2.2	1b
3. b	1.3.0	1c
4. c	1.4.0	1d
5. d	1.5.0	1e
6. c	1.6.0	1f
7. d	1.7.2	1g
8. a	1.8.1	1h

MODULE 03405

Building Management Systems

Source: AndreyPopov/iStock/Getty Images

Objectives

Successful completion of this module prepares you to do the following:

1. Identify and describe the operation of basic digital controllers.
    a. Identify the four primary control point classifications.
    b. Describe analog and discrete input and output devices.
    c. Describe closed control loops and the related algorithms.
2. Describe the architecture of a building management system.
    a. Describe a DDC peer-to-peer network.
    b. State the functions of a packaged unit digital controller.
    c. Outline BMS system control of applied HVAC systems.
3. Explain how users interface with a BMS and identify tasks that can be accomplished.
    a. Describe how users access and interface with a BMS.
    b. Identify various tasks that users can accomplish through BMS access.
4. Describe various building management system operating sequences and control strategies.
    a. Identify occupied building temperature operating sequences.
    b. Identify unoccupied building temperature operating sequences.
    c. Describe other building HVAC control strategies beyond temperature control.
5. Explain the concept of interoperability and identify common BMS protocols.
    a. Describe the primary communication protocols in use.
    b. Explain how web browsers are used to interact with systems using various protocols.

Performance Task

Under supervision, you should be able to do the following:

1. Interpret, identify, and describe HVAC system operating data viewed through building management system software.

Overview

Building management systems take the use of electronic controls and the internet to another level by integrating the control of all building systems, including HVAC, lighting, security, and fire control in a graphical environment. Most, if not all, building systems can be monitored and controlled from a central location, either locally or from thousands of miles away. HVAC service technicians may not install and service building management system hardware and software, but they must be able to understand and interact with these systems to effectively service the equipment and diagnose problems.

Digital Resources for HVACR

NCCER SCAN ME

Scan this code using the camera on your phone or mobile device to view the digital resources related to this craft.

NCCER Industry-Recognized Credentials

If you are training through an NCCER-accredited sponsor, you may be eligible for credentials from NCCER. The ID number for this module is 03405. Note that this module may have been used in other NCCER curricula and may apply to other level completions. Contact NCCER at 1.888.622.3720 or go to **www.nccer.org** for more information.

You can also show off your industry-recognized credentials online with NCCER's digital credentials. Transform your knowledge, skills, and achievements into credentials that you can share across social media platforms, send to your network, and add to your resume. For more information, visit **www.nccer.org**.

1.0.0 Basic Digital Controllers

Performance Task

1. Interpret, identify, and describe HVAC system operating data viewed through building management system software.

Objective

Identify and describe the operation of basic digital controllers.

a. Identify the four primary control point classifications.
b. Describe analog and discrete input and output devices.
c. Describe closed control loops and the related algorithms.

Building management system (BMS): A centralized, computer-controlled system for managing the various systems in a building, such as HVAC, lighting, and security. Also known as a *building automation system (BAS)*.

Direct digital control (DDC): A system that uses digital controllers to maintain precise control over system components; an alternate term for *building management system* and *building automation system*.

Most of the previous coverage of electronic controls has involved individual electronic control devices. Examples of such devices are programmable electronic thermostats and electronic ignition modules for gas furnaces. This module focuses on entire control systems used to control all essential building functions, including but not limited to the HVAC system, lighting, smoke control, and building security. The composite of these systems is known by many names, including **building management system (BMS)**, building automation system (BAS), facility management system, and **direct digital control (DDC)** system.

Many building management systems are being implemented as part of a process called *integrated building design*. Integrated building design is a collaborative decision-making process that uses a project design team with diverse knowledge and experience from a project's inception through its construction phase. The integrated building design process is, in turn, often incorporated into the overall process of creating greener buildings.

A green building is a sustainable structure that is designed, built, and operated in a manner that conserves resources while also being cognizant of its environmental impact. The green-building design process considers many areas, including the following:

- Occupant health
- Staff productivity
- Construction materials
- Building systems
- Operation and maintenance
- Water use and conservation
- Energy use and conservation

To implement many of these design elements, the building management system must be able to apply control strategies across multiple building systems.

Going Green

LEED Certification

The most energy-efficient and environmentally sound buildings require considerable preparation. In 1993, the Leadership in Energy and Environmental Design (LEED) certification was introduced. The purpose of LEED is to encourage the adoption of sustainable construction and building management standards as established by the US Green Building Council (USGBC). Today, it is the most widely used green-building rating system in the world.

On the HVAC side, meeting the LEED certification standards means selecting energy-efficient equipment and sizing it correctly for the structure and the heating/cooling loads. Preserving the indoor air quality in the most energy-efficient way possible is essential.

The NCCER office building, located in Alachua, FL, and constructed in 2011, was designed and built to the LEED Silver level of certification.

The heart of any building management system is a series of digital controllers (*Figure 1*) used to monitor and control various building functions. The typical **digital controller** is comprised of three major components: an *input module*, a *microprocessor*, and an *output module*.

Digital controller: An electronic device that uses an input module, a microprocessor to determine appropriate responses to sensed conditions, and an output module to perform control functions.

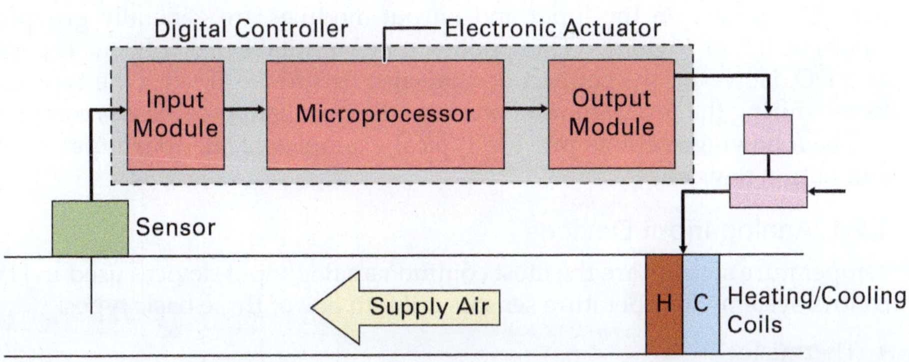

Figure 1 Typical digital controller application.

Digital controllers fall into two basic groups: product-specific controllers, better known as **product-integrated controllers (PICs)**, and general-purpose controllers. PICs are digital controllers designed to control specific equipment and are often factory installed in the equipment. The controllers found in many chillers are examples of PICs.

Product-integrated controllers (PICs): Digital controllers designed to control specific equipment, typically installed by the equipment manufacturer at the factory level; also known as a *product-specific controller*.

General-purpose controllers, on the other hand, are more generic and can be configured to control a variety of equipment. These controllers are typically field installed. Such controllers are used to control generic, less complex devices with simple operating schemes, such as exhaust fans, pumps, and similar equipment.

1.1.0 Control Point Classification

Algorithm: A set of mathematical rules and logic statements used in a computer or microprocessor to solve a specific problem. In HVAC applications, algorithms are typically used to control a wide range of system functions based on varying conditions.

Control point: The term for each input and output device wired to a digital controller.

In *Figure 1*, a sensor located in the supply air duct provides an input signal to the controller. The controller translates the input to a temperature. That temperature is compared to a programmed setpoint. The microprocessor, using an **algorithm**, determines the appropriate response and requests action from system components through its output module.

In this example, a hot-water or chilled-water valve is being controlled, so the valve opens or closes accordingly. Each input and output device wired to a controller is called a **control point**, or simply, a point.

Control points are either *analog* or *binary*. Binary points, also referred to as *digital* or *discrete* points, are those that only have one of two values, such as open or closed. A more detailed explanation of each point type follows:

- *Analog input (AI)* — A sensor used to measure variables that have a range of values, such as temperature or pressure. The sensor signal varies over the predefined range of values. A duct temperature sensor is an example.
- *Discrete input (DI)* — A sensor used to measure the status of binary devices, such as the position of a pressure switch. Only one of two possible conditions exists, e.g., switch open or switch closed.
- *Analog output (AO)* — An output signal used to modulate the position of a device, such as a valve or damper. The output signal from the controller to the AO point often varies from 4 to 20mA or from 0 to 10VDC, and the valve responds by moving to a position determined by the intensity of the signal.
- *Discrete output (DO)* — An output signal used to change the position of a device from one state to another. Examples of discrete outputs are the starting and stopping of a pump or fan motor. Discrete outputs can only provide one of two possible conditions, such as on or off.

1.2.0 Input and Output Devices

Analog or discrete devices are wired to the digital controller as inputs or outputs. The points on the input and output modules are generally grouped in blocks of 4, 8, or 16 points. These points may be suitable for only one type, such as a DO. However, the connection may also be universal, with the type established during the programming phase as the installation process is completed.

The following sections provide typical examples of the four types of input and output devices.

1.2.1 Analog Input Devices

Temperature sensors are the most common analog input devices used in HVAC control systems. Temperature sensors fall into one of three basic types:

- Thermistor
- Resistance temperature detector (RTD)
- Transmitter

Thermistors (*Figure 2*) change the resistance in proportion to changes in temperature. Thermistors can be used to sense the temperature of liquids or gases, such as air. These thermistors are the same as those used in heat pumps, digital thermostats, and similar products.

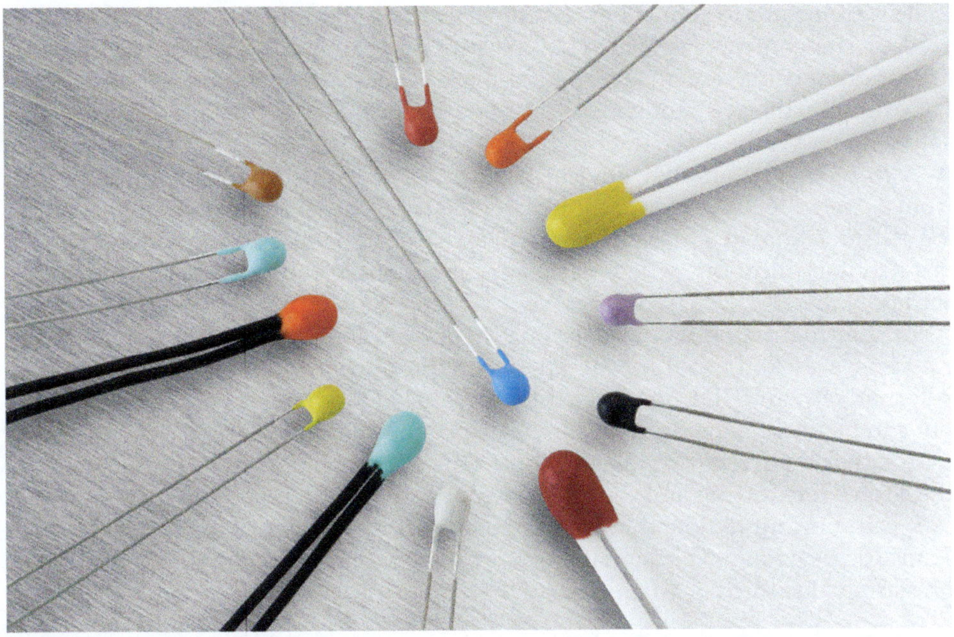

Figure 2 Examples of thermistors.
Source: U.S. Sensor Corp.

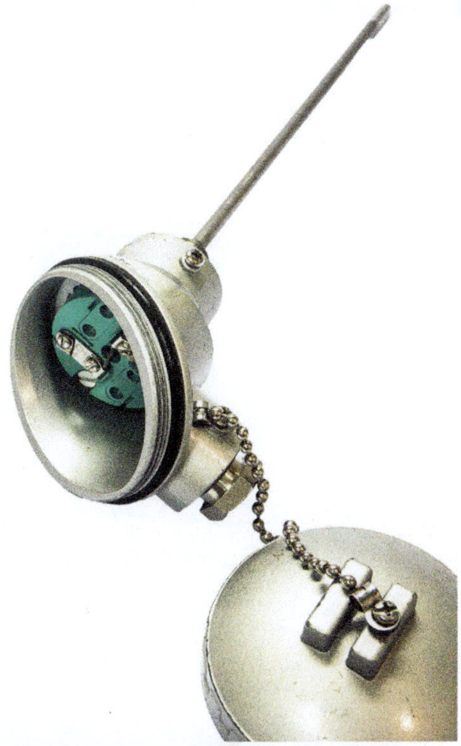

Figure 3 Resistance temperature detector (RTD).
Source: blueskyline/iStock/Getty Images

An RTD (*Figure 3*) uses a longer resistance-sensing element as opposed to measuring temperature at a single point. Thus, RTDs are commonly used to sense an average temperature along the length of the element.

When the medium being measured requires a sophisticated sensing element, or when there is a long distance between the point and the digital controller, a *transmitter* is used. A transmitter senses the specific range of a variable (like temperature) and then translates the variable into a current or voltage signal. It transmits a linear current (4–20mA) or voltage (0–10VDC) signal back to the controller. However, do not confuse the term *transmitter* to mean that the device is wireless.

On Site

Control Signals

Most control signals to actuators today are specified as 0–10VDC instead of 4–20mA. The reason is that it is easier to monitor and measure a voltage than a current, and the signal is equally precise. A simple voltmeter can be used to measure the voltage output signal.

NOTE

Actuators that accept voltage signals are not confined to the 0–10VDC range. Many of today's actuators operate at 2–10VDC instead.

1.2.2 Discrete Input Devices

DI devices are typically *dry contacts* in switches or relays. A dry contact is powered from an external source, rather than the voltage being provided by the switching device itself. Thus, most common HVAC-related switches are considered dry. Devices that might provide a DI to a digital controller include auxiliary starter contacts, differential pressure switches, and flow switches.

A unique type of discrete input device is a meter that can send a pulsed signal to a digital controller. The frequency and duration of the pulses usually represent power consumption or flow rates to a BAS. These meters have a pulse conversion rate that is programmed into the digital controller to correctly interpret the signal.

1.2.3 Analog Output Devices

Analog outputs control components like actuators that position valves or dampers (*Figure 4*). Most actuators accept Industry-standard output control signals of either 4–20mA or 0–10VDC. Although a small valve is shown in *Figure 4*, modulating actuators can be used on very large valves as well.

Figure 4 Control valve with modulating actuator.
Source: BELIMO Aircontrols (USA), Inc.

It is important to remember that devices like actuators are not usually powered by the controller signal. They require a separate power supply to enable operation and movement. The signal merely determines the device's position. For example, the actuator shown in *Figure 4* requires a 24VAC or 24VDC power source to move, but a separate signal for instructions.

A *transducer* is another device that can be controlled by an analog output from a controller. A pressure transducer (*Figure 5*), for example, converts an analog output signal to a pressure value. The use of transducers is common in large or retrofit applications where valves or dampers are equipped with pneumatic actuators.

To control a pneumatic actuator, a current-to-pressure (I/P) transducer converts a 4–20mA output signal into a proportional 3–15 psig air output. Alternatively, an E/P transducer that converts a DC voltage to a pneumatic output can be used. Some transducers, like the one shown in *Figure 5*, can be used with either a current or voltage input. This approach can significantly reduce the cost of retrofitting an existing pneumatic installation.

It is important to note that some transducers can also serve as input devices. Transducers can convert signals to a pressure value, for example, or convert a pressure value into a signal.

1.2.4 Discrete Output Devices

Discrete outputs typically control low-voltage (24VDC) control relays with the contacts wired into low- or line-voltage circuits to control loads like motors or heating elements, or to enable a piece of equipment. A discrete output can also control valves that do not need to modulate position, but simply need to be opened or closed.

Pulse-width modulation (PWM): A common control technique based on the conversion of an analog signal into discrete parts. The signal produced appears as a chain of pulses that form square waves.

In some cases, an analog output can position actuators using **pulse-width modulation (PWM)**. As shown in *Figure 6*, a PWM signal stays on for a given time and stays off for a certain time. The percentage of time for which the signal remains on is known as the *duty cycle*.

The output voltage of a PWM signal is represented by the percentage of the duty cycle. For example, for a 100% duty cycle, if the operating voltage is 10VDC, then the output voltage will also be 10VDC. If the duty cycle is 50%, then the output voltage will be 5VDC. One discrete output is used to pulse the device in one direction, and another output is used to pulse the device in the other direction.

On Site

The Use of Points

In the controls industry, each project is measured by the number of points to be installed. For example, a job may be referred to as a "100-point job" or a "1,000-point job," indicating the number of input or output devices in a system. This provides a basis for the amount of labor required to install a system.

Figure 5 Transducer (I/P or E/P).
Source: CONTROLAIR

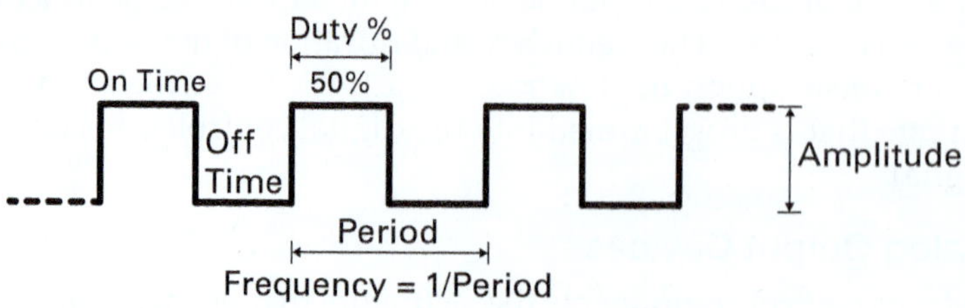

Figure 6 Pulse-width modulating square wave output.

1.3.0 Closed Control Loops

The basis for most digital control actions is a *closed control loop*. *Figure 7* shows a closed control loop being used to control the air temperature in a hydronic heating application. A sensor in the duct measures the supply air temperature. The input module monitors the sensor and transmits the air temperature to the microprocessor.

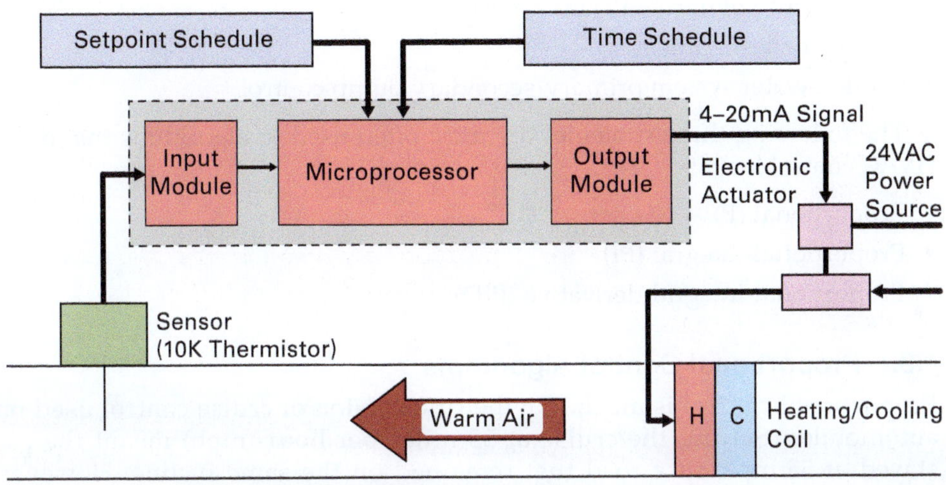

Figure 7 Closed control loop example.

The microprocessor compares the duct temperature, for example 95°F (35°C), to a programmed setpoint, perhaps 110°F (43°C), and calculates a milliamp signal to send to the hot-water valve actuator. Since the air temperature is well below the setpoint, the hot-water valve must be driven further open. The output module sends the signal to the valve actuator, and the valve opens further to increase flow through the coil.

A greater difference between the setpoint and the measured value causes a higher signal. In other words, the signal to the valve is proportional to the difference between the setpoint and the actual temperature. As the values come closer together, meaning the air temperature rises and approaches the setpoint, the signal weakens, and the valve starts to modulate toward the closed position.

Of course, this is just one example of many possible closed control loops. The automatic and consistent feedback created by this loop makes the process a closed control loop. The sensor automatically measures the impact of any change ordered by the microprocessor. As shown in *Figure 7*, both the time and setpoint schedule also play a significant role in the loop.

Closed control loops are executed by running multiple algorithms at the same time within the microprocessor. The algorithms interpret the inputs and decide what output(s) to send in response. Algorithms are a series of logic statements, repeated continuously at high speed. For example, the heating coil algorithm for *Figure 7* might consist of code that represents the following sequence:

Step 1 Read the duct temperature.

Step 2 Compare it to the setpoint. If the duct temperature is below the setpoint, calculate the difference and apply a multiplier to determine the intensity of the output signal to the valve actuator.

Step 3 Send *X* milliamps to the hot-water valve actuator.

Step 4 Wait for *Y* time and then go back to Step 1 to determine the current conditions and recalculate.

In a microprocessor, these steps may be repeated many times in just a few seconds.

Many control loops require time schedules to determine when specific equipment should be turned on or off. Using setpoints, time schedules, and point data, a microprocessor can use algorithms to perform a wide range of control functions. Some of the typical HVAC algorithms are as follows:

- Heating/cooling coil water flow control
- Humidification or dehumidification
- Mixed air temperature optimization based on indoor and outdoor conditions
- Variable air volume (VAV) fan control
- Electric heat staging
- Chilled-water system primary/secondary pump control

The following three strategies are most often used in algorithms that drive HVAC control loops:

- Proportional (P)
- Proportional-integral (PI)
- Proportional-integral-derivative (PID)

1.3.1 Proportional Control Algorithms

Proportional control is much like the first version of cruise control used on automobiles. Setting the cruise at 55 miles per hour (mph) meant the car stayed at 55 mph on a road that remained on the same incline. However, when the car started up a hill, it might slow to 50. Missing the setpoint on the low side is called *undershoot*. When the car traveled down a hill, it might increase to 60 mph. Missing the setpoint on the high side is called *overshoot*. Thus, when using proportional control loops, due to control inefficiencies as well as overshoot and undershoot, the speed fluctuated within a wide range.

Even though the setpoint was 55 mph, the speed of the car did not stay constant. The difference between the actual speed of the car when it settled at a fixed speed (the control point) and the desired speed of the car (setpoint) is called *offset*. Offset is a measure of the inefficiency of a proportional control loop. Despite the issues associated with proportional control, it is still a viable approach for some applications.

1.3.2 Proportional-Integral Control Algorithms

An *integral* control function can be added to the control loop of cruise control to eliminate offset over time. When the speed falls below 55 mph as you climb a hill, the integral function moves the accelerator pedal down and causes the car to increase speed until it returns to 55 mph again.

The reverse is true when the automobile travels down a hill. The accelerator is released, slowing the engine speed to try and maintain the 55 mph setpoint. Thus, the integral control function eliminates offset over time, and keeps the car at 55 mph regardless of the terrain being traveled.

One problem remains, however. Going up a very steep hill will cause the car to decelerate at a faster rate than going up a mild grade. The reverse is true for a steep downhill grade versus a mild grade. Utilizing the PI function, the car's cruise control will take longer to bring the car back to 55 mph on a steep hill than it will for a moderately inclined hill. Proportional-integral (PI) control does not respond to the rate at which the speed change occurs (the acceleration of the error); it simply eliminates the error over time.

PI control is used in most HVAC applications where the controlled variable changes slowly over time. The control of room temperature is one such example because room temperature tends to change gradually over time. PI control is the most common function used in HVACR control systems.

1.3.3 Proportional-Integral-Derivative Algorithms

Adding a derivative term to the cruise control function allows the control to respond to the car's rate of acceleration or deceleration when it encounters hills. The derivative function measures the rate of change of the variable (speed) and causes the fuel flow to increase its compensation as the rate of change increases. The net result is that the cruise control can rapidly bring the car back to 55 mph, regardless of the incline of the hill, up or down.

Proportional-integral-derivative (PID) control is used when the controlled variable is changing rapidly over time and/or when the rate of change varies significantly. The control of duct pressure in a VAV system is one such example. *Figure 8* summarizes the three functions involved in PID control and shows the benefits of each function.

Element	Deals With	Control Concept	Added Benefit	Response to Deviations
Proportional (P)	Error	Signal Proportional to Error	Control Signal Proportional to Error (Demand-Based)	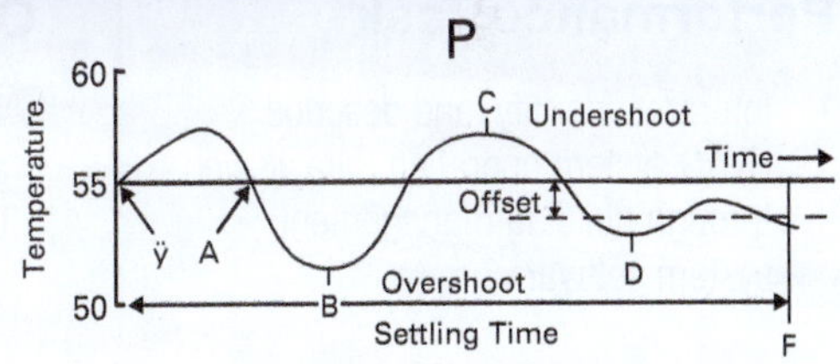
Integral (I)	Time	Integrates Error Over Time to Respond to Small Error	Eliminates Offset Over Time (Accuracy)	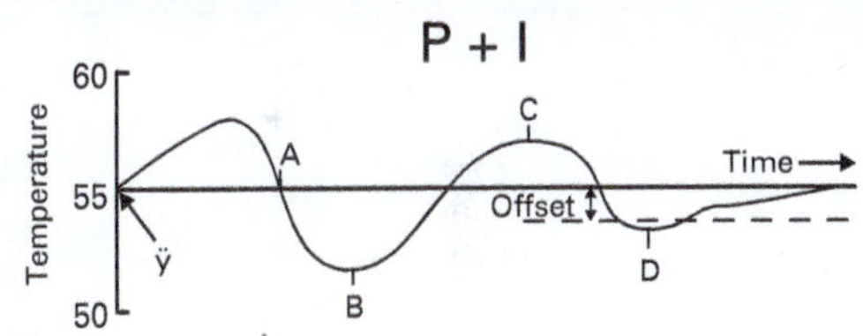
Derivative (D)	Rate	Responds to Rate at which Setpoint Is Approached	Shortens Settling Time (Quick Response) and Less Overshoot/Undershoot (Precise)	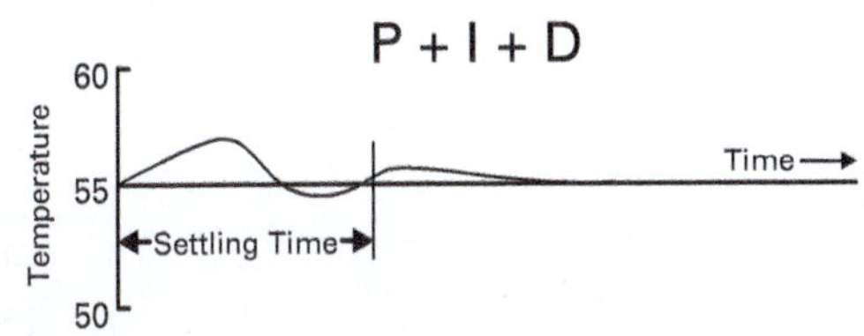

Figure 8 PID control summary.
Source: Carrier Corporation

Most HVAC control applications will use the PI function and not the full PID function because the controlled variable changes slowly over time. In fact, using the derivative function on a slowly changing process can actually make the control response worse. In most HVAC applications, the derivative function is not necessary to maintain the conditions needed, making the PI function the most common function used.

1.0.0 Section Review

1. Which of the following is another term used to describe binary control points?
   a. Dual points
   b. Single points
   c. Analog points
   d. Discrete points

2. Which of the following is the most common analog input in HVAC control systems?
   a. Voltage
   b. Pressure
   c. Resistance
   d. Temperature

3. Which closed control loop function is most often used in HVAC control functions?
   a. Proportional
   b. Proportional-integral (PI)
   c. Proportional-integral-derivative (PID)

2.0.0 Building Management System Architecture

Performance Task

1. Interpret, identify, and describe HVAC system operating data viewed through building management system software.

Objective

Describe the architecture of a building management system.

a. Describe a DDC peer-to-peer network.
b. State the functions of a packaged unit digital controller.
c. Outline BMS system control of applied HVAC systems.

Building management often starts with the automation of a building's HVAC systems, and then expands to include the building's lighting, security, and fire/smoke control systems. The primary objective of the BMS is to provide energy management routines to save energy; provide monitoring and reporting capability for planned maintenance; and permit human access and system intervention from the building or from a remote location. *Figure 9* shows one example of building management system architecture.

The automation of the building's HVAC systems is divided into two broad categories:

- Systems using packaged equipment, such as split systems, packaged rooftop units, and indoor packaged units
- Applied systems that incorporate equipment such as chillers, boilers, cooling towers, air handlers, and air terminals

The two lower sections of *Figure 9* show the controllers the BMS will interface with to control the individual pieces of HVAC equipment. The middle section above them shows the next level of building control—the interface of the BMS management functions. Finally, the top two sections of *Figure 9* show the system's local and remote access capability. All the system's digital controllers are connected through the **network** communication **bus**. The bus is a simple shielded cable that is physically wired to each controller in the network. Fiber optic cable may also be used to enable significantly faster data transfer.

Network: A structured linking of computers and other devices to allow communication and data sharing, which may rely on either wired or wireless connections.

Bus: A data communication path, often relying on a multiwire cable, that links all the components in a computer or between computers.

The scope of building management system functions is twofold. First is the use of digital controllers to control equipment. Second is the use of a network to link all controllers in the building to a centralized point for the software to reside and enable local human access and monitoring.

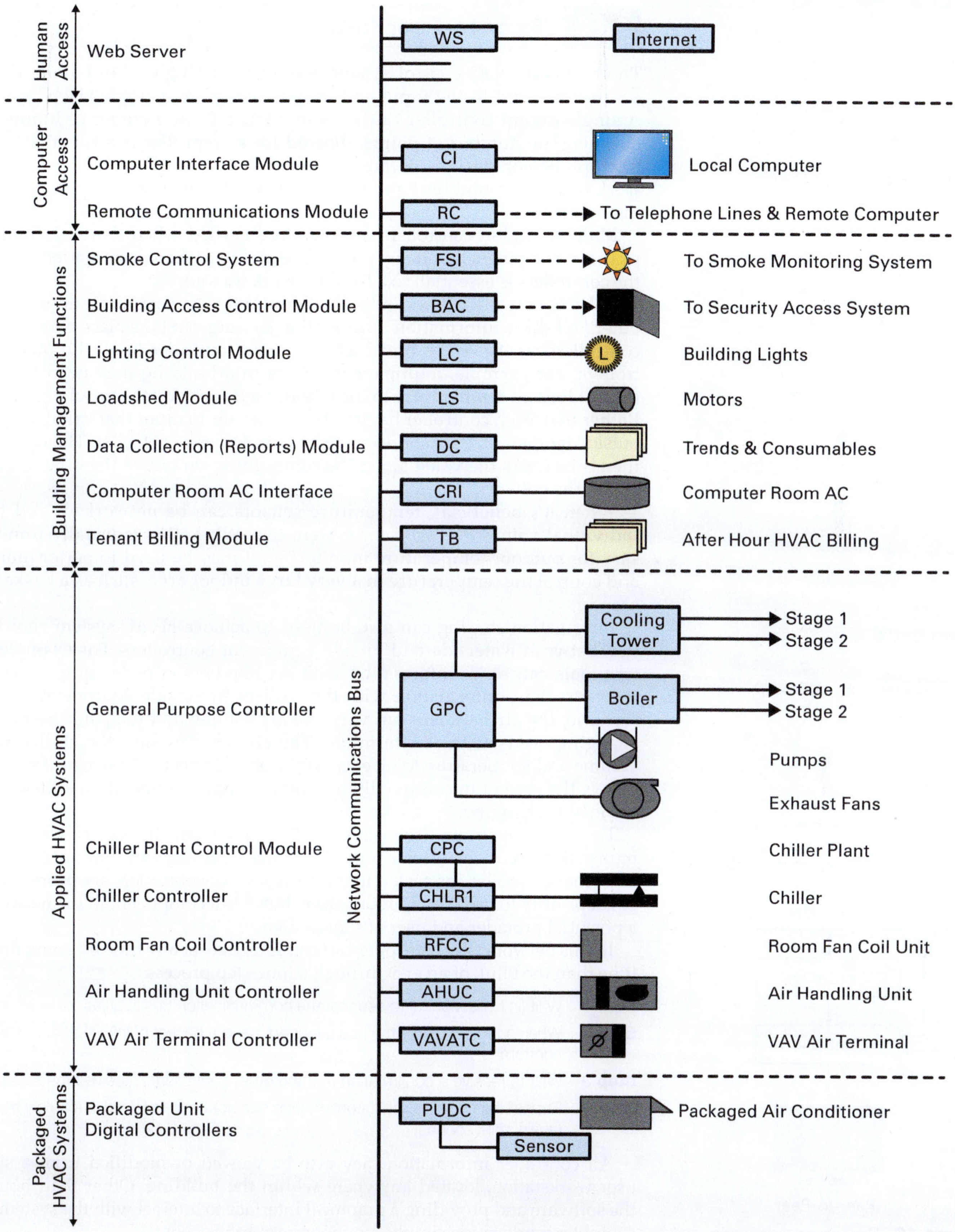

Figure 9 An example of BMS architecture.

2.1.0 Peer-to-Peer Networks

The simplest DDC control system would be a single, stand-alone digital controller wired to the input and output devices of a single HVAC system. A single digital controller can accommodate a finite number of inputs and outputs. The number of points allowed for a controller is a function of the microprocessor design. Because a stand-alone digital controller can only work within its point limitation, it would only be sufficient for a small building on its own.

Multiple digital controllers are therefore needed when there are multiple pieces of equipment and many points. The ability to share information between the controllers is essential, so a DDC network is required.

A DDC network allows individual controllers to communicate with each other and share information. Information sharing can take place between two controllers, or it may be broadcast from one controller to all the other controllers. For example, multiple controllers in a building may need to know the outdoor air temperature to include in their individual control algorithms. Rather than each controller having its own sensor to input that value, only one outside temperature sensor needs to be wired to one controller. The controller then broadcasts the value along the communication bus to the other controllers in the network.

When it's beneficial, temperature sensors can be networked, and their individual values averaged for system use. While this is usually unnecessary for outdoor temperature monitoring, it may be used to better monitor and control the temperature in a very large indoor area, such as a basketball arena.

Information sharing can also be used to achieve HVAC system coordination between waterside and airside equipment controllers. For example, air terminals can share information with air handlers (airside equipment) that then can share information with the chillers (waterside equipment). In this scenario, the air handlers can vary the air volume and temperature to meet changing needs at the air terminal. The chillers can vary the chilled-water volume and temperature to meet changing airside needs. Sharing information allows the system to continually optimize equipment operation and see how the load is changing.

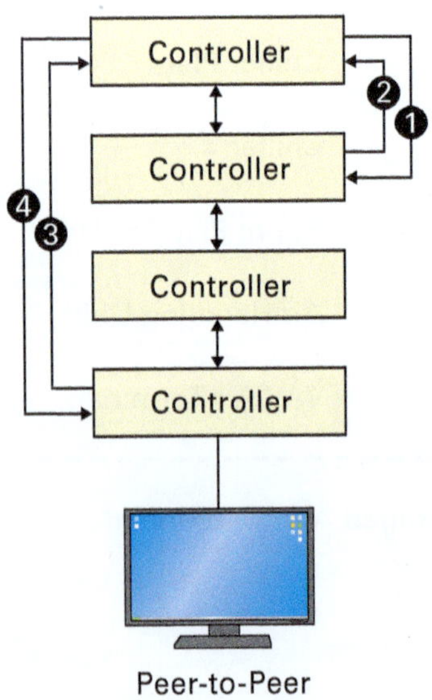

Figure 10 Peer-to-peer network.

Peer-to-peer networks (*Figure 10*) get the name from the way in which communication occurs between controllers. Each controller can communicate with any other controller connected to the network. All controllers are peers, having equal priority to use the communication bus. Each component in the network is a potential provider and user of shared data.

In this network, any controller can communicate at any time by going (in less time than the blink of an eye) through a four-step process:

Step 1 Wait (in milliseconds) to see that the communication bus is quiet.

Step 2 When sufficient quiet time has occurred, send a packet of information to another controller(s).

Step 3 Wait to receive a confirmation that the other controller(s) received the message.

Step 4 Repeat the transmission if confirmation has not been received in a timely manner (again, milliseconds).

All controller information may also be viewed or modified from a single user workstation, located anywhere within the building. Other than housing the software and providing a graphical interface to interact with the system, the computer workstation simply functions as another peer.

On Site

Polling Networks

Polling networks work differently than peer-to-peer systems. A master polling controller looks for votes from the members of the network and makes a decision as to what to provide in terms of heating or cooling. Polling networks are typically used to control variable air volume and temperature (VVT) zoning systems, as introduced in NCCER Module 03315, *Zoning, Ductless, and Variable Refrigerant Flow Systems.*

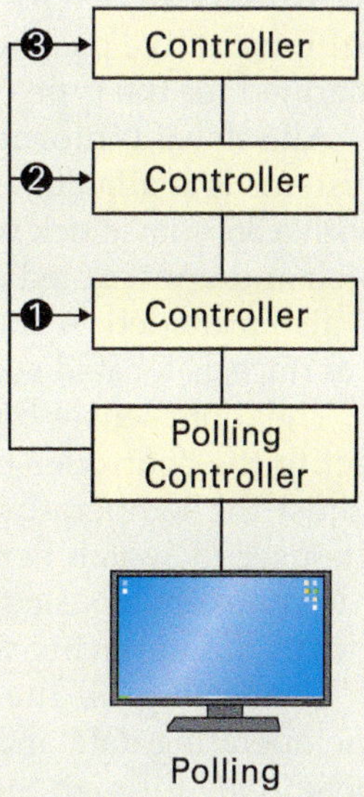

The controllers can share information because a common language or set of rules known as a **protocol** is used. The information traveling along the bus must be configured in a specific way to be recognized. The protocol is designed to check the data for errors and make sure that only one device transmits at a time. DDC control system manufacturers choose their own protocol that governs communication along the communication bus.

The speed at which information moves on the network is known as the **baud rate**. The baud rate is stated in **bits** per second (bps). The higher the baud rate, the faster the data moves. Fiber optic cable is required to reach the highest baud rates and transmit data over a longer distance.

Protocol: An established set of rules that specify how to format and exchange data transmitted between devices in a network, allowing connected devices to communicate with each other, regardless of differences in their internal structure and design.

Baud rate: The rate at which information is transmitted across communication lines.

Bits: Abbreviation for binary digits; the smallest element of data that a computer typically handles. A bit represents an off or on state (zero or one) in a binary system.

2.2.0 Packaged Unit Digital Controllers

Manufacturers of HVAC equipment often install a digital controller on equipment at the factory, selected to meet the job specifications and chosen communication protocol. This type of controller is also referred to as a *product-integrated controller (PIC)* or an *application-specific controller* (*Figure 11*). The controller

Figure 11 Carrier product-integrated controller (PIC) for a packaged unit.
Source: Carrier Corporation

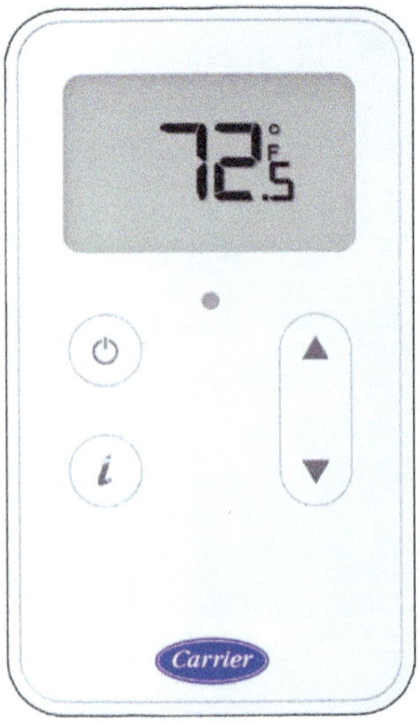

Figure 12 Carrier ZS zone temperature sensor and interface.
Source: Carrier Corporation

shown is designed for a packaged rooftop unit. Most PICs are found on unitary equipment like packaged units. The inputs and outputs of a PIC are specifically planned for the type of equipment to which it is attached.

All relevant information regarding the packaged unit's operating and safety systems is available within the controller. Such a controller is typically provided with a 365-day clock with up to eight occupancy periods and setpoints for occupied and unoccupied periods. Holiday and daylight savings time programming is also provided. A room temperature sensor (*Figure 12*), with or without a user interface, is located somewhere in the zone.

The room sensor buttons (when present) provide limited system control, subject to the controller programming. For example, the user may only be able to adjust the temperature within a 5°F (2.8°C) range. An override button allows the user to switch between the occupied and unoccupied modes of operation. Alternatively, a basic room temperature sensor is used, and the occupant has no means of interfacing with the HVAC equipment.

Connecting all the PICs to the building's peer-to-peer network bus makes the operating data and scheduling accessible through the local workstation, and potentially through the internet. *Figure 13* shows how multiple pieces in this family of BMS controls work together to control a single piece of equipment. Of course, many units and their related components can all be connected to the same bus.

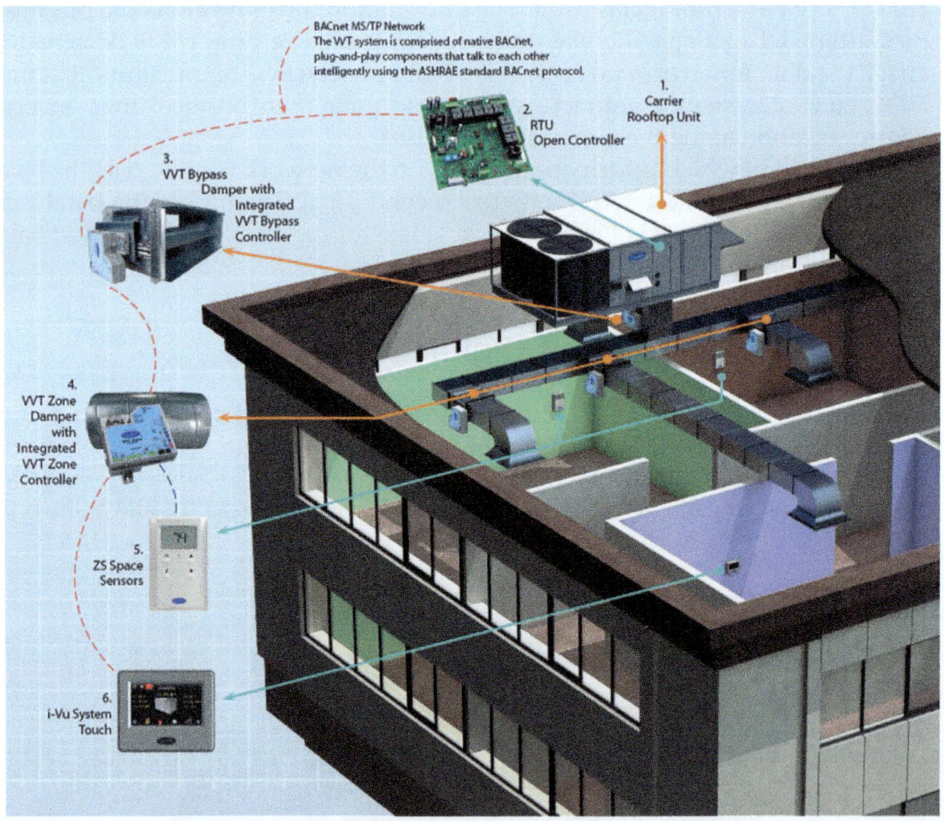

Figure 13 Carrier i-Vu® system overview.
Source: Carrier Corporation

2.3.0 Applied HVAC Systems

Applied HVAC systems typically provide comfort to a specialized market of HVAC users, such as hospitals, schools, manufacturing facilities, office complexes, malls, and hotels. Applied equipment includes boilers, chillers, cooling towers, large air handling units, and similar components.

General-purpose controllers are typically used to integrate the boiler, cooling tower, pumps, and exhaust fans into the BMS. These controllers may be installed in the field by the BMS controls contractor or the mechanical contractor. General-purpose controllers are more versatile in design than PICs, because these controllers must be capable of interfacing with a diverse selection of equipment.

The following sections describe the use of digital controllers in a variable air volume (VAV) system served by chillers.

Around the World

Shanghai Science and Technology Museum

In the midst of Shanghai's Pudong financial district, the Shanghai Science and Technology Museum provides an awe-inspiring architectural monument. The world-renowned project showcases the very best in every field of science, encompassing over 732,000 square feet, or 17 acres.

The HVAC equipment includes over 100 large air handling units, 300 exhaust fans, 7 heat recovery systems, and roughly 3,200 control points associated with the BMS. Over 300 control points are dedicated to the chilled-water plant alone, which includes thermal storage for off-peak chilled-water generation. The BMS brings the HVAC, lighting, security, and life safety systems together for consistent monitoring and control.

Source: Weiming Xie/Shutterstock

2.3.1 Chilled-Water Systems

Figure 14 provides a schematic for a typical chilled-water system. Two parallel water-cooled chillers supply chilled water to three air handling units (loads) equipped with two-way control valves. Two chilled-water pumps and two condenser water pumps serve the chillers and distribute water to the loads. Each chiller has a leaving chilled-water temperature sensor (LCHW). Another sensor (CHWST) monitors the mixed chilled-water supply temperature as it enters the loads. The chilled-water return temperature sensor (CHWRT) monitors the water temperature as it leaves the loads.

Each chiller is sized to handle 50% of the building load. In many cases, a third chiller, condenser water pump, cooling tower, and chilled-water pump is installed to provide redundancy.

The chiller condensers, condenser water pumps, and cooling towers are piped in parallel. Sensors are installed in the condenser water piping entering (CWST) and leaving (CWRT) the tower.

Digital controllers are mounted on each chiller and on each air handling unit. In addition, a general-purpose controller (GPC) is used to control the cooling tower fan(s). Each chiller controller contains its own operating system, safety

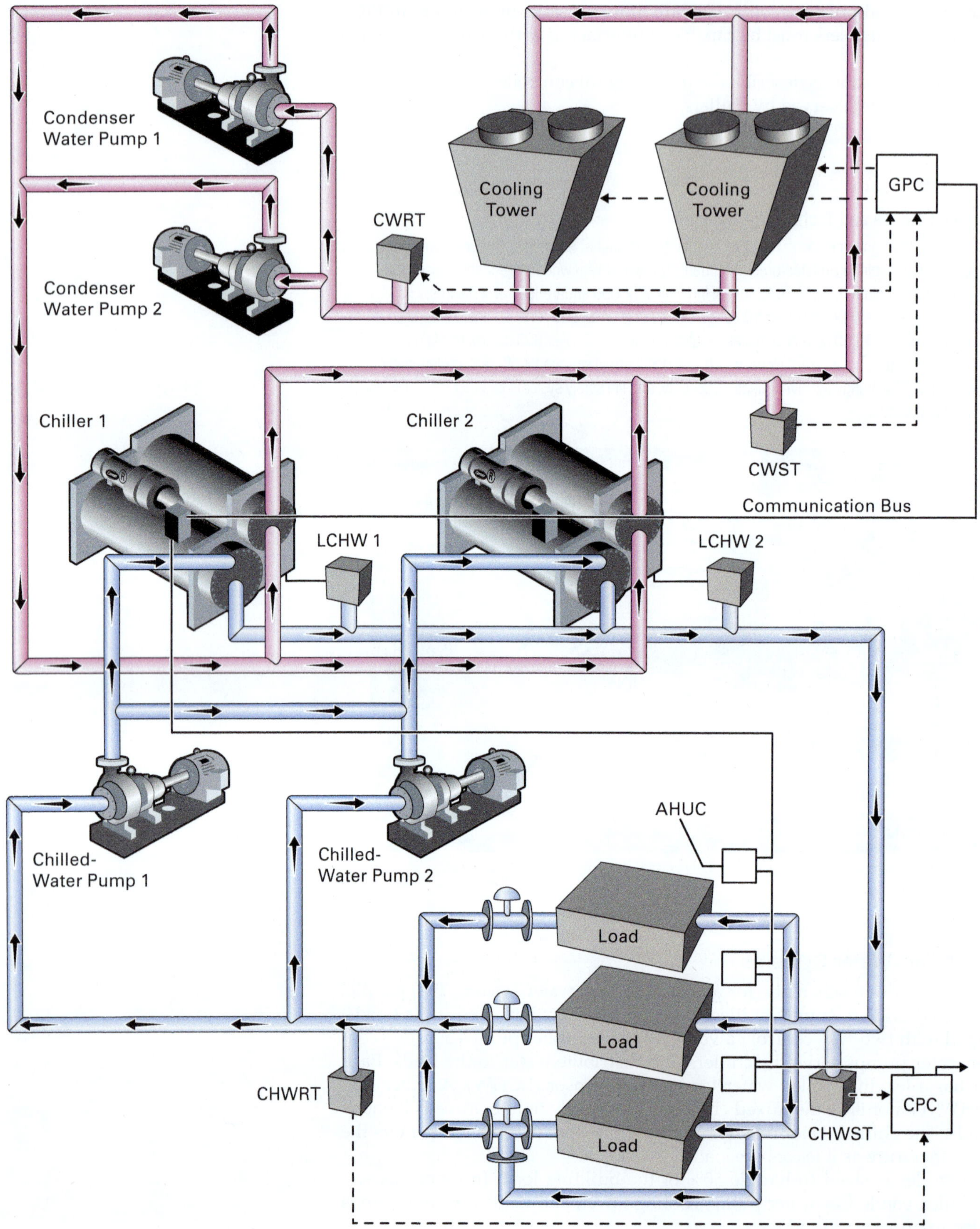

Figure 14 Typical chilled-water system.
Source: Carrier Corporation

circuits, time clock, occupancy schedule, and chilled-water setpoint. A chiller plant controller (CPC) is provided to manage the two chillers as a single chilled-water plant to modulate the capacity as needed. When requested to start by the CPC, each chiller can start the chilled-water and condenser water pumps. All the digital controllers are connected to a communications bus.

Digital controllers mounted on each air handling unit control the units based on zone needs and time scheduling. The chilled-water valves modulate to maintain a supply air temperature of about 55°F (13°C). The air is supplied to a VAV system.

On Site

Chillers

Most commercial buildings equipped with chilled-water systems have one or two chillers in the plant room. Large facilities may have several in the mechanical room. Air-cooled chillers, like the one shown here, are also common.

The chiller plant control module (referred to as the CPC in *Figure 14*) provided by manufacturers can typically control up to eight chillers, and the chillers can be of a mixed type and capacity, such as two water-cooled centrifugal chillers and an air-cooled screw chiller that might have been added during a building expansion.

Source: Sakan Piriyapongsak/iStock/Getty Images

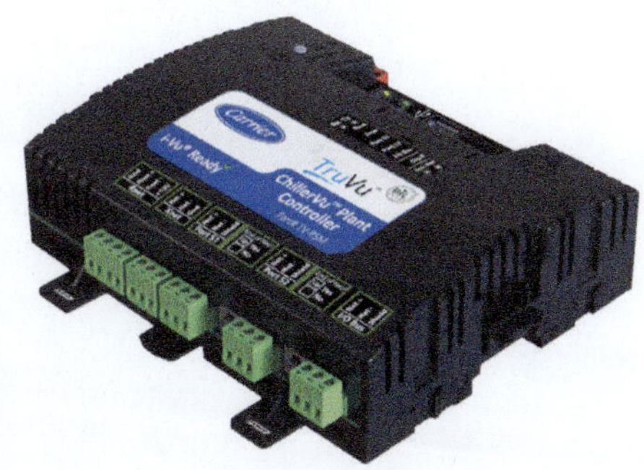

Figure 15 Chiller plant control module.
Source: Carrier Corporation

2.3.2 Chiller Plant Control Module

Because most buildings will have more than one chiller, a chiller plant control module (*Figure 15*) is required to coordinate chiller operation. This controller monitors the needs of the building and starts the chillers based on various inputs. The controller then uses both the chiller plant's LCHW and CHWRT sensors to determine how many chillers need to be operating and what the individual chilled-water setpoints need to be.

Besides starting and stopping the chillers, the chiller control module also equalizes their run hours, keeps the chillers equally loaded when more than one is operating, starts another chiller when one fails, and performs similar chiller plant management functions.

2.3.3 General-Purpose Controller Functions

Other components associated with an applied HVAC system are the boiler, cooling tower, and exhaust fans. A general-purpose digital controller (GPC) is generally used to integrate these pieces of equipment into the system. For example, the GPC shown in *Figure 16* can handle the aforementioned equipment as well as pumps, air handling units, lighting systems, and similar equipment.

Typically, the boiler for such a larger building comes with its own factory-mounted control panel. The controller contains the operating system and safety circuits related to the boiler. The GPC integrates the cycling of the

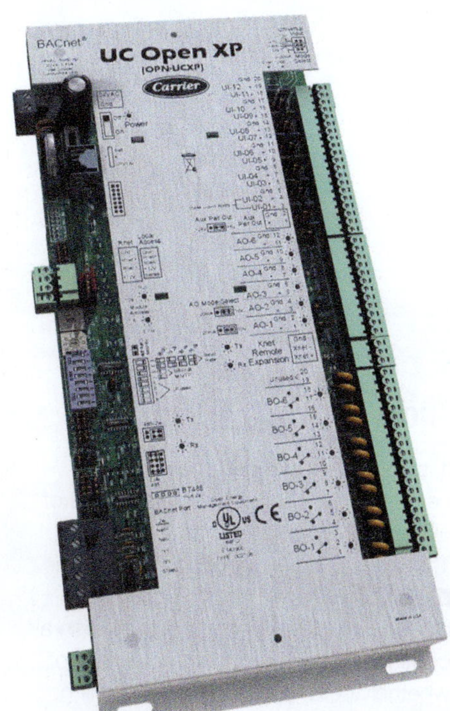

Figure 16 General-purpose controller.
Source: Carrier Corporation

boiler with the building's heating needs and resets the boiler's hot-water temperature setpoint as necessary to accommodate the outdoor conditions and current building temperature.

2.3.4 VAV Systems

VAV systems (*Figure 17*) are often used in buildings with loads exceeding 25 tons that require simultaneous cooling and heating at the zone level. An air handling unit provides conditioned air to the individual VAV boxes that then control the airflow into their respective zones.

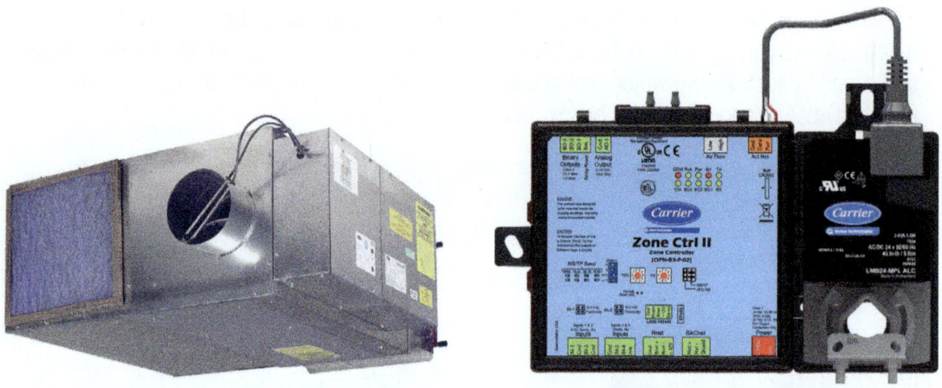

(A) Fan-Powered VAV Terminal **(B) VAV Terminal PIC and Integrated Damper Actuator**

Figure 17 Fan-powered VAV terminal and product-integrated controller.
Source: Carrier Corporation

The air handling unit has its own controller and can modulate the supply air volume as needed downstream, using a variable frequency drive (VFD) for the blower motor. Each VAV box is fitted with a PIC (or general-purpose controller) and a zone sensor. A VAV box PIC typically has an integrated damper actuator that controls the flow of conditioned air through the box. VAV boxes around the perimeter of the building may also be equipped with electric heat or a hot-water coil; boxes in the core of the building rarely require additional heat. All the controllers are connected to the same bus for communication.

During occupied periods, the air handling unit supplies cool air in the supply duct. VAV systems typically remain in the cooling mode year-round during occupied periods. The supply air temperature sent to the VAV boxes is typically maintained between 50°F and 65°F (10°C and 15°C). Most of today's systems monitor the present load demand in the building and/or the outdoor temperature, then determine the best supply air temperature setpoint. This is referred to as *supply air reset*. Both the air handling unit's blower and cooling resources are used minimally during unoccupied hours.

Refer to *Figure 18*. Individual VAV boxes work to satisfy zone needs as follows:

- When the zone sensor requires cooling and is several degrees warmer than desired, the VAV box damper is wide open, allowing the full volume of conditioned air supplied by the air handling unit to enter the zone.
- As the zone temperature gets closer to the setpoint, the damper begins to close.
- When the zone setpoint is reached and there is no demand for cooling, the damper moves to its minimum position (not closed), allowing a small amount of conditioned air to continue flowing through.

NOTE

It is important to remember that the central air handling is also the source of all fresh air introduced into the building. Therefore, the supply of conditioned air passing through a VAV box cannot be closed completely, ensuring that fresh air continues to enter a zone regardless of whether there is a demand for cooling.

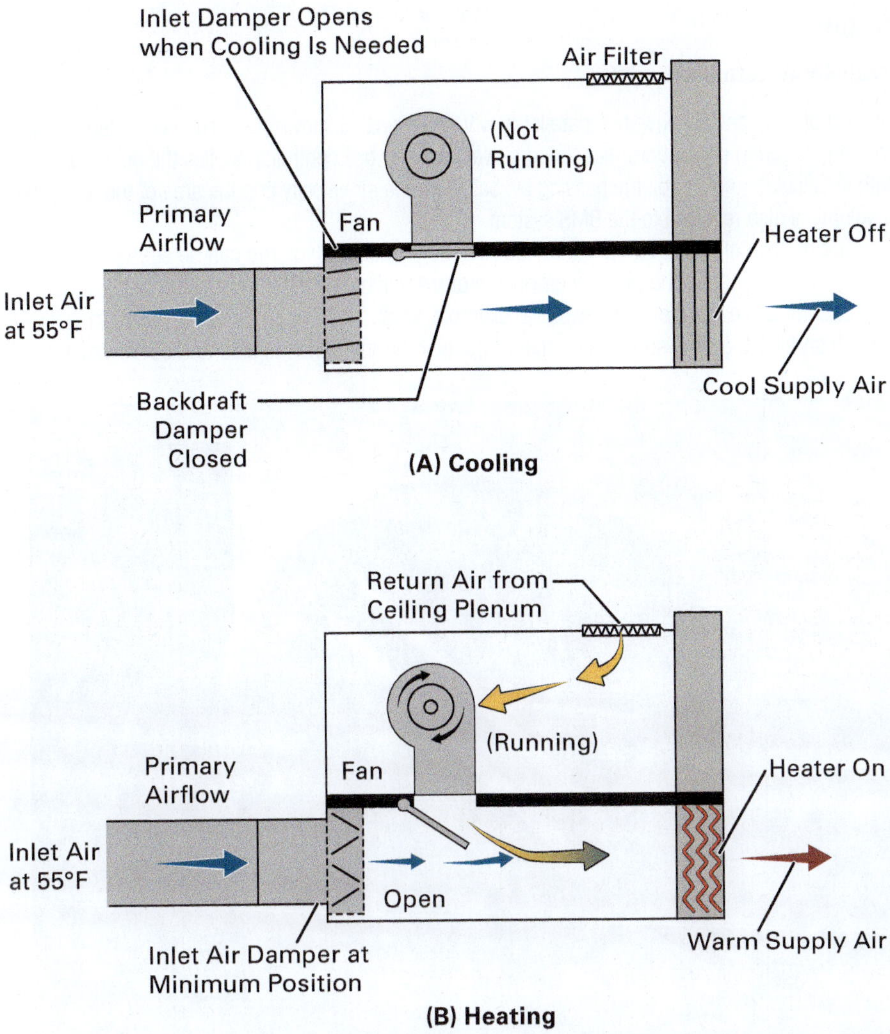

Figure 18 Parallel VAV box heating and cooling operating modes.

- If the zone temperature falls below the setpoint, the VAV box integral blower and heater (if equipped) cycle on. Return air that is typically drawn from the ceiling plenum is mixed with the conditioned air passing through. The mixed air passes over the heater and into the supply ductwork for the zone. The VAV box damper remains at its minimum open position, with most of the air for the zone being recirculated.

Note that operating sequences vary based on system design, facility use, and geographic region, as well as other factors.

The primary air handling unit operates independently of the VAV boxes it serves, with little need for them to share information. The air handling unit only needs to be aware of the VAV box operating schedules and then maintain the duct pressure setpoint downstream when the boxes are active. As VAV boxes open and close to satisfy the demands of the space, the air pressure in the primary conditioned air duct rises and falls. The VFD for the air handling unit blower responds to the static pressure sensor and adjusts the blower speed as noted to maintain a constant pressure. Very precise system air pressures can be maintained this way.

The mechanical portion of the system, such as the chilled- and hot-water systems, operate in much the same way. While data such as an operating schedule is shared on the bus, direct operating inputs come from water temperatures and flow rates, which relate to demand. Chiller controllers modulate system capacity based primarily on these factors, with other factors often added to the algorithms.

On Site

Parallel VAV Terminal

The inlet of a typical fan-powered, parallel-flow VAV terminal is shown here. The perforated tubes crossing the round air inlet are measuring air velocity. The box controller applies the necessary math to calculate the air volume passing by, based on the air velocity and the area of the duct. The air volume is also reported to the BMS system.

The horizontal round damper that controls the primary air volume can be seen just beyond the two tubes. Because the primary air duct is at a significantly higher pressure than the duct downstream of the damper (2.5" w.c., for example, versus 0.5" w.c.), the inlet duct connection is much smaller than the supply air outlet of the box (on the opposite side and out of view).

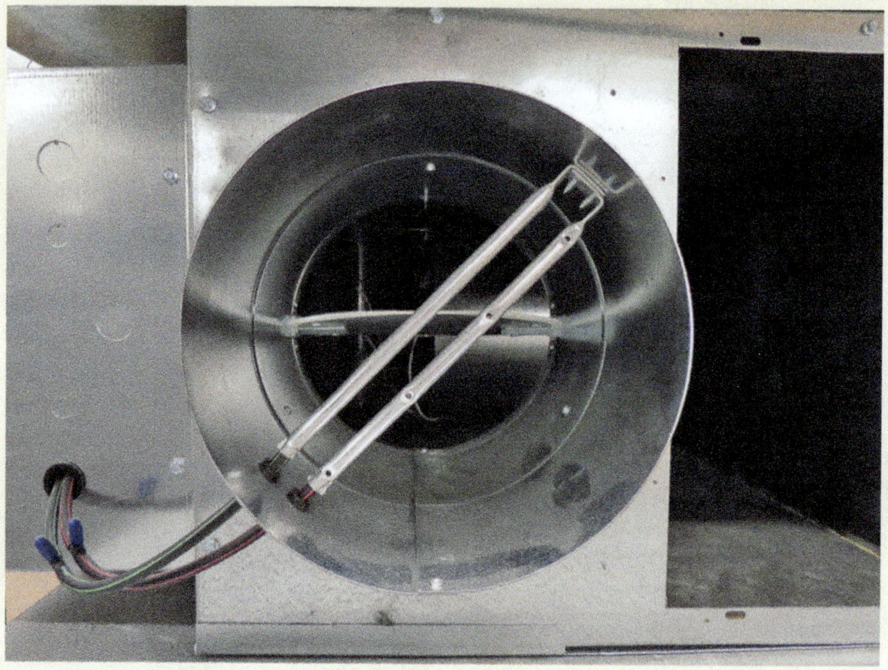

2.0.0 Section Review

1. Which term refers to the speed at which information moves between network components?
   a. Bit rate
   b. Byte rate
   c. Baud rate
   d. Protocol rate

2. Where are you most likely to find a PIC?
   a. Cooling tower
   b. Pump package
   c. Hot-water boiler
   d. Packaged rooftop unit

3. In a VAV control system, what is the most important information that is shared directly between the air handling unit and the VAV box controllers?
   a. The supply air pressure
   b. The operating schedule
   c. The outdoor temperature
   d. The supply air temperature

3.0.0 BMS User Functions

Objective

Explain how users interface with a BMS and identify tasks that can be accomplished.

a. Describe how users access and interface with a BMS.
b. Identify various tasks that users can accomplish through BMS access.

Performance Task

1. Interpret, identify, and describe HVAC system operating data viewed through building management system software.

Digital control modules and software must be accessed to achieve building subsystem management functions. The following sections describe methods used to interface with building management systems and some of the tasks that can be accomplished.

3.1.0 User Interfaces

Users can interface with today's building management systems in a variety of ways:

- A portable **local interface device**, connected directly to a digital controller
- A locally connected (to the facility network) computer workstation
- A device on the **local area network (LAN)** or **wide area network (WAN)**
- A remote computer workstation, tablet, or smartphone via the internet

A local interface device (*Figure 19*) is typically a portable device used by technicians to interface directly with a system controller. It is connected to a digital controller with a cable that usually provides both power and a communication path. Portable local interface devices can be moved from one controller to another as needed for viewing, diagnosing, and changing controller information. Many controller manufacturers provide such a device designed specifically for technicians to interface with their products. These devices are generally considered to be technician tools.

Local interface device: A portable device with a keypad and display for data entry, used to connect directly to a digital controller via cable.

Local area network (LAN): A server/client network connecting multiple computers and devices within a building or building complex.

Wide area network (WAN): A server/client network spread over a larger geographical area, generally with a radius of half a mile or more.

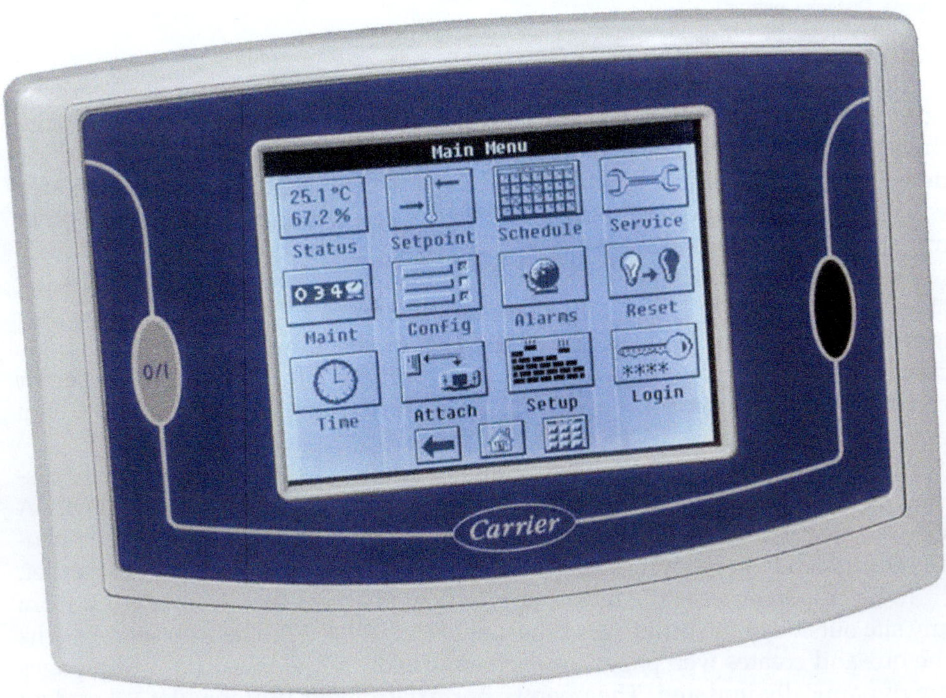

Figure 19 Digital controller local interface device.
Source: Carrier Corporation

NOTE

The term *local interface device* may have a different meaning in other topic areas.

In the past, when more unique protocols were in use among BMS manufacturers, branded local interface devices were essential tools. Today, however, open protocols are widely used in BMS. This allows a wider variety of common tools to connect and access controller and system information regardless of manufacturer.

Some BMS controllers today have a built-in display and interface (*Figure 20*). Such controllers do not require a local interface device, making on-site setup more convenient.

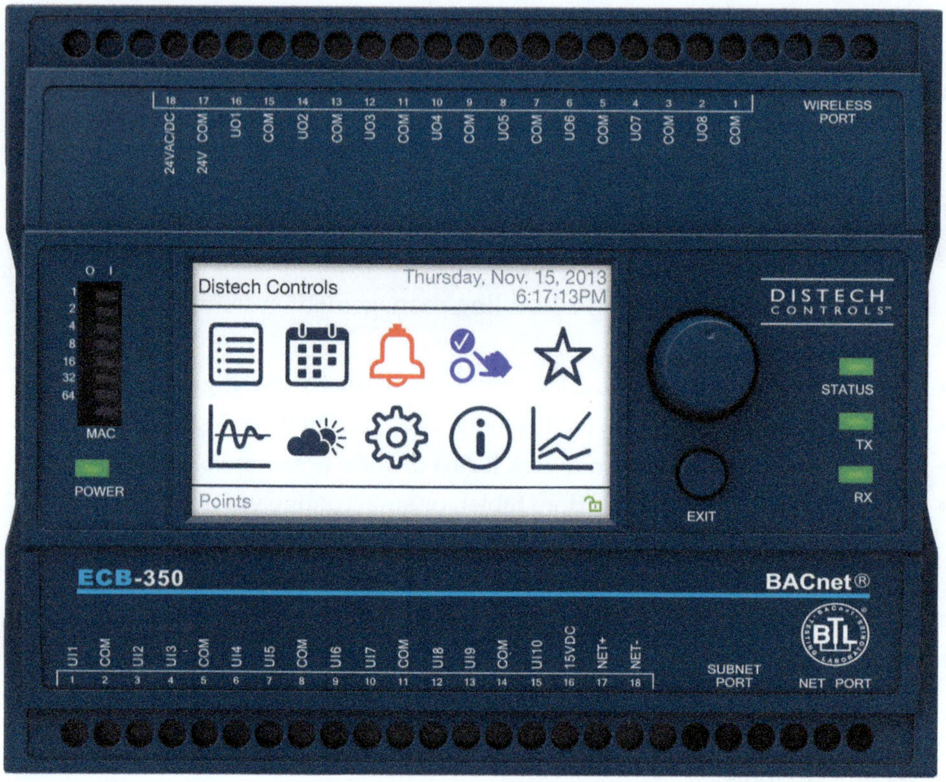

Figure 20 Digital controller with integrated graphical interface.
Source: Distech Controls

Most building management systems have at least one computer workstation in the facility (*Figure 21*). Through the workstation, which is loaded with the appropriate control system software, the operator can interact with all the devices on the connected network through a graphical interface. Dedicated workstations usually offer more options and accessibility than tablets and other mobile devices.

Despite the power of dedicated workstations, many users benefit from being able to access the BMS from anywhere on the network. In most cases, the user is only interested in completing a single task or checking the status of a single component. Networks link devices into a server/client format, where the server computer provides central data storage for all network controller data. Client devices access the network through the server.

A LAN is typically used within a single building. All devices capable of displaying the information and offering a keyboard can interface with the BMS. A WAN is a network that incorporates a larger geographical area and can be virtually unlimited in geographical size. A WAN may connect several LANs together.

Today, the internet is the most popular way to access the BMS network from any site outside of the building's **Ethernet** LAN (*Figure 22*). The web server scans the bus and creates web pages for each controller on the bus. These web pages are dynamically updated. The open **Hypertext Transfer Protocol (HTTP)** and an **Internet Protocol address (IP address)** are used to interface with the web pages. A device's IP address is used to identify the device with which you want to

Ethernet: A system of wired connections used to create a local area network, including the communication protocol to be used and the method of avoiding simultaneous information transfer by two or more network devices.

Hypertext Transfer Protocol (HTTP): The base protocol used for transferring data between clients and servers on the internet.

Internet Protocol address (IP address): A unique address that electronic devices use to identify themselves and communicate with each other on the internet.

Figure 21 Local BMS workstation with a graphical interface.
Source: Distech Controls

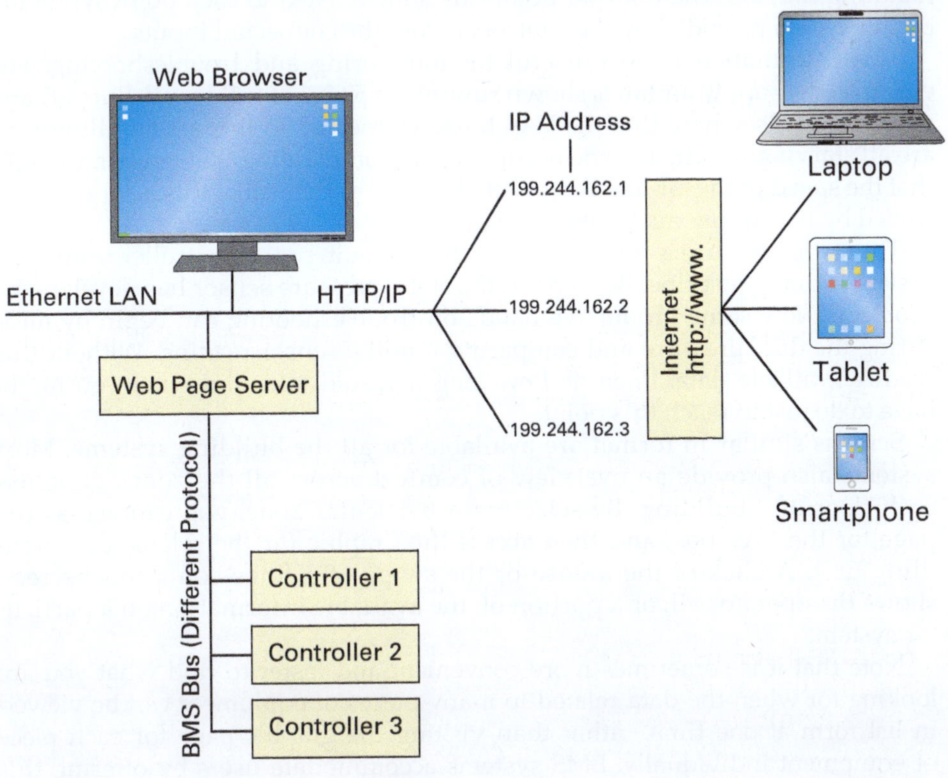

Figure 22 Internet access to a BMS.

connect. One advantage to this approach is that the device used to access the system does not require the BMS software to be installed; any device with internet access will work.

Smartphones and tablets are now used to access web-based information and sites regularly. This capability is especially valuable to companies that have technicians on the road, moving constantly from place to place. Accessing and working within a BMS is simply a matter of the system software being optimized for access by these devices, so that the web pages fit neatly on the small screens.

3.1.1 Secure Access and Operator Management

Security controls, including passwords, are used to limit access to a BMS. In the operator management portion, operators are given differing levels of authority. Some can only look at data, some can run diagnostic tests, and others can change the operating parameters or programming. The higher the level of authority, the more control an individual has over the system.

BMS system security goes beyond concern for the system's integrity. Hackers that gain access to these systems can take full control of the building's numerous utilities, power systems, and building access. Two-factor authentication is often necessary to help ensure only authorized users gain access. Individuals who are given access but no longer require it, perhaps due to a change of employment, should be immediately removed from the system. Some systems allow a user to be entered for a limited time.

3.1.2 Graphical Interface

A graphical interface uses line art or photographs to represent various subsystems and equipment. Dynamic, live information about the operation of the unit is superimposed on these images.

Figure 21 provides an example of a graphical interface. In this case, an air handling unit and the relevant points are shown. Next to each point symbol is dynamic data provided by the controller from the connected inputs.

This information is very useful for monitoring and troubleshooting. For example, the supply air fan is shown running at 52.2% of full speed. But perhaps you've seen elsewhere that the VAV boxes connected to this air handling unit are all starving for air, and the occupants are complaining. Remember, though, that the speed of the air handling unit blower in systems like this is largely controlled by the supply duct static pressure.

The static pressure is within 0.02" of the setpoint, so the controller thinks the system is doing just fine. But what if the static pressure sensor has developed a problem? Now some productive hands-on troubleshooting can begin by measuring the duct pressure and comparing it to the sensor reading. Without this readily available data, imagine how long it would take and what you might have to do just to reach this point.

Screens similar in format are available for all the building systems. Most systems also provide an overview of comfort across all the zones or across each floor of a building. By selecting a particular zone, you can access the page for the VAV box, and then access the graphic for the relevant air handling unit. A click of the mouse or the swipe of a finger on a touchscreen shows the operator all, or a portion of, the available information on a particular system.

Note that it is sometimes more convenient and faster to find what you are looking for when the data related to many pieces of equipment can be viewed in list form at one time, rather than visiting the graphic page for each piece of equipment individually. BMS systems accommodate users by offering this option as well.

3.2.0 Other Building Management System Functions

HVAC systems controlled by a BMS have features and functions that go far beyond simply maintaining the desired temperature in a space. The most common features include the following:

- Occupancy scheduling
- Point management
- Alarm management
- **Tenant billing**
- Lighting control
- Building access and security
- Fire and smoke control

Tenant billing: The process used to track resource usage and charge a building tenant for specified uses of the building's HVAC system, as well as other systems.

3.2.1 Occupancy Scheduling

Figure 23 shows a typical occupancy schedule. In this example, the Weekly Work Schedule for the building is being edited. Once the operator makes a change, the change is sent down to all controllers that are set to operate on that same schedule.

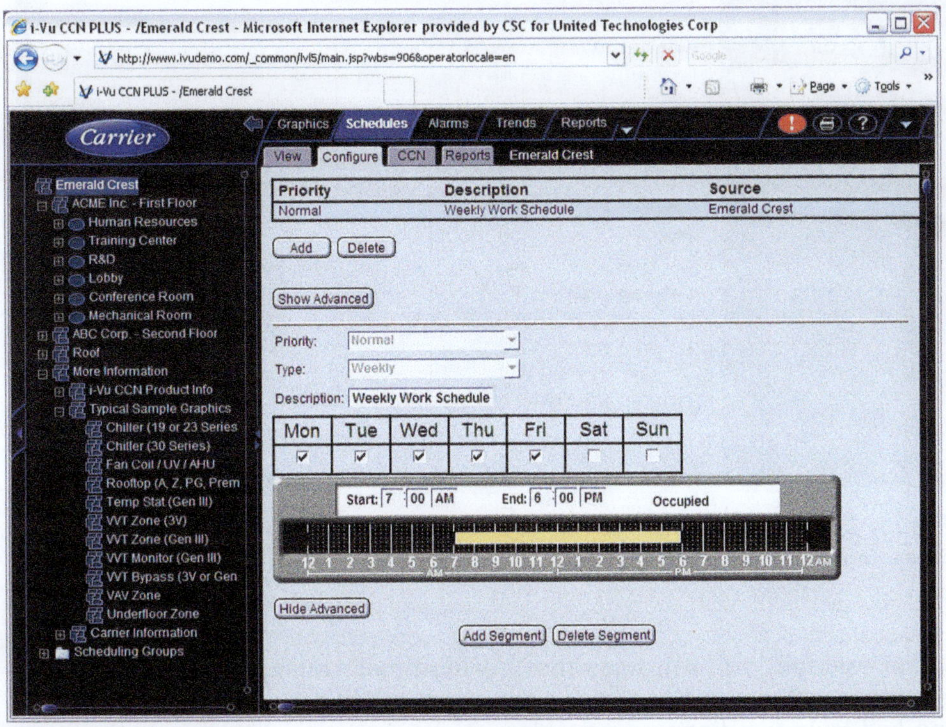

Figure 23 BMS occupancy scheduling screen.
Source: Carrier Corporation

Once the controllers have that information, a temporary loss of system connection won't interfere with the programmed operating schedule. The controllers will continue to operate on that schedule.

Time schedules can be specific to one controller, applied to a specified group of controllers, or they can be global and apply to all controllers. Many different schedules can be created to accommodate any facility's operational plan.

3.2.2 Point Management

The operator can view and modify the algorithms and points connecting to any controller. This information is displayed on dynamic screens that interrogate the highlighted controller through the network bus. In *Figure 24*, the points shown are associated with the first-floor human resources offices of ACME Inc., a tenant in the Emerald Crest building. Any changes made here are automatically stored at the controller level and on the server that stores a mirror image of the controller's file.

Note that there are both inputs and outputs shown here. Space Temperature, for example, is clearly an input, while Heat is an output that is currently disabled. From this screen, you cannot only view current information, but also force some things to change despite the controller's programming. Forcing a change can be very helpful during testing and troubleshooting.

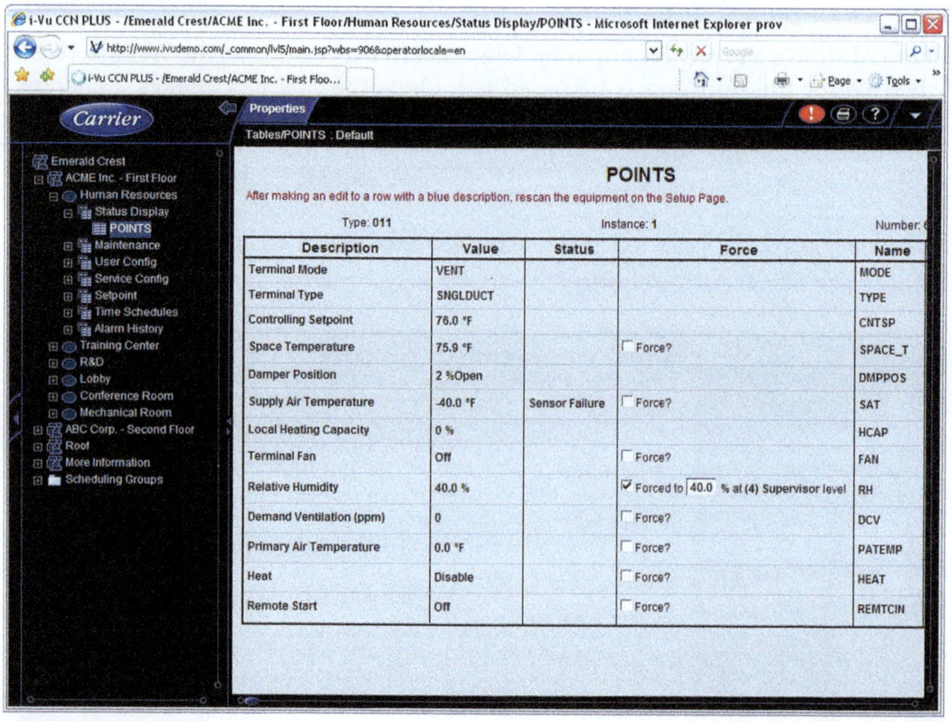

Figure 24 Point list control screen for a single controller.
Source: Carrier Corporation

For example, you can force the VAV terminal's integral fan to start. Even the space temperature can be forced, which might seem odd. But when testing and trying to diagnose a problem, perhaps you need to change the Space Temperature value temporarily and see how the VAV box responds. You can temporarily force the space temperature higher or lower as needed. This is the equivalent of physically going to the zone sensor and changing the setting manually.

A wide variety of reports can also be configured and generated from various points. *Figure 25* shows a report of space temperatures from various parts of the facility, with a snapshot of the conditions taken every 20 minutes. At the top of the screen, note that these reports can be exported to a variety of formats.

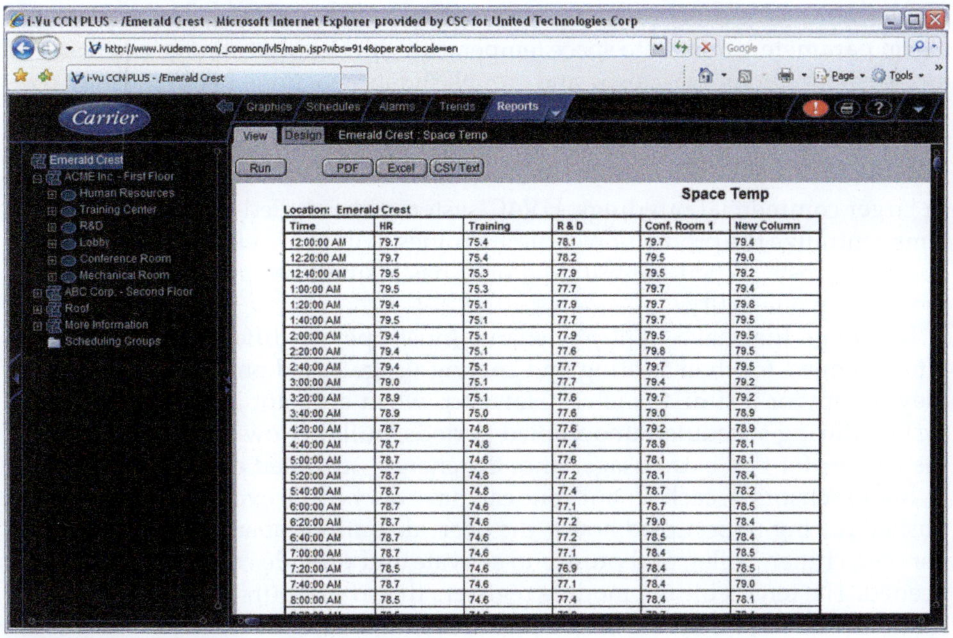

Figure 25 Space temperature report.
Source: Carrier Corporation

3.2.3 Alarm Management

Whenever an alarm condition occurs in a controller, the alarm is transmitted to the central computer and stored in an alarm file (*Figure 26*). The operator can be notified of an alarm visually and/or audibly, then address the alarm condition by reviewing the alarm file and acting accordingly. How specific alarms are reported is a user option.

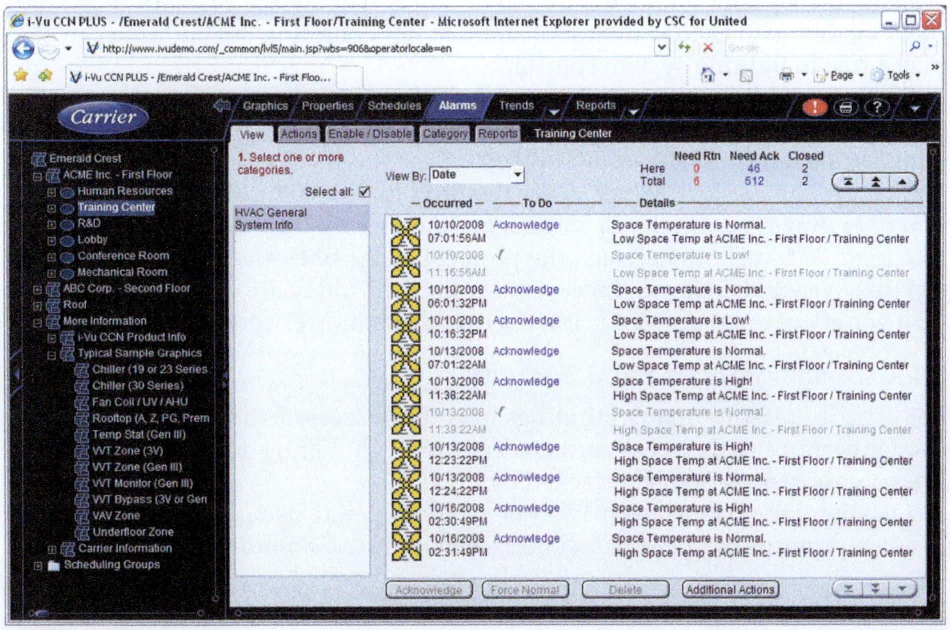

Figure 26 Alarm file for a single controller.
Source: Carrier Corporation

The alarm history is stored so operators or technicians can spot trends. For example, a particular alarm may repeatedly occur at certain times of day or under specific environmental conditions. This information can often help track down the problem.

Note that you can often cause an alarm entry as well. In the previous section, forcing the Space Temperature value higher or lower than the actual temperature as a diagnostic aid was discussed. However, depending on the alarm parameters related to space temperature, forcing that change can result in an alarm. Previous diagnostic and troubleshooting actions taken must be considered when one examines an alarm file.

3.2.4 Tenant Billing

In larger commercial buildings, HVAC systems like chilled- and hot-water systems centralize the production of these utilities to improve efficiency and reduce operating costs. When the building only has a single tenant, all the resources are used by one entity.

Buildings that have multiple tenants must operate differently. Tenants are often charged for their heating and cooling usage based on the square footage they occupy or a similar metric. However, when a tenant activates the HVAC system during scheduled unoccupied periods, building owners want to charge the tenants for that extra usage, since the energy expended can be substantial.

Room sensors in each tenant zone can have a schedule override button. When pushed during unoccupied hours, the override button causes the associated air handler, chiller, boiler, and pumps to activate and provide cooling or heating as needed. The tenant billing module registers this after-hours demand and stores the zone that requested the extra HVAC usage and how many hours were used. At the end of the month, the tenant billing module works with the system software to produce a bill for each tenant.

Tenants can also be billed separately for the electrical power used monthly. Current transformers (CTs) can be placed on each power lead that feeds power to the space. The CTs become inputs to a controller that manages the information and feeds it into the tenant billing module of the software.

3.2.5 Lighting Control

Lighting, both internal and external, uses a substantial portion of the energy in a commercial building. The BMS is often tasked with controlling the lighting systems throughout the facility. Time of day is usually the primary factor, but other factors such as the amount of ambient light entering the building can also play a role in lighting system control.

Commercial lighting systems are controlled by their own digital controller system and may use one of several unique communication protocols like digital addressable lighting interface (DALI), ZigBee Wireless, or building automation and control network (BACnet®). For the BMS to access information from a system operating on a different protocol, an interface module is needed. The interface module translates between the protocol of the BMS and the protocol of the lighting system. However, since so many systems today are designed to operate on a variety of open protocols, interface modules aren't needed as often.

3.2.6 Building Access and Security

For security reasons, many buildings must limit access to authorized individuals. Only those with access cards or keys may gain entry to the building, and their access history is easily tracked by the BMS.

Like lighting systems, security systems may operate using a different communication protocol and require an interface module to communicate with the BMS.

3.2.7 Fire and Smoke Control

In larger commercial buildings, the ability to control smoke evacuation from the building is a matter of life safety. Redundancy in control is often required, and smoke evacuation fans and dampers often require stand-alone controls that allow the equipment to operate (assuming power is present) despite the failure of the BMS.

Remember that a ready source of fresh oxygen is an essential component of a growing fire. Once a fire begins, the introduction of fresh outside air must be stopped immediately. Building air handling units of all sizes and types must shut down to prevent the circulation of fire and smoke through the air distribution system. Fire dampers in air distribution systems are either closed automatically by the fire control system or by the melting of the thermal links that hold the damper blades open.

For these reasons, smoke evacuation systems do not play a role until the fire is extinguished. Smoke evacuation fans are powerful fans that exhaust large volumes of air (smoke) from the building. The exhausted air is rapidly replaced by fresh outside air. Thus, evacuating smoke from the building too quickly can feed or help re-ignite a fire.

Figure 27 is a schematic for a typical smoke evacuation control system including the smoke detectors, detection and alarm panels, and digital controllers communicating with the fans and smoke dampers. Smoke dampers are normally closed (NC), and the fans are dedicated to the sole purpose of smoke evacuation. These fans are not used for any other purpose.

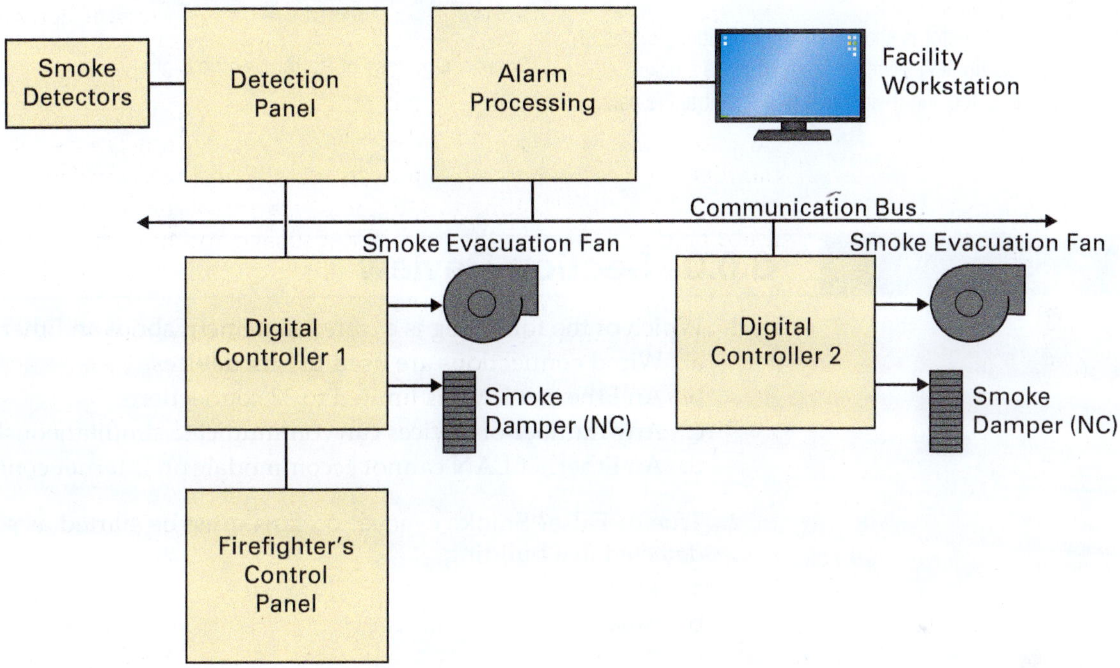

Figure 27 Smoke evacuation control system schematic.

An operator's control panel allows building personnel to monitor the status of the equipment and perform periodic testing. The firefighter's control panel allows the on-site fire marshal to view and control fire- and smoke-related equipment. The fire marshal can also initiate the smoke evacuation system when the time is right.

Not every building must be equipped with a dedicated smoke evacuation system. In this case, the HVAC system can be placed in the following modes from the fire marshal's panel:

- *Building Pressurization* — The HVAC system provides the maximum amount of air to the VAV zones by opening the outdoor air dampers 100%. All exhaust fans are turned off and no return air is circulated to minimize smoke entry into unaffected zones. All VAV dampers are positioned to the maximum cooling position. This also forces the air handling unit to provide the maximum volume of air that the VAV boxes collectively demand.
- *Evacuation* — This mode removes smoke from the zones by running the central powered exhaust fan only. This fan is normally used to help control building pressure, but it is now used to help evacuate the building. The return dampers remain closed to avoid recirculating any smoky air. The VAV dampers now move to the closed position.
- *Smoke purge* — This mode flushes smoke from the zones by running the supply fan operating on 100% outside air and the powered exhaust fans. The return air dampers remain closed. The VAV dampers again move to the maximum cooling position.

On Site

Smoke Control

The control of fire and smoke is a more sophisticated process than you might imagine. A limited number of manufacturers specialize in this product market. The fire system is typically on a separate network from that used for the HVAC equipment, and the two systems cannot be controlled through a single software package. This is a requirement of NFPA 101®, *Life Safety Code*®. The smoke control system interacts with the HVAC system through the control of safety switch and relay contacts wired into the HVAC equipment control circuit. In other words, a great deal of the interactivity relies on wired circuits.

A typical duct smoke detector is shown here. It is designed to detect smoke in all types of air distribution systems. Many of these smoke detectors can be networked, feeding their status to the central fire panel.

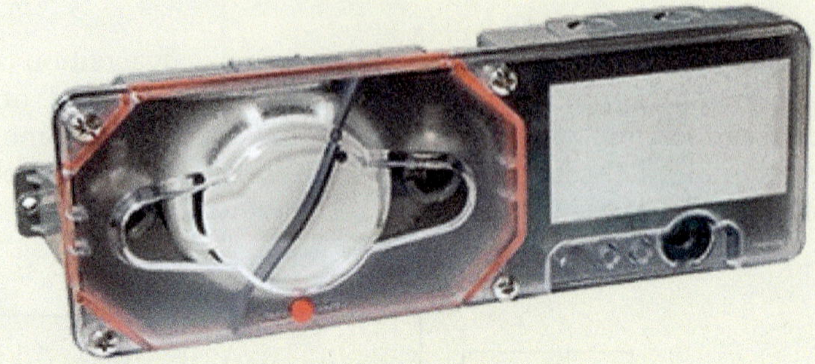

Source: Courtesy of Potter Electric Signal

3.0.0 Section Review

1. Which of the following is a *correct* statement about an Ethernet LAN?
   a. Wired connections are used to link devices.
   b. An Ethernet LAN is limited to 32 connections.
   c. Any number of devices can communicate simultaneously.
   d. An Ethernet LAN cannot accommodate an internet connection.

2. True or False? Smoke evacuation fans must be started as soon as a fire is detected in a building.
   a. True
   b. False

4.0.0 System Control Strategies

Performance Task

1. Interpret, identify, and describe HVAC system operating data viewed through building management system software.

Objective

Describe various building management system operating sequences and control strategies.

a. Identify occupied building temperature operating sequences.
b. Identify unoccupied building temperature operating sequences.
c. Describe other building HVAC control strategies beyond temperature control.

This section builds on discussions related to VAV systems that are often managed with BMS software, because VAV systems are found in larger buildings.

4.1.0 Occupied Building Temperature Control

This section covers the sequence of operation for what is known as a single-duct VAV system (*Figure 28*). Slight variations in this sequence do occur when fan-powered VAV boxes are used.

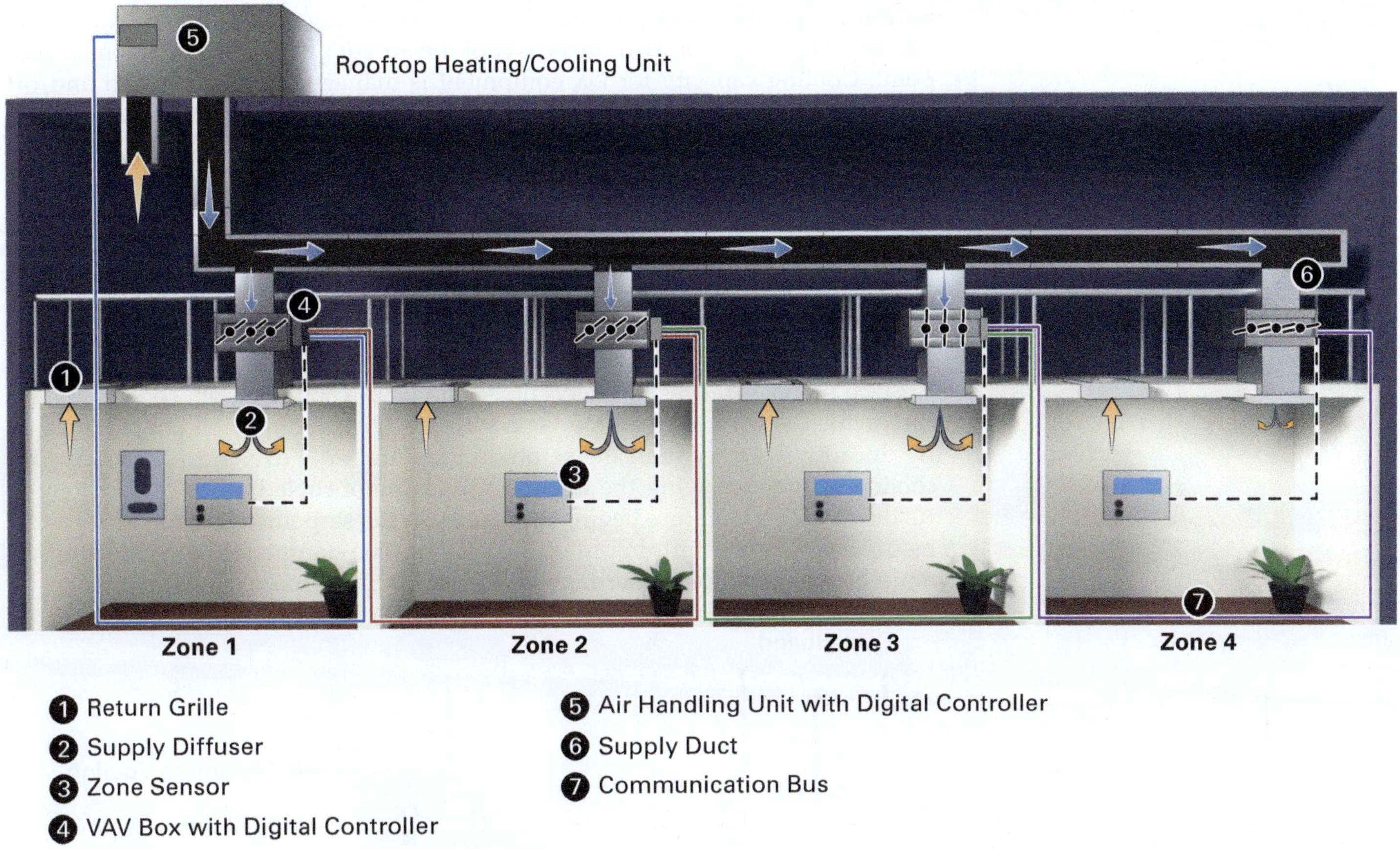

Figure 28 Basic single-duct VAV system.

In a VAV system, a central air handling unit provides air to single-duct VAV boxes, usually at a medium to high pressure, such as 2.5" w.c. The pressure is reduced downstream of the VAV boxes so the system operates quietly. The VAV boxes are equipped with electric or hot-water reheat coils, and each VAV box has a digital controller connected to the communication bus.

4.1.1 Occupied Period Scheduling

In the BMS, the occupancy schedule typically resides in the VAV box controllers. The schedules for the individual zones can vary; they do not need to be the same. When all the zones are operating in the unoccupied mode, the air handling unit also moves to the unoccupied mode and then stops. However, it is needed if even one VAV box is still operating in the occupied mode.

As mentioned previously, each zone sensor can be provided with a manual override button. Pushing the button allows an occupant to manually start the system when the zone is unoccupied. The system will start and remain occupied for a preconfigured period, such as two hours, and then return to its unoccupied mode. If the occupant wishes to remain longer, the override button would need to be pressed again after the time has elapsed.

4.1.2 Occupied Mode

As an occupied period begins, the central air handling unit blower starts. A variable-frequency drive (VFD) adjusts the blower motor speed as necessary to maintain a supply-duct static pressure setpoint. The pressure in the main supply duct fluctuates as VAV dampers open and close, and the VFD responds accordingly.

The air handling unit's outdoor air dampers also open to a preconfigured minimum position, after being 100% closed during the unoccupied period. The cooling apparatus also starts to cool the air supply to the programmed setpoint. Cooling capacity for DX equipment is managed through staged and/or inverter-driven compressors. Chilled water flow through a coil in the air handling unit is usually modulated by a three-way valve, but a standard two-way valve is used in some designs.

If the unit is equipped with an economizer and the outdoor air is suitable, it will be used as the first stage of cooling by driving the outside air damper open. If the economizer is not enough to satisfy the demand, supplemental cooling is provided by the DX equipment or chiller. Also, when the economizer is active, relief air fans and/or dampers must be activated to relieve the increasing building pressure.

Once the VAV moves to the occupied mode, it responds to its zone temperature, which is compared to the zone setpoint (*Figure 29*). If the temperature is significantly warmer than the setpoint, the damper is driven to its maximum cooling setting, allowing the maximum volume of cooled air to flow through. As the room temperature begins to close in on the setpoint, the damper starts to modulate toward its minimum open position. The damper remains at its minimum position, allowing conditioned fresh air into the space.

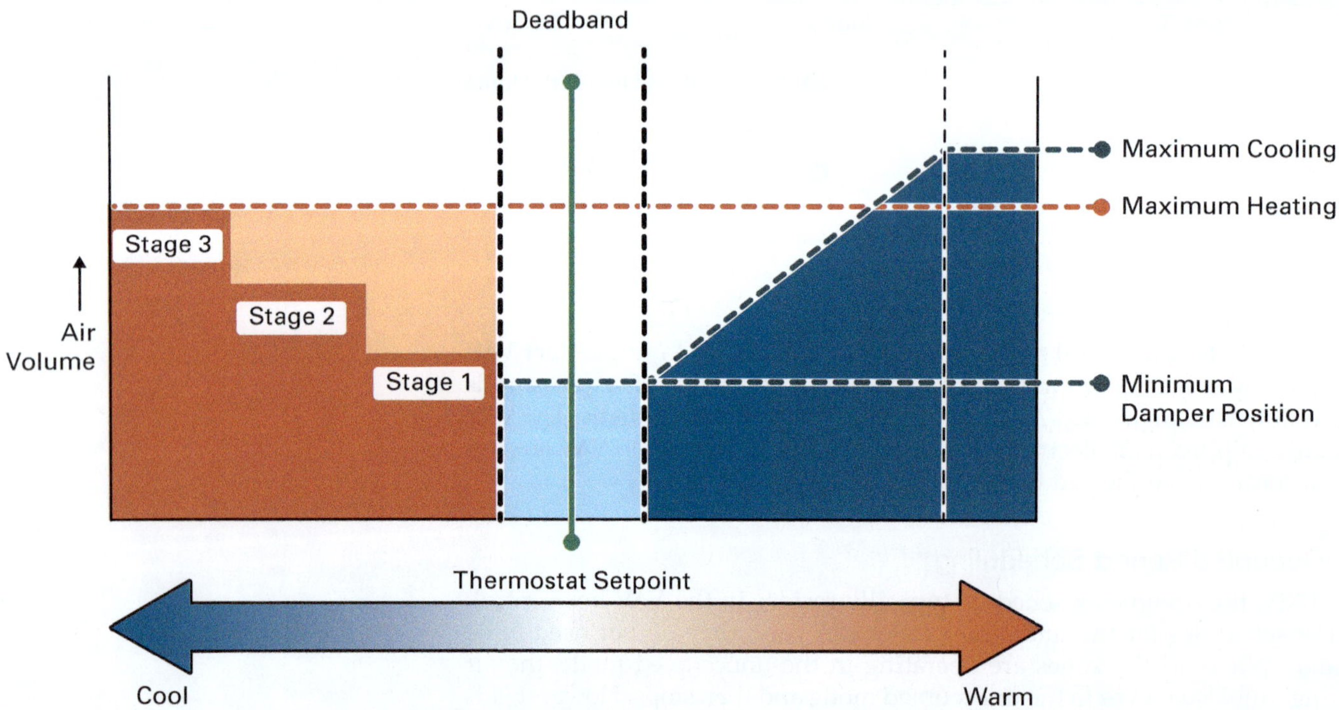

Figure 29 Single-duct VAV box with three stages of electric heat.

If the zone is too cold, then the VAV box heater (electric or hot-water coil) is energized. *Figure 29* shows a typical arrangement for a VAV box with three stages of electric heat. To ensure there is sufficient air passing over the heating elements to prevent overheating, the damper must open in increments as heating stages are energized. In cold weather, when the building is likely cold as the occupied period begins, all three stages of heat may be energized, while the damper is driven to its maximum position for the heating mode.

4.2.0 Unoccupied Building Temperature Control

The unoccupied mode provides an opportunity to save energy, as there is no reason to provide fresh air and the same comfort conditions when the building is empty.

When the schedule calls for an unoccupied period to begin, the central air handling unit shuts down and the outside air dampers close. The zone cooling setpoints are typically raised and the heating setpoints lowered to save energy.

In some cases, the air handling unit and cooling equipment may operate just like they do during an occupied period, except that the outside air dampers remain closed. A more common approach, though, is to have the air handling unit respond only to the demands of a single, representative zone. When that zone requests cooling, the air handling unit starts, and most, if not all, VAV boxes open to provide conditioned air to every zone until the representative zone is satisfied.

When heating is needed during an unoccupied period, the air handling unit starts and allows each VAV box to energize its heat source and open its dampers accordingly. The outside air dampers serving the air handling unit remain closed. If the central air handling unit is equipped with some form of heat, that heat will be used instead to keep the building warm.

It is important to note that operating strategies in both the occupied and unoccupied modes can vary significantly. Every building is unique, but the biggest difference is found in the location. A strategy that works well in Houston will likely not be effective in Minneapolis, and vice versa.

4.2.1 Morning Warmup

Before the system moves to the occupied mode in the winter, a morning warmup cycle is initiated to bring the temperature to or near the occupied setpoint. The goal is to reach those setpoints before occupancy begins, rather than restart the system at the time of scheduled occupancy.

During morning warmup, if the air handling unit is equipped with heat, the air handling unit controller modulates the heating coil to match the demand, which is often substantial in the morning. Heat from the VAV boxes is used if the air handling unit has no heating capability. During morning warmup, the outside air dampers remain closed to minimize the heating load. Remember that these dampers have been closed throughout the building's unoccupied mode.

Morning warmup may end before the time allotted has elapsed. This largely depends on the outdoor conditions. Like the unoccupied mode, a single zone thermostat may be used to monitor morning warmup progress.

Building management systems often use adaptive recovery algorithms to determine when to begin morning warmup, based on how hard the system had to work to maintain the unoccupied conditions.

4.2.2 Night-Time Free Cooling

VAV systems can use an energy-saving routine called night-time free cooling (NTFC). During the unoccupied cycle, the central air handler's controller can initiate the NTFC routine if a sufficient difference exists between the outside air and the building temperature. When activated, the controller opens the outdoor air and exhaust/relief air dampers 100%, starts the supply fan, and locks out the mechanical cooling equipment.

When activated by the NTFC mode, the zone controllers open the VAV dampers to the maximum cooling position until the NTFC setpoint is reached. Once the NTFC setpoint is reached, the VAV dampers return to the minimum cooling position. Each VAV damper cycles between these two positions until the building becomes occupied or all zone NTFC setpoints have been satisfied.

4.3.0 Other Control Strategies

Figure 30 Wall-mounted CO_2 sensor with display.
Source: Image courtesy Dwyer Instruments, LLC

In addition to controlling temperatures in a building, a BMS is also capable of controlling carbon dioxide (CO_2), humidity, and other characteristics that may affect indoor air quality (IAQ).

4.3.1 Demand-Controlled Ventilation

Building management systems can improve IAQ by controlling the CO_2 levels in the building using a control strategy called *demand-controlled ventilation* (DCV). A CO_2 sensor (*Figure 30*) is mounted in one or more zones and connected as an input to the zone VAV box controller(s).

When occupied CO_2 levels exceed the setpoint (measured in parts per million), the zone controller modulates the primary air damper between the minimum and maximum position limits to admit more air and reduce the CO_2 content. More primary air from the air handling unit, which is partially comprised of fresh outdoor air, is pushed into the space to dilute the CO_2 concentration.

If overcooling of the zone results, the VAV damper is temporarily repositioned as needed to support reheat. The electric heaters or hot-water coil is activated to warm the space. Although this consumes additional energy, reducing the CO_2 content is a system priority.

The air handling unit controller also responds to elevated CO_2 levels and adjusts the outdoor air damper accordingly. When multiple zones report elevated CO_2 levels, the outside air damper is opened beyond the minimum position to increase the ratio of outside air to return air.

4.3.2 Humidity Control

When a zone is occupied, the operation of the central air handling unit normally prevents excess humidity by dehumidifying the supply air. This is a natural part of the cooling process. However, this may be insufficient when the cooling equipment is operating at reduced capacity.

Humidity control when the cooling load is low can be improved by sensing humidity at the zone level. An optional relative humidity sensor can be wired to the VAV controller. VAV box controllers can send the humidity information back to the air handling unit controller. In response, the air handling unit adds cooling capacity to lower the supply air temperature, creating a drier supply airstream.

4.3.3 Building Pressurization

Many buildings require pressure control, primarily because of the consistent introduction of outside air into the building. As a higher volume of outside air is drawn into the building, the building pressure increases. However, there are many other factors in building pressure, including exhaust and makeup air fans in kitchens and restrooms. While some equipment is increasing the pressure, other equipment is reducing the pressure.

The central air handling unit can remove building air to reduce the pressure by either letting it flow out naturally through relief louvers or vents, or it can be fitted with a powered exhaust system. The louvers and vents have dampers that are opened and modulate to find a position that results in the target building pressure. However, in some cases, this arrangement doesn't get enough air out in a timely manner.

A powered exhaust system responds to a pressure sensor located in a key building space. A variable-speed exhaust blower starts on an increase in pressure beyond the setpoint. Its speed is modulated to bring the building pressure down to the desired pressure.

4.0.0 Section Review

1. What is the override button on a VAV system zone sensor used for?
   a. Occupant request to decrease the temperature setpoint
   b. Occupant request to bring 100% outside air into the zone
   c. Occupant request to change the system from unoccupied to occupied
   d. Occupant request to maximize airflow despite the temperature setpoint
2. When are the outside air dampers for the air handling unit in a VAV system typically closed?
   a. Only during morning warmup
   b. Only when the system is in the unoccupied mode
   c. When the building is occupied and cooling is needed
   d. When the building is unoccupied and during morning warmup
3. Demand-controlled ventilation is related to the control of _____.
   a. CO_2
   b. Smoke
   c. Temperature
   d. Building pressure

5.0.0 Interoperability

Objective

Explain the concept of interoperability and identify common BMS protocols.

a. Describe the primary communication protocols in use.
b. Explain how web browsers are used to interact with systems using various protocols.

Performance Tasks

There are no Performance Tasks in this section.

Facility managers and maintenance personnel like to have real-time access to all the buildings under their care, with all the details at their fingertips. They want to know what building space each tenant occupies, when and how it is being used, the phone numbers of tenant leadership, and what kind of special assistance might be required in case of an emergency. They also want the ability to be able to optimize the building owner's utility costs. Of course, they also want easy access to the HVAC systems in each building and the ability to control those systems from a remote location.

Ideally, facility managers can also determine who manufactured each piece of equipment, when its warranty period ends, what periodic maintenance has been performed, and when the next maintenance activity is due.

The accurate transfer of information required to meet the facility manager's needs across buildings and building systems is called *interoperability* (*Figure 31*). Meeting all these needs is indeed possible today, with proper planning by a qualified system integration company. It is not only possible to integrate an HVAC control system with other systems such as security, lighting, and fire/life safety, but multiple buildings can also be connected.

Years ago, manufacturers developed their own protocols. For end users, it meant that they were locked into a company's products and devices, since the devices of other companies were not compatible. As you might guess, the market refused to accept that long term. The development of an **open protocol** was both inevitable and necessary.

Open protocol: A communication standard that is not owned by a particular entity and is not limited to specific products, allowing interoperability between different devices and manufacturers.

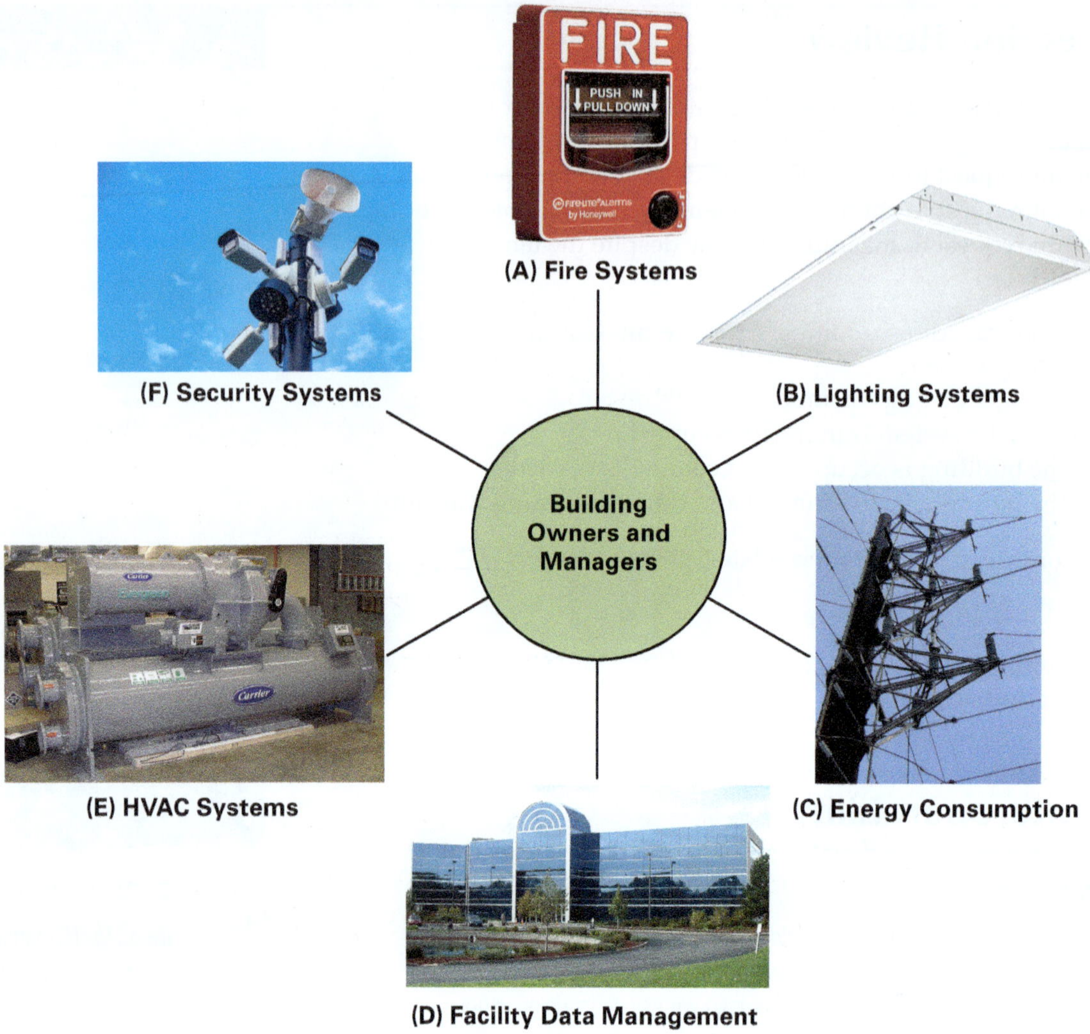

Figure 31 Interoperability.
Sources: Courtesy of Potter Electric Signal (A); Used with permission, Acuity Brands Lighting (B); Gri-spb/iStock/Getty Images (F)

Building owners and managers want flexibility when it comes to selecting equipment. Thus, open protocols quickly became the norm. Any company that sought to continue the use of a unique or proprietary protocol was quickly deleted from the list of welcome bidders on construction projects.

Unfortunately, none of the leading protocols are perfect, and none can satisfy the needs of every system type. Each protocol has advantages and disadvantages. However, with these limitations in mind, several open protocols have reached a high level of popularity.

5.1.0 Building Management System Protocols

The following sections describe the most popular protocols being used in building systems today. Note that the list is not exhaustive, and it does not include specific industries that may continue to operate using unique protocols that suit the application.

5.1.1 BACnet®

BACnet® was developed by ASHRAE and originally approved as a standard in 1995. The latest version, at this writing, of ASHRAE Standard 135-2020, *A Data Communication Protocol for Building Automation and Control Networks*, was released in 2020. This building automation and control network (BACnet®) protocol provides a comprehensive set of message rules that provide interoperability between various building systems.

As *Figure 32* shows, each manufacturer can design and provide a BACnet® interface module that would serve as a **gateway** to their own proprietary protocol bus. The gateway enables two-way communication between the manufacturer's network and the BACnet® Ethernet.

Gateway: A network device that allows two or more applications or networks operating under different protocols to share information.

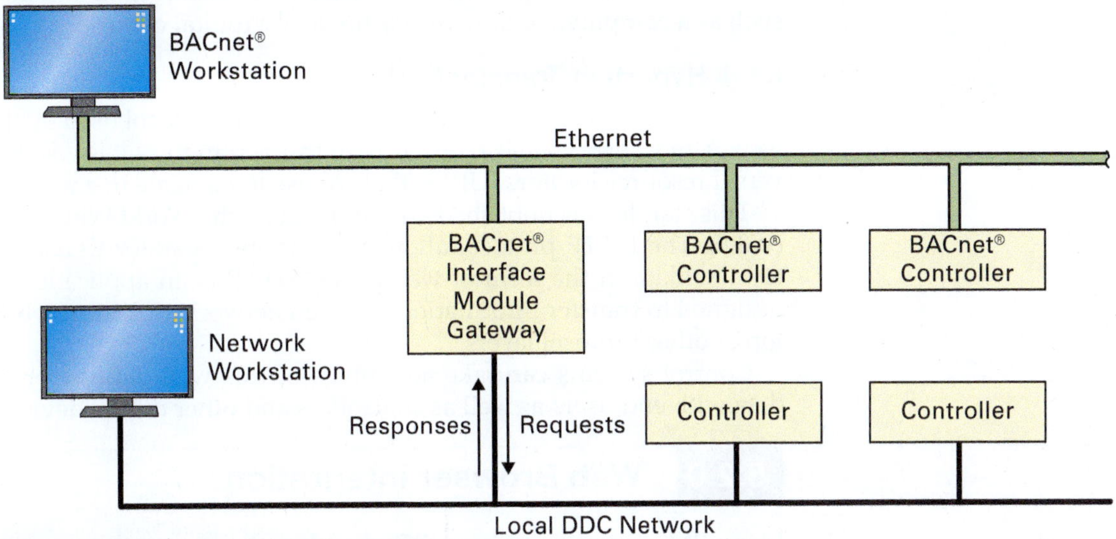

Figure 32 BACnet® overview.

Each BACnet® module would provide a limited amount of information transfer as determined by the module's manufacturer. Depending on the volume of information to be exchanged, multiple BACnet® modules might be required. However, as time passed, with manufacturers slowly doing away with their proprietary protocols, the need for an interface has been significantly reduced.

5.1.2 LON Technologies

Another developer of an open protocol is the Echelon Company, with its local operating network (LON) technologies, including LonTalk™, LonWorks™, and LonMark™.

Echelon has developed LON technology based on a neuron chip microprocessor. The neuron chip contains **firmware** that operates on a seven-layer communication protocol. LonTalk™ is the seven-layer protocol that allows controllers (also referred to as *nodes*) to operate using an efficient, reliable communication structure. Each node in the network contains a chip with embedded intelligence that implements the protocol and performs control functions.

Firmware: Software permanently stored on a device's read-only memory (ROM) to guide basic operation.

LonWorks™ is Echelon's DDC network that uses LonTalk™ protocols. It is also part of the American National Standards Institute (ANSI) approved version of ASHRAE's BACnet® standard.

Echelon's LON technology products were designed to provide solutions for a general manufacturing environment; however, the technology has been equally applied to the building and HVAC management industry. Because each node sold by Echelon has a LON chip in it, any product with this chip has the necessary protocols to communicate with similar products. Likewise, any manufacturer that uses LON chips in its products ensures interoperability with any other manufacturer's devices that include LON chips.

One advantage of the LON product line is that data can be transferred using a twisted wiring pair, Ethernet wiring, or even through power lines and electrical systems.

5.1.3 Modbus

The Modbus standard was originally developed by the Modicon Company and later became the trademark of Schneider Electric. Schneider made Modbus available for free and gave the Modbus protocol specifications to the Internet Engineering Task Force. Thus, Modbus has become widely used and is popular in the industrial, metering, power generation, and electrical equipment industries.

Modbus is designed around a master-subordinate relationship. It is not a peer-to-peer arrangement as described earlier in this module. In this relationship, communication always occurs in pairs—a master device initiates a request and then waits for a response. The master device is responsible for initiating every interaction. Typically, the master is a human machine interface (HMI), such as a computer, and the subordinate is a digital controller.

5.1.4 Hypertext Transfer Protocol

Hypertext Transfer Protocol (HTTP) is the base protocol used by the world wide web. Almost everyone is familiar with this acronym as it is visible in many universal resource locators (URLs) that we use to navigate in a web browser.

This standard is published and updated by the World Wide Web Consortium (W3C). The HTTP protocol allows a computer operator to use a web browser to view data in the form of web pages. HTTP is an application-layer protocol designed to transfer information between networked devices while running on top of other protocol layers.

Control systems can take advantage of this well-defined protocol to share data with end users as well as controllers and other digital devices.

5.2.0 Web Browser Integration

Many BMS projects can be satisfied with an integration system like that shown in *Figure 22*, but others require more complex interoperability to achieve the operational needs. To meet this greater complexity, many manufacturers now provide a web-based connection to their control systems. Thus, building owners and operators can use the common HTTP protocol through a standard web browser to interface with building systems.

Figure 33 shows the network architecture of such a system. In this example, a unique protocol can be used at a lower level when its qualities best suit the

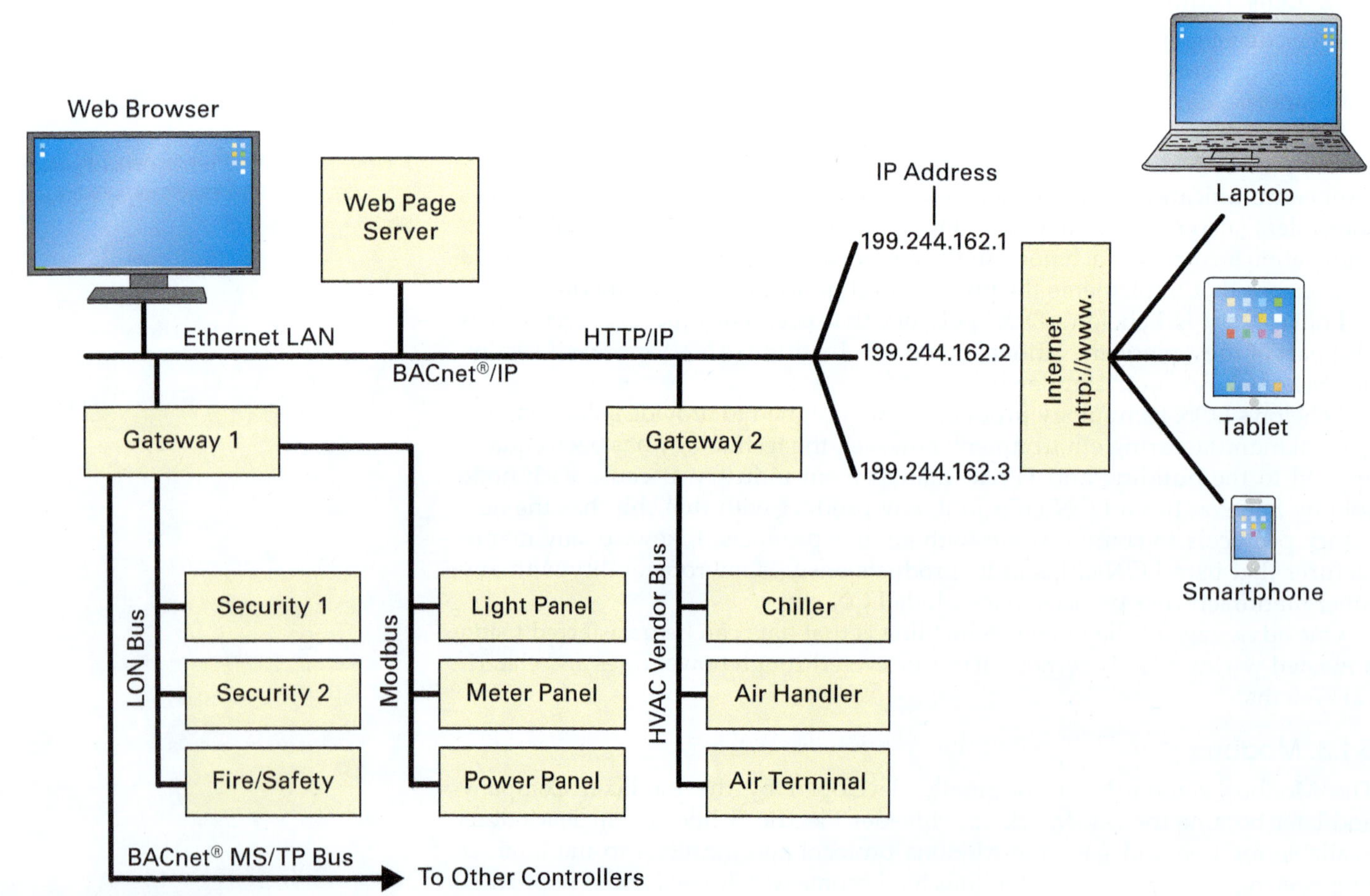

Figure 33 Web browser integration schematic.

control needs. Thus, separate control protocols for lighting, HVAC, building security, and other systems may still exist, but it doesn't appear that way to the user.

System integration is achieved using gateways, located on the building's Ethernet LAN. The gateways do not have keyboards or monitors, though the gateways have powerful central processing units with multiple cable connections that match the transport medium requirements of the common open protocols. In addition, the gateways have a bus connection that matches the downstream protocol.

The gateways are programmed to regularly extract data from controllers located in the building. This extracted data is then compiled and temporarily stored in two-way information tables recognized by the common protocol used on the Ethernet LAN. BACnet® is becoming increasingly popular for this purpose.

Also located on the LAN is a web page server. The web page server uses the common protocol to regularly extract table information from the gateways and to create web pages for viewing. The data can be viewed in a tabular or graphical format by an operator through a web browser on a computer, tablet, or smartphone. The communication is two-way, and any operator (with the proper security level) can manually intervene in the operation of a building system.

Integrated control strategies across building systems can also be achieved through special routines written into the web page server. For example, data from an ID card reader can be sent to the lighting and HVAC system controllers when a specific employee enters a building, activating the VAV box and lighting system serving their workspace. When that employee leaves the building, again scanning their ID card, those systems can instantly return to the unoccupied mode.

On Site

Lifechanging Connectivity and Information

Today's service managers that work with larger commercial buildings are still very busy people, but what they do and how they do it has significantly changed.

When a problem is reported from a building with an accessible BMS, the service manager often begins by logging in and investigating the problem system or area, before a technician is dispatched. Perhaps a controller has stopped communicating, or a temperature sensor is reading out of range, creating a secondary problem. With this knowledge, the service manager can have the dispatched technician go directly to the problem component and begin troubleshooting immediately. Depending on how they prefer to work, the technician may log into the system to see if their effort has solved the problem, or the manager can monitor the system from afar and let them know the status. In some cases, a quick workaround can be developed while parts are ordered.

5.0.0 Section Review

1. The BACnet® protocol provides ______.
   a. priority to HVAC data travel
   b. the gateway to all other protocols
   c. a comprehensive set of message rules
   d. wireless access to building management systems

2. True or False? The HTTP protocol allows building owners to access building management systems through a web browser.
   a. True
   b. False

Module 03405 Review Questions

1. A typical digital controller is comprised of which major components?
   a. Sensors and an input module
   b. Microprocessor and an output module
   c. Input module, gateway, output module
   d. Input module, output module, and microprocessor

2. A general-purpose controller would most likely be chosen to control which of the following?
   a. A pump
   b. A centrifugal chiller
   c. A 40-ton VAV HVAC rooftop unit
   d. A water-cooled, 30-ton indoor packaged unit

3. A discrete input or output is one that ______.
   a. has two states (binary)
   b. is hidden on the network
   c. varies within a wide range
   d. is proportional to a sensed value

4. A microprocessor gathers information and prepares a response based on ______.
   a. a gateway
   b. algorithms
   c. a point type
   d. previous experience

5. Which type of sensor uses a longer resistance-sensing element to sense an average temperature along its length?
   a. PID
   b. RTD
   c. Analog thermistor
   d. Binary thermocouple

6. Which of the following is a common analog output used for valves and actuators?
   a. 24VAC
   b. 20VDC
   c. 120VAC
   d. 4–20mA

7. Which of the following would be the *most* accurate algorithm style used for information processing when the controlled variable is changing rapidly over time?
   a. Proportional
   b. Proportional-integral
   c. Proportional-derivative
   d. Proportional-integral-derivative

8. The rate at which information travels on a network is known as the ______.
   a. bit rate
   b. bus rate
   c. baud rate
   d. LAN rate

9. In a VAV system, why do VAV box dampers consistently remain open to some degree?
   a. To add white noise and improve acoustics
   b. To ensure that some outside air is flowing into the space
   c. Because some level of cooling is always needed in large buildings
   d. To prevent the central air handling unit from over-pressurizing the duct

10. A VAV box PIC typically has an integrated ______.
   a. bus
   b. gateway
   c. power supply
   d. damper actuator

11. Which protocol allows the user to interact with a BMS through a web browser?
   a. LON
   b. HTTP
   c. Modbus
   d. BACnet®

12. To help ensure a BMS remains secure, access to the system for a registered user may still require ______.
   a. editorial privileges
   b. two-factor authentication
   c. your Social Security number
   d. two users to sign in together

13. True or False? When a controller temporarily loses its connection to the rest of the system, it immediately shuts down the affected equipment.
   a. True
   b. False

14. When a fire starts in a building, ______.
   a. all the air handling units must be shut down
   b. smoke evacuation must begin immediately
   c. the outside air dampers must be driven open
   d. the VAV box dampers must be driven open

15. What controls the position of a VAV box damper during the occupied mode?
   a. Zone temperature
   b. Building pressure
   c. Supply duct pressure
   d. Differential pressure

16. A building morning warmup mode begins ______.
   a. at the scheduled time of occupancy
   b. anytime a selected zone needs heat
   c. when all the zones are in need of heat
   d. before the scheduled time of occupancy

17. Which of the following occurs during night-time free cooling?
   a. The air handling unit blower shuts down.
   b. The exhaust/relief air dampers are closed.
   c. The mechanical cooling equipment remains off.
   d. The outdoor dampers open to the minimum position.

18. True or False? Equipment that communicates with using proprietary protocol can communicate with a BACnet® system using a gateway.
    a. True
    b. False

19. Which of the following protocols does *not* operate using a peer-to-peer relationship?
    a. HTTP
    b. Modbus
    c. LonTalk™
    d. BACnet®

20. Which protocol is used to communicate through the internet using a web browser?
    a. HTTP
    b. Modbus
    c. BACnet®
    d. LonTalk™

Answers to odd-numbered questions are found in the Module Review Answer Key at the back of this book.

Answers to Section Review Questions

Answer	Section	Objective
Section One		
1. d	1.1.0	1a
2. d	1.2.1	1b
3. c	1.3.3	1c
Section Two		
1. c	2.1.0	2a
2. d	2.2.0	2b
3. b	2.3.4	2c
Section Three		
1. a	3.1.0	3a
2. b	3.2.7	3b
Section Four		
1. c	4.1.1	4a
2. d	4.2.1	4b
3. a	4.3.1	4c
Section Five		
1. c	5.1.1	5a
2. a	5.2.0	5b

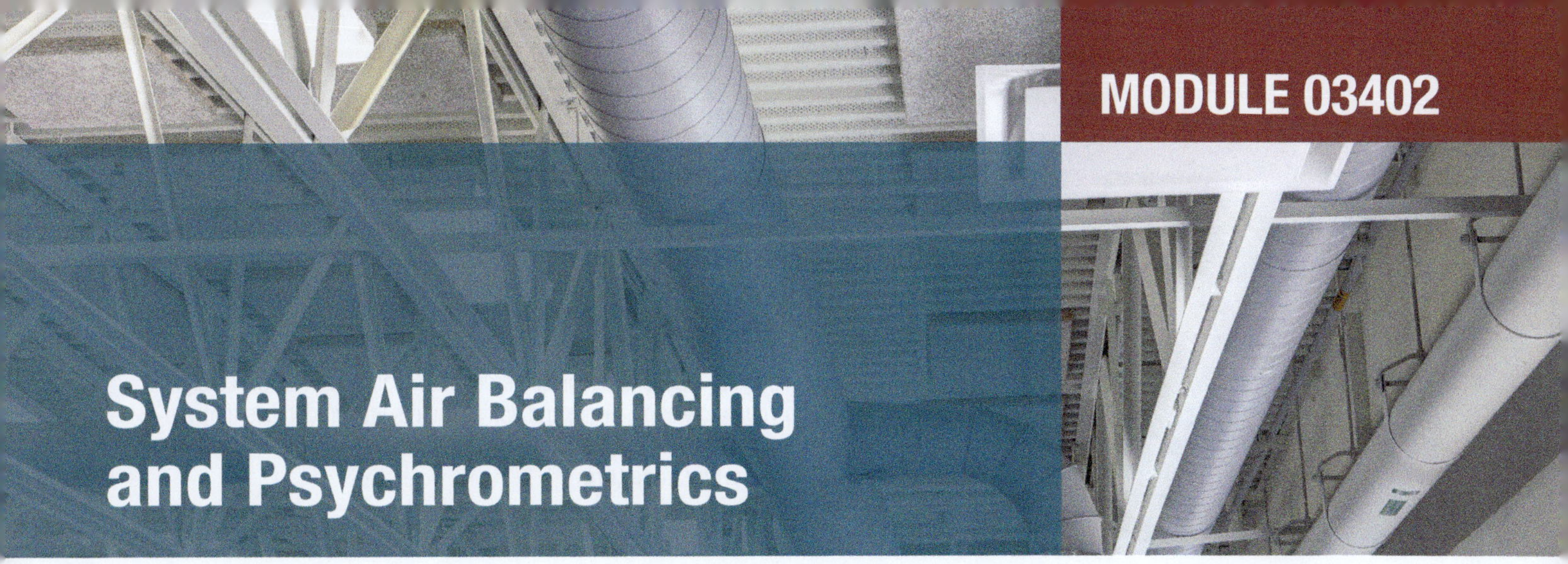

MODULE 03402

System Air Balancing and Psychrometrics

Source: Brian Brown/iStock/Getty Images

Objectives

Successful completion of this module prepares you to do the following:

1. Describe the properties of air as related to temperature, pressure, and volume.
   a. Identify the properties of air related to airflow and air balancing.
   b. State Dalton's, Boyle's, and Charles' laws.
2. Describe the study of psychrometrics and explain how it is used to determine specific air properties.
   a. Describe psychrometrics and the related properties of air.
   b. Describe the structure of the psychrometric chart.
   c. Explain how to use the psychrometric chart to determine specific air properties.
3. Describe the air balancing process and explain how to balance an air distribution system.
   a. Describe air balancing and define common terminology.
   b. Identify the tools and instruments used in air balancing.
   c. Describe the steps to take prior to beginning an air balancing task.
   d. Explain how to measure system and terminal airflow and make adjustments.
   e. Explain how to balance a system using a thermometer.

Performance Tasks

Under supervision, you should be able to do the following:

1. Demonstrate the use of air pressure and velocity measurement instruments on a duct system.
2. Measure or calculate the air volume at supply and return air grilles and registers.
3. Adjust a damper in a branch supply duct or takeoff to achieve a target air volume.

Overview

Even when forced-air HVAC equipment is properly installed and operating, the air must be distributed to the conditioned spaces in a volume consistent with the load. Balancing the airflow requires an understanding of the characteristics of a moving airstream and the ability to use specialized test instruments and air balancing devices.

Digital Resources for HVACR

Scan this code using the camera on your phone or mobile device to view the digital resources related to this craft.

SCAN ME

NCCER Industry-Recognized Credentials

If you are training through an NCCER-accredited sponsor, you may be eligible for credentials from NCCER. The ID number for this module is 03402. Note that this module may have been used in other NCCER curricula and may apply to other level completions. Contact NCCER at 1.888.622.3720 or go to **www.nccer.org** for more information.

You can also show off your industry-recognized credentials online with NCCER's digital credentials. Transform your knowledge, skills, and achievements into credentials that you can share across social media platforms, send to your network, and add to your resume. For more information, visit **www.nccer.org**.

1.0.0 Air Properties

Performance Tasks

1. Demonstrate the use of air pressure and velocity measurement instruments on a duct system.
2. Measure or calculate the air volume at supply and return air grilles and registers.

Objective

Describe the properties of air as related to temperature, pressure, and volume.

a. Identify the properties of air related to airflow and air balancing.
b. State Dalton's, Boyle's, and Charles' laws.

The efficient operation of an HVAC system requires more than just properly operating heating, cooling, and electrical equipment. Of equal importance is the delivery of the correct volume of conditioned air to the occupied space. Adjusting system airflow to satisfy this requirement is referred to as *air balancing*.

Air testing and balancing includes measuring, adjusting, and documenting the performance of a building's air distribution system to ensure it meets design specifications. Air balancing is necessary because comfort depends on having the right volume of conditioned air delivered to each space. Poor air balance reduces comfort. A balanced air system produces the right volume of air and delivers it to specific areas in a manner consistent with the heating/cooling load.

Before learning how to balance a system, you need a basic understanding of the properties of air. Technicians must also become familiar with basic laws of physics that apply to air systems. This section covers these topics at an introductory level.

1.1.0 Air Properties

Earth is surrounded by a mixture of gases composed of 78% nitrogen, 21% oxygen, and 1% other gases. Together, these gases form our atmosphere, which extends about 6,200 miles above Earth's surface and is held close to the planet by gravity.

Atmospheric gases have weight. That weight is measured in pounds per square inch (psi). At sea level, a column of air with a base of 1 square inch (in^2), extending to the outer edge of the atmosphere, exerts a weight and pressure of about 14.7 psi. The pressure exerted on Earth's surface from the weight of our atmosphere is called *atmospheric pressure* (*Figure 1*).

Atmospheric pressure can be measured with a barometer. *Figure 2* shows a basic mercury-tube barometer. It consists of an open dish of mercury, with a mercury-filled tube sealed at one end and inverted vertically into the dish.

When this column of air is applied to a mercury-tube barometer at sea level and 70°F, it will cause the mercury in the tube to rise to a height of 29.92", or

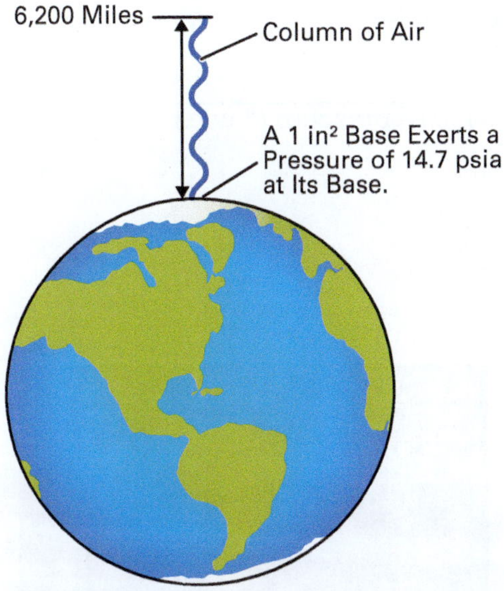

Figure 1 Atmospheric pressure.

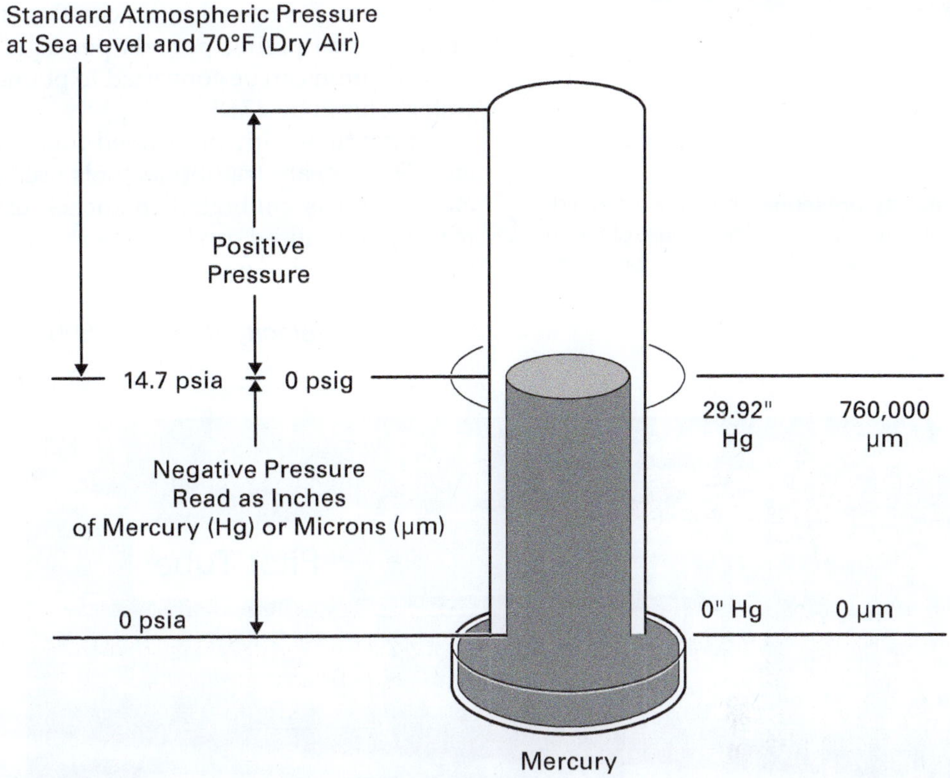

Figure 2 Atmospheric, absolute, and gauge pressure relationships.

29.92" Hg. This is referred to as *standard atmospheric pressure*. The actual atmospheric pressure at any location differs to some degree due to altitude, temperature, moisture content, and weather conditions.

These values of 14.7 psia and 29.92" Hg for dry air at sea level and a temperature of 70°F, are standards that are used frequently in the HVACR craft. One pressure scale, called the *absolute pressure* scale, is based on these described barometer values. On this scale, pressures are expressed in pounds per square inch absolute (psia), starting from 0 psia. At standard conditions, normal atmospheric pressure is 14.7 psia. Any space pressure below 14.7 psia, such as an evacuated refrigerant circuit, is said to be in a vacuum. A perfect vacuum, rarely encountered, is found at 0 psia.

Another scale, called the *gauge pressure* scale, is more frequently used to quantify air pressure. Gauge pressure uses the atmospheric pressure as the zero point. In other words, 0 psig and 14.7 psia represent the same pressure. Positive gauge pressures are expressed in pounds per square inch gauge (psig). Negative pressures, those below 0 psig, are usually expressed in inches of mercury vacuum (inHg vacuum) or in microns (µm). The micron vacuum scale ranges from 760,000 µm at 0 psig / 14.7 psia to 0 µm, which represents a perfect vacuum.

Gauge pressure can easily be converted to absolute pressure by adding 14.7 to the gauge pressure value. For example, a gauge pressure of 10 psig equals an absolute pressure of 24.7 psia (10 + 14.7). Conversion between the absolute and gauge pressure scales is often necessary in calculations.

In HVACR work, inches of water column (" w.c.) is the frequently used unit to quantify low pressures. One example is when measuring the air pressure in a duct. Inches of water column express the height that an applied pressure will lift a column of water. The atmospheric pressure of 14.7 psia at sea level and a temperature of 70°F will support a column of water 406.9" (33.9') high. Therefore, for every pound per square inch of pressure, a column of water will rise to a height of 27.68", or about 2.3'.

Pressures expressed in psi can therefore be converted to inches of water column by multiplying pounds per square inch by 27.68. Conversely, inches of water column can be converted to pounds per square inch by dividing inches of water column by 27.68.

Velocity pressure: The pressure in a duct that results from the movement of the air; total pressure minus the static pressure equals velocity pressure.

A pitot tube with an inclined-tube manometer (*Figure 3*) or a digital manometer (*Figure 4*) are traditional tools used to measure duct pressure. The manometer shown is calibrated in inches of water column. The **velocity pressure** readings measured can be converted to velocity (speed) in feet per minute (fpm) by using the following formula:

$$Velocity\ (fpm) = 4{,}500 \times \sqrt{velocity\ pressure\ (''\ w.c.)}$$

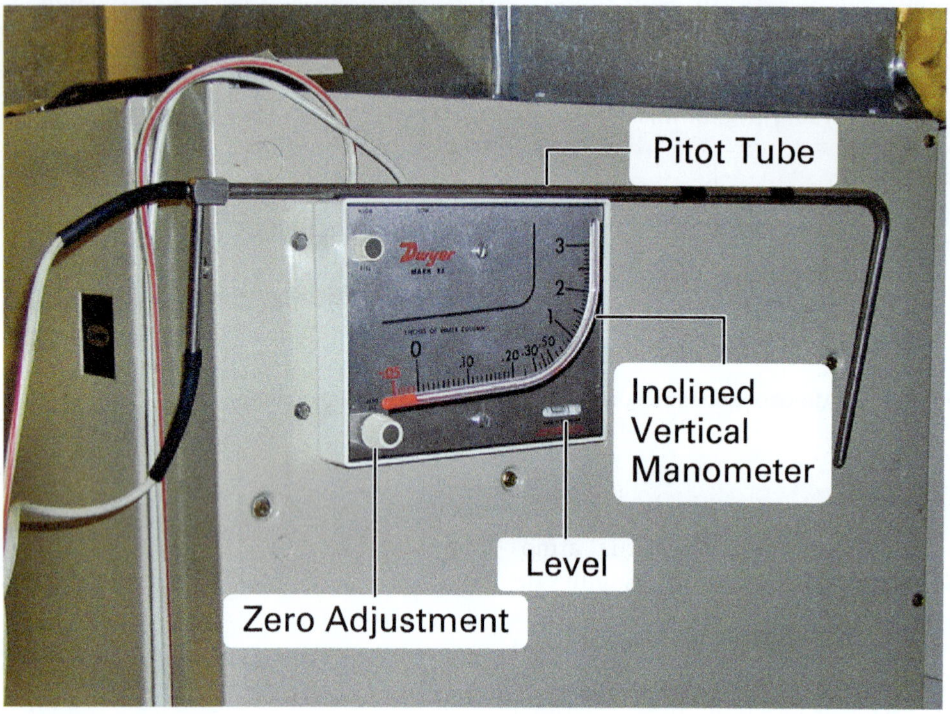

Figure 3 Pitot tube and inclined tube manometer.

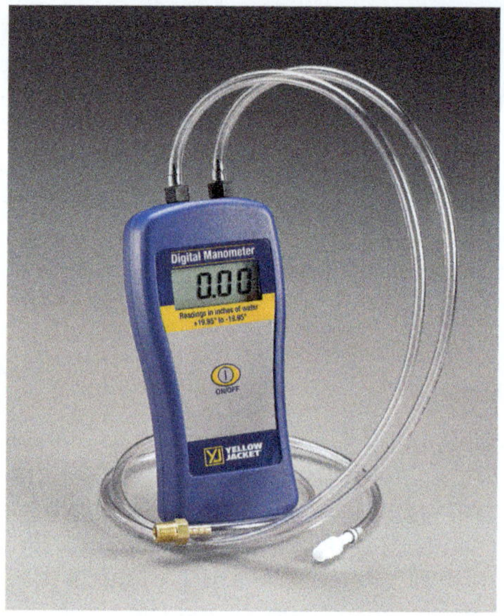

Figure 4 Digital manometer.
Source: Ritchie Engineering Co., Inc. / YELLOW JACKET

For example, a velocity pressure reading of 0.04" w.c. is equal to a velocity of 801 fpm, calculated as follows:

$$Velocity\ (fpm) = 4{,}500 \times \sqrt{0.04''\ w.c.}$$
$$Velocity\ (fpm) = 4{,}500 \times 0.02$$
$$Velocity\ (fpm) = 801\ fpm$$

1.1.1 Weight of Air

NOTE

Note the use of the term *dry air* when referring to standard air. Although it is rare to encounter air that is completely dry, moisture content is a major factor in the characteristics of air and the amount of heat that it contains. Dry air is used to provide a standard reference condition.

Air is compressible and has weight. Its density and weight vary with the elevation above Earth's surface. At higher elevations, air weighs less than at lower elevations. For example, the layer of air over Denver, CO (mean elevation 5,280') is not as heavy as it is over New Orleans, LA (mean elevation 100').

Per the American Society of Heating, Refrigerating and Air-Conditioning Engineers (ASHRAE), *standard air* is dry air with a density of 0.075 pounds per cubic foot (lb/ft^3) at 70°F and a barometric pressure of 29.92" Hg (sea level). Using standard air in many calculations provides a consistent baseline, but the value must be changed to maintain accuracy when working at a significantly different altitude.

Heating and cooling loads and equipment capacities are based on British thermal units per hour (Btuh). Some calculations used in HVACR work are based on pounds of air rather than the volume of air. Therefore, it is also important to know how to convert the volume of air flowing in a system from cubic feet per minute (cfm) to pounds per hour.

At standard conditions, a cubic foot of air weighs about 0.08 lb. If a system is moving 1 cfm then, over the course of an hour it is moving 60 ft^3 of air. That 60 ft^3 of air weighs 4.8 lb.

With that in mind, consider an example that is more closely related to an HVAC system. If a three-ton cooling system is handling 450 cfm/ton of capacity, then the total system airflow is 1,350 cfm. Over the course of an hour, the system is moving 60 times as much, or 81,000 ft^3. At 0.08 lb/ft^3, the weight of the air moved each hour is an astonishing 6,480 lb. Although it may seem hard to believe, air distribution systems move far more weight than we might guess.

Since air is lighter at higher elevations, the weight carried each hour would be less. The air's ability to absorb heat is less as well, due to the lower density. At sea level, it takes 1.08 Btuh to raise the temperature of 1 cfm by 1°F; at 5,000' of elevation, it takes only 0.892 Btuh to raise the temperature of the lighter, lower density air by 1°F. To summarize, at higher elevations, more air must be moved to transfer the same amount of heat that can be transferred at sea level.

It is helpful to understand these characteristics of air at different altitudes so that you can apply it to the work in your specific service area. In addition, understanding the relationship between the volume and weight of air will be helpful as we explore the topic of psychrometrics.

1.1.2 Air Pressure, Velocity, and Volume

In an air distribution system, air flows from an area of higher pressure to an area at a lower pressure (*Figure 5*). The blower in an air handler or fan coil unit creates the pressure differential necessary for air to move. In a typical system, the highest air pressure in the system is found at the blower outlet. The pressure gradually decreases as it passes through the supply duct, due to friction losses. As the air leaves the grilles and enters the space, it eventually equalizes with the pressure in the space itself.

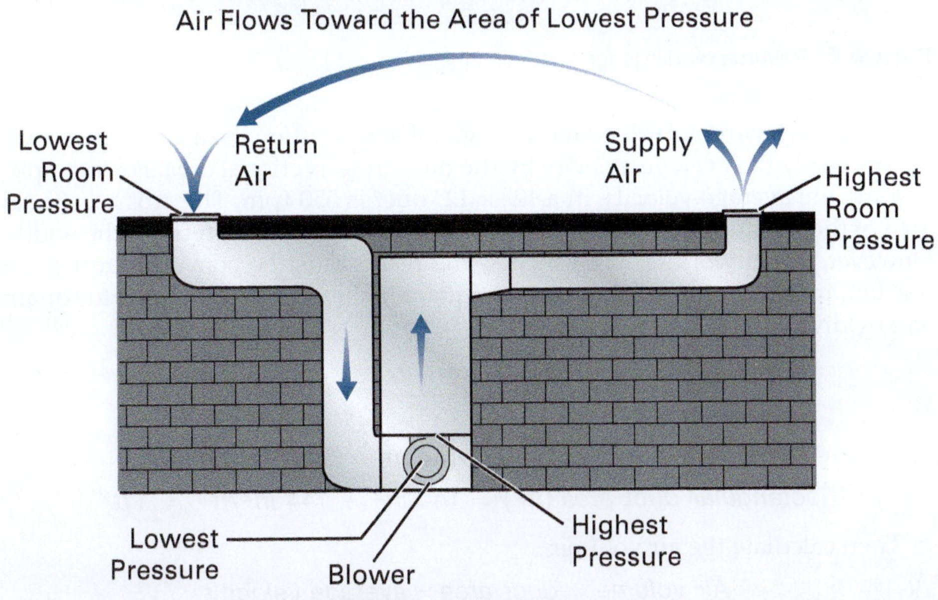

Figure 5 Air movement in an HVAC system.

The air is then attracted to any area that is at a lower pressure, which is found at the return air grille. The air pressure in the return duct gradually decreases as it approaches the blower wheel inlet, where the lowest pressure in the system is found.

Pressure losses in the ducts result from several factors that create resistance to airflow in the system. Most of the pressure losses result from friction as the air molecules rub against the walls of the duct. Along the duct surface, the molecules travel slower than the air molecules moving through the center of the duct.

Losses also occur when air molecules bump into each other as the molecules move through the duct. This happens because the air molecules tumble and mix rather than flow in a straight line. Note that friction losses increase with an increase in air speed and at any point where the air must change direction.

Static pressure: The pressure exerted uniformly in all directions by air within a duct system. Total pressure minus velocity pressure equals static pressure.

The blower must apply enough pressure to overcome friction losses and keep the air moving. This pressure, called the **static pressure**, is exerted equally in all directions inside the duct.

Air Velocity and Volume

The volume of air moving in a duct system is measured in cubic feet per minute. The volume can be determined if the velocity and the cross-sectional area of the duct are known. However, the air does not move through the duct at the same speed from side to side and top to bottom. For example, the air moving along the duct walls moves slower, due to the friction.

Traverse readings: A series of velocity readings taken at several points over the cross-sectional area of a duct or grille.

To get an accurate picture of the air velocity in the duct, the velocity is measured at numerous points across the duct (*Figure 6*). These are called **traverse readings**. To measure the velocity in a duct, a velometer or *anemometer probe* is inserted through the wall and stopped at specific points to take readings. Then the readings are averaged by adding together and dividing by the number of readings. The probe has markings along its length to ensure consistent dimensions. This topic will be explored with more detailed instructions in *Section 3.4.1, Duct Readings and Balancing.*

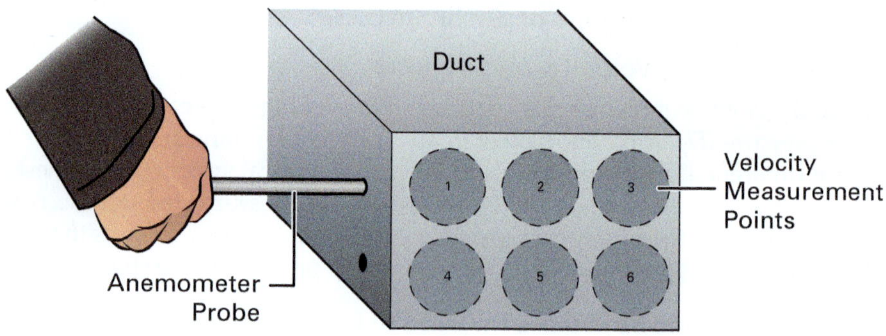

Figure 6 Traverse readings for duct velocity.

Once you have determined the average velocity, finding the air volume is simple–multiply the average velocity by the duct cross-sectional area. For example, assume the average velocity in a 12" × 12" duct is 550 fpm. The cross-sectional area of the rectangular duct is found by multiplying the length times the width. However, when working with inches, the result must be converted to square feet before calculating the air volume. Square inches can be converted to square feet by dividing by 144:

$$\text{Rectangular duct area} = L \times W$$

$$\text{Rectangular duct area} = 12" \times 12"$$

$$\text{Rectangular duct area} = 144\ in^2$$

$$\text{Rectangular duct area } (ft^2) = 144\ in^2 \div 144\ in^2/ft^2 = 1\ ft^2$$

Then calculate the air volume:

$$\text{Air volume} = \text{duct area} \times \text{average velocity}$$

$$\text{Air volume} = 1\ ft^2 \times 550\ fpm$$

$$\text{Air volume} = 550\ cfm$$

A_k factor: The net free area of an air distribution outlet or an inlet stated in square feet. The terminal device manufacturer specifies this value to aid in airflow calculations.

Velocity measurements can also be taken at the face of return and supply grilles and diffusers. It is important to remember that the cross-sectional area of a terminal is not equal to its actual measured area because the vanes restrict the opening. Grille manufacturers always provide data pertaining to the equivalent open area rather than the actual grille measurement. This factor is known as the **A_k factor**. The *A* stands for area.

The concept for calculating the air volume from a grille or diffuser face is much the same as that for duct—the average velocity across the face is multiplied by the terminal's A_k factor to determine the volume. To do so, the velocity measurements must be taken in multiple locations around the perimeter of the terminal and averaged. However, the task is somewhat tedious, and an easier way to determine the air volume flowing into or out of a single terminal exists. The flow hood, or *balometer*, is introduced in *Section 3.2.4, Velometers, Anemometers, and Balometers*.

Calculating Total Air Volume

Several methods can determine the total air volume flowing through a system. Sometimes it is best to use more than one method and compare the results for improved accuracy and an understanding of where problems may lie.

For example, you can determine the total supply air output by taking traverse measurements downstream of the blower before any takeoffs. You can do the same thing at the return duct, and the two values should be the same; insignificant differences generally result from measurement errors.

You can also measure or calculate the air volume from all the supply registers and diffusers and add them together. Repeat the process for all the return air grilles. The supply and return values should again be very close, but this method introduces more opportunities for inaccuracy in measurement. However, significant differences in the supply and return volumes calculated this way typically indicate duct leakage somewhere.

Total system airflow can also be determined using the manufacturer's performance data pertaining to a specific piece of equipment. This method of determining total system airflow requires measurement of system static pressure (*Figure 7*). The manufacturer's service manuals provide data indicating the air handling capabilities of their units, in cubic feet per minute, for a specific equipment configuration. The data is usually presented on a chart or a graph. By comparing static pressure readings to the chart, the volume of air being moved can be determined.

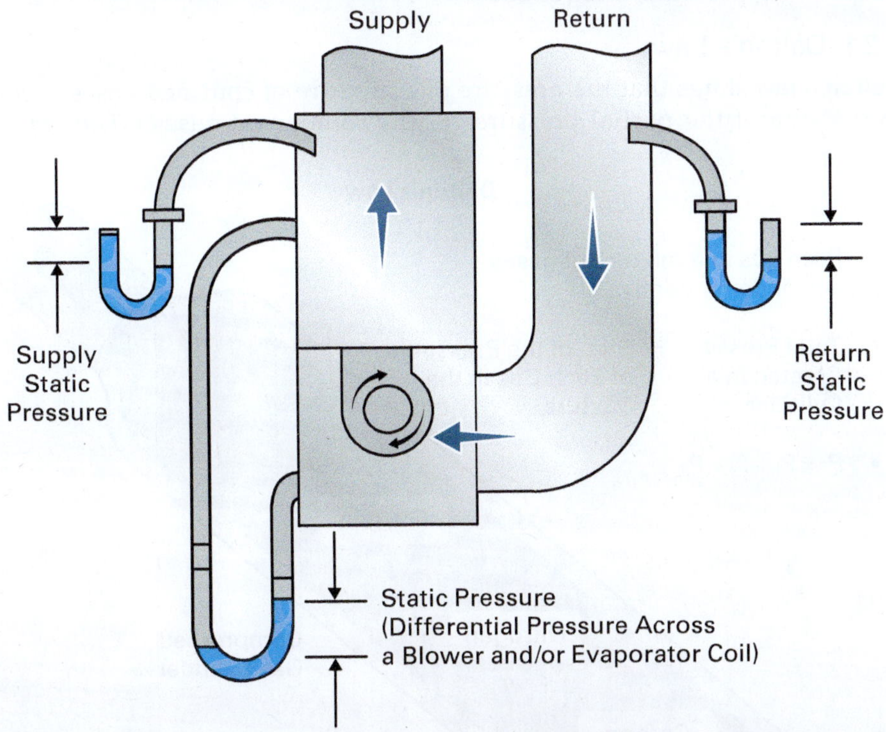

Figure 7 Measuring static pressure.

The total system pressure can be determined by calculating the difference between the supply and return static pressures; the return duct will have a negative pressure while the supply duct will have a positive pressure. Static pressure measurements across a blower or evaporator coil may also be required for performance charts.

Did You Know?

Air Balance History

In the early 1980s, there were a few books about air balancing and fewer than 1,000 people were certified to conduct a test and balance. Rob "Doc" Falke was one individual who believed in the principles of air balancing and the measurement of system performance. He was also excited to implement a new device called a Flowhood® from Shortridge Instruments, as he was certain it could help his contracting firm solve some problems.

Mr. Falke quickly learned some valuable lessons using the Flowhood® on the job. With the help of magazine editor Dominick Guarino, who was originally sent to interview Mr. Falke, he devised a plan to bring fresh training on this and related topics to the forefront. That training company was named the National Comfort Institute, and its first training event was held in 1993.

Today, the National Comfort Institute (NCI) remains "an organization that provides heating, air conditioning, plumbing, and electrical contractors with a focused offering of services and tools to help them improve their businesses, differentiate themselves, grow, and become more profitable."

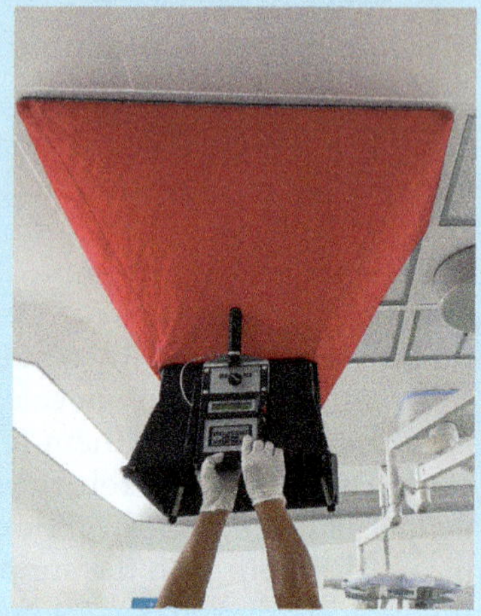

Source: Sucharas wongpeth/iStock/Getty Images

1.2.0 Relevant Air Laws

Several physical laws apply to air. These laws pertain to the characteristics of temperature, pressure, and volume.

1.2.1 Dalton's Law

Dalton's law: States that the pressure of a mixture of confined gases is equal to the sum of the partial pressures of the individual component gases. The partial pressure is the pressure that each gas would exert if it alone occupied the same volume of the mixture at the same temperature.

Dalton's law states that the pressure of a mixture of confined gases is equal to the sum of the partial pressures of the component gases (*Figure 8*). The

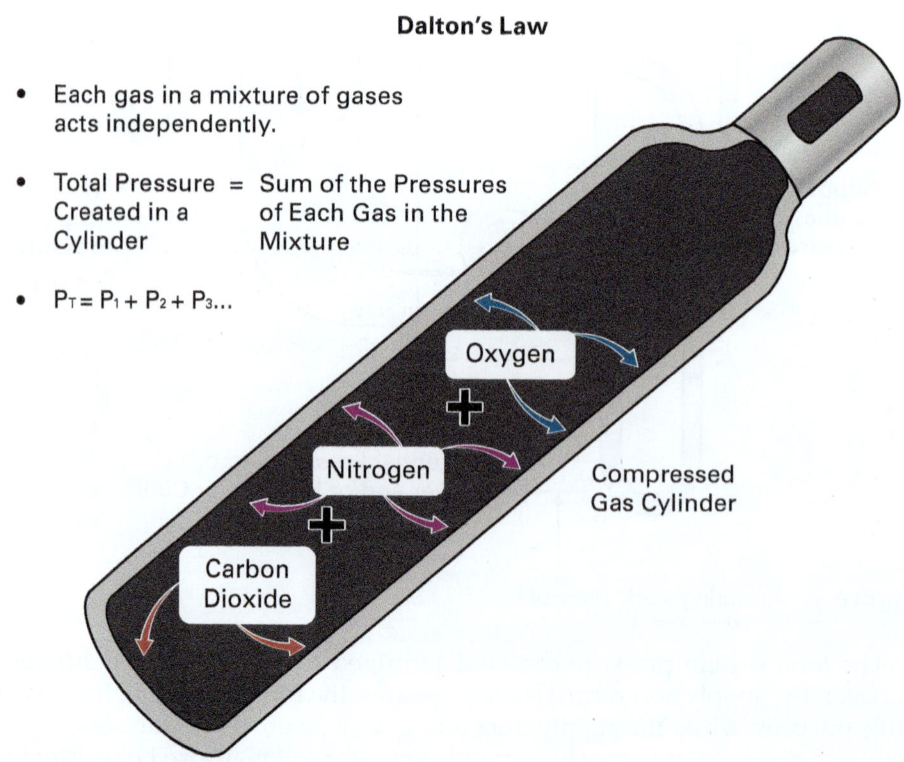

Figure 8 Dalton's law.

partial pressure refers to the pressure that each gas would exert if it alone occupied the volume of the confined area, at the same temperature. The equation is as follows:

$$P_T = P_1 + P_2 + P_3\ldots$$

1.2.2 Boyle's Law

Boyle's law states that, given a constant temperature, the pressure on a confined gas varies inversely with the volume it occupies. Similarly, at a constant temperature, the volume of a given quantity of gas varies inversely with the pressure applied. In other words, more pressure causes a reduction in the volume occupied, while providing a gas more space (volume) results in a reduction in pressure.

These three formulas outline this law mathematically:

$$\text{Formula 1}: \; P_O \times V_O = P_N \times V_N$$
$$\text{Formula 2}: \; V_N = (V_O \times P_O) \div P_N$$
$$\text{Formula 3}: \; P_N = (P_O \times V_O) \div V_N$$

Where:

P_O = original pressure (psia)
P_N = new pressure (psia)
V_O = original volume (cubic feet)
V_N = new volume (cubic feet)
Absolute pressure (psia) = psig + 14.7

Study Example:
What is the new volume of 2 ft^3 of gas at 30 psig if it is compressed to 60 psig, providing the temperature remains constant?

To find the new volume, the second equation is used. However, to convert the gauge pressure to absolute, we must add 14.7 psi to the pressure values:

$$\text{Formula 2}: \; V_N = (V_O \times P_O) \div P_N$$
$$V_N = (2\text{ ft}^3 \times 44.7\text{ psia}) \div 74.7\text{ psia}$$
$$V_N = 89.4 \div 74.7\text{ psia}$$
$$V_N = 1.20\text{ ft}^3\text{ (rounded)}$$

To summarize this example, doubling the pressure required pushing the gas into a much smaller space.

Boyle's law: States that, given a constant temperature, the pressure on a confined gas varies inversely with the volume the gas occupies. Similarly, at a constant temperature, the volume a confined gas occupies varies inversely with the pressure applied.

1.2.3 Charles' Law

Charles' law states that, given a constant pressure, the volume of a confined gas varies directly with a change in temperature. Similarly, given a constant volume of gas, the pressure varies directly with a change in temperature.

Formulas 4 through 6 mathematically demonstrate Charles' law when working with a constant pressure:

$$\text{Formula 4}: \; V_O \times T_N = V_N \times T_O$$
$$\text{Formula 5}: \; V_N = (V_O \times T_N) \div T_O$$
$$\text{Formula 6}: \; T_N = (V_N \times T_O) \div V_O$$

Where:

T_O = original absolute temperature (Kelvin)
T_N = new absolute temperature (Kelvin)
V_O = original volume (cubic feet)
V_N = new volume (cubic feet)
Absolute temperature = (°F + 459.67) ÷ 1.8

Charles' law: States that, given a constant pressure, the volume of a confined gas varies directly with a change in temperature. Similarly, given a constant volume of gas, the pressure varies directly with a change in temperature.

Formulas 7 through 9 demonstrate Charles' law when working with a constant volume:

$$\text{Formula 7}: P_O \times T_N = P_N \times T_O$$
$$\text{Formula 8}: T_N = (T_O \times P_N) \div P_O$$
$$\text{Formula 9}: P_N = (T_N \times P_O) \div T_O$$

Where:

T_O = original absolute temperature
T_N = new absolute temperature
P_O = original absolute pressure (psia)
P_N = new absolute pressure (psia)
Absolute temperature = (°F + 459.67) ÷ 1.8
Absolute pressure (psia) = psig + 14.7

Study Example:
What is the new volume of 8 ft^3 of gas at 40°F if the temperature is raised to 80°F while the pressure remains constant? To find the new volume, use Formula 5. But first, you must convert the temperature in °F to Kelvin:

$$K = (°F + 459.67) \div 1.8$$

Convert 40°F to K:

$$K = (40°F + 459.67) \div 1.8$$
$$K = 277.6K$$

Convert 80°F to K:

$$K = (80°F + 459.67) \div 1.8$$
$$K = 299.8K$$

$$\text{Formula 5}: V_N = (V_O \times T_N) \div T_O$$
$$V_N = (8\ ft^3 \times 299.8K) \div 277.6K$$
$$V_N = 2{,}398.4 \div 277.6$$
$$V_N = 8.64\ ft^3$$

To summarize, increasing the temperature of a gas from 40°F to 80°F, without a change in pressure, requires the space it occupies to be increased from 8 ft^3 to 8.64 ft^3. If the temperature is increased without providing additional space, the pressure would increase.

1.0.0 Section Review

1. Which of the following is a characteristic of what is referred to as *standard air*?
   a. 70°F
   b. 0 µm
   c. 0" Hg
   d. 14.7 psig

2. True or False? Boyle's law states that, given a constant temperature, the volume of a given quantity of gas varies inversely with the pressure applied.
   a. True
   b. False

2.0.0 Psychrometrics

Objective

Describe the study of psychrometrics and explain how it is used to determine specific air properties.

a. Describe psychrometrics and the related properties of air.
b. Describe the structure of the psychrometric chart.
c. Explain how to use the psychrometric chart to determine specific air properties.

Performance Tasks

There are no Performance Tasks in this section.

Psychrometrics is the study of air and water vapor mixtures. Understanding the relationship between the properties of air is critical to comfort, as well as many manufacturing processes. To advance your skills as an HVACR craft professional, you must understand psychrometrics. This knowledge will enable you to solve some problems that might not be diagnosed any other way.

Psychrometrics: The study of the physical and thermodynamic properties of air-vapor mixtures.

2.1.0 Psychrometric-Related Air Properties

Air has weight, density, temperature, heat, and can also conduct heat. When in motion, air also has momentum and inertia. Air can hold substances in suspension and in solution. As mentioned previously, dry air is a mixture of gases composed of about 78% nitrogen, 21% oxygen, and 1% other gases.

Specific volume describes how much space one pound of standard dry air occupies (*Figure 9*). At standard conditions, 1 lb of dry air occupies a volume of 13.33 ft^3. The **specific density** of dry air at standard conditions is 0.075 lb/ft^3.

Specific volume: The space occupied by one pound of standard air. Per ASHRAE standards, one pound of standard air occupies a volume of 13.33 ft^3.

Specific density: The weight (mass) of air per cubic foot. Per ASHRAE standards, standard air weighs 0.075 lb/ft^3.

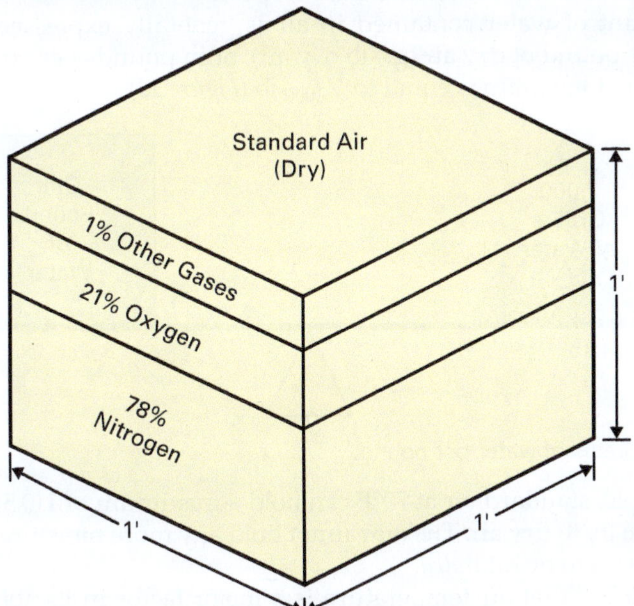

Specific Volume = 13.33 ft^3/lb

Specific Density = 0.075 lb/ft^3

Specific Heat = 0.24 Btu/lb/°F

Figure 9 Standard dry air composition and characteristics.

The **specific heat** of a substance refers to the amount of heat required to change the temperature of one unit of mass by 1° (the pound and °F, respectively, for our purposes). Standard dry air has a specific heat of 0.24 Btu/lb/°F. The *sensible heat formula* can be used to calculate how many Btus are needed to raise the temperature of dry air. It is called the sensible heat formula because it is related to dry air, thus discounting the presence of water vapor and the latent heat it contains.

Specific heat: The amount of heat required to change the temperature of one mass unit of a substance by one degree.

NOTE

Specific heat can also be defined using other units of mass as well as other units of temperature.

The basic sensible heat formula for a volume of standard dry air can be simplified as follows:

Sensible heat (Btuh) = specific heat × specific density × 60 min/hr × air volume (cfm) × ΔT

Sensible heat (Btuh) = (0.24 × 0.075 × 60 min/hr) × air volume (cfm) × ΔT

Sensible heat (Btuh) = 1.08 × air volume (cfm) × ΔT

Where:

cfm = air volume in cubic feet per minute
ΔT = change in temperature (°F)

Study example:

How much heat is required to raise the temperature of 450 cfm of air by 30°F?

Sensible heat (Btuh) = 1.08 × air volume (cfm) × ΔT

Sensible heat (Btuh) = 1.08 × 450 cfm × 30°F

Sensible heat (Btuh) = 14,580 Btuh

2.1.1 Water in Air

Rarely is a mass of air encountered that is completely free of water vapor. If the air we worked with in the HVACR craft were completely dry, it would certainly simplify many of the calculations! However, water vapor is always present in the air, and the volume of it can vary dramatically from one sample to another.

The water vapor entrained in air has many sources. Outdoors, it comes from the evaporation of seawater and other bodies of water, both small and large. Inside buildings, water vapor comes from cooking, showers, and similar activities. The amount of water vapor that air can hold depends on its temperature. Warm air can hold more moisture than cold air.

The amount of water contained in air is generally expressed in grains of moisture per pound of dry air (gr/lb dry air), or in pounds per pound of dry air (lb/lb dry air). One grain is equal to $^1/_{7,000}$ lb (*Figure 10*).

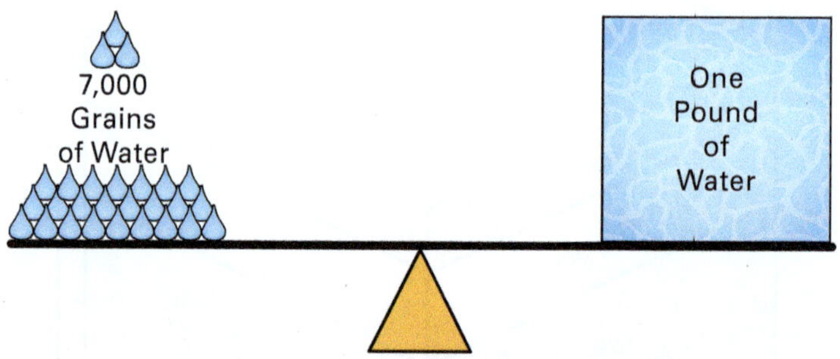

Figure 10 Grains of water per pound.

For example, standard air at 70°F can hold a maximum of 110.5 gr/lb dry air, or about 0.016 lb/lb dry air. The air cannot hold any more moisture at this condition, so it is said to be *saturated*.

But remember that air temperature is a major factor in its ability to absorb and hold water vapor. At 50°F, air is saturated when it contains about 53 gr/lb dry air, or about 0.0075 lb/lb dry air. The point at which air is saturated is also referred to as its **dew point**.

Dew point: The temperature at which air becomes saturated and water vapor begins to condense and fall freely from the air.

2.1.2 Relative Humidity

Relative humidity (RH) represents the ratio of the moisture present in air to the volume it can hold at saturation. If a sample of air is totally saturated with moisture, its relative humidity is 100%. If it contains only half of the moisture it can hold at its current temperature, the relative humidity is 50%. *Figure 11* shows how the same volume of moisture in a pound of dry air affects the relative humidity as the air temperature changes.

The amount of humidity in the air affects the rate at which perspiration evaporates from the skin. Dry air enables faster evaporation, which makes the skin feel cooler. Moist air prevents rapid evaporation, making the body feel warmer.

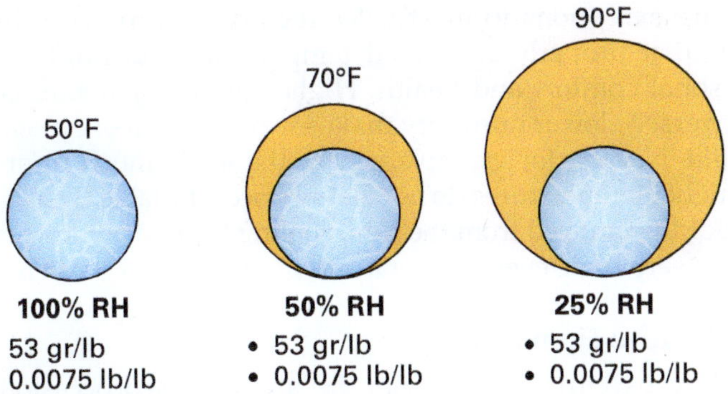

Figure 11 Relative humidity.

The proper control of relative humidity is therefore critical to comfort. It is also very important in many manufacturing processes, product storage, and similar applications. Most people feel comfortable when the indoor temperature and humidity conditions fall within a certain range, called the *comfort zone.*

The comfort zone is generally considered to be between 68°F and 78°F, with a relative humidity of 30% to 60% (*Figure 12*). Of course, not every person will

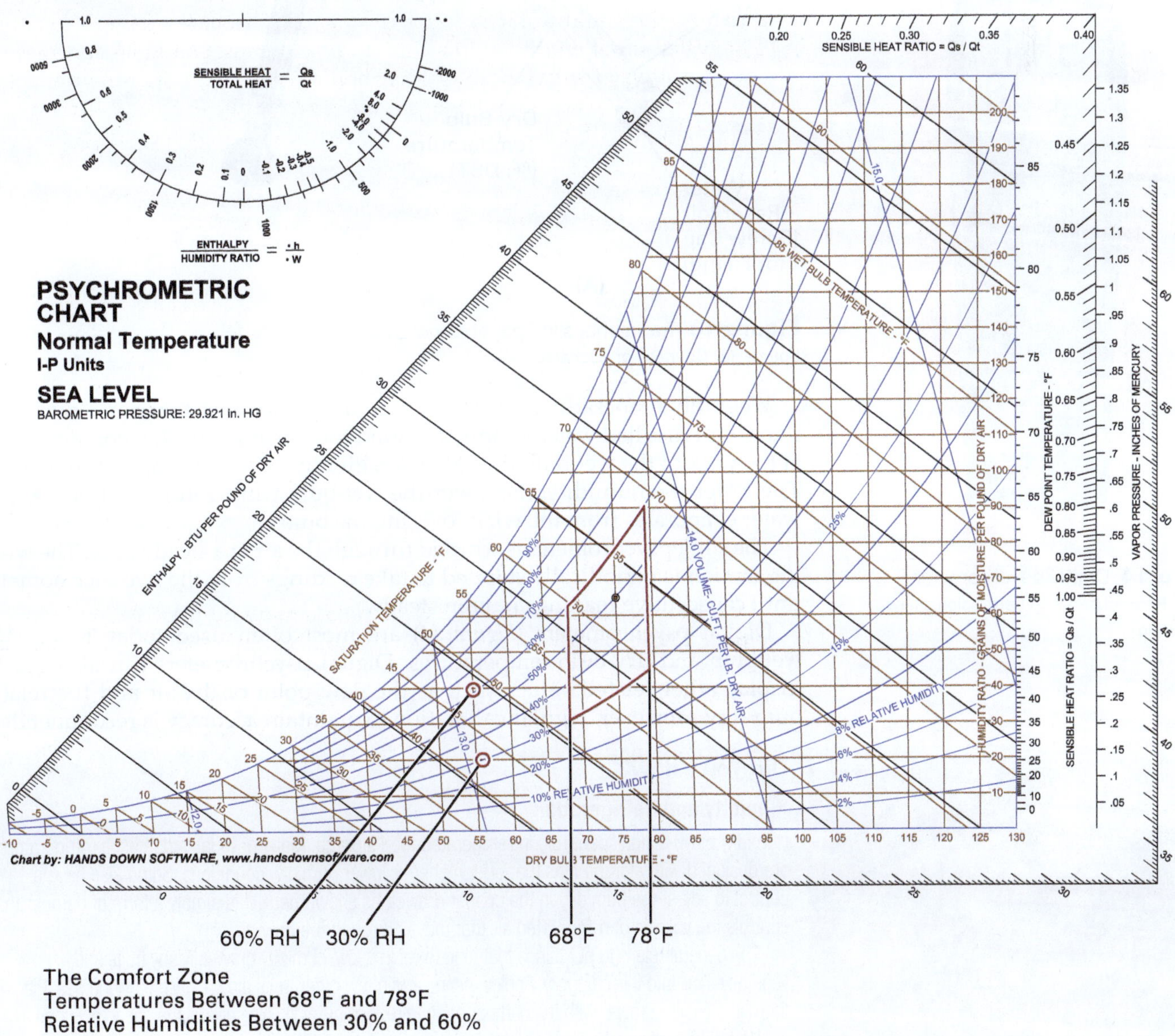

Figure 12 The comfort zone plotted on a psychrometric chart.
Source: CHART BY: Hands Down Software, www.handsdownsoftware.com

agree; there are exceptions to most rules regarding comfort. What people do agree with is that properly controlled temperature and humidity conditions improve personal comfort and health. Higher humidity makes cooler air feel warmer. Conversely, lower humidity makes warmer air feel cooler.

In states like Nevada for example, where the air is much drier than that of Florida, the higher temperatures do not feel so uncomfortable. This is because perspiration is rapidly removed from the skin, cooling the body. But when the humidity in Florida is very high, even temperatures in the upper 80s can be miserable.

Measuring Relative Humidity

Traditional sling psychrometers (*Figure 13*) are still used occasionally to measure air temperature and determine the relative humidity. Sling psychrometers contain two identical thermometers—one to measure dry-bulb temperatures and the other to measure wet-bulb temperatures. Dry-bulb temperatures represent the sensible heat in the air. The reading from a wet-bulb thermometer accounts for the moisture content of the air.

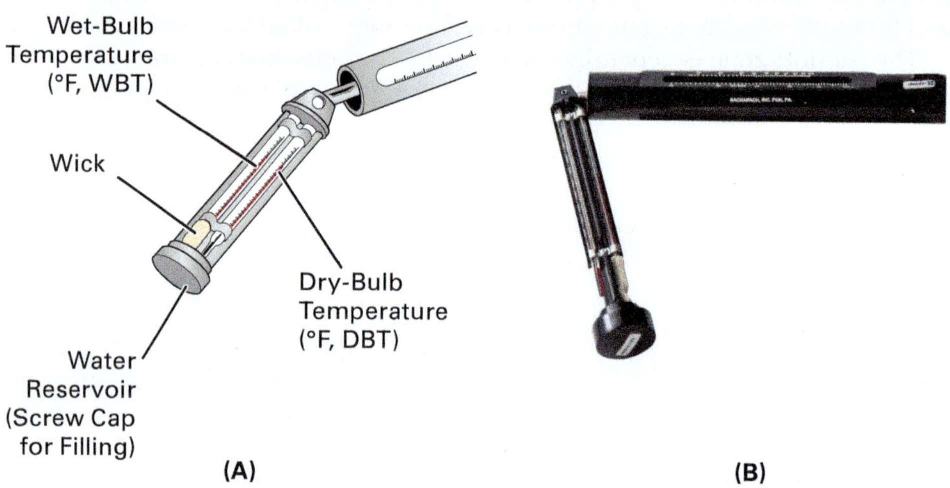

Figure 13 Traditional sling psychrometer.
Source: (B) Courtesy of Bacharach, Inc.

Wet-bulb temperatures are taken with a thermometer that has a wick saturated with distilled water wrapped around the sensing bulb. Evaporation occurs at the wick of the wet-bulb thermometer, giving it a lower temperature reading. The lower the humidity, the lower the wet-bulb temperature will be, as more water evaporates from the wick covering the bulb.

The sling psychrometer is whirled through the air at a steady pace. The whirling motion is periodically stopped to take readings from the two thermometers until consecutive readings remain steady.

Digital psychrometers (*Figure 14*) are most often used today to measure wet-bulb and dry-bulb temperatures. Digital psychrometers can also quickly display other related values such as the dew point of the air and the relative humidity. However, regular calibration to maintain accuracy is recommended.

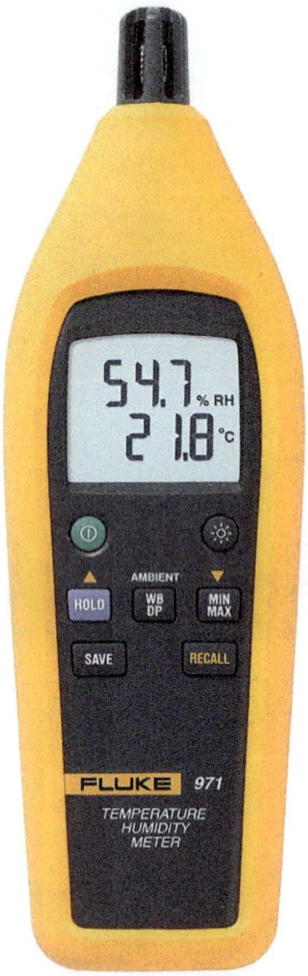

Figure 14 Digital psychrometer.
Source: Reproduced with Permission, Fluke Corporation

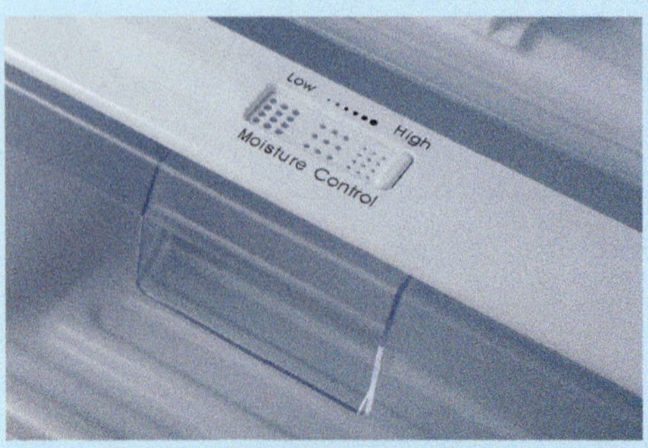

Source: Tohid Hashemkhani/iStock/Getty Images

Did You Know?

Humidity and Refrigeration

Humidity plays a significant role in refrigerated food storage. Many of today's domestic refrigerators are designed with storage bins that offer higher or lower humidity conditions. Some also have a control to adjust the humidity in the crisper drawer, as shown here. It is simply a form of damper that controls the volume of refrigerated air that moves through the drawer.

The general rule is to put things that tend to rot in a low-humidity drawer. Many fruits and vegetables fall into this category. Apples, pears, and a few other fruits also release ethylene gas. Placing them in a crisper drawer with the damper open to reduce humidity also allows the gas to escape.

Vegetables that wilt, such as leafy greens, are best kept in a high-humidity drawer. The increased humidity helps keep the vegetables fresh and hydrated.

2.1.3 Dew Point

The dew point is the temperature at which air becomes saturated with water vapor and the water begins to form water droplets that fall freely out of the air. When the relative humidity is 100% (saturated air), the dew point, wet-bulb temperature, and dry-bulb temperature are all the same. This is true because the air cannot hold any more moisture. At this condition, dewfall and fog typically occur. Rain forms at much higher altitudes before falling to the ground.

2.1.4 Enthalpy

Water vapor in the air contains heat. The latent heat it contains added to the sensible heat yields the total heat of the air. The latent heat referred to here is the heat the water absorbed as it vaporized.

The total heat content of an air and water vapor mixture is called *enthalpy*. Enthalpy is measured using a wet-bulb thermometer and is expressed in Btus per pound (Btu/lb). The change in enthalpy after air passes through an HVAC system can be used to determine the capacity of a typical air conditioning system, using the following equation:

$$\text{Total Heat Transferred (Btuh)} = (H1 - H2) \times 4.5 \times \text{cfm}$$

Where:

H# = enthalpy of the entering and leaving airstreams
cfm = system airflow

NOTE

This equation with a practice problem was presented in NCCER Module 03215, *Basic Maintenance*.

2.2.0 Psychrometric Chart Structure

The psychrometric chart provides a graphical representation of air properties and their relationship. It is used to determine how air properties vary with the amount of moisture in the air.

The psychrometric chart shown in *Figure 15* is based on normal temperatures at sea level, so it is considered the normal chart. Charts are also available for low temperatures, high temperatures, and for various elevations other than sea level. Metric psychrometric charts are also available. With the dry-bulb and wet-bulb temperatures of an air sample, the psychrometric chart can be used to find the value for one or more other properties.

Did You Know?

Psychrometric Chart Development

In 1911, Dr. Willis Carrier presented his work called the "Rational Psychrometric Formulas" to what was then known as the American Society of Mechanical Engineers (ASME). This led to the development of the psychrometric chart. A normal-temperature psychrometric chart is based on a barometric pressure of 29.92 inHg. Though based on atmospheric pressure at sea level, the normal temperature chart can be used for field work up to 2,000' in elevation. If working at higher elevations, technicians should use a chart for that elevation. Dr. Carrier is widely considered to be the father of modern air conditioning.

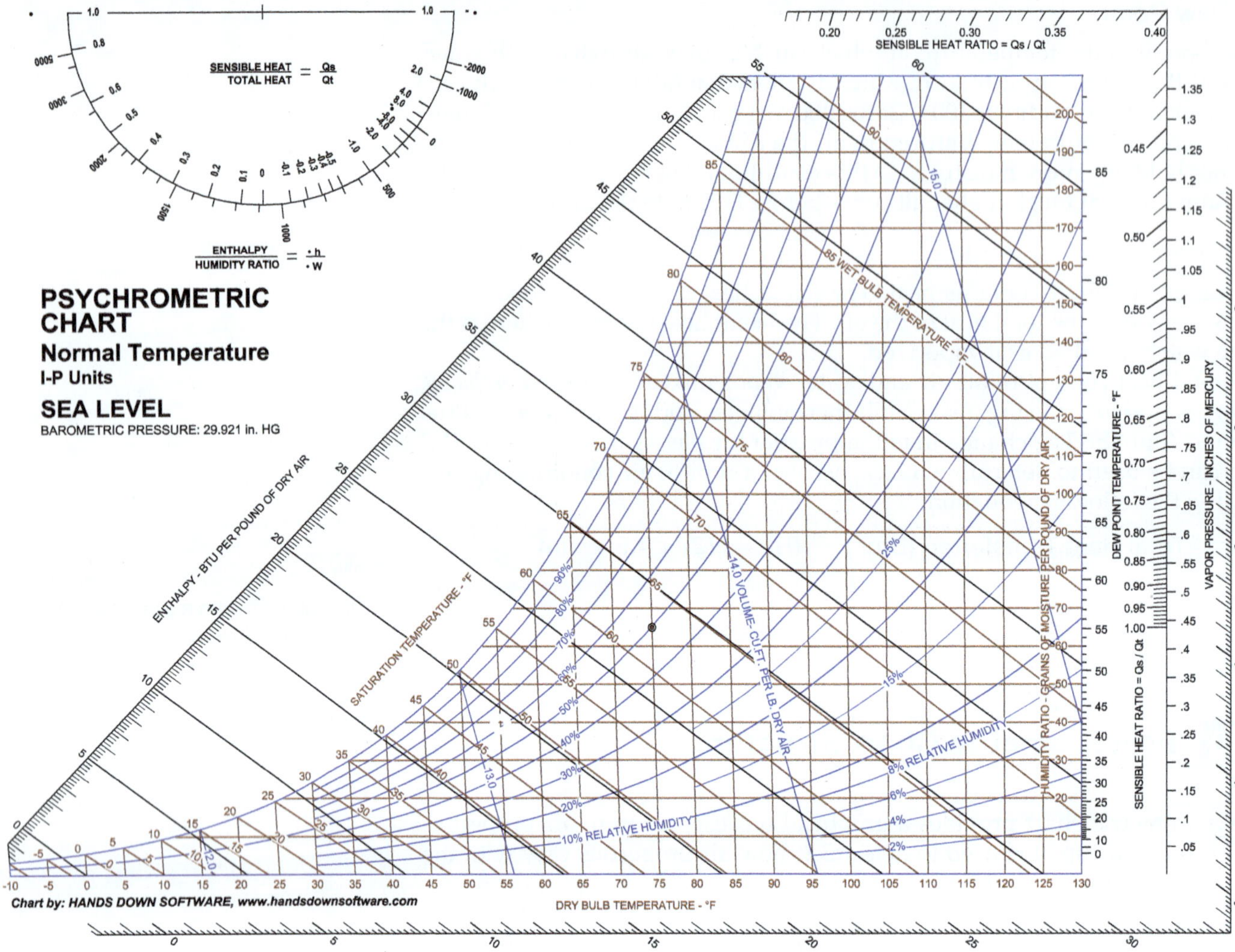

Figure 15 Example psychrometric chart.
Source: CHART BY: Hands Down Software, www.handsdownsoftware.com

Refer to *Figure 16*. The scales typically shown on a psychrometric chart are as follows:

- Dry-bulb temperature (°F)
- Grains of moisture per pound of dry air
- Relative humidity
- Dew point (°F)
- Wet-bulb temperature (°F)
- Enthalpy
- Specific volume
- Sensible heat factor

2.2.1 Dry-Bulb Temperature

The dry-bulb temperature scale, also referred to as the *sensible-heat temperature scale*, is laid out horizontally across the bottom, with increment lines extended vertically, as shown in *Figure 16A*.

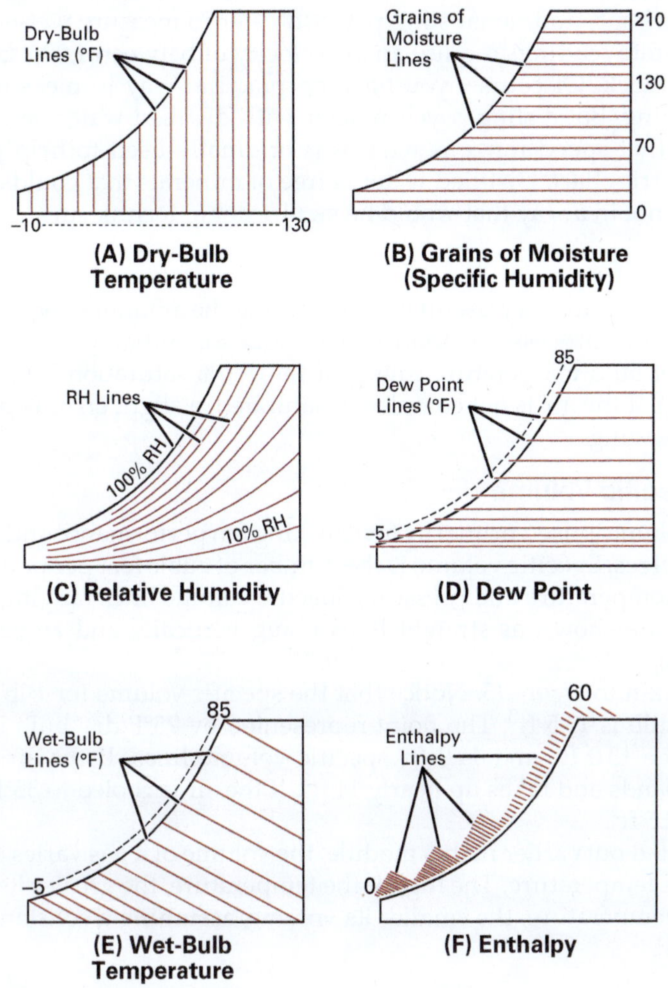

Figure 16 Psychrometric chart scales.

2.2.2 Grains of Moisture (Specific Humidity)

The grains scale (*Figure 16B*) represents the volume of water vapor per pound of air. The scale may also be referred to as the *specific humidity scale*. This scale runs vertically with increment lines extended horizontally.

Another scale on a full-sized chart, like the one shown in *Figure 15*, located just to the right of the grains scale, can be used to determine the humidity ratio.

2.2.3 Relative Humidity

Relative humidity is shown as curving lines from 0% to 100% RH (*Figure 16C*). At the 100% RH line, the air is fully saturated. Any reduction in air temperature results in condensation (dew).

2.2.4 Dew Point Temperature

The temperature at which the relative humidity has reached 100% (saturation) is also the dew point. The dew point temperature lines (*Figure 16D*) run horizontally like the grains of moisture lines. However, the lines are not at consistent intervals, spreading further apart as the scale rises from bottom to top.

2.2.5 Wet-Bulb Temperature

The wet-bulb temperature lines slant to the left at about a 45-degree angle (*Figure 16E*). The scale is the same one that is used to show dew point and saturation.

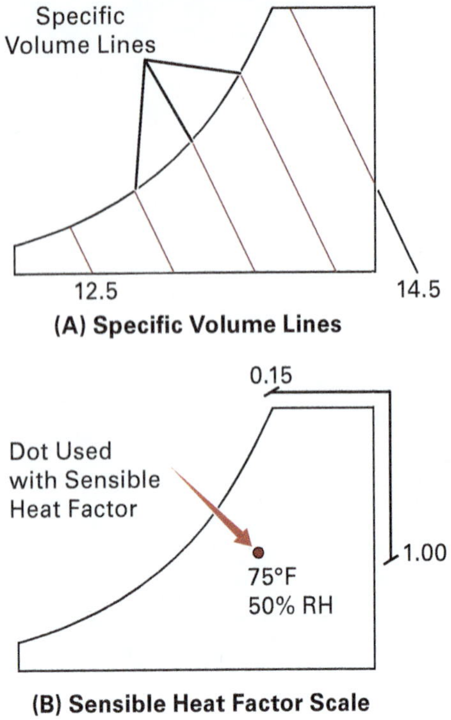

Figure 17 Specific volume and sensible heat factor.

When using a traditional sling psychrometer to measure wet-bulb temperatures, accurate readings require an air velocity of between 1,000 and 1,500 fpm across the wick. Of course, you have no practical way to measure the speed of your swing. But a sling psychrometer with distilled water on the wick and rotated at that speed range is what was originally used to help populate the psychrometric chart. Distilled water is free of minerals that could affect evaporation, but not in a way that would be significant to technicians.

2.2.6 Enthalpy

Enthalpy values are very useful in determining the amount of heat that is added or removed in a process. An approximate value for enthalpy is found by extending the wet-bulb temperature line past the 100% saturation line on the chart (*Figure 16F*). If the air is not completely saturated, a slight error is present in the enthalpy reading.

2.2.7 Specific Volume

Specific volume is used primarily for checking fan performance and determining fan motor sizes. Specific volume is the number of cubic feet per pound of dry air at a given temperature and pressure. Specific volume lines ranging from 12.5 to 14.5 ft^3/lb are shown as straight lines rising vertically and angled to the left (*Figure 17A*).

Refer again to *Figure 15*. Notice that the specific volume for 1 lb of dry air at 75°F dry bulb is 13.5 ft^3. The point represented by 75°F dry bulb is positioned between the 13.0 ft^3 and 14.0 ft^3 specific volume lines. If the air is heated to 95°F, it expands and takes up nearly 14 ft^3. If the air is cooled to 35°F, it occupies roughly 12.5 ft^3.

As pointed out earlier in this module, the volume of a gas varies directly with its absolute temperature. The higher the temperature, the greater its volume; the lower the temperature, the smaller its volume, assuming a constant pressure is maintained.

2.2.8 Sensible Heat Factor

The sensible-heat factor scale provides a useful method for determining sensible heat. This scale, located to the right of the chart, provides the sensible heat ratio (*Figure 17B*). The scale ranges from about 0.15 to 1.00, or 15% to 100%, representing the percentage of sensible heat in the air.

This scale is used when plotting processes such as cooling or dehumidification, allowing determination of the ratio of sensible heat to total heat. Whatever remains represents latent heat. For example, if the sensible heat ratio is 0.60, or 60%, 40% of the heat the air contains is latent heat that can be attributed to its water content.

The dot located at the intersection of the 75°F dry-bulb temperature (DBT) and the 50% RH lines on the chart is provided for use in conjunction with the sensible-heat factor scale. It is used when finding the dew point of a device when the device's sensible heat factor is known. Note that not every psychrometric chart has this dot plotted on the chart, and many may not include the sensible-heat factor scale.

2.3.0 Using the Psychrometric Chart

The remainder of this section focuses on the use of a psychrometric chart for servicing HVACR systems. For more advanced study, see the *Additional Resources* section for this module.

2.3.1 Using Known Properties to Find Unknown Properties

Figure 18 shows a psychrometric chart with lines plotted for dry-bulb temperature, wet-bulb temperature, dew point, specific humidity, and relative humidity. When any two of these values are known, the exact condition of the air can be located on the chart and the values for all the other properties can be found from this one point.

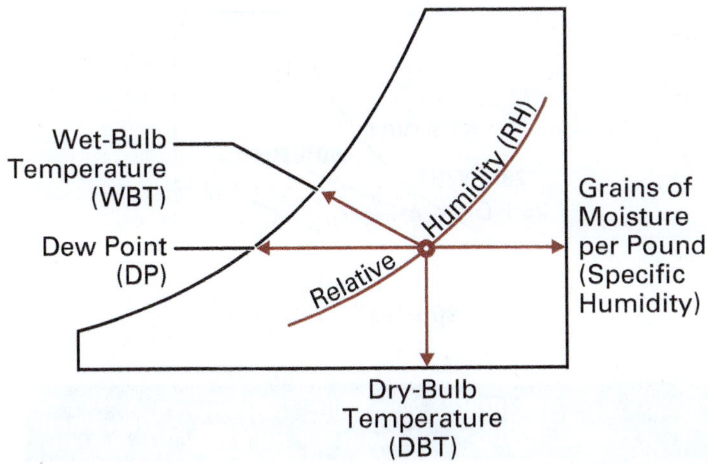

Figure 18 Example of air properties plotted on a psychrometric chart.

For example, assume that the dry-bulb temperature is 75°F and the wet-bulb temperature is 65°F for a given sample of air. Plotting the dry-bulb and wet-bulb temperatures on the psychrometric chart to determine their intersection, you can see that the relative humidity is about 58%, the dew point is 59.5°F, and the specific humidity is about 76 gr/lb dry air. A simplified plot of this example is shown in *Figure 19*.

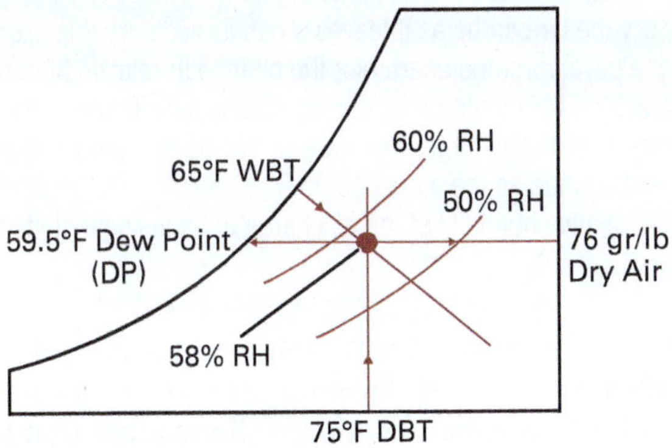

Figure 19 Finding unknown air properties.

Practice finding these values on a psychrometric chart for different combinations of dry-bulb and wet-bulb temperatures measured in your classroom, home, or even outdoors. A straightedge or ruler helps.

2.3.2 Working with Changes in Sensible Heat

The psychrometric chart can be used to evaluate air conditioning processes, such as changes in the sensible heat of air. Changes in dry-bulb temperature and is indicated by a horizontal line on the chart.

An example of a change in sensible heat is plotted in *Figure 20*. Assume that air at 30°F DBT and 80% RH is heated to 75°F DBT. If the air is heated with no moisture added or removed (common), the process results in a horizontal line from point 1 to point 2 as shown in *Figure 20*. With the information provided, you can determine the change in wet-bulb temperature, relative humidity, and dew point resulting from the change in dry-bulb temperature.

NOTE

Your results may differ slightly, depending on the size and composition of the psychrometric chart in use. Larger charts are more easily examined and allow for more accurate interpolation between values.

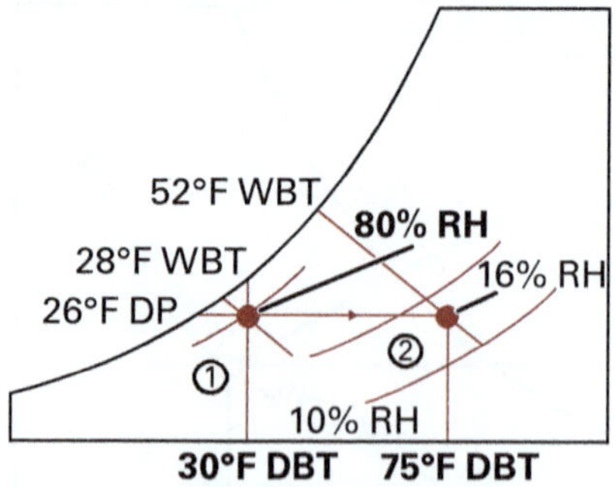

	Dry-Bulb Temp	Wet-Bulb Temp	% Relative Humidity	Dew Point
Air at	30°F	28°F	80%	26°F
Is Heated to	75°F	52°F	16%	26°F

(Given Values Are Shown in Bold)

Figure 20 Psychrometric chart plot of sensible heat increase.

On Site

Relative Humidity During the Winter Months

Many people complain of dry conditions in their home or business during the winter, and for good reason. Humidity plays a very important role in winter comfort.

To see just how dry the air can be as it leaves a gas furnace, try this exercise with very common values. On the psychrometric chart, plot the change in relative humidity that occurs when air at 72°F and a 30% RH (common indoor conditions for a home in the winter) is heated by a gas furnace to a discharge temperature of 120°F. Remember that the volume of water in the air does not change as the air is heated.

You'll find that the relative humidity of the discharge air is a single digit, and dry air is thirsty air!

Note that the dew point temperature remains the same, because no water vapor has been added or removed. Heating air reduces its relative humidity, but not the actual volume of water it contains. Remember that the term *relative humidity* is so named because it refers to the volume of water vapor in the air relative to its temperature.

An example of changes that occur when air cools is plotted in *Figure 21*. Assume that outdoor air at 95°F DBT and 75°F WBT cools overnight to 80°F DBT. The change results in a horizontal line from point 1 to point 2 as marked on the chart. With the information provided, you can determine the new wet-bulb temperature, relative humidity, and dew point.

Again, the dew point remains the same because no water has been added or removed. What this exercise reveals is that, on this particular night, dew will not fall outdoors unless it happens at higher altitudes and falls through the air below.

Remember, however, that air passing across a cold evaporator coil is nearly always cooled below its dew point, and water condenses on the coil surface. This clearly represents a change in moisture content. Measure the return air and discharge air dry-bulb and wet-bulb temperatures separately, then plot those points on the chart independently to see how the characteristics change.

2.3.3 Working With Enthalpy

Enthalpy—the air's heat content—is found on the chart by following along a wet-bulb temperature line, past the saturation line, and out to the enthalpy scale. Changes in enthalpy can be used to determine the total heat added to or removed from air.

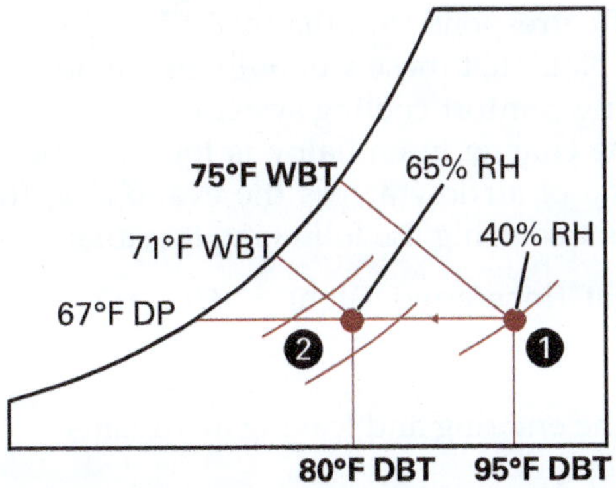

	Dry-Bulb Temp	Wet-Bulb Temp	% Relative Humidity	Dew Point
Air at	**95°F**	**75°F**	40%	67°F
Is Cooled to	**80°F**	71°F	65%	67°F

(Given Values Are Shown in Bold)

Figure 21 Psychrometric chart plot of sensible heat decrease.

For example, *Figure 22* shows a plot of the dry-bulb and wet-bulb temperatures measured across an evaporator coil. The measured wet-bulb temperatures can be used to check the cooling capacity delivered by the evaporator.

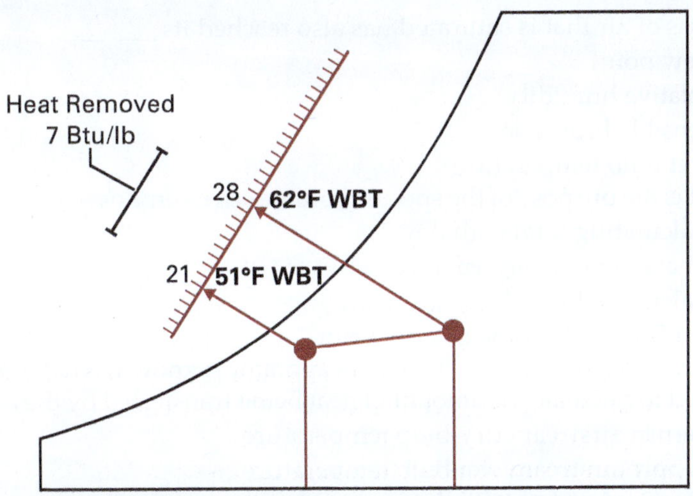

	Wet-Bulb Temp (WBT)	Enthalpy	Airflow Volume
Air Entering Evaporator	62°F	28 Btu/lb	1,000 cfm
Air Leaving Evaporator	51°F	21 Btu/lb	1,000 cfm

(Given Values Are Shown in Bold)

Figure 22 Using enthalpy values to find cooling capacity.

As shown, the wet-bulb temperatures for the air entering and leaving the evaporator are extended past the saturation line, and out to the enthalpy scale. In this example, the measured temperatures of 62°F wet-bulb temperature

(WBT) and 51°F WBT correspond to enthalpy readings of about 28 Btu/lb and 21 Btu/lb, respectively. Note that these wet-bulb temperatures represent common values for an operating comfort-cooling system.

By subtraction, the change in enthalpy is found to be 7 Btu/lb. Using this value and the volume of airflow across the evaporator, the delivered cooling capacity can be calculated using the following formula:

$$\text{Total Heat Transferred (Btuh)} = (H1 - H2) \times 4.5 \times \text{cfm}$$

Where:

H# = enthalpy of the entering and leaving airstreams
cfm = system airflow

As shown in *Figure 22*, the system airflow was determined to be 1,000 cfm. Now the amount of heat transferred can be calculated:

$$\text{Total Heat Transferred (Btuh)} = (H1 - H2) \times 4.5 \times \text{cfm}$$
$$\text{Total Heat Transferred (Btuh)} = (28\ \text{Btuh/lb} - 21\ \text{Btuh/lb}) \times 4.5 \times 1{,}000\ \text{cfm}$$
$$\text{Total Heat Transferred (Btuh)} = 7\ \text{Btuh/lb} \times 4.5 \times 1{,}000\ \text{cfm}$$
$$\text{Total Heat Transferred (Btuh)} = 31.5 \times 1{,}000\ \text{cfm}$$
$$\text{Total Heat Transferred (Btuh)} = 31{,}500\ \text{Btuh}$$

2.0.0 Section Review

1. A mass of air that is saturated has also reached its ______.
   a. dew point
   b. relative humidity
   c. sensible heat ratio
   d. wet-bulb temperature
2. What is the purpose of the specific volume scales on a psychrometric chart?
   a. Calculating total enthalpy
   b. Checking fan performance
   c. Calculating heat loss/heat gain
   d. Finding the volume of a duct run
3. If the volume of airflow across an evaporator is known, what other value is needed to calculate the amount of heat being transferred by the evaporator?
   a. Return airstream dry-bulb temperature
   b. Supply airstream wet-bulb temperature
   c. Grains of moisture in the supply airstream
   d. Change in enthalpy as the air crosses the evaporator

3.0.0 The Air Balancing Process

Performance Tasks

1. Demonstrate the use of air pressure and velocity measurement instruments on a duct system.
2. Measure or calculate the air volume at supply and return air grilles and registers.
3. Adjust a damper in a branch supply duct or takeoff to achieve a target air volume.

Objective

Describe the air balancing process and explain how to balance an air distribution system.

a. Describe air balancing and define common terminology.
b. Identify the tools and instruments used in air balancing.
c. Describe the steps to take prior to beginning an air balancing task.
d. Explain how to measure system and terminal airflow and make adjustments.
e. Explain how to balance a system using a thermometer.

Balancing an air distribution system means delivering conditioned air in the correct volume to each of the areas in the structure being conditioned, resulting in consistent conditions throughout. The satisfactory distribution of the air depends upon a well-designed duct system and a properly chosen fan.

After a system is installed, it should be balanced to ensure the volume of air delivered to each terminal is as designed. Balancing the system will often require adjustments to the fan speed as well as to the position of strategically placed dampers.

3.1.0 Room Air Movement Terminology

The air balancing process begins with understanding some common terms associated with air terminals. These terms can apply to any air system. While some of these terms may have been introduced in previous modules, many will likely be new to you in this context.

Note that the terms have been placed in alphabetic order, but some of the terms may be used in the definition of others.

- *A_k factor* — This term was introduced in *Section 1.1.2, Air Pressure, Velocity, and Volume.* It represents the net free area of a terminal, stated in square feet. The terminal device manufacturer specifies this value to aid in airflow calculations. For example, a square ceiling diffuser with an 8" neck isn't just an 8" opening from which the air emerges—louvers, blades, and perforated faces partially block the opening. The A_k factor compensates for these obstructions.
- *Aspect ratio* — The ratio of width to height of a duct or air terminal.
- *Drop* — The vertical distance that the lower edge of a horizontally projected airstream falls or rises over a specified distance (*Figure 23*). It is the result of the combined effects of gravity, the air's density, and the vertical expansion of the air path that results from entraining room air. The reported drop for a supply air terminal is generally related to its throw.

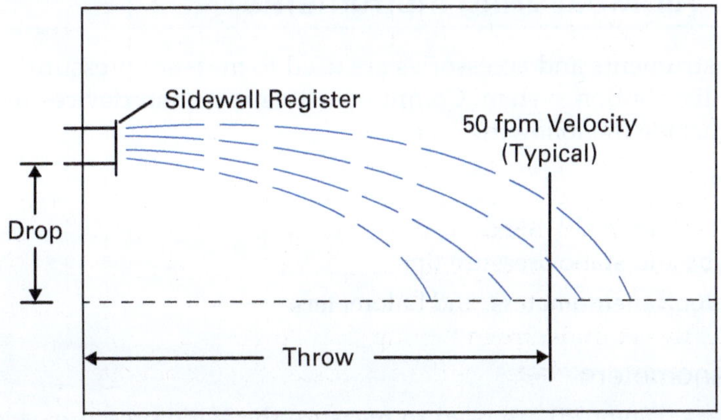

Figure 23 Air drop and throw.

- *Entrained air* — Room air that is drawn into and mixed with the supply airstream leaving a terminal. Also known as *secondary air,* the resulting turbulence in room air causes eddy currents that are referred to as *secondary air motion.*
- *Primary air* — In the context of air distribution and balancing, the air delivered to the room or conditioned space directly from a supply duct or terminal.
- *Spread* — The horizontal divergence of a supply airstream after it exits a terminal (*Figure 24*).
- *Terminal* — A device placed at a duct termination to discharge or admit air, often chosen to influence how the air is discharged. Examples include grilles, registers, and diffusers.

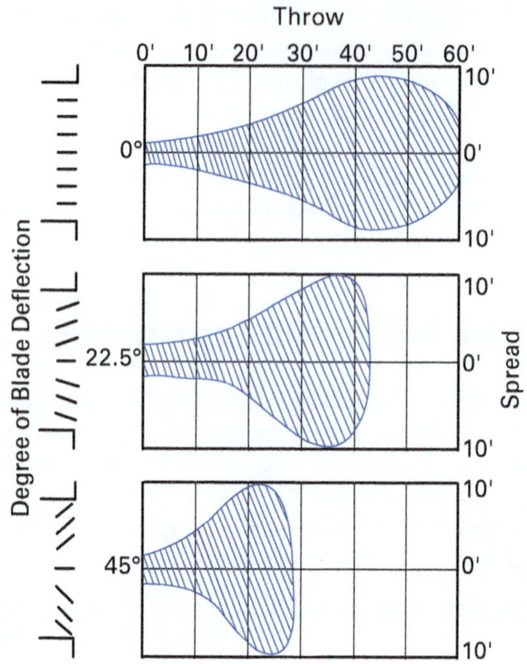

Figure 24 Air throw and spread patterns.

- *Throw* — The distance from the center of a supply air terminal to a point in the airstream where the highest sustained velocity has been reduced to a specified level, usually 50 fpm, as shown in *Figure 23*. The specified velocity for the throw measurement is called the *terminal velocity* (V_t). In *Figure 24*, note how the throw is affected by the angle of the blades, or *deflection*, on the register's face.

3.2.0 Air Measuring Instruments

Several instruments and accessories are used to measure pressure and velocity in an air distribution system. Common instruments and devices used for this purpose include the following:

- Manometer
- Differential pressure gauge
- Pitot tube and static pressure tips
- Velometers, anemometers, and balometers

3.2.1 Manometers

Manometers (*Figure 25*) are used to measure the static pressure, velocity pressure, and total pressure in air distribution systems. Manometers used for this purpose in the United States are calibrated in inches of water column (" w.c.).

Traditional manometers may use water or a special oil. Most models use a type of oil that has a specific gravity of 0.826 as the measuring fluid. The manufacturer of the gauge specifies the type of oil to be used; therefore, substitution for a specified oil is not recommended, as it will likely affect the accuracy.

Traditional manometers come in many styles, including U-tube, inclined, and combined U-inclined. These manometers work on the principle that air pressure affects the level of a sloped or vertical liquid column, and the change in column height could be measured. The concept of measuring low pressures in inches of water column was indeed based on these instruments.

Individual U-tube and inclined manometers are available in many pressure ranges. Inclined manometers are usually calibrated in the lower pressure ranges and are more sensitive than U-tube manometers.

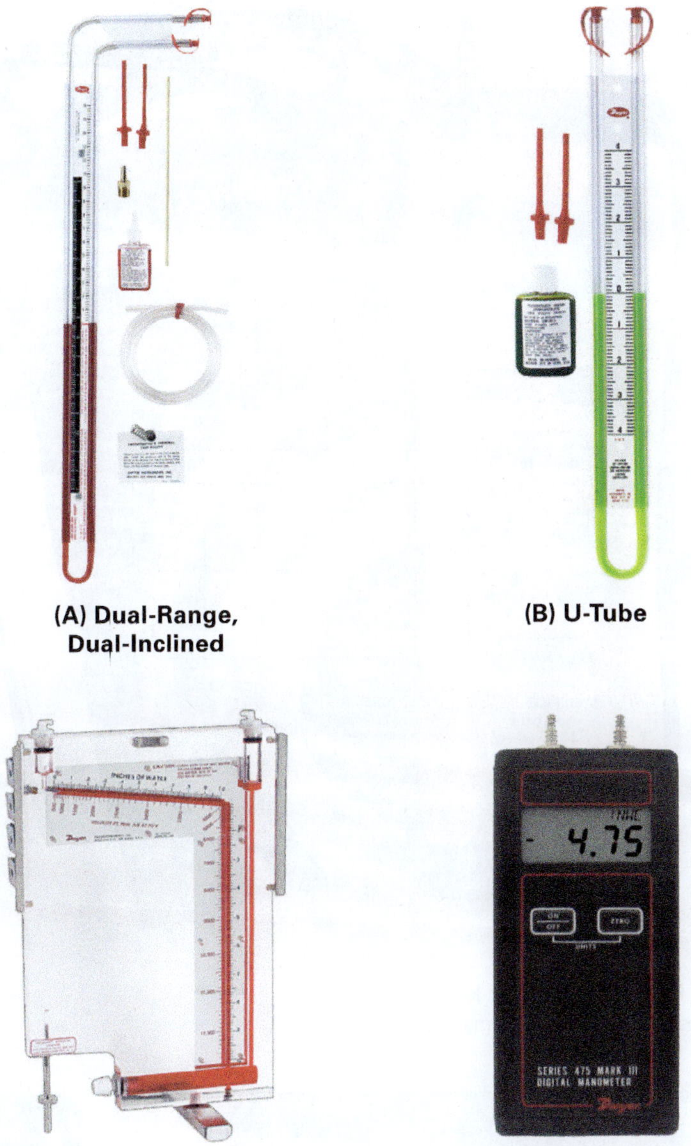

(A) Dual-Range, Dual-Inclined

(B) U-Tube

(C) Inclined-Vertical

(D) Digital

Figure 25 Manometers.
Source: Images courtesy of Dwyer Instruments, LLC

U-inclined manometers combine both the sensitivity of the inclined manometer with the higher-pressure-range capability of the U-tube manometer in one instrument.

Inclined-vertical manometers combine an inclined section for high accuracy and a vertical manometer section for extended range. The manometers also have an additional scale that can measure air velocity in feet per minute.

Digital manometers are a better choice when speed, efficiency, and flexibility are important. Digital models can often measure pressure across a much wider range than traditional models. The manometer shown in *Figure 26* has a range of ±2" to ±60" w.c. It can measure pressure in millimeters of water column, millibars, and pounds per square inch, and has a tiny pump for testing furnace draft pressure switches. Manometer kits provide you with most, if not all, of the needed accessories. Wireless manometers with a compatible app (*Figure 27*) can provide updates while you work or make adjustments elsewhere in the system.

3.2.2 Differential Pressure Gauge

Differential pressure gauges (*Figure 28*) are typically used to measure duct pressure, filter resistance, and furnace draft. Some are capable of measuring both pressure and velocity. Dual-scale gauges are calibrated for pressure in inches of

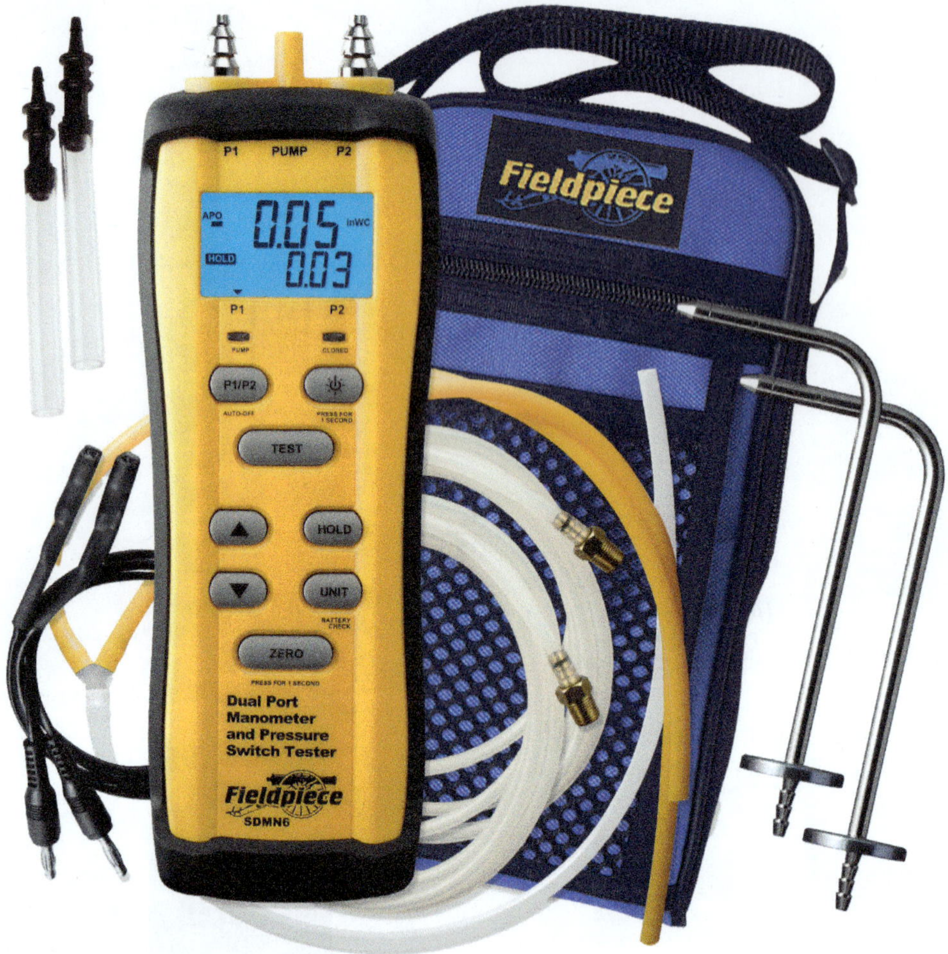

Figure 26 Digital manometer kit.
Source: Courtesy of Fieldpiece Instruments

(A) Wireless Manometer **(B) Job Link® App**

Figure 27 Wireless manometer and compatible mobile app.
Source: Courtesy of Fieldpiece Instruments

water column and for air velocity in feet per minute. Like manometers, the type of gauge used in air balancing is designed exclusively for very small pressures.

Differential pressure gauges are sometimes permanently installed on large air handling units to monitor the pressure drop across the filters. Readings when the filters are fresh are compared to future readings to determine when filter replacement is necessary. Pitot tubes and/or static pressure tips are often needed to make air pressure and velocity measurements in ductwork.

Figure 28 Portable differential pressure gauge.
Source: Image courtesy Dwyer Instruments, LLC

3.2.3 Pitot Tubes and Static Pressure Tips

Pitot tubes (*Figure 29*) and static pressure tips are used with manometers and pressure gauges to measure duct pressure. The standard pitot tube used for measurements in ducts 8" and larger has a $^{5}/_{16}$" diameter outer tube with eight equally

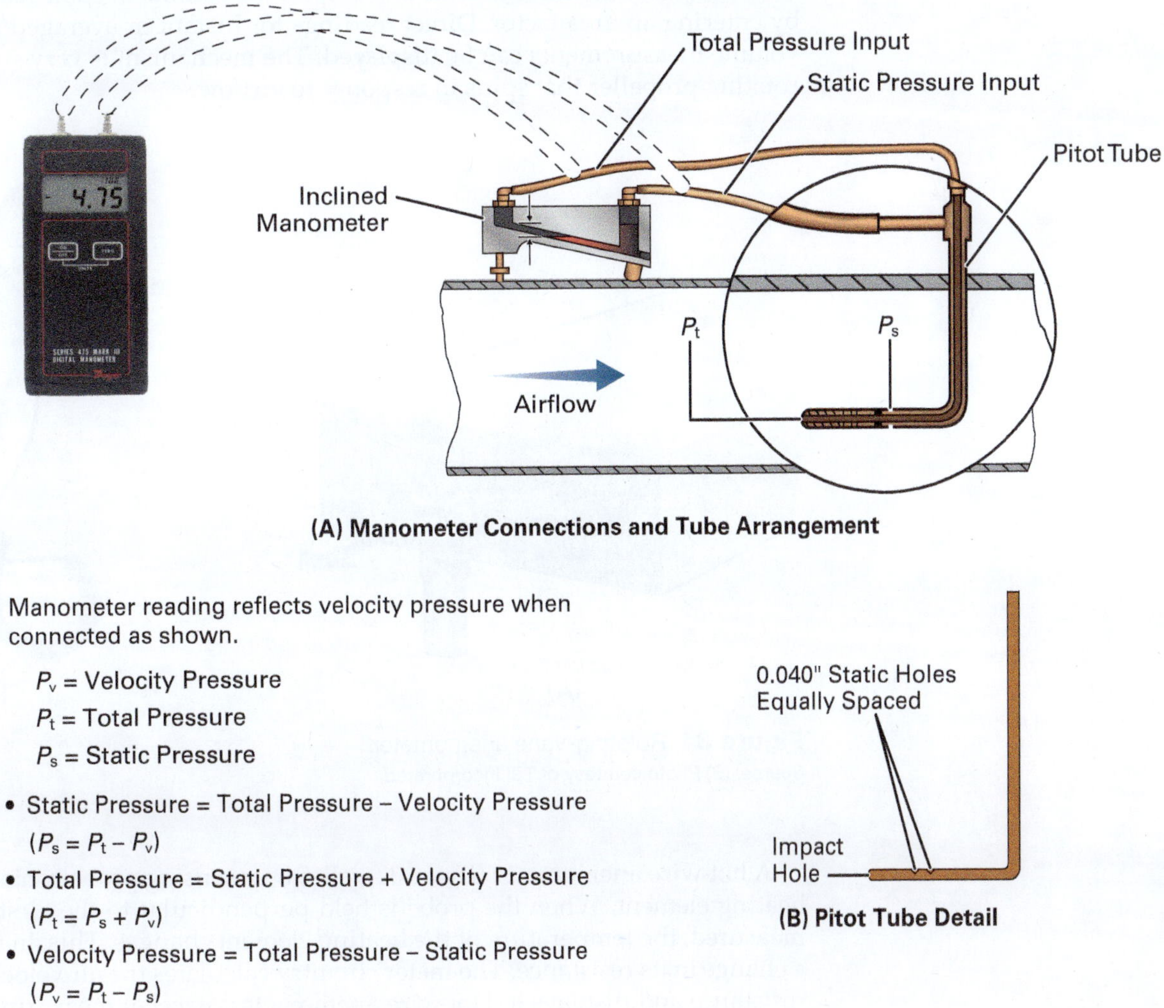

Figure 29 Using a pitot tube with manometers.
Source: (A photo) Image courtesy Dwyer Instruments, LLC

Total pressure: The sum of the static pressure and the velocity pressure in an air duct; the pressure produced by the supply air fan or blower.

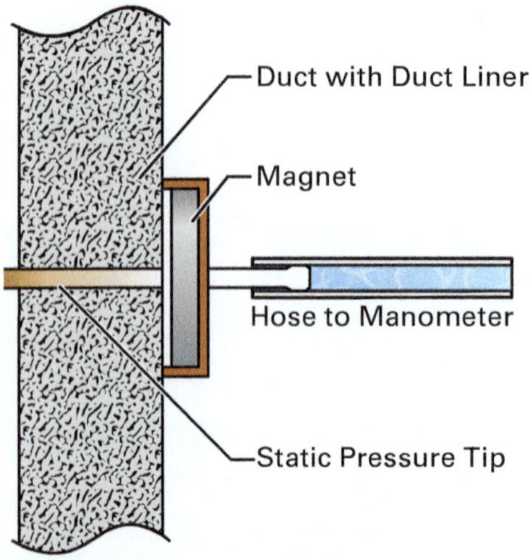

Figure 30 Static pressure tip.

spaced 0.04" diameter holes used to sense static pressure. Pocket-size pitot tubes with a smaller outer tube are recommended for ducts less than 8" diameter. The tiny holes in the side connect to a separate passage that surrounds the inner passage. In other words, a pitot tube assembly consists of a tube inside a tube (concentric tubes).

The passage between the inner and outer tubes transfers pressure from the lateral sensing holes to the static pressure connection of the pitot. When the **total pressure** input is connected to the high-pressure port of the manometer, and the static pressure tube is connected to the low-pressure port of the manometer, velocity pressure is indicated directly. *Figure 29* also shows the relationship between static, velocity, and total pressures.

To ensure accurate velocity pressure readings, the pitot tube tip must be pointed directly into the airstream. Pitot tubes come in various lengths ranging from 6" to 60", with graduation marks along their length to show the insertion depth. Simple static pressure tips (*Figure 30*) are used to measure static pressure alone.

3.2.4 Velometers, Anemometers, and Balometers

Velometers and anemometers are used to measure air speed. Velometers typically provide readings in feet per minute. Although the two terms once identified two different devices, today velometers and anemometers are largely considered the same thing in the HVACR world. For simplicity, devices that measure air speed (velocity) will be referred to as *anemometers* for the remainder of this module.

The two most common anemometers in use today are the rotating-vane anemometer and the hot-wire anemometer. The rotating-vane anemometer (*Figure 31*) can be used to measure air velocity at grilles and in ducts. The model shown here can calculate and display air volume in cubic feet per minute by entering an area factor. Direct readings for instant or averaged velocity and volume measurements can be displayed. The mechanism is very simple, with a rotating propeller that spins in response to airflow.

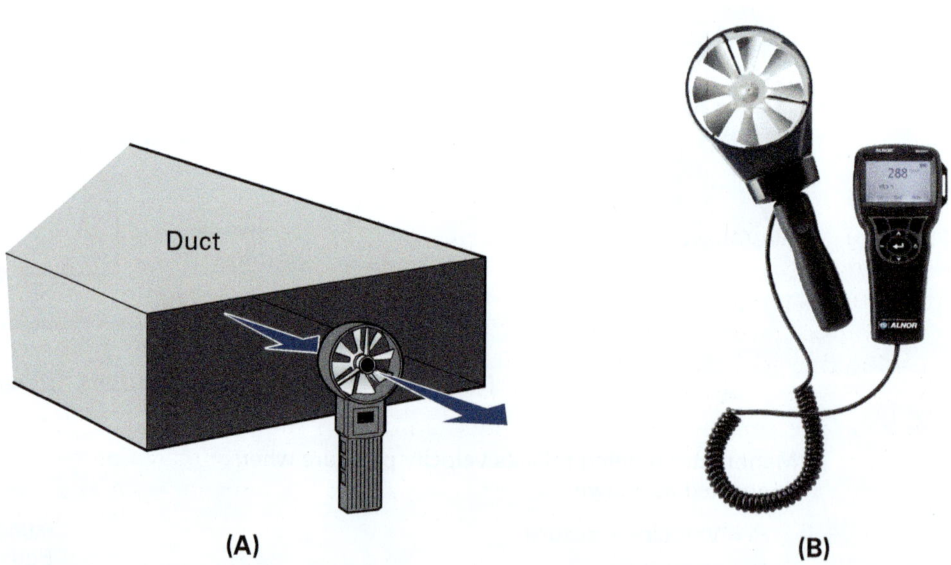

Figure 31 Rotating-vane anemometer.
Source: (B) Photo courtesy of TSI Incorporated

A hot-wire anemometer (*Figure 32*) relies on a sensing probe containing a tiny heating element. When the probe is held perpendicular to the airstream being measured, the temperature of the heating element changes. This, in turn, causes a change in its resistance. The meter circuitry calculates the air velocity from the resistance and displays it. Hot-wire anemometers excel at measuring very light drafts that may not be able to force a rotating vane to spin. However, the model shown in *Figure 32* can also measure velocity up to 3,960 fpm.

Figure 32 Hot-wire anemometer.
Source: Courtesy of Fieldpiece Instruments

Figure 33 Flow hood, or balometer.
Source: Photo courtesy of TSI Incorporated

Another flow measurement instrument, called a flow hood or balometer, is frequently used to measure air volume at terminals directly (*Figure 33*). At the heart of the balometer is an anemometer. Since the area inside the hood where velocity measurements are taken is fixed in size, the anemometer can easily make the calculation to convert air velocity to air volume. This is the most common device used to measure airflow from terminals today. With the hood assembled, you need only place it into position over a terminal and read the display.

3.3.0 Air Balancing Preparations

Every air distribution system should be balanced before it is considered complete and ready for action. Balancing involves making measurements and setting dampers to deliver the required volume of air where it is needed.

When system characteristics prevent an air balancing technician from accomplishing this goal, the technician will often be asked for suggestions to resolve the problems. Conducting the air balance typically provides the technician with a unique perspective from which valid and practical suggestions can be made. Thus, it's a good idea to consider the potential reasons for encountered problems.

Some factors should be considered before making certain adjustments in an air distribution system, including the following:

- Air entering or exiting return and supply grilles at a high velocity can cause objectionable noise.
- Complaints about drafts can result from excessively cold air leaving terminals at a significant velocity.
- A decrease in system airflow can increase the temperature rise across a furnace, stressing the heat exchanger and possibly causing the limit switch to

open. Comfort is clearly important, but the equipment also needs to be considered before changes are made.

- **Stratification** can result from reduced airflow. Stratification results in warmer air remaining at the ceiling, while colder air collects at the floor. Air distributions systems are typically designed to reduce or eliminate stratification, although stratification has a purpose in some applications.

Stratification: The natural layering of undisturbed room air due to temperature differences. Warmer air is concentrated at the ceiling while cooler air is concentrated at the floor. Mechanical ventilation interrupts or eliminates natural stratification.

A satisfactory compromise must often be made between temperature rise or drop through the HVAC equipment, total system airflow, terminal velocities, and occupant comfort. The ideal compromise can usually be achieved with a properly designed duct system and with properly sized heating and cooling equipment. However, ideal systems and conditions rarely exist, and the air balancing technician must work within the limitations of the system at hand. For example, many systems are equipped with balancing dampers at the branch duct takeoffs, but some systems have dampers only in select branches. This is less than ideal, but it may also be adequate for some applications.

Balancing an HVAC system requires time and patience. It is important to follow a systematic process to achieve the best results in the time allotted.

3.3.1 Pre- and Post-Balance Tasks

A systematic balancing procedure begins with preparing a schematic of the air distribution system. This is relatively easy when plans are available; make a copy that you are free to highlight and make notes on.

The layout should provide the following information:

- The location of all dampers, turning vanes, and terminals.
- The size of all main and branch ducts, and the volume of air (in cubic feet per minute) expected to flow through each section. This includes the fresh air and return air volumes. Although duct sizes should be shown on the plans, you will likely need to compile some of the air volume information and note it on the plans where appropriate.
- You may also wish to convert the air volume in duct sections to the appropriate velocity for the duct size. You will typically need to determine the average velocity in a duct to determine the air volume.
- Number all the terminals, both supply and return, using a simple system that makes it easier to find a specific terminal number.

As needed, prepare forms needed to record system measurements, such as pitot tube traverse readings for ducts (*Figure 34*). Forms available from the

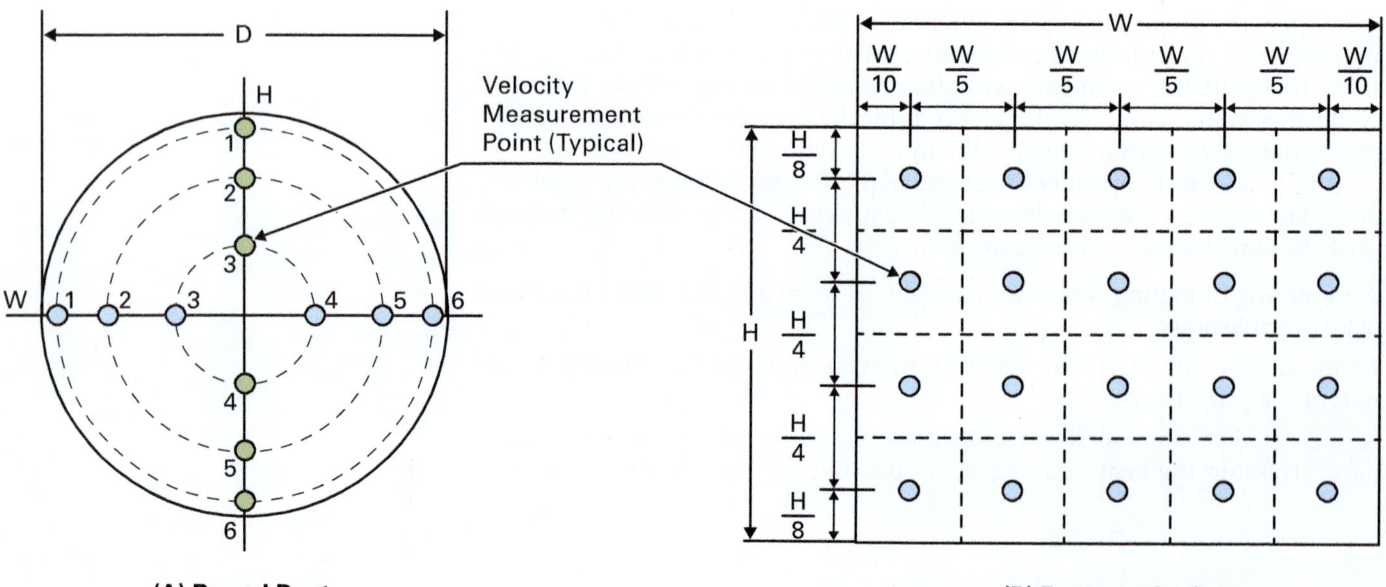

(A) Round Duct

(B) Rectangular Duct

Figure 34 Duct traverse readings form.

National Environmental Balancing Bureau (NEBB) can be used for this purpose. Examples of completed forms are provided in *Appendix 03402A*.

Next, ensure that the equipment and air distribution system are clean and fully operational:

Step 1 Check the blower and its components. Make sure that the blower wheel is free to rotate and the housing is clear of obstructions.

Step 2 Examine the blower drive for proper belt tension and pulley alignment. Record the model numbers of both the drive and driven pulleys if readable. Otherwise, document the appropriate measurements of the pulleys and shaft sizes in case changes need to be made.

Step 3 Ensure any vibration eliminators are in place and serviceable.

Step 4 Examine the duct system to make sure that all branch dampers and terminal dampers are in the engineered, marked positions or fully open. In most cases, the dampers will be fully open, awaiting final adjustment by the air balance team.

Step 5 Make sure that all fire dampers are open.

Step 6 Ensure the air filters and coils are clean.

Step 7 Start the blower and ensure it is rotating in the correct direction. Also document the motor voltage and current at the present conditions and ensure both are within the motor's normal operating parameters.

WARNING!

Watch out for rotating, pressurized, or hot components. Follow all safety instructions labeled on the equipment and provided in the manufacturer's service manual.

Step 8 Energize all exhaust and makeup air equipment that can affect the HVAC system and are expected to operate during occupied hours.

Step 9 Close all windows and outside doors. Open any interior doors that may significantly affect air movement. However, this situation should be rare, as system designers can rarely depend on doors being in any given position.

Remember that doors are usually undercut to allow air to flow under them toward a central return, unless the space has a dedicated return air grille or a transfer grille arrangement. The same volume of air coming into a room needs to flow freely from the room.

Transfer grilles, when installed, may be installed in doors or through walls, but these locations often interfere with private conversation. For this reason, transfer grilles are most often installed in the ceiling of commercial buildings (*Figure 35*). However, as you move through a home or facility, ensure that provisions have been made for air to easily leave each conditioned space.

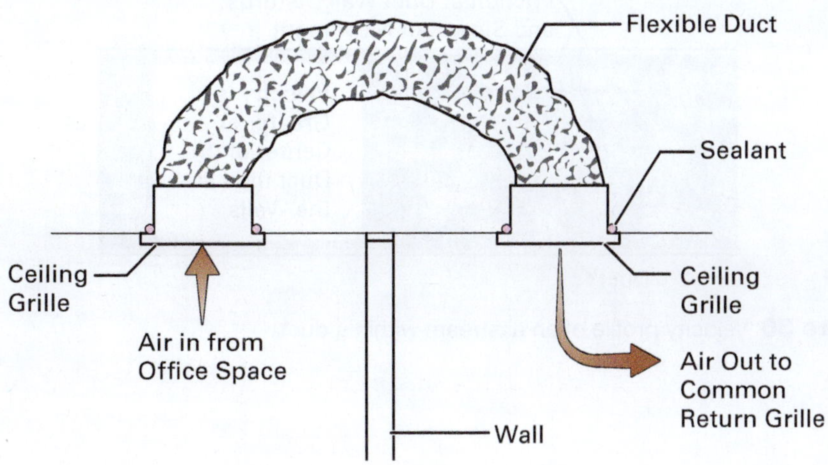

Figure 35 Overhead transfer grille arrangement.

On Site

NEBB

The National Environmental Balancing Bureau (NEBB) was founded in 1971 and is currently headquartered in Gaithersburg, MD. NEBB provides topical education and various certifications in testing, adjusting, and balancing all types of HVAC systems. NEBB's primary function is to "establish, promote, and maintain high quality standards through certification of Firms, Professionals and Technicians."

During new construction projects, check with other workers and crafts to ensure that they are aware that the system will be running. Ideally, airborne dust is minimized while the system is running. In existing buildings, let the building occupants know what you are doing and what to expect before you begin. In larger buildings, it often helps to encourage the leadership team to inform all staff members about the balancing process in advance, and that you will need to access all supply and return terminals and various points in the duct system. This usually requires occupants to be temporarily displaced, and, in some cases, to move furnishings. Conducting an air balancing procedure in a heavily occupied building is much more challenging and time consuming, but it is essential that you work with the occupants professionally and with patience.

After the air balance is complete, be prepared to check the operation of the mechanical equipment. Ensure that the temperature rise across furnaces is still within the furnace specifications and check the cooling system capacity and airflow. Also ensure that blower motors and compressors are still operating at the expected current.

3.4.0 Measuring and Adjusting Airflow

An air distribution system can be broken into several areas for balancing. A complete air balance begins with measuring the airflow in the duct system, then moves to the individual terminals. The volume of outside air volume must also be addressed.

Select the instruments that will do the best job and ensure the instruments are calibrated and in good condition. Whenever possible, use the same instrument for the entire job to minimize calibration errors. If more than one instrument of a similar type will be used, compare the devices by taking measurements at the same point. The variation between the instruments should be limited to ±5%.

The air balancing process can be very repetitive. Since adjustments in one place can affect the airflow in other places, it may require duct and terminal readings be taken a number of times as adjustments are made.

3.4.1 Duct Readings and Balancing

Measuring and adjusting the air volume in the main ducts begins with traverse readings to measure velocity, then calculating the air volume using the velocity and area of the measured duct.

When taking traverse readings, remember that the velocity of the airstream is not uniform across the duct. Friction slows the air along the duct walls, resulting in a higher velocity in the center of the duct (*Figure 36*). Elbows and other fittings also significantly influence the velocity in various parts of the duct.

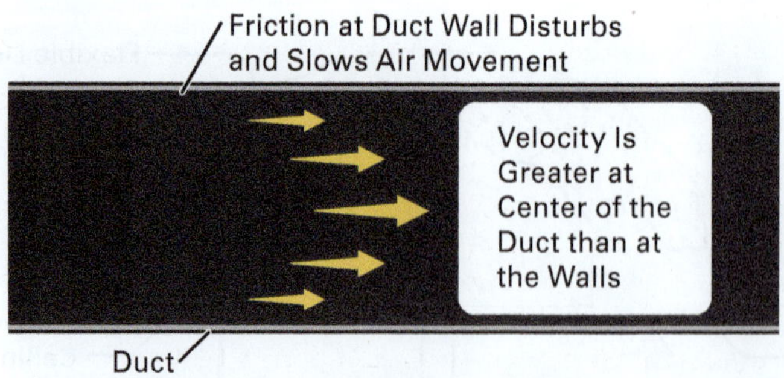

Figure 36 Velocity profile of an airstream within a duct.

For that reason, a traverse pattern such as shown in *Figure 34* allows readings taken at different points in the duct to be averaged. The goal is to break the duct into specific areas, and then measure the velocity in each of the areas. Several different methods are available to approach this mathematically.

The *equal area method* is commonly used, but it often results in the calculated average velocity being higher than reality. *Figure 37* shows a duct broken into equal areas. Readings are taken at the center of each square.

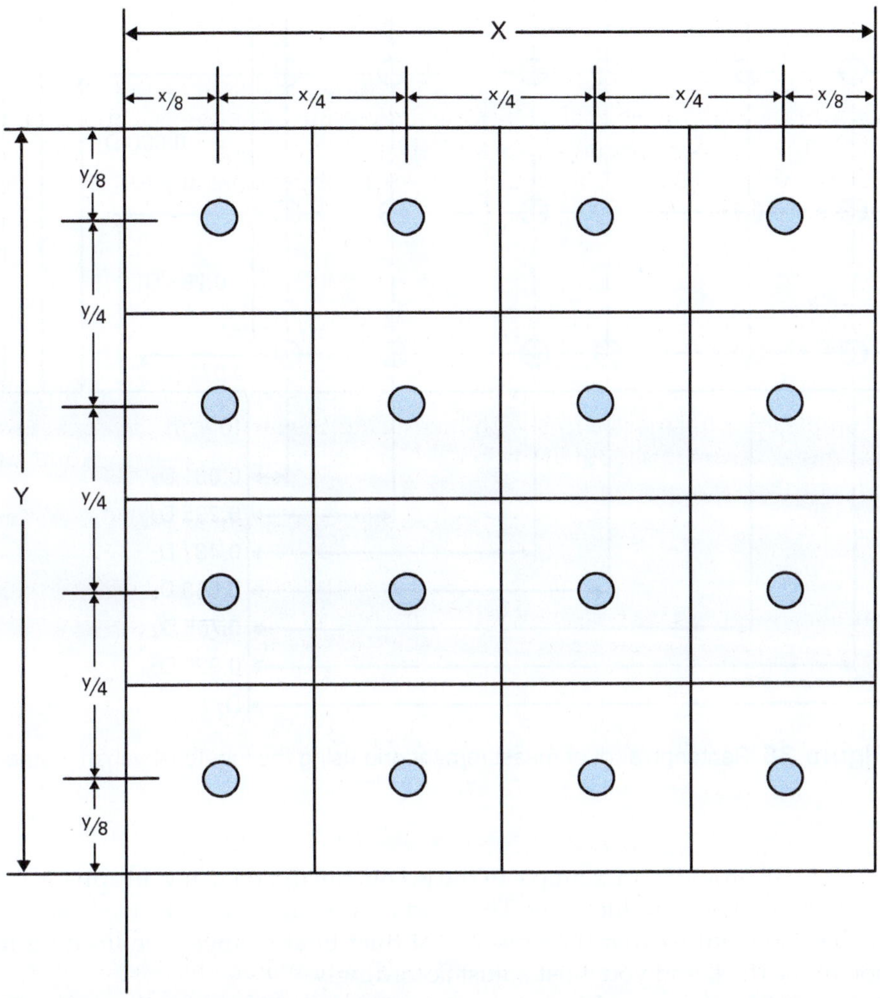

Figure 37 Rectangular duct measurement grid using the equal area method.

Figure 38 shows the same duct broken into zones using a spacing algorithm known as the *log-Tchebycheff method*. Using this approach, more readings are generally required, but it has been proven to provide more accurate measurements.

Note various apps and websites are readily available to make this step painless. For example, the Associated Air Balance Council (AABC) offers a mobile app that provides a wealth of information, as well as a calculator for determining measurement points in round or rectangular ducts. On the internet, visit *www.tabcalcs.com* for a free calculator and other balancing guidance.

NOTE

The method used to plot duct measurements points will typically be specified in construction documents for new construction. For service-related testing, where measurements may only be taken in a few key locations, the method used to plot duct measurement points is a matter of choice.

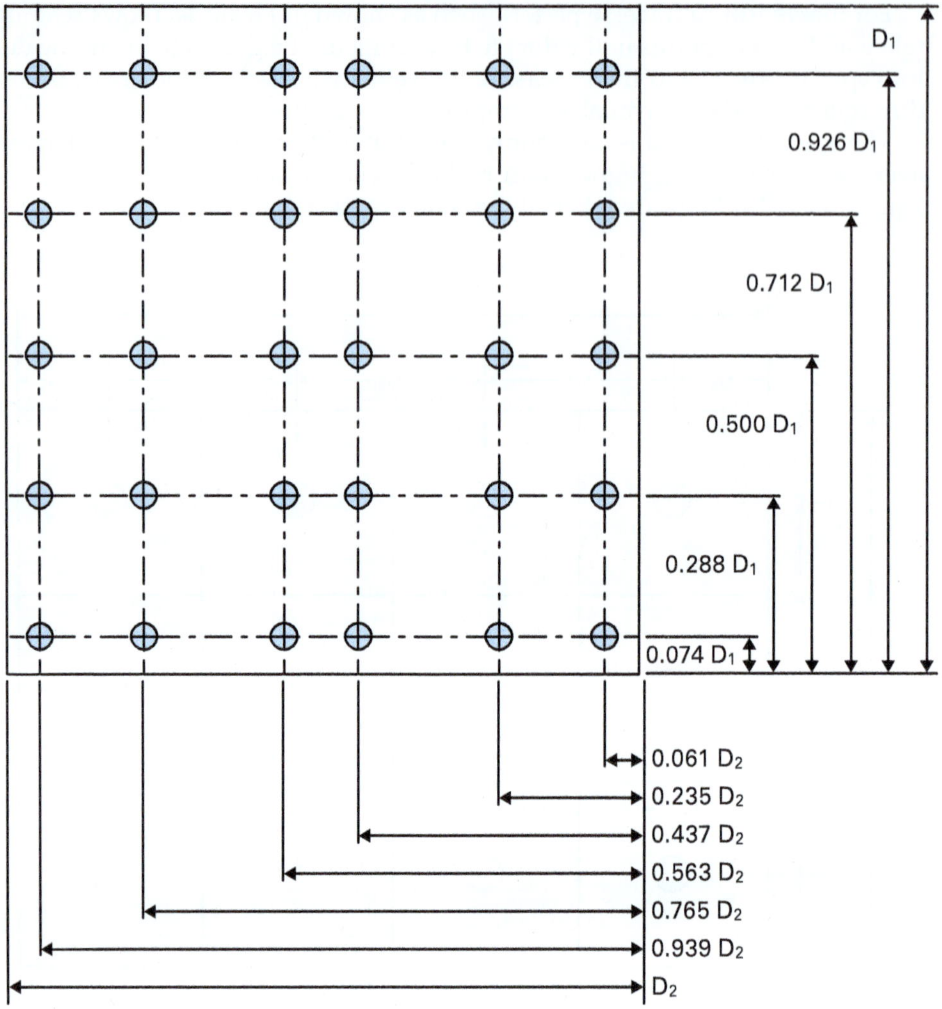

Figure 38 Rectangular duct measurement grid using the log-Tchebycheff method.

Figure 39 provides an example of the points calculated using an app for a 24" × 12" duct fitted with 1" duct liner. This grid was created using the log-Tchebycheff method. Remember that the presence of duct liner changes the inside dimension of the duct, and you must adjust accordingly.

For the duct shown, 30 measurements are needed. Mark the anemometer or velometer probe with the insertion depth graduations indicated by the *W* dimensions to place the pitot tube in the duct at the proper location for each measurement. The *H* dimensions shown indicate where the holes will be drilled on one side of the duct. Alternatively, the holes can be drilled from the bottom of the duct at the *W* dimensions, and the probe inserted through each hole to the *H* dimensions.

Round ducts are handled in much the same way. *Figure 40* shows the measurement points plotted for a 10" round duct with no duct liner. Using the log-Tchebycheff method, a total of 18 readings are needed, and the points are plotted along three lines across the duct.

The location of measurements taken along the length of the duct is a very important consideration. Obstructions and changes of direction affect airflow dramatically. Measurements should be taken at least 7.5 times the duct diameter downstream from turns or obstructions, and at least 3 duct diameters upstream from these features (*Figure 41*). For example, in a 6" round duct, the

NOTE

Taking duct measurements at the perfect location relative to turns and obstructions is not always possible. In those cases, document the exception in your notes and consider the data gathered accordingly.

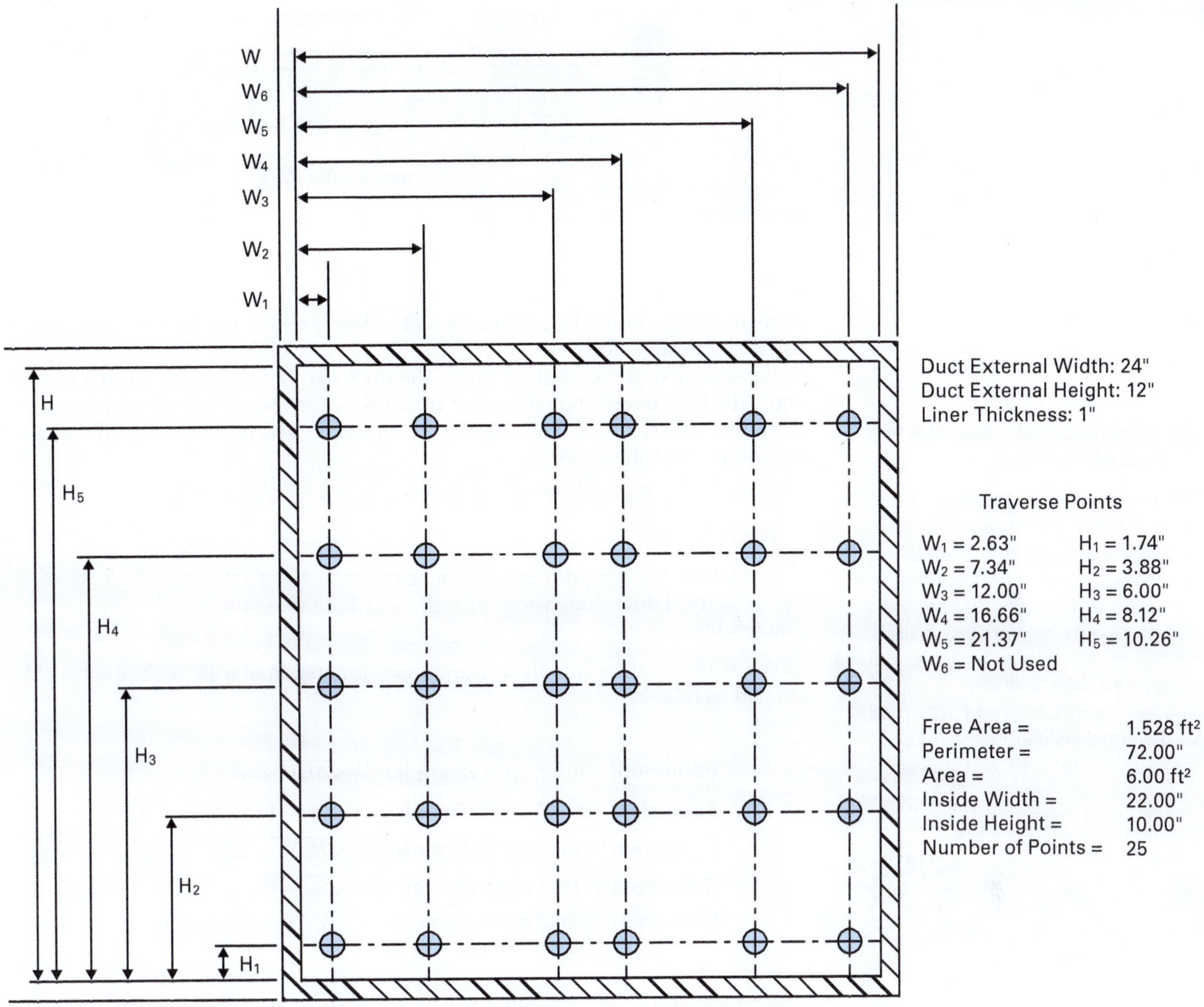

Figure 39 Example traverse points for a 24" × 12" duct using the log-Tchebycheff method.

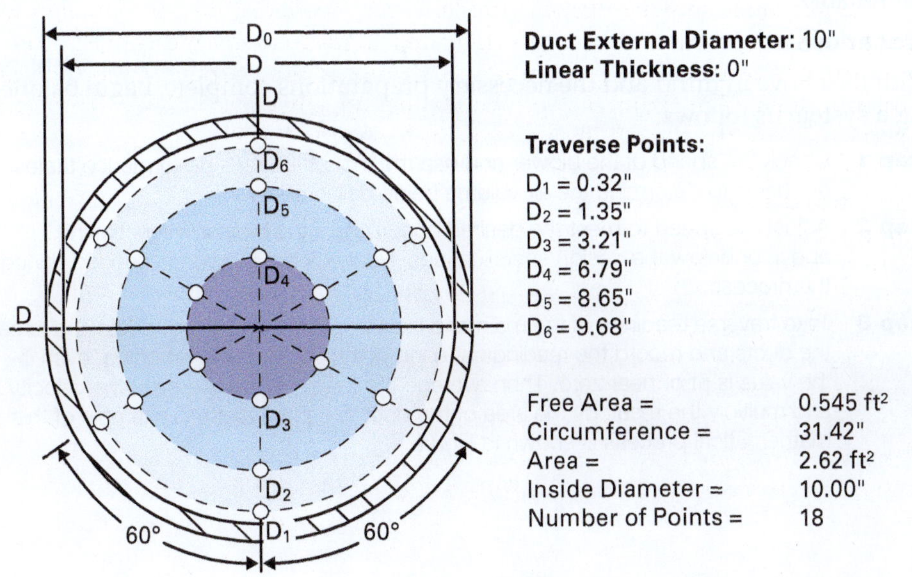

Figure 40 Example traverse points for a 10" round duct using the equal area method.

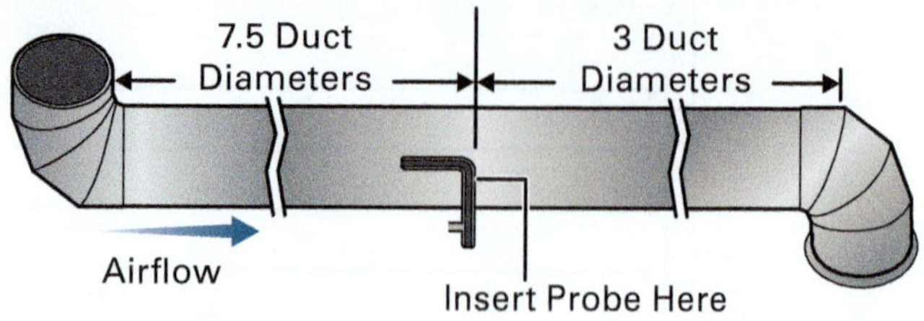

Figure 41 Recommended location for measurements in round duct.

measurements should be taken at least 45" downstream and 18" upstream from turns and obstructions.

For square and rectangular ducts, the duct size can be converted to a round equivalent to determine where the measurements should be taken relative to turns and obstructions. The following formula is used to determine an equivalent round duct diameter:

$$\text{Equivalent round duct diameter} = \sqrt{4HV \div \pi}$$

Where:

H = horizontal duct dimension (inches)
V = vertical duct dimension (inches)
π = 3.14

For example, the equivalent round duct dimension of a 15" × 12" duct is calculated as follows:

$$\text{Equivalent round duct diameter} = \sqrt{4HV \div \pi}$$
$$\text{Equivalent round duct diameter} = \sqrt{(4 \times 15'' \times 12'') \div 3.14}$$
$$\text{Equivalent round duct diameter} = \sqrt{720'' \div 3.14}$$
$$\text{Equivalent round duct diameter} = \sqrt{229.3''}$$
$$\text{Equivalent round duct diameter} = 15.1'' \text{ (round to 15'')}$$

Before drilling holes in the duct to make traverse measurements, be sure you have plugs on hand to fit the size of the drilled holes. Every hole must be securely closed with a plug. Do not use tape alone to cover the openings, as it is not reliable.

On Site

Changing Rectangular Duct Size to a Round Duct Equivalent

There is a quick way to determine the equivalent round duct size of a rectangular duct without doing any math at all. Using a Ductulator® or similar tool that shows the relationship of velocity, static pressure, and volume in various duct sizes, set the rectangular duct size in the window as shown in the figure. Then simply look at the equivalent round duct diameter in the upper window. In this example, the round duct equivalent of a 15" × 12" duct is shown.

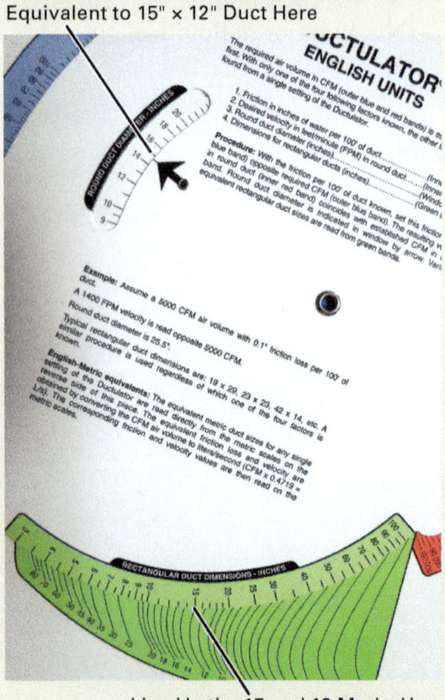

Procedure

With the above in mind and the necessary preparations complete, begin balancing a system as follows:

Step 1 Check the speed of the blower and use the manufacturer's performance tables or curves to determine the air volume being delivered by the unit.

Step 2 Adjust the speed to meet the desired conditions if the blower is functioning above or below the design specifications. Recheck the blower as needed during the process.

Step 3 Take traverse readings of the main and branch supply and return ducts. Traverse the ducts and record the readings at each point. Record every reading, even if the value is at or near zero. Then average the readings to determine the velocity and multiply the result by the area of the duct in square feet. An example of this mathematical process is shown in *Figure 42*.

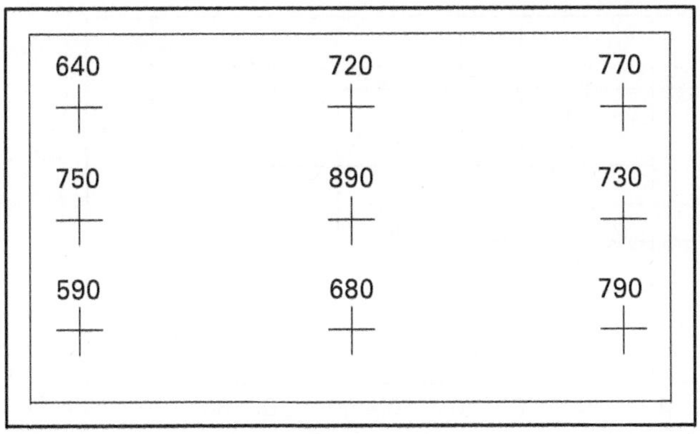

12" × 15" Duct Area = 1.25 ft²

640 + 720 + 770 + 750 + 890 + 730 + 590 + 680 + 790 = 6,560 fpm

Average Velocity = 6,560 fpm ÷ 9 readings = 729 fpm

Air Volume = 729 fpm × 1.25 ft²

Air Volume = 911 cfm

Figure 42 Calculating air volume from velocity readings.

Step 4 Once the readings have been taken in the main ducts, adjust any volume and/or splitter dampers to deliver the design airflow in each section of duct. Some branches will likely be receiving too much airflow, while others are receiving too little. Start by reducing the airflow in areas that are receiving too much air, and then take new readings on areas that were short of air earlier.

Step 5 Secure plugs in all holes made in the ducts to take measurements.

When a damper is used to reduce the airflow in one section, the air volume in other sections should increase. However, dampers increase friction, so the total system air volume may decrease as dampers intrude into the airstream.

Continue the process throughout the system. Reconcile the readings taken in the supply and return air ducts with each other, and then against the system fan curves. If the entire system is short of airflow, this can only be corrected by making adjustments in blower speed.

The damper is the primary component available to control airflow in ducts. Closing a damper to any degree increases the resistance to airflow. However, the reduction in airflow that occurs is rarely proportional to the percentage of closure. In other words, closing a fully open damper until it is 50% open does not mean that the air volume will be reduced by 50%. This relationship between the position of the damper and the air volume is called the *flow characteristic* or the *mean effective airflow* of the damper.

Figure 43 provides an example of a flow characteristic chart for both opposed-blade and parallel-blade dampers. The curve O (opposed-blade damper) shows that, to obtain 25% of the maximum airflow, the damper should be about 23% open. To achieve 50% of the maximum airflow, the damper only needs to be open about 39%. Note that the actual air volume passing through depends on the pressure applied by the fan.

The curve P demonstrates the flow characteristic of parallel-blade dampers. Notice that the parallel-blade damper only needs to be about 20% open to achieve 50% airflow. It is important to remember that these percentages are approximate for these families of dampers. The flow characteristic varies by damper size, blade profile, static pressure, and other factors.

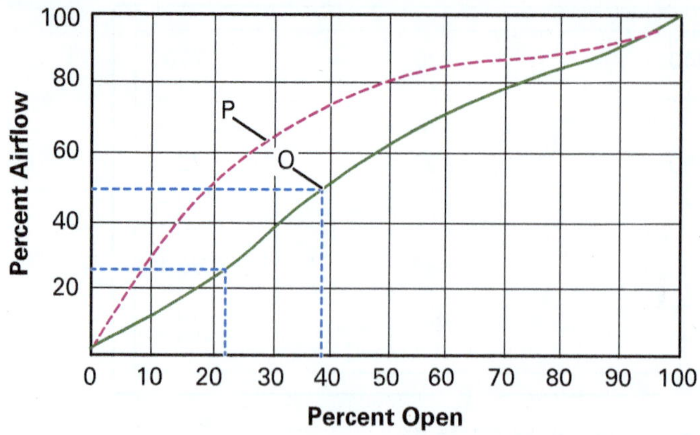

Figure 43 Damper position-to-airflow ratio, or flow characteristic.

NOTE

Completed air balance documents for a system in a commercial building are provided for review in *Appendix 03402A*.

Although it is helpful to understand how different dampers respond to changes in position, you won't need such charts to conduct an air balance. In the end, finding the right damper position to receive the desired airflow is largely a trial-and-error process.

3.4.2 Terminal Readings and Balancing

As described earlier, the air volume entering or leaving a terminal can be determined by taking a series of velocity readings across the face, averaging the velocity, then multiplying the average velocity by the terminal's A_k factor to determine the volume.

This process requires a specified velometer probe positioned at various points along the face of the terminal. Each manufacturer provides this information for the many styles of terminals it offers. *Figure 44* shows the guidance offered by one manufacturer for a stamped-steel ceiling diffuser. The A_k factors are provided along with guidance in taking the velocity readings.

Figure 45 shows how the velometer probe is typically held for a louvered sidewall register. Each manufacturer will specify the equipment to be used and precisely how the velometer probe must be placed to acquire accurate readings.

Today, the flow hood, or balometer, is most often used to measure the air volume at a terminal (*Figure 46*). Flow hoods typically read directly in cubic feet per minute, eliminating the many velocity measurements and related math. However, they are not practical for every situation.

For example, if the grille or register is located on a narrow wall or on the side of a duct, the hood will not be sealed on all four sides, allowing air to escape without passing through the meter. The entire perimeter of the hood must seal against the surface around the grille. You may be able to effectively block off portions of the hood to compensate (*Figure 47*), but this is also time-consuming. However, these hoods are well-suited for ceiling diffusers and most sidewall terminals, which are very common in commercial buildings.

Now, with the understanding of how to determine the air volume from the terminals, the balancing process can begin. Note that this process assumes that the airflow in the main ducts has been tested and balanced. However, it is also important to note that very small systems, such as the average residential

⑤ CARNES FIELD BALANCING DATA ⑤

STAMPED STEEL LOUVERED DIFFUSERS

$$CFM = V_K \times A_K$$

AREA FACTOR (A_K) TABLE – MODEL SFA

NECK SIZE	NOMINAL LOUVERED AREA		
	12 3 12	18 3 18	24 3 24
	Horizontal Throw		
5	12	–	–
6	14	0.22	0.29
7	17	0.25	0.28
8		0.27	0.28
10		0.38	0.42
12		0.48	0.50
14		–	0.62

AREA FACTOR (A_K) TABLE – MODEL SAA

NECK SIZE	NOMINAL LOUVERED AREA					
	Horizontal Throw			Vertical Throw		
	12 3 12	18 3 18	24 3 24	12 3 12	18 3 18	24 3 24
	0.09	–	–	0.10	–	–
	0.12	0.20	0.35	0.11	0.21	0.30
	0.15	0.23	0.31	0.12	0.23	0.27
	–	0.26	0.33	–	0.24	0.36
	–	0.29	0.32	–	0.23	0.30
	–	0.38	0.39	–	0.30	0.33
	–	–	0.60	–	–	0.48

FIELD BALANCING

The actual volume of air being discharged from an outlet can be determined by measuring the outlet velocity in feet per minute (fpm) and multiplying by an area factor (A_K).

$$CFM = V_K \times A_K$$

The Alnor velometer, with the 2220-A jet is the recommended equipment for balancing Carnes stamped diffusers.

The Alnor Model 6000P with 6070P probe can be used with the same A_K factors.

Place the Alnor jet into the correct louvered space as shown in the sketches below. Point the jet as directly as possible into the air stream and move jet slowly along the lip of the cone to obtain the highest reading. Average the readings from all four sides to obtain V_K. Select the correct A_K from the tables and apply the formula to obtain the CFM.

ALNOR JET POSITION – MODEL SFA

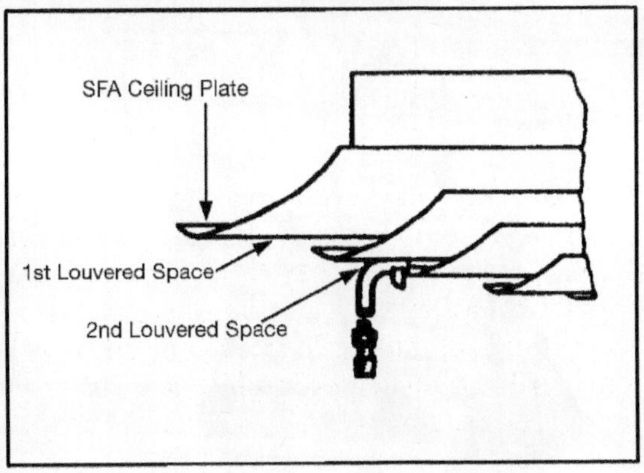

ALNOR JET POSITION – MODEL SAA

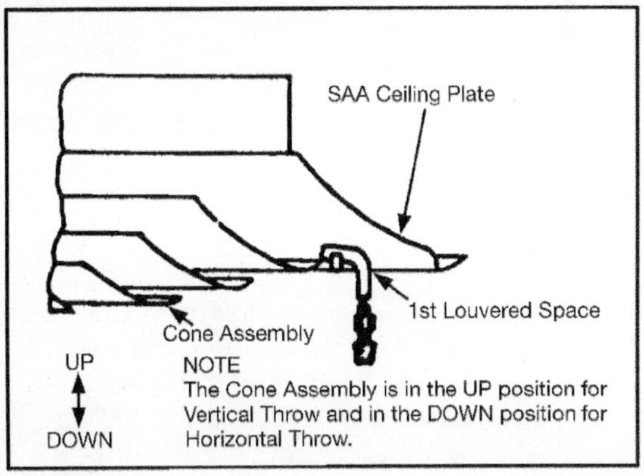

CARNES

Figure 44 Field balancing data for a ceiling diffuser.
Source: Carnes Company

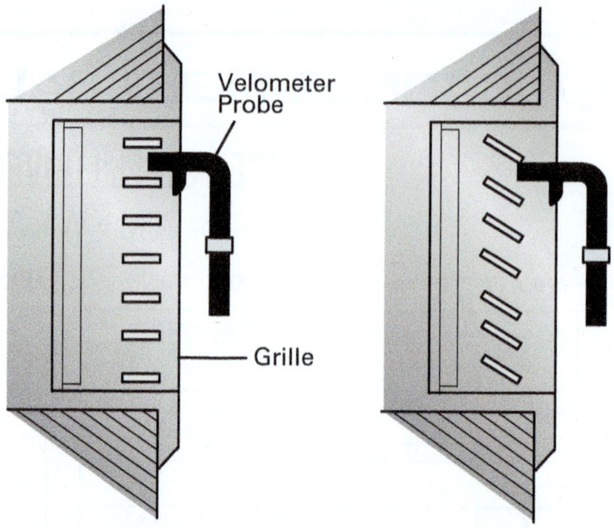

Figure 45 Velometer probe placement for a sidewall register.

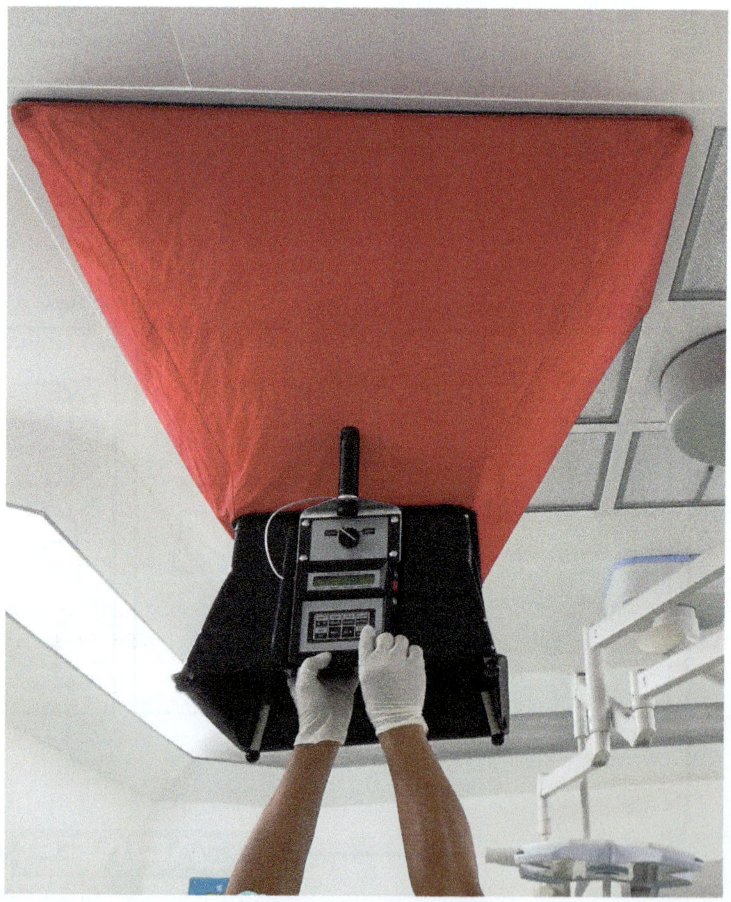

Figure 46 Measuring terminal airflow with a flow hood.
Source: Sucharas wongpeth/iStock/Getty Images

Figure 47 Blocking a portion of the flow hood inlet.
Source: BanksPhotos/iStock/Getty Images

system, may be measured entirely at the terminals. These systems usually have dampers to work with only at branch line connections, and even those may be absent.

To measure terminals for air balancing, proceed as follows:

Step 1 Start by measuring the airflow at the farthest outlet in each branch, or the farthest from the air handling unit. Work your way back toward the air handling unit. Some branch lines may only have a single terminal attached. Before taking the measurements, ensure any damper that may be incorporated into the damper is fully open.

Step 2 If the farthest terminal measures below the design volume, leave its damper fully open and go on to read the next terminal upstream. If the air volume at the farthest terminal is above the design volume, use the branch-line damper to throttle the air as necessary, and then proceed to the next terminal. It is best to avoid using a damper that is provided at the terminal face for air balancing, for two reasons: Terminal dampers are best used by occupants to make small or temporary adjustments; closing the damper at the terminal often increases noise.

Step 3 Repeat the procedure one or more times (some experts recommend at least three rounds). This is because the adjustment of one damper typically affects the other branches to some degree.

Step 4 Ensure that all the final readings are documented along with the initial readings.

NOTE

Completed air balance documents for a system in a commercial building are provided for review in *Appendix 03402A*.

3.4.3 Mixed-Air System Adjustment

When outside air is being introduced into the space through the HVAC system, it is necessary to measure and adjust the volume to meet the specifications. Motorized outside-air dampers are typically available for this purpose. Mixed-air temperature measurements can be used to confirm that the correct amount of outside air is being added to the return airstream.

Step 1 Loosen the outside-air damper drive rod from the damper and actuator.

Step 2 Ensure the damper actuator is stroked to the fully closed position.

Step 3 Manually move the damper to the fully closed position.

Step 4 Install and secure the drive rod to the damper and actuator.

Step 5 Determine the minimum percentage of outside air required from the plans or specifications. If the specifications are given as a specific volume rather than as a percentage, divide the outside air volume requirement by the total system air volume to obtain the outside air percentage.

Step 6 Set the fresh-air control potentiometer on the actuator to the proper value (30%, for example). Remember, however, that this relates to damper position and does not reflect the volume of airflow admitted through the damper.

Step 7 With the system operating and stabilized in the current mode, measure the return air temperature in the return duct, upstream from where the outside air is admitted (*Figure 48*).

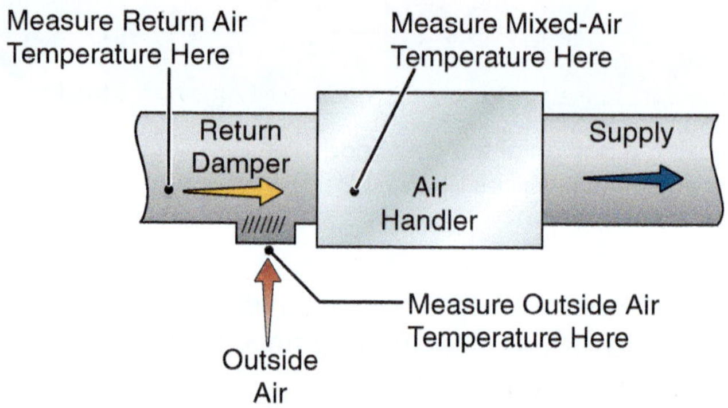

Figure 48 Measuring return and outside air temperatures.

Step 8 Measure the outside air temperature where it enters the system.

Step 9 Determine what the mixed-air temperature should be based on the percentage of outside air required, the return air dry-bulb temperature, and the outside air dry-bulb temperature, using the following formula. Assume that 30% is the required percentage of outside air, the outside air temperature is 95°F, and the return air temperature is 78°F. Note that the percentage of outside air must be converted to a decimal value:

$$\textit{Mixed Air Temperature} = RA_{db} + (\%\ OA)(OA_{db} - RA_{db})$$

Where:
RA_{db} = return air dry-bulb temperature
$\%OA$ = percentage of outside air required
OA_{db} = outdoor air dry-bulb temperature

$$\textit{Mixed Air Temperature} = RA_{db} + [(\%\ OA)(OA_{db} - RA_{db})]$$
$$\textit{Mixed Air Temperature} = 78°F + [0.3 \times (95°F - 78°F)]$$
$$\textit{Mixed Air Temperature} = 78°F + [0.3 \times 17°F]$$
$$\textit{Mixed Air Temperature} = 78°F + 5.1°F$$
$$\textit{Mixed Air Temperature} = 83°F \text{ (rounded)}$$

Step 10 Now measure the mixed-air temperature downstream from the fresh-air inlet. Ensure the measurement is taken where the outside and return airstreams have thoroughly mixed. Compare the temperature to the target mixed air temperature.

Step 11 Adjust the position of the outside air damper using the potentiometer or, if necessary, by adjusting the linkage. If the mixed air temperature is too low, the outside air damper must be opened further. If the mixed air temperature is too high, the outside air damper must be closed some. Adjust the damper until the mixed air temperature is correct.

NOTE

If the mixed air temperature can't be increased to the proper temperature using the outside air damper, the return air damper may need to be partially closed to get the correct mix.

3.5.0 Air Balancing by Thermometer

Small systems can often be balanced by what is known as the *thermometer method*. The thermometer method is based on producing consistent temperatures throughout the space, so it can be quite effective in practice. An anemometer or balometer may still be used to document the air volume from each terminal. This can be valuable information for future reference.

Some factors in residential systems can make system balancing difficult. While similar problems are often found in commercial systems, commercial systems are usually designed and installed to a more stringent specification. A project engineer often oversees the work as well.

Common residential system issues include the following:

- *Grossly oversized equipment* — In the past, oversized equipment was often sold as a cure-all for residential comfort problems. Oversizing typically creates a whole new set of problems. Fortunately, this practice is not as common as it once was.
- *Inadequate duct systems* — Ducts are often undersized, oversized, and/or poorly insulated. Terminals may also be poorly placed, and the ducts on both the supply and return sides of the system may leak.
- *Poor temperature control* — The room thermostat may be of low quality, out of calibration, or in a poor location.
- *Equipment maintenance* — The HVAC equipment may be poorly maintained, if at all.

Regardless of whether the system is residential or commercial, these factors are best addressed before air balancing is done. A well-balanced system can't accommodate or cover up these problems, making any effort to balance the system largely useless. Address the many other possible causes of discomfort first.

Use an accurate and dependable digital thermometer for balancing, along with a proper probe for air measurement. Use the same device throughout the process, eliminating potential errors between multiple devices. For thermometer balancing, prepare as follows:

Step 1 Check all registers and grilles for blockage by furniture, drapes, or appliances. Open any partially or fully closed dampers located at the terminals. For supply registers and diffusers equipped with adjustable vanes, position the vanes to provide the best air pattern for the space.

Step 2 Ensure that the wire entry hole behind the thermostat is plugged to prevent drafts from affecting the internal sensor.

Step 3 Check the air filters and the blower wheel for dirt and obstructions. Replace dirty air filters.

Step 4 Check the duct system for balancing dampers and ensure they are fully open. It is also helpful to be aware of the damper location in the system and consider how to resolve imbalances as you continue to work.

Step 5 Check the duct system for obvious leaks. Many systems will benefit from pressure testing before balancing begins to ensure all leaks are found and repaired. Duct leaks are far more prevalent than most owners realize.

3.5.1 Thermometer Balancing in the Heating Mode

As you begin the balancing process, remember to take measurements at 4' to 5' above the floor in each space. Make sure that your reading is not being influenced by the supply airstream from a terminal.

Proceed as follows to balance a system by thermometer in the heating mode:

Step 1 Adjust the thermostat setpoint about 3°F above the room temperature. Thermometer balancing is best done when the space temperature is within 5°F of the normal comfort setpoint.

Step 2 Allow the system to operate for 10–15 minutes.

Step 3 Move to each conditioned space and record the temperature in each area.

Step 4 Adjust the balancing damper(s) to reduce airflow into the spaces that have the highest temperature. Do not reduce the airflow too much at once though, as the airflow in the remaining areas will likely increase. If a space remains significantly cooler than the others, look for obstructions in the duct or other causes before proceeding further. Clearing an obstruction in one place can have a substantial effect on the entire system.

Step 5 Allow the space to cool somewhat and repeat the process until the temperature in each area is as close as possible to being equal.

Step 6 If a fossil fuel furnace is in use, complete the task by checking the temperature rise across the heat exchanger. For other systems, make sure system performance has not been negatively affected by any changes.

On Site

Thermometer Balancing

A digital thermometer with a proper probe should be used for thermometer balancing. Inexpensive dial and stick thermometers typically lack the accuracy needed. Infrared thermometers that measure surface temperatures should also be avoided. These other types of thermometers are often convenient and have a place in the craft, but air balancing is not one of those places.

Source: Reproduced with Permission, Fluke Corporation

3.5.2 Thermometer Balancing in the Cooling Mode

Thermometer balancing in the cooling mode is completed in largely the same manner as in the heating mode. Of course, in the cooling mode, you are identifying spaces that are cooler than others and adjusting the dampers to reduce airflow into those spaces. If a space remains significantly warmer than the others, look for obstructions in the duct or other causes before proceeding further.

When the process is complete, ensure the equipment is still working properly and has not been adversely affected by significant changes in airflow. This is especially important if you have changed the blower speed or otherwise significantly changed the total airflow through the system.

As mentioned previously, an anemometer or balometer may still be used to document the air volume from each terminal. The environment in commercial buildings often changes, with copy machines, people, and other loads being relocated. Having a snapshot of the system airflow as a reference can be very helpful when determining how to accommodate changes in the environment that affect comfort.

3.0.0 Section Review

1. As air leaves a supply terminal, it mixes with room air that is referred to as ______.
   a. fresh air
   b. primary air
   c. terminal air
   d. entrained air

2. Standard pitot tubes are best used for pressure measurements in ducts ______.
   a. of all sizes
   b. 4" and larger
   c. 8" and larger
   d. that are rectangular only

3. True or False? Transfer grilles in commercial buildings are usually installed through the walls of offices.
   a. True
   b. False

4. If you must use more than one instrument, such as an anemometer, on a balancing job, how much variation is allowed between the instruments for readings taken at the same location?
   a. 2%
   b. 3%
   c. 5%
   d. 10%

5. Thermometer balancing is best done when the space temperature is within how many degrees of the normal comfort setpoint?
   a. 1°F
   b. 2°F
   c. 5°F
   d. 8°F

Module 03402 Review Questions

1. An atmospheric pressure of 0 psia is equal to a pressure of ______.
   a. 0 µm
   b. 0 psig
   c. 29.92" Hg
   d. 760,000 µm

2. Per ASHRAE, which of the following is a characteristic of standard air?
   a. 70°F temperature
   b. 0.150 lb/ft^3 density
   c. 0" Hg barometric pressure
   d. gr of water per pound of dry air

3. Per Boyle's law, assuming a constant temperature, how does the pressure of gas behave relative to the space (volume) it occupies?
   a. The pressure increases if the volume increases.
   b. The pressure increases if the volume decreases.
   c. The pressure decreases if the volume decreases.
   d. The pressure stays the same regardless of a change in volume.

4. Per Charles' law, assuming the volume remains constant, how does the pressure of a contained gas behave relative to the temperature?
   a. The pressure decreases if the temperature is increased.
   b. The pressure increases if the temperature is decreased.
   c. The pressure decreases if the temperature is decreased.
   d. The pressure remains the same regardless of a change in temperature.

5. The condition in which the dry-bulb, wet-bulb, and dew point temperatures of an air mass are all equal occurs when the ______.
   a. relative humidity is 50%
   b. relative humidity is 100%
   c. air is considered standard air
   d. enthalpy is at its lowest value

6. On a psychrometric chart, what is the relative humidity of air at 95°F DBT and 75°F WBT?
   a. 30%
   b. 40%
   c. 50%
   d. 60%

7. Plotting 95°F DBT and 75°F WBT on a psychrometric chart shows that the dew point is ______.
   a. 57°F
   b. 61°F
   c. 64°F
   d. 67°F

8. Plotting 95°F DBT and a 75°F WBT on a psychrometric chart shows that the enthalpy of the air is ______.
   a. 24 Btu/lb dry air
   b. 38 Btu/lb dry air
   c. 41 Btu/lb dry air
   d. 47 Btu/lb dry air

9. Which of the following *best* describes primary air?
   a. Any air that hasn't been heated or cooled
   b. Air delivered to a room from supply terminals
   c. It is another term used to describe entrained air
   d. System return air before it mixes with outside air

10. The term that identifies the distance from the center of an outlet to a point in the room where the terminal velocity has been reduced to a specified level is called the ______.
    a. drop
    b. throw
    c. spread
    d. terminal velocity

11. True or False? To measure the velocity of an extremely light draft, a rotating-vane anemometer is the best choice.
    a. True
    b. False

12. You can't rely solely on the outside dimensions of a duct in airflow calculations due to the possible presence of ______.
    a. a duct liner
    b. turning vanes
    c. balancing dampers
    d. elbows and takeoffs

13. When taking velocity measurements inside a duct, how far downstream should you be from a 90-degree turn in the duct?
   a. 3 duct diameters
   b. 5 duct diameters
   c. 7.5 duct diameters
   d. 10 duct diameters

14. If the average velocity measured across a grille is 800 fpm, and the A_k factor for the grille is 1.5, what is the air volume discharged?
   a. 533 cfm
   b. 801 cfm
   c. 950 cfm
   d. 1,200 cfm

15. You can calculate the resulting mixed air temperature for a system using the percentage of outside air required and the ______.
   a. return air and outdoor air dry-bulb temperatures
   b. return air and outdoor air wet-bulb temperatures
   c. return air dry-bulb and outdoor air wet-bulb temperatures
   d. return air wet-bulb and outdoor air dry-bulb temperatures

Answers to odd-numbered questions are found in the Module Review Answer Key at the back of this book.

Answers to Section Review Questions

Answer	Section	Objective
Section One		
1. a	1.1.1	1a
2. a	1.2.2	1b
Section Two		
1. a	2.1.1	2a
2. b	2.2.7	2b
3. d	2.3.3	2c
Section Three		
1. d	3.1.0	3a
2. c	3.2.3	3b
3. b	3.3.1	3c
4. c	3.4.0	3d
5. c	3.5.1	3e

MODULE 03406

System Startup and Shutdown

Source: Baloncici/iStock/Getty Images

Objectives

Successful completion of this module prepares you to do the following:

1. Explain how to store, start up, and shut down boiler systems.
   a. Explain how to shut down and prepare boilers for storage.
   b. Explain how to prepare and start up steam and hot-water boilers.
2. Explain how to start up and shut down chillers and cooling towers.
   a. Explain how to start up and shut down reciprocating chillers.
   b. Explain how to start up and shut down centrifugal chillers.
   c. Explain how to start up and shut down cooling towers and clean water-cooled condensers.
3. Explain how to start up air handling units and packaged rooftop units.
   a. Explain how to start up an air handling unit.
   b. Explain how to start up packaged rooftop units.

Performance Tasks

Under supervision, you should be able to do the following:

1. Prepare and complete the startup or shutdown procedures for at least one of the following:
   - Steam or hot-water boiler
   - Chiller
   - Cooling tower or evaporative condenser
2. Prepare and complete the startup procedures for an air handling unit or packaged HVACR unit.

Overview

When boilers, chillers, and other commercial HVAC systems are shut down for extended periods and then returned to service, specific routines must be followed to prepare the system for operation. This helps to ensure things go smoothly when very large, expensive, and critical equipment is at stake. This module provides an overview of the startup and shutdown requirements for various types of equipment.

Digital Resources for HVACR

SCAN ME

Scan this code using the camera on your phone or mobile device to view the digital resources related to this craft.

NCCER Industry-Recognized Credentials

If you are training through an NCCER-accredited sponsor, you may be eligible for credentials from NCCER. The ID number for this module is 03406. Note that this module may have been used in other NCCER curricula and may apply to other level completions. Contact NCCER at 1.888.622.3720 or go to **www.nccer.org** for more information.

You can also show off your industry-recognized credentials online with NCCER's digital credentials. Transform your knowledge, skills, and achievements into credentials that you can share across social media platforms, send to your network, and add to your resume. For more information, visit **www.nccer.org**.

1.0.0 Boiler Startup and Shutdown

Performance Task

1. Prepare and complete the startup or shutdown procedures for at least one of the following:
   - Steam or hot-water boiler
   - Chiller
   - Cooling tower or evaporative condenser

Objective

Explain how to store, start up, and shut down boiler systems.

a. Explain how to shut down and prepare boilers for storage.
b. Explain how to prepare and start up steam and hot-water boilers.

CAUTION

A boiler should be started and shut down as directed in the boiler manufacturer's instructions. Do not operate, repair, or dismantle a boiler before reviewing the manufacturer's service literature.

This section describes common tasks required to prepare larger commercial/industrial steam and hot-water heating boilers for startup and shutdown. However, these procedures are generic in nature. Due to the unique characteristics of boilers and boiler systems, it is essential that you follow the guidance of the equipment manufacturer.

1.1.0 Boiler Storage

Many large boilers experience extended periods of time shut down. Boilers require special attention to ensure that both the waterside and fireside components are protected from corrosion and deterioration. When not properly prepared for a lengthy shutdown, boilers can suffer irreversible damage. More damage can result from a month of sitting idle than during an entire year of operation.

The maintenance actions described in this module are general in nature. Remember that the startup or shutdown of a specific piece of equipment should always be performed per the manufacturer's guidance. The specific shutdown procedures for boilers vary, but the tasks and sequence of performance are essentially the same. The shutdown procedure typically followed for a steam boiler (*Figure 1*) is outlined in this section.

One area of concern during the prolonged shutdown, or **layup**, of a boiler is the damage from corrosion. Sulfuric and other acids that form during the combustion of oil and gas react with moisture and oxygen. During layup, even the fireside can be exposed to moisture from water vapor and/or condensation. In high-humidity conditions, the corrosive effect can erode and weaken boiler tubes and combustion chambers.

Layup: Refers to the status of a boiler when it is shut down for a significant period and not considered immediately operable.

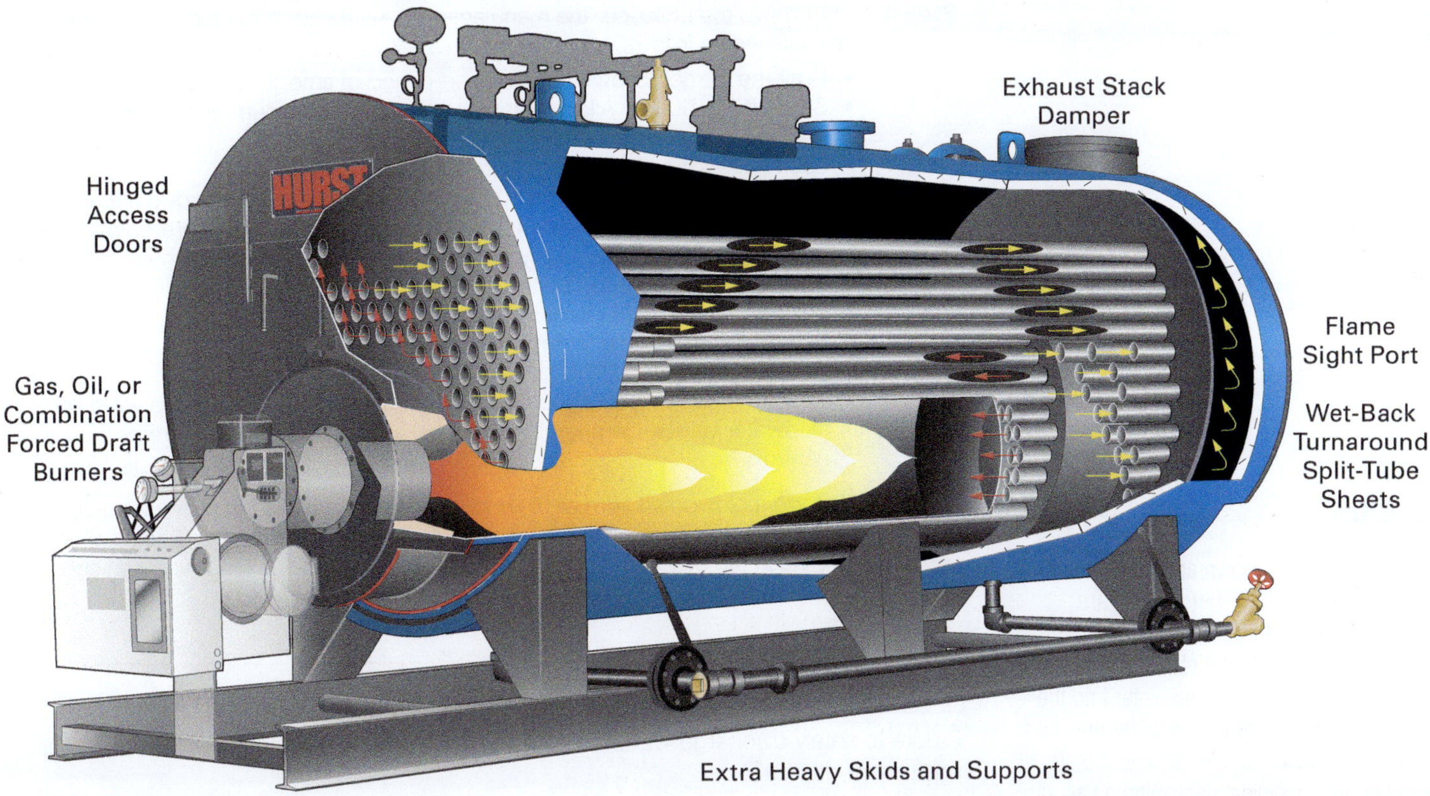

Figure 1 Horizontal firetube steam boiler.
Source: Image courtesy of Hurst Boiler Welding & Co., Inc. of Coolidge, GA

Corrosion during the layup period can be greatly reduced by thoroughly cleaning soot and other byproducts of combustion from the boiler's fireside components immediately after it cools down enough for maintenance. The condition of the combustion chamber and fireside of the boiler also provides insight into how well the burner is operating. After a thorough cleaning, a heater placed in the combustion chamber temporarily can help drive out any remaining moisture.

When preparing to shut down a boiler, it is best to reduce the load slowly so that the boiler cools at a slow and consistent rate, avoiding the potential for **thermal shock**. Boiler tube sheets, where dissimilar metals often meet, are particularly susceptible to damage from thermal shock, as are cast iron components.

Thermal shock: A potentially powerful and destructive response to major and rapid changes in temperature due to expansion or contraction, resulting in metal cracking or distorting.

1.1.1 Boiler Preparation for Dry Storage

Dry storage of a boiler is typically required when the boiler will be exposed to freezing temperatures or a humid environment, or when it will be out of service for more than 30 days. Dry storage involves draining all the water and carefully cleaning the waterside and fireside surfaces to remove all deposits such as scale and soot.

After cleaning, heaters are used to vaporize any remaining moisture, and the fireside surfaces may be coated with an anticorrosive product. **Desiccants** are then placed inside to absorb residual moisture, and the boiler is sealed to prevent fresh air and water vapor from entering.

Desiccants: Substances such as silica gel or calcium chloride ($CaCl_2$) that readily attract and hold water and water vapor, limited by the water-holding capacity of the material.

The following procedure outlines the preparation of a typical steam boiler for dry storage:

WARNING!

Preparing a boiler for storage involves working with energized equipment containing pressurized and/or hot components. Follow all manufacturer's instructions, OSHA regulations, and jobsite safety requirements.

Step 1 Shut down the boiler per the manufacturer's instructions. General instructions are outlined as follows:

- Limit the burner to low-fire for a short period of time.
- Shut off the burner. Allow feedwater pumps to continue running as needed.
- Lock out and tag the boiler power and fuel supply.
- Reduce the draft through the burner manually if the draft control is not automated and doesn't close when the burner is shut down. This prevents the boiler from cooling too quickly.
- Remember that a large boiler will retain a massive amount of heat for quite some time. The boiler will continue to make steam for a short time. Ensure the feedwater pumps continue to maintain the correct water level.
- After the boiler has stopped steaming, close the main steam stop valve and feedwater valves. De-energize, lock out, and tag pump power supplies and controls.

Step 2 Perform a blowdown from all blowdown outlets. After the pressure has dropped to no more than 1 psig, carefully open any air cocks to prevent a vacuum from forming in the boiler as it continues cooling. In extreme cases, a vacuum can cause boiler tubes or shells to collapse.

CAUTION

Remember that blowdown and drain water must be allowed to cool before they are discharged into a sewage system. Follow all federal, state, and local regulations that govern the discharge of wastewater into the environment. Do not drain the boiler until all the pressure has been relieved and the air cocks opened to avoid establishing a vacuum in the boiler.

WARNING!

Failure to safely discharge water or steam from a boiler can result in serious burns or equipment damage.

Step 3 After the boiler has cooled below 140°F (60°C), drain the boiler as directed in the manufacturer's service literature. Note that cooling to this temperature can take hours. Draining the water from a hot boiler causes sludge and other suspended solids to quickly bake and harden on the tubes and other surfaces.

Step 4 Begin cleaning and flushing out the waterside with a high-pressure water stream as soon as practical. Clean and flush feedwater and control lines. All scale and silt must be removed, as these materials will hold moisture.

WARNING!

Make sure that all steam and other system valves, blowdown valves, and electrical switches are turned off before opening boiler handholes and manhole covers. Always adequately vent the boiler before entering and apply lockout/tagout procedures.

CAUTION

Due to the extremely fine nature of soot particles, use only vacuums and collection bags specifically designed for soot. An inappropriate vacuum cleaner will be quickly ruined.

Step 5 While the waterside dries, inspect all gaskets and replace as necessary. Use a fan to help dry the waterside if necessary.

Step 6 Thoroughly clean the fireside surfaces to remove soot and combustion deposits. Clean the firetubes with a rotary brush kit and vacuum the loosened material with an appropriate vacuum for soot (*Figure 2*).

Step 7 Inspect all the fireside metal surfaces for damage or corrosion.

Step 8 Brush and vacuum the refractories clean and inspect for damage. Clean and fill any cracks over $\frac{1}{4}$" with high-temperature bonding mortar. Small fractures in the refractory are expected and acceptable.

Step 9 Remove scale or deposits from the waterside surfaces. Check for signs of internal leakage or corrosion.

Figure 2 Firetube boiler cleaning kit with vacuum.
Source: Goodway Technologies Corp.

Step 10 Check and clean the following:

- Low-water cutoff piping and control
- Water level controls and piping connections
- Blowdown valves and piping

Step 11 Wash and coat the refractories using the material recommended by the boiler manufacturer. This may be a high-temperature, air-dry mortar diluted with water to the thickness of light cream.

Step 12 Coat the fireside surfaces with an anticorrosive material recommended by the boiler manufacturer.

Step 13 Once the boiler is dry and the fireside has been treated with an anticorrosive, place a desiccant inside as recommended by the boiler manufacturer. Some manufacturers recommend specific types of lime, as well as silica gel. Pay close attention to the volume of material that should be placed in the waterside and/or fireside as directed by the manufacturer. The volume will be based on the boiler's internal volume. Put the desiccant in half-filled trays to allow room for expansion as moisture is absorbed by the desiccant.

Step 14 Close and seal all boiler openings, including handholes and manholes. Use new gaskets as needed. Close the burner air dampers and all vents to prevent moisture-laden air from reaching the fireside surfaces of the boiler.

Step 15 Prominently display a tag on the boiler declaring that it must not be operated until the desiccants are removed and the boiler is refilled with water.

Step 16 At six-week intervals, open the boiler handholes and manholes and inspect the desiccant in the boiler. Renew any desiccant that is saturated or regenerate the desiccant as the manufacturer recommends.

On Site

Silica Gel

Silica gel is one of the safest and most popular materials to use for boilers in dry storage. About 5 lb of silica gel is generally recommended per 30 ft^3 of boiler volume.

Silica gel does not visibly change as it absorbs moisture. Some products, however, are modified to change color when the gel needs to be replaced or dehydrated. The product shown here changes to green when it is saturated.

Silica gel can absorb up to 40% of its weight in water vapor and has an almost indefinite shelf life when stored airtight. Gentle heating drives the absorbed moisture out, making it reusable.

Source: Dry & Dry

1.1.2 Boiler Preparation for Wet Storage

A boiler is typically kept in wet storage when it is held in standby, when storage temperatures will remain above freezing, or when dry storage is otherwise impractical. A boiler in wet storage, meaning that the waterside remains filled, can be restarted much more quickly than a boiler in dry storage.

The following procedure outlines the preparation for wet storage of a typical steam boiler:

CAUTION

The water charged into the boiler for wet storage must be properly treated before it is added to prevent corrosion. Using untreated water can cause serious damage while it is in storage. De-aerating the boiler water is an equally important step.

Step 1 Shut down, drain, and clean the boiler as outlined in the dry storage procedure.

Step 2 With the boiler still valved off from the rest of the system, refill it to overflowing from the top with treated water. If de-aerated water is not available, fire the boiler for a short time until the water reaches a temperature of about 200°F (about 90°C) to drive dissolved gases and oxygen from the water.

Step 3 Let the boiler cool back down to room temperature, and then add enough water until it again overflows from the top. If the amount of water added is significant, add additional chemical treatment to compensate. Circulate or stir the boiler water to ensure that the chemicals are properly dispersed.

Step 4 With the boiler flooded (filled to the top, beyond the normal operating water line), close all the boiler openings.

Step 5 Maintain a slight pressure (1–3 psig) on the waterside of the boiler. Nitrogen is typically used for this purpose because it is inert, does not support corrosion, and is readily available.

Step 6 Protect the exterior surfaces and components from rust by coating them with a corrosion inhibitor or paint. Cover all tube and firebox surfaces with a coating of mineral oil or similar corrosion inhibitor.

Step 7 Leave the flue and fireside/combustion chamber doors open during the shutdown. This helps to keep the doors dry. Keep the boiler room dry and well ventilated.

Step 8 Prominently display a tag on the boiler declaring that it must not be operated until the combustion chamber is closed and the boiler is drained down to the normal waterline.

1.2.0 Boiler System Startup

A steam boiler must be started up as directed by its manufacturer. The specific startup procedures for boilers vary, but the tasks and sequences are essentially the same. A typical startup procedure for a steam boiler is outlined here.

Step 1 If not previously done, inspect the steam piping and ensure all the strainers have been cleaned in preparation. Ensure that all control and safety relief valves are functional.

Step 2 Check that all tools and equipment have been removed from the waterside and fireside of the boiler. If the boiler has been in dry storage, remove all desiccant material and trays. Make sure that all boiler openings, including handholes and manholes, are properly secured.

Step 3 Ensure the boiler is filled with properly treated water up to the normal operating level. Vent the boiler while it is filling so that air can escape as it fills. If the boiler has been flooded with treated water in wet storage, drain the water down to the proper operating level.

CAUTION

Fill the boiler with water that is at least 70°F (21°C); cold water should not be used to fill a steam boiler.

Step 4 Examine the linkages of all dampers, metering valves, and similar devices to ensure the linkage is free to move throughout the full stroke of the actuator.

Step 5 Ensure the floats in the low-water cutoff and pump controls can move freely. Also, make sure these controls are level and the piping is plumb.

Step 6 Ensure all drains and discharge piping are intact and routed to the appropriate location. Remember that hot wastewater cannot be drained directly into the local sewer system until it is cooled.

Step 7 Inspect all room openings that provide combustion air and ensure they are open and clear.

Step 8 Examine any feedwater treatment equipment, such as filters, chemical feeders, demineralizers, softeners, and de-aerators, and ensure all are operational and ready for use.

Step 9 Ensure all pressure safety control settings are set slightly above the highest steam pressure required. The safety control setting must be at least 10% below the setting of the steam pressure relief valve(s). Also ensure that all safety control settings and pressure relief valve settings do not exceed the maximum pressure rating of the boiler or its attached components.

Step 10 Make sure that the main steam stop valve is closed.

Step 11 Follow the proper procedures for energy-source lock and tag removal. Provide electrical power to the burner and the operating controls and provide fuel to the burner.

Step 12 Bleed the air from the main burner and pilot burner fuel lines.

Step 13 If practical and safe to do so, momentarily energize all three-phase motor loads associated with the boiler and ensure the direction of rotation is correct. If this is not practical, monitor these motors for the proper rotation as they start automatically.

WARNING!

Never stand directly in front of a burner or boiler during startup. Accumulated fuel vapors present in the combustion chamber or flue passages could explode. If the burners fail to ignite, allow enough time for the boiler's burner and control system to purge accumulated vapors from the combustion chamber before reattempting ignition.

Step 14 Set the boiler controls for a cold startup (if this burner mode exists), and then fire the boiler as directed by the manufacturer's operating instructions. On a cold start, it may be necessary to attempt ignition several times before the last of the air has been forced out of the fuel lines.

Step 15 Monitor the operation of the burner and controls while the boiler warms slowly. If the burner controls do not offer a cold-start mode, limit the burner to low-fire operation as it warms up. As the boiler warms up, watch the water level carefully to make sure that it remains at the proper level. It will likely rise in the early stages as the water expands, before the boiler begins to steam.

Step 16 After the boiler water is thoroughly warmed up and approaching steaming temperatures, enable high-fire operation on the burner. This will happen automatically if the burner is operating in a cold-start mode.

Step 17 When the steam gauge just begins to register pressure building in the boiler, blowdown the gauge glass, water column, and low-water cutoff.

Carryover: Refers to any contaminant that is transferred along with steam, typically from a boiler to a system, but contaminants can also be transferred into a boiler that is at a lower pressure than the steam header. Carryover can be comprised of contaminants in solid, liquid, or vapor form.

NOTE

If above-normal amounts of water treatment chemicals were added to a boiler during a period of wet storage, it may be necessary to blow down the boiler more frequently during the first few days of operation. This is required to reduce the chemicals to normal levels.

Step 18 Perform a combustion analysis on the burner. Make this test and any other burner-related adjustments as instructed by the test instrument and boiler manufacturer's service literature.

Step 19 When the boiler pressure is very close to the pressure on the steam header, very slowly open the main steam stop valve until it is fully open. If the valve is opened too quickly, the header can rapidly lose pressure. Also, by opening the valve slowly, **carryover** from either system can be prevented.

Step 20 With the burner now operating automatically between low-fire and high-fire, ensure that it shuts down completely when the steam pressure reaches the control setpoint.

Step 21 Monitor the boiler through enough operating cycles to make sure it and the various subsystems are operating correctly.

Step 22 Visually examine the steam piping system to verify the following:

- The system is free of leaks.
- All pressure-reducing valves and stations are operating properly.
- Steam traps serving both loads and steam mains are operating properly and condensate is returning to the feedwater system as expected.

Step 23 Record the boiler and system operating characteristics.

On Site

Burner Management Systems

In days gone by, monitoring the operation of a big boiler could be a full-time job, requiring an operator to be consistently in motion to monitor gauges, thermometers, and other indicators. Although the process of heating water to make steam has not changed much over the years, the level of control sophistication has certainly changed for the better.

Also known as supervisory control systems, burner management systems allow an operator to monitor many points and control multiple burners and the steam system from a comfortable control room.

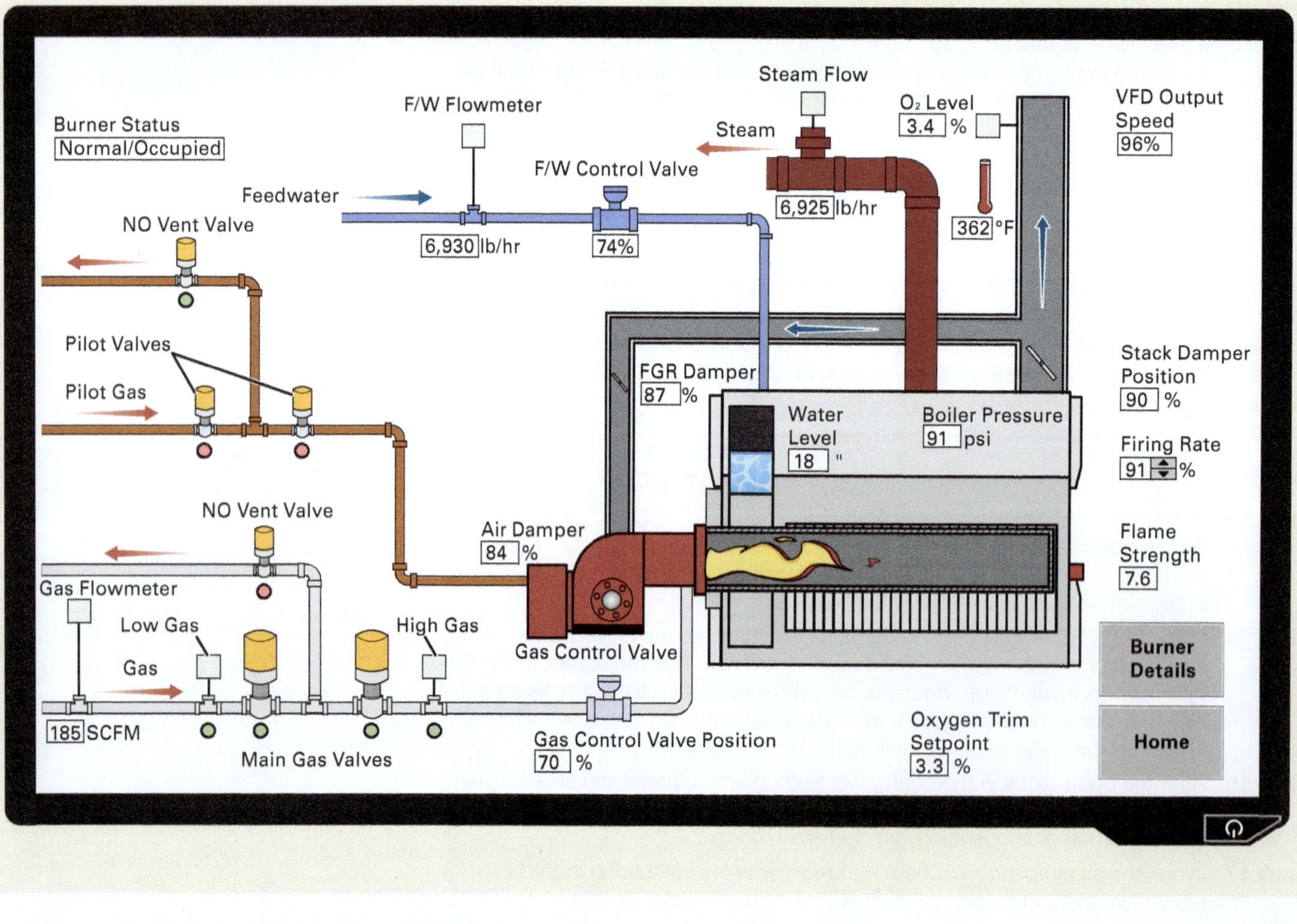

1.2.1 Hot-Water Boiler Startup

Hot-water boilers (*Figure 3*) are generally built and operated like steam boilers. However, hot-water boilers are flooded with water, because no space for steam to form is needed. Hot-water boilers also present less stored energy and fewer hazards, unless operated at high pressures.

Figure 3 Hot-water firebox boiler.
Source: Image courtesy of Hurst Boiler Welding & Co., Inc. of Coolidge, GA

High-temperature hot-water (HTHW) boilers and plants are typically designed to operate at temperatures ranging from 350°F to 420°F (175°C to 215°C). The system pressure must remain at least 25 psig above the saturation pressure of the high-temperature water's maximum temperature to prevent pump cavitation and flashing of superheated water to steam.

WARNING!

High-temperature hot-water (HTHW) systems present extreme hazards. Minute cracks and leaks can allow high-temperature, high-pressure water to escape and flash to steam, resulting in a virtually invisible high-pressure steam plume that can easily cut flesh and cause extreme injuries or death. This training program does not prepare trainees for work with HTHW systems and the surrounding environment. Do not work on or with HTHW systems without the proper training to ensure your safety.

The sections that follow present some considerations related to hot-water boiler startup as opposed to steam systems. Remember that all boilers should be started up as directed in the manufacturer's service instructions and plant policy.

Filling and Venting a Hot-Water System

When filling a new or dry hot-water system with treated water, the entire boiler, piping system, and terminals must be filled. Manual air vents must be opened (if used) to expel any air trapped in the system. Automatic air vents may also be found at strategic locations. Manual air vents are usually placed at the top of coils and other hot-water loads, as well as at the highest points of the system.

Because a hot-water boiler is filled with water, eliminating air, the hot-water outlet usually includes a dip tube that extends 2" to 3" into the boiler. This dip tube traps any air at the top of the boiler shell that is released from the water during heating. The collected air is piped to the expansion tank.

On Site

Dual-Rated Pressure Vessels

Boiler manufacturers will often build a single boiler pressure vessel that is pressure-rated for both steam and hot-water applications. When a boiler is built, different operating and safety controls can be added to make a single pressure vessel suitable for either steam or hot-water use.

System Operating Temperatures

The minimum water temperature recommended for many industrial hot-water boilers is 140°F (60°C). At lower temperatures, combustion gases may condense in the flue piping and fireside of the boiler and cause corrosion. This problem is more severe in systems that are operated intermittently or are significantly oversized for the heating load. The boiler water-temperature limit switches are typically set 5°F to 10°F (2.8°C to 5.5°C) above the boiler's aquastat setting.

Water Circulation Considerations

The system piping and controls are arranged to prevent the possibility of pumping large volumes of cold water into a hot boiler. For example, if 160°F (60°C) boiler water is suddenly replaced with 70°F (21°C) water returning from a cold system, it could cause thermal shock. This is avoided by gradually warming the water in the boiler and the water in circulation.

As with steam boilers, allow the boiler and system to come up to temperature slowly. Limit the burner to its low-fire mode or use cold-start controls if equipped. The threat of thermal shock can also be reduced by having the burners interlocked with the circulating pump(s), so the burners cannot operate unless the pump is running.

During initial startup and subsequent operation, it is important to maintain the pressure/temperature relationships in a hot-water boiler per the boiler manufacturer. *Figure 4* shows the recommended pressure/temperature relationships for one manufacturer's hot-water boilers. The chart shows the minimum pressure that should be maintained on the boiler based on the water temperature. For example, if the boiler must provide 260°F (127°C) water, the pressure should be maintained at a minimum of 55 psig.

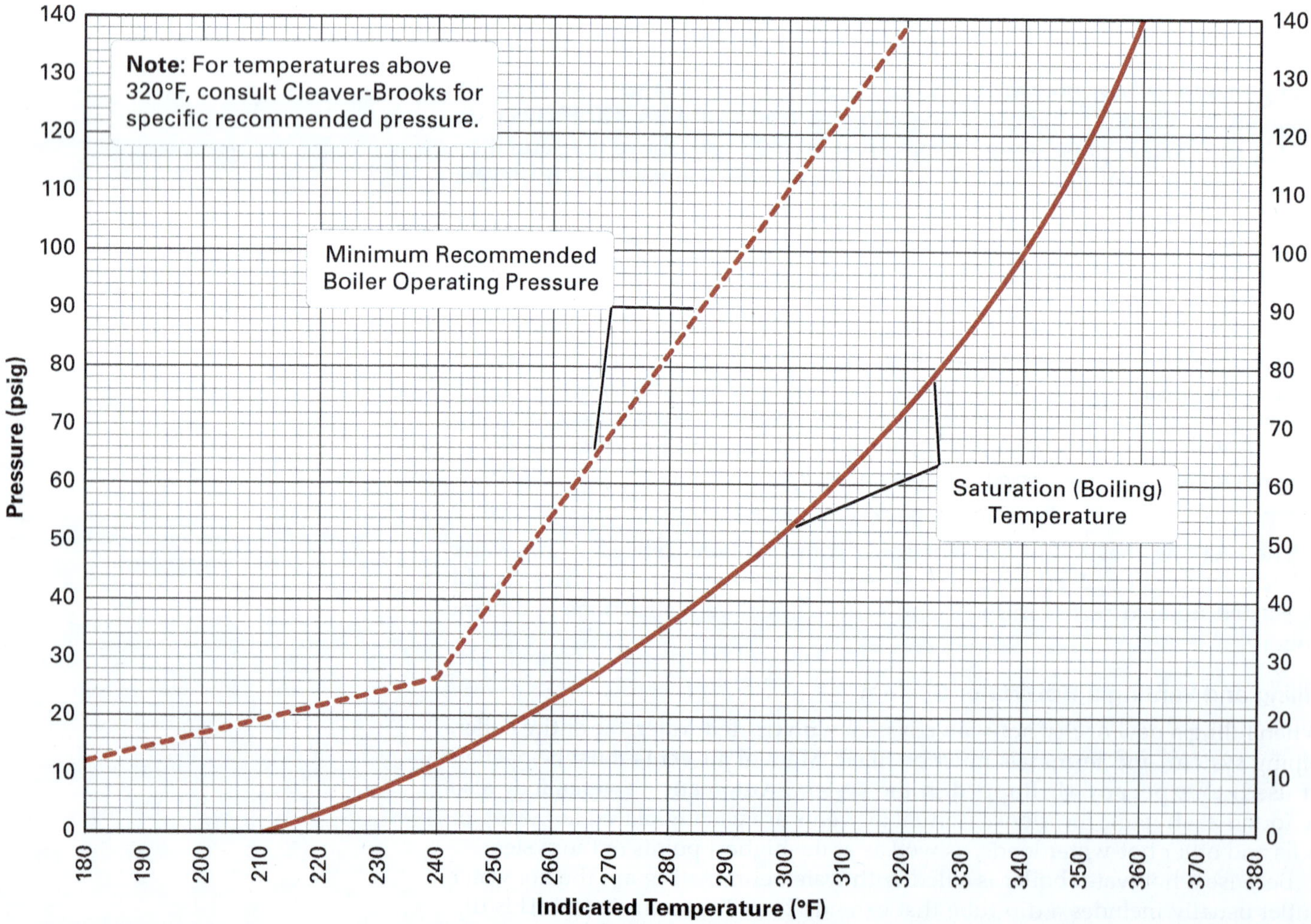

Figure 4 Typical hot-water boiler pressure/temperature relationship.
Source: Cleaver-Brooks, Inc.

At pressures below those shown on the chart, the boiler could potentially begin producing steam. A safety factor is reflected in the pressures shown, as this is a dangerous condition for a hot-water boiler. Of course, not all hot-water boilers can operate at the pressures and temperatures shown. The average residential boiler, for example, may just make the chart at 180°F (82°C) and a minimum operating pressure of 12 psig.

Constant water flow through a hot-water boiler helps eliminate water stratification inside and results in more consistent water temperatures leaving the boiler. Boiler manufacturers usually recommend minimum and maximum flow rates based on the water temperature drop as it moves through the system. *Figure 5* shows a chart used by one manufacturer to determine the circulation rate in a system based on temperature drop.

Boiler Size (BHP)	Boiler Output (1,000) Btuh	System Temperature Drop – °F									
		10	20	30	40	50	60	70	80	90	100
		Maximum Circulating Rate – gpm									
15	500	100	50	33	25	20	17	14	12	11	10
20	670	134	67	45	33	27	22	19	17	15	13
30	1,005	200	100	67	50	40	33	29	25	22	20
40	1,340	268	134	89	67	54	45	38	33	30	27
50	1,675	335	168	112	84	67	56	48	42	37	33
60	2,010	402	201	134	101	80	67	58	50	45	40
70	2,345	470	235	157	118	94	78	67	59	52	47
80	2,680	536	268	179	134	107	90	77	67	60	54
100	3,350	670	335	223	168	134	112	96	84	75	67
125	4,185	836	418	279	209	168	140	120	105	93	84
150	5,025	1,005	503	335	251	201	168	144	126	112	100
200	6,695	1,340	670	447	335	268	224	192	168	149	134
250	8,370	1,675	838	558	419	335	280	240	210	186	167
300	10,045	2,010	1,005	670	503	402	335	287	251	223	201
350	11,720	2,350	1,175	784	587	470	392	336	294	261	235
400	13,400	2,680	1,340	895	670	535	447	383	335	298	268
500	16,740	3,350	1,675	1,120	838	670	558	479	419	372	335
600	20,080	4,020	2,010	1,340	1,005	805	670	575	502	448	402
700	23,430	4,690	2,345	1,565	1,175	940	785	670	585	520	470
800	26,780	5,360	2,680	1,785	1,340	1,075	895	765	670	595	535

Figure 5 Typical hot-water boiler flow rate chart.
Source: Cleaver-Brooks, Inc.

To establish an accurate water circulation rate for the system, measure the temperature drop between the supply outlet and return inlet of the boiler. Then enter the chart on the row appropriate for the boiler size or horsepower.

Did You Know?

Boiler Sizing

One boiler horsepower (BHP) is equal to 9.803 kilowatts (kW) of energy and a heat output of 33,475 Btuh.

1.0.0 Section Review

1. Why are desiccants used inside the fireside of a boiler during dry storage?
   a. To absorb any residual moisture inside
   b. To help prevent soot buildup in the future
   c. To prevent the combustion chamber from drying out
   d. To seal any cracks in the combustion chamber lining

2. What happens if the main boiler stop valve is opened too quickly after a steam boiler heats up, but its pressure is lower than the steam header?
   a. Steam will recondense in the boiler.
   b. Thermal shock to the steam header is likely.
   c. The boiler burner will shut off due to overload.
   d. The boiler pressure rapidly rises and carryover may occur.

2.0.0 Chiller System Startup and Shutdown

Performance Task

1. Prepare and complete the startup or shutdown procedures for at least one of the following:
   - Steam or hot-water boiler
   - Chiller
   - Cooling tower or evaporative condenser

Objective

Explain how to start up and shut down chillers and cooling towers.

a. Explain how to start up and shut down reciprocating chillers.
b. Explain how to start up and shut down centrifugal chillers.
c. Explain how to start up and shut down cooling towers and clean water-cooled condensers.

Chillers produce cold water that is routed to chilled-water coils or other heat exchangers. Chilled water is not only used for larger comfort cooling applications, but also for many industrial processes. Like boilers, chillers must be started up and shut down properly to avoid damage and unexpected problems. Chillers also have unique maintenance and cleaning requirements.

2.1.0 Reciprocating and Scroll Air-Cooled Chillers

Packaged scroll and reciprocating chillers (*Figure 6*) are typically used on jobs requiring chiller capacities ranging from 2 to 200 tons. Large reciprocating chillers usually have more than one compressor. The condenser may be water-cooled

Figure 6 Packaged air-cooled chiller with scroll compressors.
Source: Carrier Corporation

or air-cooled. The evaporator, sometimes referred to as the *cooler,* is often a direct-expansion, shell and tube heat exchanger in which water flows through the tubes and refrigerant occupies the shell (*Figure 7*). However, evaporator designs differ among chiller types and manufacturers.

NOTE

Throughout this section, to avoid redundancy, the term "reciprocating" is intended to refer to chillers that use reciprocating hermetic or semi-hermetic compressors, as well as those that use hermetic scroll compressors.

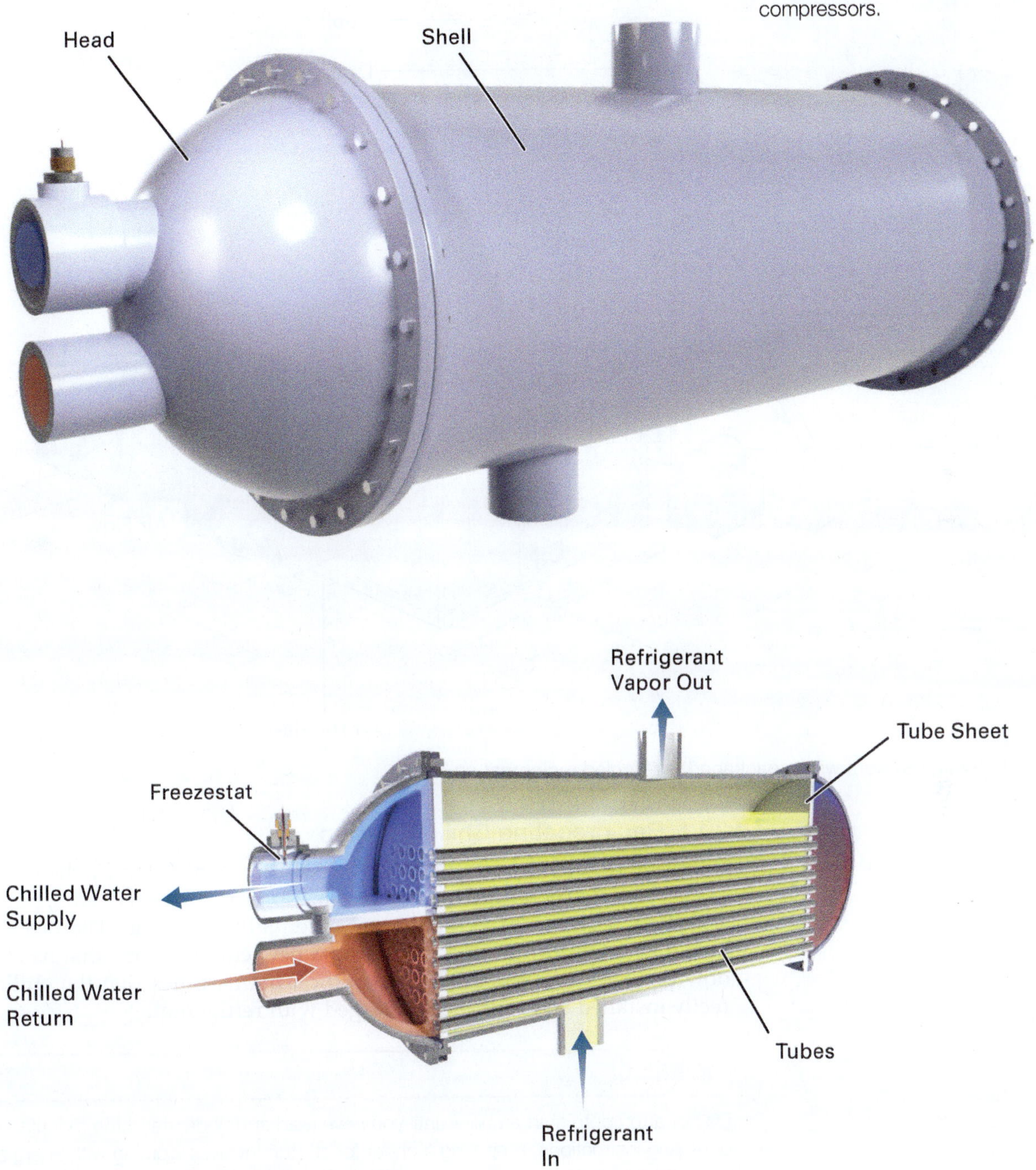

Figure 7 Shell and tube chiller evaporator (cooler).

Figure 8 shows a simplified view of a packaged, roof-mounted reciprocating chiller with an air-cooled condenser. The chiller shown is supplying chilled water to the coil of an air handling unit for comfort cooling. Multiple air handling units are usually served by a single chiller. The following sections describe the startup and shutdown of a reciprocating chiller and related chilled-water system.

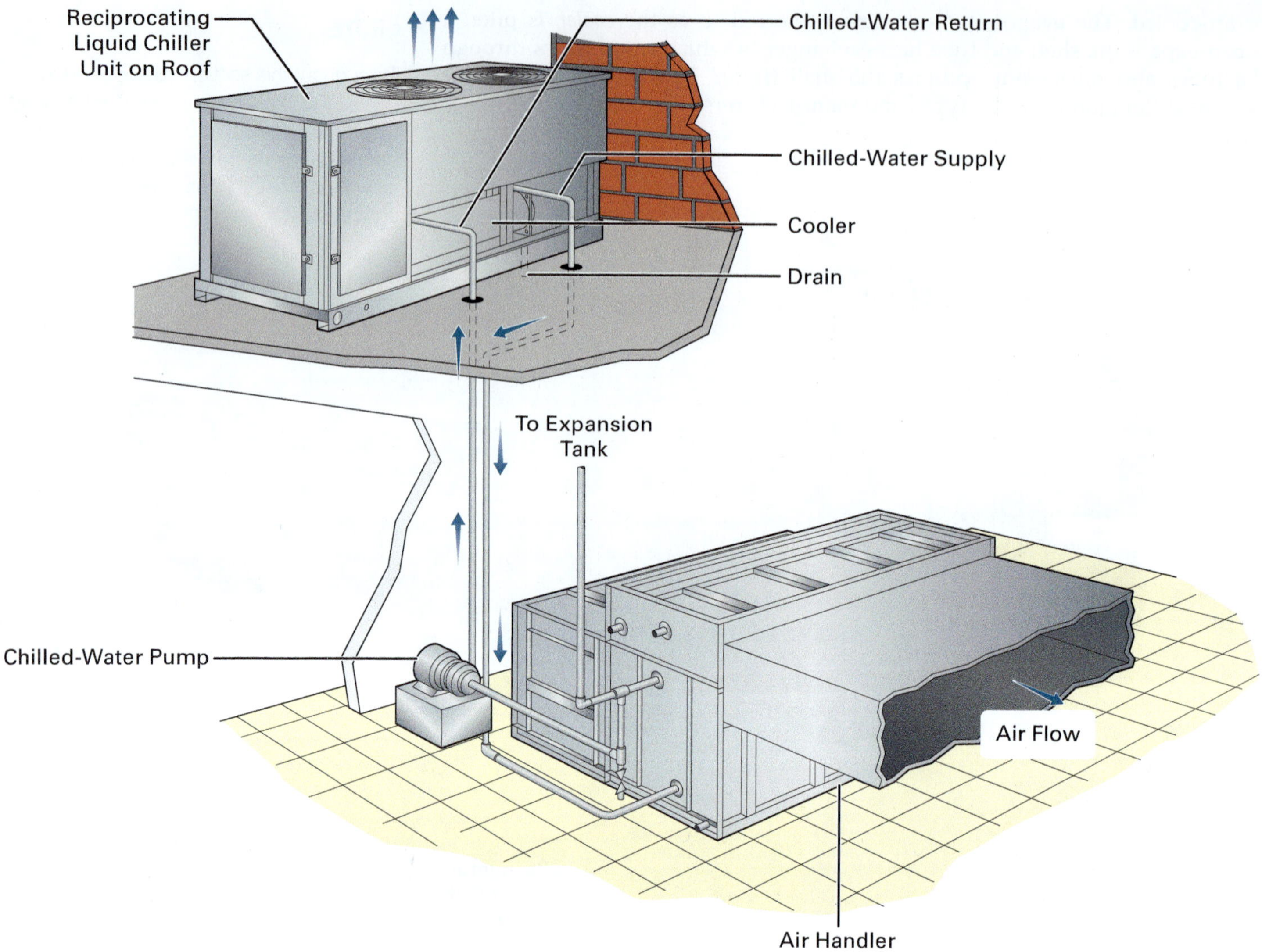

Figure 8 Chilled-water system with a packaged, air-cooled reciprocating chiller.

2.1.1 Reciprocating Chiller Startup

Complete the startup of a reciprocating chiller as directed by the manufacturer's service instructions. The specific startup procedures will vary, but the tasks and their sequence of execution are essentially the same. The tasks outlined herein are typically related to starting a new chiller or one that has been shut down for an extended period. The procedure assumes that the chiller is correctly installed and is properly charged with refrigerant.

WARNING!

Do not attempt to start a chiller until you have read and understand the manufacturer's operating instructions. Preparing a chiller for startup involves working with energized equipment containing pressurized and/or hot components. Follow all manufacturer's instructions, OSHA regulations, and jobsite safety requirements.

CAUTION

Manufacturers usually require the chiller to be powered at least 24 hours prior to startup to energize the crankcase heaters, allowing time to warm the oil and vaporize condensed refrigerant.

Step 1 Ensure all tools, equipment, shipping protection, and debris are removed from the chiller and the work area. Ensure the compressor crankcase heater(s) is energized and the crankcase is warm.

Step 2 Perform an inspection of the chilled-water circuit piping to verify the following:

- All system strainers are clean and installed.
- All piping connections are secure, and piping is properly mounted to accommodate the weight of the water when added.

- Coil fins and tubing are undamaged, and the coils are free of debris and obstructions.
- Pressure-relief valves are set to the correct setpoint. The setpoint should be visible on the valve's tag.
- All balancing and isolating valves are set per the job specifications or opened to the required position.
- Refrigerant and water stop valves for the evaporator (cooler) circuit are open.

Step 3 Fill the system with clean water that has been treated with corrosion and biological growth inhibitors as prescribed.

- When starting a system that has been shut down through the previous heating season and charged with a glycol solution to prevent freezing, first drain any antifreeze solution that may have been added for protection. Then flush and refill the system with appropriately treated water.

Step 4 Purge the air from the piping, coils, and/or heat exchangers using vents positioned at the highest points of the circuit. Once the air is removed, confirm that the water flow, water level, and systems pressures are within the specifications.

Step 5 With the system under pressure, visually inspect the piping system and connected equipment for signs of leaks.

Step 6 Perform a preliminary inspection of the chiller to verify the following:

- The provided voltage meets the unit nameplate requirements.
- The compressor crankcase is warm, indicating that the crankcase heaters are operating.
- Oil is visible in the compressor sight glass (if equipped).
- The compressor(s) floats freely on its mounting springs or isolators.
- The chilled-water temperature setpoint and the freezestat (low-temperature safety control) setpoint is correct.
- The condenser coils are free of debris and obstructions.
- The condenser fans are free to turn, and the blades are secure. Also ensure the fans are properly positioned in the venturi (discharge orifice).
- All refrigerant stop valves are open, and the unit is charged with refrigerant.

Step 7 Start the chilled-water pump(s) and check for the following. Note that chilled-water pump control may be tied directly to the chiller to automate the process:

- Pump rotation is in the proper direction.
- Pump is quiet and lubrication has been provided.
- Motor and pump shaft are properly aligned. Misalignment is often accompanied by vibration.
- Pump motor current is within the motor's rated load amps (RLA).

Step 8 Start the air handling equipment.

- In some cases, the air handling units may be left off during chiller startup to reduce the initial load on the chiller.

Step 9 Turn on the chiller. The chiller compressor(s) should start after a short delay. Remember that one or more condenser fans may be controlled by pressure or temperature and may not start immediately.

NOTE

Most chillers have compressor short-cycle protection that provides a delay-on-break of about five minutes. The delay is activated when the compressor stops at the end of a normal cooling cycle, if a safety device opens, or if there is a power interruption. Be sure to allow enough time for the compressors to start. There may also be other delays related to a connected control system.

Step 10 Once the compressors have started and the refrigerant circuit has stabilized, check the superheat, subcooling, and refrigerant charge. With packaged chillers free of external refrigerant piping, these values should be very close to the manufacturer's documented values.

Step 11 Monitor the leaving chilled water temperature and the temperature drop through the chiller. Also ensure that the chiller shuts down as expected when the chilled-water temperature reaches the setpoint. This may take some time, however, depending on the current chilled-water load.

Step 12 Confirm that the chiller's freezestat is functioning properly per the manufacturer's instructions.

Step 13 Check the operation of the cooler heater cable, if the chiller is so equipped. This is an important accessory if the chiller is exposed to freezing temperatures, which is common for an outdoor, air-cooled chiller.

Step 14 Record all chiller and chilled-water pump operating data using suitable forms. Records should generally include the information summarized in *Figure 9* and *Figure 10*.

Chiller Data

Location: ______ Manufacturer: ______

Model Number: ______ Serial Number: ______

Compressor Operating Data

Suction Pressure/Temperature: ______ Crankcase Heater Amperes: ______

Discharge Pressure/Temperature: ______ Chilled-Water Control Setting: ______

Oil Pressure/Temperature: ______ Condenser Water Control Setting: ______

Voltage: ______ Low-Pressure Cutout Setting: ______

Amperage T1/T2/T3: ______ High-Pressure Cutout Setting: ______

kW Input: ______ Safety Control Check: ______

Evaporator (Cooler) Operating Data

Entering/Leaving Water Pressure: ______ Flow in Gallons per Minute (gpm): ______

Water Pressure Difference: ______ Refrigerant Temperature: ______

Entering/Leaving Water Temperature: ______ Refrigerant Pressure: ______

Water Temperature Difference: ______

Condenser Operating Data

Water-Cooled Condenser

Entering/Leaving Water Pressure: ______

Water Pressure Difference: ______

Entering/Leaving Water Temperature: ______

Water Temperature Difference: ______

Flow in Gallons per Minute (gpm): ______

Refrigerant Temperature: ______

Refrigerant Pressure: ______

Air-Cooled Condenser

Entering/Leaving Dry-Bulb Temperature: ______

Air Temperature Difference: ______

Refrigerant Pressure: ______

Refrigerant Temperature: ______

Figure 9 Packaged reciprocating chiller operating data.

Chiller Data

Location: ______ Manufacturer: ______

Model Number: ______ Serial Number: ______

Pump Operating Data

Pump Off Pressure: ______ Final Suction Pressure: ______

Valve Shutoff Differential: ______ Final Pressure Difference: ______

Valve Open Differential: ______ Final gpm: ______

Valve Open gpm: ______ Voltage: ______

Final Discharge Pressure: ______ Amperage T1/T2/T3: ______

Figure 10 Chilled-water pump operating data.

2.1.2 Reciprocating Chiller Shutdown

A few tasks are required to prepare an air-cooled reciprocating chiller for an extended shutdown. If the chiller is being shut down for the winter season, some manufacturers recommend the evaporator be isolated from the chilled-water circuit by closing the valves on the supply and return piping. With the evaporator isolated, remove the drain plug and drain the water from the shell. Some also recommend adding an approved glycol solution cooler to prevent any residual water from freezing.

If the water is routed through the tubes of the evaporator, isolate the evaporator from the chilled-water circuit and drain the water from the heads.

Regardless of which type of evaporator is used, the evaporator heat cable should continue to operate. If the power supply for the chiller must be disabled, ensure the heat cables are powered from a separate, dependable source. The compressor crankcase heaters may also be left energized throughout a seasonal shutdown.

Some manufacturers further recommend that the bulk of the refrigerant charge be isolated in the condenser or receiver during a lengthy shutdown. This minimizes the amount of refrigerant that might be lost due to a small leak on the low-pressure side of the system. Isolating the refrigerant in the condenser or receiver usually involves bypassing the low-pressure protective switch temporarily and closing the refrigerant valve leaving the condenser or receiver. The compressor(s) is then used to pump the refrigerant into the high-pressure side of the system. This should be done only as directed in the chiller manufacturer's instructions.

CAUTION

Pumping down a chiller can cause the water in the evaporator to freeze, seriously damaging the evaporator. Take the necessary precautions and follow the manufacturer's guidance carefully. Ensure that the chilled-water pump remains running during this process to help vaporize the refrigerant.

2.2.0 Centrifugal Chillers

Centrifugal chillers are normally water-cooled. *Figure 11* shows a typical water-cooled centrifugal chiller. Condenser water is cooled by a cooling tower. The following sections describe the tasks needed to start up and shut down a centrifugal chiller and related chilled-water system.

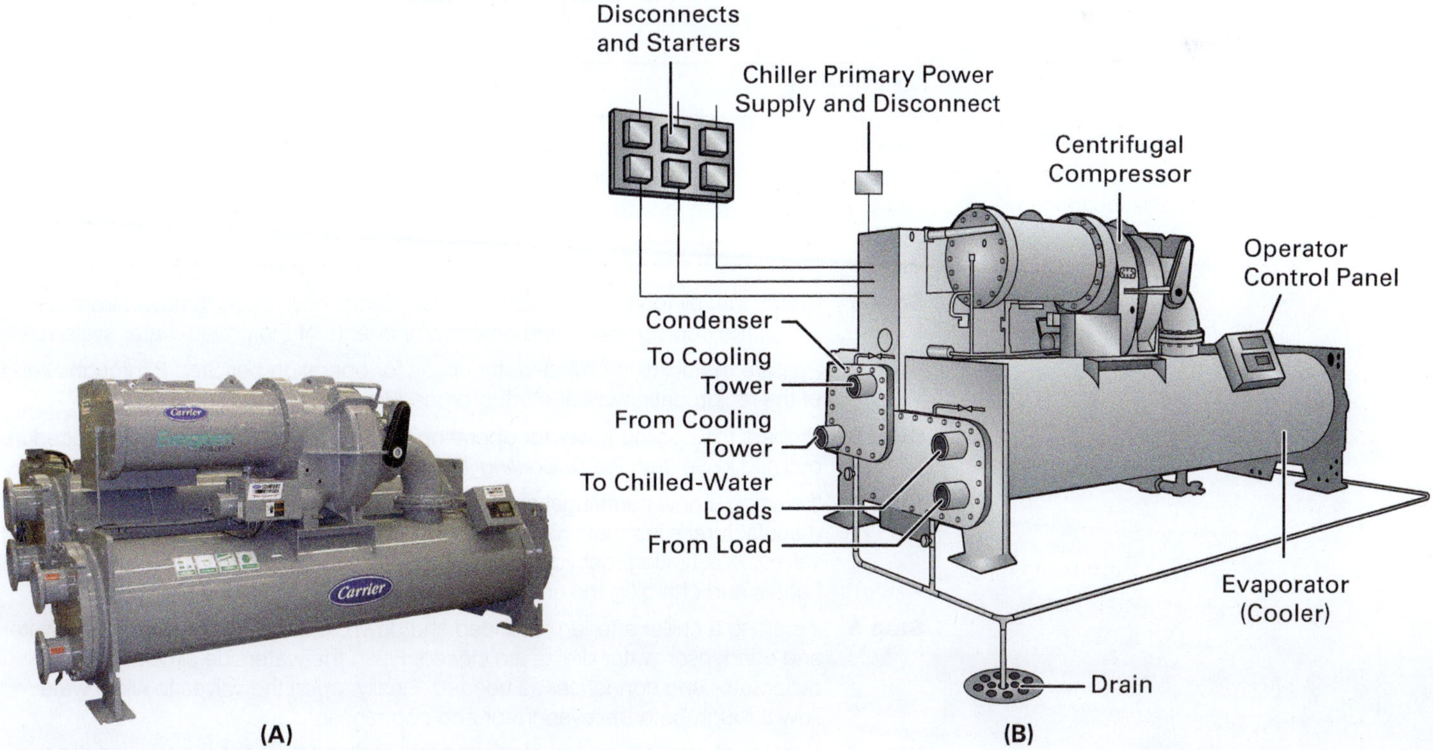

Figure 11 Typical water-cooled centrifugal chiller.
Source: Carrier Corporation (A)

Going Green

Impact of LEED on System Commissioning

The Leadership in Energy and Environmental Design (LEED) green-building rating system was introduced earlier in the HVACR craft training program. LEED requirements have an impact on every phase of building construction, from design through commissioning. The HVACR commissioning process is now more extensive than ever before.

In many cases, a third-party commissioning agent is employed by the building owner for the task, in addition to the traditional system testing and balancing contractor. Filling these new positions requires both experience and higher levels of training. Today's commissioning agents are often on the go and travel to every corner of the country.

2.2.1 Centrifugal Chiller Startup

The startup procedures for a centrifugal chiller are often more automated than those that apply to reciprocating chillers. Centrifugal units have a microprocessor-based controller that monitors and controls equipment operation. The microprocessor matches the cooling capacity of the chiller to the cooling load. It also monitors system operating conditions to execute capacity overrides or safety shutdowns.

Operator interface with the chiller is usually done at a local interface panel on the chiller. Much of the same information can be accessed and edited from a connected building management system (BMS or BAS), but a new or awakened chiller must be started with a technician at the machine. There, an operator can input setpoints, schedules, and setup options. Typically, these inputs are entered in response to software-driven prompts shown on the control panel display.

Because the operational software and the format of the operator interface vary widely among manufacturers, it is essential to complete the startup of a centrifugal chiller as directed in the manufacturer's instructions.

The following procedure briefly outlines the tasks required to start up a centrifugal chiller with a water-cooled condenser. The tasks described are those that are commonly performed when starting a new chiller after installation or when it has been shut down for an extended period.

CAUTION

Do not attempt to operate a chiller until you have read and understand the manufacturer's operating instructions. Different chillers often have unique requirements.

WARNING!

Preparing a chiller for startup involves working with energized equipment containing pressurized and/or hot components. Follow all applicable manufacturer's instructions, OSHA regulations, and jobsite safety requirements.

Step 1 Check that all tools, equipment, or other debris have been removed from the chiller, the cooling tower, and other components of the chilled-water system.

Step 2 Prepare the building chilled-water circuit for operation per *Step 2* through *Step 5* of the reciprocating chiller startup procedure.

Step 3 Prepare the cooling tower for operation per the cooling tower startup procedure outlined in *Section 2.3.0, Cooling Towers* of this module.

Step 4 If starting a new centrifugal chiller, perform the necessary prestart tasks per the manufacturer's instructions. These instructions typically include opening specific valves, leak testing, dehydration and evacuation, energizing and checking the oil heater, and charging the unit with refrigerant.

Step 5 If starting a chiller after an extended shutdown, make sure that the chilled-water and condenser water drains are closed. Flush the waterside circuits of the evaporator and condenser as needed. Finally, open the valves to allow water flow through both the evaporator and condenser.

Step 6 Restore power to the main starter panel, cooling tower fan starter, and the machine control center. Check the oil heater to ensure it is operating and warming the oil.

Step 7 Ensure the cooling tower water sump is at the appropriate level and makeup water is available. Start the condenser water pump and monitor the flow until all the air has been forced out of the circuit and the flow is stable.

Step 8 Start the chilled-water pump and ensure all the air has been purged from the water system.

Step 9 Ensure the oil temperature has reached the minimum specified temperature for operation. The acceptable temperature range is usually between 140°F to 160°F (60°C to 71°C).

Step 10 Perform the following tasks per the manufacturer's instructions:

- Dry run the unit to test the startup sequence.
- Perform a compressor motor rotation check.
- Conduct oil pressure and compressor stop checks.
- Calibrate the motor current demand setting.

Step 11 Enable and start the chiller per the manufacturer's instructions. If the chiller fails to start, check for conditions that prohibit starting, such as activated machine safety devices, no occupied mode schedule selected for the current period, or short-cycle time delays have not yet timed out. The reason for the delay may be visible on the chiller's display, along with a countdown to start.

An example of an automated sequence for centrifugal chiller startup is shown in *Figure 12* and is described as a series of events that follow.

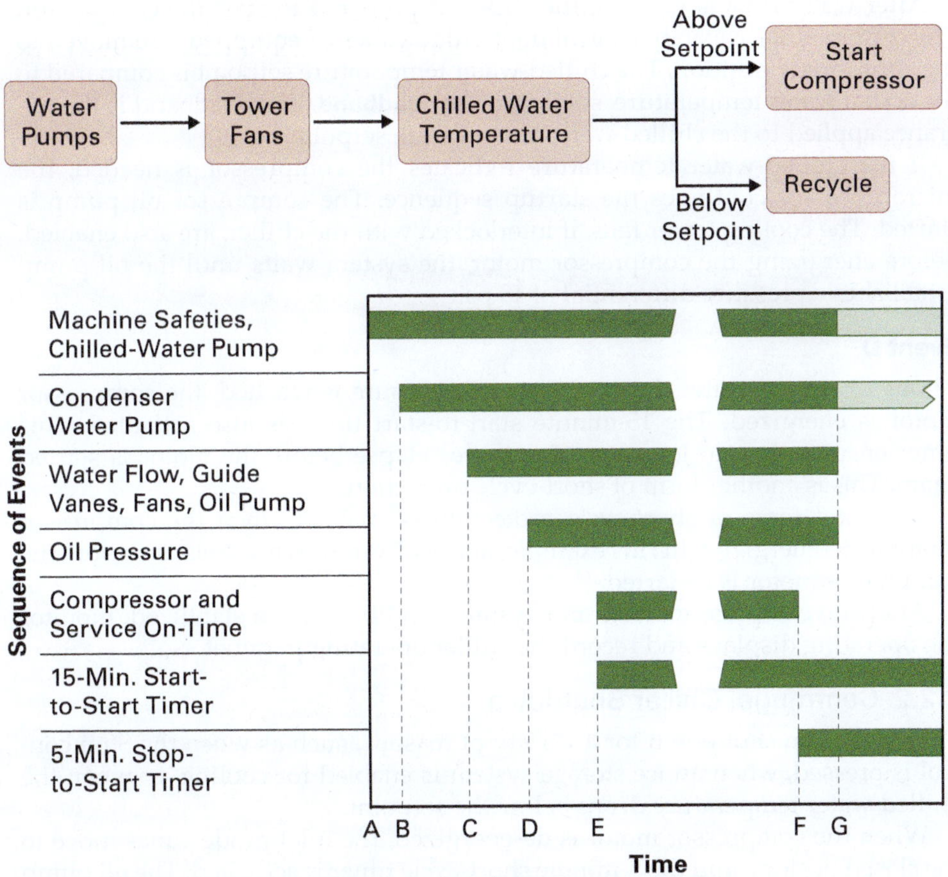

Figure 12 Typical centrifugal chiller startup sequence.

Event A

When chilled-water pump control is linked to the chiller, enabling the chiller starts the pump, assuming the equipment safety controls are all satisfied. The safety controls typically look for the following conditions:

- Temperature sensors out of range
- Pressure transducers out of range
- Compressor discharge temperature excessive

NOTE

Note that the times and other values provided in the sequence are for example only. Any events in the sequence that fail to take place typically result in an aborted startup sequence and a fault message on the control panel display.

- Motor winding temperature excessive
- Evaporator refrigerant temperature too low
- Condenser pressure excessive
- Oil pressure out of range
- Line voltage out of range
- Compressor motor overloaded
- Motor acceleration time too slow

Event B

Assuming all the safety controls are satisfied, the condenser water pump starts after a short delay.

Event C

The system controller begins looking for proof that the chilled-water and condenser water pumps are operating and creating flow. The chiller may wait as long as five minutes to verify water flow. Chilled water and condenser water flow is usually proven by paddle-type flow switches or differential pressure switches that indicate flow directly. Note that the waiting period allocated to monitor the water flow can often be changed at the control panel, if desired.

After water flow is verified, the chiller is prepared to start the compressor. The compressor's capacity-control inlet guide vanes (if equipped) are moved to the appropriate position. The chilled-water temperature setpoint is compared to the actual water temperature, subject to the **deadband**. The deadband is the tolerance applied to the chilled-water temperature setpoint.

Deadband: For a chiller, the tolerance around the chilled-water temperature setpoint. For example, a 1°F deadband controls the water temperature to within ± 0.5°F of the control point temperature (0.5°F + 0.5°F = 1°F deadband).

If the chilled-water temperature indicates the compressor is needed, the microprocessor continues the startup sequence. The compressor oil pump is started. The cooling tower fans, if interlocked with the chiller, are also enabled. Before energizing the compressor motor, the system waits until the oil pump establishes a pressure differential of 18 psi.

Event D

About 15 seconds after the minimum oil pressure is reached, the compressor motor is energized. The 15-minute start-to-start timer is also activated. This timer ensures that no less than 15 minutes elapse before the motor is started again. This is another form of short-cycle protection.

A second form of short-cycle protection is enabled when the compressor motor is de-energized. In this example, at least 5 minutes must elapse before the compressor motor is restarted.

After the compressor starts and system conditions have stabilized, monitor the operating displays and record the chiller operating parameters.

2.2.2 Centrifugal Chiller Shutdown

The chiller can shut down for a variety of reasons, such as when the Stop control is pressed, when an ice storage system is enabled for cooling, or when the chilled-water temperature drops below the setpoint.

When the compressor motor is de-energized, the inlet guide vanes move to the closed position and the 5-minute short-cycle timer is activated. The oil pump is turned off and the chilled-water and condenser water pumps may or may not continue running. If the condenser water temperature is still high, the tower fan(s) may continue to run until the water temperature falls below the setpoint.

During normal operation, the chiller may automatically cycle off and wait until the load increases significantly before it restarts. This mode of operation is called the **recycle shutdown mode**. It occurs when the compressor is running at minimum capacity and the chilled-water temperature drops 5°F (2.8°C) below the chilled-water control setpoint (*Figure 13*). The chilled-water pump remains running. The chiller control system automatically restarts the chiller when it senses that the chilled-water temperature has increased to 5°F above the

Recycle shutdown mode: A chiller mode of operation in which automatic chiller shutdown occurs when the compressor is operating at minimum capacity, but the chilled-water temperature drops below the chilled-water temperature setpoint. The chilled-water pump(s) continues running.

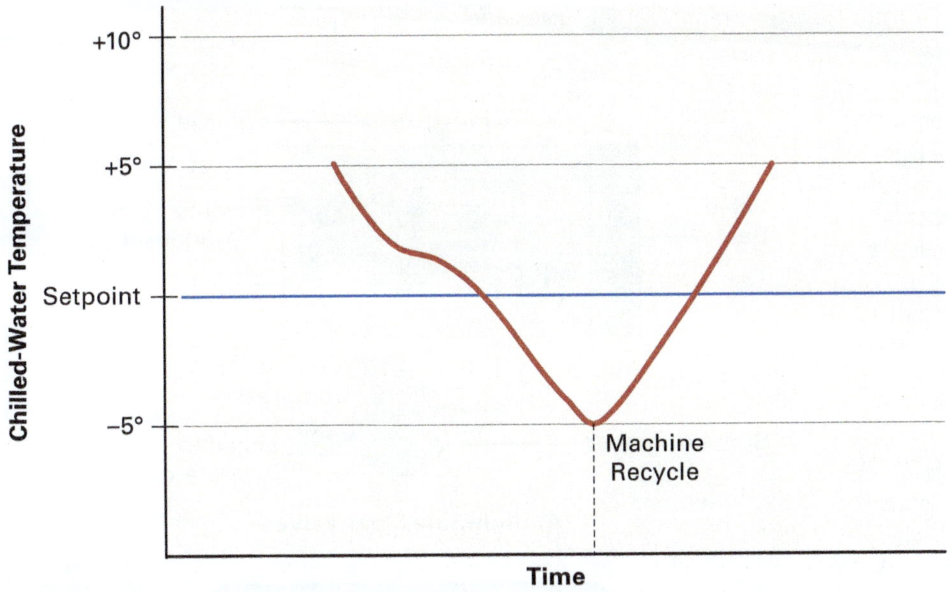

Figure 13 Recycle shutdown mode setpoints.

chilled-water control setpoint. Normally, the chiller would operate inside a tighter deadband. But the conditions indicate the cooling load is very light, so rushing to restart repeatedly is not necessary or productive.

Centrifugal chillers are better left running than cycling on and off often. These motors are large, and a lot of energy is required to get the impeller turning at a high speed. In most cases, assuming there is any significant load on the chiller, it will run consistently through an occupied period while controlling its capacity across a broad range to match the load.

2.2.3 Centrifugal Chiller Extended Shutdown

When the centrifugal chiller will be shut down for an extended period, the refrigerant may be recovered and stored in a tank. This prevents refrigerant loss if leaks develop. Additionally, after a long storage period, the machine should be leak tested before it is recharged. A holding charge, usually dry nitrogen at a low pressure, is added to prevent moisture-laden air from leaking into the chiller. Some chillers are equipped with a pump-out system and isolation valves to make the refrigerant transfer to storage more convenient.

If the machine may be exposed to freezing temperatures, drain the water from the condenser and evaporator. Leave the oil charge in the machine with the oil heater and controls energized to maintain the minimum oil reservoir temperature. This prevents moisture from condensing in the oil.

Extended shutdowns offer an opportunity to perform scheduled maintenance activities, which can be too time-consuming when the compressor needs to remain available. These activities may include the following:

- Oil sampling and lab testing for contamination of all types. Wear metals in the oil are monitored to evaluate and track system health.
- Checking the calibration of safety and operating controls.
- Checking the calibration of all pressure transducers.
- Checking the calibration of all temperature sensors.
- Inspecting and calibrating pressure relief valves.
- Changing the oil and oil filters.
- Changing the refrigerant filter-driers.
- Changing the oil reclaim system filters and/or strainers.
- Inspecting and cleaning the refrigerant float valve or other metering device.
- Inspecting and cleaning the evaporator and condenser tubes.
- Inspecting and cleaning the contacts of high-current contactors and starters.

On Site

High- and Low-Side Float Valves

High- and low-side float valves are the most common metering devices used on low-pressure chillers. The high-side float valve reacts to the level of liquid in the condenser, opening and closing the orifice in response to changes in the liquid level. The amount of liquid leaving the condenser is driven by the amount of refrigerant vaporizing in the evaporator and flowing into the suction line. This, in turn, is a function of the heat load applied to the evaporator. As the cooling load increases, more liquid must leave the condenser to keep up. This extra liquid raises the float, lifting the orifice to send more refrigerant to the evaporator.

The low-side float valve is the best control device for use with flooded evaporators. It may be installed directly inside the evaporator or in a separate float chamber outside the evaporator. Note that it is installed on the low-pressure side of the orifice, while the high-side float valve is installed on the high-pressure side. As the load on the evaporator increases, more liquid boils off, reducing the liquid level in the float chamber. The reduced liquid level lowers the float, opens the orifice, and supplies more refrigerant to the evaporator.

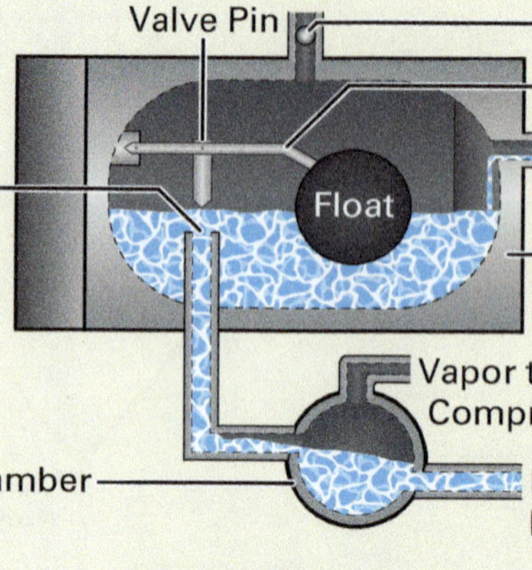

(A) High-Side Float Valve

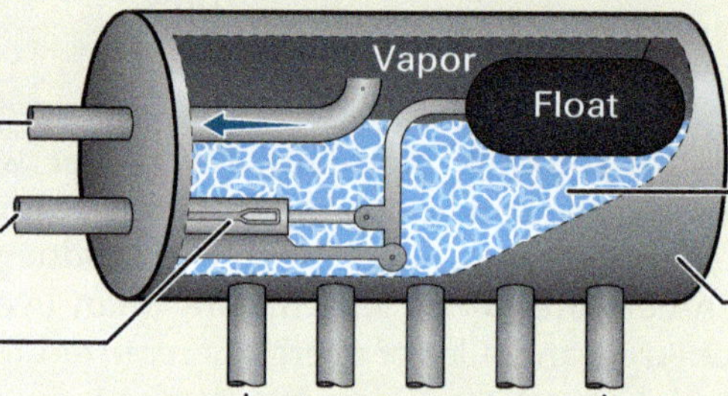

(B) Low-Side Float Valve

2.2.4 Rotary Screw Chillers

Rotary screw chillers are typically used on jobs requiring chiller capacities over 75 tons. The chiller may be water-cooled and located indoors, or air-cooled and located outdoors (*Figure 14*). The evaporator is typically a direct-expansion design with thermostatic or electronic expansion valves.

(A) Air-Cooled Rotary Screw Chiller

(B) Water-Cooled Rotary Screw Chiller

Figure 14 Rotary screw chillers.
Source: Carrier Corporation (A-B)

Like centrifugal chillers, the startup, operation, and shutdown of most screw chillers is highly automated. A microprocessor-based control system monitors and controls all machine operations. The sequence and tasks involved in the startup and shutdown of a screw chiller are basically the same as those described for a centrifugal chiller.

2.3.0 Cooling Towers

Preparing a cooling tower (*Figure 15*) for operation is essential for efficient and trouble-free system operation. Important health concerns related to cooling-tower water systems must also be considered.

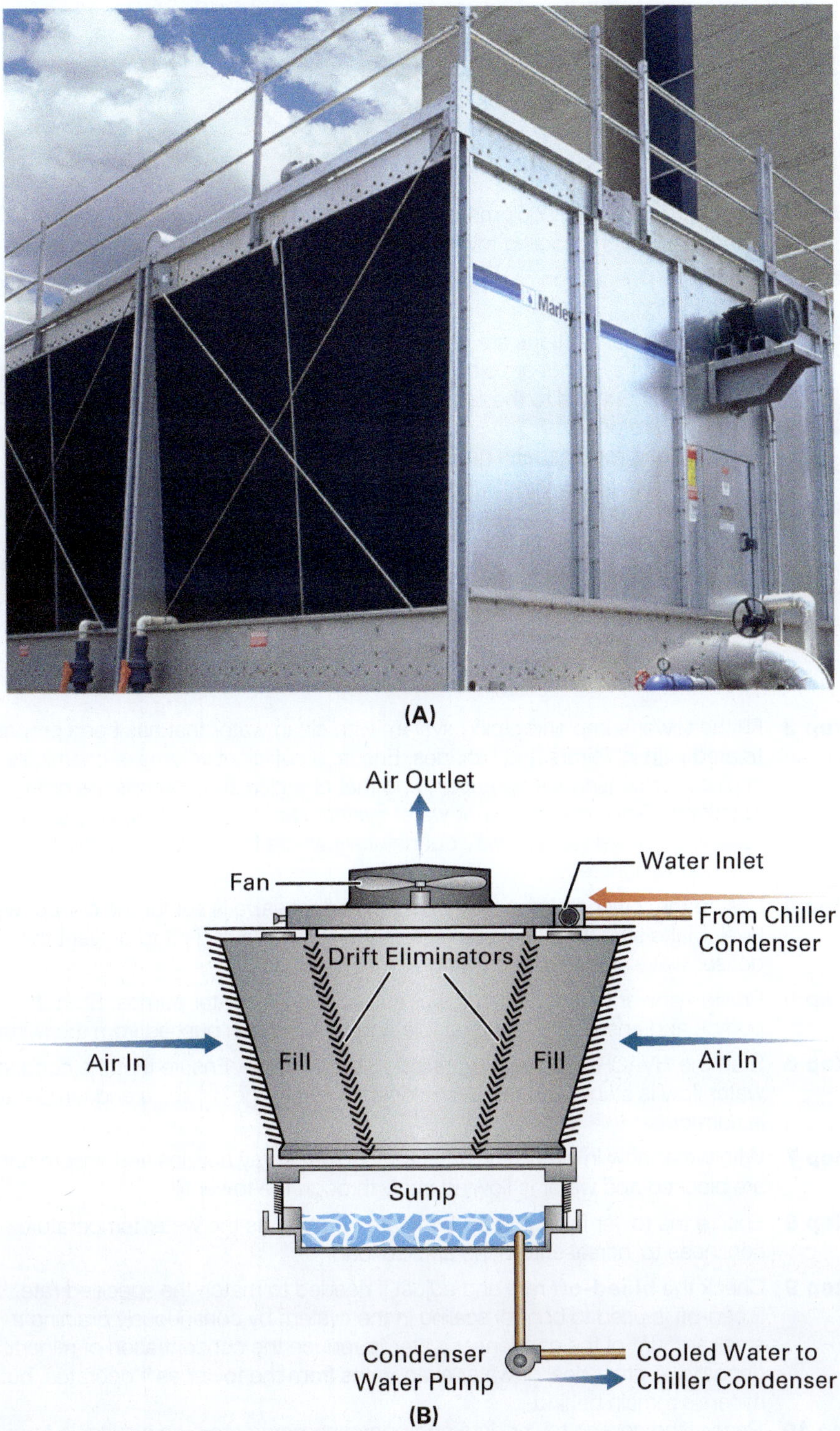

Figure 15 Typical induced-draft cooling tower.

Poorly prepared and maintained cooling towers are known to support the growth of *Legionella* bacteria, the cause of Legionnaires' disease. The aerosol produced by a cooling tower has the potential of infecting not only the people near the equipment, but also the surrounding area. Although HVACR service technicians rarely have direct responsibility for water treatment, they must be vigilant and ensure that cooling towers are being treated and the equipment is functional.

WARNING!

Preparing a cooling tower for startup involves working with energized equipment, including pumps and large fans or blowers. Follow all applicable manufacturer's instructions, OSHA regulations, and jobsite requirements.

2.3.1 Cooling Tower Startup

Step 1 Check that all tools, equipment, packaging, and shipping components are removed from the cooling tower.

Step 2 Perform an inspection of the cooling tower and condenser water piping to verify the following:

- All piping connections are complete and secure. System strainers and/or filtration systems are clean and installed. Balancing and isolating valves are set or opened to the required positions. Shutoff valves at the unit condenser are open.
- Pump and motor shafts have been aligned, and bearings have been lubricated. Ensure the direction of rotation is correct.
- Tower fan motors and drives are ready for operation, and drive belts are at the proper tension. Fan blade assemblies are secure and free to rotate. All drive guards are in place and secure. Ensure the direction of rotation is correct.
- There is no apparent mechanical damage to the tower. Tower components, such as the sump and drift eliminators, are free of visible biological growth.
- The sump heater is in place and functional. Test the thermostat.

CAUTION

Thoroughly clean the piping system of protective oils, grease, welding flux, dirt, and other debris following a new installation. This work must be done before the system is filled with water that will remain in circulation.

Step 3 Fill the tower sump and piping system with clean water that has been properly treated with inhibitors and biocides. Ensure a sufficient volume of chemicals is in place for continuous operation, and that chemical feed pumps are online and functional. Since cooling tower water systems are open, air bleeding is not necessary, as air will work its way out naturally where the water is directly exposed to the atmosphere.

Step 4 Ensure the tower makeup water control or float valve is set for the correct water level. Make sure that the water level in the sump is sufficient to prevent the condenser water pump from drawing air into the water.

Step 5 Provide power to the cooling tower and condenser water pumps. Start the pumps and ensure the flow is stable and air has been pushed from the system.

Step 6 Start the HVACR equipment connected to the tower. Ensure that the condenser water flow is stable and that the refrigerant discharge pressure and temperature is correct.

Step 7 With water now in circulation, check the tower spray nozzles and ensure none are blocked and water is flowing freely through the tower fill.

Step 8 Ensure the tower fans start and operate normally as the water temperature continues to increase from the applied load.

Step 9 Check the **bleed-off** rate and adjust if needed to match the specified rate. Bleed-off is used to control scaling in the system by continuously draining a small amount of the condenser water to reduce the concentration of mineral deposits. A great deal of water evaporates from the tower as it operates, but the minerals remain behind.

Step 10 Record the cooling tower data and operating parameters on a suitable form. The records should include the information summarized in *Figure 16*.

Bleed-off: Consistently draining (bleeding off) a controlled amount of water from a water circuit to carry away minerals and other impurities that remain as water evaporates from a cooling tower or evaporative condenser, reducing the concentration of the minerals in the remaining water.

Cooling Tower Data

Location: ______________________ Manufacturer: ______________________

Model Number: ______________________ Serial Number: ______________________

Pump Data

Make/Model Number: ______________________ Motor HP/rpm: ______________________

Serial Number: ______________________ Volts/Phase/Hertz: ______________________

Motor Make/Frame: ______________________ Gallons per Minute (gpm): ______________________

Fan Data

Number of Fan Motors: ______________________ Motor Sheave Diameter/Bore: ______________________

Motor Make/Frame: ______________________ Fan Sheave Diameter/Bore: ______________________

Motor HP/rpm: ______________________ Sheave C/L Distance: ______________________

Volts/Phase/Hertz: ______________________ Number Belts/Size: ______________________

Operating Water Flow Parameters

Entering/Leaving Water Pressure: ______________________ Water Temperature Difference: ______________________

Water Pressure Difference: ______________________ Flow in Gallons per Minute (gpm): ______________________

Entering/Leaving Water Temperature: ______________________ Bleed-Off gpm: ______________________

Figure 16 Cooling tower operating data.

Going Green

Bleed-Off Water Control

Older cooling tower systems were often fitted with a simple manual valve to allow for bleed-off. In many cases, the valve was set by HVACR technicians to a flow rate far greater than that required. In some cases, calculations were never made to determine the correct volume, nor was the volume of water being bled off ever measured. If the objective of reducing impurities was reached, this method seemed sufficient, but it undoubtedly wasted a great deal of water.

Today, automated controls coupled with precise methods of monitoring the water conductivity allow the bleed-off rate and volume to be accurately controlled. The devices control a solenoid valve in the bleed-off line, opening it only when impurity levels indicate the need for dilution. The result is that millions of gallons of precious water are saved each year.

Because of the conditions under which cooling towers operate, periodic maintenance must be completed faithfully to reduce corrosion and scale buildup and to eliminate the growth of algae, bacteria, and other living organisms. The tower manufacturer furnishes service manuals that recommend scheduled maintenance tasks and intervals.

2.3.2 Cooling Tower Shutdown

The cooling tower and condenser water system can be prepared for an extended shutdown as follows:

WARNING!

Use a face shield, goggles, rubber gloves, and other PPE appropriate for the work and chemicals involved when shutting down and cleaning cooling towers. Biological growth and water treatment chemicals both represent hazards.

Step 1 De-energize the cooling tower equipment, including the water treatment pumps and controllers.

Step 2 If the makeup water line is exposed to freezing temperatures, turn off the water supply and drain the supply line. Drain all recirculating condenser water lines. Be sure to follow all local, state, and federal laws pertaining to the discharge and disposal of the drained water, based on its contents.

Step 3 Remove and clean strainers and filtration equipment (if applicable).

Step 4 Clean and service the tower makeup water controls or float valve.

Step 5 Drain the condenser water pumps. Inspect and perform any scheduled maintenance on the pumps and motors specified in the pump manufacturer's service literature. Manufacturers often recommend that the pump be filled with a glycol solution that also contains corrosion inhibitors.

Step 6 Inspect and perform any periodic maintenance recommended by the manufacturer for the tower fans and fan motors.

Step 7 Inspect and clean the tower distribution basin, spray nozzles, fill, drift eliminators, and sump.

Step 8 As needed, paint metal parts and components to prevent corrosion. Use a paint that is suitable for consistent water contact.

Step 9 When practical, to reduce the airborne dirt and debris carried into the tower during shut down, cover or screen the fan and louver openings.

2.3.3 Cleaning Water-Cooled Condensers

Water-cooled shell and tube condensers served by open cooling towers are subject to contamination and scaling. Condensers and similar heat exchangers should be periodically inspected and cleaned per the manufacturer's instructions.

The amount of fouling and scale buildup encountered depends largely on the water temperature and the concentration of minerals contained in the water. Mineral concentrations in condenser water are often higher than in normal water, despite using bleed-off to reduce the concentration. Some of the entrained minerals break down into insoluble compounds when warmed. The minerals precipitate out of the water and stick to the tubing walls. The higher the temperature to which the water is heated, the more rapidly the scale develops.

Scale on condenser tubes does not form in a uniform manner. Heavier scaling takes place on the tubing immediately adjacent to the cold-water inlet, where the temperature difference is highest. As the scale builds up at the entry of a tube and insulates it, the higher temperature difference moves further back into the condenser and heavier scale begins to develop. This is referred to as *progressive scaling.* An inspection of the tubing at the cold-water inlet generally provides a good indication of how much scale has developed. The rest of the condenser will usually have less scale in comparison.

Typically, shell and tube condensers and evaporators are cleaned at least once a year. The environmental conditions may require the condenser tubes to be cleaned more often. The tubes are cleaned using rotating brushes designed to remove scale without damaging the tube surface.

Figure 17 shows a popular tube cleaning kit. Several different types of brushes are required, depending on the debris to be removed. The brush is mounted to

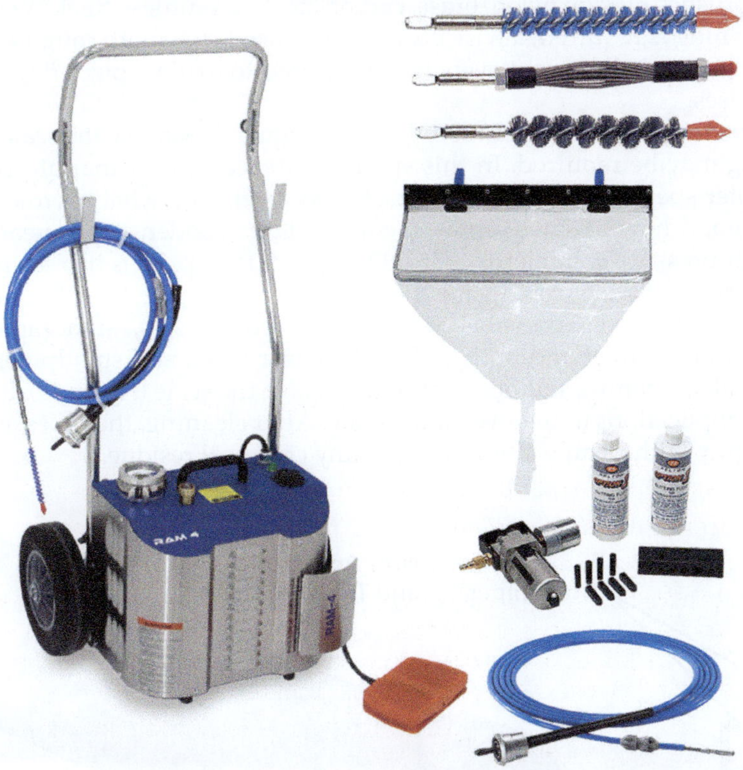

Figure 17 Contractor's tube cleaning kit.
Source: Goodway Technologies Corp.

the flexible shaft on one end, while the other engages a square drive mounted on the machine. A water source is also attached to the machine. The machine rotates the brush while also pumping water through the shaft casing, flushing the debris out as it breaks free of the tube (*Figure 18*). The operator controls the machine with a foot pedal.

Figure 18 Cleaning chiller condenser tubes.
Source: Goodway Technologies Corp.

A wide variety of brushes and other driven tool attachments complete the setup. Brushes may be nylon, brass, carbon steel, or stainless steel. When aggressive cleaning is required, a wire buffer can be used to smooth rough spots in the tubes. Special drill bits are even available, used to drill through deposits in the worst cases.

If brushing the condenser tubes fails to adequately remove the scale, chemical cleaning may be required. In this situation, the condenser manufacturer and/or a water specialist should be contacted to determine what chemicals should be used and how to safely apply them. For large condensers or heat exchangers, high-pressure water jetting (10,000 psig to 20,000 psig) is increasingly being used to remove scale from inside the tubes.

For smaller, non-accessible water-cooled condensers, standpipes can be installed in the inlet and outlet. The condenser tubes and standpipes are then filled with a cleaning solution that will dissolve the scale (*Figure 19*). It may take several applications to remove all the scale. After cleaning, the tubes are flushed thoroughly with clean water to remove any chemical residue.

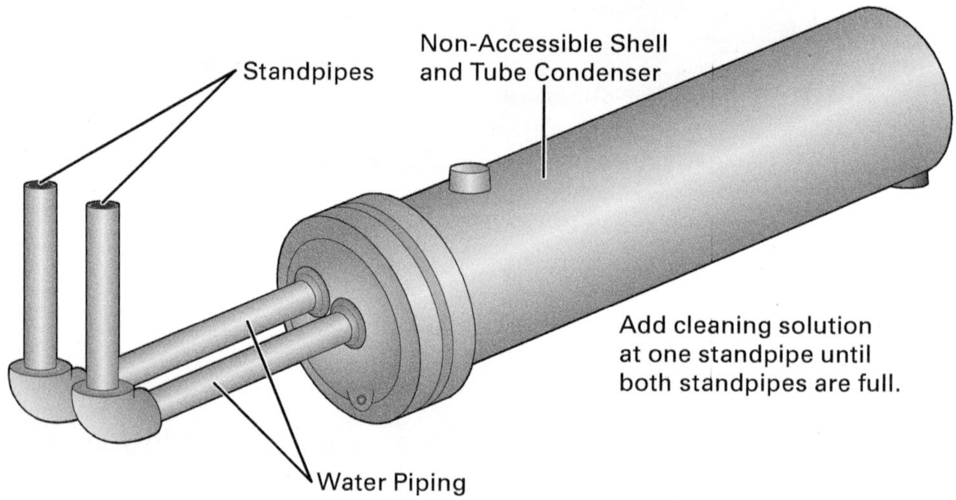

Figure 19 Non-accessible condenser cleaning setup.

2.0.0 Section Review

1. For what purpose must power be provided to reciprocating and scroll chillers well in advance of startup?
   a. To energize the crankcase heaters
   b. To help humid air escape from the control panel
   c. To ensure the electronic components work reliably
   d. To increase the pressure of the charged refrigerant

2. When a recycle shutdown mode is initiated, which of the following best describes the status of the chiller?
   a. It is under a light cooling load.
   b. It is under a heavy cooling load.
   c. The chilled-water temperature is excessively low.
   d. The chilled-water temperature has risen well above the setpoint.

3. What will cause the condenser water pump to draw air into the water from a cooling tower sump?
   a. Scale on the fill
   b. Blocked spray nozzles
   c. Incorrect float valve setting
   d. Misaligned drift eliminators

3.0.0 Air Handling and Rooftop System Startup

Objective

Explain how to start up air handling units and packaged rooftop units.

a. Explain how to start up an air handling unit.
b. Explain how to start up packaged rooftop units.

Performance Task

2. Prepare and complete the startup procedures for an air handling unit or packaged HVAC unit.

Larger air handling units are used in commercial applications and may be connected to a variety of heating and cooling equipment. The blowers in larger air handling units are typically belt driven, although modular panel fans driven by electronically commutated motors (ECMs) are now offered on the market. Packaged rooftop units are self-contained and generally provide heating and cooling. Like other HVACR products, these units must be started up properly to ensure reliability and good performance.

3.1.0 Air Handling Units

The air handling unit and air distribution systems must be prepared and started before condensing units, chillers, and boilers can be brought on line. Condensing units, for example, cannot be operated without some load on the evaporator coil. Although chillers and boilers can be started to some degree before the air distribution system is functional, the load imposed is minimal and a proper startup cannot be completed until a consistent load is present.

3.1.1 Air Handling Unit Startup

The following procedure describes the startup of a typical air handling unit (*Figure 20*) connected to a constant-volume, single-duct air distribution system. However, it is important to note that air handling units come in many shapes and sizes and can be equipped with a wide variety of features and integrated systems. The startup of an air handling unit should always be performed per the manufacturer's instructions, with the requirements for all connected

Figure 20 Typical air handling unit with chilled-water and hot-water coils.

Did You Know?

Air Handlers and Fan Coil Units

While you may hear these two terms used interchangeably, air handlers and fan coil units represent different types of equipment. An air handling unit is typically referred to as *applied equipment*, and is a more complex and full-featured piece of equipment than a fan coil unit. Air handlers can generally be customized by the manufacturer to provide varying amounts of air volume and heating/cooling capacity to suit the needs of an engineer for a specific job.

Fan coil units are considered *unitary equipment* and are generally smaller than air handling units, but there is some overlap with air handling units in capacity. Fan coil units are stock items that cannot be customized, with the exception of the size of the optional electric heat package.

WARNING!

Startup of an air handling unit and air distribution system involves working on energized equipment with rotating and moving parts. Follow all manufacturer's instructions, OSHA regulations, and jobsite safety requirements.

equipment relevant as well. Before starting an air handling unit, or *air handler*, perform the following preliminary checks:

- Obtain the building plans and specifications, locate all system components relevant to the air handler, and familiarize yourself with the layout.
- De-energize, lock, and tag all electrical power and other sources of energy while completing the preliminary inspection.
- Inspect all fans and blower assemblies. Verify the following:
  - All shipping restraints, protective covers, and packaging materials have been removed.
  - Blower wheels spin freely, are secured, and are correctly positioned on the shaft.
  - Drive belts are aligned and have the correct tension; drive motors are securely fastened.
  - Belt guards are in place.
- Visually inspect the complete air distribution system including the complete return air duct system. Verify the following:
  - All tools, equipment, and debris have been removed.
  - The ductwork is complete and sealed, all end caps are in place, and all access doors are closed and secured. Note that duct leak testing must also be completed for many jobs.
  - All balancing, volume, and terminal dampers are open, or they are in the specified positions.
- Verify that all the air filters are clean, in place, and are the correct size and specified type, providing complete coverage of the return air inlet.
- Ensure the air handling unit coils are undamaged and clean, and that all piping connections are complete and insulated as required. Verify that the condensate drain pan opening and installed drain line are clear.
- Ensure all automated dampers, such as return air and outside air dampers, are open or correctly set so that air travels through the system as expected. Outside air dampers typically remain closed when the blower is off.

Once the preliminary checks have been successfully completed, proceed with the startup as follows:

Step 1 Remove tags and locks as necessary, then energize and enable the air handling unit. Check the voltage supplied to the unit and ensure it is within the motor specifications.

CAUTION

Proper operation of motorized system and air handling unit dampers during startup is critical. If dampers are closed or restricting airflow, the casings, housings, and ductwork can be damaged from differential pressures created by powerful blowers. When starting a new system, a temperature control or building automation system (BAS) specialist or system designer should be available to make sure all automated dampers are properly programmed and positioned.

Step 2 Select the mode of operation that calls for maximum airflow. Normally, this is the cooling mode with wetted coils, meaning that the associated chiller or condensing unit must also be operational to allow condensate to develop.

Step 3 Ensure the blower starts, and the direction of rotation is correct. Note that some systems also have a return air blower to overcome the resistance of lengthy return air ducts (*Figure 21*). Ensure it is functioning properly as well.

Figure 21 Return air blower.

Step 4 Check that all automatic dampers are under control and are in the proper position.

Step 5 Measure and record the operating voltage and current of the blower motor(s). Calculate both current and voltage imbalance. If the amperage exceeds the motor nameplate full load amperage, stop the fan to determine the cause and make the necessary adjustments. After any adjustment in supply fan speed, always recheck the current draw to ensure the motor is not overloaded. Remember that all blower motor electrical-current measurements must be made with the access doors closed.

Step 6 Measure and record the rotational speed of the blower(s). If the fan is not running at the specified speed, or as close as standard pulleys will allow, the problem must be corrected before proceeding further. Adjustments in blower speed and/or variable frequency drive (VFD) settings may also be made when the system is air balanced.

3.1.2 Air Handling Unit Shutdown

The shutdown of an air handler is relatively simple. The outdoor air dampers are driven closed when the air handler is off to prevent cold air from leaking into the building, and to protect chilled- and hot-water coils from freezing temperatures.

During extended periods of shutdown, perform scheduled maintenance and cleaning as recommended by the manufacturer.

3.2.0 Packaged HVAC Units

Packaged air conditioning units provide heating, cooling, and airflow from a single unit. The unit shown in *Figure 22* is a 20-ton unit with 235,000-Btuh heating input. It is equipped with an optional economizer. When in the economizer mode, one or more compressors do not operate, thus saving energy. When the economizer is active, the building pressure rises, so relief air dampers or a powered exhaust fan is needed to relieve excess pressure, unless building pressure is managed some other way.

Figure 22 Typical packaged unit.

Packaged units have traditionally used standard direct-drive permanent split capacitor (PSC) or three-phase blower motors with larger blowers using belt drives. However, larger ECM motors are now available, and more packaged units are being delivered with ECM-driven blowers. As shown in *Figure 23*, the blower assembly is significantly different than a traditional blower assembly, as the blower uses a vane axial design as opposed to a centrifugal blower wheel.

Figure 23 Packaged unit with ECM-driven blower.
Source: Carrier Corporation

3.2.1 Packaged Unit Preparation

The startup procedure outlined here is relevant for many packaged units connected to an air distribution system. A startup in both the cooling and heating modes is presented. The process begins with a preliminary inspection as follows:

WARNING!

Startup of a packaged unit involves working on energized equipment with rotating, pressurized, and moving parts. Follow all manufacturer's instructions, OSHA regulations, and jobsite safety requirements.

- Obtain the building plans and specifications, locate all system components relevant to the unit, and familiarize yourself with the layout.
- De-energize, lock, and tag all electrical power and other sources of energy while completing the preliminary inspection.
- Inspect the condenser fan(s) and blower assembly. Verify the following:
  - All shipping restraints, protective covers, and packaging materials have been removed.
  - Blower wheels spin freely, are secured, and are correctly positioned on the shafts. The blower wheels must not be making contact with the housing. Remember that ECMs typically do not spin freely, due to the permanent magnets in the motor.
  - Drive belts are aligned and have the correct tension; drive motors are securely fastened.
  - Belt guards are in place.
- Inspect the compressors and ensure that the mount bolts have been loosened to allow flexibility in the vibration isolators.
- Connect a set of gauges and ensure the system has a refrigerant charge.
- Check all accessible electrical connections throughout the unit.
- Visually inspect the complete air distribution system including the return air duct. Verify the following:
  - All tools, equipment, and debris have been removed.
  - The ductwork is complete and sealed, all end caps are in place, and all access doors are closed and secured. Note that duct leak testing must also be completed for many jobs.
  - All balancing, volume, and terminal dampers are open or are in the specified positions.
- Verify that all the air filters are clean, in place, and are the correct size and specified type, providing complete coverage of the return air inlet. For packaged units, air filters are often found indoors at return air grilles.
- Ensure the coils are undamaged and clean. Verify that the condensate drain pan opening and installed drain line are clear.
- Ensure all automated dampers, such as the outside air dampers, are open or correctly positioned. Outside air dampers typically remain closed when the blower is off.
- For gas-fired units, open the main fuel supply valves and ensure the gas valves are switched on. Measure the gas pressure and ensure it is within the specifications shown on the data plate. Most gas units are designed for a maximum inlet pressure of $\frac{1}{2}$ psig. Check the gas piping inside the unit for gas leaks.
- Inspect the hot surface ignitor (HSI) and ensure it is intact. If the unit uses spark ignition, ensure the igniter and flame sensor position is correct and the spark gap measurement is as specified.

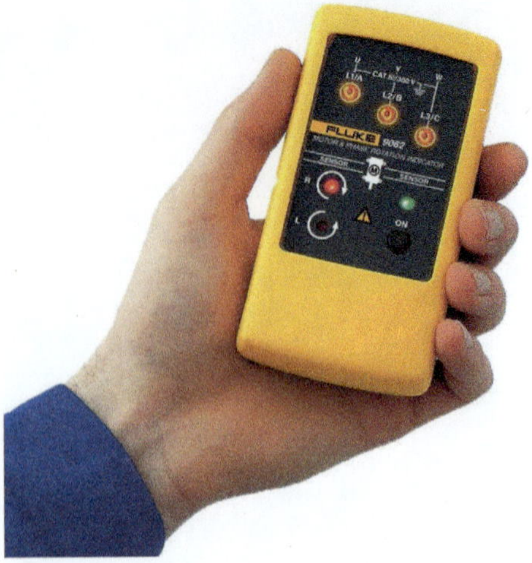

Figure 24 Motor and phase rotation indicator.
Source: Reproduced with Permission, Fluke Corporation

3.2.2 Packaged Unit Startup – Cooling Mode

Once the preliminary checks have been successfully completed, proceed with the startup as follows:

Step 1 Restore power to the unit and check the voltage supplied. Prepare one or more clamp-on ammeters to read the compressor current.

Step 2 Set the external control(s) to start only the indoor blower (mode switch Off and fan switch On or Continuous). The unit blower should start almost immediately, although some units may delay the blower for a short period. Ensure the blower is operating as expected and rotating in the correct direction (three-phase motors). A phasing meter can also be used to check for correct rotation prior to startup (*Figure 24*).

- For packaged heat pumps, the blower will usually operate at the same speed for both the heating and cooling modes. In units equipped with fossil-fuel heating, the blower may operate at a higher speed for cooling. This is less common in commercial applications. It is not uncommon for commercial packaged units to operate at one blower speed for both modes. Belt-driven blowers always operate at the same speed for both modes of operation.

Step 3 Reset the fan switch to Auto on the external controls. Then change the operating mode to cooling and ensure the setpoint is well below the room temperature.

Step 4 The compressor(s) and indoor fan should start. Ensure all three-phase scroll compressors are rotating in the correct direction by monitoring system suction and discharge pressures. Normal suction and discharge pressures usually indicate correct phasing. Also ensure the condenser fan motor(s) start and run. Measure the operating current of each compressor.

Step 5 Ensure the indoor blower started as expected, and the speed and operating current are correct.

Step 6 Check the proper refrigerant charge per the manufacturer's instructions. Check and record the superheat and subcooling after the unit has stabilized.

Step 7 Ensure any motorized dampers have moved to the correct position. Outdoor air dampers, for example, should be at their minimum outside air position per the job specifications.

Step 8 If practical, check the operation of the economizer and relief air dampers, if equipped. Unless it is cool outdoors, some resetting and manipulation of the economizer controls will be needed to activate it.

Step 9 Measure and record all relevant unit information and operating characteristics for the cooling mode.

3.2.3 Packaged Unit Startup – Heating Mode

Many packaged units contain gas heating sections that have similar operating sequences, regardless of the manufacturer. The following is a typical sequence of operation for a packaged unit with electronic ignition and single-stage heat.

1. Thermostat or control system calls for heat.
2. Burner induced draft motor starts and is proven by a pressure switch.
3. Pilot gas valve opens and igniter arcs.
4. Pilot ignites, flame is proved by the flame sensor current, and arcing ceases.
5. Main gas valve opens and burners ignite.
6. Indoor blower starts after a short delay to allow heat exchanger to warm up.

Start up the gas furnace in a packaged unit as follows:

Step 1 Set the external controls to the heating mode.

Step 2 Observe the burner firing sequence and ensure the burner flame is stable.

Step 3 With the burner operating, check for the proper gas manifold pressure using a manometer. Note that the manifold pressure may differ when a multistage gas valve changes from low-fire to high-fire.

Step 4 Ensure the indoor blower started as expected and the direction of rotation is correct. Check the operating current and speed.

Step 5 Measure the temperature rise across the heat exchanger.

Step 6 Ensure all motorized dampers are under control and in the proper position.

Step 7 Measure and record all relevant unit information and operating characteristics for the heating mode.

3.0.0 Section Review

1. True or False? When an air handling unit shuts down, the outside air dampers typically move to the full open position.
   a. True
   b. False

2. When starting a new packaged rooftop unit today, what type of blower might you find in place of a traditional centrifugal blower wheel?
   a. Radial
   b. Spherical
   c. Vane axial
   d. Backward inclined

Module 03406 Review Questions

1. Thermal shock tends to have a more significant effect on areas where dissimilar metals meet and on which of these materials?
   a. Cast iron
   b. All gaskets
   c. Carbon steel
   d. Flue piping

2. When a boiler is prepared for wet storage, a slight pressure should be maintained on the waterside using ______.
   a. oxygen
   b. nitrogen
   c. compressed air
   d. carbon dioxide (CO_2)

3. When preparing a steam boiler for startup, the main steam stop valve should be ______.
   a. fully open
   b. 10% open
   c. 50% open
   d. fully closed

4. When initially firing a steam boiler at startup, the burners should be operated at ______.
   a. low-fire
   b. mid-fire
   c. high-fire
   d. standby

5. When the steam gauge just begins to register pressure during the startup of a steam boiler, ______.
   a. manually test the pressure relief valve
   b. vent the boiler so that no air is trapped
   c. quickly open the main steam stop valve
   d. blow down the gauge glass, water column, and low-water cutoff

6. What is the minimum number of hours that manufacturers usually require the compressor crankcase heaters in a chiller to be energized before startup?
   a. 8 hours
   b. 12 hours
   c. 24 hours
   d. 36 hours

7. When a centrifugal chiller is shutdown and may be exposed to temperatures below 32°F, ______.
   a. the oil charge should be drained from the compressor
   b. the refrigerant should be transferred to a storage tank
   c. a holding charge of high-pressure nitrogen should be added
   d. the water should be drained from the evaporator and condenser

8. Which of the following statements about centrifugal chillers prepared for an extended shutdown is *correct*?
   a. The refrigerant charge should be transferred to a storage tank.
   b. The chiller refrigerant circuit should be maintained in a vacuum.
   c. The entire refrigerant charge should be transferred into the condenser section.
   d. The chiller refrigerant circuit should be filled with oxygen as a holding charge.

9. A small amount of water is consistently bled from a cooling-tower condenser water circuit to reduce ______.
   a. algae growth
   b. the water temperature
   c. mineral concentrations
   d. *Legionella* bacteria growth

10. What accessory do packaged HVAC units often need when they are equipped with an economizer?
   a. Dehumidifier
   b. Relief air dampers
   c. Oversized filter-drier
   d. Electric heating package

Answers to odd-numbered questions are found in the Module Review Answer Key at the back of this book.

Answers to Section Review Questions

Answer	Section	Objective
Section One		
1. a	1.1.1	1a
2. d	1.2.0	1b
Section Two		
1. a	2.1.1	2a
2. a	2.2.2	2b
3. d	2.3.1	2c
Section Three		
1. b	3.1.2	3a
2. c	3.2.0	3b

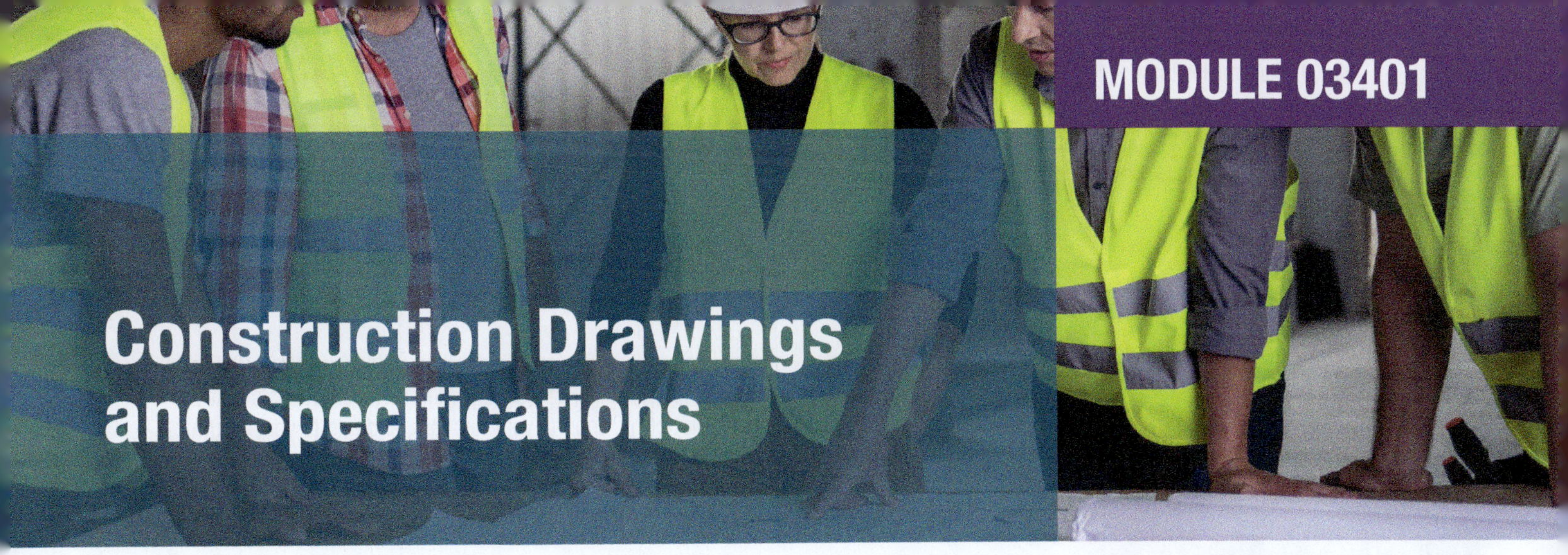

MODULE 03401

Construction Drawings and Specifications

Source: Tashi-Delek/istock/Getty Images

Objectives

Successful completion of this module prepares you to do the following:

1. Describe the types of construction drawings HVACR technicians encounter and use.
   a. Explain how to approach a set of construction drawings for an initial review.
   b. Describe site plans.
   c. Describe plan views, elevations, detail drawings, and section drawings.
   d. Describe mechanical, electrical, and plumbing drawings.
   e. Describe shop drawings.
   f. Describe as-built drawings.
   g. Identify and explain the importance of building codes to the design and construction process.
2. Describe the importance of specifications, submittals, and takeoffs in construction projects.
   a. Describe the contents and role of project specifications in construction.
   b. Describe the purpose of equipment and material submittals.
   c. Describe a takeoff and explain how a simple takeoff is conducted.

Performance Tasks

Under supervision, you should be able to do the following:

1. Identify and interpret information found on architectural drawings such as elevations and floor plans.
2. Identify and interpret information found on mechanical drawings such as air distribution systems and piping drawings.
3. Identify and interpret information found on HVACR-related shop drawings.

Overview

Everyone involved in the installation of HVACR systems must be able to interpret the project drawings and specifications. The drawings show the location of all relevant components, dimensions, and many of the details needed to execute the work. Other important details are found in the project specifications.

During the estimating and planning processes, the drawings are used to identify the amount and types of equipment, accessories, and materials needed for the job. Accurate interpretation of the drawings is essential to price the job, and then to acquire all the needed materials. As a job gets started, everyone is holding a set of plans that are the basis for every move made from start to finish.

Digital Resources for HVACR

SCAN ME

Scan this code using the camera on your phone or mobile device to view the digital resources related to this craft.

NCCER Industry-Recognized Credentials

If you are training through an NCCER-accredited sponsor, you may be eligible for credentials from NCCER. The ID number for this module is 03401. Note that this module may have been used in other NCCER curricula and may apply to other level completions. Contact NCCER at 1.888.622.3720 or go to **www.nccer.org** for more information.

You can also show off your industry-recognized credentials online with NCCER's digital credentials. Transform your knowledge, skills, and achievements into credentials that you can share across social media platforms, send to your network, and add to your resume. For more information, visit **www.nccer.org**.

1.0.0 Construction Drawings

Performance Tasks

1. Identify and interpret information found on architectural drawings such as elevations and floor plans.
2. Identify and interpret information found on mechanical drawings such as air distribution systems and piping drawings.
3. Identify and interpret information found on HVACR-related shop drawings.

Objective

Describe the types of construction drawings HVACR technicians encounter and use.

a. Explain how to approach a set of construction drawings for an initial review.
b. Describe site plans.
c. Describe plan views, elevations, detail drawings, and section drawings.
d. Describe mechanical, electrical, and plumbing drawings.
e. Describe shop drawings.
f. Describe as-built drawings.
g. Identify and explain the importance of building codes to the design and construction process.

This module reviews and builds on the information previously studied in NCCER Module 00105, *Introduction to Construction Drawings*. The module focuses on the complete drawing package and techniques for reading various drawings and project specifications.

Did You Know?

Blueprints

The term *blueprint* is derived from a method of plan reproduction used in the past. A true blueprint shows the details of the structure as white lines on a blue background.

The method involved coating a paper with a specific chemical. After the coating dried, an original hand drawing was placed on top of the paper. Both papers were then covered with a piece of glass and set in the sunlight for about an hour. The coated paper was developed much like a photograph. After a cold-water wash, the coated paper turned blue and the lines from the drawing remained white.

Today, most drawing reproduction methods produce a black line on a white background and are made on larger printers and copiers. However, the term *blueprint* is still widely used, especially by craftworkers with years of experience in the construction industry.

Source: Michael Burrell/iStock/Getty Images

Construction drawings and specifications tell the HVACR technician and installer, as well as other skilled craftworkers, how to build a specific building or structure. The construction specifications are part of the contract documents and are used in conjunction with the drawings. The specifications contain detailed written information and instructions that supplement the information on the drawings. No room is available for this information on the actual drawings.

As an HVACR technician, you must be able to analyze and interpret drawings and specifications. Failure to do so accurately often results in costly rework. Depending on the severity of an error, it can also expose your employer to liability for its effect on the job and the work of other crafts.

1.1.0 Viewing a Set of Drawings

The following general procedure is suggested as a method for reading a set of drawings. Use this procedure to familiarize yourself with a set of drawings with which you are not familiar:

Step 1 Locate and read the title block. The title block identifies the drawing and provides some important information. It contains critical information such as the scale, date of last revision, drawing number, and the name of the architect or engineer. If you must remove and fold a sheet from a set of drawings for portability, be sure to fold the sheet with the title block facing up.

Step 2 Find the North arrow. Always orient yourself to the structure on its site from above. This helps you to visualize and describe the location and direction of other building elements more accurately.

Step 3 Check the list of drawings in the set. Note the sequence of the drawings. Many drawing packages have an index on the cover page. Notice that the prints are broken into several categories, such as the order shown in *Figure 1*. However, drawing sets do not have to contain all the drawing categories shown, nor are the sets always in precisely the same order.

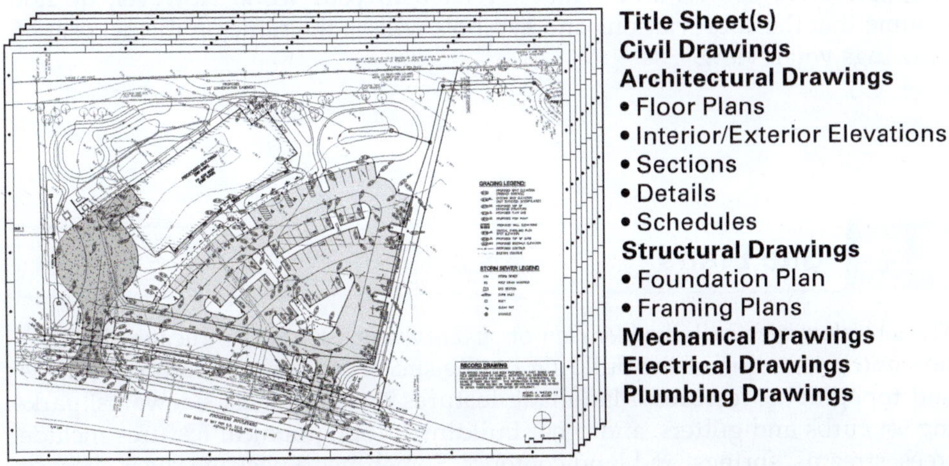

Figure 1 Typical categories of drawings in a drawing package.

Step 4 Study the **site plan** to evaluate the location of the building on the site in relation to other structures or landmarks. Also note that the geographic location of the building may be indicated on the site plan, including its map coordinates.

Step 5 Check the **floor plan** to get an idea of the layout. Try to visualize yourself walking through the area shown. Consider details such as entrances, corridors, and any special features.

Step 6 Study the features that may extend vertically through floors, such as plumbing, stairways, elevator shafts, HVAC ductwork, or chilled-water piping. Determine the location of important elements such as where utilities enter the building and where the mechanical rooms are located.

Site plan: A civil drawing that indicates the location and position of a structure on its construction site.

Floor plan: A plan view showing a horizontal section of the floor arrangement, usually conceived as if the viewer is positioned just beneath the ceiling of the same floor.

Step 7 Review the floor and wall construction details in relation to your area of interest. For example, wall construction has a significant effect on how duct penetrations are created and installed for fire protection.

Step 8 Review the foundation plans and become familiar with how the building is supported and whether any portion of it is below grade or underground.

Step 9 Study the mechanical plans for relevant HVAC and plumbing details.

Step 10 Review the electrical distribution panels and how power is routed to the equipment, as well as the installation of the lighting and other significant power supplies.

Step 11 Check the notes on relevant pages and review the specifications related to relevant construction details. Look for any discrepancies.

Step 12 Thumb through the sheets of drawings until you are familiar with the order and where you can find important details as the project progresses.

1.1.1 Symbology and Abbreviations

Many years ago, when construction drawings were drawn by hand, the use of symbols and abbreviations on the drawings were more standardized. Over the years, computer-assisted drafting/computer-assisted manufacturing (CAD/CAM) software packages were developed and are now as common as computers on desks.

However, the result is that most standardization has been lost as software developers created their own symbology libraries to suit their work and customer needs. The issue is further complicated by the massive growth over the years in the types of components and materials that must be shown in construction drawings. In other words, many symbols used today were unnecessary years ago.

Today, it is essential to review the abbreviations and symbology shown on the drawing set with which you are working. No standardized resource is available on which you can confidently rely. As you review a new set of drawings and initiate the work, you will quickly become familiar with the most common abbreviations and symbols related to your work. However, do not assume that the abbreviations and symbols will be the same on the next set of drawings you review.

1.2.0 Site Plans

The site plan, also called a *plot plan* or *civil drawing* (*Figure* 2), indicates the location and orientation of the building on the site. It can include both man-made and topographic features. Man-made features include roads, sidewalks, parking lot curbs and gutters, and other buildings. Topographical features include trees, streams, springs, and land contours. Sometimes a package cover sheet or site plan contains a large-scale map of the overall area that indicates where the project is located (*Figure* 3).

A *roof plan*, when available, may also be part of the site plan. More complex roof plans, such as those that show the exact positions of HVACR equipment, generally have a place in the drawing package. General notes pertaining to grading and shrubbery may also be included on the site plan, although a separate landscaping plan may also exist.

A utility site plan may also be included in the drawing set. It shows the locations for underground utilities such as gas and water pipelines, sanitary sewers, power transformers and feeds, and communication system cables.

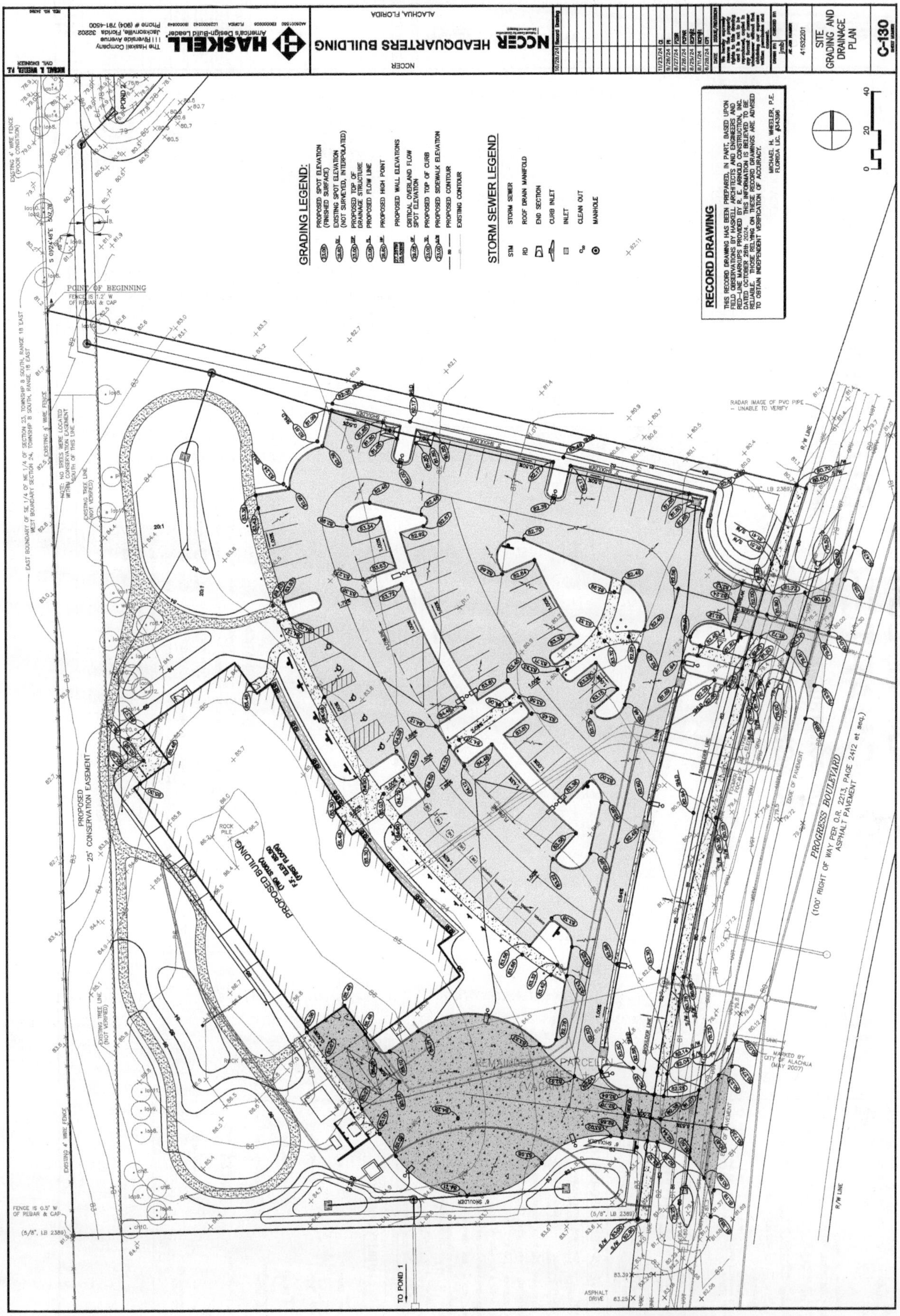

Figure 2 Site plan.

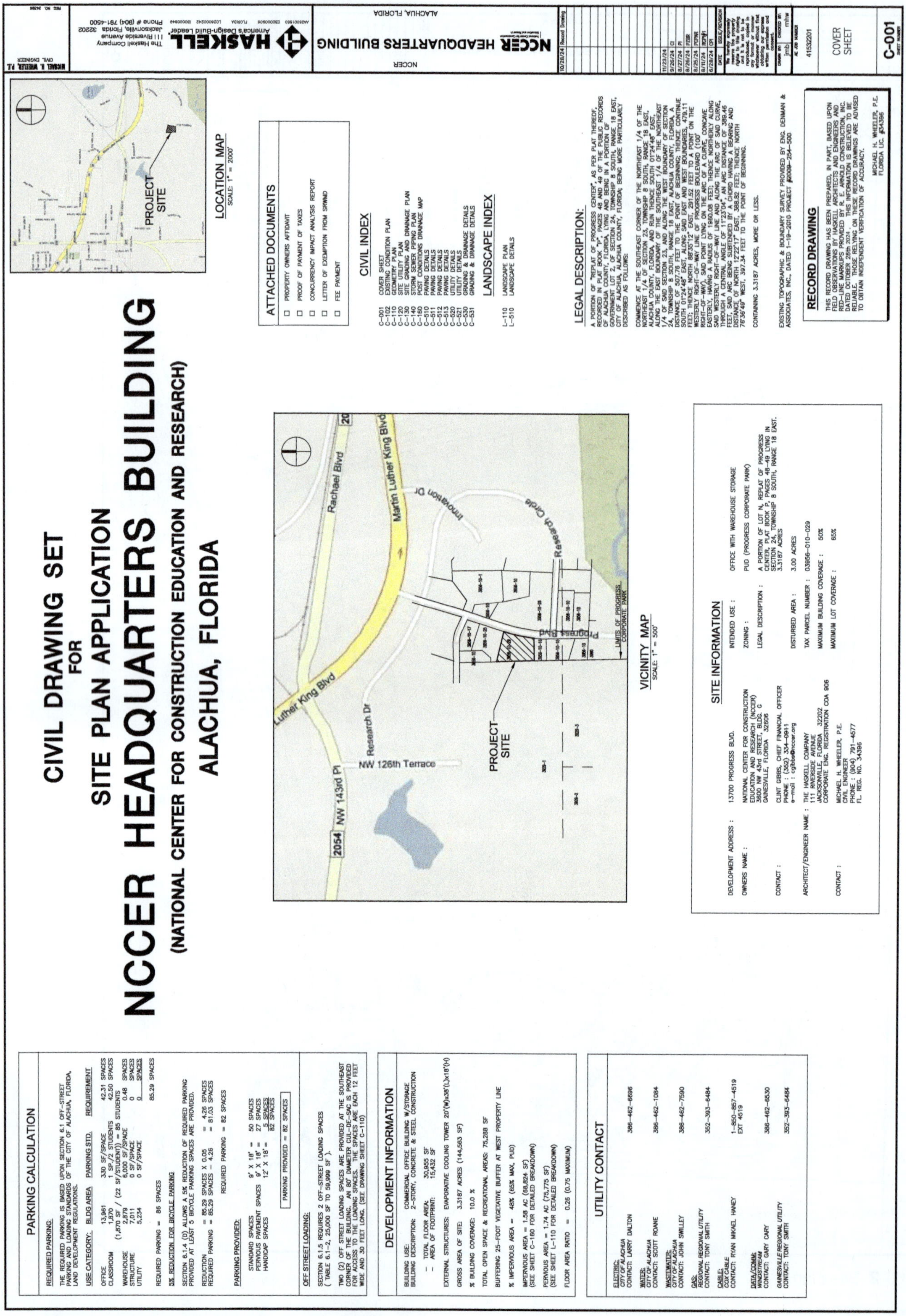

Figure 3 Plan package cover sheet with large-scale site plan.

1.3.0 Plan Views, Elevations, Detail Drawings, and Section Drawings

Several types of drawings are used to show different structural aspects and perspectives of a building. Taken together, the drawings show the full picture of how the building is constructed and provide important details.

1.3.1 Plan Views

Plan views: Views of an object or structure from overhead. A floor plan is a common example of a plan view.

Plan views typically include drawings such as the floor plan (*Figure 4*), roof plan, and ceiling plan. The floor plan is one of the most important working drawings for many crafts. It is often the first drawing done by the architect. It shows the

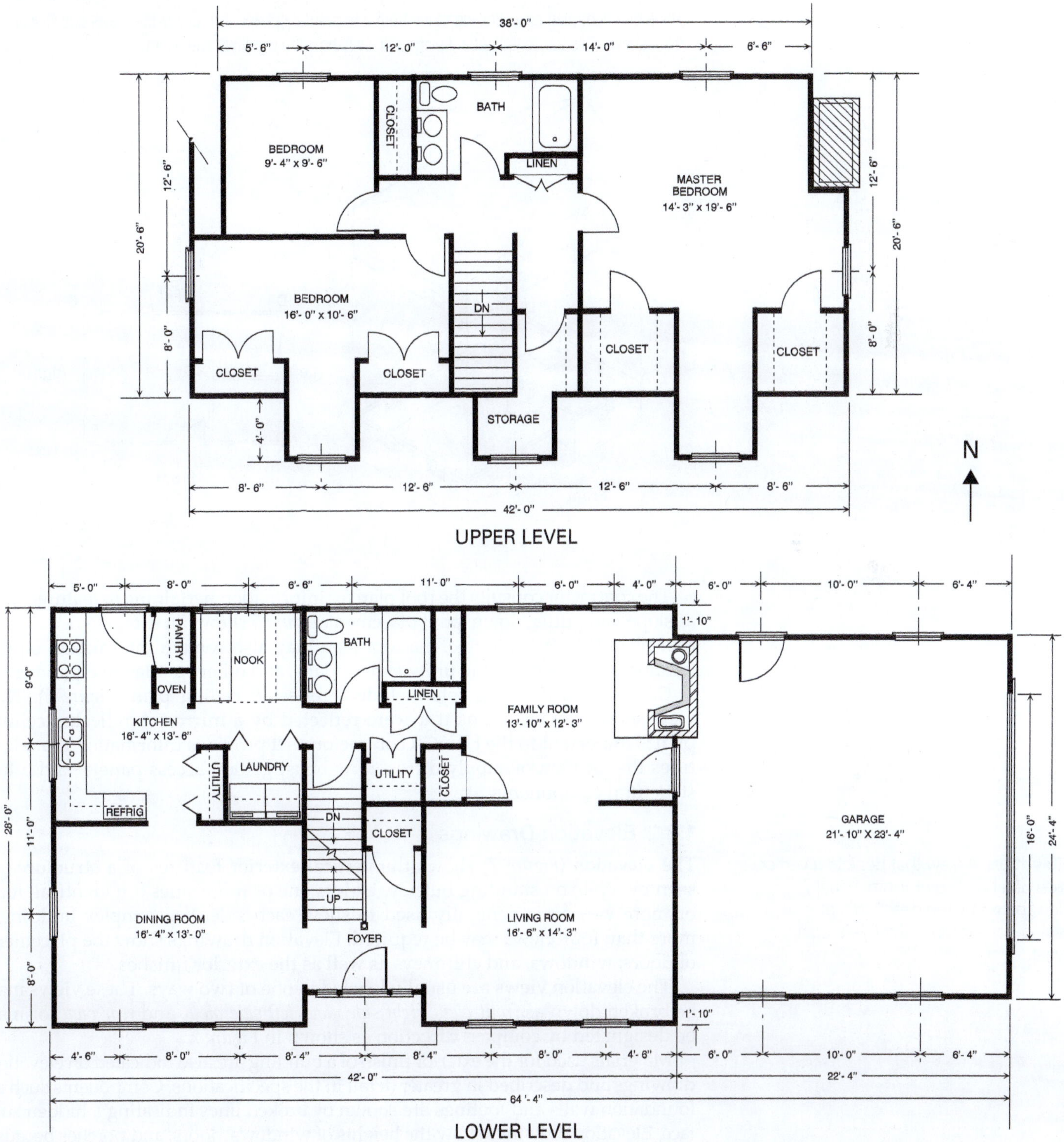

Figure 4 Floor plans for a home.

length and width of the building and the location of the rooms and other spaces within the building. Each floor of a structure has its own floor plan.

A floor plan is a plan view of a horizontal section taken at some distance above the floor, usually midway between the floor and ceiling. However, the cutting plane of the sectional drawing may be offset to a higher or lower level so that it cuts through the desired features such as windows and doors. The cut crosses all openings for that floor or story. This view gives the dimensions of the window and door openings and indicates which way the doors must swing.

On Site

Floor Plans

Think of a floor plan as a drawing made from a horizontal cut about 4' above floor level and viewed from above. This horizontal cutting plane will also show the windows and doors.

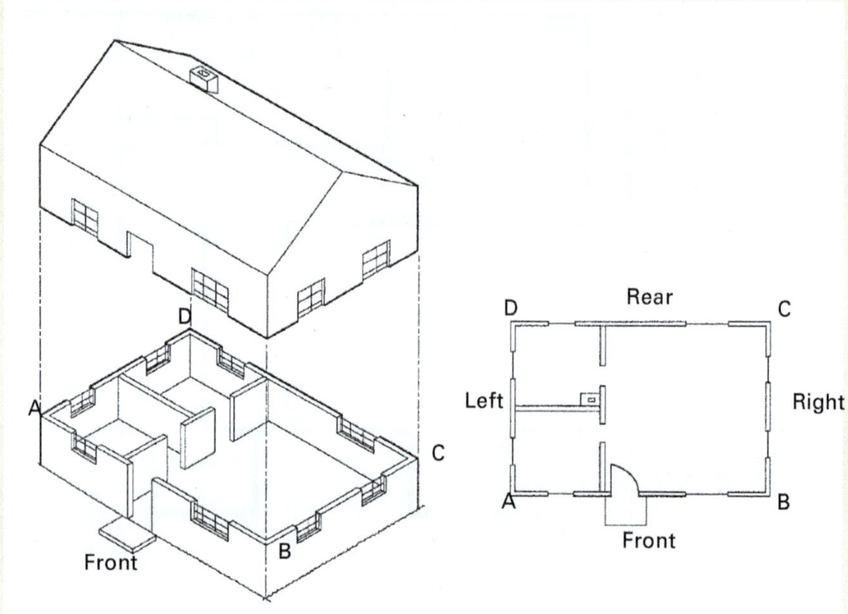

The contractor consults the roof plan for information pertaining to features such as slope and gutter/downspout placement. *Figure 5* shows an example of a typical roof plan. Where applicable, the roof plan may also contain information on the location of air conditioning units, exhaust fans, and other ventilation equipment.

Commercial drawing sets include a reflected ceiling plan (*Figure 6*). This view shows the ceiling as if it were reflected by a mirror. A reflected ceiling plan is important to the HVACR contractor, as it provides information that identifies the location of supply diffusers, exhaust grilles, access panels, and other structural components.

1.3.2 Elevation Drawings

Elevation: A view that depicts a vertical side of a building or room, usually designated by the direction that side is facing; for example, right, left, east, or west.

The **elevation** (*Figure 7*) shows the vertical exterior features of a structure, as seen by a person standing outside. Unless one or more sides are identical, four or more views are generally used to show each side. For complex buildings, more than four views may be required. Elevation drawings show the placement of doors, windows, and chimneys as well as the exterior finishes.

The elevation views are usually labeled in one of two ways. These views may be broken down as *front view, right side view, left side view,* and *rear view,* or may be designated by compass direction as shown in *Figure 7.*

Materials used for the exterior finish of a building are also indicated on elevation drawings and described in greater detail in the specifications. Components such as foundation walls and footings are shown by broken lines indicating a hidden surface. Elevation drawings show the heights of windows, doors, and porches because these measurements cannot always be shown in a practical way on floor plans.

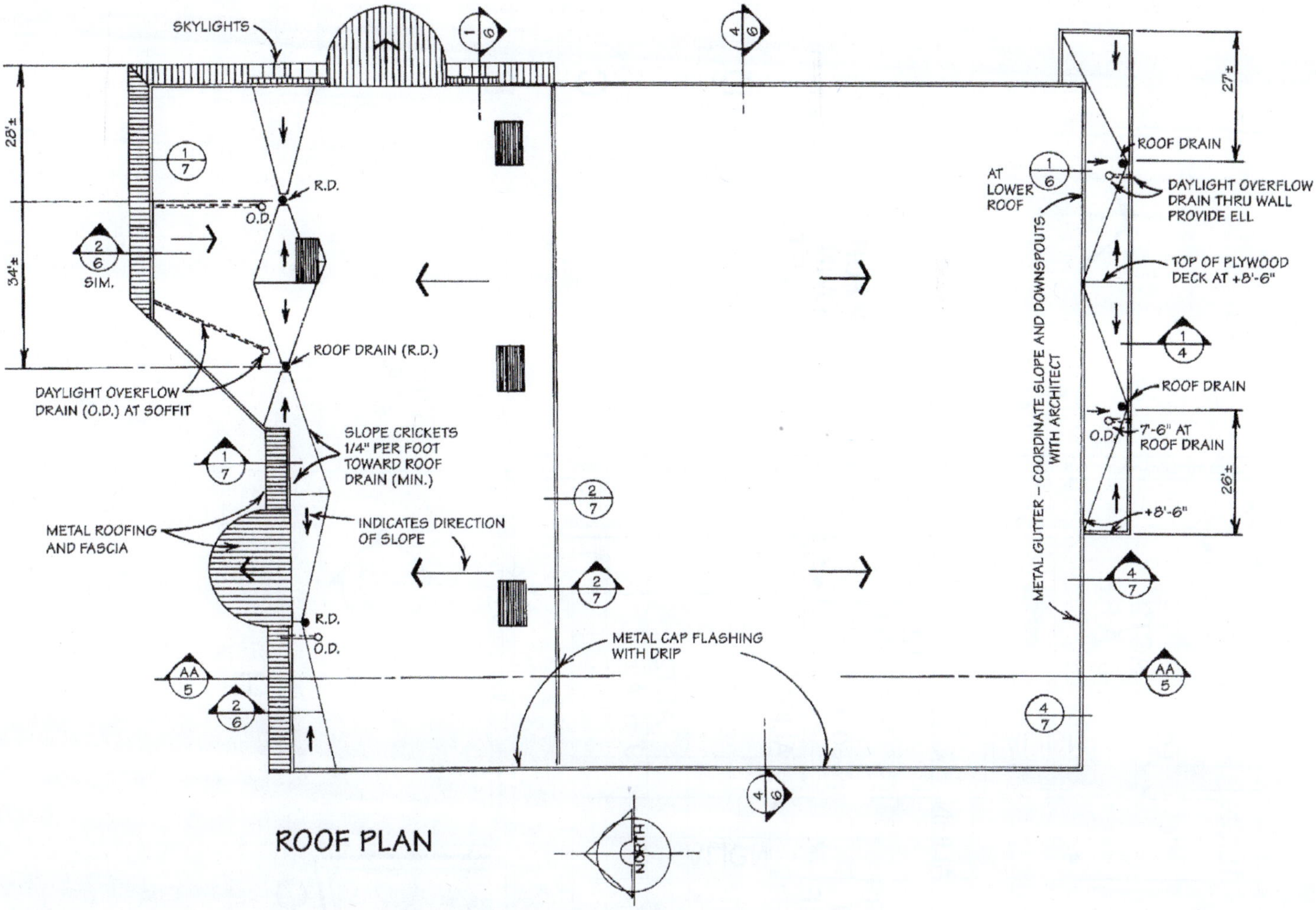

Figure 5 Example roof plan.

On Site

Protection of Underground Utilities

To avoid damage or interruptions to underground utilities, it is mandatory to contact the various companies for utility location and marking prior to any digging. All states have what is known as a *one-call notification center* that makes digging notification easy by calling 811 or contacting the center through a website. The phone number 811, referred to as the *Call Before You Dig* number, is now reserved for this purpose nationally. Typically, a request to mark the location of utilities must be made at least three working days before the digging is to begin.

Pocket cards and similar materials are readily available from state centers to remind contractors and excavators of the requirement and provide other handy information. All utility locators must use the same color code to identify specific utilities. The American Public Works Association (APWA) establishes the universal color codes used to mark underground utilities.

APWA UNIFORM COLOR CODE
FOR MARKING UNDERGROUND UTILITY LINES

WHITE	Proposed Excavation
PINK	Temporary Survey Markings
RED	Electric Powerlines, Cables, Conduit, and Lighting Cables
YELLOW	Gas, Oil, Steam, Petroleum, or Gaseous Materials
ORANGE	Communication, Alarm or Signal Lines, Cables, or Conduit
BLUE	Potable Water
PURPLE	Reclaimed Water, Irrigation, and Slurry Lines
GREEN	Sewers and Drain Lines

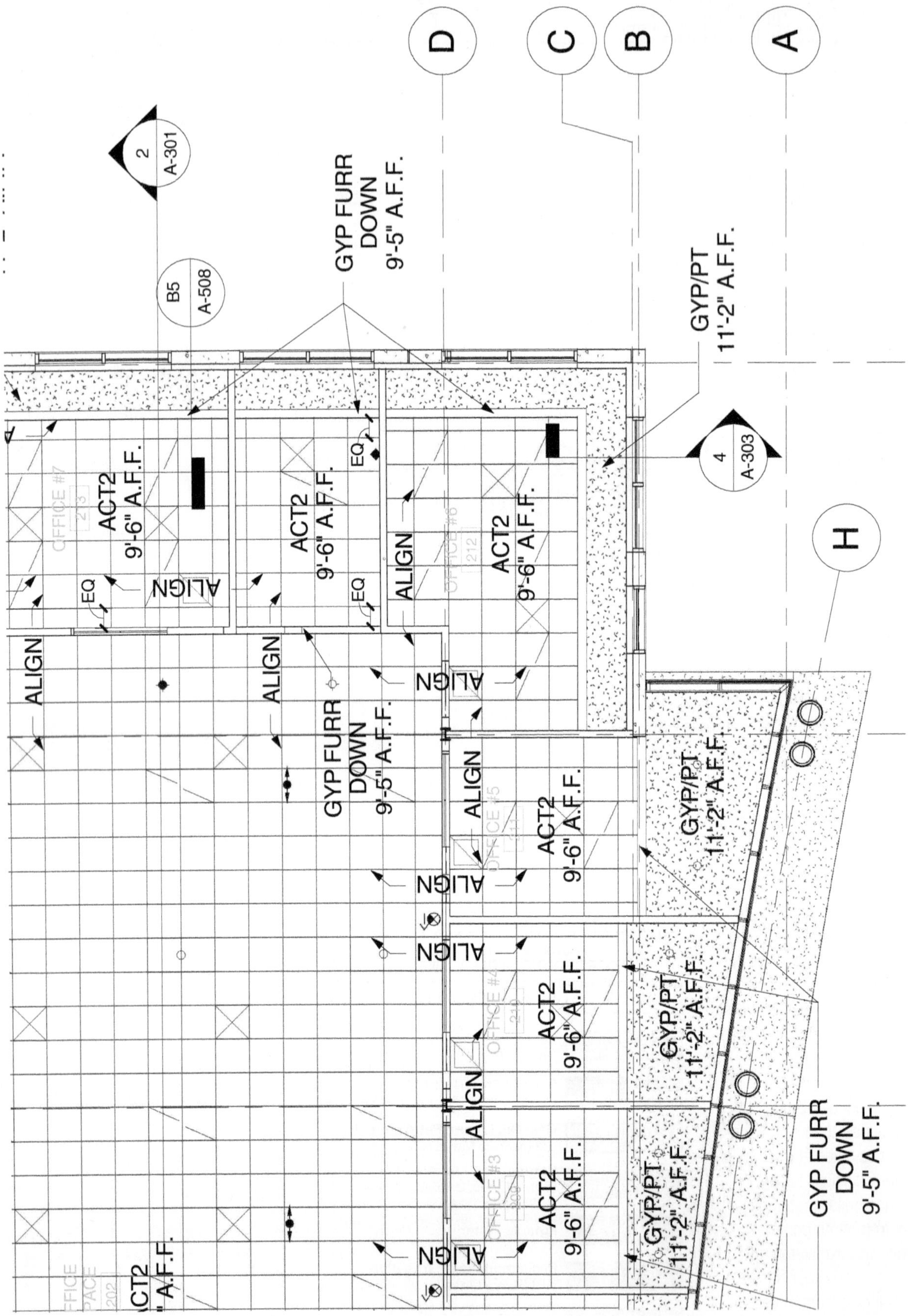

Figure 6 Example reflected ceiling plan.

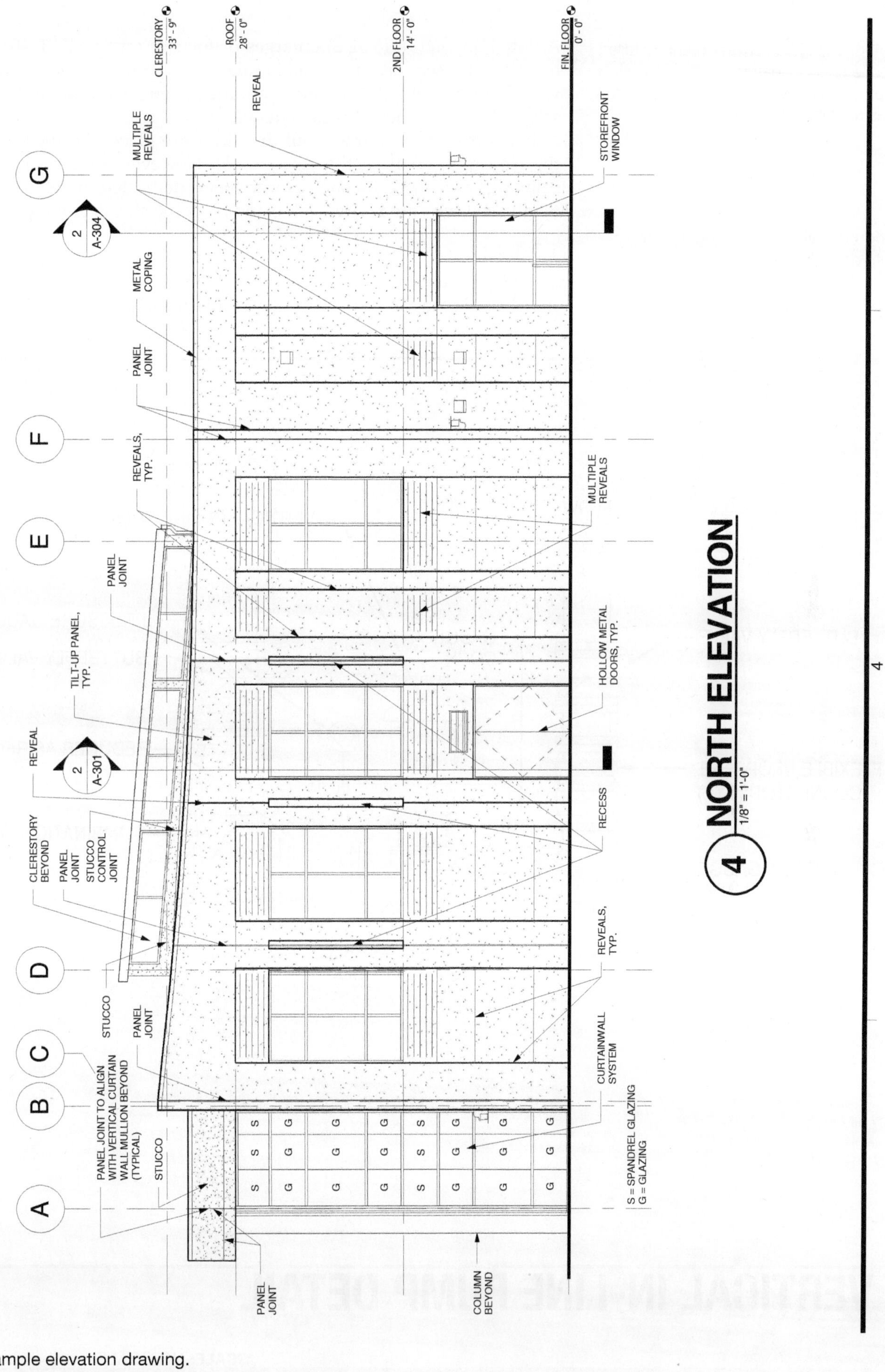

Figure 7 Example elevation drawing.

Detail drawing: An enlarged and detailed view of a specific feature or area of work that provides more elaborate information than is available in a large format.

Section drawing: A drawing that depicts a feature of a building as if it were cut through the middle, exposing otherwise unseen details.

1.3.3 Detail Drawings

A **detail drawing** (*Figure 8*) enlarges or focuses on special features and details that can't be seen at the normal scale of the drawing. Note that the scale can vary from one detail drawing to another, even on the same sheet. The scale is selected to fit the need for each detail.

For some projects, the detail drawings are placed on the same sheet where the feature appears at the larger scale. However, for complex projects, these drawings are placed on one or more sheets dedicated to details. In this case, the detail drawings are referenced to and from the sheets to which they apply.

1.3.4 Section Drawings

A **section drawing** (*Figure 9*) is a cut-away view that allows the viewer to see the inside of a structure or how something is put together internally. A

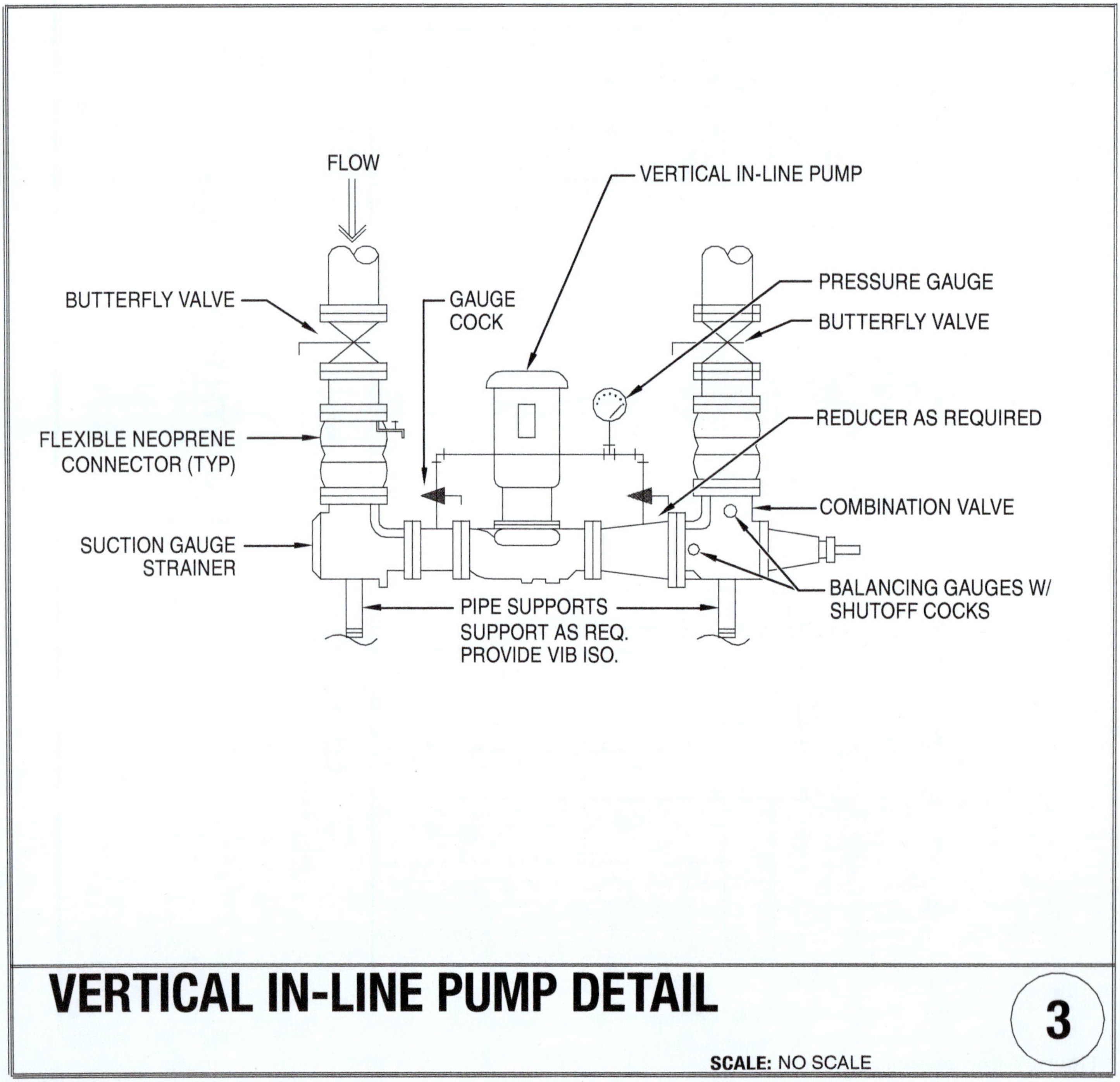

Figure 8 Example vertical inline pump detail.

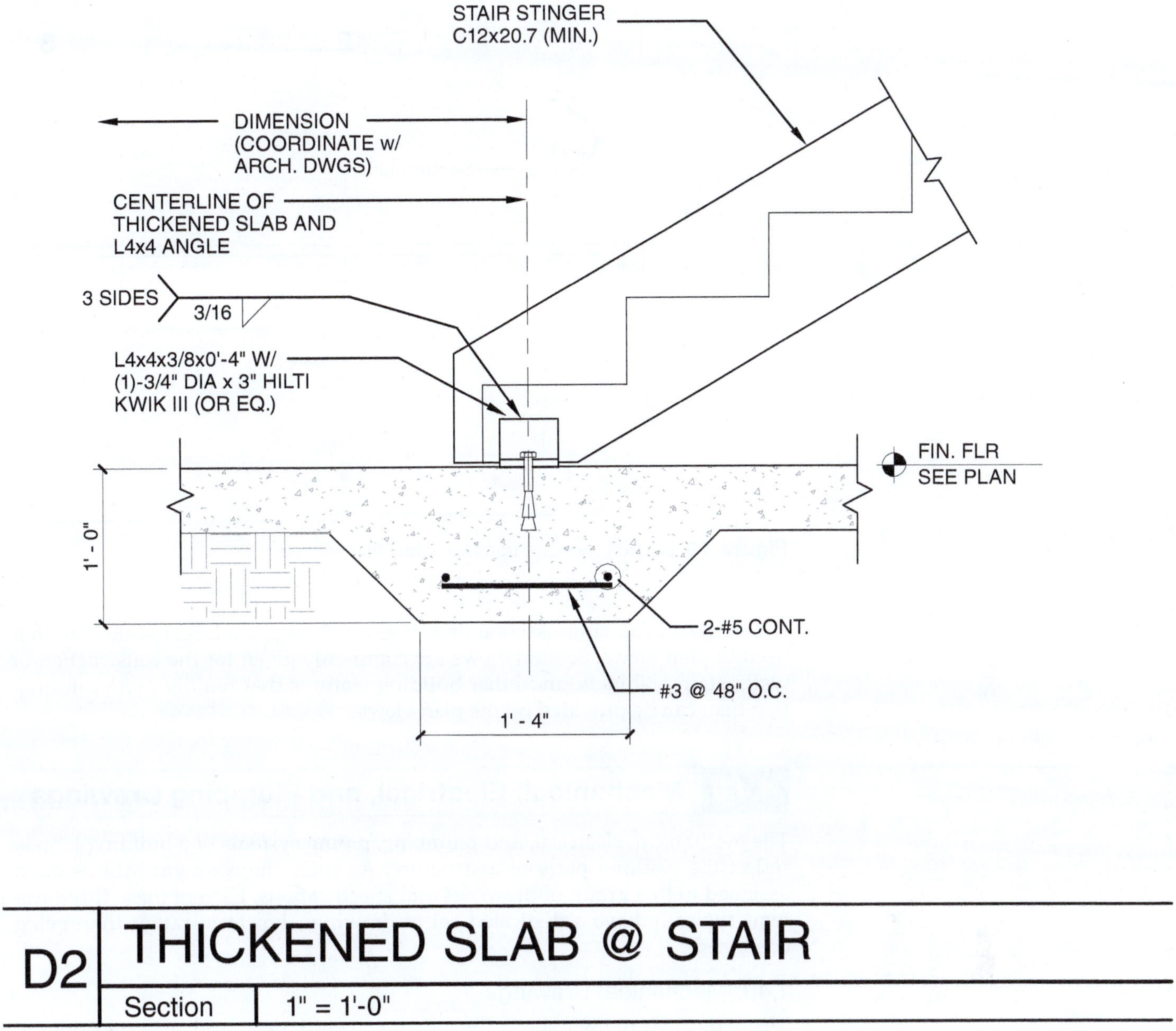

Figure 9 Section drawing showing the thickened and reinforced slab at the base of stairs.

specific area or feature is drawn as if a cut has been made through the middle of it. Note that many section drawings are also detail drawings, and vice versa.

When a sectional cut is made along the long dimension of a building, it is called a **longitudinal section**. When it is made through the short dimension of a building, it is called a **transverse section**. The point on the original drawing that the section was taken from, showing where the imaginary cut has been made, is indicated by a *section line*. A section line is a special line with arrows on each end to indicate the direction of view. Letters and numbers identifying the section drawing are placed on the arrows or adjacent to the line. *Figure 10* shows a transverse section designated as Section 2 that is found on drawing A303.

Longitudinal section: A section drawing on which the cut is made parallel to the longest dimension of a structure.

Transverse section: A section drawing on which the cut is made along the short dimension of the structure.

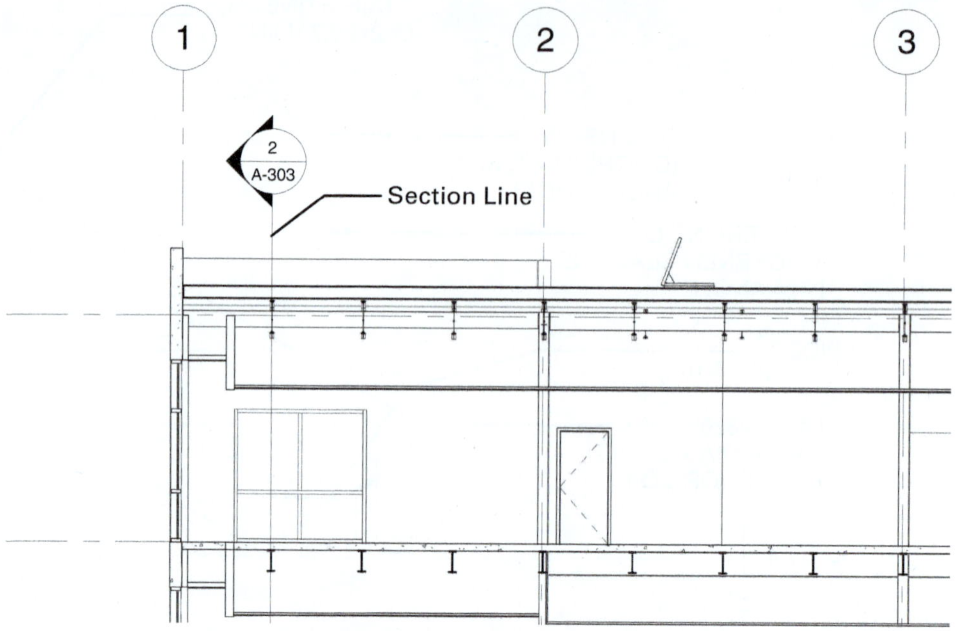

Figure 10 Section line for a transverse building section.

Like detail drawings, section drawings are drawn to a larger scale than that used in plan views. Section views are commonly given for the construction of walls, stairs, cabinets, and other building features that require more information than can be provided on the plan views.

1.4.0 Mechanical, Electrical, and Plumbing Drawings

The mechanical, electrical, and plumbing/piping systems of a building largely reflect the working parts of a structure. As such, the acronym MEP is often assigned to this group of drawings and specifications. Each of these three systems typically have a dedicated set of drawings, however, rather than being combined.

1.4.1 Mechanical Drawings

Mechanical drawings show the heating, ventilation, and air conditioning systems, as well as any other mechanical systems. For some residential jobs, the mechanical drawing may be combined with the plumbing drawing and show very few details other than the locations of the main components. This is because the details are often determined on these projects by the HVACR contractor who must also be responsible for the performance. For most commercial projects, however, the mechanical drawings usually show detailed information about the HVAC equipment, air distribution system, and/or hydronic system.

HVAC system drawings are usually overlaid on tracings of the floor plan to allow the location and layout of the HVAC system to dominate. *Figure 11* shows an example of the chilled- and hot-water piping to four air handling units, overlaid on a floor plan. *Figure 12* shows the piping routed to three of the four air handling units with some additional detail. *Figure 13* shows the piping connections in even greater detail, including the position of each component of the piping connections.

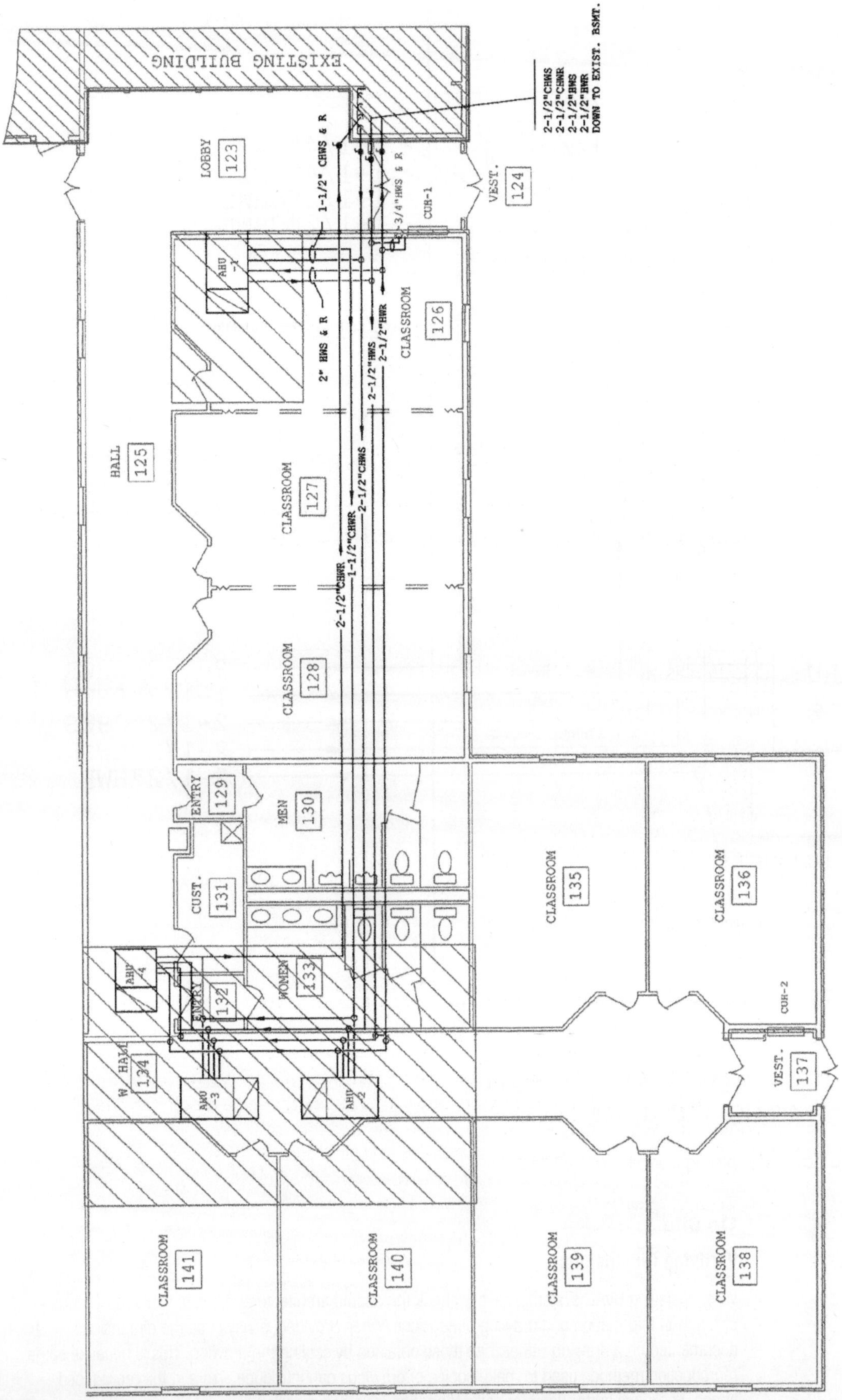

Figure 11 HVAC chilled-water and hot-water piping connections (1 of 3).

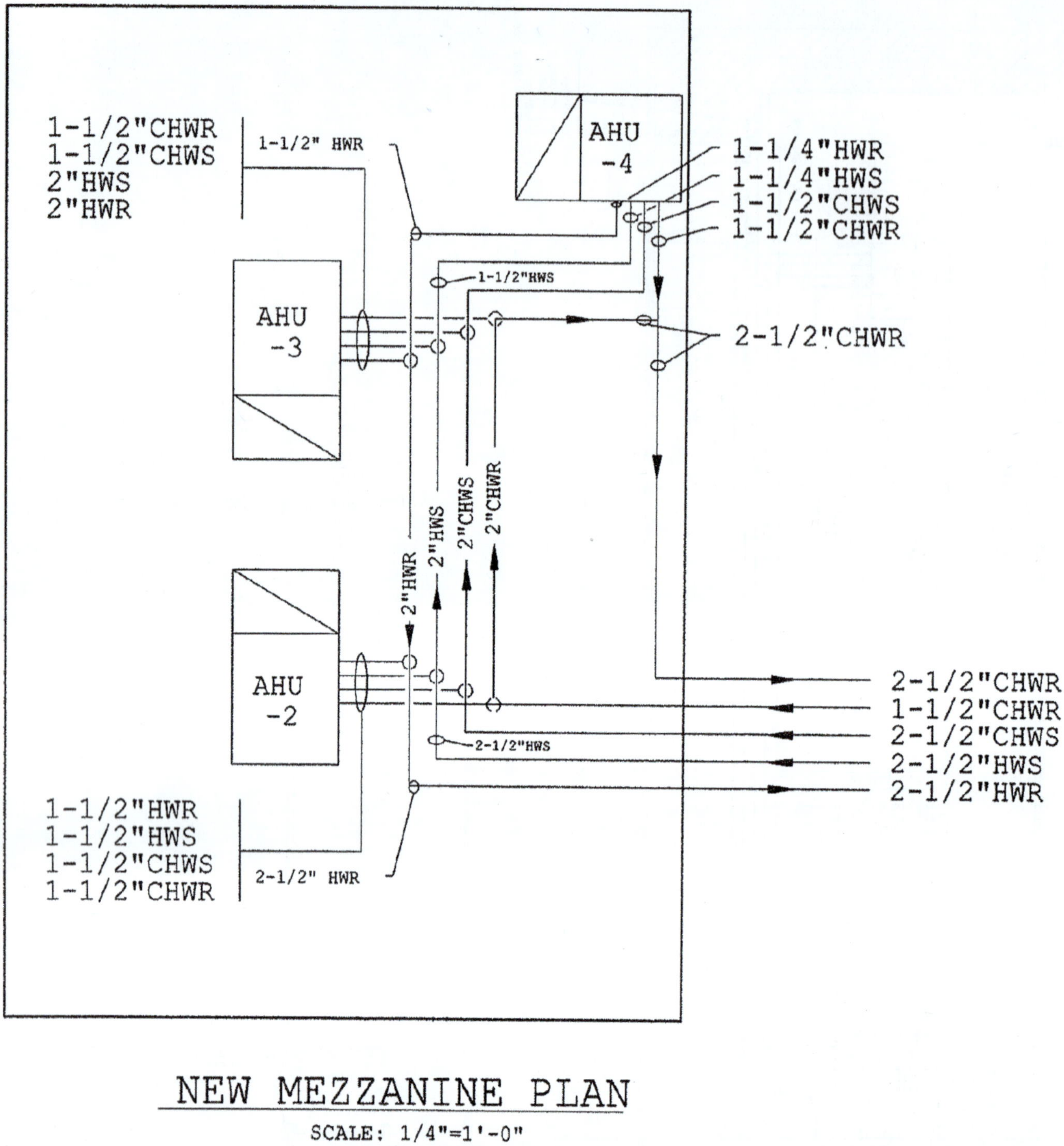

Figure 12 HVAC chilled-water and hot-water piping connections (2 of 3).

On Site

Verifying Dimensions

When installing HVAC systems, always check the related architectural plans to verify the accuracy of rough-in information and drawing dimensions. When provided, always use the dimensions documented on a drawing rather than those obtained by scaling the drawing. This is because some reproduction methods used to make copies of drawings can introduce errors in the reproduced image. Scaling is better used for estimating than for precise work.

Also, when devices are to be located at heights specified above the finished floor (AFF), be sure to determine the actual height of the flooring to be installed. Some materials such as ceramic tile and carpeting can add significantly to the height of the subfloor on which you might be working.

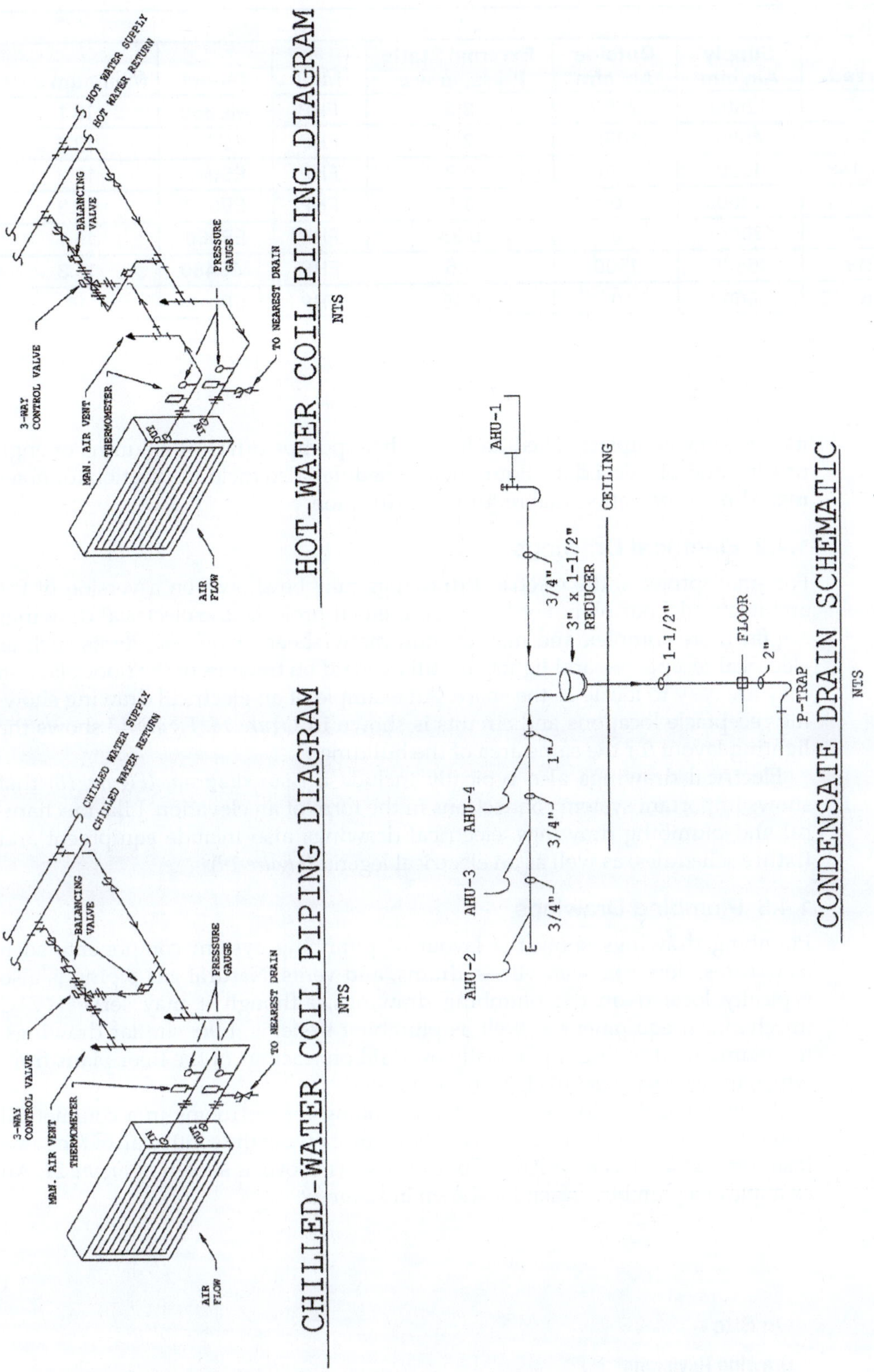

Figure 13 HVAC chilled-water and hot-water piping connections (3 of 3).

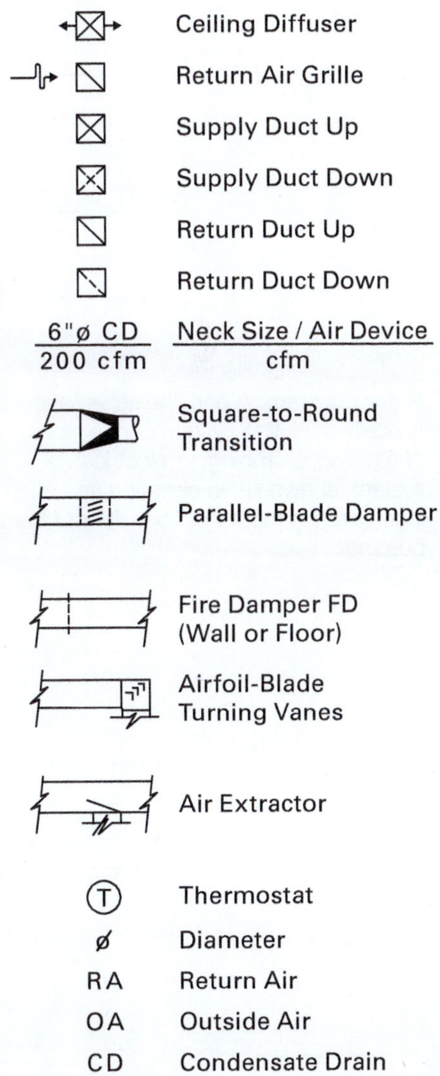

Figure 14 HVAC legend (partial).

Mechanical drawings normally include an HVAC legend listing the various symbols pertaining to the plan. *Figure 14* shows a partial HVAC legend.

Some mechanical drawings also contain **schedules** of HVACR systems (*Figure 15*). This type of schedule has no relationship to time, but instead identifies specific pieces of equipment and provides key engineering data. Note that *Figure 15* only shows a portion of the water-source heat pump schedule, due to

Schedules: Tables that identify and specify the types and sizes of components or equipment required for the construction of a building.

Water Source Heat Pump Schedule									
Unit ID	Location	Area Served	Supply Air, cfm	Outside Air, cfm	External Static Press, in w.c.	MFR	Model	Minimum EER
WSHP1-1	Floor	Mech 145	12800	2300	3.5	FHP	MC360	16.1
WSHP1-2	Floor	Mech 128	4500	1050	2.5	FHP	EC210	13.7
WSHP1-3	Ceiling	Wellness 148	1230	150	0.8	FHP	ES042	16.3
WSHP1-4	Ceiling	Main Elec 147	1200	0	0.5	FHP	EP042	16.3
WSHP1-5	Ceiling	Server 227	2000	0	0.39	FHP	EP060	15.1
WSHP2-1	Floor	Mech 219	16630	1500	3.5	FHP	MC480	16.3
WSHP2-2	Ceiling	Elev Equip 225	500	0	0.18	FHP	EP015	16

Figure 15 Schedule of HVAC equipment (partial).

its size—the complete schedule for the heat pumps offers 41 columns of engineering and physical data. Equipment schedules also include a section for notes regarding information that requires clarification.

1.4.2 Electrical Drawings

For small projects, the electrical drawings may be shown on a version of the architectural floor plan. For large commercial projects, the electrical drawings are far more complex and may contain many sheets. However, items such as electrical receptacles and lights are still overlaid on tracings of the floor plans so they are easy to locate in the space. An example of an electrical drawing showing receptacle locations and circuits is shown in *Figure 16*. *Figure 17* shows the lighting layout for the same area of the building.

Electrical drawings also typically include a **riser diagram** (*Figure 18*) that shows important system connections in the form of an elevation. Like mechanical and plumbing drawings, electrical drawings also include equipment and fixture schedules, as well as an electrical legend (*Figure 19*).

Riser diagram: A one-line schematic typically depicting system connections of a piping, plumbing, or electrical system, shown as an elevation and often incorporating more than one floor of the building.

1.4.3 Plumbing Drawings

Plumbing drawings show the layout of plumbing system components such as fixtures, domestic water lines, drains, and vents. Natural gas piping is also typically located on the plumbing drawings, although it may serve HVAC (mechanical) equipment as well as plumbing systems. Like similar drawings, the plumbing drawings are usually overlaid on tracings of the floor plans from which unnecessary details have been omitted.

A plumbing drawing showing the plumbing for restrooms in a commercial building is shown in *Figure 20*. A riser diagram showing the south wall of the same restrooms, as well as the second-floor restrooms above, is shown in *Figure 21*. An example of a plumbing legend is shown in *Figure 22*.

On Site

Drawing Revisions

When a set of construction drawings has been revised, always make certain that the active and current set is used for all future layout work. Either destroy the old, obsolete set of drawings or else clearly mark on the affected sheets Obsolete Drawing—Do Not Use. A good practice is to remove the obsolete drawings from the set and file those as historical copies for future reference. Also, when working with a set of construction drawings and written specifications for the first time, thoroughly check each page to see whether any revisions or modifications have been made to the originals. Doing so can save time and expense for everyone concerned with the project.

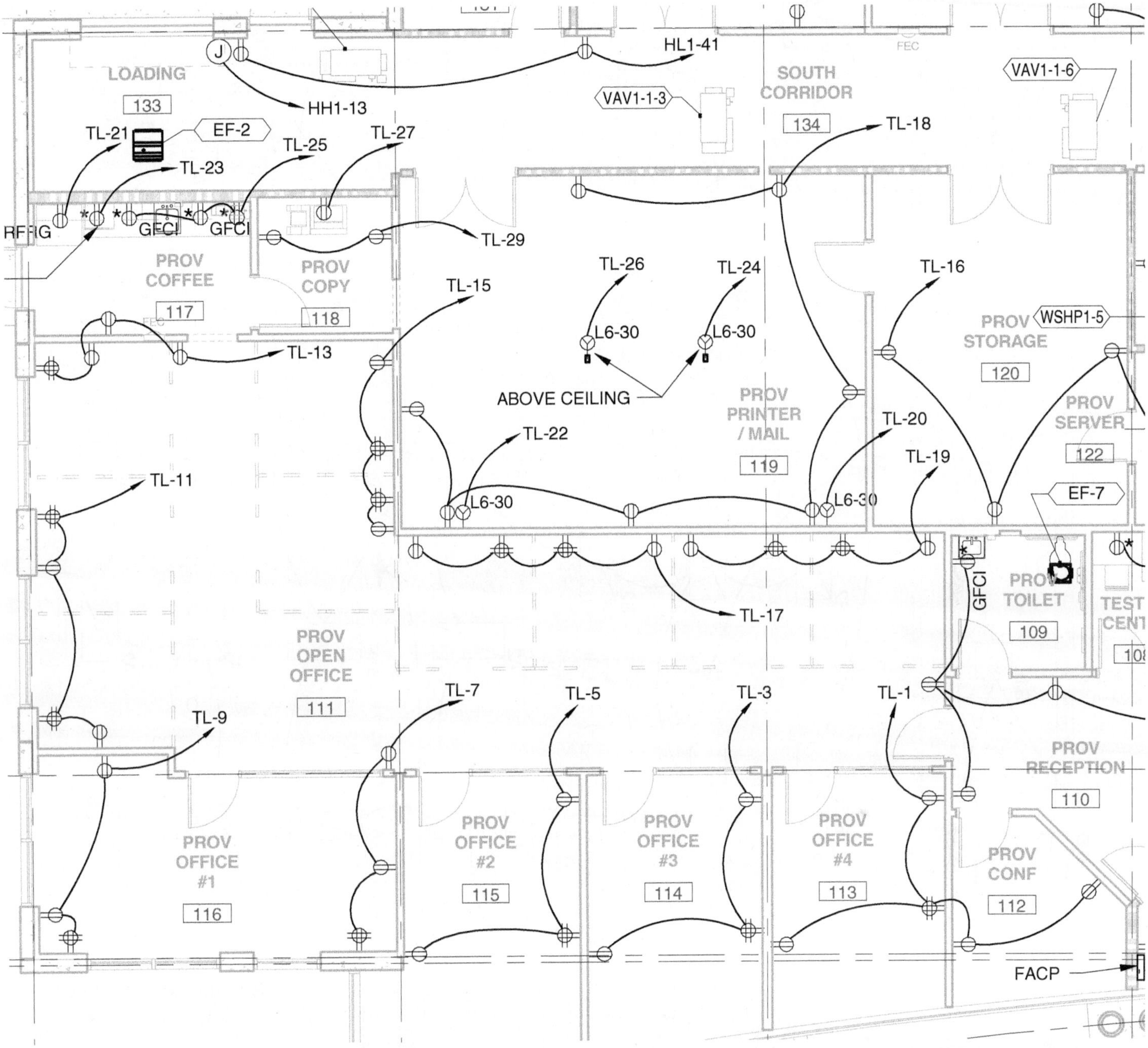

Figure 16 Electrical receptacle plan (partial).

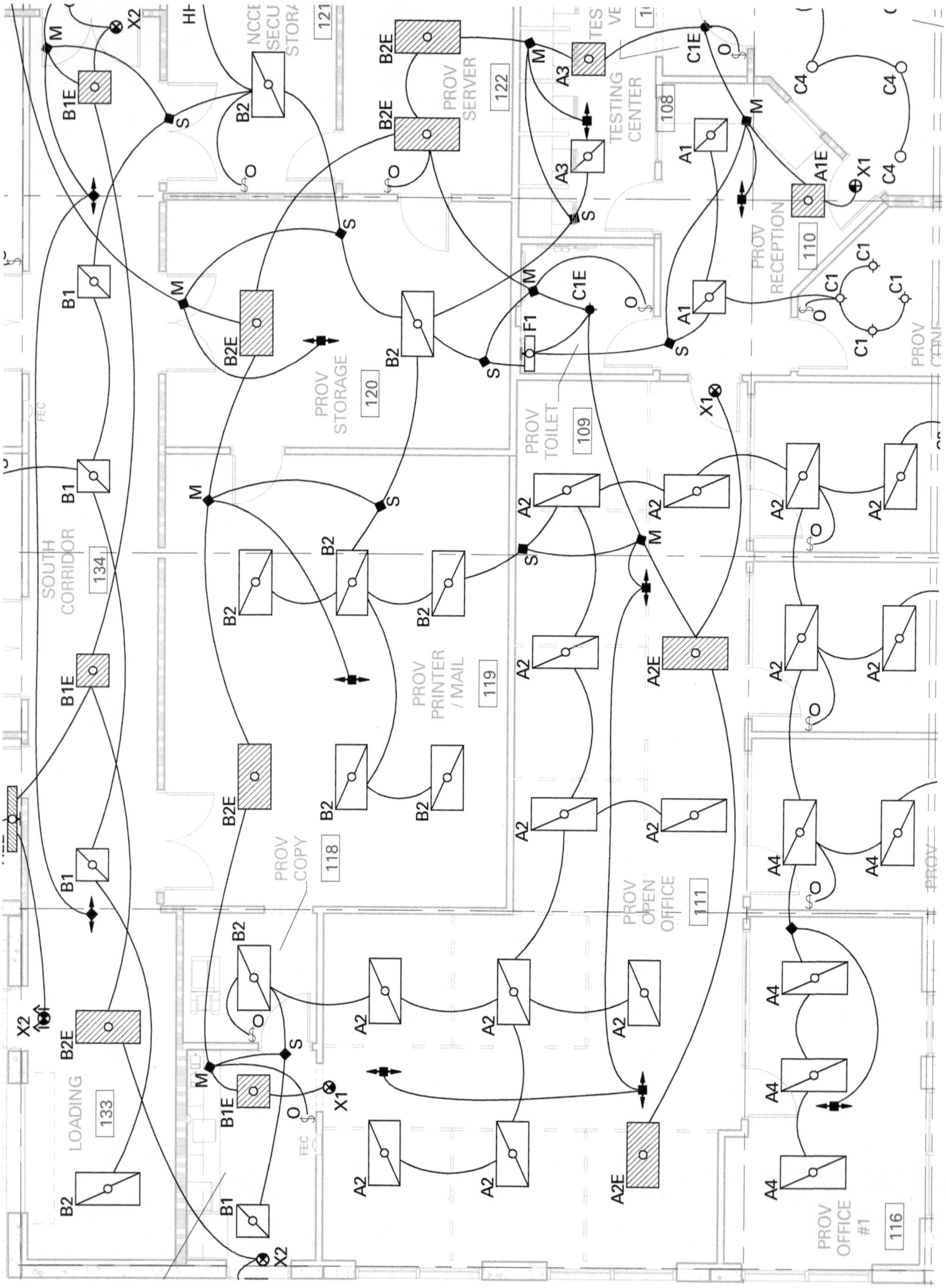

Figure 17 Electrical lighting plan (partial).

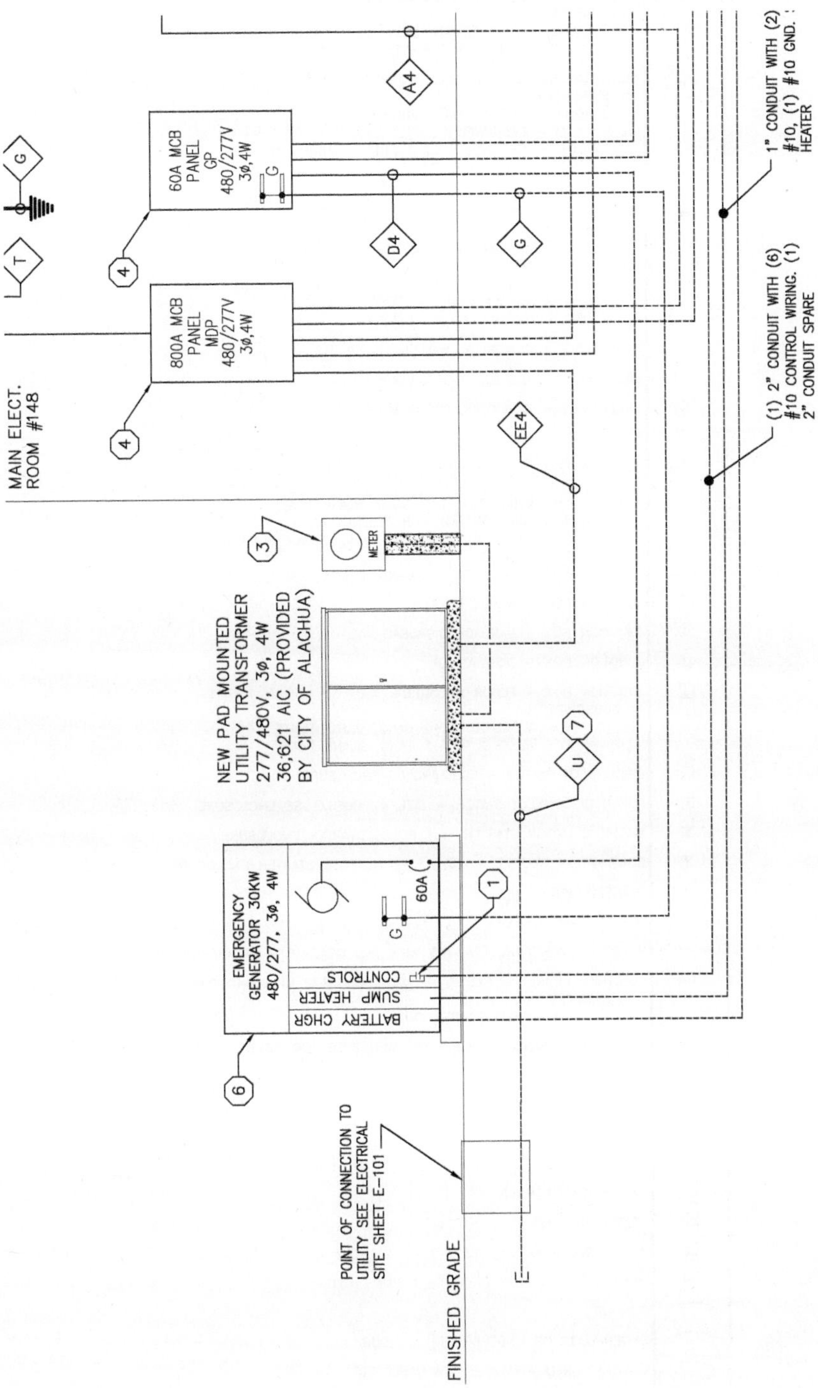

Figure 18 Electrical riser diagram (partial).

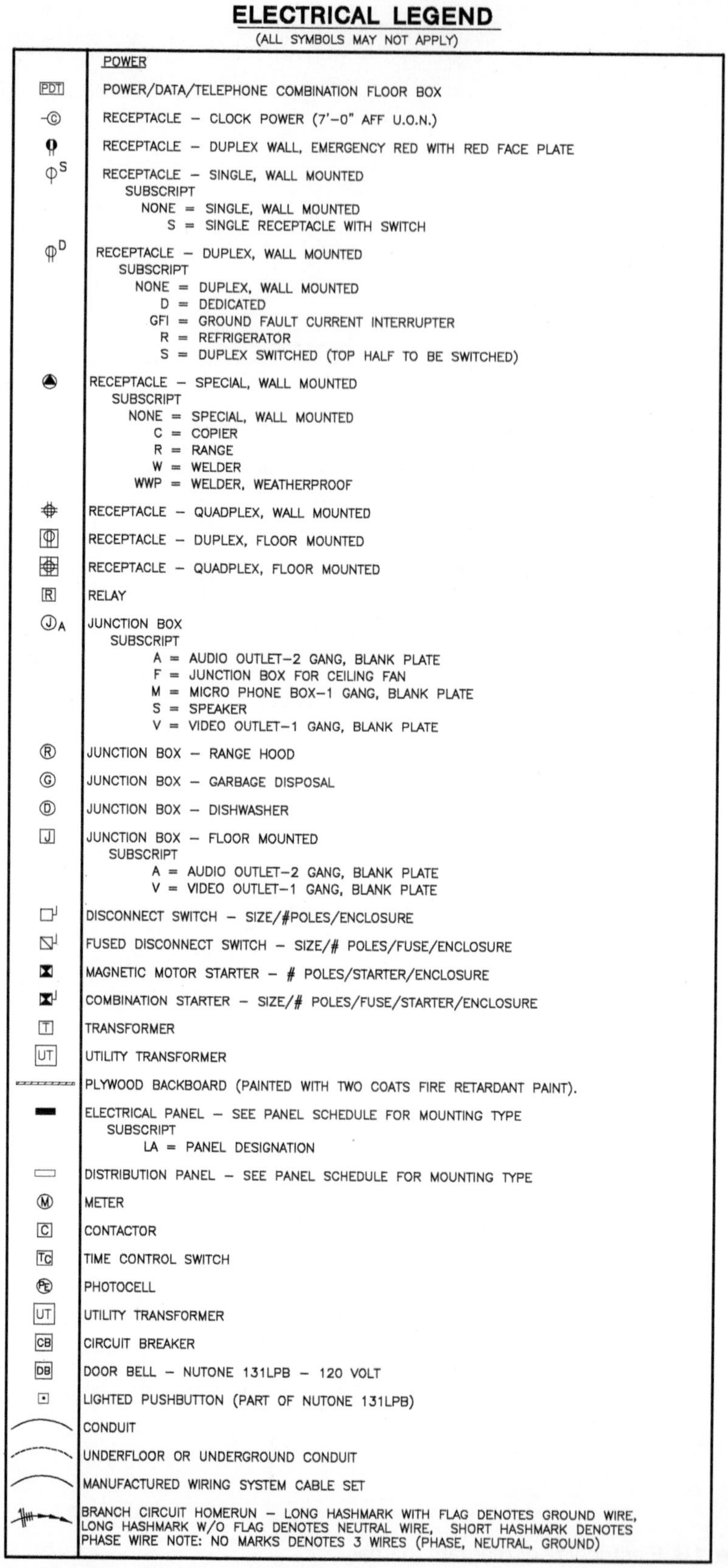

Figure 19 Electrical legend (partial).

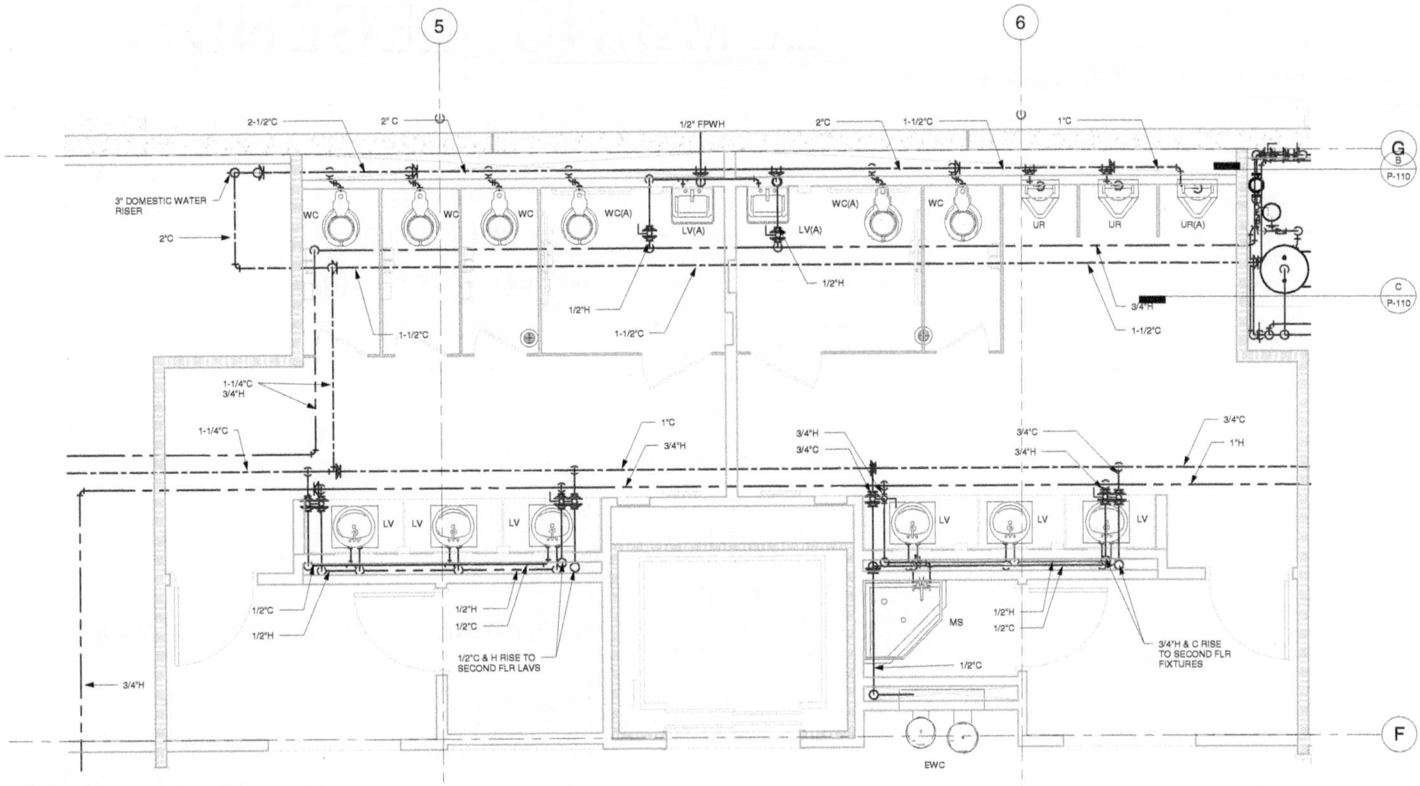

Figure 20 Plumbing drawings for commercial restrooms (partial).

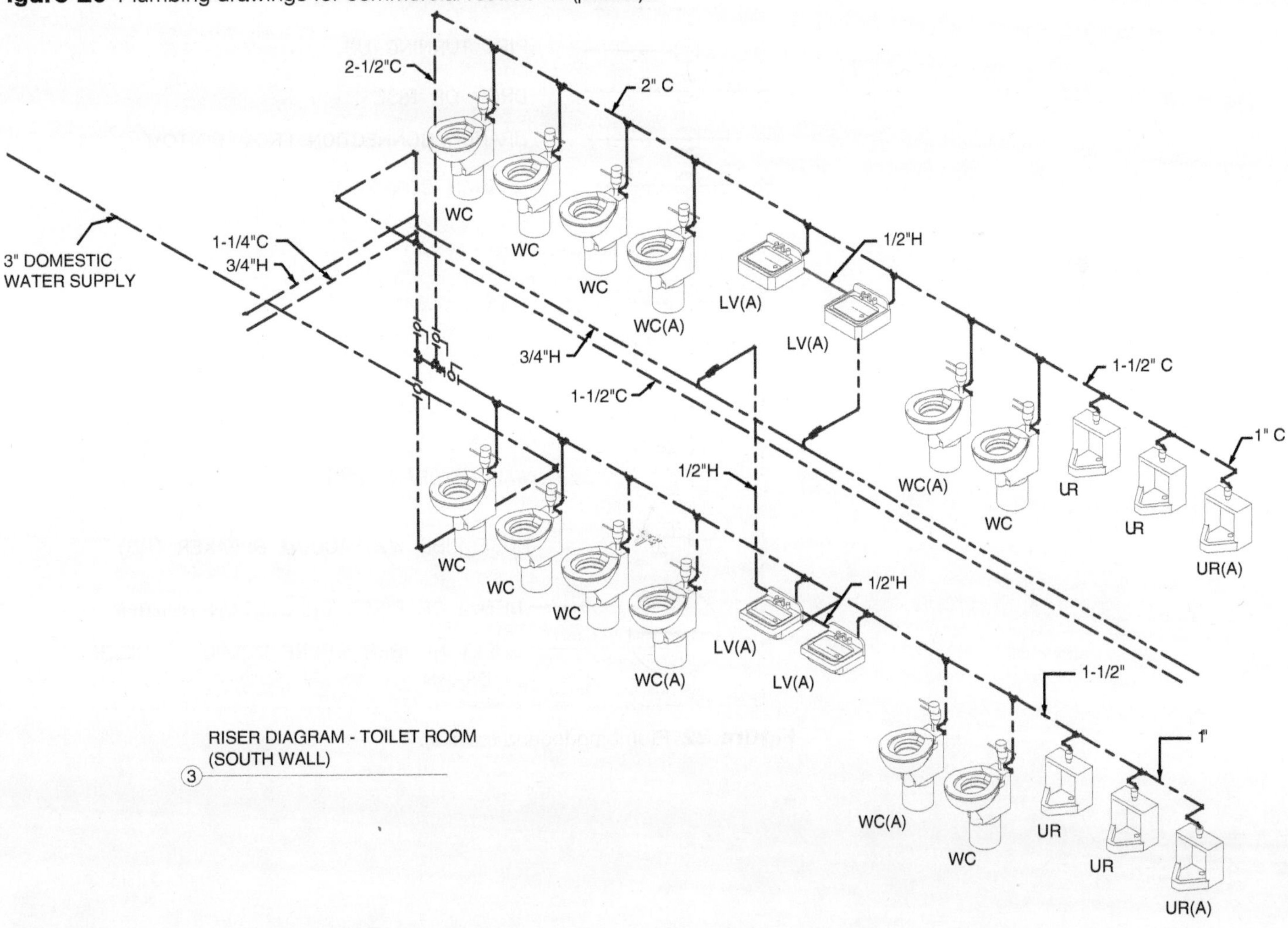

Figure 21 Plumbing riser diagram.

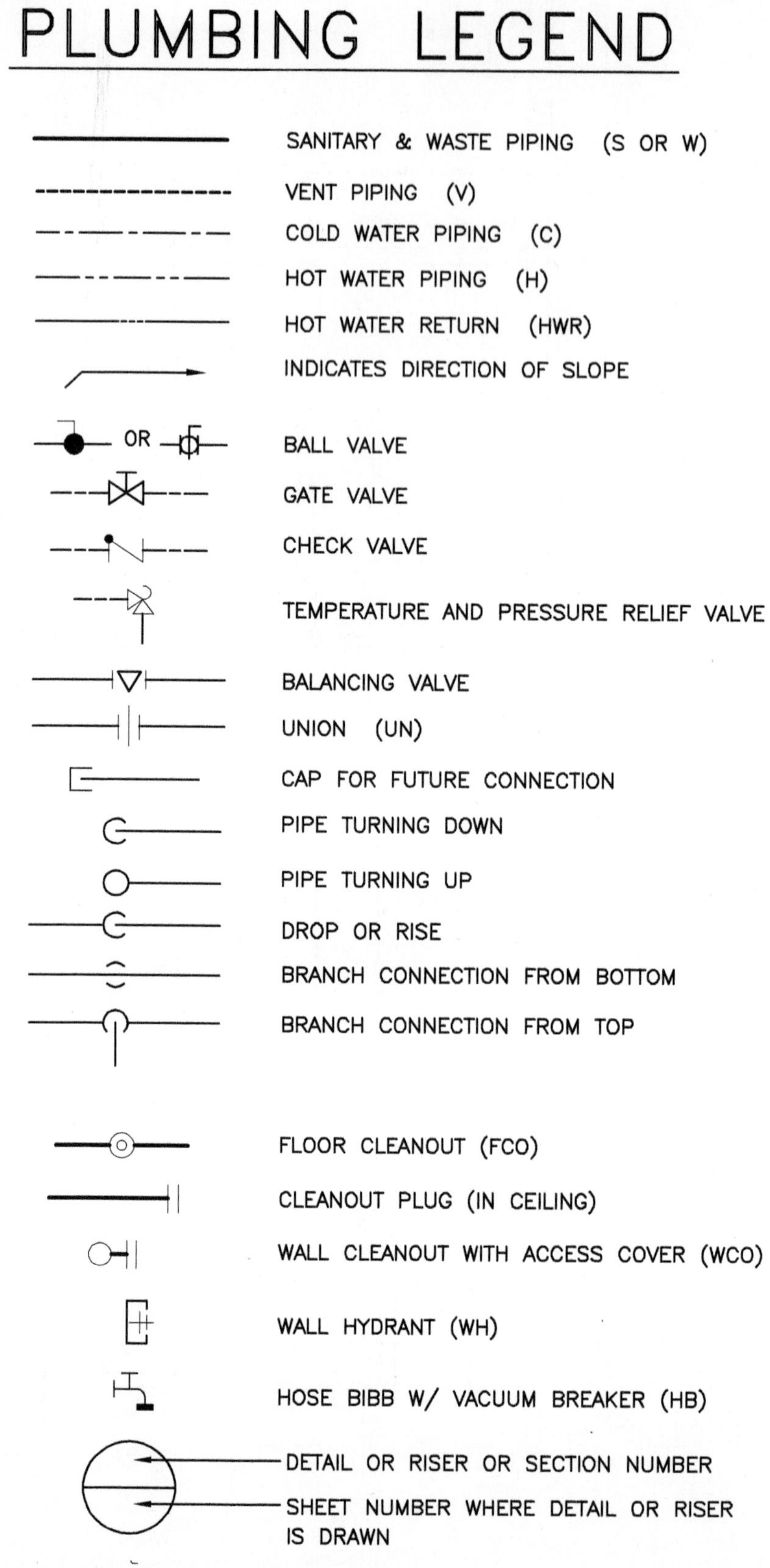

Figure 22 Plumbing legend (partial).

1.5.0 Shop Drawings

Shop drawings, sometimes called fabrication drawings, are an important piece of the construction puzzle. Several types of shop drawings may be encountered.

Shop drawings: Drawings that indicate how to fabricate and/or assemble specific components and subassemblies of a construction project. Shop drawings may be drafted from the construction drawings of a project by one or more of the subcontractors or be provided by a manufacturer.

One type of shop drawing is a detail drawing that a draftsperson creates after the engineer designs the overall structure. For example, the shop drawing for a structural steel subassembly illustrates the joints used, shows the location of all holes and openings, and provides notes specifying how each part is to be assembled.

A second type of shop drawing pertains to special items of manufactured equipment or assemblies for a project. This type is prepared by the equipment manufacturers. This drawing shows dimensions, details of construction, methods for securing the equipment to the structure, and all other pertinent data that the architect and contractor both need to know for the final placement and installation of the equipment. The contractor must check the dimensions of a submitted item to make sure it will fit into the space provided.

A third type of shop drawing is very similar in development to the structural steel shop drawing. This is the primary type used by sheet metal fabricators and installers. *Figure 23* shows an example of an HVAC air distribution system shop drawing. This shop drawing was developed from the design drawings provided

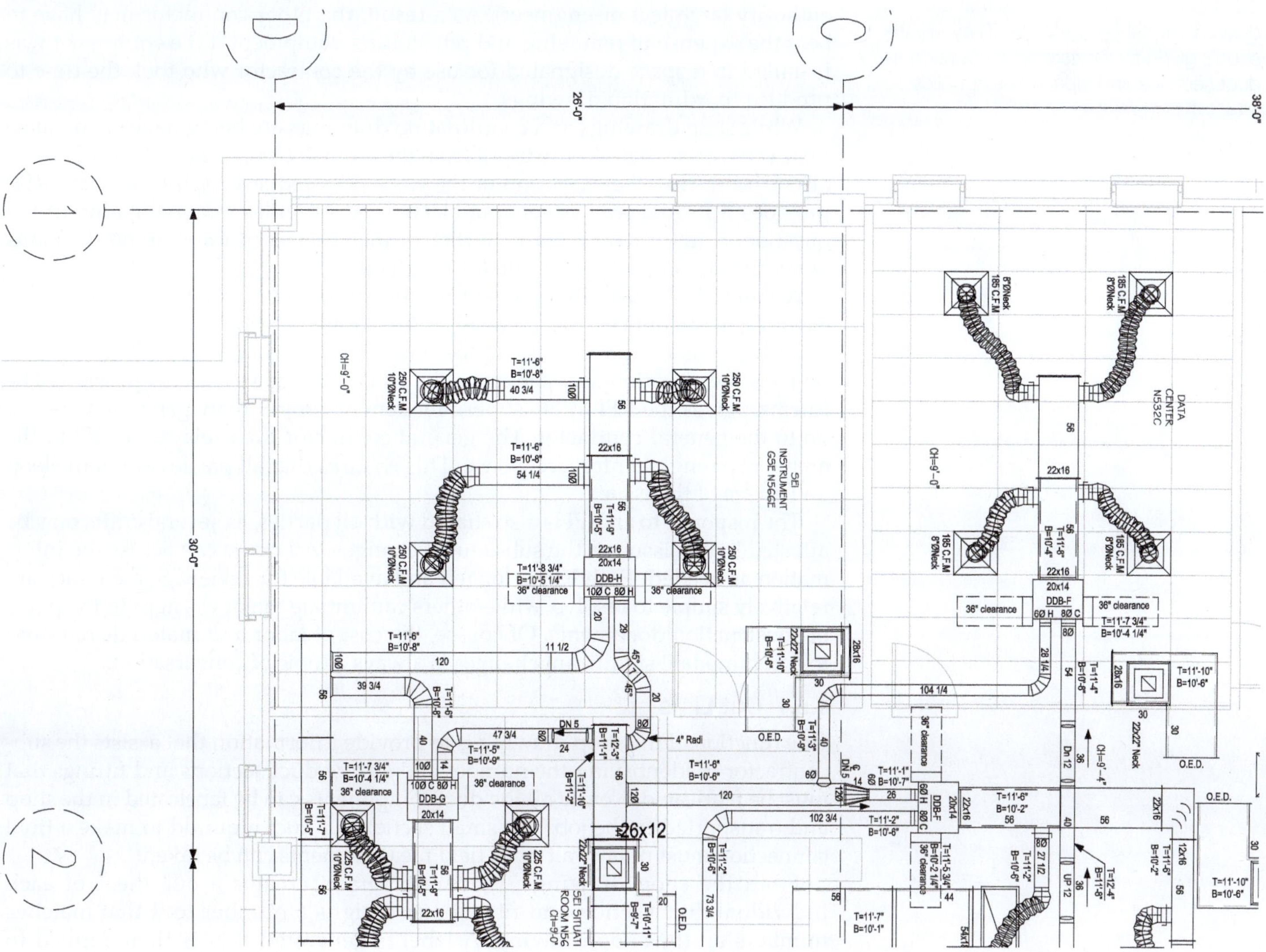

Figure 23 HVAC air distribution shop drawing.
Source: Kogok Corporation

by the architect/engineer. Shop drawings are usually drawn to a scale several times larger than the original drawing to improve clarity and accuracy when scaling the drawing. This type of shop drawing shows the exact layout of the ductwork, the sizes and types of fittings, the types of connectors and hangers required for installation, and notes that will assist in duct fabrication and installation.

Contractors commonly use the following approach for the development of shop drawings:

- When the contract is signed, a full set of drawings is given to the mechanical contractor's drafting department.
- Upon receipt of the mechanical drawings, the drafting department begins developing the shop drawings.
- The drafting department also researches the plumbing, electrical, HVAC piping, architectural, and structural prints to determine if there are any conflicts in the layout. The resulting drawings are commonly called **coordination drawings**.

Coordination drawings: Elevation, location, and other shop drawings produced for a project by the individual contractors for each craft to prevent a conflict regarding the installation of each craft's materials and equipment. Coordination drawing development evolves through a series of meetings between the various contractors.

Cut lists: Information sheets that are derived from shop drawings. They are the shop's guide for fabricating items such as duct sections and pipe subassemblies.

Coordination drawings are generally produced prior to finalizing shop drawings, which must be submitted to the architect for approval. Once approved, **cut lists** can be created from the shop drawings, and duct fabrication can begin.

Some contracts require coordination drawings to be drawn and submitted, while it is simply recommended in others. In the case where one contractor elects to make coordination drawings and another does not, the contractor who made the drawings may be given the installation right-of-way by the presiding authority (architect or engineer). As a result, the other contractor may have to bear the expense of removing and reinstalling equipment if the equipment was installed in a space designated for use by the contractor who took the time to produce coordination drawings.

While shop drawings and coordination drawings are being developed, questions often arise that require input from the architect or engineer. Questions can also arise during the execution of the work. The request for information (RFI) provides a formal way of seeking clarity and provides documentation of the question. A discrepancy, craft conflict, or incomplete information on the plans or specifications are best handled through an RFI.

A hierarchy is usually followed on the job before an RFI is prepared. For example, if you notice a discrepancy in the plans, you should notify your immediate supervisor. That will likely be your foreman on a jobsite. The foreman reviews the situation and generates the RFI, being as specific as possible. The foreman gives the RFI to the superintendent or project manager, who passes it on to the general contractor. The general contractor then relays the RFI to the architect or engineer for a response. This ensures that all parties are in the loop and aware of the issue.

The response to an RFI is also shared with all parties, as several crafts may be affected by the issue or the subsequent changes. All crafts can access the information as needed. Each RFI is logged and available for review. Some issues are relatively simple to resolve, while others can initiate major changes in the plans or construction documents. Of course, the cost of labor and materials required to accommodate significant changes is always a topic of conversation.

1.5.1 Cut Lists

One function of the shop drawing is to provide information that assists the subcontractor in identifying the number and size of duct sections and fittings that must be fabricated. Nearly all the duct components can be fabricated in the shop and transported to the job. The small sections of duct required to make a final connection must often wait until field measurements can be taken.

From the shop drawings, the draftsperson creates a cut sheet of each individual duct section and fitting and assigns a number to it that matches numbers on the shop drawing. A label is generated that is then applied to the completed piece as it is produced. The cut sheets are then given to the fabrication shop.

On Site

Building Information Modeling

Building information modeling (BIM) is becoming increasingly popular and helps do away with the tedious process of creating coordination drawings. BIM creates a digital 3D representation of the building that places components and systems into their relative positions, providing better coordination between the crafts and allowing all parties to visualize the finished product in greater detail. By streamlining workflows, avoiding costly errors and rework, and providing accurate materials takeoffs, BIM can significantly reduce both labor and material costs as well as accelerate the pace of a project.

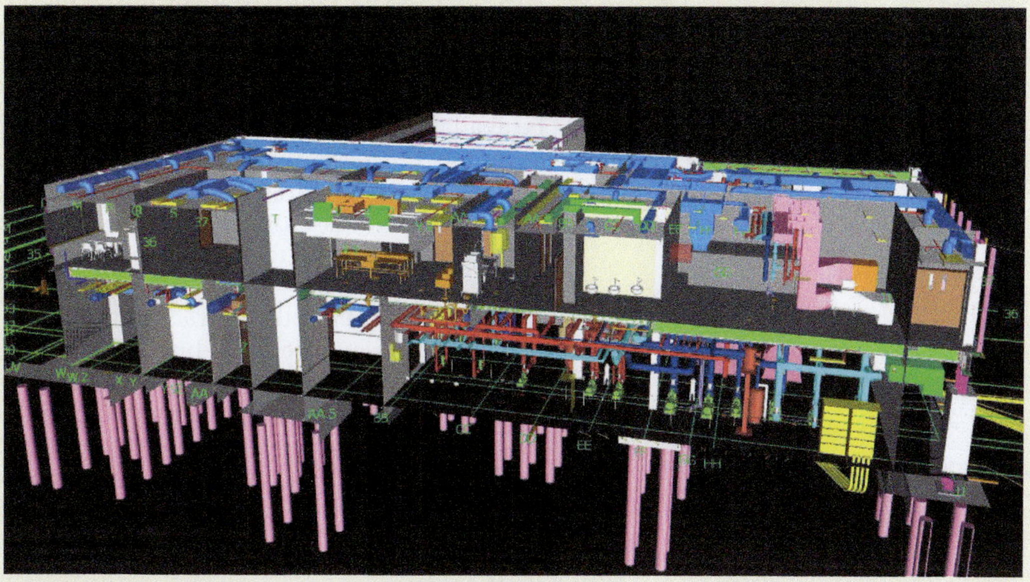

Source: U.S. Army Corps of Engineers (USACE)/Johnny Jaccuzzo

Today, cut sheets and lists are often generated by software that communicates directly with the shop equipment (*Figure 24*). The fabrication information is fed directly to a plasma cutter that cuts out a sheet metal blank that is then broken (bent) into the final shape and assembled.

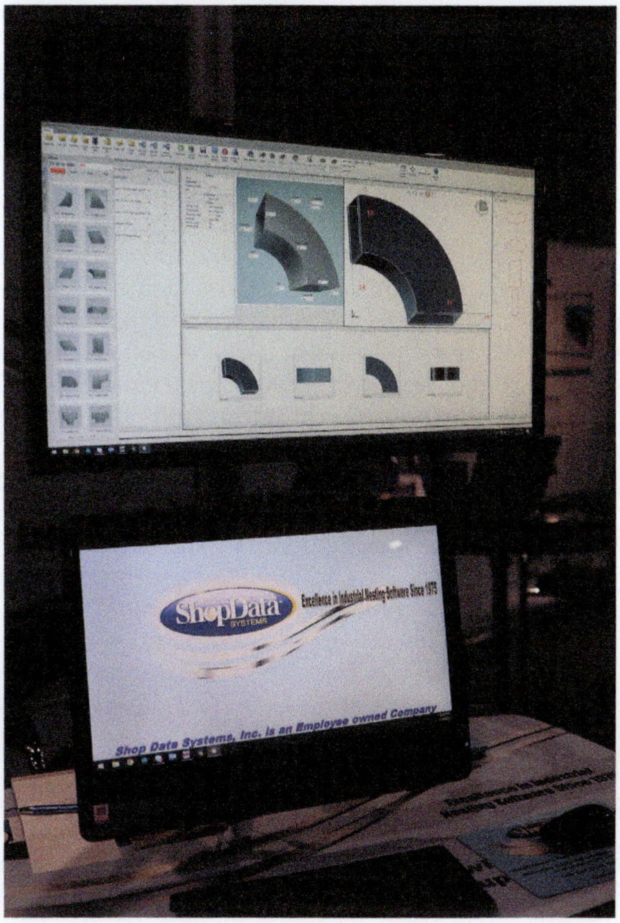

Figure 24 Software-generated cut list.

1.6.0 As-Built Drawings

As-built drawings must be made on alteration or addition projects, on jobs where design modifications must be made to make room for other various crafts, or to alter the previously identified location of a component or piece of equipment. In some cases, particularly on additions or alterations, the as-built drawings may be available from the project architect or engineer.

Simply put, as-built drawings represent the actual results of the construction process. These drawings are the final drawings prepared as a project concludes. As-built drawings must capture all the changes that occurred. This drawing set is very important to those who will occupy and maintain the facility in the future.

The project's various contractors and subcontractors must often prepare and maintain their own set of as-builts. This is because they have the most intimate knowledge of the work they performed. At the end of the project, the as-builts are combined into a comprehensive drawing set that is delivered to the project owner.

The as-built drawings are then placed on the architect's plan set. The drawings are sometimes created or printed in a contrasting color, such as red, but they are more often simply stamped clearly with the words AS BUILT.

The information provided on the as-built drawings about the location and position of ducts, wiring, and plumbing makes it much easier to perform repairs and maintenance. When something specific must be found in a building for service or maintenance, such as a valve or damper above the ceiling, the as-built drawings provide more accurate information than the original construction drawings. As-builts are also used as the basis for future remodeling and alterations to the building.

1.7.0 Building Codes

Building codes that are national in scope provide minimum standards to protect the life and safety of the public by regulating and controlling the design, construction, and quality of materials used in modern construction. These codes have also expanded to govern use and occupancy, location of a type of building, and the ongoing maintenance of all buildings and facilities. Once adopted by a local jurisdiction, these model building codes then become law.

When states, counties, and cities adopt a model code as the basis for their own code, these entities may change it to meet local conditions. These entities might add further restrictions, or might adopt only a portion of the model code. An important general rule to remember about building codes is that, in almost every case, a more stringent local code takes priority over the model code. While a local jurisdiction cannot weaken the requirements of a state-adopted model code, the local jurisdiction can and often does make it more stringent.

The building codes are the legal instruments that enforce public safety in the construction of human habitation and assembly structures. These codes are used not only in the construction industry but also by the insurance industry for compensation appraisals and claims adjustments, and by the legal industry for court litigation.

Prior to 2000, three model building codes were available. The three major code-writing groups, Building Officials and Code Administrators International (BOCA), International Conference of Building Officials (ICBO), and Southern Building Code Congress International (SBCCI), combined into one organization called the International Code Council (ICC) with the purpose of writing one nationally accepted family of building and fire codes. The first edition of the *International Building Code*® (*IBC*®) was published in 2000. At the time of this writing, the 2024 edition is the latest version. However, most states and local jurisdictions do not immediately adopt the latest release of the code. Time is needed to plan for adjustments in code requirements at the local level and implement more stringent requirements where appropriate.

In 2002, the National Fire Protection Association (NFPA) published its own building code, NFPA 5000®, *Building Construction and Safety Code*®. At the time of this writing, 2024 is the latest edition.

Thus, two nationally recognized codes compete for adoption by the 50 states—the *International Building Code*® and NFPA 5000®. The format and chapter organization of the two codes differ, but the content and topics covered are largely the same.

An HVACR technician must be familiar with the local building codes, laws, and restrictions that affect each construction project. This includes a basic understanding of other codes that govern specific types of work, such as the mechanical and plumbing codes, to ensure compliance. While architects and engineers never intend to design a building or system that does not meet the applicable codes, errors do occur, and the construction team needs your help in recognizing unforeseen problems.

1.0.0 Section Review

1. True or False? If you must remove and fold a sheet from a set of drawings for portability, fold the sheet with the title block facing up.
   a. True
   b. False

2. Site plans may also be referred to as ______.
   a. sections
   b. elevations
   c. plot plans
   d. foundation plans

3. Elements such as the heights of windows, doors, and porches are often shown on which type of drawing?
   a. Floor plan
   b. Riser diagram
   c. Detail drawing
   d. Elevation drawing

4. HVAC system drawings are usually overlaid on tracings of the ______.
   a. site plan
   b. floor plan
   c. structural drawing
   d. elevation drawing

5. HVAC duct fabricators follow cut lists to prepare and fabricate duct fittings, with the cut lists typically extracted from the details of the ______.
   a. floor plans
   b. specifications
   c. shop drawings
   d. elevation drawings

6. An HVAC technician who needs to locate a specific damper in an air distribution system should look for it on the _____.
   a. site plan
   b. specifications
   c. as-built drawings
   d. coordination drawings

7. True or False? Model building codes are legal instruments that cannot be changed in any way when adopted by local jurisdictions.
   a. True
   b. False

2.0.0 Specifications, Submittals, and Takeoffs

Performance Tasks

There are no Performance Tasks in this section.

Objective

Describe the importance of specifications, submittals, and takeoffs in construction projects.

a. Describe the contents and role of project specifications in construction.
b. Describe the purpose of equipment and material submittals.
c. Describe a takeoff and explain how a simple takeoff is conducted.

Submittals: Written and illustrated documents related to project equipment and materials provided to the general contractor for review by the subcontractor responsible for the item to ensure the submitted components meet the specifications.

Specifications and **submittals** perform interrelated roles in construction projects. Specifications are designed to document the precise standards and requirements for the work to be done. The materials and equipment to be used on the project are also clearly documented. When the project specifications call for selected pieces of equipment or accessories to be furnished and installed, a submittal document must be provided. The submittals provide important details and confirm to the engineer and contractor that the equipment or component meets the specifications.

2.1.0 Specifications

On Site

Notes on Drawings

Sometimes notes on drawings contradict written specifications or are inconsistent with other requirements. Even though the specifications usually take precedence, the notes are often closer to the true intent. In any case, initiate an RFI to clarify such a discrepancy when you see it.

Specifications are written documents provided by the architectural and engineering firm to the general contractor and subcontractors. Specifications define the quality of work to be done and describe the materials that must be used. The specifications are very important to the architect and owner because they ensure contractor compliance with the expectations of the design team. The specifications provide a great deal of information that can't be placed on the plans due to limited space.

Specifications consist of many elements that differ between projects. On the other hand, architectural and engineering firms often have certain specifications that they consistently apply to projects.

Specifications have several important purposes, including the following:

- Clarify information that cannot be shown on the drawings
- Identify the work standards, types of materials to be used, and responsibilities of various parties to the contract

- Provide information about the details of construction
- Serve as a guide for contractors bidding on a construction project
- Serve as a standard of quality for materials and workmanship
- Serve as a guide for compliance with building codes
- Serve as the basis of agreement between the owner, architect, and contractors in settling any disputes

Specifications generally contain both non-technical and technical content. The non-technical content is often referred to as *special and general conditions.*

2.1.1 Special and General Conditions

Special and general conditions cover the non-technical aspects of the contractual arrangement. Special conditions cover topics such as safety, temporary construction, shop drawings required, and similar information. General conditions include but are not limited to the following:

- Contract terms
- Responsibilities for examining the construction site
- Types and limits of contractor insurance
- Permits and the payment of fees
- Use and installation of needed utilities
- Supervision of the construction process

The general conditions section is the area where misunderstandings often occur. Therefore, these conditions are usually very explicit on large, complex construction projects. The specifications, along with the drawings, are available to bidders for review before a contract is awarded. However, during the bidding process, bidders typically have questions that require an *addendum* to the bidding documents. An addendum is used to clarify or change previously offered information. In most cases, bidding contractors must acknowledge receipt of each addendum released when submitting a bid.

Once a contract is awarded, changes put forth by the architect or engineer must be executed through a **change order** (*Figure 25*). Thus, the difference between an addendum and a change order is a matter of timing. An addendum is written before a contract is awarded, while a change order is drawn up after the contract is awarded. Addenda and change orders both take priority over the original plans and specifications.

Change order: The documentation of a change in the originally agreed-upon and contracted scope of work, usually related to additions or deletions from the scope with corresponding alterations to the completion date and/or the contract value.

Did You Know?

Permits

All required construction permits must be acquired before beginning a project. The permitting process establishes the existence of the project with the local jurisdiction and determines what types of inspections are required. Failure to acquire a permit before work begins usually results in fines being assessed.

The laws in some states give owners the right to refuse payment for any work that is done prior to receiving the proper permits. In addition, once the error has been discovered, work that has been covered up by walls and other obstructions may have to be demolished to allow for a proper inspection of the hidden work.

Construction Change Order

Project Name			
Location of Work			
Project Manager		Client	
Requesting Party		Date of Request	

Change Request Overview

Description of Changes Needed	
Reason for Change	
Support and Justification Documents	
Specifications	
Additional Information	

Change in Contract Price

Original Price	
Net Changes of Previous Change Orders	
Net Increase/Decrease	
Total Contract Price with Approved Changes	

Change in Contract Times

Original Times	
Net Changes of Previous Change Orders in Days	
Net Increase/Decrease	
Total Contract Time with Approved Changes	

Accepted by *Signature*	
Name and Title	
Date	

Approved by *Signature*	
Name and Title	
Date	

Figure 25 Example of a change order form.

2.1.2 Project Technical Information

The technical portions of the specifications cover the work to be done by the contractors and identify the material and installation standards to be met. This includes information on materials covered by standards developed by a number of organizations, such as the American Society of Testing Materials (ASTM) International.

Certain standards may be *incorporated by reference* into a contract. In other words, by referring to a specific standard, it becomes part of the contractual obligation.

2.1.3 Specification Format

The most used specification-writing format in North America is the MasterFormat®. The Construction Specifications Institute (CSI) and Construction Specifications Canada (CSC) jointly developed this standard. The MasterFormat® calls for specifications to be broken down by divisions dealing with similar construction requirements, products, and activities. The format makes it easy to write and use the specifications. Its prevalence in construction means that everyone dealing with construction drawings and specifications soon learns the basic order of the divisions and where the information relative to their task is found.

Prior to 2004, the organization of construction specifications was based on a standard comprised of 16 sections, otherwise known as *divisions*. The divisions and the subsections were individually identified by a five-digit numbering system. The first two digits represented the division number, and the next three individual numbers represented successively lower levels of breakdown. For example, the number 13213 represented Division 13, subsection 2, sub-subsection 1 and sub-sub-subsection 3.

The MasterFormat® standard has experienced significant changes since then. What had been 16 divisions was expanded to 4 major groupings and 49 divisions, with additional divisions reserved for future expansion.

The first 14 divisions are essentially the same as the previous format. Subjects previously in Division 15, *Mechanical*, have been relocated to Division 22, *Plumbing*, and Division 23, *Heating, Ventilating, and Air Conditioning (HVAC)*. The topics under Division 16, *Electrical* have been relocated to new Division 26, *Electrical*, and Division 27, *Communications*. Division 27 was largely created to cover the world of low-voltage electrical systems and work. The numbering system was also changed to six digits to allow for more subsections in each division, which allows for finer definition in the content.

In today's numbering system, the first two digits represent the division number. The next two digits represent subsections of the division, and the two remaining digits represent the third-level sub-subsection numbers. The fourth level, if required, is a decimal and number added to the end of the last two digits. For example, the MasterFormat® number 312219 represents Division 31, *Earthwork*, subsection 22, *Grading*, and sub-subsection 19, *Finish Grading*.

The current version of the MasterFormat® is shown in *Figure 26* for review.

MasterFormat Groups, Subgroups, and Divisions

Procurement and Contracting Requirements Group

Division 00 – Procurement and Contracting Requirements
- Introductory Information
- Procurement Requirements
- Contracting Requirements

Specifications Group

General Requirements

Division 01 – General Requirements

Facility Construction Subgroup

Division 02 – Existing Conditions
Division 03 – Concrete
Division 04 – Masonry
Division 05 – Metals
Division 06 – Wood, Plastics, and Composites
Division 07 – Thermal and Moisture Protection
Division 08 – Openings
Division 09 – Finishes
Division 10 – Specialties
Division 11 – Equipment
Division 12 – Furnishings
Division 13 – Special Construction
Division 14 – Conveying Equipment
Division 15 – Reserved for Future Expansion
Division 16 – Reserved for Future Expansion
Division 17 – Reserved for Future Expansion
Division 18 – Reserved for Future Expansion
Division 19 – Reserved for Future Expansion

Facility Services Subgroup

Division 20 – Reserved for Future Expansion
Division 21 – Fire Suppression
Division 22 – Plumbing
Division 23 – Heating, Ventilating, and Air Conditioning (HVAC)
Division 24 – Reserved for Future Expansion
Division 25 – Integrated Automation
Division 26 – Electrical
Division 27 – Communications
Division 28 – Electronic Safety and Security
Division 29 – Reserved for Future Expansion

Site and Infrastructure Subgroup

Division 30 – Reserved for Future Expansion
Division 31 – Earthwork
Division 32 – Exterior Improvements
Division 33 – Utilities
Division 34 – Transportation
Division 35 – Waterway and Marine Construction
Division 36 – Reserved for Future Expansion
Division 37 – Reserved for Future Expansion
Division 38 – Reserved for Future Expansion
Division 39 – Reserved for Future Expansion

Process Equipment Subgroup

Division 40 – Process Interconnections
Division 41 – Material Processing and Handling Equipment
Division 42 – Process Heating, Cooling, and Drying Equipment
Division 43 – Process Gas and Liquid Handling, Purification, and Storage Equipment
Division 44 – Pollution and Waste Control Equipment
Division 45 – Industry-Specific Manufacturing Equipment
Division 46 – Water and Wastewater Equipment
Division 47 – Reserved for Future Expansion
Division 48 – Electrical Power Generation
Division 49 – Reserved for Future Expansion

Figure 26 MasterFormat® groups, subgroups, and divisions.
Source: ©2018 The Construction Specifications Institute, Inc. (CSI). MasterFormat® excerpt and trademarks used under permission from CSI.

2.2.0 Submittals

Submittals are documents that illustrate specific pieces of equipment or accessories that are to be ordered, purchased, and installed by the subcontractor. The submittal document is provided by the manufacturer or their authorized distributor to the subcontractor. It is then submitted to the general contractor to be approved and added to the other construction documents. The submittals must be approved before a contractor can proceed with ordering and installation.

Subcontractors receive submittals for any major pieces of equipment or subassemblies that they are contractually obligated to provide. A *manufacturer's certified submittal* provides accurate dimensional information to a normal tolerance expected. However, normal submittals are more generic in nature and the dimensions are an approximation without a guarantee of accuracy.

The submittal sheet illustrates the material or piece of equipment that has been defined in the specifications and that must conform to the standards outlined in the specification. For example, the specifications for an inline centrifugal duct fan may be stated as follows:

> *Centrifugal inline duct fans shall be ACME Company, Model VIDB direct drive or Model VIBA belt drive, as shown on the plans and schedules. Fans shall be constructed of heavy-gauge steel and electro-coated acrylic enamel finish over an iron-phosphate primer. Wheels 12" in diameter and larger shall have median foil blades to assure quiet, efficient operation. The motor drive compartment shall be isolated from the airstream and externally ventilated. Bearings shall be pre-lubricated and sealed for minimum maintenance and designed for 200,000 hours of operation. Internal parts (wheels, shaft, bearings, motor, and drive) shall be accessible for inspection and repair or replacement without disturbing connecting ductwork. Fans shall be furnished with a mounted safety disconnect. Single-phase motors shall have integral overload protection. V-belt drives shall be adjustable. Horsepower and noise levels shall not exceed the values shown in the equipment schedule; oversized motors to meet air delivery performance are not acceptable. Performance ratings shall be certified for air and sound.*

The submittals related to this specified duct fan are shown in *Figure 27A* and *Figure 27B*. The general contractor and engineering team will review the submittals and ensure that the information meets the job requirements.

However, the specifications may also allow for substitutions. Substitutions, of course, must still meet the original specifications. In most cases, a contractor chooses a substitute because of better pricing. Contractors may seek pricing for the substitute equipment before preparing the bid and know that they can reduce their price using the substitute. This strategy can provide an advantage over other bidders.

One of the hazards of working with substitute equipment though, is that it could complicate the work in some way for both the contractor providing it as well as other contractors. This usually places the financial responsibility for any additional cost squarely on the shoulders of the contractor requesting and providing the substitute. Thus, providing substitute materials and equipment can be a complicated business. When possible substitutes are not listed but a contractor believes that a suitable and practical substitute exists, it requires a consultation with the owner and/or design team.

Regardless of substitution, a submittal for the equipment is acquired by the subcontractor, who submits copies of it to the general contractor, owner, architect, and code enforcement authorities, as required. The subcontractors either accept the submittal or reject it with clear cause. If the submittal is accepted by all parties, the item may then be ordered and installed by the contractor. Changes to the drawings often follow the acceptance of substituted equipment.

SUBMITTAL SHEET HV 4-8
INLINE CENTRIFUGAL DUCT FANS

MODEL VIDB – DIRECT DRIVE

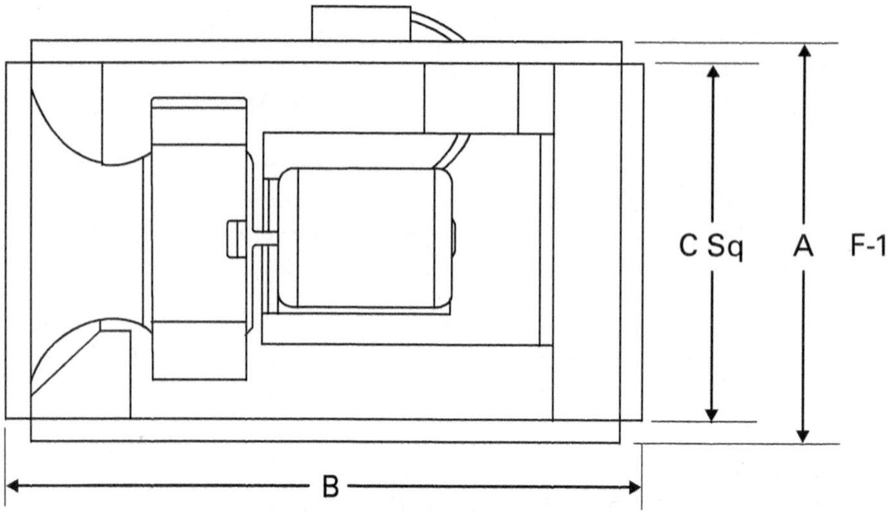

UNIT SIZE	A	B	C	INLET OR OUTLET AREA	WHEEL DIAMETER	WT (lb)
06	13⅞"	20"	12"	0.979 ft^2	10¾"	30
08	13⅞"	20"	12"	0.979 ft^2	10¾"	40
10	13⅞"	20"	12"	0.979 ft^2	10¾"	40
12	17⅞"	27⅜"	16"	1.750 ft^2	11 13/16"	75
15	*21⅞"	31"	20"	2.740 ft^2	14⅞"	90
18	*21⅞"	33⅜"	26"	4.650 ft^2	17 13/16"	140

* A-1 Larger on access door sides

JOB NAME AND LOCATION	SUBMITTED BY
SECTION 230000 2.4.1	

Figure 27A Duct fan submittal sheet (1 of 2).
Source: Carnes Company

SUBMITTAL DATA TYPE VIDB IN LINE DUCT FANS

MODEL VIDB 10 – DIRECT DRIVE UNITS

SIZE 10	MOTOR HP	RPM	TIP SPEED	CFM @STATIC PRESSURE, (In W.G.) 1/8"	1/4"	3/8"	1/2"	5/8"	3/4"	1"
10-D3S with ES60 Speed Control	35 MHP Standard Motor 100W Max.	700	1970	200						
		800	2250	295						
		900	2535	365	160					
		1000	2815	430	290					

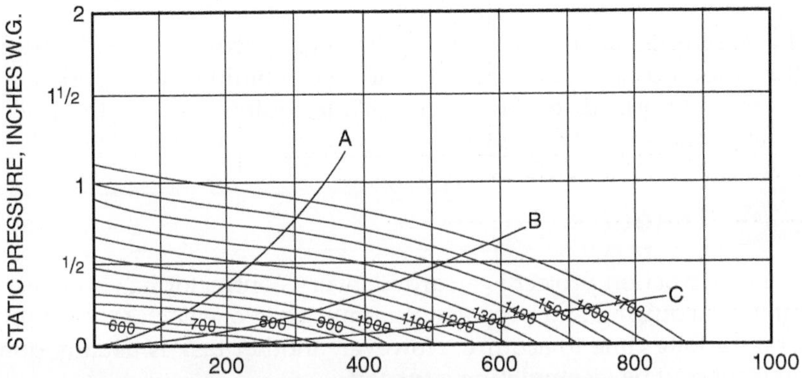

SOUND LEVEL DATA VIDB SIZE 10

SIZE RPM	*	SOUND POWER LEVEL BY OCTAVE BANDS 1	2	3	4	5	6	7	8	LWA	SONE
10-10D 1105	A	62.5	72.5	57.5	54.5	51.0	49.5	48.0	47.0	60.5	5.6
	B	62.5	69.0	58.0	56.5	53.5	50.0	48.0	47.0	60.0	5.2
	C	62.5	68.0	59.0	58.5	57.0	51.5	48.5	47.0	62.0	5.5
10-10H 1710	A	72.0	82.0	67.0	64.0	60.5	59.0	57.5	56.5	69.5	10.7
	B	72.0	78.5	67.5	66.0	63.0	59.5	57.0	56.5	69.5	9.8
	C	72.0	77.5	68.5	68.0	67.0	61.0	58.0	56.5	71.0	10.0

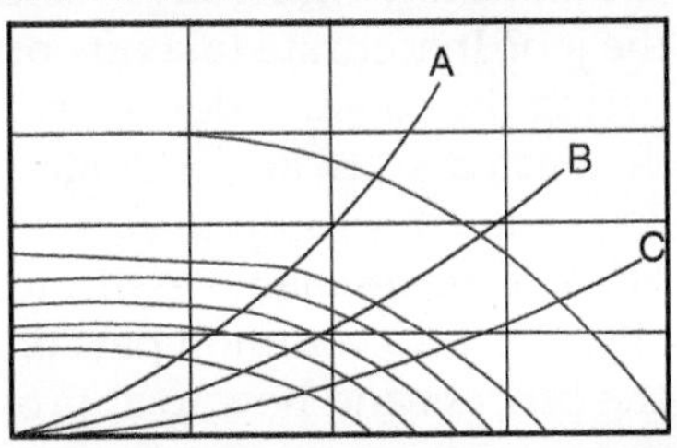

PROJECT:	AGENT:	
	SUBMITTAL DATE:	JOB NO:
The CARNES Company Manufactured Products		DRAWING NO:
		Date: Rev:

Figure 27B Duct fan submittal sheet (2 of 2).
Source: Carnes Company

On Site

Project Equipment Selection

Architects and engineers, like the rest of us, routinely develop a preference for one brand of equipment or another. They then typically design and develop construction plans around their preferred brand and type.

Rather than write every specification related to that equipment from scratch, they often turn to the manufacturer for assistance in developing the specification. In this case, the documented features and characteristics of the preferred equipment become the specification. The result is that, if the product has any significant features or characteristics that make it unique, it can be challenging for a substitute to be accepted.

Debates between the contractor and the engineering team can and do occur as a result. In most cases, the contractor must convince the others that a particular feature or characteristic is unimportant and the lack (or presence) of the feature does not make the substitute an inferior product. The owner or engineering team has no obligation to accept a substitute for an item that was clearly shown on the plans and in the specifications. However, in the spirit of competition, project specifications often list typically accepted substitutes in the specifications.

The submittal acquisition and submission process usually begins very quickly after a contract is awarded. Until the submittals are approved, contractors aren't authorized to place an order for equipment, and the lead time for some items can easily be months.

2.3.0 Takeoffs

Once a construction project is awarded, every contractor has a variety of tasks to initiate. Acquiring and preparing submittals is one such task, as some items may take a long time to acquire. However, another task is usually done before seeking submittals—completing a **takeoff**.

Takeoff: The result of reviewing, measuring, and quantifying all the materials and equipment needed for a project, as indicated by the drawings. The information from a takeoff is then used to initiate purchases.

A takeoff involves reviewing, measuring, and quantifying all the materials and equipment required to complete a project. The project drawings and specifications are used as the basis. In fact, in many cases, the takeoff was completed as part of the bidding process. It is impossible to accurately bid on a project without performing a takeoff with some level of detail.

Takeoffs cover a lot of ground. To the extent possible, every item must be accounted for. Both large items, such as chillers and packaged air conditioning units, as well as many smaller items need to be counted. How many feet of thermostat or communication wire will be needed? How many ½" 90-degree copper elbows are on the job? How much silver solder should be purchased to avoid running out on the job? Inaccurate takeoffs often result in shortages, lost labor, and problems meeting the construction schedule.

Although many sophisticated software packages are available today to help the process, those packages are typically used by professional estimators for larger projects. As an HVACR technician, small projects that require a takeoff to be completed manually may be a common part of your career. Therefore, it is helpful to understand the process and how to approach it.

2.3.1 Takeoff Tools and Materials

The following materials will make the measuring, counting, and summarizing tasks easier:

- Colored pencils for checking off items on drawings as each one is accounted for; assign a color to systems or specific types of material to avoid confusion.
- A simple mechanical counter for counting similar pieces such as 90-degree elbows as you review the drawing
- A digital scale (*Figure 28*) or similar device for measuring long duct and piping runs
- Architect's and engineer's drafting scales
- An adding machine or calculator
- A magnifying glass for examining details on the drawings (when working from paper)

Figure 28 Digital scale.
Source: Courtesy of Calculated Industries

Dedicated forms prepared for certain materials are also helpful. An example of a simple form for pipe fittings is shown in *Figure 29*. Regardless of the form

PROJECT NAME ______________________ JOB NO. ______________

MATERIAL ____________________ DRAWING NUMBER ______________

FITTING TYPE	FITTING SIZES (INCHES)						
NO. OF JOINTS							

Figure 29 Example pipe fitting takeoff sheet.

used to document your counts, it's best to also record the drawing number at the top of each sheet for future reference.

After the takeoff is complete, prepare summaries of each type of material. Mark each sheet to indicate the numbers have been transferred to the summary. The summaries can be used to begin preparing pricing requests or purchase orders for the various suppliers.

2.3.2 Takeoff Procedures

It is a good idea to develop a logical sequence for takeoffs and stick to it. Of course, some topic areas will not be relevant on every project. Developing a logical sequence helps ensure that nothing is missed. As a rule, begin with the primary equipment and work your way down to items such as the interconnecting piping and controls.

The basic takeoff procedure for items such as valves, fittings, piping, pipe hangers, and accessories is as follows:

Step 1 Read the specifications carefully. List the materials and joints specified for each system. Remember that there are many types of fittings and piping material, so differentiation is essential for accuracy.

Step 2 Prepare the required takeoff sheets.

Step 3 Take off each system as indicated on the floor plans, section details, and flow diagrams. When working with two-dimensional drawings, it is very important to visualize the installation in three dimensions as you count. Move back and forth between floor plans, elevations, and riser diagrams when necessary to form a three-dimensional mental picture of specific areas.

Step 4 Double-check the drawing scale and measure pipe lengths in feet; count the fittings, valves, hangers, for each section of pipe.

Step 5 List all together quantities of different fittings for the same size of pipe.

Step 6 After completing the takeoff, prepare the summary sheets and transfer the counts to them.

Use the same basic procedure for all the other systems and materials. While takeoffs are rarely perfect, if ever, completing the task systematically will always yield better results. If the takeoff is being completed as a basis for bidding a project, the next step will be to assign unit costs to each unique component.

2.0.0 Section Review

1. True or False? The general conditions outlined in project specifications identify the specific material and installation standards to be met.
   a. True
   b. False

2. To obtain approval for a substitute piece of equipment when the specifications do not list any possible substitutions, you must ______.
   a. request authorization from the general contractor
   b. petition the appropriate code enforcement authorities
   c. create a change order and share it with the design team
   d. consult with the owner and/or design team for the project

3. True or False? A takeoff is often accomplished as part of the bidding process.
   a. True
   b. False

Module 03401 Review Questions

1. You can find the date of the last revision for a set of drawings by looking at which portion of the drawing package?
   a. Site plans
   b. Title block
   c. Submittals
   d. Drawing index

2. Site plans may include some construction features such as _____.
   a. ductwork layout
   b. wall composition
   c. water supply lines
   d. parking lot curbs and gutters

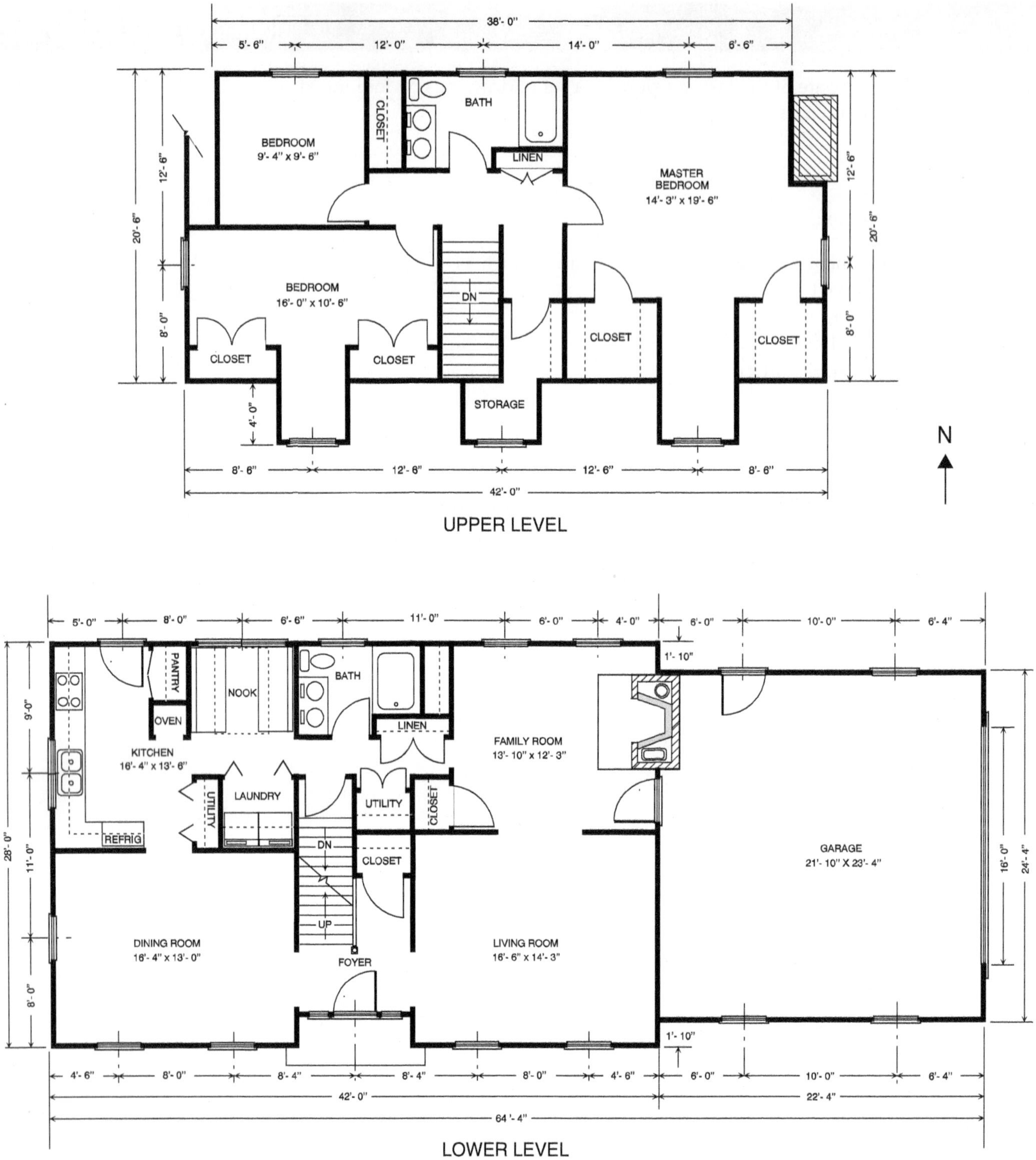

Figure RQ01

3. Refer to *Figure RQ01*. In the garage, what is the distance between the center of the north window and the center of the door on the same wall?
   a. 6'-0"
   b. 6'-4"
   c. 8'
   d. 10'

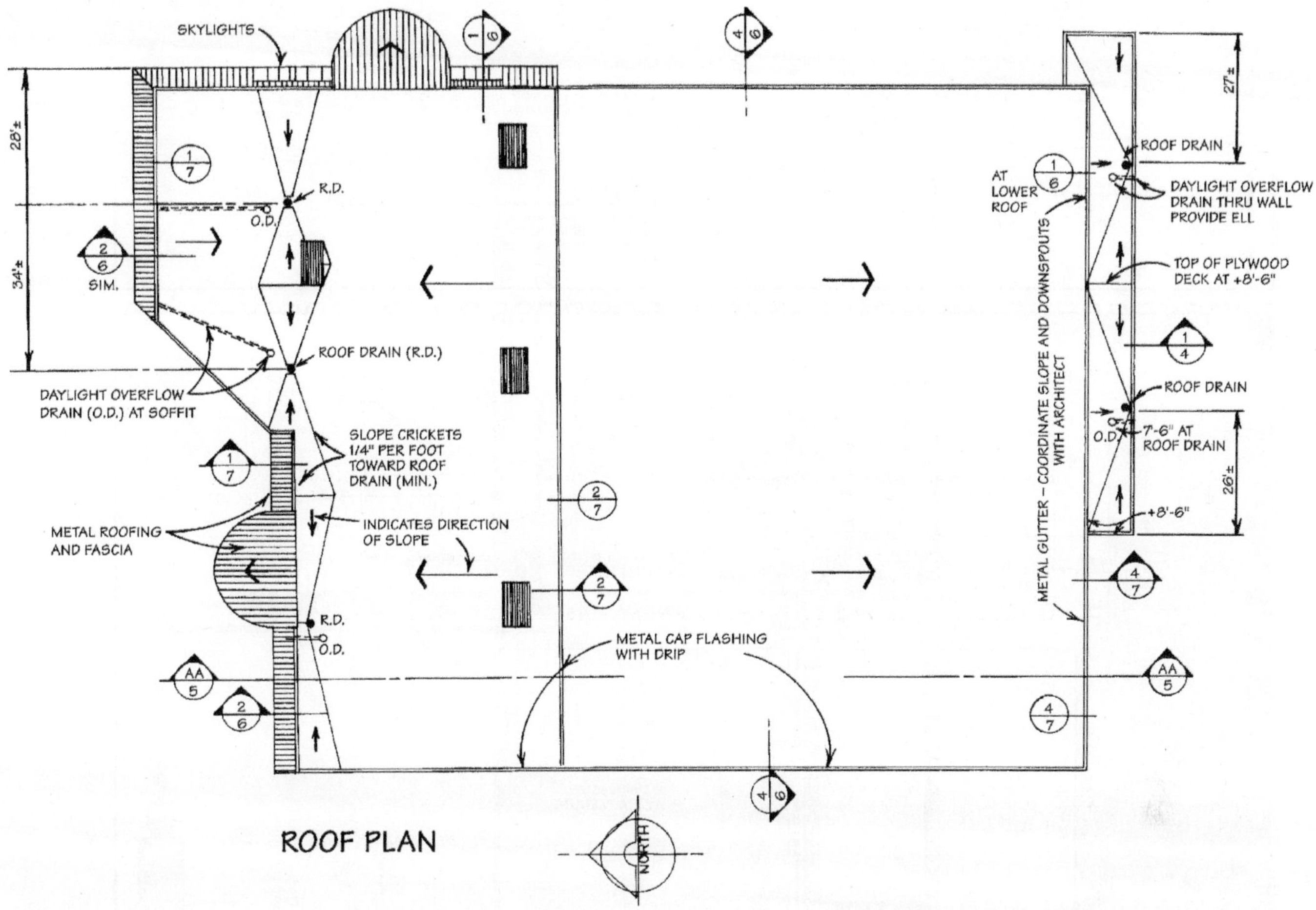

Figure RQ02

4. Refer to *Figure RQ02*. How many roof drains are called for in the roof plan?
   a. Three
   b. Four
   c. Five
   d. Six

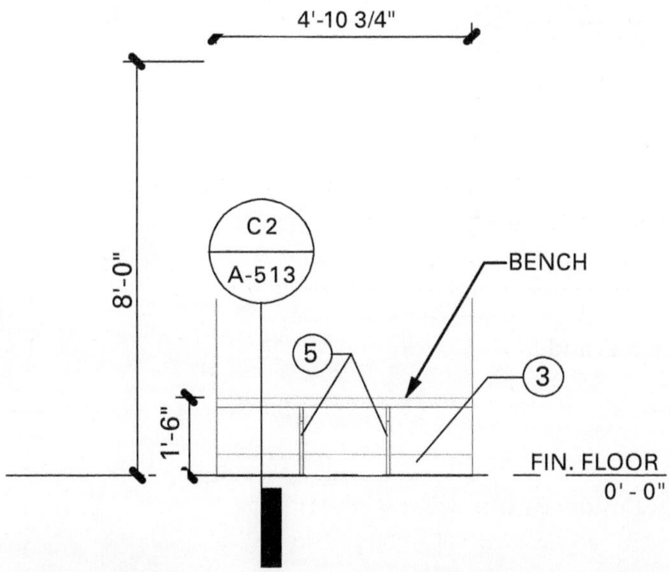

Figure RQ03

5. Refer to *Figure RQ03*. If the finished floor will be 1" above the subfloor on which you are currently working, at what height would you place the bench from the subfloor?
   a. 17"
   b. 18"
   c. 19"
   d. 20"

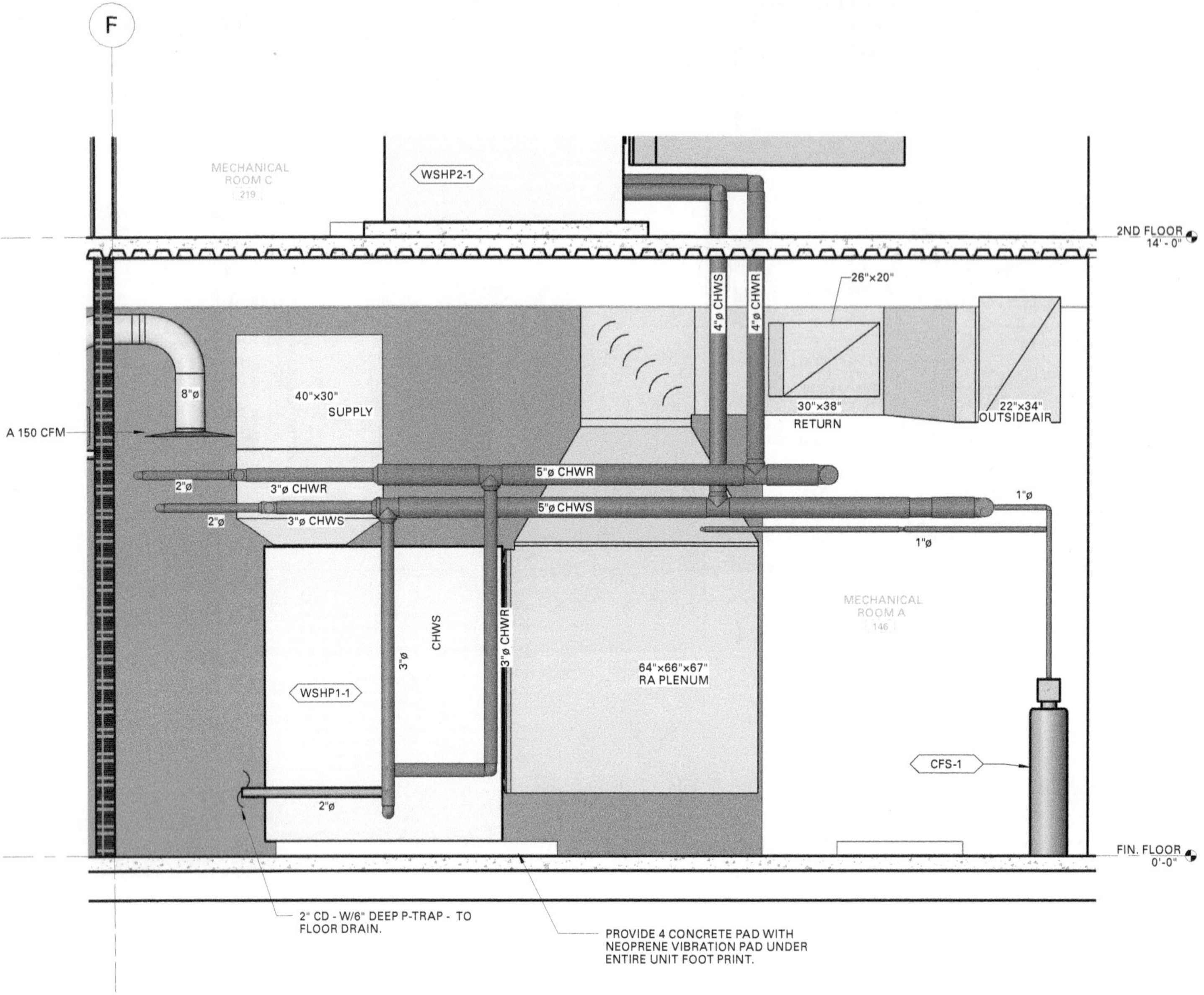

Figure RQ04

6. What size(s) of round duct is required to complete the ductwork in this room, based solely on the view shown in *Figure RQ04*?
   a. 12"
   b. 8"
   c. 3" and 5"
   d. All sizes between 1" and 5"

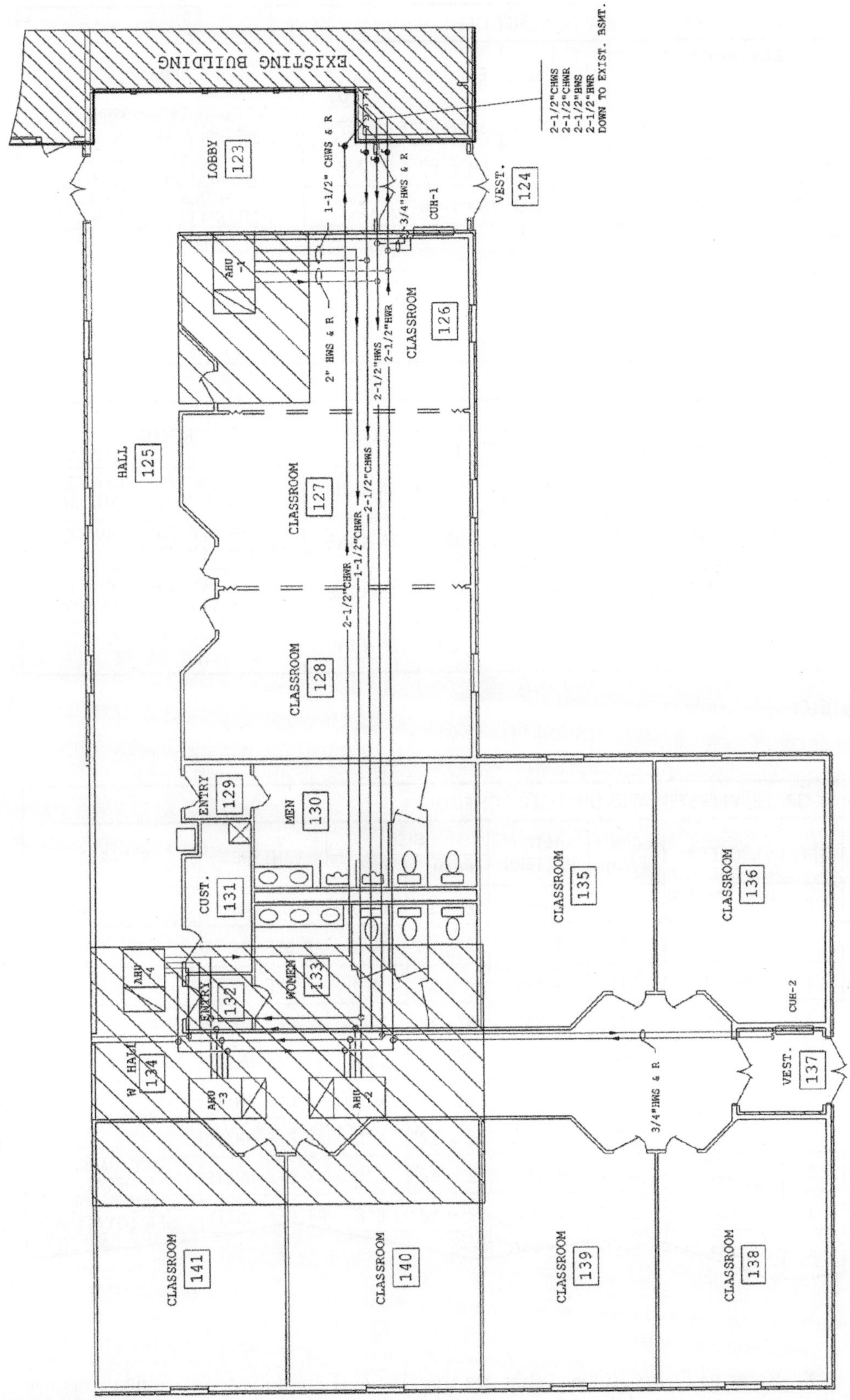

Figure RQ05

7. Refer to *Figure RQ05*. What size chilled-water piping should be connected to Air Handling Unit 1?
   a. $1\frac{1}{4}$"
   b. $1\frac{1}{2}$"
   c. 2"
   d. $2\frac{1}{2}$"

CABINET UNIT HEATER SCHEDULE												
UNIT HEATER NO.	LOCATION	C.F.M.	FAN MOTOR				MBH	GPM	EWT	EAT	MAX. WATER P.D.	REMARKS
			H.P.	VOLTS	PHASE	Hz						
CUH-1	124	400	1/12	115	1	60	23	2.3	180°F	60°F	2.7	McQUAY #CHF004 SEMI-RECESSED, R.H COIL
CUH-2	137	400	1/12	115	1	60	23	2.3	180°F	60°F	2.7	McQUAY #CHF004 SEMI-RECESSED, L.H COIL
CUH-3	143	400	1/12	115	1	60	23	2.3	180°F	60°F	2.7	McQUAY #CHF004 SEMI-RECESSED, R.H COIL

NOTES:
1. 3 Speed Control
2. Front Discharge
3. With Return Air Filters

PUMP SCHEDULE											
UNIT NO.	LOCATION	SERVICE	GPM	MBH	MOTOR					TYPE	REMARKS
					RPM	H.P.	VOLTS	PHASE	Hz		
P-1	MECH. ROOM	NAVE	45	41'	1750	1 1/2	208/230	3	60	IN-LINE	B/G #60-20T SERVICE 40% GLYCOL SOLUTION
P-2	MECH. ROOM	CHW TO AHU 1-4	65	37'	1750	1 1/2	208/230	3	60	IN-LINE	B/G #60-20T SERVICE 40% GLYCOL SOLUTION
P-3	MECH. ROOM	RECIR. TO TANK	40	17'	1750	1/2	208/230	3	60	IN-LINE	B/G #60-13T SERVICE 40% GLYCOL SOLUTION
P-4	EXIST. MECH. ROOM	HW	73	31'	1750	1 1/2	208/230	3	60	IN-LINE	B/G #60-20T HOT WATER

NOTES:
1. Starters And Disconnects By E.C.

GRILLE, REGISTER AND DIFFUSER SCHEDULE										
ITEM	MANUFACTURER	MODEL NO.	QTY.	LOCATION	CFM EACH	AIR PATTERN	SIZE		FINISHES	REMARKS
							FRAME	NECK		
A	BARBER COLMAN	SFSV	8	126,127,128 144	245	4-WAY	12"× 12"	8"Ø	#7 OFF-WHITE	
B	BARBER COLMAN	SFSV	2	142	275	4-WAY	18"× 18"	10"Ø	#7 OFF-WHITE	
C	BARBER COLMAN	SFSV	4	140, 141	240	4-WAY	12"× 12"	8"Ø	#7 OFF-WHITE	
D	BARBER COLMAN	SFSV	2	139	270	4-WAY	18"× 18"	10"Ø	#7 OFF-WHITE	
E	BARBER COLMAN	SFSV	2	138	280	4-WAY	18"× 18"	10"Ø	#7 OFF-WHITE	
F	BARBER COLMAN	SFSV	2	136	250	4-WAY	12"× 12"	6"Ø	#7 OFF-WHITE	
G	BARBER COLMAN	SFSV	2	135	235	4-WAY	12"× 12"	6"Ø	#7 OFF-WHITE	
H	BARBER COLMAN	SFSV	3	134	100	4-WAY	12"× 12"	6"Ø	#7 OFF-WHITE	FIRE DAMPER SEE DETAIL A
I	BARBER COLMAN	SFSV	1	134	190	4-WAY	12"× 12"	8"Ø	#7 OFF-WHITE	FIRE DAMPER SEE DETAIL A
	[illegible]	SFSV	1	134		[illegible]	12"[illegible]			FIRE DAMPER

Figure RQ06

8. Refer to *Figure RQ06*. What are the size and air delivery specifications for items tagged with an F on the relevant set of mechanical drawings?
   a. 12" × 12"; 240 cfm
   b. 12" × 12"; 250 cfm
   c. 18" × 18"; 235 cfm
   d. 18" × 18"; 280 cfm

9. Natural gas piping is typically shown on which of the following drawings?
   a. Floor plan
   b. Elevation drawings
   c. HVAC drawings
   d. Plumbing drawings

10. Cut lists are typically generated from ______.
    a. elevations
    b. floor plans
    c. specifications
    d. shop drawings

11. The final recipient of a request for information (RFI) form is the ______.
    a. foreman
    b. superintendent
    c. general contractor
    d. architect or engineer

12. The final drawings prepared for a construction project are called ______.
    a. submittals
    b. change orders
    c. shop drawings
    d. as-built drawings

13. Two national building codes adopted by state and local jurisdictions are the *International Building Code*® and ______.
    a. BOCA
    b. NFPA 5000®
    c. MasterFormat®
    d. AIA Standard Format

14. In the current CSI-formatted specifications, you would expect to find information on HVAC systems in ______.
    a. Division 7
    b. Division 10
    c. Division 15
    d. Division 23

15. To secure approval to furnish and install HVAC equipment such as exhaust fans, a subcontractor must acquire and provide the general contractor with a(n) ______.
    a. cut list
    b. takeoff
    c. submittal
    d. as-built drawing

Answers to odd-numbered questions are found in the Module Review Answer Key at the back of this book.

Answers to Section Review Questions

Answer	Section	Objective
Section One		
1. a	1.1.0	1a
2. c	1.2.0	1b
3. d	1.3.2	1c
4. b	1.4.1	1d
5. c	1.5.0	1e
6. c	1.6.0	1f
7. b	1.7.2	1g
Section Two		
1. b	2.1.2	2a
2. d	2.2.0	2b
3. a	2.3.0	2c

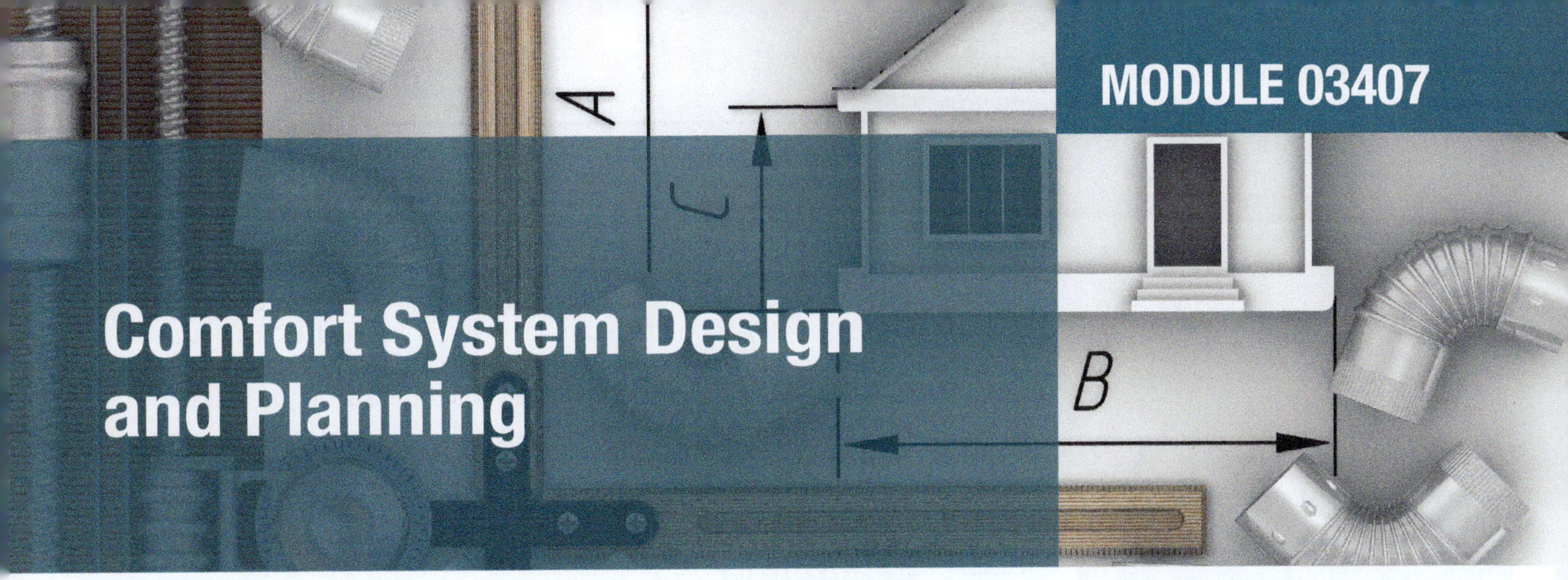

MODULE 03407

Comfort System Design and Planning

Source: Waldemarus/iStock/Getty Images

Objectives

Successful completion of this module prepares you to do the following:

1. Describe the HVAC system design process and describe the evaluation of a structure for load calculations.
   a. Describe the HVAC system design process.
   b. Describe the evaluation of a structure to gather information for load calculations.
2. Explain how heating and cooling load calculations are produced.
   a. Identify how and where heat is gained or lost through a building structure.
   b. Identify specific cooling and heating load factors.
3. Explain how to select equipment based on load calculations and plan for support systems that are part of an installation.
   a. Explain how to select cooling equipment.
   b. Explain how to select heating equipment.
   c. Explain how to select heat pumps.
   d. Identify common support systems and common planning requirements.
4. Explain how to design and size air distribution systems.
   a. Identify basic duct design considerations.
   b. Describe how to size duct systems and evaluate equipment selections.
   c. Identify heat load and system design factors unique to commercial buildings.

Performance Tasks

Under supervision, you should be able to do the following:

1. Using a simple home floor plan provided by the instructor, perform a heat load calculation using a standardized method.
2. Using a manual duct design calculator or app, size the trunk and branch ducts for an instructor-provided residential air distribution system and set of design parameters.
3. Calculate the total system friction loss (external static pressure) for a selected air distribution system.

Overview

The proper design of HVAC comfort systems, along with the selection of the proper heating and cooling equipment, are critically important to the success of an installation. Many variables influence system design. Contractors who rely solely on rule-of-thumb methods for equipment sizing and design put themselves at a competitive disadvantage. Successful contractors calculate the heating and cooling losses of the structure, then select the equipment and design an air distribution system that places conditioned air where it is needed.

Digital Resources for HVACR

Scan this code using the camera on your phone or mobile device to view the digital resources related to this craft.

SCAN ME

NCCER Industry-Recognized Credentials

If you are training through an NCCER-accredited sponsor, you may be eligible for credentials from NCCER. The ID number for this module is 03407. Note that this module may have been used in other NCCER curricula and may apply to other level completions. Contact NCCER at 1.888.622.3720 or go to **www.nccer.org** for more information.

You can also show off your industry-recognized credentials online with NCCER's digital credentials. Transform your knowledge, skills, and achievements into credentials that you can share across social media platforms, send to your network, and add to your resume. For more information, visit **www.nccer.org**.

1.0.0 HVAC Design Processes and Structure Evaluation

Performance Task

1. Using a simple home floor plan provided by the instructor, perform a heat load calculation using a standardized method.

Objective

Describe the HVAC system design process and describe the evaluation of a structure for load calculations.

a. Describe the HVAC system design process.

b. Describe the evaluation of a structure to gather information for load calculations.

NOTE

It is suggested that the following modules be reviewed to best understand the material presented here: NCCER Module 03109, *Air Distribution Systems* and NCCER Module 03213, *Sheet Metal Duct Systems*.

This module introduces HVAC comfort system design concepts. Design flaws, poor installation practices, and significant changes in the building after the equipment was placed into service are often the source of discomfort, noise, or inefficient operation.

Becoming an effective HVAC system designer requires extensive knowledge and experience. This module introduces the factors that affect the selection and design of HVAC systems and ductwork. The focus will be on residential systems, but the coverage will also touch on commercial system design. Upon completion, you should have a clear understanding of what information is required to complete a load calculation and the important factors in system design.

Material published by the Air Conditioning Contractors of America (ACCA) Association, Inc. is referenced extensively in this module. In addition to this text, it is strongly recommended that ACCA Manual J®, *Residential Load Calculation* be reviewed. ACCA Manual N®, *Commercial Load Calculation* is used for commercial system design for trainees who are interested in learning more.

The design process begins with gathering information about the structure and its environment. With that information in hand, the heat loss and heat gain of the structure can be calculated. Once these values are known, equipment can be selected, and a duct system can be designed.

1.1.0 Overview of the Design Process

The HVAC comfort-system design process has several important goals and steps (*Figure 1*):

- Surveying the building or reviewing the plans to determine the structure's characteristics
- Determining the heat gain and loss of the structure under the worst expected conditions
- Selecting the type and size of equipment needed to overcome the calculated heat gain and loss
- Sizing the distribution system to move the required amount of air through the primary equipment to ensure comfort throughout the space and reliable equipment operation
- Considering and planning other system factors and subsystems, such as the electrical power requirements and condensate disposal system

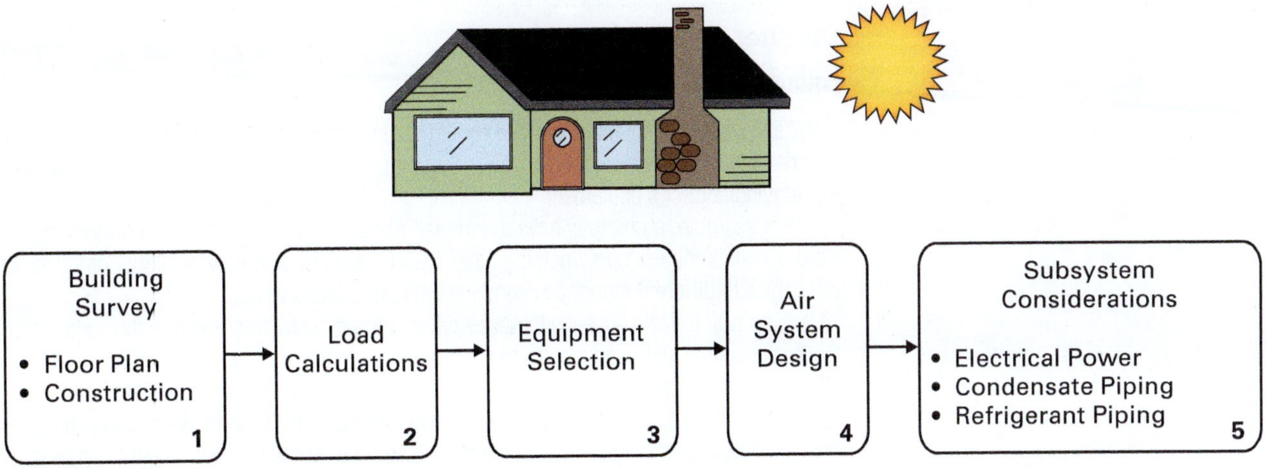

Figure 1 The HVAC comfort-system design process.

As shown in *Figure 1*, the first step in this process is to collect information about the building. When dealing with new construction, the building plans and specifications are the only reliable resource.

For existing structures, the construction drawings are still very helpful, but not often available. In this case, a field survey is needed to acquire the information to the best of your ability. A floor plan showing building dimensions, overhangs, number and type of windows and doors, and similar information is ideal, even if it must be generated through a detailed site visit.

Load calculations on paper can be time-consuming. In the past, the information was reviewed in an office where it was calculated manually using a variety of printed tables and other resources. Software can now quickly produce a load calculation in the field or in the office, especially on simpler residential jobs. What the software can't do is produce the information about the building's construction characteristics.

Regardless of how the calculation is performed, it is important that the information gathered on a site survey be as accurate as possible. In many cases, some assumptions must be made about certain characteristics, such as the wall insulation that is out of sight.

The primary purpose is to calculate the winter heat loss and summer heat gain. Once this is known, equipment can be selected to accommodate those loads. This in turn leads to the volume of airflow available to distribute in the system.

The typical load-calculation process will yield five values:

- Sensible cooling load
- Latent cooling load
- Minimum cooling airflow required
- Sensible heating load
- Minimum heating airflow required

As shown in *Figure 1*, the third step is to use the load calculations to select the equipment using the manufacturer's product data. For example, if the cooling load is 52,000 Btuh (about $4\frac{1}{2}$ tons of cooling), select equipment with that capacity or as close to it as possible without under-sizing. This will likely require a nominal 5-ton unit (60,000 Btuh) as a nominal 4-ton unit (48,000 Btuh) is insufficient.

Keep in mind that the load calculations will be based on design conditions for your area. This information includes the minimum and maximum expected temperatures. On rare occasions, conditions will exceed these values, but this doesn't render the equipment inoperative. When the equipment is properly selected using accurate load calculations, it will operate efficiently and effectively, providing comfort in the worst expected conditions.

On Site

Equipment Sizes

HVAC equipment manufacturers make equipment in nominal sizes. However, each model has its own real performance data that vary from system to system, and with different combinations of the same manufacturer's equipment.

For example, a unit may have a nominal rating of 5 tons, but its actual performance at various conditions will be something other than 60,000 Btus. It may be higher or lower, and it will vary with different indoor coil combinations and airflow values.

Even with today's advanced inverter-driven compressors that modulate the speed to track the load, nominal unit sizes are still the norm on the market. But these units are certainly more efficient and can track and respond to loads significantly smaller than the nominal capacity.

Always operating a compressor at its maximum capacity, when the load is something less than the design load roughly 98% of the time, is very inefficient. But that is how it has been done for many years. Traditionally, systems merely cycle on and off to track the heating and cooling load.

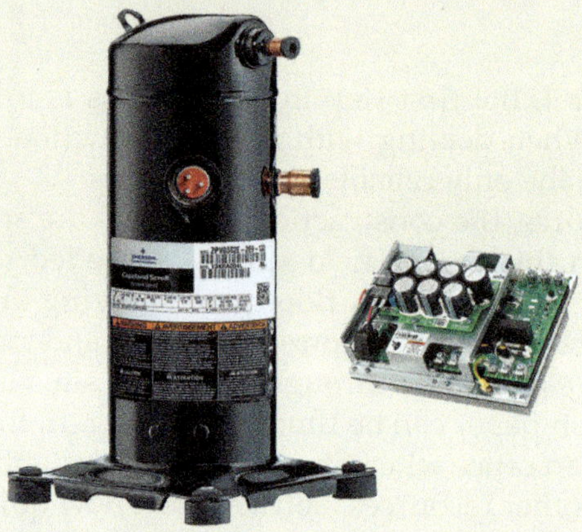

Source: Courtesy of Emerson

The heating equipment is selected using the same philosophy. The furnace or other heating appliance is selected from the manufacturer's product literature based on the heating load and the required amount of airflow. Other factors that affect the selection are covered throughout this module.

Once the equipment is selected, you will know the air volume it can produce for distribution and at what external static pressure (ESP) and motor speed. All three of these values are essential to the design process. The air distribution system is then designed to accommodate the airflow needed while ensuring the external static pressure remains within the specifications. In the case of an existing home, you can measure and sketch the duct system and ensure it can support the new equipment. Modifications are sometimes necessary.

This is a very important step. If the duct system is not properly designed, the system will not perform well, resulting in noise and comfort problems, even if the equipment selection is ideal.

With many of the primary considerations addressed, subsystems and accessories can be considered and planned. This step also contributes to quantifying and acquiring the materials needed to do the work. Here are some of the things to consider in this step:

- Will the electrical service support the load? What size wire and disconnect switch will the equipment require?
- What size gas line is needed and at what pressure must the gas be provided?
- What type of vent piping will the furnace need, and what route will it take?
- How much and what type of control wiring is needed between the units?
- Where should the room thermostat be located for best performance?
- How much and what size refrigerant and condensate piping is needed?

- What route will the piping, as well as the control wiring, need to take?
- Can the condensate be drained along the same route, or must it take a different route? Will a condensate pump be necessary to lift the condensate?
- Would a humidifier or whole-house, high-efficiency air filter be positive and practical additions to the system?

1.1.1 Oversizing Equipment

Keep in mind that comfort cooling systems both cool and dehumidify. A significant part of the evaporator's load is the latent heat transfer dedicated to condensing water vapor. If the equipment is significantly oversized, the excess capacity will bring the space temperature down too quickly in most conditions, and it will not run long enough to condense much water vapor. If the system is undersized, it will run longer and do an acceptable job of dehumidification, but it will lack the capacity needed to maintain the thermostat setpoint on the hottest days.

An oversized furnace will also short cycle, but dehumidification isn't a factor. The cycle rate may be a minor nuisance and result in some inefficiency. An undersized furnace will maintain comfort in milder weather, but it will be unable to maintain the setpoint on the coldest days.

The capacity of these systems is put to the test at different times of the day. The maximum load a cooling system usually encounters is late afternoon on a summer day. The sun has been out all day and daytime temperatures tend to peak at that time. The maximum load a heating system will encounter is during the early hours of a winter morning, just before dawn. The atmosphere has had all night to cool off, and outdoor temperatures tend to be coldest around dawn.

Oversized equipment costs more to install and to operate, so the user receives less comfort while paying more. Although the load-calculation process can be time-consuming, load-estimating software makes it faster and easier. All too often, contractors fail to run loads and select equipment by experience alone, using "rules of thumb." Some load programs, especially those provided by equipment manufacturers, also select the appropriate equipment.

Many software suppliers allow potential customers to download trial versions of their load-estimating program from the internet. A few free programs are also available. Take advantage of these offers and get to know some of the programs available and the kind of inputs needed to complete accurate load calculations. Don't guess at heating and cooling loads.

1.2.0 Building Evaluation and Survey

It is very important to obtain as much information as possible about the structure. The information is obtained from construction drawings or through a physical survey.

When possible, it is also important to talk to the current user in an existing home or business. Users can describe any current performance problems or special considerations that must be factored into the process. Residential designs are relatively straightforward unless special health requirements need consideration. Commercial projects can be significantly more complicated.

A great deal of information is needed to prepare load calculations. The factors that affect heat loss and heat gain are also summarized in *Figure* 2 and *Figure* 3. However, many details can't be placed on a drawing. Factors in heat loss and gain include, but are not limited to, the following (note that some factors only affect heat loss or heat gain, but not both):

- Building orientation to the sun
- Exterior dimensions of the building
- Construction type (frame, masonry, brick)
- Color of sun-exposed exterior roof and walls
- Roof material

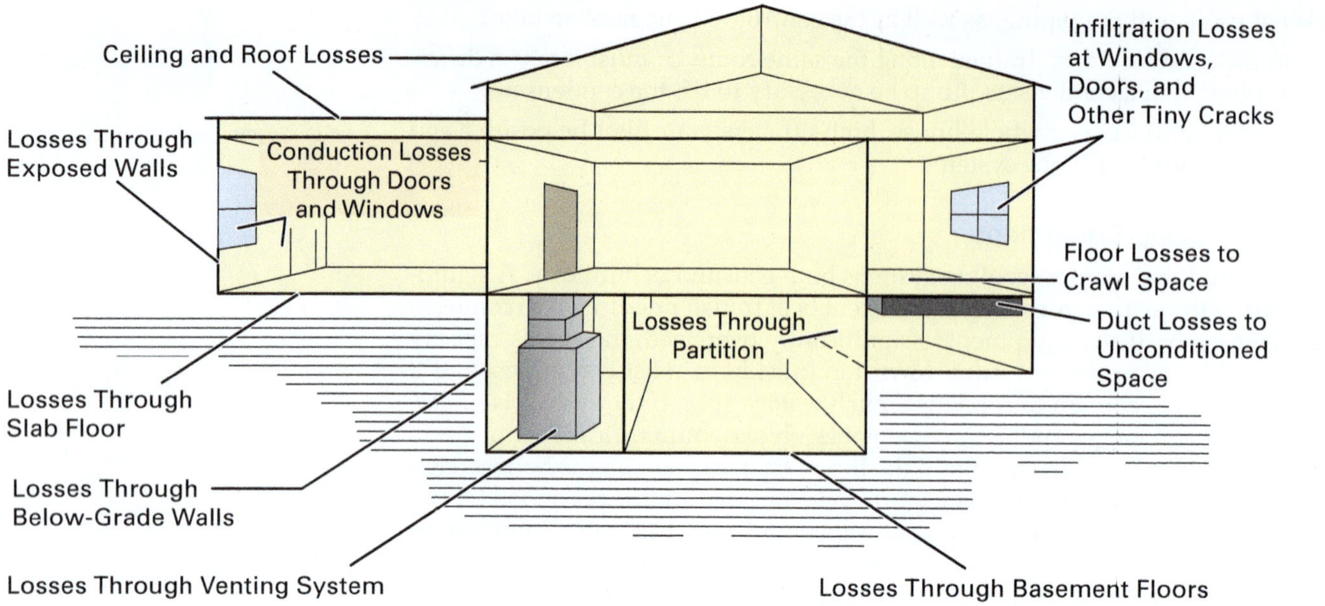

Figure 2 Winter heat losses.
Source: ACCA - Air Conditioning Contractors of America

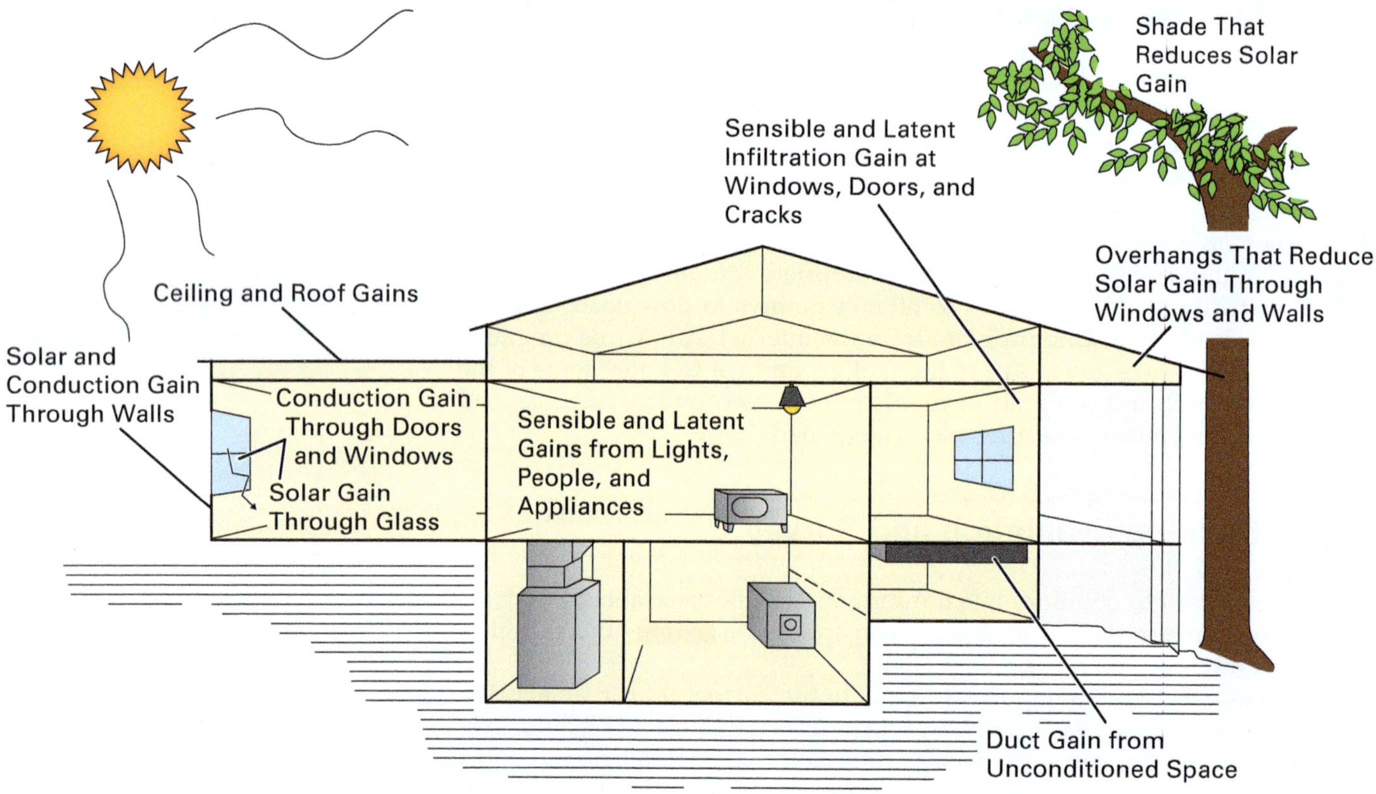

Figure 3 Summer heat gains.
Source: ACCA - Air Conditioning Contractors of America

- Thickness and type of insulation in the attic, walls, and floors
- Walls, floors, and ceilings exposed to unconditioned spaces such as garages and crawl spaces
- General tightness of the building (infiltration)
- Number, size, and type of windows and doors
- Interior window covering, such as draperies, curtains, or blinds
- Shading from trees, porch overhangs, permanent awnings, and adjacent buildings

When evaluating an equipment replacement, never assume that the capacity of the existing equipment is correct. Complete load calculations and ask the user about their experience to date. When the outdoor temperature is at or near the design temperature, properly sized heating or cooling equipment should run almost constantly to maintain the design setpoint. If the equipment cycles regularly when the outdoor temperature is at or above design temperatures, the equipment is possibly oversized.

On Site

Downsizing Equipment

You will occasionally encounter oversized systems during surveys. However, it can be difficult to convince a homeowner to downsize their system with the promise of better performance. Be sure your calculations are accurate, take the time to explore the issue with the owner carefully, and look for opportunities to solve comfort problems with efficient solutions.

For example, perhaps a homeowner likes to reduce the temperature in the master bedroom from 75°F to 70°F at night during the summer. This might only be done by reducing the temperature of the entire home or floor using a large system.

In this case, you might consider the addition of a smaller high-efficiency ductless heat pump system in the master suite. This system can be used to reduce the temperature in the master suite without affecting the rest of the home. Alternatively, consider a zoning system with an inverter-driven compressor that can more efficiently track the load when only the master bedroom needs cooling at night.

During a survey, be sure to note any remodeling that has been done since the original system was installed. If a room has been added or windows replaced, the heating and cooling loads will be affected. Also survey, measure, and sketch the existing ductwork. If a fossil-fuel furnace is present, measure and sketch the venting system. Keep in mind that old chimneys and venting systems may not be safe or adequate for use with modern furnaces.

During this survey and owner consultation, think about equipment options and configuration, possible zoning arrangements, and other aspects of the project. Buildings of all sizes may require more than one system and/or a zoning arrangement to ensure that all areas of the building are adequately and efficiently heated and cooled. For example, consider whether it is practical to replace an induced-draft furnace with a condensing model. The most important installation factor to consider here is the venting system that will have to be replaced. Is the furnace in a location that lends itself to these changes in the venting system, or will it be complicated or perhaps impractical?

A checklist for recording information about the building is an effective way to ensure that all the information needed for a load calculation is collected. *Figure 4* shows an example survey form for residential load calculations. Of course, many HVAC software packages provide forms like this that can be completed on a mobile device.

More than one type of form may be necessary to collect all the information. *Figure 5* shows a sketched floor plan with annotations. Note the lack of detail regarding interior walls and features. This form was prepared to run a **block load calculation** that considers the structure as a single room.

Block load calculation: A load calculation that considers the entire structure as one single area, essentially considering the structure to be a single room.

Between the data entered on the survey checklist and this floor plan, enough information should be available to do a block load calculation. The other option is to calculate the load on a room-by-room basis, then bring the individual loads together in a summary. The advantage of room-by-room calculations is that these calculations provide helpful insight as the air distribution system for each room is sized. However, room-by-room calculations are more time-consuming to complete.

The standard for reliable residential load-calculation guidance is provided by Manual J®, *Residential Load Calculation*, published by the Air Conditioning Contractors of America (ACCA). ACCA also publishes Manual N®, *Commercial*

Survey and Checklist (New Construction, Add-On, Replacement)

1. Design Conditions

Location: City ANYTOWN State U.S. Latitude 34°N

Temperature °F	Summer	Winter
Outside Design DB/WB	96/77°F	20°F
Daily Range	22°F	xxxxxx
Inside Design DB/%RH	75/50%	72/30%
Difference	21°F	52°F
Swing	3°F	xxxxxx

Special Internal Loads (Computers, Etc.) NONE

Frequent Entertaining ☐ Doors Opened Often ☐

2. Orientation and Type

House Faces: N NE E SE (S) SW W NW

Single Story ☐ Two Story ☑ Split Level ☐

3. Construction

Ceiling Height: Basement 7 1/2' 1st Floor 8' 2nd Floor 7'

Insulation (Inches) 0 1 2 3½ 6 8 12 R Value

Walls:
Frame ☑ Masonry Above Grade ☑ Bsmt ☐ ☐ ☐ ☐ ☐ (Frame) R-11
Masonry Below Grade: 0-5' ☑ >5' ☐ ☑ ☐ ☐ ☐ ☐ R-0

Roof:
Ceiling under Ventilated Attic ☑ ☐ ☐ ☐ ☐ ☐ ☑ ☐ R-25
Roof-Ceiling Combination ☐ ☐ ☐ ☐ ☐ ☐ ☐ R-
Roof on Exposed Beams or Rafters ☐ ☐ ☐ ☐ ☐ ☐ ☐ R-
Roof Color: Dark ☑ Light ☐

Floor:
Slab on Grade ☐ Edge Insulation: ☐ ☐ ☐
Floor over Garage or Vented Crawl space ☐
Floor Finish: Hardwood ☐ Carpeted ☐
Ceiling Below: Yes ☐ No ☐ ☐ ☐ ☐ ☐ ☐ R-
Floor over Unheated Basement or Enclosed Crawl space ☐
Ceiling Below: Yes ☐ No ☐ ☐ ☐ ☐ ☐ ☐ R-
Basement Floor >2' Below Grade ☑

4. Windows & Doors

Windows:
Type: Movable ☑ Fixed ☐ Jalousie ☐
Glass: Single ☐ Double (Single + Storm) ☑ Triple ☐
Clear ☐ Tinted ☐ Reflective ☐ Low "e" ☐
Frame: Wood ☑ Metal ☐ Metal w/TB ☐

Number	Size: Width	Height	Sq. Ft.	Type
2	3'	3'	9.0	24" O.H. - SOUTH
9	3'	4.5'	13.5	SHADED
3	3'	4.5'	13.5	24" O.H. - SOUTH
6	3'	4.5'	13.5	0" O.H.

Skylight: Wood ☐ Metal ☐ Metal w/TB ☐

Doors:
Type PANEL Storms METAL

Shading:
Internal: None ☑ Full ☐ Half ☐
Overhangs: 0" ☑ 12" ☐ 24" ☑ 36" ☐ 48" ☐ (24" - SOUTH)
Permanent External Shading NONE

5. Infiltration

Weatherstripping: Windows ☐ Doors ☑
Building Tightness: Loose ☐ Medium ☑ Tight ☐
Ventilation Fan ☐ Location CFM
Fireplaces: No. 1 Loose ☐ Medium ☑ Tight ☐

6. Present Equipment Survey

Indoor System:
Forced Air Furnace: Heat Only ☐ Heat/Cool ☑
Furnace w/Cooling ☐ Heat Pump w/Fan Coil ☐ Hydronic ☐
Fuel: Natural Gas ☑ LP Gas ☐ Oil ☐ Electricity ☐
Unit: Upflow ☑ Downflow ☐ Horizontal ☐ Package ☐
Location/Condition: W. END, BSMT. Good ☐ Avg ☑ Poor ☐ Age 15
Make: BRAND "X" Model No: XYZ
Capacity: Input 100,000 Output 76,000 BTU ☑ KW ☐
Blower: Motor HP 1/3 Direct Drive ☑ Belt Drive ☐
Multiple Speeds ✓ Blower Dia. 10½" Blower Width 10½"

Vent System: Condition: Good ☐ Avg ☑ Poor ☐ Age
Metal: Single Wall ☐ Double Wall Type "B" ☐ Dia.
Masonry: Unlined ☐ Lined ☑ Liner Size x "
PVC Plastic: ☐
Vent Connector: Dia. 6" Length ft. Corroded: Yes ☐ No ☑
Water Heater: 40,000 BTUH Common Vent: Yes ☑ No ☐ Replace: Yes ☐ No ☐

Outdoor:
Unit: Split System ☐ Package Unit ☐ Room Air ☐
Condensing Method: Air Cooled ☐ Water Cooled ☐
Water Supply: City ☐ Well ☐ Max Summer Temp °F
Location/Condition: Good ☐ Avg ☐ Poor ☐ Age
Ratings: Capacity BTUH; EER or SEER
Condensate: Gravity Drain ☐ Pump ☐ Sump ☐ Drywell ☐ Floor ☐
Emergency Overflow Pan in Attic: Yes ☐ No ☐
Refrigerant Lines: Length ft.
Diameter: Suction " Liquid "

Other Equipment:
Central Humidifier: Yes ☐ No ☑ Electronic Air Cleaner: Yes ☐ No ☑
Zoning: Yes ☐ No ☑ Heat Recovery Vent: Yes ☐ No ☑ Other:

7. Utilities

Natural Gas Meter Location N.E. CORNER, BASEMENT
LP Gas Tank Location
Oil Tank Location ; Pump Above ☐ Below ☐
Tank Size ; Distance from Pump
Electrical Service: Volts 240 Phase 1 Hz 60 Amps 100
Location of Entrance Panel N.E. CORNER, BASEMENT
Major Elec. Loads: Range/Self Clean ☐ Range/Oven ☑ Range Top ☐ Single Oven ☐
Double Oven ☐ Dryer ☐ Dish ☑ (→ FUTURE) KW Other KW

8. Controls

Zones: Single ☑ Multi ☐ Number 1
Thermostat Type: Heating ☑ Heat & Cool ☐ Continuous Fan ☑
Auto Changeover ☐ Clock-type w/Night Setback ☐ Programmable ☐
Location of Master Thermostat: LIVING ROOM - INSIDE WALL

9. Air Distribution System

Supply:
Location: Basement ☑ In Slab ☐ Crawl space ☐ Ceiling ☐ Attic ☐ Soffit ☐
Exposure: In Unconditioned Space ☐ To Outdoor Temp. ☐
Insulation: 0" ☑ 1/2" ☐ 1" ☑ 1 1/2" ☐ 2" ☐ (1" @ Wall Stacks)
Plenum: Width 16" Depth 19" Height 42 Clearance to Rafters 2"
Main Trunk Duct: Width 20" Height 8"
Runout Diameter: 6"
Outlets: Floor Perimeter ☑ Baseboard ☐ Ceiling ☐ High Sidewall ☐ Low Sidewall ☐

Return:
Location: Basement ☑ In Slab ☐ Crawl space ☐ Ceiling ☐ Attic ☐ Soffit ☐
Exposure: In Unconditioned Space ☐ To Outdoor Temp. ☐
Insulation: 0" ☑ 1/2" ☐ 1" ☐ 1 1/2" ☐ 2" ☐
Main Trunk Duct: Width 16" Height 8"

Figure 4 Example of a residential load-calculation survey form.
Source: Carrier Corporation

Load Calculation, which provides similar guidance for commercial buildings. Both Manual J® and Manual N® are available in print and digital versions. Most companies that offer competitive software for load calculations base the calculations on the ACCA publications. ACCA maintains a list of approved software vendors to ensure compliance with the rigorous standards.

Load calculation is a number-crunching process with many variables. One advantage of using software is the ability to quickly compare how

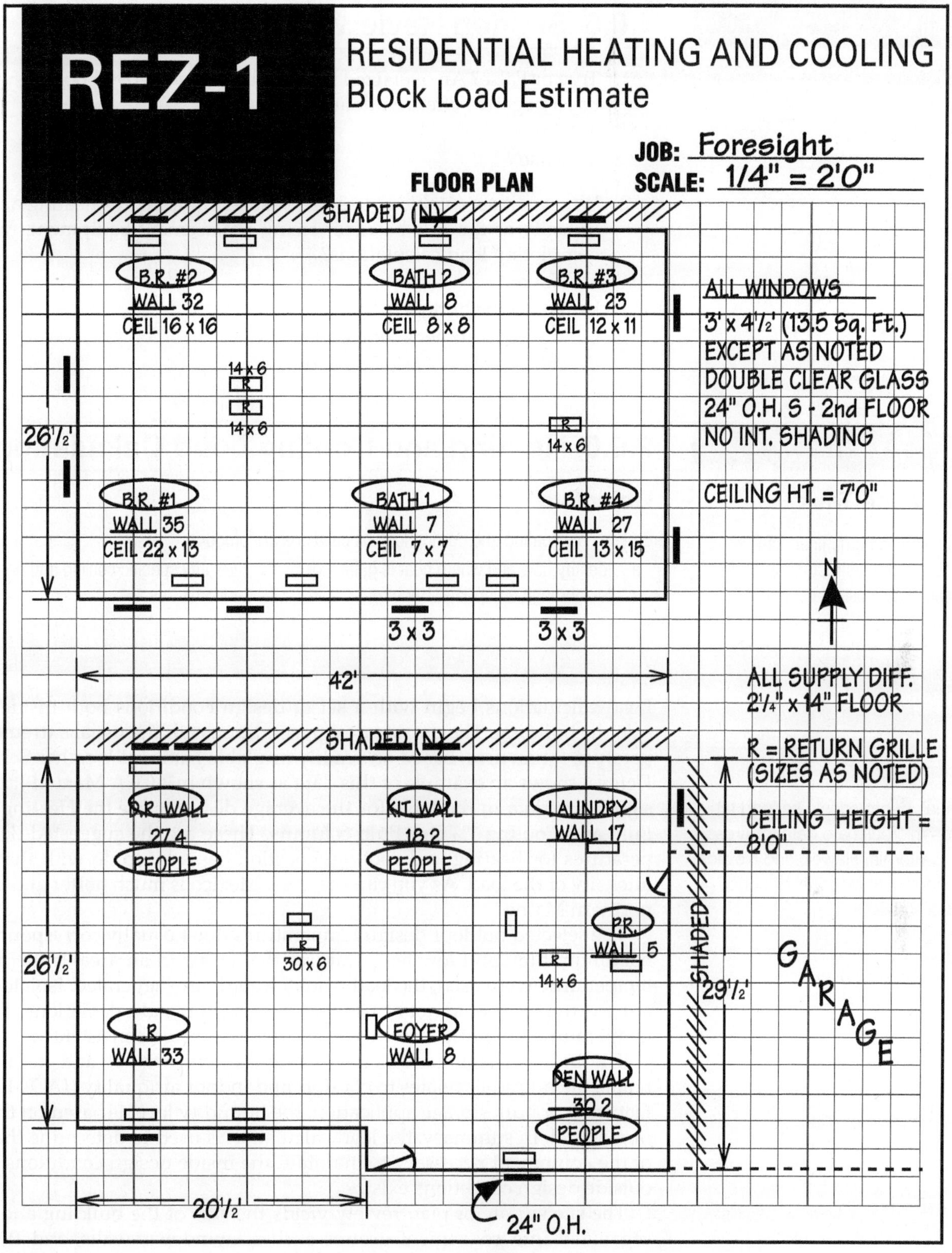

Figure 5 Example of an annotated floor plan for block load calculations.
Source: Carrier Corporation

different building construction factors might affect the load. For example, if a set of plans shows standard window glass, it is easy to change the entry to energy-efficient or triple-pane glass and see how it affects the load. Perhaps a different grade of windows will reduce the load enough to use smaller equipment and result in a significant return on investment (ROI) over a short period of time.

1.0.0 Section Review

1. True or False? An oversized furnace, like an oversized cooling system, also has a negative effect on dehumidification.
   a. True
   b. False

2. When is it acceptable to use the existing equipment's heating and cooling capacity as the sole basis for sizing replacement equipment, rather than completing load calculations?
   a. Never
   b. When the budget is tight
   c. At the customer's request
   d. When time is a critical factor

2.0.0 Heating and Cooling Load Calculations

Performance Task

1. Using a simple home floor plan provided by the instructor, perform a load calculation using a standardized method.

Objective

Explain how heating and cooling load calculations are produced.

a. Identify how and where heat is gained or lost through a building structure.
b. Identify specific cooling and heating load factors.

NOTE

ACCA Manual J® provides a second table that provides microclimate data for areas of Arizona, California, Hawaii, and Nevada, due to the significant disparity of conditions in those states.

Load calculations begin with a set of design conditions both inside and outside the building. The outside design conditions are the usual extremes of temperature for an area, based on readily available National Weather Service data. *Figure 6* shows an example of this data as shown in ACCA Manual J®. The most significant data in the table for the average designer are the Heating 99% Dry Bulb and Cooling 1% Dry Bulb columns. These are the suggested design temperatures for heating and cooling. Location has a lot to do with the type and intensity of the load. As you can see, New Mexico is much hotter and drier than northern Maine.

The selected outdoor design temperatures don't usually correspond with the maximum high and low temperatures that occur in an area. If the maximum temperature were used, the load results would be exaggerated. Isolated temperature extremes should not be used as the basis for load calculations. The design temperature is slightly closer to what can be expected consistently.

The inside design conditions include the temperature and humidity. In a home, the humidity relates to comfort and indoor air quality (IAQ). In a manufacturing or bulk storage application, the humidity level is based on the process or material. In summary, the load calculation is used to determine the capacity of the equipment necessary to maintain the inside design conditions when the outside design conditions exist.

The site survey or plan review yields the size of the building and its parts, organized into convenient groups such as windows, walls, and floors. This information is further divided into subgroups that make it easier to perform the load calculations.

These details lead to determining the heat transfer multipliers (HTMs) for each construction element. The HTM represents the heat gain and/or loss to be applied to the windows, walls, floors, and other construction elements.

To obtain the total sensible heat load for the building, the window, wall, door, floor, and ceiling areas are multiplied by the respective HTMs and then added together. Examples can be seen in the sample building load analysis reports

Table 1A
Outdoor Design Conditions for the United States

Location	Elevation Feet	Latitude Degrees North	Heating 99% Outdoor Dry Bulb	Cooling						HDD_{65} / CDD_{50} Ratio
				Outdoor Air		Design Grains			Daily Range (DR)	
				1% Dry Bulb	Coincident Wet Bulb	55% RH Indoors	50% RH Indoors	45% RH Indoors		
Alabama										
Alexander City	686	33	22	93	76	38	45	52	M	0.55
Anniston Metropolitan AP	600	34	25	92	76	36	43	50	M	0.51
Auburn-Opelika AP	774	33	28	91	74	28	35	42	M	0.43
Birmingham Municipal AP	630	34	24	93	75	32	39	45	M	0.49
Cairns AFB (Ozark)	299	31	30	93	76	37	44	51	M	0.28
Centreville WSMO	453	33	26	92	76	37	43	50	M	0.45
Decatur (Athens DD)	592	34	16	93	74	26	32	39	M	0.79
Dauphin Island	30	30	38	87	77	53	60	66	L	0.19
Dothan Municipal AP	322	31	31	93	76	35	42	49	M	0.26
Florence AP (Muscle Shoals DD)	581	34	21	94	75	30	37	44	M	0.60
Gadsden Municipal AP AWOS	568	34	22	91	75	32	38	45	M	0.68
Huntsville IAP, Jones Field	643	35	22	92	75	31	38	45	M	0.62
Maxwell AFB (Montgom)	174	32	32	95	77	37	44	51	M	0.28
Mobile Regional AP	220	31	31	92	76	42	48	55	M	0.25
Mobile CO	26	30	29	93	77	43	50	56	M	0.25
Montgomery, Dannelly Field	203	32	27	94	76	36	43	49	M	0.35
Muscle Shoals Regional AP	561	35	22	93	75	33	40	46	M	0.60
Ozark, Fort Rucker	356	31	31	94	77	42	49	56	M	0.28
Selma-Craig AFB	166	32	26	95	77	40	47	53	M	0.37
Talladega	528	33	22	94	76	36	43	50	M	0.69
Tuscaloosa Municipal AP	187	33	25	93	76	38	45	52	M	0.44
Alaska										
Adak NAS	20	52	23	57	53	-19	-13	-6	L	90
Ambler	289	67	-38	73	58	-26	-19	-12	M	24
...uk Pass	2,156	68	-35	66	53	-36	-29	-22	M	77
	131	61	-4	68	57	-19	-12	-6	L	14
	138	61	-7	70	58	-19			L	11
	194	61	-9	71	58				L	13
	42	61	–13	71	58					
		55	20	69						
			-22	71						

Figure 6 Example of ACCA design-conditions table assembled from NWS data from across the US.
Source: ACCA - Air Conditioning Contractors of America

provided in *Appendix 03407A, Sample Residential Load Calculation*. For example, under the heating calculations, the wall area is multiplied by an HTM of 3.3 Btuh/ft^2. The HTM is derived by the software from the information previously input about the wall construction.

Figure 7 shows a typical input screen from a popular load-calculation app. The software shown offers features such as the following:

- ACCA-approved app based on ACCA Manual J®
- Calculates system losses, duct sizes, and blower ESP requirements based on ACCA Manual D,® *Residential Duct Systems*
- Calculates block loads or room-by-room loads and provides room airflow requirements
- Assists in compliance with the *International Residential Code*® (IRC®) and the *International Energy Conservation Code*® (IECC®)

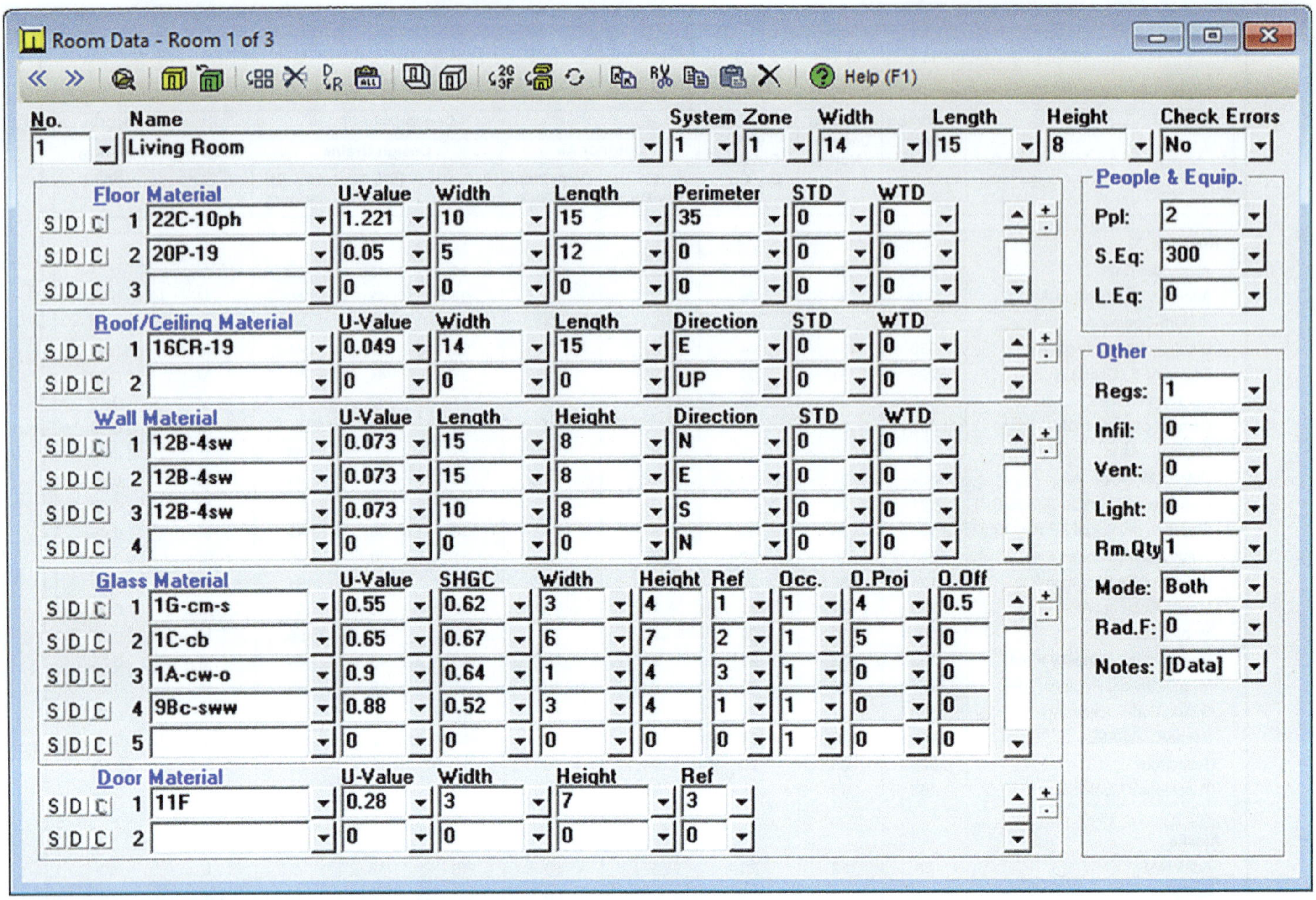

Figure 7 Example of a data input screen from a load-calculation app.
Source: Elite Software

Once all the information is entered, the summary printout (*Figure 8*) provides the details of the load calculation for heating and cooling. Note that the complete report spans seven pages and can be viewed in *Appendix 03407A, Sample Residential Load Calculation*. In addition to providing the information you need, detailed reports like this are appreciated by potential clients and demonstrate a professional approach. Load-calculation apps may also provide features such as equipment selection, operating-cost comparisons, and airflow requirements, enhancing the programs' versatility.

On Site

Hydronic System Design

Although load calculations for a home are completed in the same manner regardless of system type, the design process for a hydronic system differs significantly. Instead of calculating air volumes and duct sizes, the designer must determine water temperatures, waterflow rates, pipe sizes, and pump capacity. For a hydronic floor heating system, the designer must also determine how many lineal feet of tubing is needed to heat each area.

wrightsoft

Project Summary
Entire House
Bertie Heating and Air Conditioning

Job:
Date: [redacted]
By: Ryan Johnston

1730 NE 23rd Ave, Gainesville, FL 32609 Phone: 352-331-2005 Email: ryanj@bertieair.com Web: www.bertieair.com License: CAC058522

Project Information

For: [redacted]
Gainesville, FL

Notes:

Design Information

Weather: Gainesville Rgnl, FL, US

Winter Design Conditions

Outside db	33	°F
Inside db	70	°F
Design TD	37	°F
Ventilation Method	MJ8	

Summer Design Conditions

Outside db	92	°F
Inside db	75	°F
Design TD	17	°F
Daily range	M	
Relative humidity	50	%
Moisture difference	47	gr/lb

Heating Summary

Structure	27548	Btuh
Ducts (R-6.0)	7212	Btuh
Central vent (0 cfm)	0	Btuh
Humidification	0	Btuh
Piping	0	Btuh
Equipment load	34759	Btuh

Sensible Cooling Equipment Load Sizing

Structure	17701	Btuh
Ducts (R-6.0)	7812	Btuh
Central vent (0 cfm)	0	Btuh
Blower	0	Btuh
Use manufacturer's data	n	
Rate/swing multiplier	0.97	
Equipment sensible load	24722	Btuh

Infiltration

Method	Simplified
Construction quality	Semi-tight
Fireplaces	0

	Heating	Cooling
Area (ft²)	2079	2079
Volume (ft³)	20790	20790
Air changes/hour	0.22	0.11
Equiv. AVF (cfm)	76	38

Latent Cooling Equipment Load Sizing

Structure	2209	Btuh
Ducts	1471	Btuh
Central vent (0 cfm)	0	Btuh
Equipment latent load	3680	Btuh
Equipment Total Load (Sen+Lat)	28402	Btuh
Req. total capacity at 0.75 SHR	2.7	ton

Heating Equipment Summary

Make	Carrier	
Trade	15 SEER2 HP	
Model	GH5SAN43600AA0	
AHRI ref	210998688	
Efficiency	7.5 HSPF2	
Heating input		
Heating output	34000	Btuh @ 47°F
Temperature rise	27	°F
Actual air flow	1140	cfm
Air flow factor	0.033	cfm/Btuh
Static pressure	0.50	in H2O
Space thermostat		

Capacity balance point = 35 °F

Backup:
Input = 10 kW, Output = 34601 Btuh, 100 AFUE

Cooling Equipment Summary

Make	Carrier	
Trade	15 SEER2 HP	
Cond	GH5SAN43600AA0	
Coil	FJ4DNXB36L	
AHRI ref	210998688	
Efficiency	12.0 EER2,14.3 SEER2	
Sensible cooling	25650	Btuh
Latent cooling	8550	Btuh
Total cooling	34200	Btuh
Actual air flow	1140	cfm
Air flow factor	0.045	cfm/Btuh
Static pressure	0.50	in H2O
Load sensible heat ratio	0.87	

Calculations approved by ACCA to meet all requirements of Manual J 8th Ed.

wrightsoft ACCA Right-Suite® Universal 2024 24.0.01 RSU59482 [redacted]
[redacted] - Manual JD&S.rup Calc = MJ8 Front Door faces: W
Page 1

Figure 8 Sample report (partial) from a load calculation app.
Source: Bertie Heating and Air Conditioning Inc.

2.1.0 Heat Transfer

At this stage of your training, you know that heat can be moved from one place to another. Heat is always transferred in one direction—from a higher temperature to a lower temperature. The greater the temperature difference, the more rapid the transfer. For example, if it is 70°F inside and 0°F outside (21°C and –18°C, respectively), heat moves rapidly out of the structure because of the significant temperature difference. If it is 70°F inside and 65°F outside (21°C and 18°C), the heat loss is nearly insignificant because the temperature difference is very small.

Effective insulation installed in floors, walls, and ceilings greatly reduces heat loss and heat gain. Other construction materials, such as exterior siding and sheathing, also reduce heat loss and gain. The higher the **R-value** of a material, the lower the conductive heat transfer will be.

R-value: A number identifying the thermal resistance of a given thickness of insulating material; higher numbers reflect greater resistance to heat flow.

Conduction is just one way that heat is transferred. Heat transfer takes place through conduction, radiation, and convection.

2.1.1 Conduction

Conduction is the transfer of heat energy in a substance from particle to particle from the warmer region to the colder region. If, for example, one end of a rod is heated over a flame, heat travels by conduction from the hot end to the cool end. Heat transfer by conduction occurs not only within an object or substance but also between different substances in contact with one another.

Going Green

Roof Color

Dark-colored roofs are common in colder climates because the heat gained helps to offset the winter heat losses through the roof. In contrast, white roofs are widely used in the south and desert southwest to reflect solar radiation and reduce heat gain through the roof. The roof color can significantly affect the load calculations for a structure. The photo shows an energy-saving white foamed roof on a home in Arizona.

2.1.2 Radiation

Radiant heat does not rely on contact between substances. Radiant heat can even travel through a vacuum, although it can't pass through an opaque object. It can also be reflected from one surface to another, just as light is reflected by a mirror.

Radiant heat passing through air does not warm the air through which it passes. For example, a roof with the sun shining on it might be heated to 180°F (82°C) by the rays of the sun, but the air through which the radiant heat travels remains far cooler.

All objects and substances radiate some heat. The higher the temperature of the object, the greater the amount of heat radiating from it. The amount of radiant heat given off per unit of time depends on both the temperature of the radiating body and the type and color of its surface. A rough, dark surface, for example, radiates more heat than a smooth, light surface of the same dimensions. In addition, the darker the surface, the more solar radiation it will absorb. Thus, dark surfaces always have higher heat gains than light surfaces when exposed to radiant light from the sun.

2.1.3 Convection

Convection is the transfer of heat energy through the movement of a fluid. A fluid is roughly defined as anything that flows; thus, all gases and liquids are fluids. Air, being a mixture of several gases, is also a fluid. As air circulates, it transfers heat from one place to another.

In a gas furnace, for example, heat energy is released by the combustion process and transferred through the heat exchanger by conduction. Air from the conditioned space is forced over the heated walls of the heat exchanger by the blower. Because the air is cooler than the heat exchanger, heat flows from the heat exchanger to the air. The heat exchanger is designed such that the air passes over its surface in a wiping motion, allowing a large amount of heat to transfer to the airstream.

The warmed air then moves through the air distribution system. Through convection, the warm air warms the ductwork and eventually the objects and occupants in the conditioned space.

2.1.4 Heat Gain and Loss

In the summer, heat is transferred through the exterior walls into the building. Theoretically, a wall continually losing heat to a room would eventually cool down to the room air temperature. However, heat that the wall loses to the room air is continuously replaced with heat from outside the structure.

The rate of heat transfer through a wall separating spaces at different temperatures depends on three main factors:

- The area of the wall
- The temperatures of the two spaces (temperature differential)
- The heat-conducting properties of the wall

As the surface area of a wall increases, the amount of heat transferred also increases. For example, a wall with an area of 200 ft^2 can conduct twice as much heat as a wall with an area of 100 ft^2 exposed to the same conditions.

The temperature difference affects the rate of heat transfer and the amount of heat passing through. If the difference in temperature between the two spaces is 50°F (28°C), only half the sensible heat will flow through the wall than it would if the temperature difference were 100°F (56°C).

These heat flow principles also apply to windows, roofs, floors, and other building surfaces. In summary, the flow of heat through any surface is directly proportional to its area and the difference in temperature of the spaces separated by the surface. The third factor affecting heat transfer through building components is the wall composition and its thickness.

The terms *conductivity* and *conductance* are used to describe heat flow through building materials. **Thermal conductivity** refers to the ability of a material to conduct heat. The best conductors of heat are typically metals. Materials such as wood, cork, and expanded foam are poor thermal conductors.

Thermal conductivity can be defined as the heat flow per hour (Btuh) through one square foot of a 1" thick **homogeneous** material when the temperature difference between the two sides is 1°F (*Figure 9*).

Thermal conductivity: A number reflecting the heat flow per hour, in Btuh, through one square foot of 1" thick homogeneous material per 1°F of temperature difference between the two surfaces. The actual conductance of the material is changed by the actual temperature difference and material thickness.

Homogeneous: Describes a mixture, substance, or material that is uniform throughout its composition.

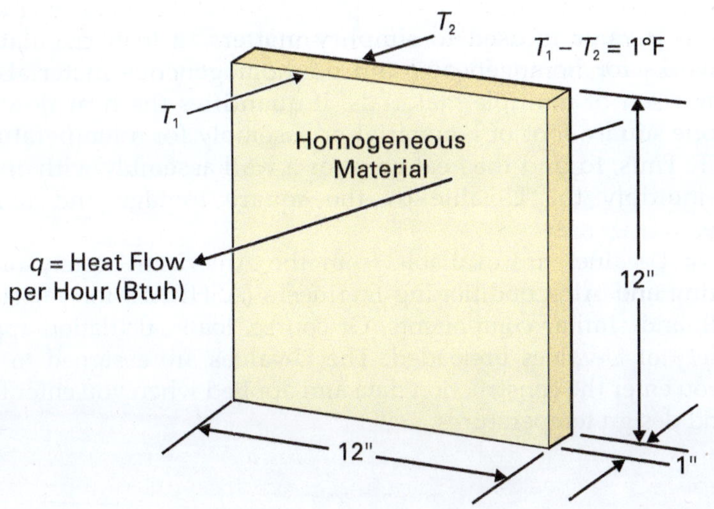

Figure 9 Concept of thermal conductivity illustrated.

A similar term, conductance, is used to quantify heat flow through materials such as glass blocks, hollow clay tile, concrete blocks, and other composite materials. These materials are not homogeneous. Thus, heat does not flow through the material at the same rate through every inch of thickness.

Thus, the conductance of a material is defined as the heat flow rate in Btuh through one square foot of a nonhomogeneous material of a specified thickness when there is a 1°F temperature difference between the two surfaces (*Figure 10*).

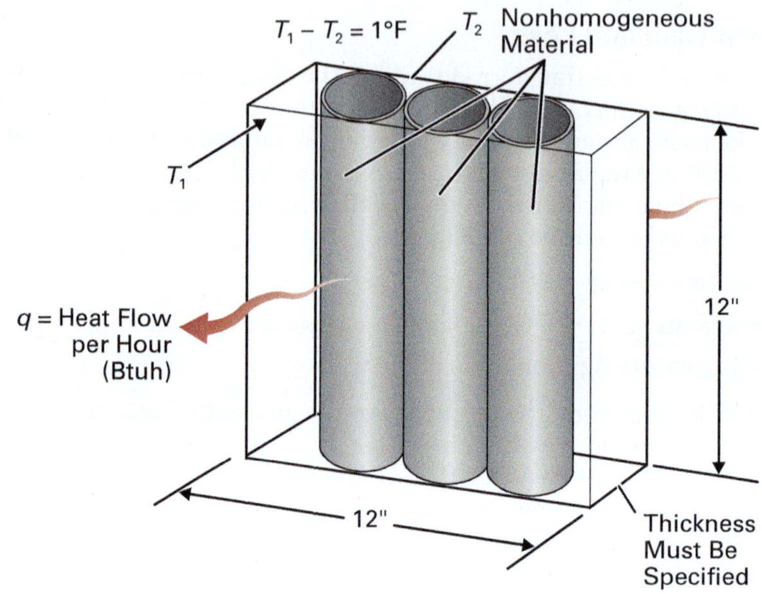

Figure 10 Conductance illustrated.

Conductivity and conductance are not interchangeable terms. Conductivity is the heat flow through 1" of a homogeneous material; conductance is the heat flow through the entire specified thickness of a nonhomogeneous material. The conductance of air may also be considered during load calculations.

It is inconvenient to measure all the surface temperatures of the materials that comprise a wall, yet it is easy to find the temperature on both sides of a wall with an ordinary thermometer. Therefore, for nonhomogeneous materials like a wall comprised of several different materials, it is much easier to use a value that reflects the conductance of a wall structure than for its individual components.

U-value: A heat-transfer coefficient rating of how well a building assembly, such as a wall consisting of multiple materials, conducts heat. The U-value is used as a multiplier in load calculations.

The term **U-value** is used to simplify matters in load calculations. The U-value works for homogeneous and nonhomogeneous materials, such as walls comprised of multiple materials. It quantifies the heat flow per hour through one square foot of a material or assembly for a temperature difference of 1°F. Thus, to find the heat gain for a wall assembly with an assigned U-value, multiply the U-value by the square footage and temperature difference.

Tables of U-values are available from the American Society of Heating, Refrigerating and Air-Conditioning Engineers (ASHRAE) for various ceiling, floor, wall, and similar components. Of course, load-calculation apps have a wide variety of U-values preloaded. The U-values are assigned to the comp onent as you enter the construction data and applied when you enter the square footage and design temperatures.

2.2.0 Other Heating and Cooling Load Factors

The materials used in constructing a building are not the only factors that affect heat gain and heat loss. The location and positioning of a building, for example, also all have an effect.

A residence shaded by trees or a building in the shadow of taller buildings will have a smaller cooling load than more exposed buildings. This is especially true if the exposed building is oriented with a lot of glass exposed to the sun. Of course, many tall buildings are largely sheathed with glass and fully exposed. Today's glazing options can provide far better thermal performance, both in terms of conductivity and solar gain, than plain glass.

The pie charts in *Figure 11* show the average impact of various construction factors on cooling and heating loads. *Table 1* lists various construction factors and the impact on heating and/or cooling loads. These factors will be explored further as the module continues.

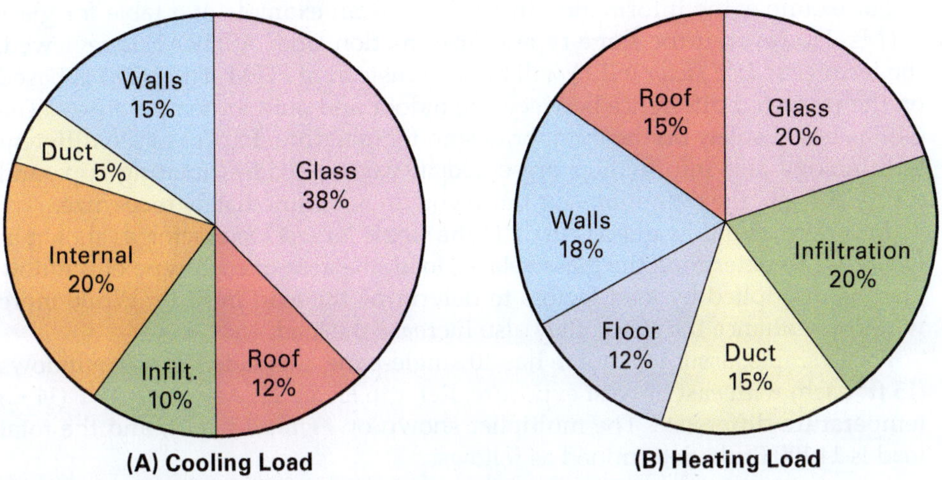

Figure 11 Ratio of effect on building loads for various structural components.

TABLE 1 Effects of Construction Factors on Cooling and Heating Loads

Factor	Cooling Load	Heating Load	Comments
Windows and Glass Doors	X	X	Major impact on cooling load; significant impact on heating load
Shaded Glass	X		Shading reduces solar gain
Wood and Metal Doors	X	X	Greater effect on heating load
Exterior Walls	X	X	Degree of impact depends on insulation value and type
Below-Grade Masonry Walls		X	Limited but consistent due to the steady soil temperatures
Floors		X	Soil or crawl space beneath has little or no impact on the cooling load
Roof	X		Darker colors increase the absorption of radiant heat
Infiltration	X	X	Greater effect on heating load due to the increased temperature difference
Partitions to Unconditioned Spaces	X	X	Some effects but far smaller than direct exposure to outdoor conditions
Duct System	X	X	Greater effect on heating load
Internal Loads	X		People, appliances; all have positive influences on the heating load

The architect as well as the HVAC system designer can improve energy efficiency by specifying the use of items such as heavier insulation, tight-fitting windows and doors, triple-pane glazing, and reflective glass. Changes in glazing alone can significantly reduce the heating and cooling load. However, a higher cost is associated with all enhanced materials.

On Site

Consider the Future

Use caution when considering shading from adjacent buildings or trees for a cooling load calculation. Buildings may be torn down, but far more often, it's the trees that end up being removed. Be sure any buildings or trees you are counting on for shade will be around for a while and think about the effect on your calculation if that factor were removed. Although you cannot plan and design around every future event that could theoretically happen, be sure you are at least aware of the possibilities.

2.2.1 Window Glass

Before discussing the topic of window glass, it is important to point out that the terminology and process often differ between load-estimating packages. The results are comparable, but the factors used to arrive at the answer to each calculation may differ. Most of the examples used in this module were taken from ACCA Manual J® for consistency.

For example, the information in *Figure 12* is an example of a table for glass HTMs. However, if the same type of information from ASHRAE is reviewed, the term *glass load factor* (GLF) will be seen instead of HTM. The HTM is based on the projected difference between the indoor and outdoor temperatures. The GLF is based solely on the outdoor design temperature. Expect to see different terminology and information presented in various load-estimating apps and forms. Review the definitions for those you do not immediately recognize.

In cooling, window glass is usually the single largest load factor, as shown in *Figure 11*. To determine the glass-related load, the area of each type of window glass is multiplied by load factors to determine the load in Btuh. While more windows enhance the space, they also increase the load.

For example, assume a home has 10 single-pane, unshaded 3' × 5' windows (15 ft^2 each) with east or west exposure. Refer to *Figure 12*. Assume a 25°F (14°C) temperature difference. The multiplier shown on *Figure 12* is 94, and the total load is 14,100 Btuh, determined as follows:

$$\text{Glass load} = (\text{number of windows} \times \text{area}) \times \text{HTM}$$
$$\text{Glass load} = (10 \text{ windows} \times 15 \text{ ft}^2) \times 94 \text{ Btuh/ft}^2$$
$$\text{Glass load} = 150 \text{ ft}^2 \times 94 \text{ Btuh/ft}^2$$
$$\text{Glass load} = 14{,}100 \text{ Btuh}$$

Thus, the 10 windows alone create a cooling load exceeding 1 ton. Note that if the same windows have internal shading (refer to the lower tables of *Figure 12*), the load would be reduced significantly. Half-open blinds, for example, drop the HTM to 66, which reduces the heat load by about 30%.

These are some of the tradeoffs to consider when working on load calculations. If this is an existing home needing a new system, can you rely on the occupant to keep the blinds no more than half open when the windows are exposed to sunlight? Perhaps not. If this is new construction, the owner could be offered the option of installing more energy-efficient windows. The initial cost would be higher, but the savings in the size of the cooling system, combined with lower operating costs, would eventually pay off.

If a project is sized for windows with full, insulated draperies and the owner later switches to light, sheer curtains, it can affect the ability of the system to meet the cooling demand, especially if there is a large amount of glass exposed to the east and west. It may not be enough of a difference to cause discomfort throughout the home, but local discomfort is certainly possible.

Windows that are fully or partially shaded by roof overhangs, awnings, or shade screens, must also be taken into account when calculating the cooling load. A fully shaded east, west, or south window is basically treated like a north window, meaning it will not be exposed to direct sunlight. If a window is only partially shaded, the percentage of shading must be factored in. Manual J® provides a special calculation in which the shaded and unshaded portions are

Table 3A

$MJ8_{AE}$ Procedure: Cooling HTM for Windows and Glass Doors

No External Sun Screen, Clear Glass

Cooling Load (Btuh) = HTM × Reference Area
Bay Window HTM = 1.15 × Table HTM
Garden Window HTM = 2.00 × Table HTM
French Door HTM = 0.70 × Table HTM
Use Single Pane, Clear Glass for Jalousie Window

Default indoor design temperature = 75°F.
Outdoor design temperature provided by Table 1.
Load area appears above HTM values.
Table 3, note 2 specifies the order of application of the HTM adjustment procedures.

Recommended Adjustments

1) Full outdoor insect screen = 0.80 × Table HTM
   Half outdoor insect screen = 0.90 × Table HTM
   Full indoor insect screen = 0.90 × Table HTM
   Half indoor insect screen = 0.95 × Table HTM
2) Shade by external overhang (See Table 3E-1)

No Optional Adjustments

Table 3A — Clear Glass

No Internal Shade

Default Assembly Performance	Single Pane						Double Pane						Triple Pane					
	U-Value 0.98		SC 0.85		SHGC 0.74		U-Value 0.56		SC 0.75		SHGC 0.65		U-Value 0.42		SC 0.70		SHGC 0.61	
Design CTD	10	15	20	25	30	35	10	15	20	25	30	35	10	15	20	25	30	35
Exposure	HTM for Rough Opening						HTM for Rough Opening						HTM for Rough Opening					
North	24	29	34	39	44	49	18	21	24	27	29	32	16	18	20	22	24	26
NE or NW	56	61	66	70	75	80	46	49	52	55	57	60	42	44	46	48	50	53
East or West	80	84	89	94	99	104	67	70	73	76	78	81	62	64	66	68	70	72
SE or SW	68	73	78	83	88	93	57	60	63	65	68	71	52	54	56	59	61	63
South	40	45	50	55	60	65	32	35	38	41	44	46	29	31	33	36	38	40

Vertical or Horizontal Blinds with Slats At 45 Degrees

Default Assembly Performance	Single Pane						Double Pane						Triple Pane					
	U-Value 0.98		SC 0.60		SHGC 0.52		U-Value 0.56		SC 0.50		SHGC 0.44		U-Value 0.42		SC 0.45		SHGC 0.39	
Design CTD	10	15	20	25	30	35	10	15	20	25	30	35	10	15	20	25	30	35
Exposure	HTM for Rough Opening						HTM for Rough Opening						HTM for Rough Opening					
North	16	21	26	31	35	40	11	13	16	19	22	25	9	11	13	15	17	19
NE or NW	36	41	46	50	55	60	27	30	33	36	38	41	24	26	28	30	32	34
East or West	51	56	61	66	71	76	40	43	46	49	51	54	35	37	40	42	44	46
SE or SW	44	49	54	59	64	68	34	37	40	42	45	48	30	32	34	36	38	40
South	25	30	35	40	45	49	18	21	24	27	29	32	16	18	20	22	24	26

Drape or Roller Shade Half Drawn

Default Assembly Performance	Single Pane						Double Pane						Triple Pane					
	U-Value 0.98		SC 0.70		SHGC 0.61		U-Value 0.56		SC 0.60		SHGC 0.52		U-Value 0.42		SC 0.55		SHGC 0.48	
Design CTD	10	15	20	25	30	35	10	15	20	25	30	35	10	15	20	25	30	35
Exposure	HTM for Rough Opening						HTM for Rough Opening						HTM for Rough Opening					
North	17	22	27	32	37	41	12	14	17	20	23	26	10	12	14	16	18	20
NE or NW	40	45	50	55	60	65	32	34	37	40	43	46	28	30	32	34	36	38
East or West	58	63	68	73	78	83	47	50	53	55	58	61	42	44	46	49	51	53
SE or SW	50	54	59	64	69	74	40	43	45	48	51	54	35	38	40	42	44	46
South	28	32	37	42	47	52	21	24	26	29	32	35	18	20	22	24	27	29

Drape or Roller Shade Fully Drawn

Default Assembly Performance	Single Pane						Double Pane						Triple Pane					
	U-Value 0.98		SC 0.50		SHGC 0.44		U-Value 0.56		SC 0.45		SHGC 0.39		U-Value 0.42		SC 0.40		SHGC 0.35	
Design CTD	10	15	20	25	30	35	10	15	20	25	30	35	10	15	20	25	30	35
Exposure	HTM for Rough Opening						HTM for Rough Opening						HTM for Rough Opening					
North	15	20	25	30	34	39	10	13	16	19	21	24	8	10	12	15	17	19
NE or NW	31	36	41	46	51	56	25	28	31	33	36	39	21	24	26	28	30	32
East or West	44	49	54	59	64	69	37	40	42	45	48	51	32	34	35	38	40	42
SE or SW	38	43	48	53	58	63	31	34	37	40	42	45	27	29	31	33	35	37
South	22	27	32	37	42	47	17	20	23	25	28	31	14	16	19	21	23	25

Figure 12 Example of Manual J® glass HTMs.
Source: ACCA - Air Conditioning Contractors of America

treated as separate entities, and then the results are added to determine the load for the entire window. Remember that French doors, sliding glass doors, and other construction features are also treated as windows in load calculations.

Going Green

Energy-Efficient Windows

A single pane of glass provides very little insulation. A single pane of glass has an R-value less than 1. Remember that the greater the R-value, the greater the insulating value. Adding another pane with a ½" air space between the panes more than doubles the R-value. The air space between the panes of glass also acts as an insulator, slowing heat transfer. Windows are commonly designed with two or three layers of glass separated by an air space to improve performance.

To obtain even more insulating value, the space between the panes can be filled with argon gas, which conducts heat more slowly than air.

A special type of glass known as *low emissivity*, or low-e, glass provides even greater insulating quality. *Emissivity* refers to the ability of a material to absorb or radiate heat. Low-e glass is coated with a very thin metallic substance on the inside of a double-pane window. In cold weather, radiated heat from walls, floors, and furniture is reflected back into the room by the low-e coating. This reduces the heat loss, which in turn reduces heating costs. In the summer months, radiated heat from outdoor sources, such as the sun, roads, and parking lots, is reflected away from the building by the low-e coating.

Although windows with low-e glass are considerably more expensive than standard windows, these windows usually return the investment within three or four years through reduced heating and cooling costs.

Source: Coprid/iStock/Getty Images

2.2.2 Walls, Roofs, Ceilings, and Floors

Any surface exposed to the outdoors must be figured into the load calculation. These surfaces gain and lose heat as described earlier. The amount of load a surface represents depends on the construction material used. The load is calculated in a similar way to the window-glass load—by applying a factor. In Manual J®, tables are provided showing the U-value for each type of panel (wall, roof, ceiling, or floor) assembly. *Figure 13* shows a portion of a Manual J® table for frame walls and partitions. On the single page shown, all values are for brick siding.

Table 4A
Heating and Cooling Performance for Opaque Panels
U-Values and Group Numbers or CLTD Values

Construction Number 12
Frame Walls and Partitions

Wall or partition with brick veneer, plus interior finish (40 to 50 Lb / SqFt)
Wall with siding or stucco, or light partition, plus interior finish (7 to 20 Lb / SqFt)
Exterior finish code: b = brick veneer; s = stucco or siding
Framing code: w = wood, m = metal (studs 16 Inches on center, 75% cavity, 25% framing)
Reference Area = Gross Wall Area - Area of Window and Door Openings

Construction Number	Insulation R-Values	Description of Construction	Exterior Finish	U-Value with Wood Studs	U-Value with Metal Studs	Group Number
12A — No Insulation In Stud Cavity						
12A-0b w/m 12A-0s w/m	Cavity: None Board: None	Frame construction, no cavity insulation, no board insulation, wood sheathing	Brick Siding	0.253 0.240	0.315 0.295	E A
12A-2b w/m 12A-2s w/m	Cavity: None Board: R-2	Frame construction, no cavity insulation, R-2 board insulation	Brick Siding	0.194 0.186	0.230 0.219	E A
12A-3b w/m 12A-3s w/m	Cavity: None Board: R-3	Frame construction, no cavity insulation, R-3 board insulation	Brick Siding	0.162 0.157	0.187 0.180	F B
12A-4b w/m 12A-4s w/m	Cavity: None Board: R-4	Frame construction, no cavity insulation, R-4 board insulation	Brick Siding	0.139 0.135	0.157 0.152	F B
12A-5b w/m 12A-5s w/m	Cavity: None Board: R-5	Frame construction, no cavity insulation, R-5 board insulation	Brick Siding	0.122 0.119	0.136 0.132	G C
12A-6b w/m 12A-6s w/m	Cavity: None Board: R-6	Frame construction, no cavity insulation, R-6 board insulation	Brick Siding	0.109 0.106	0.120 0.117	G C
12B — R-11 Insulation In 2 x 4 Stud Cavity						
12B-0b w/m 12B-0s w/m	Cavity: R-11 Board: None	Frame construction, R-11 cavity insulation, no board insulation, wood sheathing	Brick Siding	0.097	0.122	H B
12B-2b w/m 12B-2s w/m	Cavity: R-11 Board: R-2	Frame construction, R-11 cavity insulation, R-2 board insulation	Brick Siding	0.086	0.106	I C
12B-3b w/m 12B-3s w/m	Cavity: R-11 Board: R-3	Frame construction, R-11 cavity insulation, R-3 board insulation	Brick Siding	0.079	0.096	J D
12B-4b w/m 12B-4s w/m	Cavity: R-11 Board: R-4	Frame construction, R-11 cavity insulation, R-4 board insulation	Brick Siding	0.073	0.088	J D
12B-5b w/m 12B-5s w/m	Cavity: R-11 Board: R-5	Frame construction, R-11 cavity insulation, R-5 board insulation	Brick Siding	0.068	0.081	K E
12B-6b w/m 12B-6s w/m	Cavity: R-11 Board: R-6	Frame construction, R-11 cavity insulation, R-6 board insulation	Brick Siding	0.064	0.075	K F
12C — R-13 Insulation In 2 x 4 Stud Cavity						
12C-0b w/m 12C-0s w/m	Cavity: R-13 Board: None	Frame construction, R-13 cavity insulation, no board insulation, wood sheathing	Brick Siding	0.091	0.115	I C
12C-2b w/m 12C-2s w/m	Cavity: R-13 Board: R-2	Frame construction, R-13 cavity insulation, R-2 board insulation	Brick Siding	0.081	0.101	J D
12C-3b w/m 12C-3s w/m	Cavity: R-13 Board: R-3	Frame construction, R-13 cavity insulation, R-3 board insulation	Brick Siding	0.075	0.092	K E
12C-4b w/m 12C-4s w/m	Cavity: R-13 Board: R-4	Frame construction, R-13 cavity insulation, R-4 board insulation	Brick Siding	0.069	0.084	K E
12C-5b w/m 12C-5s w/m	Cavity: R-13 Board: R-5	Frame construction, R-13 cavity insulation, R-5 board insulation	Brick Siding	0.064	0.078	K F
12C-6b w/m 12C-6s w/m	Cavity: R-13 Board: R-6	Frame construction, R-13 cavity insulation, R-6 board insulation	Brick Siding	0.060	0.072	K G

Figure 13 Manual J® U-values for frame walls and partitions (partial).
Source: ACCA - Air Conditioning Contractors of America

Going Green

Energy Performance Ratings

The National Fenestration Rating Council (NFRC) is a nonprofit organization created by the window, door, and skylight industry that includes manufacturers, architects, code officials, and government agencies among their members. The primary goal of the organization is to provide accurate information to measure and compare the energy performance of windows, doors, and skylights. NFRC has established a voluntary energy performance rating and labeling system for fenestration (window and door) products.

The NFRC label contains vital information such as U-factor, solar heat gain coefficient, visible transmittance, and air leakage. Manufacturers can also ask that a condensation rating be determined for their product labels.

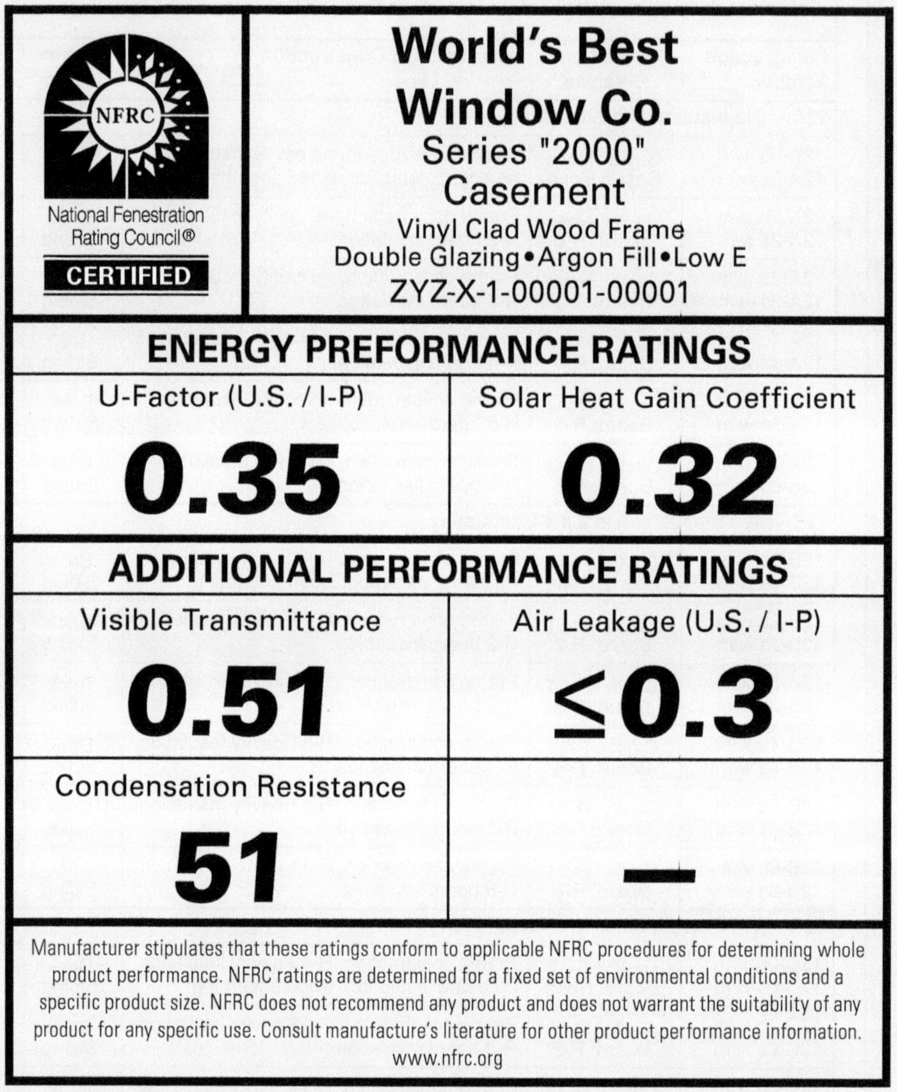

Source: National Fenestration Rating Council (NFRC)

A separate table containing similar data for heating load factors is also provided. Note that, when determining the load represented by exposed walls, the square footage for applicable windows and doors is subtracted from the wall area. Otherwise, those areas will be considered twice.

As shown in *Figure 13,* insulation thickness plays a significant role in the load presented by exposed surfaces. Money invested in insulation will generally pay for itself over time.

When calculating heating loads, consideration must also be given to walls, floors, and ceilings that adjoin unconditioned spaces such as garages, crawl spaces, and attics. Although these surfaces aren't directly adjacent to the outdoors, the unconditioned space still represents a heating and/or cooling load, even with insulation in the partition.

Basements with walls below grade also require special consideration when calculating a heating load. Even with insulation, the load can be substantial. The portion of the wall that is below grade has a very consistent heat loss, although the temperature difference of exposed walls is typically much higher. Therefore, the portions above and below grade must be treated separately, like partially shaded windows. Basement floors must also be calculated separately. If the first floor of the structure is above an unheated basement or crawl space, it can represent a significant heating load. Below-grade basement walls aren't usually a factor in the sensible cooling load, although dehumidification may be needed.

Roof area has a major effect on the cooling load. In Manual J®, the reflective qualities of the roof are also important. As discussed earlier, light-colored roofing material will reflect radiant heat better than dark material. The insulation provided is also an important factor in the roof load.

2.2.3 Infiltration

Infiltration affects heating and cooling loads. Air enters a building through cracks around windows, doors, and other construction joints. Fireplaces and vents without dampers are another source of infiltration.

Infiltration generally contributes more to the heating load than the cooling load. This is because of the greater temperature difference between the conditioned inside air and the infiltrating outside air. Additional caulking and weather stripping can make a dramatic difference, especially in cold climates.

The volume of infiltration is a function of the building's *tightness*. Tightness is a function of several factors, as stated in ACCA Manual J® where it is broken down into classifications:

- *Tight* — Continuous infiltration barrier with all cracks and penetrations sealed; tested leakage of windows and doors less than 0.25 cfm per running foot of crack; vents and exhaust fans have dampers; recessed ceiling lights gasketed or taped; no combustion air required or combustion air from outdoors; no duct leakage.
- *Average* — Plastic vapor barrier, major cracks and penetrations sealed; tested leakage of windows and doors between 0.25 and 0.50 cfm per running foot of crack; electrical fixtures that penetrate the envelope not taped or gasketed; vents and exhaust fans have backdraft dampers; combustion air from indoors; intermittent ignition and flue damper; minimal duct leakage to unconditioned space.
- *Loose* — No infiltration barrier or plastic vapor barrier; no attempt to seal cracks and penetrations; tested leakage of windows and doors greater than 0.50 cfm per running foot of crack; vents and exhaust fans do not have backdraft dampers; combustion air from indoors; no flue damper; significant duct leakage to unconditioned space.

In addition, two other categories have now been inserted between the others to provide greater accuracy: *semi-tight* and *semi-loose.*

Figure 14 shows how the effect of infiltration is determined. Infiltration is apportioned to the load calculation based on the square footage of the home and the chosen category of tightness. The Manual J® table provides the number of air changes per hour (ACH) that result from infiltration. One ACH means that infiltration replaces all the air within the structure once per hour. Of course, when it is very hot or cold outdoors, replacing even 20% of the air in a home every hour with outside air represents a major load.

You must convert the ACH provided in the Manual J® tables to infiltration cubic feet per minute (ICFM), which is used to determine the volume of outside air that must be heated and cooled. This is done by multiplying the ACH value by the actual cubic footage of the area, then dividing by 60 to convert from

Table 5A
Infiltration Air Change Values for Three or Four Exposures

Default Air Change Values for Single Story Construction

Construction	Air Changes per Hour — Heating					Infiltration [1] Cfm for One Fireplace
	Floor Area of Heated Space (SqFt)					
	900 Or Less	901 to 1500	1501 to 2000	2001 to 3000	3001 or More	
Tight	0.21	0.16	0.14	0.11	0.10	0
Semi-Tight	0.41	0.31	0.26	0.22	0.19	13
Average	0.61	0.45	0.38	0.32	0.28	20
Semi-Loose	0.95	0.70	0.59	0.49	0.43	27
Loose	1.29	0.94	0.80	0.66	0.58	33

1) For one additional fireplace, add 7 Cfm to the above fireplace values. For two or more additional fireplaces, add 10 Cfm (total) to the above.

Construction	Air Changes per Hour — Cooling					Fireplace Infiltration Cfm
	Floor Area of Air Conditioned Space (SqFt)					
	900 Or Less	901 to 1500	1501 to 2000	2001 to 3000	More than 3000	
Tight	0.11	0.08	0.07	0.06	0.05	0
Semi-Tight	0.22	0.16	0.14	0.11	0.10	
Average	0.32	0.23	0.20	0.16	0.15	
Semi-Loose	0.50	0.36	0.31	0.25	0.23	
Loose	0.67	0.49	0.42	0.34	0.30	

Default Air Change Values for Two Story Construction

Construction	Air Changes per Hour — Heating					Infiltration [1] Cfm for One Fireplace
	Floor Area of Heated Space (SqFt)					
	900 Or Less	901 to 1500	1501 to 2000	2001 to 3000	3001 or More	
Tight	0.27	0.20	0.18	0.15	0.13	0
Semi-Tight	0.53	0.39	0.34	0.28	0.25	13
Average	0.79	0.58	0.50	0.41	0.37	20
Semi-Loose	1.23	0.90	0.77	0.63	0.56	27
Loose	1.67	1.22	1.04	0.85	0.75	33

1) For one additional fireplace, add 7 Cfm to the above fireplace values. For two or more additional fireplaces, add 10 Cfm (total) to the above.

Construction	Air Changes per Hour — Cooling					Fireplace Infiltration Cfm
	Floor Area of Air Conditioned Space (SqFt)					
	900 Or Less	901 to 1500	1501 to 2000	2001 to 3000	More than 3000	
Tight	0.14	0.11	0.09	0.08	0.07	0
Semi-Tight	0.28	0.21	0.18	0.15	0.13	
Average	0.41	0.30	0.26	0.21	0.19	
Semi-Loose	0.64	0.47	0.40	0.33	0.29	
Loose	0.87	0.64	0.54	0.44	0.39	

Figure 14 Infiltration ACH values.
Source: ACCA - Air Conditioning Contractors of America

hours to minutes. For example, assume the infiltration for a 1,600 ft^2 home with an 8' ceiling is found to be 0.31 ACH:

$$ICFM = [(square\ footage \times ceiling\ height) \times ACH] \div 60\ \text{minutes}$$
$$ICFM = [(1{,}600\ \text{ft}^2 \times 8') \times 0.31] \div 60\ \text{minutes}$$
$$ICFM = [12{,}800\ \text{ft}^3 \times 0.31] \div 60\ \text{minutes}$$
$$ICFM = 3{,}968\ \text{cfh} \div 60\ \text{minutes}$$
$$ICFM = 66\ \text{cfm}$$

Note that the Manual J® tables offer different values for single-story and multistory homes.

2.2.4 Duct Losses

Duct losses affect both heating and cooling loads, although the losses have a greater effect on heating loads. The US Department of Energy states that residential duct systems in poor condition can lose 25% to 40% of the heating or cooling energy passing through, due to leakage and heat transfer through the duct walls. Adding insulation and sealing leaks can significantly reduce those losses.

Figure 15 shows Manual J® heat loss, sensible heat gain, and latent heat gain factors assigned to duct systems based on outside air temperature (OAT), the square footage of the home, and R6 insulation. R6 is the insulating value assigned to most 2" thick ductwrap. Correction factors are provided for different insulation values. There are a number of these tables in Manual J®, each offered for specific duct configurations and locations (i.e., attic, crawl space, or basement).

2.2.5 Cooling Load Factors

Some load factors only affect the cooling load. For example, internal factors such as people and appliances only contribute to the cooling load.

Indeed, internal factors typically reduce the heating load, because internal factors are items that produce heat. People emit heat, so the more occupants there are in the building, the greater the cooling load. If the occupants entertain large groups, that must also be factored into the load calculation. In the absence of the specific number of occupants provided by a homeowner, a rule of thumb is to consider the load generated by one person per bedroom, then add one person to the total.

Figure 16 shows default scenarios and load values for both appliances and people. A separate Manual J® table provides more specific information for a variety of appliances. Note that the table uses 230 Btuh for the sensible load from people, plus 200 Btuh for the latent load (evaporation of moisture from the skin). Of course, if people are exercising or working hard, their heat output rises sharply. In a room-by-room load calculation, the people are apportioned to the rooms they would normally occupy during the peak load hours.

Appliances, baths, showers, and lighting also add heat indoors. In *Figure 16*, a default appliance load of 1,200 Btuh is shown to cover the sensible load of a refrigerator and a vented range hood. However, very few homes with kitchens are this sparsely equipped. You'll likely need to look at the other options to find the scenario for today's average kitchen.

Table 7C-T — Trunk and Branch Supply and Return System in 16C, 16CR or 16CF Attic

7C-T
Ambient dry-bulb temperature = Outdoor db + 11 (heating) and Outdoor db + 25 (cooling) Supply location = Center of room ceilings Nominal return Cfm = 400 Return location = Close to air handler
Duct leakage Cfm per SqFt of duct surface area (supply / return) = 0.06/0.06; 0.09/0.15; 0.12/0.24; 0.24/0.47; 0.35/0.70 Duct wall insulation R-value = 2, 4, 6 and 8

Base Case Heat Loss Factor (BHLF)
R6 Insulation, ASHRAE Sealed (Supply = 0.12, Return = 0.24)

Table 1 OAT	1,000	1,500	2,000	2,500	3,000
	Square Feet of Floor Area				
–10	0.220	0.240	0.260	0.280	0.310
0	0.200	0.220	0.240	0.260	0.290
10	0.180	0.200	0.220	0.240	0.260
20	0.170	0.180	0.200	0.220	0.240
30	0.140	0.170	0.180	0.200	0.220
40	0.110	0.140	0.160	0.180	0.200

Interpolate or extrapolate for OAT and Floor Area.

R-Value Correction (WIF—Heat Loss)	R2	R4	R6	R8
	1.85	1.24	1.00	0.85

Base Case Sensible Gain Factor (BSGF)
R6 Insulation, ASHRAE Sealed (Supply = 0.12, Return = 0.24)

Table 1 OAT	1,000	1,500	2,000	2,500	3,000
	Square Feet of Floor Area				
85	0.190	0.240	0.290	0.320	0.350
90	0.210	0.260	0.290	0.320	0.350
95	0.230	0.260	0.290	0.320	0.350
100	0.240	0.260	0.290	0.320	0.360
105	0.240	0.260	0.300	0.330	0.370

Interpolate or extrapolate for OAT and Floor Area.

R-Value Correction (WIF—Sensible Gain)	R2	R4	R6	R8
	2.00	1.28	1.00	0.86

Base Case Latent Gain (BLG)
R6 Insulation, ASHRAE Sealed (Supply = 0.12, Return = 0.24)

Table 1 Grains	1,000	1,500	2,000	2,500	3,000
	Square Feet of Floor Area				
10	191	265	317	363	387
20	303	421	503	576	614
30	420	583	697	798	850
40	541	751	897	1,027	1,094
50	666	925	1,105	1,265	1,347
60	795	1,104	1,319	1,510	1,608
70	929	1,290	1,541	1,764	1,879

1) The latent gain value is zero if the Table 1 grains value is zero.
2) Interpolate or extrapolate for Grains.
3) Negative Table 1 grains values produce negative latent loads.

Leakage Correction (LCF) for Heat Loss

Leakage	R2	R4	R6	R8
0.06 / 0.06	0.84	0.77	0.71	0.67
0.09 / 0.15	0.92	0.89	0.85	0.83
0.12 / 0.24	1.00	1.00	1.00	1.00
0.24 / 0.47	1.27	1.37	1.44	1.52
0.35 / 0.70	1.52	1.76	1.90	2.04

Leakage Correction (LCF) for Sensible Gain

Leakage	R2	R4	R6	R8
0.06 / 0.06	0.80	0.74	0.69	0.64
0.09 / 0.15	0.90	0.87	0.84	0.82
0.12 / 0.24	1.00	1.00	1.00	1.00
0.24 / 0.47	1.34	1.45	1.52	1.59
0.35 / 0.70	1.71	1.97	2.14	2.27

Leakage Correction (LCF) for Latent Gain

Leakage	Any R-Value
0.06 / 0.06	0.28
0.09 / 0.15	0.63
0.12 / 0.24	1.00
0.24 / 0.47	2.17
0.35 / 0.70	3.58

Default Duct Wall Surface Area (SqFt)

Floor Area Look-Up Value									
1,000 SqFt		1,500 SqFt		2,000 SqFt		2,500 SqFt		3,000 SqFt	
Supply	Return	Supply	Return	Supply	Return	Supply	Return	Supply	Return
202	91	302	132	388	168	471	200	527	227

See Sections 6-8 and 23-6 for instruction for determining the floor area look-up value.

Surface Area Factors

Leakage	Ks	Kr
0.06 / 0.06	0.694	0.306
0.09 / 0.15	0.660	0.340
0.12 / 0.24	0.643	0.357
0.24 / 0.47	0.624	0.376
0.35 / 0.70	0.606	0.394

Surface Area Adjustment Factor (SAA) for Heat Loss or Sensible Gain

SAA = Ks x (Actual supply area / Default supply area) + Kr x (Actual return area / Default return area)
See Table 7A-AE, 7B-AE, 7C-AE or 7D-AE for an example application of this equation.

Surface Area Adjustment (LGA) for Latent Gain

LGA = Actual return-side area / Default return-side area
See Table 7A-AE, 7B-AE, 7C-AE or 7D-AE for an example application of this equation.

Procedure for Heat Loss and Sensible Gain Factor Adjustment
Step 1: Select the base case heat loss factor or base case sensible gain factor.
Step 2: Apply R-value correction to default value.
Step 3: Apply leakage correction to Step 2 value.
Step 4: Apply the surface area adjustment to the Step 3 value.

Procedure for Latent Gain Adjustment
Step 1: Select the base case latent gain factor.
Step 2: Apply leakage correction to Step 1 value.
Step 3: Apply surface area adjustment to the Step 2 value.

Notes
1) This table provides load factors for systems that feature multiple returns (approximately 400 Cfm per return).
2) Multiple returns improve comfort and room air motion, stabilize room pressures and blower Cfm (as interior doors open and close) and reduce the noise level in the conditioned space.
3) The load factors in this table are compatible with duct systems that are designed according to ***Manual J***, ***Manual S*** and ***Manual D*** procedures.
4) Duct systems designed by other procedures may not provide adequate air distribution (surface area adjustment does produce an acceptable duct load estimate for such systems).
5) ACCA recommends sealing duct systems that have leakage rates greater than the 0.12 / 0.24 scenario. Use the data for leakier scenarios to evaluate the benefit of the sealing work.

Figure 15 Manual J® duct heat gain/loss factors.
Source: ACCA - Air Conditioning Contractors of America

Table 6A
Default Scenarios and Values for Internal Loads

Appliance, Equipment and Lighting Loads			
Default Appliance Load	**Sensible Btuh**	**Latent Btuh**	**Notes**
Refrigerator and range with vented hood.	1,200	~	1,200 Btuh applied to the kitchen.
Scenario Options			
1) Refrigerator, range with vented hood, dish washer, clothes washer and vented clothes dryer, electronic equipment and lighting allowance.	2,400	~	1,000 Btuh for the kitchen, 500 Btuh for the utility room, 900 Btuh allowance for a TV or computer and a few lighting fixtures.
2) Two refrigerators or one refrigerator and one freezer, dish washer, range with vented hood, clothes washer and vented clothes dryer, electronic equipment and lighting allowance.	3,400	~	2,000 Btuh for the room or rooms equipped with a refrigerator, 500 Btuh for the utility room, 900 Btuh allowance for a TV or computer and a few lighting fixtures.
Adjustment Options			
A) Cooking range not equipped with a hood that is vented to outdoors, or an unvented dishwasher operating during the late afternoon in mid summer, or simultaneous use of unvented range and dishwasher.	+ 850	+ 600	Light duty cooking, 25 percent of the available range capacity used for 15 to 20 minutes. One dishwasher cycle load spread over a two hour recovery period.
B) Water bed heater (400 Watts, 33 percent duty cycle).	+ 450	~	Apply to each bedroom equipped with a water bed .
C) Ceiling fan (75 Watts).	+ 250	~	Apply to the room where the fan is located.
D) Large family using TV's, stereos, computers and laundry room during the late afternoon in mid summer.	+ 1,400	~	This is an additional 410 Watt allowance for A/V or computer equipment.
E) Allowance for above average lighting load.	+ 1,705	~	Five 100 watt lights.
F) Unvented clothes dryer.	Clothes dryers must be vented for air quality and efficiency.		
Assignment			
For room cooling load estimates, correlate the equipment loads with the rooms that contain the equipment (existing construction) or which will logically contain the equipment (working from drawings).			

Internal Loads for Full Time Occupants, Guests and Plants			
Item	**Sensible Btuh**	**Latent Btuh**	**Notes**
Full Time Occupant	230	200	Occupancy = Number Bedrooms + 1
Guest	230	180	Number of guests specified by owner
Each Small Plant	~	10	Less than 12 Inches high
Each Medium Plant	~	20	12 inches to 24 inches high
Each Large Plant	~	30	More than 24 inches high

Figure 16 Default scenarios and load values for internal loads.
Source: ACCA - Air Conditioning Contractors of America

2.0.0 Section Review

1. Which of the following describes the effect of radiant energy on materials?
   a. Radiant energy cannot pass through a vacuum.
   b. Radiant energy passes through opaque objects.
   c. Radiant energy warms the air, which then warms an object.
   d. Radiant energy warms dark objects more readily than light ones.

2. Per Manual J® tables for infiltration-related air changes, what happens to the air changes per hour (ACH) as the square footage of a home increases?
   a. The ACH remains stable.
   b. The ACH increases as the square footage increases.
   c. The ACH decreases as the square footage increases.
   d. The ACH decreases for cooling and increases for heating.

3.0.0 Equipment Selection and Planning

Performance Tasks

There are no Performance Tasks in this section.

Objective

Explain how to select equipment based on load calculations and plan for support systems that are part of an installation.

a. Explain how to select cooling equipment.
b. Explain how to select heating equipment.
c. Explain how to select heat pumps.
d. Identify common support systems and common planning requirements.

The following major items of information are necessary to select the equipment for a particular application:

- The type of system desired or required
- The sensible heating and/or cooling load
- The latent heat load (cooling only)
- The required heating and/or cooling air volume

Detailed load calculations are performed to identify the equipment that comes as close as possible to matching the true load. Assuming you have included all the factors in your load calculations, you should be able to select the equipment size and capacity that is equal to the load or the next size larger that is available. For example, you can't purchase a system with a nominal cooling rating of 2.3 tons, so you select a nominal 2.5-ton unit, which is a common trade size.

Undersized equipment will not be able to handle peak loads and will take longer to return the space to the comfort level following temperature setbacks/setups. However, slight under-sizing is not generally frowned upon nor represents a comfort issue. Oversized equipment will have higher initial costs and operating costs. It will also cycle on and off more often. Oversized cooling equipment will take control of the load more quickly, which is a slight benefit when owners save energy through temperature setup or setback. However, an oversized cooling system may not run long enough to sufficiently reduce the humidity. This is arguably the most significant issue presented by oversized equipment.

Oversized heating equipment will not be as efficient. Oversizing an induced-draft, non-condensing furnace may cause additional condensation to form in the vent system, which can cause corrosion and failure.

In summary, HVAC equipment should be selected carefully relative to the calculated loads as well as any unique aspects of the application.

Residential cooling and heat pump systems are generally sized in half-ton increments in the smaller sizes, and then in 1-ton increments in larger sizes. For example, most manufacturers provide sizes of $1\frac{1}{2}$, 2, $2\frac{1}{2}$, 3, $3\frac{1}{2}$, 4, and 5 tons. Again, select the capacity that is closest to the load, but not less.

Furnace size increments often vary in increments of 20,000 Btuh. For example, a manufacturer might offer 40,000, 60,000, 80,000, and 100,000 Btuh sizes. However, this is not considered a standard among all manufacturers. After determining the heat loss, select an output size that matches or is slightly higher than the calculated load.

The airflow requirement must be calculated separately for cooling and heating. To make a preliminary equipment selection, you must approximate the heating and/or cooling air volume. However, the final determination on fan size and speed is based on the duct design process, which is discussed later in this module. Cooling air volume is generally higher than the heating air volume required. One method of estimating air volume uses the following formula:

$$\text{Cooling air volume} = \frac{\text{sensible heat gain (Btuh)}}{1.08 \times (T_1 - T_2)}$$

$$\text{Heating air volume} = \frac{\text{sensible heat loss (Btuh)}}{1.08 \times (T_2 - T_1)}$$

Where:

T_1 = outdoor dry-bulb temperature
T_2 = indoor dry-bulb temperature

3.1.0 Cooling Equipment Selection

In many parts of the country, split systems are used to satisfy residential cooling requirements. The condensing unit is placed outside the building, as close to the evaporator as possible, but generally not more than about 50' away. Consult the manufacturer for any refrigerant charge increases and changes in the tubing size for long runs of refrigerant piping. This information is generally provided in the installation guide.

The condensing unit contains the compressor, outdoor coil, and outdoor fan, as well as a contactor, safety switches, and other electrical components. The cost of the condensing unit is a function of its efficiency and features. A high-SEER cooling unit may contain a two-speed or inverter-driven, variable-speed compressor and a larger outdoor coil. A deluxe model, which is also associated with a high SEER rating, may also contain high-pressure and low-pressure safety switches, sound-deadening shields around the compressor, start-assist gear, crankcase heater, filter-drier, and similar features. High-efficiency units are more expensive, but these units will usually return the investment over time.

The indoor portion of a split cooling system includes an indoor coil, blower, and metering device. When the blower is part of a forced-air furnace, an indoor coil is added to the furnace (*Figure 17*). Otherwise, a fan coil unit is used.

Selection of the outdoor unit is made using the load calculations and the equipment manufacturer's product data. Cooling equipment must be selected to handle both the sensible and latent heat loads. The indoor unit is selected from a physically compatible group based on its capacity, ability to meet the air volume requirement, and efficiency rating.

The components of a split system must be properly matched, but multiple combinations of indoor and outdoor equipment are available. When fixed-restrictor metering devices are used, the correct piston for the indoor coil is often shipped with the condensing unit. However, due to increasing efficiency standards, an increasing number of systems are being installed with expansion valves instead of fixed restrictors.

In areas with low humidity and high dry-bulb temperatures, such as the southwest US, it is not uncommon to see an indoor coil with a higher capacity rating than the outdoor unit. Using an indoor coil with a higher capacity than

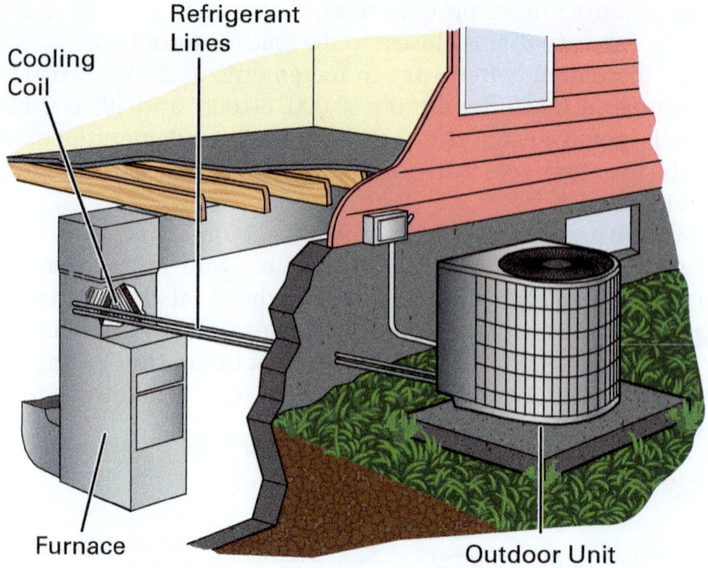

Figure 17 Split system with cooling coil on an upflow furnace.

the condensing unit trades some dehumidification performance for increased sensible cooling performance and energy efficiency.

In these cases, the nominal indoor unit capacity selected can be about a half-ton greater than that of the condensing unit. In desert-like conditions, the selected indoor coil capacity may exceed the outdoor unit by a full ton. However, it is essential to use a manufacturer-approved combination to avoid overloading the compressor and outdoor unit by selecting an indoor coil that is too large.

Did You Know?

Indoor Coil and Condensing Unit Combinations

Using an indoor coil with a higher-rated capacity than the condensing unit can have some positive benefits. The total cooling capacity of the system increases, due to the additional surface area of the indoor coil. However, the extra load from the larger indoor coil tends to raise the saturated suction temperature and pressure. Since the coil is slightly warmer, less dehumidification (latent cooling) occurs but more sensible cooling is delivered.

Although these combinations can work using an indoor coil with a fixed metering device, the positive results are limited because a fixed device cannot react to changing conditions. These coil combinations perform far better with a thermostatic expansion valve (TXV) or an electronic expansion valve (EEV). All systems perform better with a TXV or EEV.

Packaged units are installed where indoor space is limited, or the structure is built on a slab or over a crawlspace. Packaged units can be mounted on the ground with the ducts penetrating a sidewall. They are also commonly installed on the roof in commercial as well as residential applications (*Figure 18*).

3.2.0 Heating Equipment Selection

Many options are available to produce heat, including natural gas, propane, oil, heat pumps, and electric furnaces. The selection of a heating fuel depends on the local climate as well as the availability and cost of the available forms of energy. Factors that influence heating fuel selection include the following:

- Local traditions and experiences
- Construction characteristics
- Fuel availability
- Fuel costs

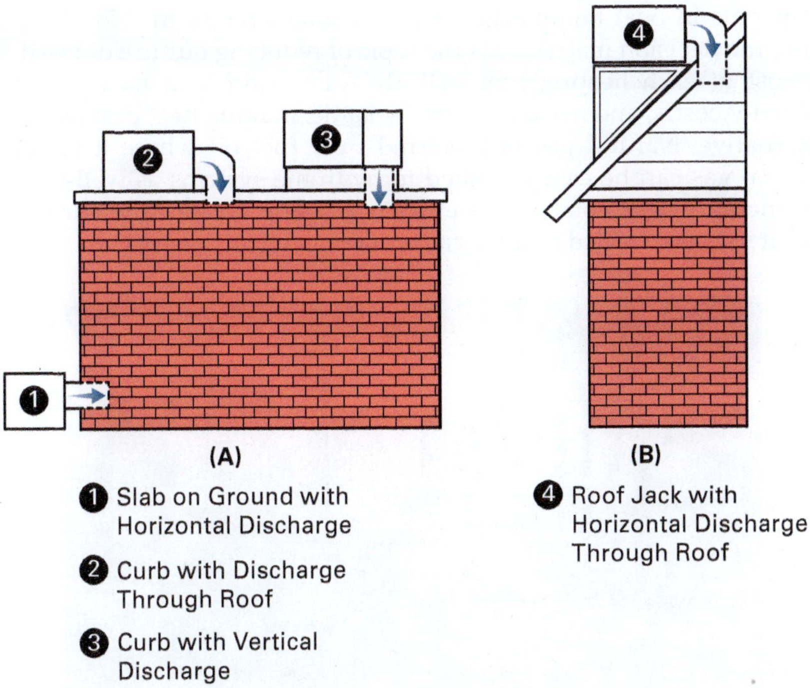

Figure 18 Common packaged unit installation locations.

Local traditions and a client's experiences often have a bearing on fuel and equipment selection. For example, although heat pumps for all climates are now widely available, some users have had many years of experience with gas or even fuel oil. As a result, they often resist a change in heating fuel.

For many years, packaged rooftop units were installed on top of homes in the southwest. Part of the reason was that evaporative coolers were traditionally installed on the roof (*Figure 19*), so the duct connections were already there. Today in the southwest US, more split systems are being installed in new homes and fewer pieces of equipment are placed on the roof. Manufactured homes with a crawlspace often use packaged units installed on a slab at ground level and ducted through the foundation wall. In homes with basements, upflow forced-air furnaces are widely used.

Figure 19 Evaporative cooler mounted on a roof jack.
Source: xshot/Shutterstock

The availability of a fuel also helps determine the selection. In areas served by gas pipelines, natural gas is often preferred. Fuel oil and propane are often used in rural areas, where natural gas may not be readily available. However, improvements in heat pump efficiency and functionality in cold climates are changing the decision matrix, as is the topic of reducing our use of fossil fuels.

The cost of each heating fuel will always remain as a factor. Fuel oil is undoubtedly costly, and pricing is often volatile, making heat pumps an attractive alternative. Natural gas is preferred over fuel oil where it is available, and natural gas can be easily applied to hydronic systems as well (*Figure 20*). Improvements in gas combustion efficiency, using condensing furnaces and water heaters, have helped natural gas remain an attractive fuel.

Figure 20 Wall-mounted, gas-fired condensing boiler installation.
Source: matteogirelli/iStock/Getty Images

However, the efficiency and improvements found in today's heat pumps have made them the most desirable alternative for many users. At this writing, tax incentives are also available for those who choose to replace fossil-fuel equipment with a heat pump. The electrification of heating systems as an alternative to fossil fuels is currently considered to be the most climate-friendly approach. In the end, the selection of a heating fuel requires careful thought as well as thoughtful discussions with the client. It is important to understand the client's preferences and goals, and it is equally important for the client to understand any functional limitations and operating characteristics that their preferences might involve.

Furnace manufacturers offer furnaces with standardized cabinet sizes to accommodate various indoor coils and installation configurations. Some also offer *high-air models*, which are used in areas where the heating load is lower than the cooling load. The desert southwest serves as an example. A home in that area may have a heat loss as low as 40,000 Btuh but have a heat gain of 48,000 Btuh or more. High-air furnaces are characterized by a larger blower and blower motor and are usually wider than the standard model of the same heating capacity.

3.3.0 Heat Pump Selection

A heat pump is an excellent choice for many applications. For design heating days, a heat pump can be supplemented with auxiliary electric heaters, although cold-climate heat pumps can produce more heat at lower temperatures than older systems. A dual-fuel system that takes advantage of multiple heat sources is also an option where a fossil-fuel source is also available.

The heat loss of the structure drives the furnace selection process, while the cooling equipment is selected based on the heat gain. With a heat pump, only a single primary piece of equipment is being selected. If the heat pump is sized to the heating load in a cold northern application, it will likely result in excessive cooling capacity. This will result in inefficiency and poor dehumidification.

To prevent this problem, a heat pump has traditionally been sized to the cooling load, and any additional heating capacity needed is made up with auxiliary heat. Again, a dual-fuel system might be an option if another fuel source is available. When a dual-fuel system is used, the furnace takes over all heating responsibilities below the balance point, since the heat pump alone cannot keep up. However, cold-climate heat pumps have changed the selection criteria for northern climates, as these units can produce more heat at colder outdoor temperatures than traditional heat pumps (*Figure 21*). Cold-climate heat pumps can often handle both the heating and cooling needs without being oversized for the cooling load. However, some supplemental heat may still be needed in the most challenging conditions.

Figure 21 Cold-climate heat pump at work.
Source: Fire-n/Shutterstock

When electric resistance heaters are added to a heat pump system, the combined capacity of the heat pump and the supplemental electric heaters should not exceed 115% of the calculated heating load. Electric resistance heaters are also used to provide emergency heat in the event of a heat pump compressor failure. In such cases, the electric heaters may be sized to provide as much as 80% of the heating load when the heat pump fails. However, only a portion of the electric heating capacity should be made available to the system during normal operation.

3.4.0 Support Systems

This section discusses various support systems associated with heating and cooling systems that must be considered during the design and installation planning process.

3.4.1 Refrigerant Piping

The refrigerant piping for a comfort air conditioning system has three major sections:

- The suction line conveys low-temperature, low-pressure, refrigerant vapor from the evaporator outlet to the compressor inlet. It is referred to more accurately as the *vapor line* in heat pump systems.
- The hot-gas (discharge) line conveys hot, high-pressure vapor refrigerant from the compressor to the condenser.
- The liquid line conveys high-pressure liquid refrigerant from the condenser outlet to the expansion device.

In split-system applications, only the suction and liquid lines, which connect the indoor and outdoor units, are installed at the site. The hot-gas line is confined to the condensing unit and is very short. The proper suction and liquid line sizes are specified in the manufacturer's product data.

The diameter and length of the refrigerant lines affect the refrigerant pressure drop. The smaller the diameter and the longer the line, the greater the pressure drop. In their product data, manufacturers will specify larger refrigerant line sizes for long runs. Larger pipe sizes reduce the pressure drop.

Remember that pipe sizes also vary for different refrigerants. Do not deviate from the recommended sizes. Undersized lines can cause noise, evaporator starvation, and excessive pressure drop. Oversized lines require a larger refrigerant charge and can create oil return problems. The compressor discharges a small amount of oil with the refrigerant. During cold startups, a greater volume of oil may be discharged. This oil must return to the compressor to maintain proper lubrication.

Oil movement through the liquid line is not a problem because the oil mixes easily with liquid refrigerant. In the vapor line, however, it can be a major problem. With no liquid to mix with, the oil must move along by maintaining a reasonable refrigerant velocity and relying on gravity whenever possible. If the vapor line is too large, sufficient velocity will not be available to move the oil up vertical piping sections (risers), and the oil will settle in low spots. Vapor lines should always be pitched in the direction of flow when possible.

The elevation of the indoor coil relative to the compressor must be considered when routing refrigerant lines. If the indoor coil is located above the outdoor unit, the suction line should loop up to the height of the indoor coil, as shown in *Figure 22A*. This prevents liquid refrigerant from migrating from the indoor coil to the compressor during the off cycle.

When the indoor coil and the outdoor unit are at the same level, the suction line should pitch toward the compressor in the outdoor unit (*Figure 22B*) with no sags or low spots in the run. The pitch allows gravity to help move the oil along. When the indoor unit is below the outdoor unit, oil must be returned to the compressor via a vertical riser in the suction line (*Figure 22C*).

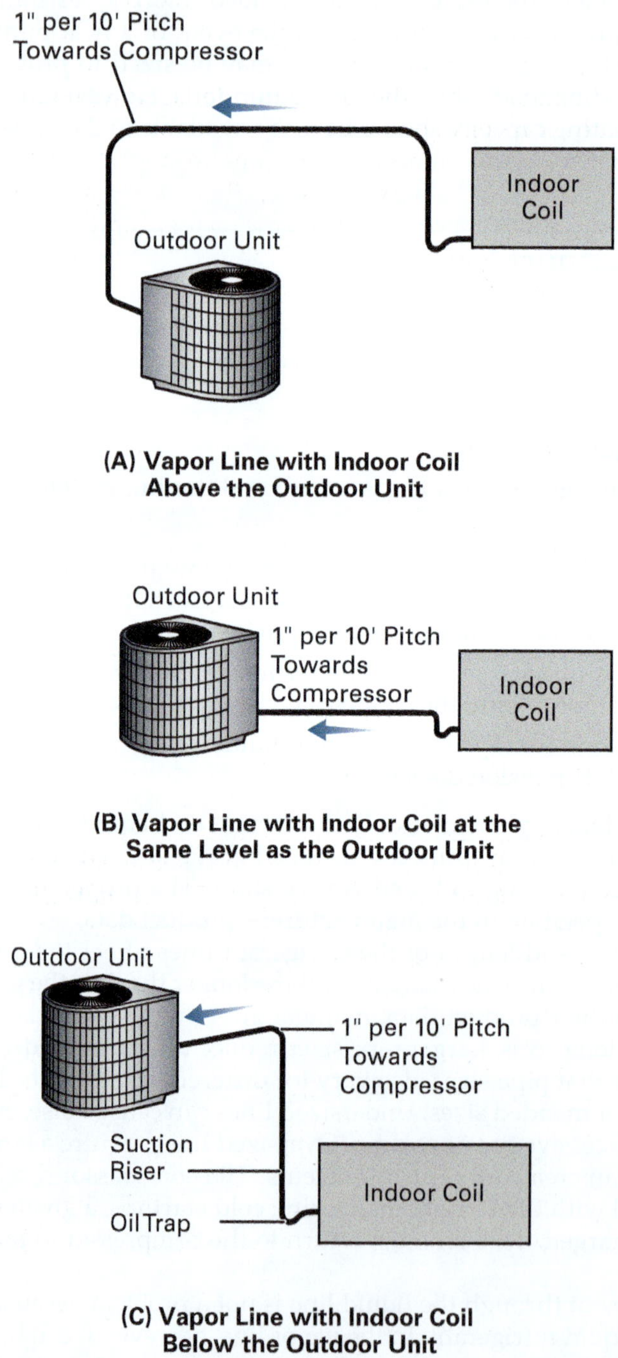

Figure 22 Suction (vapor) line routing scenarios.

An oil trap should be installed at the base of the riser to collect oil that might fall back down during the off cycle. In the trap, turbulence helps entrain the liquid oil in the vapor, helping it move up a riser. The suction line must be sized to maintain a minimum velocity to ensure the oil can travel uphill. A minimum velocity of 500 feet per minute (fpm) is valid for some refrigerants, but not necessarily all. A safer target to ensure oil return is a minimum of 750 fpm. Making the tubing smaller increases velocity but also increases pressure drop. A comfortable compromise between these two important factors is required.

Note that refrigerant oil traps are best kept shallow. Pre-formed P-traps are generally too deep and will collect too much oil. Running traps or those constructed using a 90-degree ell with two 45-degree ells are best. To further enhance oil return, horizontal runs should be pitched at a rate of 1" in 10' toward the outdoor unit. Remember to consider the thickness of the insulation when creating penetrations and planning hangers and supports.

Again, oil movement through the liquid line is not a consideration. However, an undersized liquid line causes excessive pressure drop. If there is little subcooling in the refrigerant, the reduction in pressure will cause *flash gas*—pockets of vapor—to form in the line. Essentially, an undersized liquid line becomes one big capillary tube—a metering device.

An oversized liquid line is rarely a functional problem, but it does create the need for more refrigerant to fill the system. This can make it challenging or impossible to pump the entire charge into the condenser to support repairs.

Vertical risers in the liquid line create additional pressure drop. Refer to the manufacturer's guidance for liquid line sizing when significant risers are required.

On Site

Refrigerant Lines

Refrigerant lines used with HVACR systems should always be constructed using Type ACR copper tubing. Never use copper tubing meant for use in plumbing applications, such as Type M. Equipment manufacturers offer soft-copper line sets of different lengths that are intended for use with their equipment. These line sets are factory-evacuated and contain a holding charge of dry nitrogen to prevent copper oxide formation (oxidation).

3.4.2 Condensate Piping

The condensate collected in the pan beneath the indoor coil must be efficiently drained away. A pipe connected to the pan routes the condensate to an indoor drain or outside of the building. Condensate drainage is especially important where units are installed in attics or above drop ceilings. Water leaks in these areas can cause significant damage.

Most mechanical codes require a separate secondary drain pan (*Figure* 23) under the coil or fan coil unit. This pan catches the condensate overflow from the unit in the event the primary drain is blocked. The secondary drain pan has its own drain line as shown. The outlet should be placed in a location that will be immediately noticed if water begins to drain from it. You made need to explain to the user that water draining from that specific line indicates a problem exists with the primary drain that needs to be investigated and resolved immediately.

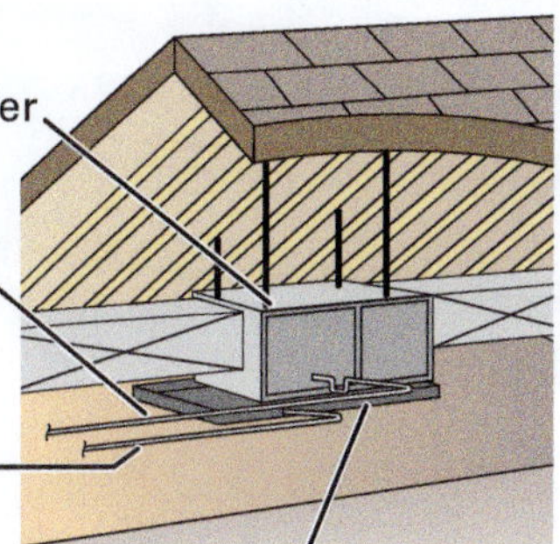

Figure 23 Secondary drain pan installation.

Drain pan safety switches that shut down the cooling equipment when water is present are also needed when the system is located where water leakage can cause damage (*Figure* 24). Some rely on magnetic floats that rise with the water level, opening a switch. Others react more quickly by sensing the presence of water beneath the switch with continuity between probes. A third type also relies on a float switch, but it is placed in the primary drain line above a trap (*Figure* 25).

Condensate safety switches must be wired into the control circuit to shut off the heating and cooling equipment if a drain problem occurs. The switch is wired in series in the system control circuit. The following shows how the overflow switch is typically connected in different systems:

- *Cooling-only fan coil unit* — Wire the switch in series with the Y circuit.
- *Condensing furnace with no cooling* — Wire the switch in series with the W circuit.

NOTE

Condensate safety switches that use probes to detect the presence of water typically must be powered from the control circuit to energize the electronic components. Float switches do not require a power source for operation.

(A) Magnetic Float Switch **(B) Water Sensor**

Figure 24 Condensate pan safety switches.
Source: RectorSeal, LLC

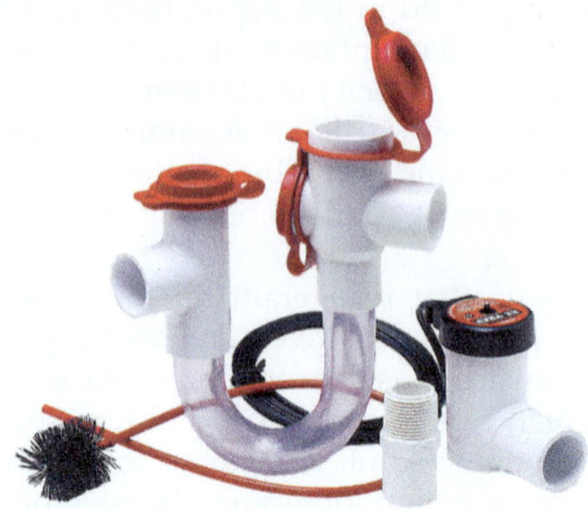

Figure 25 Line-mounted condensate safety switch.
Source: RectorSeal, LLC

- *Heating and cooling system with condensing furnace* — Although the switch can be wired into the R circuit, remember that breaking the R circuit will also result in the system blower being shut down. This can have a significant effect on a drainage situation. The blower generally creates a negative pressure inside the air handling unit cabinet. If the cabinet fills with water, which eventually overflows into the secondary pan, a great deal of water may suddenly drain from the equipment when the fan shuts down. To avoid this, wire the switch to energize a normally closed (NC) DPDT relay. The two separate sets of NC contacts are then used to break the Y and W circuits when the relay is energized. This maintains control power on the rest of the system and allows the blower to continue running.

Condensate drain lines typically require a trap to prevent water from being held in the drain pan by the pressure difference created by the blower. Traps are especially important when the drain line is under a negative pressure (upstream of the blower). Ductless indoor units do not usually require a drain trap. Always check the manufacturer's literature for guidance. Drain lines, regardless of the presence of a trap, must be pitched toward the drain at a minimum of $\frac{1}{4}$" per foot.

Condensate disposal is also required for condensing furnaces. Where condensate cannot be disposed of by gravity flow, a condensate pump is used (*Figure 26*). The same condensate pumps used for cooling applications can be used for furnace condensate disposal. However, it is important to ensure the pump is rated for furnace condensate disposal, due to the acidic nature of furnace condensate.

Most condensate pumps also have a float-actuated safety switch integrated into the pump. If the pump fails, the float rises and breaks the system control circuit. Pump safety switches should be wired into the system in the same manner as other condensate safety switches.

3.4.3 Electrical Service

Electrical service is not usually an issue in new construction because the service should be designed to handle the load created by the heating and cooling equipment. It is more likely to be an issue in cases where air conditioning is being added or replaced.

The addition of air conditioning or heat pump equipment can add anywhere from 15 to 60 amps of load per system, depending on the capacity and type of system. When selecting heat pumps for retrofit, especially those that will use electric resistance heat for supplemental heat and during the defrost cycle, it is important to evaluate the structure's electrical service before proceeding. A 5-ton heat pump with supplemental electric heaters, for example, will add more than 50 amps to a home's electrical load.

In some homes, the additional electrical load will exceed the capacity of the existing electrical service. In these situations, always consult a qualified, licensed

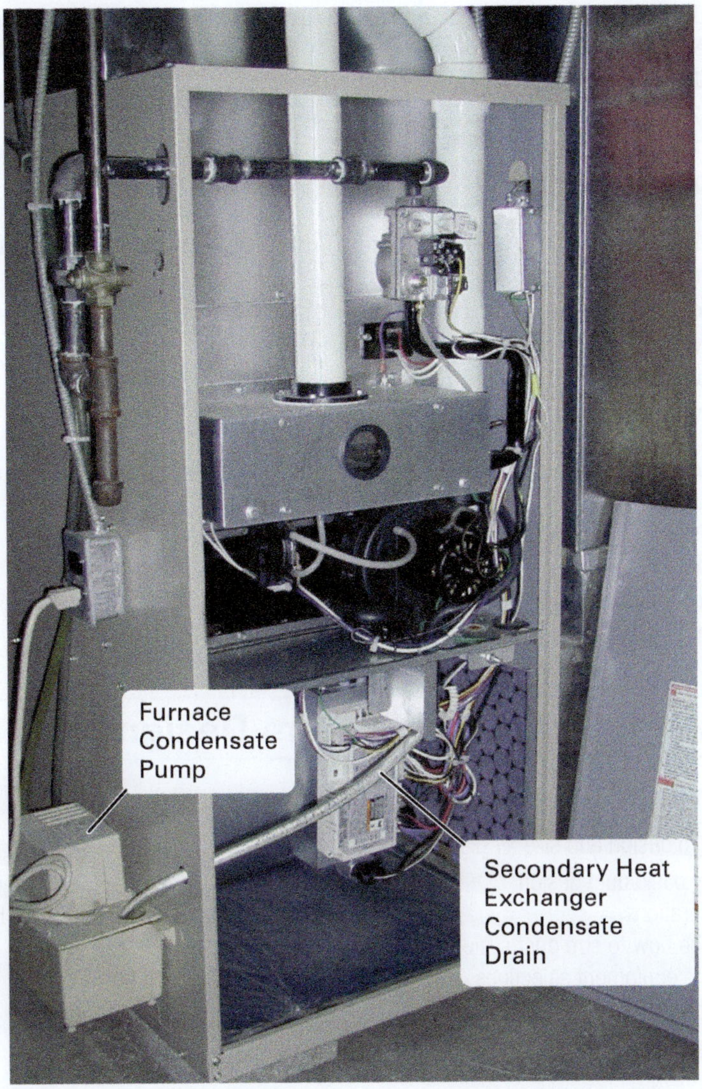

Figure 26 Condensate pump installed on a condensing furnace.

master electrician to evaluate the electrical service. Homes with emergency generators that were selected and installed to support the total electrical needs of a home or business require an even deeper evaluation. These are not topics on which HVACR technicians should typically consult. Although HVACR craftworkers should understand these electrical issues, they should not make decisions regarding the size of the electrical service, nor should they execute any electrical changes.

Note that electrical upgrades can potentially be avoided by relying on dual fuel systems, using an existing fossil-fuel source for supplemental heating with a heat pump. An upgrade may also be avoided by selecting one of today's cold-climate heat pumps that can produce far more heat at colder temperatures, potentially avoiding the need for supplemental heat altogether.

3.0.0 Section Review

1. In a desert setting, the indoor coil of an approved split-system condensing unit and indoor coil combination unit may be oversized by as much as ______.
   a. 1,000 Btuh
   b. ½ ton
   c. 1 ton
   d. 1½ tons

2. Which of the following is *not* typically a factor that affects the selection of a heating fuel?
   a. Fuel costs
   b. Allergies
   c. Local traditions
   d. Fuel availability

3. With a standard dual-fuel system and a traditional heat pump, the heat pump primarily provides heat ______.
   a. during all daylight hours, but not at night
   b. when the thermostat is operating in setback mode
   c. when the outdoor temperature is below the balance point
   d. when the outdoor temperature is above the balance point

4. When installing a condensate pump on a fan coil unit, you should wire the overflow switch in series with the ______.
   a. Y control circuit
   b. W control circuit
   c. G control circuit
   d. R control circuit

4.0.0 Air Distribution System Design

Performance Tasks

2. Using a manual duct design calculator or app, size the trunk and branch ducts for an instructor-provided residential air distribution system and set of design parameters.
3. Calculate the total system friction loss (external static pressure) for a selected air distribution system.

Objective

Explain how to design and size air distribution systems.

a. Identify basic duct design considerations.
b. Describe how to size duct systems and evaluate equipment selections.
c. Identify heat load and system design factors unique to commercial buildings.

Understanding basic air distribution system design is a benefit when installing and servicing HVAC systems. The knowledge helps a technician recognize and solve existing duct system problems, such as noise, vibration, and insufficient airflow.

The first part of this section introduces some basic system operating principles. The second part covers factors directly related to air system design. The design goals for all air distribution systems include the following:

- Supply the right volume of air to each conditioned space.
- Supply the air to each space so that stratification is minimized, while air movement is adequate for mixing but not drafty.
- Provide for the return of air from all conditioned spaces to the air handling unit.
- Operate efficiently without excessive power consumption.
- Operate quietly and with minimum maintenance required.

4.1.0 Duct Design Basics

The resistance to airflow caused by the duct system must be overcome by the system fan. The fan supplies the energy needed to overcome the resistance and maintain the necessary airflow. Air is moved by creating areas of differing pressures, causing air to move from the higher pressure toward the lower pressure. All fans produce both high- and low-pressure areas. The air inlet to a fan is below atmospheric pressure, while the outlet of the fan is above atmospheric pressure.

As shown in *Figure 27*, normal atmospheric pressure exists at the return grille and supply diffusers before the blower starts. Points A through H identify places where obstructions or other factors cause pressure losses called *dynamic losses*. With the blower running, the pressure at the face of the return air grille drops below atmospheric pressure, causing air to move into the duct. The pressure decreases through the duct where it reaches its lowest point at the blower wheel inlet.

NOTE

Dynamic losses are explored in *Section 4.1.3*.

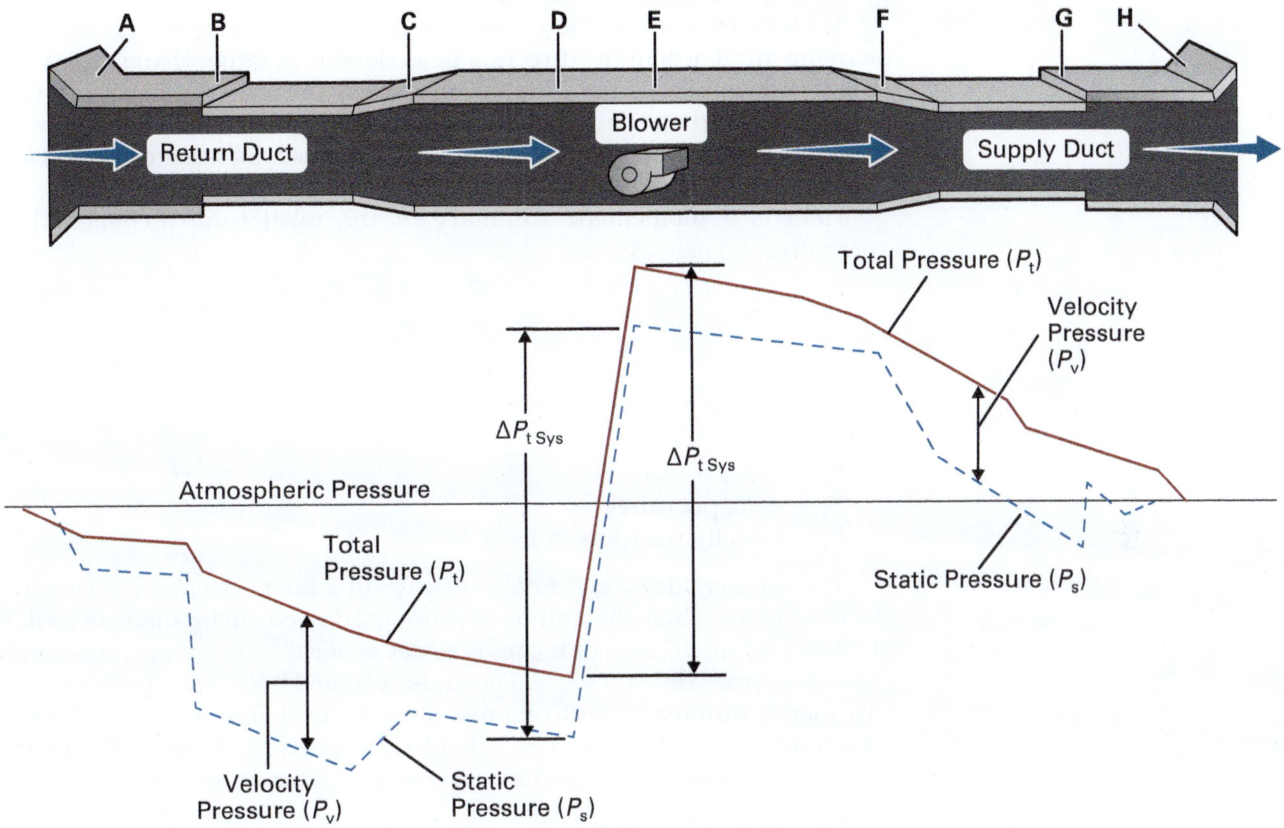

Figure 27 Pressure relationships in a duct system.

Through the action of the fan, the air pressure increases to its highest level as it leaves the blower wheel. From there, the air follows its tendency to flow from an area of higher pressure to an area of lower pressure.

The pressure difference needed to move air through a duct system depends on the air velocity, air volume, the cross-sectional area of the duct, and the length of the duct. Velocity reflects how fast the air is moving and is measured in feet per minute (fpm). Air volume is traditionally measured in cubic feet per minute (cfm). The air volume moving through a duct, in cfm, is calculated by multiplying the air velocity, in fpm, by the area, in square feet, of the duct as follows:

$$\text{Air volume (cfm)} = \text{duct area (ft}^2\text{)} \times \text{velocity (fpm)}$$

This formula can also be rearranged to find other values:

$$\text{Velocity (fpm)} = \frac{\text{air volume (cfm)}}{\text{duct area (ft}^2\text{)}}$$

$$\text{Duct area (ft}^2\text{)} = \frac{\text{air volume (cfm)}}{\text{velocity (fpm)}}$$

4.1.1 Pressure Relationships Within a Duct

Three pressures exist in an operating duct system: *total pressure, static pressure,* and *velocity pressure*. Total pressure determines how much energy must be supplied by the blower to maintain airflow. The total pressure always decreases as you progress in the direction of airflow. For any cross section of the duct, the total pressure is the sum of the static pressure and the velocity pressure.

Static pressure and velocity pressure are present when the blower is operating. Static pressure represents air pressure that is exerted uniformly in all directions within the duct. Think of it as the pressure applied equally on the internal walls of a balloon or an inflated tire.

Velocity pressure is the pressure caused by the velocity and weight of the moving air. It acts in the direction of airflow only, rather than in all directions. Mathematically, it is the difference between the total pressure and the static pressure. In other words, static and velocity pressure together comprise total pressure. Static and velocity pressures can either increase or decrease in the direction of airflow.

Here is a mathematic summary of the relationship between the three pressures:

$$P_t = P_s + P_v$$
$$P_s = P_t - P_v$$
$$P_v = P_t - P_s$$

Where:

P_t = total pressure
P_s = static pressure
P_v = velocity pressure

The static, velocity, and total pressures in a duct system are relatively small in most cases. Thus, the unit of measure used needs to be small as well. Inches of water column (" w.c.) or inches of water gauge (" w.g.) are the units commonly used to express duct pressure. The units " w.c. and " w.g. are interchangeable and are merely different ways to express the same unit. Both represent the height of a column of water that can be lifted by the pressure applied. The instrument used to measure pressures in a duct system is the manometer.

4.1.2 Friction Losses

The inside surface of the duct offers resistance to the flow of air. Because of the friction along the duct walls, the air velocity in the duct is not uniform. The velocity varies from a very low speed at the duct walls, approaching zero at a molecular level, to its maximum velocity found at or near the center of the duct. This resistance to airflow is referred to as *friction loss,* as it works against the blower. The amount of friction loss in straight duct is normally found by using air friction charts. These charts are presented later in this section.

As the cross-sectional area of a straight duct section becomes smaller, as shown in duct sections BC and FG in *Figure 27,* the air velocity and velocity pressure increases. As a result, the static and total pressure lines drop more rapidly than in the larger cross-sectional area ducts. This drop occurs because the pressure losses increase as the square of the velocity increases.

As the air moves through a straight duct section with a consistent area, the static pressure and total pressure losses increase at the same rate. This is because the velocity and velocity pressure (P_v) of the air flowing through the consistent duct are constant.

4.1.3 Dynamic Losses

Dynamic losses occur where there are changes in the direction or velocity of the air, such as transitions, elbows, and similar fittings. Dynamic losses also occur where the air encounters an obstruction, such as a damper. Turbulence in the airstream is typically the cause of dynamic losses.

When ducts transition to a smaller physical size, whether abruptly or gradually, some turbulence results. This is because of the change in velocity and the air's impact against the angled surface. The result is that the total pressure loss increases when compared with air traveling through an equal length of straight duct of a uniform size. Anything encountered by air moving through a duct that affects the status quo of moving straight ahead results in additional friction loss.

The additional loss resulting from these situations is called *dynamic loss.* In *Figure 27,* points A, B, C, F, G, and H represent changes that will cause dynamic losses. Dynamic pressure losses can be expressed as a loss coefficient value using what is known as a *C-factor.* A loss coefficient value is a value assigned to each type of duct fitting, such as elbows and transitions.

The pressure loss for a specific type of fitting is more often expressed by its *equivalent length of straight duct.* This is a more convenient value with which to work. For example, a radius elbow has the same resistance as 25' of straight duct of the same size. While C-factors more accurately represent pressure loss, rarely are C-factors used in traditional residential and commercial HVAC designs.

Both C-factors and values for the equivalent length of straight duct for fittings are found in published tables. These tables are available from the Sheet Metal and Air Conditioning Contractors' National Association (SMACNA), ASHRAE, and duct component manufacturers. *Figure 28* shows an example of the loss coefficient data and equivalent length data given in tables for a rectangular elbow. Note that 90-degree elbows fitted with turning vanes have an equivalent length of only 15'—half of what is shown for the style of elbow in *Figure 28.* The equivalent length of a 90-degree elbow can also be reduced by using a longer radius.

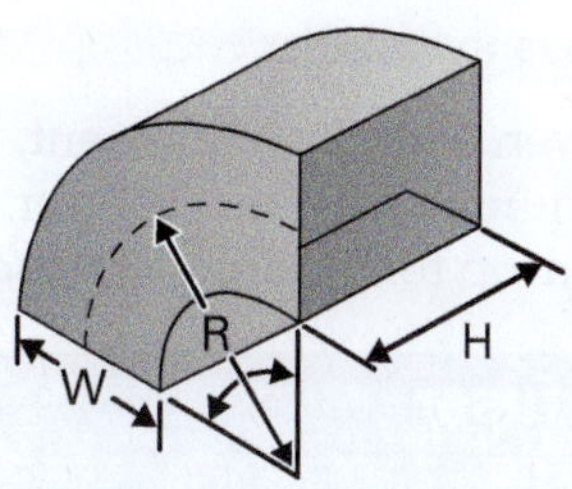

Coefficients (C-Factors) for a Rectangular 90° Ell:

Coefficients C									
R × *W*	*H* × *W*								
	0.25	0.5	0.75	1.0	1.5	2.0	4.0	6.0	8.0
0.5	1.5	1.4	1.3	1.2	1.0	1.0	1.1	1.2	1.2
0.75	0.57	0.52	0.48	0.44	0.40	0.39	0.40	0.43	0.44
1.0	0.27	0.25	0.23	0.21	0.19	0.18	0.19	0.27	0.21
1.5	0.22	0.20	0.19	0.17	0.15	0.14	0.15	0.17	0.17
2.0	0.20	0.18	0.16	0.15	0.14	0.13	0.14	0.15	0.15

Where:
H = Fitting Height
W = Fitting Width
R = Radius of Turn

(A) Loss Coefficient (C-Factor)

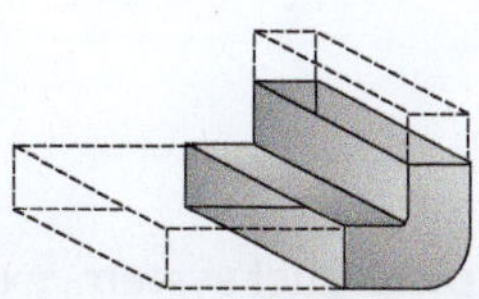

G = 30'

(B) Equivalent Length of Duct

Figure 28 Examples of fitting loss coefficient and equivalent length data.

4.1.4 Blower and Duct Static Pressure Relationship

The total pressure that the system blower must overcome is the sum of the friction losses in the supply duct system, return duct system, and any other components other than the blower itself (*Figure 29*). This includes all straight duct sections; the dynamic losses of each duct fitting or obstruction, such as registers and grilles; and the pressure loss of each duct component, such as coils, filters, and dampers. Although the conditioned space is technically part of the air distribution system too, it doesn't represent significant resistance to airflow.

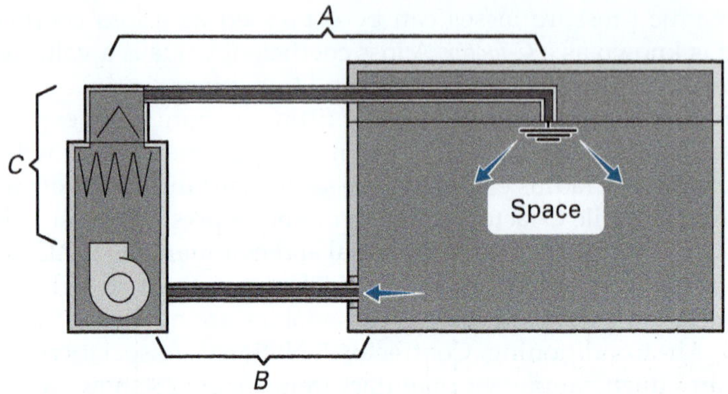

Total Friction Loss = *A* + *B* + *C*
A = Supply System Loss
B = Return System Loss
C = Components Not Included in Fan Rating

Figure 29 System friction losses to consider.

The pressure loss for a given system component, such as an indoor coil or grille, is available from the equipment manufacturer. For example, the chart in *Figure 30* shows the pressure drop for a ceiling diffuser by size and air volume.

24" Square Ceiling Diffuser									
Face Velocity		300	400	500	600	700	800	900	1000
Pressure Loss		0.006	0.010	0.016	0.022	0.031	0.040	0.050	0.062
6	cfm	50	65	85	100	115	130	150	165
Ak 0.165	Throw	3.5	4.5	5.5	6.5	8	9	10	11
8	cfm	85	110	140	170	195	225	250	280
Ak 0.280	Throw	4.5	5.5	7	8.5	10	11	12	14
10	cfm	125	170	210	250	295	335	380	420
Ak 0.420	Throw	5	6.5	8	9.5	11.5	13	15	16
12	cfm	180	240	300	355	415	475	535	595
Ak 0.595	Throw	6	8	10	11.5	13.5	15.5	17.5	19
14	cfm	245	330	410	490	575	655	740	820
Ak 0.820	Throw	7	9	11.5	13.5	16	18	20	22.5
16	cfm	310	410	515	620	720	825	925	1030
Ak 1.03	Throw	7.5	10	12.5	15	18	20	22	25
18	cfm	400	530	665	800	930	1065	1200	1330
Ak 1.33	Throw	8.5	11	14	17	20	23	26	28
20	cfm	480	640	800	960	1120	12801	1440	1600
Ak 1.60	Throw	9.5	12	16	18	22	25	28	31
22	cfm	570	760	950	1140	1330	1520	1710	1900
Ak 1.90	Throw	10.5	13.5	17	19	24	27	30	33
24	cfm	690	920	1150	1380	1610	1840	2070	2300
Ak 2.30	Throw	11	14.5	18.5	22	26	30	33	36

Terminal Velocity of 50 fpm

Figure 30 Example of a ceiling diffuser pressure-drop chart.
Source: Hart & Cooley Inc.

The static pressure of the complete duct system, external to the blower assembly, is called the external static pressure (ESP). The sizing of the system blower is based on the external pressure loss resulting from the system ductwork and components. Any resistance presented by the blower assembly itself has been accounted for in the manufacturer's performance tables.

4.1.5 Airflow in a Typical System

Figure 31 shows a simplified building air distribution system. It will be used to discuss the airflow through a system, and to review concepts and pressure relationships. The volume of air that an air distribution system must deliver is normally based on the mode of operation that requires the most airflow. Generally, this is the cooling mode, but exceptions exist.

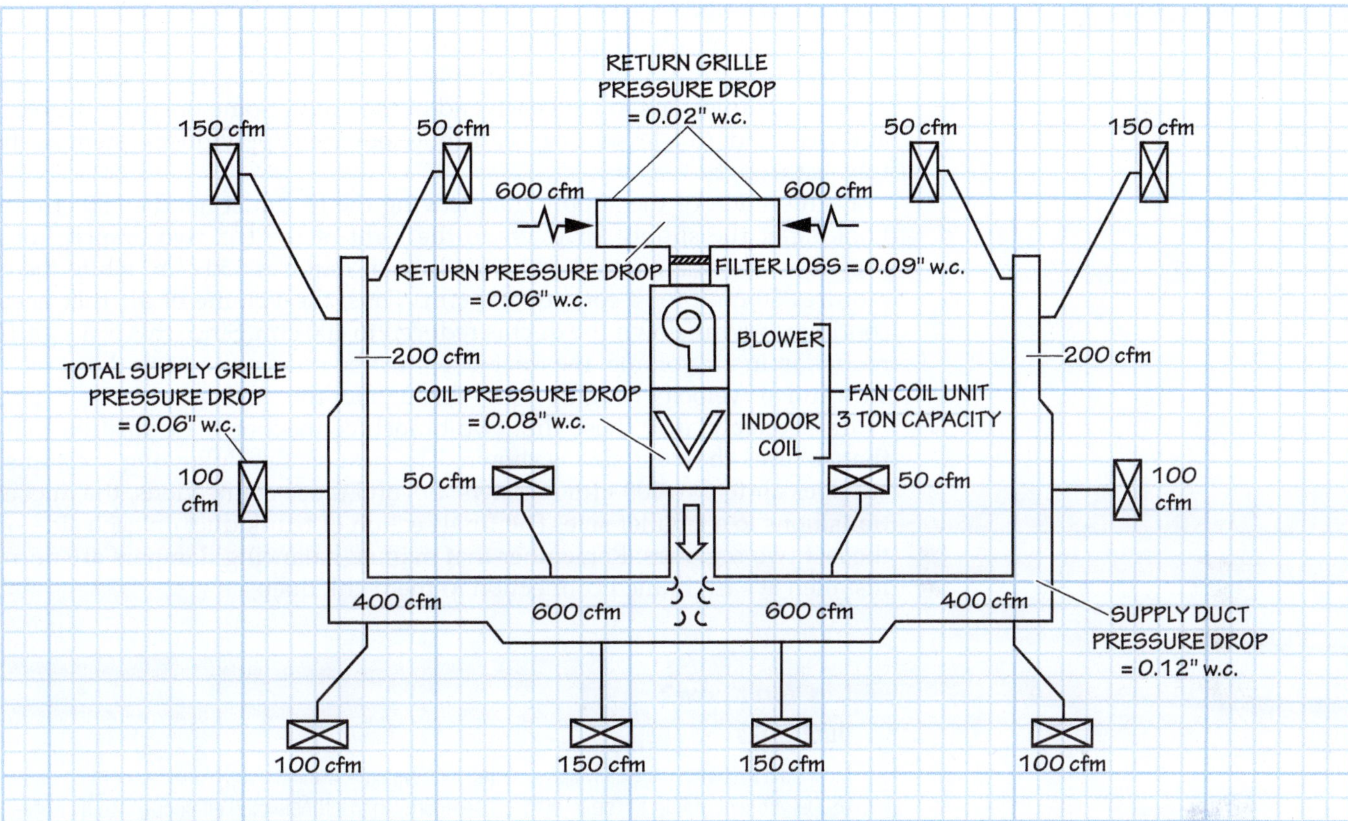

- System Cooling Capacity = 3 Tons
- At 400 cfm per Ton, Total Airflow Required = 1,200 cfm
- Total Return Duct Static Pressure Loss at 1,200 cfm = 0.19" w.c.
  - 0.06" w.c. Return Duct Static Pressure Loss
  - 0.09" w.c. Filter Static Pressure Loss
  - 0.04" w.c. Total Return Air Grille Loss
- Total Supply Duct System Static Pressure Loss at 1,200 cfm = 0.26" w.c.
  - 0.08" w.c. Evaporator Coil Static Pressure Loss
  - 0.12" w.c. Supply Duct Static Pressure Loss
  - 0.06" w.c. Total Supply Grille Static Pressure Loss
- Total External Static Pressure Loss = 0.45" w.c.

Figure 31 Simplified air distribution system.

In *Figure 31*, the airflow shown is for the cooling mode related to a 3-ton cooling system. Therefore, the system fan must be able to supply the volume of air needed for 3 tons of cooling. Since cooling requires roughly 400 cfm of air per ton, the blower in the example system must supply at least 1,200 cfm (3 tons × 400 cfm) of air.

Further, the external static pressure of the duct system that the fan must overcome is 0.45" w.c. This is derived by adding the absolute values of the supply and return external static pressures of 0.26" w.c. and 0.19" w.c., respectively.

The system shown has 12 supply registers and diffusers. Four need to deliver 150 cfm, four need to deliver 100 cfm, and four need to deliver 50 cfm. The return air enters the system through two centrally located return air grilles.

While reviewing *Figure 31*, consider the home as part of the system. Most of the supply air is delivered around the perimeter of the building. The air travels through the conditioned spaces as it flows back toward the return air grilles. A negative pressure at the return grilles causes air to move in that direction.

As the air flows down the return duct toward the blower wheel, the pressure continues to decrease slightly. The single largest pressure drop on the return side of the system occurs at the filter. At the inlet to the blower wheel, the system pressure is at its lowest. The air leaving the blower is where the highest pressure is found.

As the air moves through the indoor coil, it encounters a pressure drop of 0.08" w.c. After the air enters the supply duct, a tee routes the supply air into two reducing trunks, one to feed each end of the home. Some pressure drop occurs at the tee and, to a lesser degree, throughout the supply duct system as the air travels through.

Each section of the supply duct leaving the tee must handle 600 cfm of air. Looking at the supply air ducts to the right of the tee, two branch lines are shown as the first connections. One branch requires 150 cfm, while the other requires 50 cfm, for a total of 200 cfm. Thus, the volume of air that needs to continue moving down the duct is reduced to 400 cfm. Since this is a significant reduction in air volume, the duct size must be reduced to keep the static pressure and air velocity roughly equal throughout the system.

Immediately after these first two branch connections, a transition that further reduces the duct size is shown. The next section following the transition supplies air to two 100-cfm registers. Beyond those connections, the duct again transitions to a smaller size. The final section of trunk duct on each side serves the last two registers. Remember that manual balancing dampers are normally installed at each branch connection to the trunk duct.

Did You Know?

Duct Leaks

Most duct systems, especially sheet metal duct systems, are inherently leaky. According to the US Department of Energy, up to 40% of the energy in the conditioned air supply can be lost through leakage and inferior or damaged insulation. Much of the leakage can be attributed to poor workmanship when assembling the duct components.

To avoid these unacceptable losses, all sheet metal duct joints and seams must be properly sealed with mastic. Minor flaws in fiberglass ductboard can be sealed with aluminum foil tape, but mastic is often required as well.

Most codes and job specifications require that duct systems be sealed per SMACNA standards or a similar standard.

4.2.0 Duct System Sizing and Design

The duct design process typically proceeds as follows:

Step 1 Study the building plans or layout and determine the best location for the supply registers and return grilles to provide proper air distribution and circulation within each space.

Step 2 Determine the air volume to be delivered by each supply register and returned through each return grille. Size the supply registers and return grilles based on the air volume and the manufacturer's performance data.

Step 3 Sketch the duct system, connecting the supply registers and return grilles with the system. The sketch does not have to be elaborate; show the duct as a single line. Note that the physical layout of the duct system is often driven by the building construction and significant obstructions that may be encountered. These obstructions need to be considered in the early stages because the addition of more fittings to reroute the duct create additional resistance that must be considered when sizing the duct.

NOTE

When working with duct drawings, the duct dimensions shown are always written with the dimension for the visible side of the duct shown first. For example, when the view is looking down on the duct, as viewed on a floor plan, a duct size shown as 18" × 12" or as 18/12 means that the duct is 18" wide and 12" deep.

Step 4 Divide the duct system layout into sections and identify each of the sections. The duct system should be divided at all points where the direction of airflow and cut size needs to change.

Step 5 Size the main and branch ducts. Several methods can be used for duct sizing, but the **equal-friction method** is commonly used for residential design. A variety of apps are available for this purpose, in addition to manual duct calculators produced by SMACNA and major HVACR equipment manufacturers.

Step 6 Calculate total system pressure loss and then ensure the selected blower or fan coil unit can provide the needed airflow against this calculated resistance.

Step 7 Revisit the layout and sizing if significant field changes are required during the installation. It may be necessary to recalculate the pressure losses and resize some portions of the system.

Equal-friction method: A method of sizing air distribution systems by designing for a consistent amount of pressure loss per unit length of duct (usually 100').

Duct-system design apps can simplify the process. *Figure 32* shows a screen from a popular load-estimating program that includes duct design.

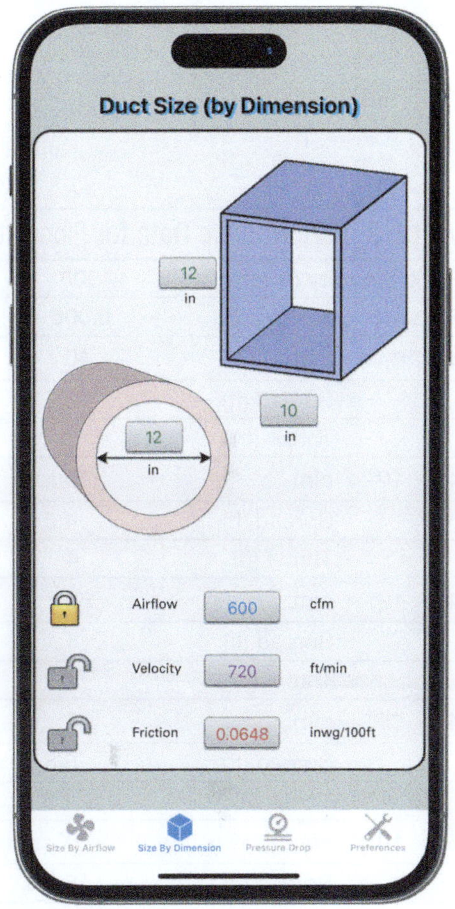

Figure 32 Screenshot from a duct design app.
Source: CyberProdigy LLC

4.2.1 Selecting Supply Outlets and Return Inlets

Locating registers and diffusers and selecting them for a reasonable volume of airflow is an important part of the process. In the average residential setting, it is best to plan for a maximum of about 150 cfm per register. If a space needs more air, use more than one register. Although larger registers and diffusers can handle more than 150 cfm, too much air from a single point can feel drafty and create temperature imbalances in the space. One diffuser manufacturer recommends that at least one outlet be used for each 4,000 Btuh heat gain in any space. The 4,000 Btuh value equates to roughly 130 cfm of airflow.

A good system design provides a clear path for air to move from the conditioned space to the return air grille(s). Individual return grilles in each room provide ideal performance, as opposed to a single return air grille in a central location. However, it is also significantly more expensive and adds a great deal of additional resistance to the system. More resistance requires more fan energy to move the air. As a result, residential systems are generally designed with one or two return grilles placed in central locations such as hallways where air can be collected from multiple rooms.

Return air grilles should not be located close to supply registers. Doing so encourages the supply air to quickly enter the return grille before it is thoroughly mixed with room air. Return grilles should also not be installed in areas such as bathrooms and kitchens. Odors from these areas will spread throughout the building.

When return grilles are placed in areas such as hallways, where doors to adjoining rooms can be closed, the doors must be undercut to allow air to escape the rooms. Alternatively, transfer grilles and ducts must be installed to provide a path for air.

4.2.2 Determining Supply Register Volume and Size

To determine the size of the supply registers for each space, first calculate the air volume needed for each room. Determine the air volume for both cooling and heating, as you'll need to work with the larger of the two air volumes. As discussed earlier in this module, the air volumes for cooling and heating are calculated using the following formulas:

$$\text{Cooling air volume} = \frac{\text{sensible heat gain (Btuh)}}{1.08 \times (T_1 - T_2)}$$

$$\text{Heating air volume} = \frac{\text{sensible heat loss (Btuh)}}{1.08 \times (T_2 - T_1)}$$

Where:

Sensible heat gain or loss = as determined by the heating and cooling load calculations

T_1 = outdoor dry-bulb temperature

T_2 = indoor dry-bulb temperature

Once the heating and cooling air volumes for each space have been calculated, the larger of the two values is used to size the supply outlets for that room. The required flow rate for cooling is commonly the larger of the two values, but the heating requirement may be larger in some regions of the country.

If more than one outlet is being used to serve a space, the total room air volume must be divided by the number of outlets to determine the size of the individual outlets. For example, if a room requiring 220 cfm is supplied by two outlets, each outlet is sized to deliver 110 cfm.

Once the air volume for each supply outlet is known, the types and sizes of diffusers can be selected using the manufacturer's product data. *Table 2* shows performance data for a series of floor registers. The top row shows the face velocity, which refers to the speed of the air as it exits the register. For residential systems, it is best to select registers and diffusers for a maximum face velocity of 500 fpm. Higher speeds tend to create drafts and noise.

TABLE 2 Performance Data for Floor Registers

Face Velocity (fpm)		300	400	500	600	700	800	900	1,000
Pressure Loss (" w.c.)		0.006	0.011	0.016	0.022	0.031	0.041	0.050	0.062
4" × 8"	cfm	40	50	65	80	90	105	115	130
	Spread (ft)	3	4	5	6.5	7.5	8.5	9.5	11
	Throw (ft)	4	4.5	6	7.5	8.5	10	11	13
4" × 10"	cfm	50	70	85	100	120	135	155	170
	Spread (ft)	4.5	5	6.5	7.5	9	10	11	13
	Throw (ft)	4	6	8	10	11	12.5	14	15.5
4" × 12"	cfm	60	80	100	120	140	160	175	195
	Spread (ft)	5	6.5	8	9.5	11.5	13	14.5	16
	Throw (ft)	4	5.5	7	8	9.5	11	12	13
4" × 14"	cfm	70	90	115	140	160	185	205	230
	Spread (ft)	5.5	7	8.5	10	12	13.5	15.5	17
	Throw (ft)	4.5	5.5	7	8.5	10	11.5	12.5	14
6" × 10"	cfm	70	95	120	145	170	190	215	240
	Spread (ft)	5.5	7	8	10	12	14	15	17
	Throw (ft)	4	5.5	7	8.5	10	11	12.5	14
6" × 12"	cfm	85	115	140	170	200	230	255	285
	Spread (ft)	6	7.5	9	11	13	15	17	19
	Throw (ft)	4.5	6	7.5	9	10	12	14	16
6" × 14"	cfm	100	130	165	200	230	265	300	330
	Spread (ft)	6.5	8	9	12	14	16.5	18	20
	Throw (ft)	4.5	6.5	8	9.5	11	13	15	17

For example, assume a space requires 130 cfm of air. A 6" × 12" register can provide that volume of air with face velocity between 400 and 500 fpm. Selecting at or below this velocity also keeps the pressure loss low. In *Table 2*, note how the pressure loss increases across the top as the face velocity increases. The spread of the air also increases with velocity. The throw represents the distance the air flows before the speed slows to 50 fpm.

The throw and spread are not necessarily a factor in sizing, but the information helps determine how far apart the registers should be placed to ensure good airflow patterns in the space. This is another reason to select floor register sizes for a velocity of 500 fpm. As you glance down the 500-fpm column in

Table 2, note that the throw and spread largely remain below 8'. Considering the general size of rooms and 8' ceilings, more spread and a longer throw may create too much air movement, as the air would still be travelling at more than 50 fpm as it struck the walls and ceiling.

4.2.3 Determining Return Grille Volume and Size

You must pay as much attention to the return duct as you do the supply duct. Sizing return grilles is done in basically the same way as sizing supply outlets. However, the parameters are a bit different.

Return air grilles should be sized for a velocity no higher than 450 fpm. For best results, use 300 fpm. This ensures the grille will remain quiet, so the occupant doesn't hear air rushing in.

Because air is entering the grille rather than exiting, the tables have no spread and throw data. *Table 3* provides an example of a table for several different sizes of return air grilles. Note that these tables may not go beyond 600 fpm. Again, maintain the velocity at the return grilles low to ensure quiet operation and avoid creating large drafts.

TABLE 3 Performance Data for Return Grilles

Face Velocity (fpm)		300	400	500	600
Pressure Loss (" w.c.)		0.010	0.019	0.029	0.041
20" × 14"	cfm	401	534	668	801
20" × 20"	cfm	575	767	959	1,150
20" × 25"	cfm	721	962	1,202	1,442

When the air handler is installed in a crawlspace, attic, or other limited-access location, it is common to include filters in the return air grilles. This replaces the need for a filter at the return side of the air handler and helps keep the return ductwork clean. If you are using filter grilles, then grille sizing must be based on the limitations of the air filter.

Filter grilles are best sized based on a maximum air volume of 2 cfm per square inch of nominal filter area. A 20" × 20" filter, for example, offers 400 in^2 of surface area. Thus, the maximum volume of air that should flow through it is 800 cfm. Exceeding this maximum will likely result in noise and buckled filters.

4.2.4 Sizing the Ducts

Several acceptable methods are used to size ductwork. This module is based on the most common method, referred to as the equal-friction method. Most duct design apps are also based on this method due to its popularity and reliable results.

The term *equal friction* does not mean that the friction losses remain constant throughout the system. It means that a specific friction loss or static pressure loss per 100' of duct is selected before the ductwork is laid out. The selected friction per 100' is used to size all sections of the duct. The equal-friction method also results in reduced velocities as air progresses through the system. Note that, in the end, the actual friction loss must be calculated for the equivalent length of the duct.

However, when using the equal-friction method for duct sizing, velocity should not be forgotten. After sizing each section of duct, it is best to also check the velocity that results from the selection and ensure it is within a reasonable range. The velocity will typically be acceptable if you design to a reasonable friction loss from the beginning.

On Site

Other Approaches to Duct Sizing

Other approaches to duct sizing include the *velocity-reduction method* and the *static-regain method.* The velocity-reduction method is based on maintaining the velocity within a specified range for certain sections of the system. For example, comfort systems might be designed to meet these parameters. These velocities generally result in quiet systems that perform well:

- *Main trunk ducts* — 750 to 1,400 fpm
- *Main branch ducts* — 600 to 1,000 fpm
- *Branch ducts* — 200 to 600 fpm

The static-regain method sizes the supply duct system to obtain uniform static pressure at all branches and outlets. It focuses on minimizing pressure losses in duct systems and ensuring that airflow remains balanced throughout a system. Although this method can result in a system that is well balanced from the beginning, it is far more complex than the other methods. Creating uniform static pressures at branches and outlets is easier on paper than in practice, as odd duct sizes may be required to make it work. The static-regain method is better suited for commercial/industrial applications, especially ventilation and exhaust systems where balancing dampers need to be minimized or eliminated.

Total Duct Length

Figure 33 shows a simple duct system. In the accompanying table, the outlets are listed in the far-left column. The method used to find the equivalent length of fittings in the duct run to outlet S5, which is highlighted on the drawing, is explained in this section.

To aid in understanding the example, the types of fittings in the duct run to outlet S5 have been marked on the figure. For each duct fitting, the capital letter identifies the type of fitting, and the number indicates the equivalent length of straight duct in feet for that fitting. For example, the E5 transition in the main supply trunk, a fitting used to reduce the size of the duct, has an equivalent length of 5'.

The fittings in the path from the air handling unit to outlet S5 have an equivalent length of 90', derived as follows:

- (1) B10 plenum takeoff = 10'
- (1) E5 reduction = 5'
- (3) E10 90-degree adjustable elbows = 30'
- (1) G30 floor diffuser boot = 30'
- (1) B15 takeoff = 15'

$$10' + 5' + 30' + 30' + 15' = 90'$$

To begin sizing the ducts, identify the supply and return runs that have the longest total length. The total length of a run is the measured straight lengths of duct in the path, plus the equivalent length of all fittings in the path. Noting the table shown in *Figure 33,* the supply run for outlet S5 has a total length of 132' (90' + 42'). Note that, although S7 may appear to be the longest run since it is at the end of the run, the total length of S7 is only 102'. Size the duct along the path to the longest run, and then drop back to size the branch ducts.

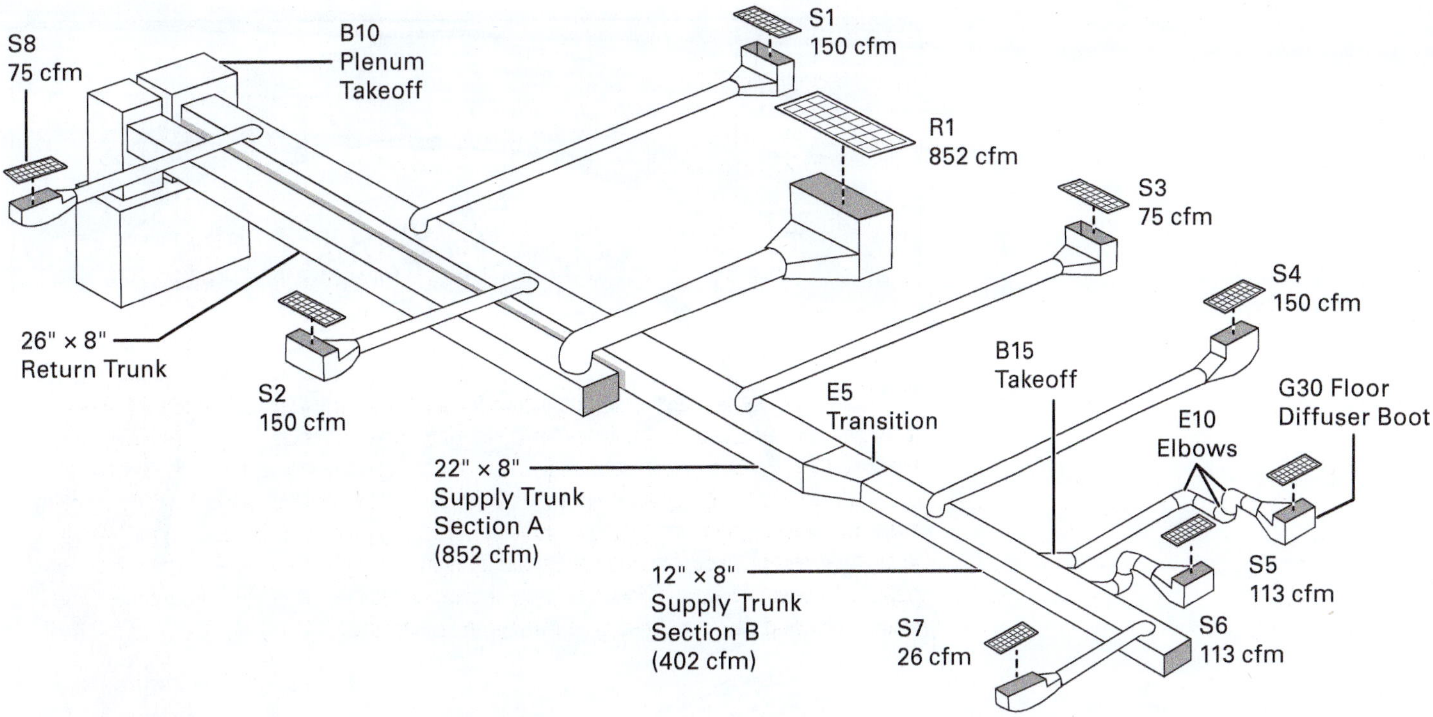

Outlet	Equivalent Length of Fittings (ft)	Measured Length of Duct (ft)	Total Equivalent Length of Run (ft)	Round Branch Duct Size
S1	55	29	84	8"
S2	55	19	74	8"
S3	55	43	98	6"
S4	60	49	109	8"
S5	90	42	132	6"
S6	80	42	122	6"
S7	60	42	102	6"
S8	55	12	67	6"
Outlet R1	**35**	**32**	**67**	**14"**

Figure 33 Example duct system.

This comparison between the S5 and S7 duct runs is presented to emphasize that the duct run with the greatest measured length is not always the run with the greatest total length. It is often necessary to calculate the total length (measured length + equivalent length of fittings) of all the runs to make sure that the longest equivalent length is used in the design process.

Choosing Duct Sizes

Duct sizing can be done using different methods. Various manual calculators are available to choose from (*Figure 34*) as well as apps. All methods include guidance in their use. Most apps, with the appropriate knowledge of the process, can be used to design a system without additional references. The apps are quite versatile and are available for smartphones and tablets as well as for computers.

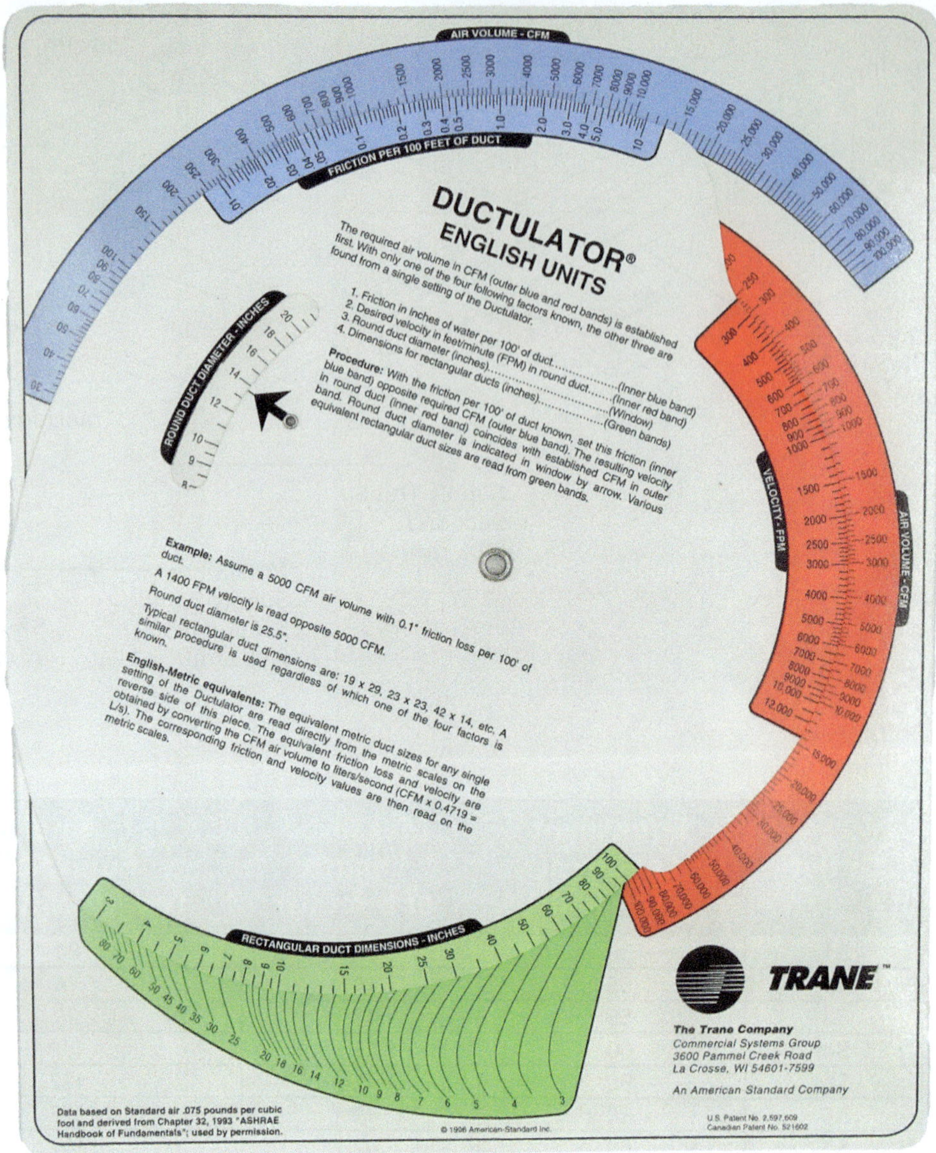

Figure 34 Manual duct-sizing calculator.

The app allows the user to input the desired air volume or velocity, then use sliders or a keyboard to change the duct size. As the duct size changes, the resulting information is displayed for the chosen air volume or velocity.

Refer to *Figure 35*. An air volume of 800 cfm has been entered and a target friction loss of 0.08" w.c. per 100' of duct. Subject matter experts (SMEs) indicate that 0.08" w.c. typically yields good results in residential systems. The *aspect ratio*, set at 1.5, is the desired ratio of duct width to height. You can increase the aspect ratio but remember that higher aspect ratios increase the friction loss, meaning that the duct must be larger to stay within the desired friction loss. This increases the cost of the duct. In this example, the app has chosen a final duct size of 10" × 16", which results in a friction loss of 0.064" w.c. per 100' of duct.

In *Figure 36, Sizing By Dimension* has been selected. This is a good feature for checking the performance of an existing duct size. Assume you are replacing a system and encounter an 18" × 14". Using the same 800 cfm value, you can see that the friction loss per 100' is much lower than our target of 0.08" w.c. This also means the velocity in the duct is very low, even for a residential application. It may be best to replace this section of duct with a smaller size for improved performance.

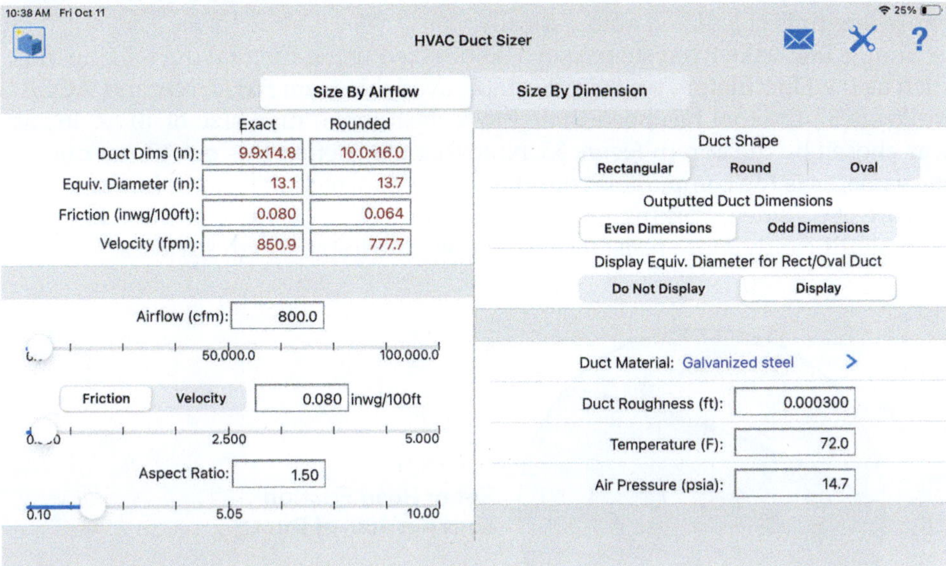

Figure 35 Screenshot of duct-sizing app.

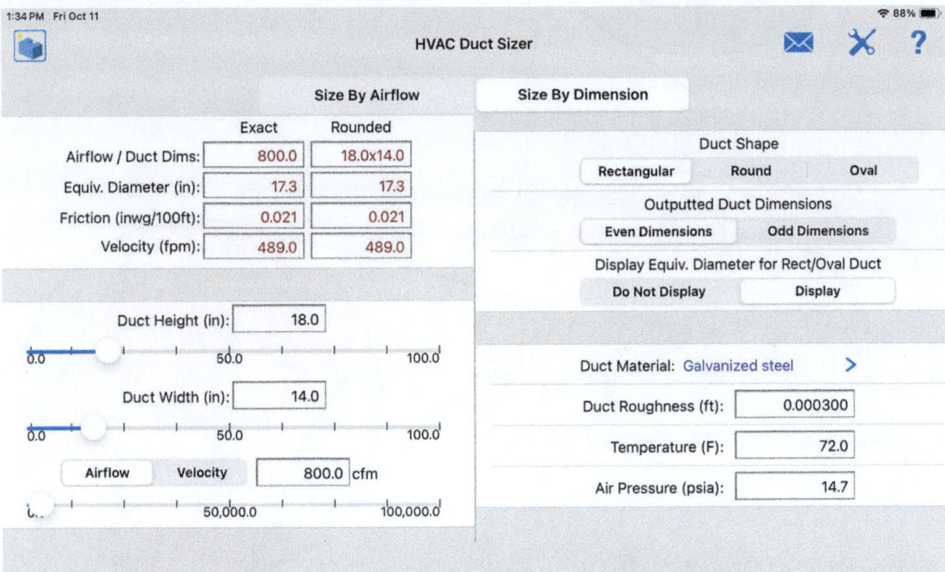

Figure 36 Screenshot of app using the Sizing By Dimension feature.

One important point to remember is that the friction values are based on losses per 100' of duct. When using the app or a manual duct-sizing calculator to size the duct for a run with a total length that is greater or less than 100', the friction loss must be corrected for the total length of the duct, which includes the equivalent length of fittings. The corrected value is easily calculated as follows:

$$\textit{Corrected friction loss} = \textit{total duct length} \times \left(\frac{\textit{friction per } 100'}{100}\right)$$

For example, assume a duct has a total length of 35', sized at a friction loss of 0.08" w.c. per 100':

$$\textit{Corrected friction loss} = 35' \times \left(\frac{0.08\text{" w.c. per } 100'}{100}\right)$$

$$\textit{Corrected friction loss} = 35' \times 0.0008\text{" w.c.}$$

$$\textit{Corrected friction loss} = 0.028\text{" w.c.}$$

The actual friction loss of 0.028" w.c. for this 35' section of duct is then added to the rest of the losses as part of the total system friction loss. Failing to correct the friction for the actual length of duct can lead to an incorrect blower selection. Many apps allow for the entry of the duct length, and the total loss for the entire length is calculated automatically.

Simple low-velocity systems can also be sized using manual duct-sizing aids such as the Ductulator® and similar tools available from SMACNA and ACCA. In *Figure 37*, the tool has been dialed to a sheet metal duct size of 10" × 16", as was chosen by the app in *Figure 35*. Note that the friction loss per 100' of duct is the same, as is the comparable round duct size. Correction factors shown on the tool are used when working with fiberglass and concrete duct, as these features offer significantly more resistance to airflow due to the rough surfaces.

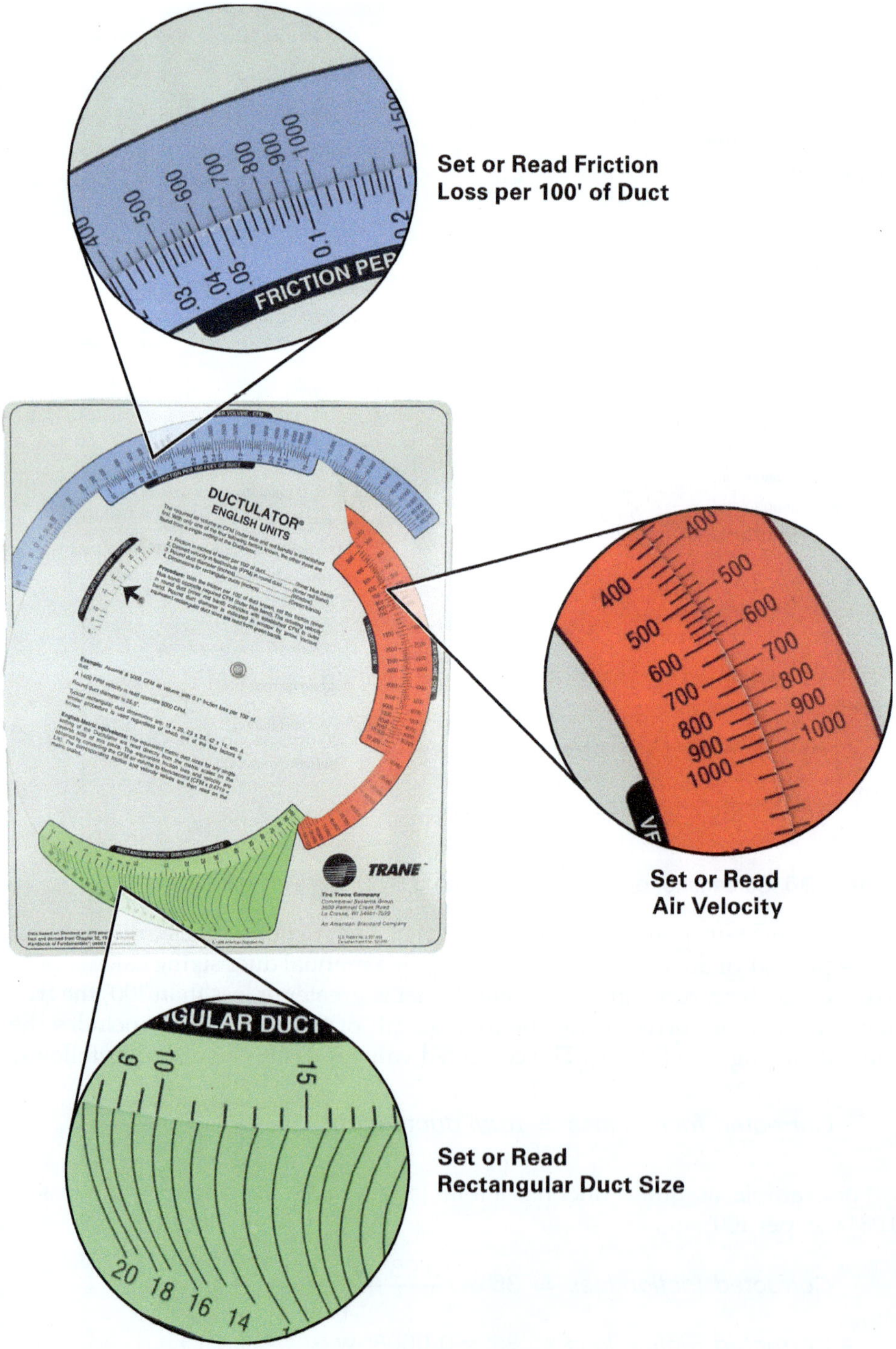

Figure 37 Duct-sizing calculator set for 10" × 16" duct.

When designing a new system using a manual duct-sizing aid, set the air volume, in cfm, to the desired friction loss. Then review the various duct sizes that would work. Because you can quickly see a variety of duct sizes instantly, these aids are often more convenient to work with than an app for an experienced craftworker.

Ductwork is generally sized starting at the air handler or furnace and working toward the longest run of the system. Refer to the duct system example in *Figure 33*. The first section of the supply trunk (Section A) must carry the total system air quantity of 852 cfm to feed outlets S1 through S8. The selected duct size is 22" × 8". Using the app, you can see the result of that selection (*Figure 38*). The friction loss per 100' of duct is 0.064" w.c., and the velocity is 790 fpm. Both are acceptable values.

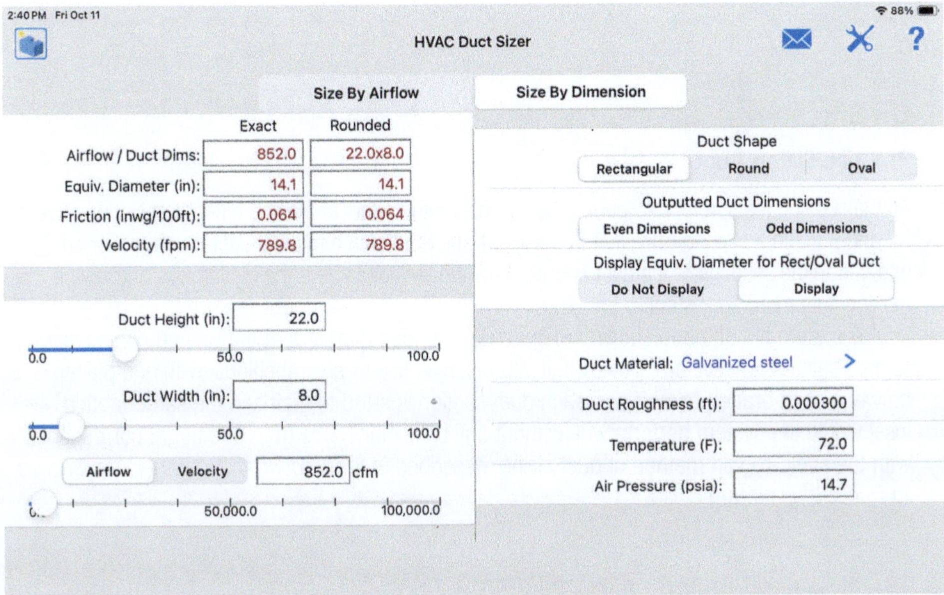

Figure 38 Performance of a 22" × 8" duct at 852 cfm.

However, consider this: As soon as the first branch duct draws its air from the trunk, the air volume in that section is reduced by 75 cfm. By the time you reach the transition, 450 cfm of air has ideally left the main duct, leaving only 402 cfm flowing. Using the app to examine this new condition (*Figure 39*), the friction loss per 100' of duct has fallen to 0.016" w.c. and velocity is below 400 fpm. An experienced designer would say that the air is now "wandering" through the duct.

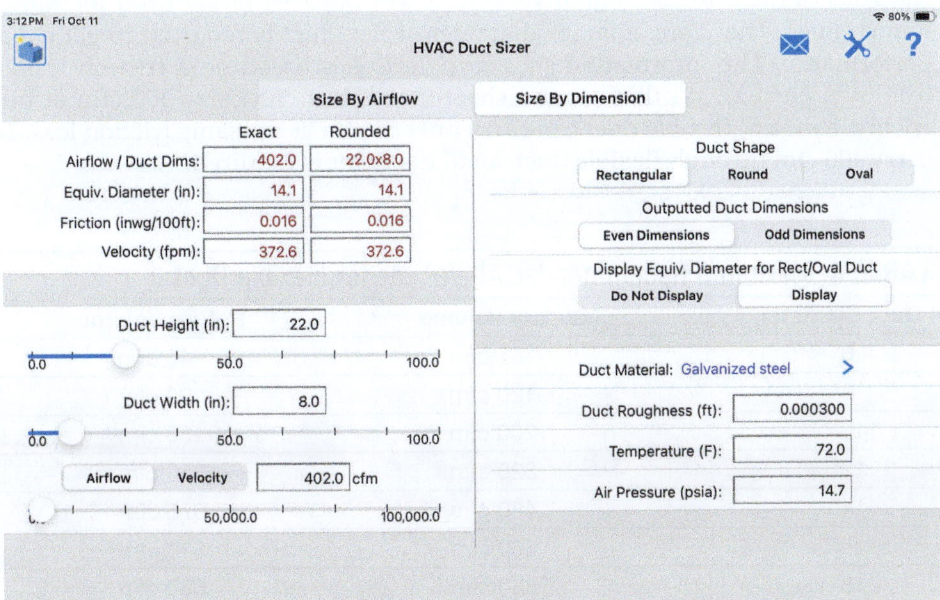

Figure 39 Performance of a 22" × 8" duct at 402 cfm.

This is the reason for transitions to reduce the size of the duct as air is delivered to the connected branches. The transition shown in *Figure 33* is designed to equalize the static pressure and restore the velocity to an acceptable level.

The section of duct following the transition (Section B) is sized to 12" × 8". This is based on the need to deliver the 402 cfm of air used to feed outlets S4 through S7. This results in a friction loss per 100' of 0.063" w.c. and a velocity of 649 fpm. These are acceptable values.

In reality, the system in *Figure 33* would perform better if the main duct were transitioned twice rather than once. Consider the effect of placing a transition between S2 and S3, then a second transition between S4 and S5. With those sections properly sized, the friction loss and velocity throughout the main trunk will remain in a narrower range.

On Site

Practice Sizing Duct

As the text suggests, refer to *Figure 33* and consider the effect of placing a duct transition between S2 and S3, and a second transition between S4 and S5. Using a manual duct-sizing tool or an app, resize the sections of the duct and see how the changes affect the friction loss and velocity, making both more consistent.

A system like the one shown in *Figure 33* will need dampers at the branch connections to help balance the system. However, if you installed five or six transitions, with one placed between each branch takeoff and each duct section is sized to maintain the same friction loss, would the system perform well without any balancing? This train of thought corresponds with the static-regain method of duct sizing, described in the *Section 4.2.4, Other Approaches to Duct Sizing* feature.

Compromises are sometimes made when sizing ducts. Some engineers and fabricators prefer to work only in even numbered sizes of duct, while others prefer working with odd-numbered sizes, such as 17" × 9". When sizing branches, a duct-sizing app may recommend a 7" round duct to carry 150 cfm of air at 0.08" w.c. friction loss per 100'. However, 7" round duct is an odd size and rarely carried by HVACR distributors. If you begin to size duct regularly for your employer, you'll find these factors affect the designer's choices.

Also remember that flexible duct offers far more resistance to airflow than sheet metal duct. In residential systems, sheet metal is rarely used for round branch lines. Therefore, a larger diameter of flex duct is required to get equal performance. The information shown in *Table 4* is based on a friction loss of 0.08" w.c. per 100'. While 6" round sheet metal duct can carry 100 cfm at this friction loss, a 6" flexible duct can carry only 55 cfm at the same friction loss. To carry 100 cfm through flexible duct, an 8" diameter is required.

NOTE

To minimize the friction loss with flexible duct, elongate the duct until it's taut, without putting unnecessary stress on the jacket.

TABLE 4 Maximum Recommended Air Volume for Flexible Duct

Duct Diameter	Supply Volume	Return Volume
6"	55 cfm	40 cfm
8"	120 cfm	90 cfm
10"	200 cfm	160 cfm
12"	320 cfm	250 cfm
14"	480 cfm	375 cfm
16"	660 cfm	530 cfm
18"	880 cfm	680 cfm
20"	1,200 cfm	900 cfm

Total Friction Loss and Blower Selection

The total friction loss of the system needs to be calculated to ensure the system blower can provide the needed performance. As previously shown, the total friction loss for an air system includes the supply system loss, return system loss, and any components not included in the system fan rating.

The friction loss for the supply system ductwork is found by taking the total length of the longest run and multiplying it by the system design friction rate. In the example system in *Figure 33*, the run feeding outlet S5 has the longest total length at 132'. This was determined in *Section 4.2.4*. The design friction loss used for duct sizing was 0.08" w.c. per 100'. Therefore, the supply-duct friction loss for the example system is determined as follows:

NOTE

If the actual friction loss of the duct selected is significantly different than the original design loss, use the actual friction loss value for each section of duct.

$$\textit{Friction loss for longest run} = \textit{total duct length} \times \left(\frac{\textit{friction per 100' duct}}{100}\right)$$

$$\textit{Friction loss for longest run} = 132' \times \left(\frac{0.08\text{" w.c.}}{100}\right)$$

$$\textit{Friction loss for longest run} = 132' \times 0.0008\text{" w.c.}$$

$$\textit{Friction loss for longest run} = 132' \times 0.0008\text{" w.c.}$$

$$\textit{Friction loss for longest run} = 0.1056\text{" w.c.}$$

The pressure loss of the floor register at the end of the run must also be added to the supply duct loss. This value can be found using the manufacturer's product data sheets for the grille. Run S5 in *Figure 33* does not show the register size, so we will size it based on *Table 2*. A 4" × 14" floor register can provide the 113 cfm of air at 500 fpm velocity and a friction loss of 0.016". Add that value to the 0.1056" friction loss for the rest of the run:

$$\textit{Total friction loss of the longest supply run} = 0.1056\text{" w.c.} + 0.016\text{" w.c.}$$

$$\textit{Total friction loss of the longest supply run} = 0.1216\text{" w.c.}$$

The friction loss for the return duct is found the same way. Take the total length of the longest return run and multiply it by the return system design friction rate. In the example system in *Figure 33*, only one return run to a single return grille is used. It has a total equivalent length of 67'. The design friction rate used for the return duct system was 0.06" w.c. per 100' of duct. Therefore, the friction loss for the return duct is calculated as follows:

$$\textit{Friction loss for longest run} = \textit{total duct length} \times \left(\frac{\textit{friction per 100' duct}}{100}\right)$$

$$\textit{Friction loss for longest run} = 67' \times \left(\frac{0.06\text{" w.c.}}{100}\right)$$

$$\textit{Friction loss for longest run} = 67' \times 0.0006\text{" w.c.}$$

$$\textit{Friction loss for longest run} = 0.0402\text{" w.c.}$$

Remember that the pressure loss of the return grille must be added to the return duct loss. This information can be found using the manufacturer's product data sheets for the specific grille. Referring to *Table 3*, we will select the 20" × 25" return grille to ensure the system remains quiet. However, we will have to interpolate the friction loss for accuracy, since 852 cfm is not shown on the chart. You can probably come up with a sufficiently accurate value with some mental math, but interpolation is the way to do it accurately.

To begin, we must assign some variables, as follows:

- $x = 852$ cfm (our example system air volume)
- $x_1 = 721$ cfm (the lower airflow value in *Table 3*)
- $x_2 = 962$ cfm (the higher airflow value in *Table 3*)
- $y =$ the unknown friction loss
- $y_1 = 0.010$" (the friction loss for 721 cfm from *Table 3*)
- $y_2 = 0.019$" (the friction loss for 962 cfm from *Table 3*)

Now we can set up the equation and find the corresponding friction loss for the grille at 852 cfm. Note that some multiplication products have been rounded:

$$y = y_1 + \left[\left(\frac{x - x_1}{x_2 - x_1}\right)(y_2 - y_1)\right]$$

$$y = 0.010\text{" w.c.} + \left[\left(\frac{852\text{ cm} - 721\text{ cfm}}{962\text{ cfm} - 721\text{ cfm}}\right)(0.019\text{" w.c.} - 0.010\text{" w.c.})\right]$$

$$y = 0.010\text{" w.c.} + \left[\left(\frac{131\text{ cfm}}{241\text{ cfm}}\right) \times 0.009\text{" w.c.}\right]$$

$$y = 0.010\text{" w.c.} + [0.543 \times 0.009\text{" w.c.}]$$

$$y = 0.010\text{" w.c.} + 0.0049\text{" w.c.}$$

$$y = 0.0149\text{" w.c.}$$

The friction loss of the return grille can now be added to the friction loss of the duct:

$$\textit{Total friction loss of the return run} = 0.0402\text{"} + 0.0149\text{"}$$

$$\textit{Total friction loss of the return run} = 0.055\text{"}$$

The last component of the system friction loss accounts for the losses for components not included in the system fan rating, such as a wet evaporator coil. The friction loss for such components is provided in the equipment product literature. However, it is important to remember the following:

- For most fan coil units that house the blower and coil in the same cabinet, the wet coil is accounted for in the performance tables.
- For furnaces and similar installations where the indoor coil is a separate component, you must add the friction loss of a wet coil from the product literature to the system friction loss.

Table 5 shows some performance data for two sizes of coils nominally rated at 2 tons. The air volume is shown across the top row and the table shows the corresponding friction loss for both a wet and dry coil. Since the friction loss is always higher for a wet coil, use the value for a wet coil.

TABLE 5 Indoor Coil Pressure Drop (" w.c.)

Unit Size	Standard cfm								
	400	**500**	**600**	**700**	**800**	**900**	**1,000**	**1,100**	**1,200**
					Dry				
2414	0.065"	0.091"	0.120"	0.154"	0.194"	0.237"	0.284"		
					Wet				
	0.066"	0.094"	0.124"	0.161"	0.203"	0.250"	0.301"		
					Dry				
2417	0.056"	0.076"	0.097"	0.123"	0.151"	0.182"	0.215"		
					Wet				
	0.060"	0.082"	0.105"	0.132"	0.163"	0.195"	0.231"		

To summarize, if this were a furnace with an add-on coil, the friction loss for the selected coil must be added to the calculated friction loss for the rest of the system. Assume that the Model 2414 coil is selected for the system. Since our 852-cfm system falls neatly between 800 and 900 cfm, we can easily interpolate by taking half the difference between the two friction loss values shown:

$$\textit{Coil friction loss} = (0.203\text{" w.c.} + 0.250\text{" w.c.}) \div 2$$

$$\textit{Coil friction loss} = 0.453\text{" w.c.} \div 2$$

$$\textit{Coil friction loss} = 0.227\text{" w.c.}$$

We can now determine the total system friction loss:

Total system friction loss = *supply duct loss* + *return duct loss* + *coil loss*

Total system friction loss = 0.1216" w.c. + 0.055" w.c. + 0.227" w.c.

Total system friction loss = 0.404" w.c.

Once the total friction loss is known, the next step is to refer to the manufacturer's product data sheets for the selected equipment to verify that its blower can deliver the needed air volume at this resistance, and what fan speed may be required to do so.

Figure 40 shows an example of a blower performance chart for a furnace. Furnace size 40-A has been selected to offset the heat loss. Even at low speed, the selected furnace can manage 850 cfm of air at a friction loss of 0.6" w.c. During system startup, the blower can be set to operate at low speed and should have no problem developing the required airflow.

NOTE

In *Figure 40*, note that the furnace blower performance includes the friction loss through the air filter. If the air filters are placed in filter grilles, the friction loss would need to be added to the total.

Fan Performance — High-Efficiency Furnace Air Delivery in cfm (with Filter)

Furnace	External Static Pressure (" w.c.)							
Size	Speed	0.1	0.2	0.3	0.4	0.5	0.6	0.7
40-A	High	1425	1370	1320	1265	1195	1125	1060
	Med-High	1315	1275	1230	1180	1125	1070	1000
	Med-Low	1145	1115	1090	1045	1010	955	880
	Low	1035	1000	970	930	900	850	800
40-B	High	1515	1465	1405	1345	1275	1210	1115
	Med-High	1350	1310	1275	1220	1160	1100	995
	Med-Low	1155	1140	1110	1070	1020	945	825
	Low	975	960	945	915	870	780	710
60	High	1510	1450	1380	1320	1230	1150	1045
	Med-High	1350	1300	1245	1190	1115	1030	945
	Med-Low	1165	1125	1090	1045	980	910	775
	Low	965	945	915	880	830	735	640
80	High	1590	1530	1470	1390	1315	1225	1140
	Med-High	1425	1390	1345	1285	1200	1135	1040
	Med-Low	1250	1220	1185	1135	1085	1010	930
	Low	1060	1055	1030	980	930	870	805

Figure 40 Furnace blower performance chart example.

Note that there is another source of friction loss that is rarely factored into duct sizing. Balancing dampers are often installed in systems to help balance the volume flowing from each supply register and diffuser. A damper limits flow to a branch by partially closing it. This adds friction loss, causing less air to enter the branch.

Remember that dampers can only reduce airflow where too much is flowing, forcing it to find an easier way forward. Systems are designed without consideration for the dampers, because a perfect design would eliminate the need for them altogether. However, in reality, the actual friction loss in a system often exceeds the calculated loss. Required field changes may also add to the friction loss. For this reason, it is never a good idea to select a system with a blower that is on the edge of its maximum capacity at the calculated static pressure.

While the process of sizing duct may seem tedious, practice helps. An experienced craftworker can lay out, design, and select the equipment for a system in a relatively short period of time. One good way to practice your skills is to evaluate the system serving your own home or the home of a friend.

Survey and document the system layout and duct sizes, then apply what you have learned to determine the system friction loss and air velocity in various duct sections to see how the system performs and how changes might affect performance.

4.3.0 Commercial Building Loads and Design Factors

ACCA Manual N®, *Commercial Load Calculation*, provides heat load-estimating data and instructions for small to medium commercial buildings using packaged or split systems, including heat pumps. Although the load factors and calculations for commercial systems are similar to those for residential systems, some important differences exist. Here are some examples of how commercial calculations differ from residential calculations:

- Commercial buildings typically use different construction materials, such as metal studs and a wider variety of exterior siding materials. While homes lean toward wood frame or masonry construction, commercial buildings often have facades of metal, masonry, and glass.
- Because of the effect of radiation and infiltration on the perimeter of a building, the loads at the perimeter and the core can vary widely. It is often necessary to deal individually with core and exterior zones as a result. For example, it is not unusual during occupied periods for perimeter zones to need cooling while the core zone is overcooled or just needs ventilation. The air distribution system for unitary equipment in large commercial buildings is typically laid out so that a unit takes care of perimeter zone or core zones, but not both, due to the diversity in the loads.
- A commercial building is more likely to have a large, open area, such as a lobby or atrium with vaulted ceilings, and a lot of traffic in and out of the building with its resulting infiltration. These represent unique heat load factors.
- The load often varies significantly as the day progresses. Solar gain can be very high on a large structure with a great deal of glass. Since the sun moves throughout the day, the load imposed by it moves as well. For this reason, the calculations in Manual N® are offered for different times of day.
- The heat load from occupants is a critical factor in commercial system design. Shopping centers, theaters, and office buildings have very dense occupancy at times, while there is virtually no one is present at other times. A lot of human traffic also means many door openings and closings. However, systems must be sized to manage the heaviest loads. For this reason, commercial buildings need diversity in capacity, based on multistage and modular systems.
- The interiors of commercial buildings are often rearranged to accommodate tenants. The system design for these buildings must be flexible enough to deal with the changes with reasonable changes to the air distribution system.

Commercial buildings may also present other load considerations such as the following:

- Large latent loads where a lot of water is present
- Large lighting loads
- Vending machines
- Refrigeration equipment
- Computers, servers, and workstations
- Printers and copiers

4.0.0 Section Review

1. What is a loss coefficient, or C-factor, used for?
   a. Determining the throw of a register or diffuser
   b. Determining the dynamic pressure loss in a duct fitting
   c. Determining the dynamic pressure loss in a straight duct section
   d. Determining the size of a rectangular duct based on pressure loss

2. Which of the following is the most common approach to sizing residential duct systems?
   a. Static-regain method
   b. Equal-friction method
   c. Friction-recovery method
   d. Velocity-reduction method

3. Which statement about commercial building HVAC system design is correct?
   a. ACCA Manual J® is used for commercial building load calculations.
   b. Radiant heat affects the interior zones more than those on the perimeter.
   c. The heat load of a commercial building is very consistent throughout the day.
   d. Unitary systems should be arranged so that each one handles either perimeter or core space.

Module 03407 Review Questions

1. An oversized cooling system will likely result in ______.
   a. poor air filtration
   b. reduced energy costs
   c. excess indoor humidity
   d. an oversized heating system

2. A block load calculation ______.
   a. is only used for commercial buildings
   b. considers the structure as a single room
   c. is the most time-consuming approach to load calculations
   d. must be done before a room-to-room load calculation can begin

3. True or False? The historical maximum temperature in an area should be used as the HVAC system design outdoor temperature.
   a. True
   b. False

4. The term *convection* refers to heat that is transferred by ______.
   a. the movement of a fluid
   b. physical contact between materials
   c. the flow of heat through a substance
   d. the light bouncing off reflective objects

5. All the factors below affect heat flow through a wall *except* ______.
   a. the area of the wall
   b. outdoor relative humidity
   c. the heat-conducting properties of the wall
   d. the temperature difference across the wall

6. A heat transfer multiplier (HTM) differs from a glass load factor (GLF) because a GLF considers ______.
   a. the relative humidity
   b. only the indoor design temperature
   c. only the outdoor design temperature
   d. the indoor and outdoor temperature difference

7. Below-grade basement walls are a significant factor in calculating ______.
   a. heating loads only
   b. cooling loads only
   c. humidification loads
   d. heating and cooling loads

8. Infiltration is apportioned to the load calculation based on the chosen category of tightness and what other factor?
   a. Number of occupants
   b. The total heating load
   c. Square footage of the home
   d. The sensible and latent cooling load

9. All the following factors must be considered in the selection of a furnace for a heating-only application *except* ______.
   a. air volume
   b. type of system
   c. latent heat load
   d. sensible heat load

10. Using an indoor coil with a higher nominal cooling capacity than the condensing unit increases the sensible cooling performance but the practice ______.
    a. interferes with oil return
    b. decreases energy efficiency
    c. decreases dehumidification performance
    d. requires a fixed-restrictor metering device

11. In a residential application where the cooling load significantly exceeds the heating load, the furnace may need to be ______.
    a. electric
    b. gas-fired
    c. undersized
    d. a high-air model

12. When the indoor coil is at a higher elevation than the outdoor unit, the suction line should ______.
    a. be no more than 25' in length
    b. be pitched back toward the indoor coil
    c. be increased in diameter to the next available size
    d. have a loop that rises to the height of the indoor coil

13. For residential systems, it is best to select registers and diffusers for a maximum face velocity of ______.
    a. 300 fpm
    b. 500 fpm
    c. 700 fpm
    d. 900 fpm

14. Duct-sizing tools and apps typically provide the friction loss for a section of duct at a given air volume based on _____.
    a. a duct length of 100'
    b. a reduction in velocity
    c. proper duct insulation
    d. a 20°F temperature difference

15. True or False? For a manufacturer-approved combination of furnace and indoor coil, the friction loss of air passing through the coil has been included in the furnace blower performance chart.
    a. True
    b. False

Answers to odd-numbered questions are found in the Module Review Answer Key at the back of this book.

Answers to Section Review Questions

Answer	Section	Objective
Section One		
1. b	1.1.1	1a
2. a	1.2.0	1b
Section Two		
1. d	2.1.2	2a
2. c	2.2.3; *Figure 14*	2b
Section Three		
1. c	3.1.0	3a
2. b	3.2.0	3b
3. d	3.3.0	3c
4. a	3.4.2	3d
Section Four		
1. b	4.1.3	4a
2. b	4.2.0	4b
3. d	4.3.0	4c

MODULE 03408

Commercial/Industrial Refrigeration

Source: Berg Chilling Systems Inc.

Objectives

Successful completion of this module prepares you to do the following:

1. Describe methods used to freeze, store, and transport food products.
   a. Identify and describe common methods used to freeze food products.
   b. Describe approaches to short-term and long-term refrigerated storage.
   c. Identify various ways to transport refrigerated food products.
2. Identify and describe various commercial and industrial refrigeration system components.
   a. Identify and describe refrigeration compressor configurations.
   b. Describe the application, control, and installation of air-cooled condensers.
   c. Identify and describe various evaporator and display case configurations.
   d. Identify and describe refrigeration system accessories, specialties, and support systems.
3. Identify and describe various types of defrost systems.
   a. Describe an off-cycle defrost system.
   b. Explain how electric defrost systems operate.
   c. Identify and describe hot-gas defrost systems.
4. Describe ammonia-based refrigeration systems.
   a. Describe the properties of ammonia as a refrigerant and identify safety considerations.
   b. Describe common components used in ammonia-based refrigeration systems.

Performance Task

Under supervision, you should be able to do the following:

1. Install or troubleshoot a commercial refrigeration condensing unit and unit cooler system.

Overview

Supermarkets, cold storage warehouses, commercial food preparation facilities, and similar businesses require much larger and more complex refrigeration systems than those used in small retail operations. Commercial and industrial refrigeration applications are often custom engineered for the structure, products, and processes required. Many work areas, such as areas where tons of fish and seafood are cleaned and prepared for sale, must also be refrigerated, and can involve highly complex control systems. This module examines the equipment and systems used in commercial and industrial refrigeration applications.

Digital Resources for HVACR

Scan this code using the camera on your phone or mobile device to view the digital resources related to this craft.

SCAN ME

1.0.0 Food Product Refrigeration

Performance Tasks

There are no Performance Tasks in this section.

Objective

Describe methods used to freeze, store, and transport food products.

a. Identify and describe common methods used to freeze food products.
b. Describe approaches to short-term and long-term refrigerated storage.
c. Identify various ways to transport refrigerated food products.

The common purpose of commercial and industrial refrigeration systems and comfort cooling systems are to remove heat. Refrigeration systems are largely used to remove heat from substances for the food, chemical, and manufacturing industries. These systems are basically constructed from the same components as air conditioning systems. Mechanical refrigeration systems can reach surprisingly low temperatures, but a totally different approach is required to reach the lowest possible temperatures.

Most commercial refrigeration involves the freezing and storage of food and other perishable **commodities**. This module focuses on the refrigeration equipment used for this purpose. The first part of this module briefly describes the methods used to preserve food and presents various kinds of commercial and industrial refrigeration equipment.

Commodities: Raw materials or agricultural products, such as iron and grain, that are commercially bought, sold, and traded.

Later in the module, refrigeration system characteristics and components that are unique to these applications are presented. You'll also learn about some special refrigeration applications and systems.

The concept of food processing starts at the point of harvest and extends to consumption (*Figure 1*). Throughout this process, the food must be properly prepared and stored to maintain its freshness and quality. *Refrigerated storage* is defined as any space where refrigeration is used to provide controlled storage conditions. It can be an entire building such as a warehouse, a section of a building dedicated to refrigerated goods, or a small cabinet or compartment. Refrigerated storage is often divided into two or more sections or areas, with each section maintained at a different temperature to suit the product.

1.1.0 Commercial Freezing Methods

The process of freezing reduces the temperature of a food product to the planned storage temperature or below, changing any water the product contains to ice. As shown in *Figure 2*, the freezing process has three phases. The first phase removes the product's sensible heat, reducing the temperature to the freezing point. During the second phase, the product's latent heat of fusion is removed and the water in the product is changed to ice crystals. During the phase change, no sensible temperature change occurs. In the final phase, sensible heat is again removed to subcool the product down to or below the storage temperature.

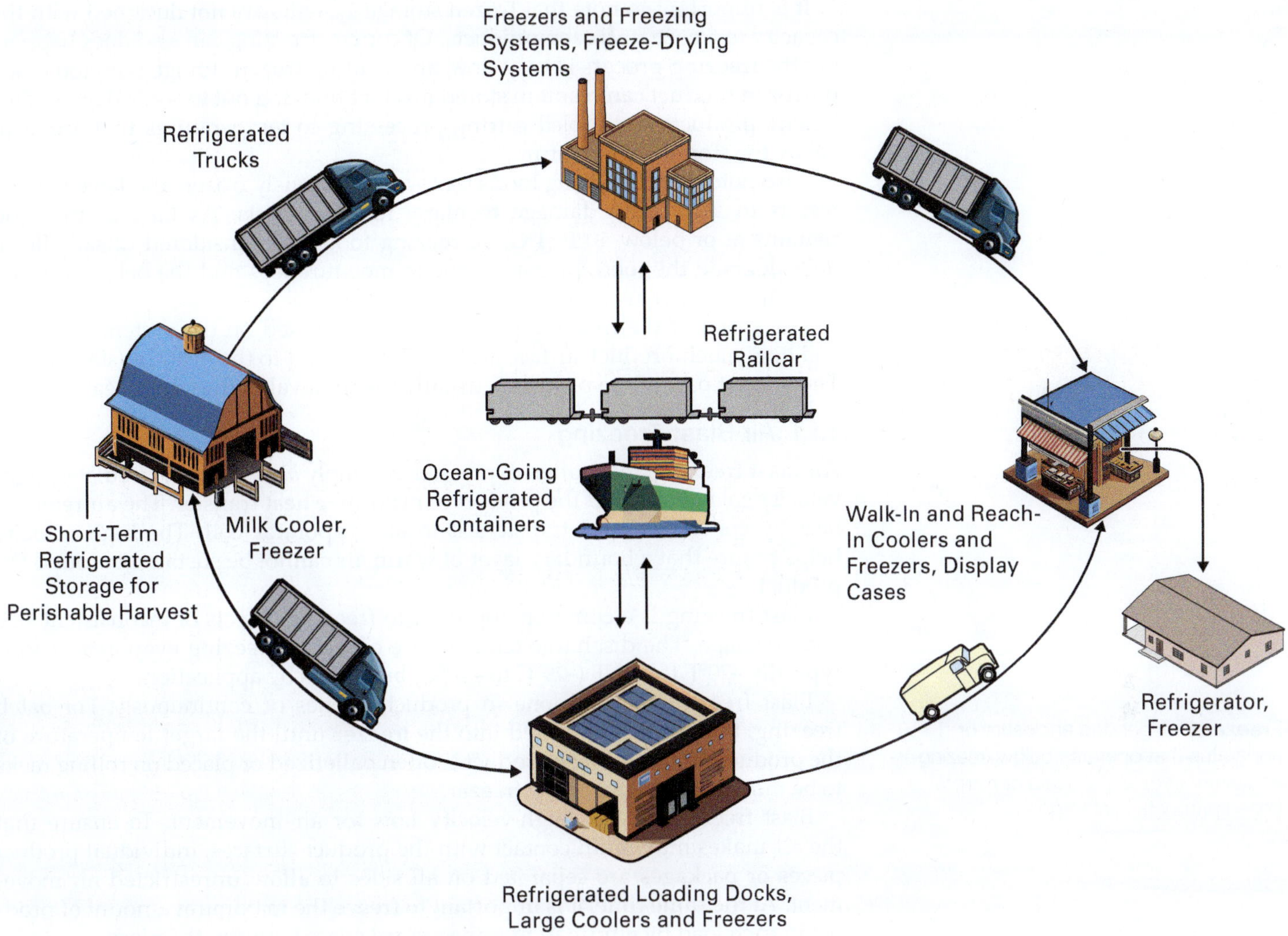

Figure 1 The movement of food from harvest to consumer.

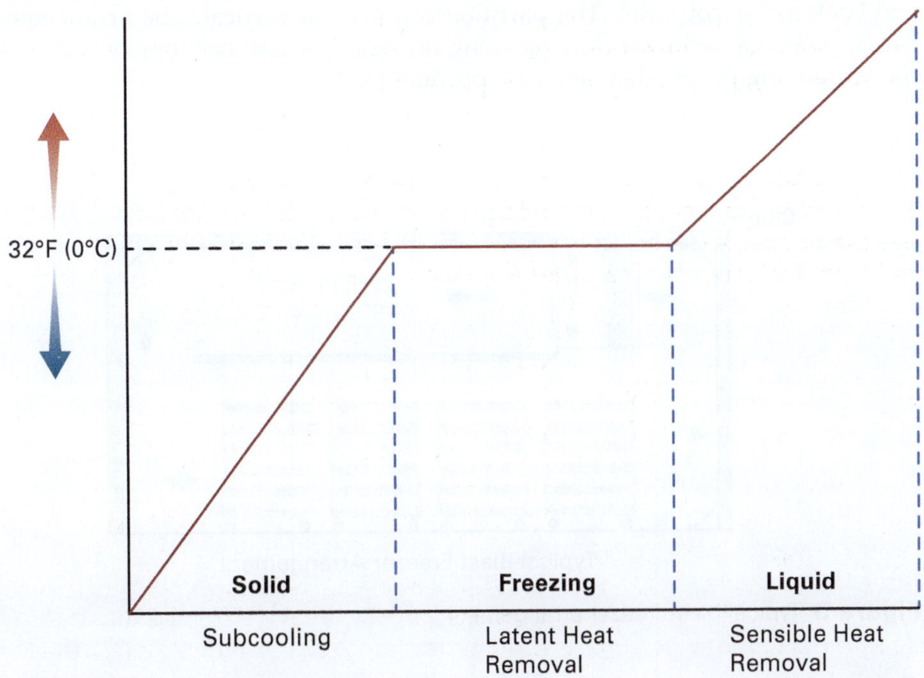

Figure 2 Typical freezing curve.

It is important to note that frozen storage systems are not designed with the capacity to freeze unfrozen products. Of course, freezing can and does happen, but the freezing process is very slow, and loading frozen storage with too much unfrozen product can result in stored product thawing out to some degree. This is why products are cooled during processing to temperatures that are at or below the storage temperature.

Also note that refreezing foods that were previously frozen and have thawed results in significant damage to many food products. As long as the food remains at or below 40°F (4°C), refreezing food isn't considered unsafe. But it does degrade the food, primarily due to moisture loss and the behavior of ice crystals as they thaw and refreeze.

The speed of the freezing process is largely based on the system's capacity and how much product surface area can be exposed to the heat transfer surface. The longest part of the process is usually the removal of the latent heat.

1.1.1 Air Blast Freezing

Air blast freezing, most often expressed as simply *blast freezing*, circulates high-velocity cold air around the product to maximize heat transfer. The air removes heat by convection and transfers it to an evaporator coil. The high velocity helps ensure that a boundary layer of warm air cannot be sustained around the product.

Blast freezing is a common approach to freezing objects of inconsistent size and/or shape. The discharge temperature of a blast-freezing evaporator coil is typically –20°F to –40°F (–29°C to –40°C) but varies by application.

Freezer: Refrigerated enclosure or apparatus that operates below freezing and either freezes products or maintains frozen products in storage.

Blast freezing can be done in product batches or continuously. For batch freezing, the product is loaded into the **freezer** until the target temperature of the product is reached. The product is often palletized or placed on rolling racks to be moved in and out of the freezer.

Blast freezers rely on high-velocity fans for air movement. To ensure that the air makes maximum contact with the product surfaces, individual product pieces or packages are separated on all sides to allow unrestricted air movement. At the same time, it is important to freeze the maximum amount of product in each load by eliminating unnecessary space between the pieces.

As shown in *Figure 3*, the evaporator coils are often separated from the product by a horizontal partition, directing the cold air through the product racks and back to the coil inlet. The partition can also be vertical. The arrangement can be *series flow*, with air flowing along the length of the load, or *crossflow*, with the air moving perpendicular to the product load.

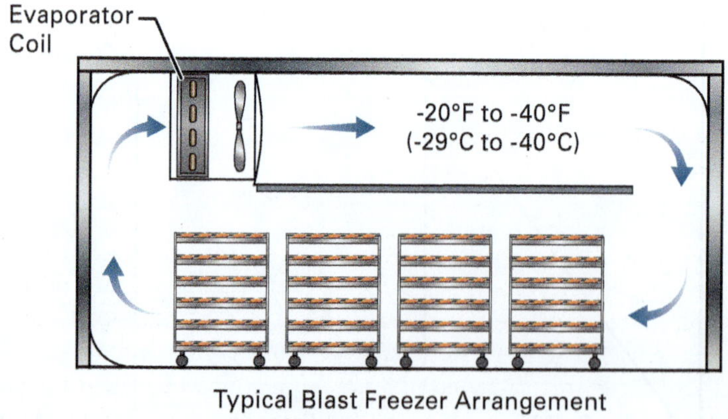

Figure 3 Typical blast freezer arrangement.

Continuous systems rely on conveyors or other means to move product slowly through a freezing zone over a predetermined period. As frozen product exits, fresh product enters at the opposite end.

1.1.2 Contact Freezing

Contact freezing produces frozen products by lightly pressing the products between cold metal plates or by contact with a moving belt exposed to the refrigerated surface. Contact freezing is typically used for products of consistent size and shape, ensuring that proper contact is made with individual pieces. Heat is extracted by conduction through the plate surfaces and into the refrigerant circulating through tubes or passages integrated into the plate.

Contact freezing is one of the most efficient approaches because conduction transfers heat more rapidly. One other advantage is that there is little or no loss of moisture from the product, as opposed to freezing methods using air. Moisture loss means product weight loss and often a reduction in product quality, significantly affecting the value of the product.

Figure 4 shows an example of a horizontal plate freezer. The plates can move vertically, allowing space for products such as fish filets to be loaded between the plates. Once loaded, the plates are lowered to make contact and the freezing process begins. The plates can also be placed vertically, and the foods can be packaged or unpackaged.

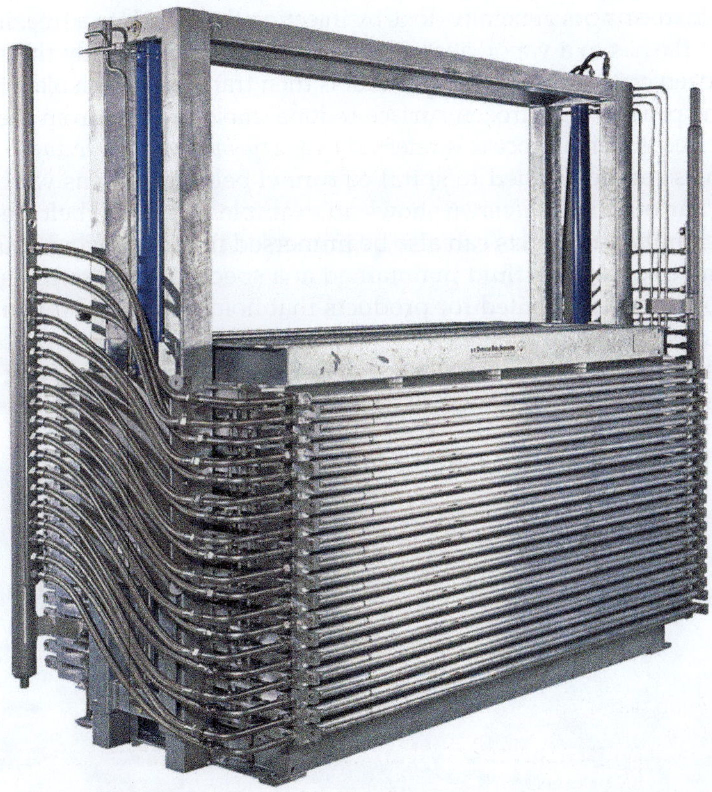

Figure 4 Horizontal plate freezer.
Source: A/S Dybvad Stal Industri

Belt freezers (*Figure 5*) are generally used for unpackaged products. A thin film serves as the belt and is driven with rollers, moving the product along the surface. Used film is disposed of to ensure good hygiene. Soft, sticky, and wet products, such as scallops, can be a problem for contact freezers as these products can adhere to the refrigerated surface. The special thin film used with belt freezers does not allow product adhesion while allowing rapid heat transfer.

Cryogenics: Field of engineering and refrigeration related to extremely low temperature applications, typically below −250°F (−157°C).

1.1.3 Cryogenic Freezing

Cryogenic fluids: Fluids that exist as a liquid or gas at temperatures below −250°F (−157°C).

In the refrigeration industry, **cryogenics** refers to the freezing of product using fluid temperatures below −250°F (157°C). The boiling points of common **cryogenic fluids**, such as carbon dioxide (CO_2), nitrogen, and helium that are normally in a gaseous state, are well suited for cryogenic freezing. Liquid

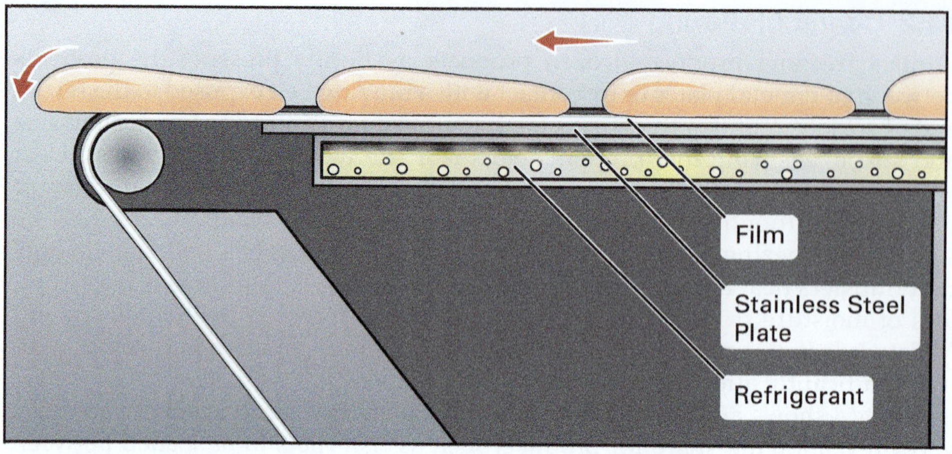

Figure 5 Belt freezer operation.

nitrogen is most often used, while liquid helium provides the lowest temperature of the three popular fluids.

Cryogenic freezing is generally done by injecting the fluid into a freezing chamber where it flashes to a vapor, absorbing heat. In some cases, only the surface of a food is frozen cryogenically. The product is then transferred to a blast freezer to complete the process. The frozen surface reduces moisture loss from the product in the blast freezer. This process is referred to as *cryomechanical freezing*.

Cryogenics can be applied to spiral or tunnel belt freezers, as well as batch freezers of various sizes. *Figure 6* shows an example of a spiral belt freezer in a cryogenic chamber. Products can also be immersed in the cryogenic fluid, floating in a tank of cryogenic fluid maintained at a specific temperature and pressure. Immersion is best suited for products that hold shape and integrity, such as shrimp, meat cubes, and berries.

In all cases, the cryogenic fluid absorbs heat from the product, causing the fluid to vaporize. The vapors are drawn away and condensed back to a liquid state.

Figure 6 Cryogenic spiral freezer.
Source: Linde, Inc.

1.1.4 Freeze Drying

Freeze drying, also known as **lyophilization**, is generally done cryogenically, relying on the rapid freezing action of this method. The product is then placed in a deep vacuum with sufficient heat applied to dehydrate the product, removing the moisture through **sublimation**.

The result is a product that no longer requires refrigeration for long-term preservation. The product is also much lighter without the water, and it can be returned to its original state by simply adding water. *Figure 7* is an example of a typical commercial freeze-drying chamber.

Lyophilization: A dehydration process that begins with freezing a substance, then reducing the pressure while adding heat to drive out moisture by sublimation; also known as *freeze drying*.

Sublimation: A change of state that progresses directly from a solid to a gas, such as the changing of ice to water vapor, without passing through the liquid state at any point.

Figure 7 Freeze-drying chamber.
Source: Cuddon Freeze Dry

Did You Know?

Other Uses of Freeze Drying

In addition to foods, freeze drying is used to manufacture pharmaceuticals and biological samples, as well as in the recovery of valuable water-damaged documents. Special freeze-drying chambers have also been developed for cannabis crops that, according to manufacturer testing, retain significantly more cannabinoids.

1.2.0 Food Storage

Cold storage is used to preserve perishable food for extended periods. Cold storage conditions are determined by the type of food or material being stored. The length of time the food is to be held in storage is also a factor. Remember that frozen storage systems are not designed with the capacity to intentionally freeze unfrozen goods.

Cold storage systems can be divided into two broad groups based on the temperature range: chilled (above freezing) and frozen.

Chilled storage applies to commodities that must be refrigerated but not frozen. For some products in this category, such as fruits, vegetables, and flowers, natural biological processes (such as the growth of shoots or ripening) continue during storage. Storing them at cool temperatures only retards these processes, with the degree of retardation depending on the temperature and the specific product.

Airflow is crucial to removing the heat and gases generated during the respiratory cycle of perishable produce. However, air movement also increases the potential for dehydrating many products. Avoiding significant dehydration requires the humidity to be maintained at the proper level.

For example, celery is best stored at temperatures between 32°F and 36°F (0°C to 2°C), with a relative humidity of 98% to 100%. Pumpkins, on the other hand, are best stored at temperatures between 50°F and 55°F (10°C and 13°C) and a relative humidity of 50% to 70%. The required storage temperature and humidity of products vary so widely that no generalizations can be made. Some individual products, such as certain kinds of potatoes, require different storage conditions depending on the intended use.

Frozen storage applies to products whose temperature must remain below freezing, such as ice cream, frozen prepared foods, meat products, and fish. Airflow is not as serious a problem for frozen products as it is for chilled products. If the product is packaged, dehydration is not a factor at all.

Cold storage facilities are used to package and store large quantities of bulk foods and materials intended for distribution to supermarkets and restaurants. The food handling and storage requirements determine the building design. An example of a frozen storage facility is shown in *Figure 8*. Cold storage facilities that handle a wide variety of foods are divided into multiple **coolers** and freezers that operate at different conditions to suit the product.

Coolers: Enclosed refrigerated storage systems maintained at temperatures above freezing. Coolers range in size from reach-in cabinets to drive-in warehouses.

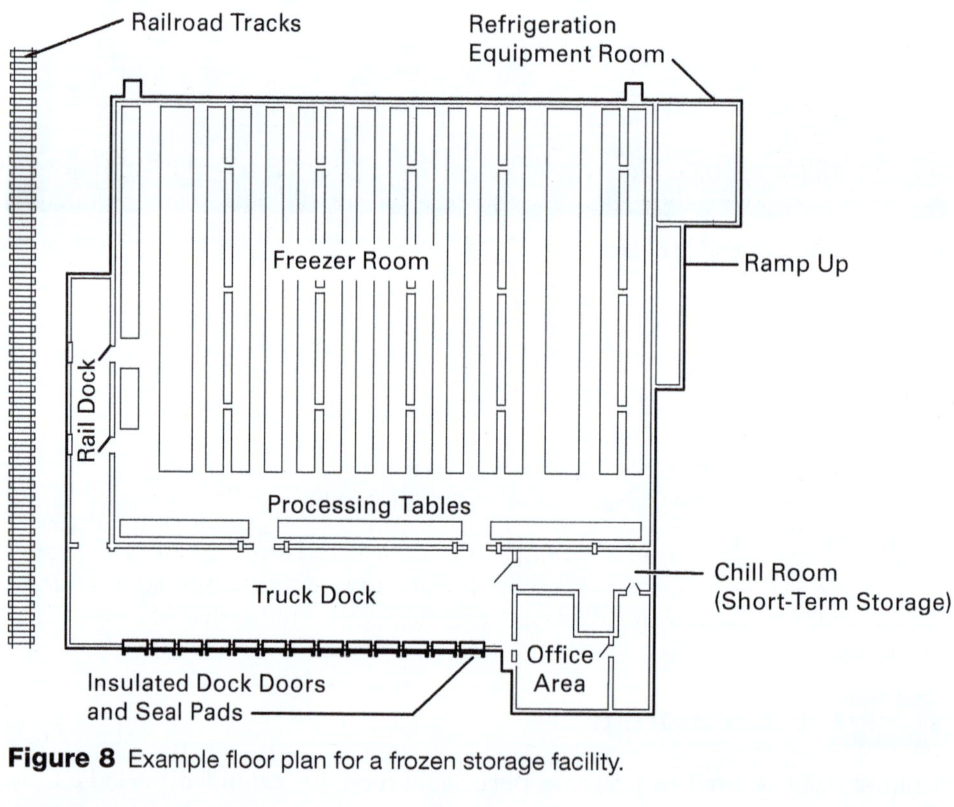

Figure 8 Example floor plan for a frozen storage facility.

The temperatures in a cold storage facility can range from 35°F to 60°F (2°C to 16°C) with humidity control possible, and down to –20°F (–29°C) without humidity control. Several large refrigeration systems are used to cool the entire facility, using evaporator pressure regulators (EPRs) and other components to maintain specific conditions in designated areas. Multiple systems also provide redundancy.

On Site

Food Processing and Distribution

Refrigerated warehouses often function as both storage and distribution points for produce and foods supplied from many sources. Many warehouses are highly automated to support efficient handling, storage, and movement of the food within the building and to the building loading docks. Chilled water can be used in refrigerated areas to both wash produce and move it through the process without damage.

Source: RGtimeline/iStock/Getty Images

1.2.1 Chill Rooms

Chill rooms are used for temporary storage of products such as fruits and vegetables. The product is normally placed in chill rooms fresh from harvest, after the necessary sorting and cleaning. In the chill room, the products are cooled to an appropriate temperature for long-term storage or further processing.

Chill rooms require temperatures ranging from 33°F to 60°F (1°C to 16°C). The specific temperature and humidity maintained in a room is determined by the product. The list of temperatures and humidity levels required for various fruits and vegetables is quite extensive. In some cases, variations as small as 2°F to 3°F (1.1°C to 1.6°C) can negatively affect product quality.

Most chill rooms also have a means of controlling the relative humidity to within 3% to 5% of the setpoint. Humidification is often done using atomizing spray nozzles as the volume of water required can be substantial.

Did You Know?

Produce Breathes!

Many products, such as apples and potatoes, continue to ripen long after harvest. The respiration associated with the ripening process creates heat and ethylene gas. If proper airflow isn't maintained, the buildup of heat and ethylene gas accelerates the ripening process, causing the produce to spoil prematurely.

1.2.2 Frozen Storage

Held at temperatures below freezing, freezer rooms are used for long-term storage. Freezing allows the storage time to be lengthened from days or weeks to months. Commercial/industrial frozen storage is typically maintained between 10°F and –10°F (–12°C to –23°C). Ice cream, however, is usually maintained at –20°F (–29°C) for long-term storage.

Frozen foods can deteriorate while in storage. Freezing does not maintain product quality forever. The most important factors contributing to the deterioration of frozen foods are the storage temperature and storage time. As frozen foods move to the consumer's household where the storage temperature is often higher and interrupted by consistent defrost cycles, the usable life is further shortened.

On Site

Food Safety

The amount of protection provided by frozen food packaging is very important. Some of the bacteria in frozen foods is killed during freezing and storage, but some types of bacteria do survive. Once defrosted, foods are immediately subject to bacterial decomposition. This is why it is wise to defrost foods slowly in the refrigerator, not on the countertop.

1.3.0 Transporting Refrigerated Food Products

The movement of refrigerated products from one point in the processing chain to the next requires refrigerated transportation methods. Exported products can spend several weeks in some form of refrigerated transportation unit.

Refrigerated transportation units must maintain the product temperature during shipment until reaching distant markets. The movement of the product from storage into the truck or cargo container must be efficient and conducted in a refrigerated environment.

Without product temperature control in transit, consumers would not be able to enjoy many of the foods found in local stores. There are three primary categories of refrigerated transportation:

- Refrigerated truck and trailer units
- Refrigerated cargo containers
- Refrigerated railcars

1.3.1 Refrigerated Truck and Trailer Units

At some point in the supply chain, refrigerated products will be transported by refrigerated trucks. Trucks used to haul perishable and frozen loads come in a variety of sizes and shapes. The most common are over-the-road refrigerated trailers, trucks, and direct-drive delivery vans.

Refrigerated trailers are used to haul almost every type of temperature-sensitive cargo. These units are used to transport products over both long and short distances. Unlike standard trailers, refrigerated trailers are insulated, have doors with seals, and have a refrigeration unit mounted on the front of the trailer.

The refrigeration equipment is a unique packaged unit containing all the necessary components (*Figure 9*). These components include the compressor, condenser, metering device, evaporator, and all applicable controls. The

Figure 9 Trailer refrigeration unit with an engine-driven, open-drive compressor.

refrigeration unit is heavily constructed, allowing it to endure the constant bumps and vibration present during travel.

Power is provided to an open-drive compressor driven by a small diesel or gasoline engine connected via a driveshaft. The engine also drives an alternator to supply DC voltage to the unit's electrical components. Fans may be belt driven or direct drive. The unit includes temperature and defrost controls, as well as the engine controls for fully self-contained operation.

Cargo placement within the trailer plays an important role in maintaining consistent product temperatures. The refrigeration unit typically discharges cold air across the top of the trailer, just under the roof (*Figure 10*). Sometimes a chute or duct is used to evenly distribute the air along the length of the trailer. The air picks up heat from the product and is drawn back toward the front of the trailer along the floor. If the cargo is loaded too tightly, the air will not be able to circulate through the load. For produce with a high rate of respiration, tight loading can cause the cargo to spoil.

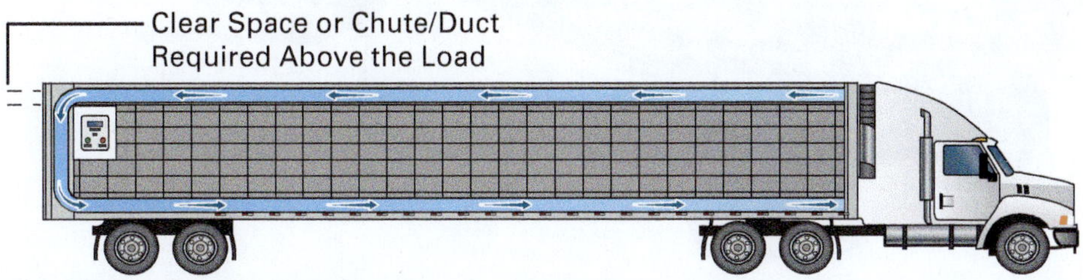

Figure 10 Typical airflow pattern in a refrigerated trailer.

Like other freezers, trailers transporting frozen products must defrost the evaporator regularly. Hot-gas defrost is fast and doesn't require the electrical current that electric heat demands. Thus, it is the most popular approach to defrosting the coil.

The primary difference between a refrigerated trailer and a refrigerated truck (*Figure 11*) is that a trailer requires a tractor to tow it from place to place. However, the refrigeration unit, since it is self-contained, can operate regardless of its connection to a tractor. Refrigerated trucks are self-contained and built on a rigid chassis. Trucks have less load capacity and are typically used for local deliveries.

Figure 11 Refrigerated truck.
Source: vitpho/iStock/Getty Images

Both refrigerated trailers and truck boxes can be configured for multiple temperature zones. A multi-temperature truck or trailer relies on a single refrigeration unit supplying two or more evaporators, or multiple systems. Lightweight, moisture-resistant bulkheads are used to separate the zones (*Figure 12*).

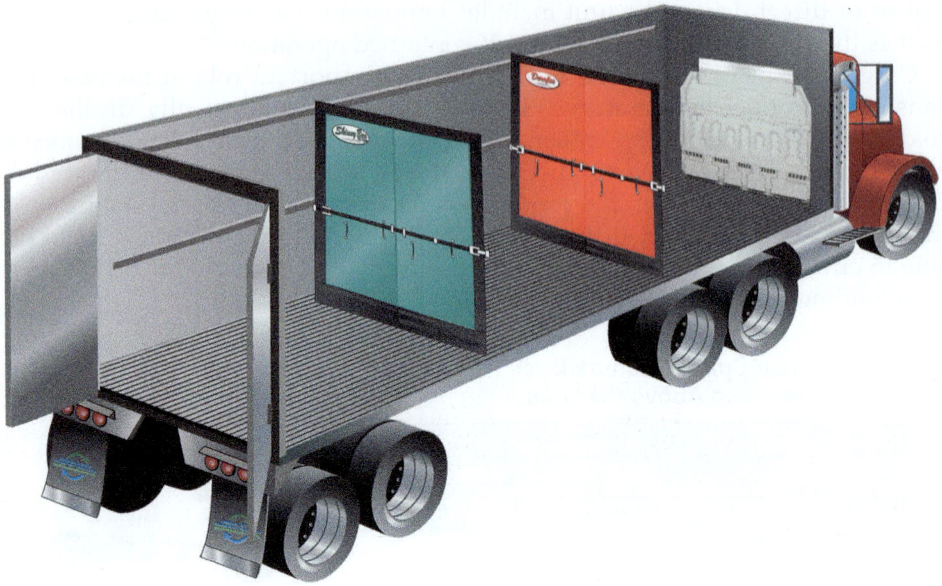

Figure 12 Trailer bulkheads.
Source: Courtesy of Signode

The smallest refrigerated trucks use compressors that are merely larger versions of the compressor in a standard automobile. The compressor is in the engine compartment and belt driven. Electrical power for controls, valves, and the fan(s) is supplied by the vehicle's alternator.

1.3.2 Refrigerated Cargo Containers

Many products are transported in refrigerated cargo containers. The refrigeration unit must be capable of operating across a wide range of temperatures, both inside and outside the container. The unit must also be heavily constructed to withstand the abuse of traveling on a flatbed truck, being moved by a forklift, and being lifted on and off a container ship by a crane. Some refrigerated containers can be placed directly on a fitted chassis that is then connected to a tractor (*Figure 13*).

Corrosion resistance is also important to the unit's design as saltwater will be present for much of its life cycle. Exported commodities are usually shipped this way.

The heavily insulated steel or aluminum boxes are typically 8' wide × 8' high × 20' or 40' long, with a refrigeration unit installed at one end. The uniform size and design of the containers allow stacking aboard ships and on the docks. The loaded containers are transported to and from the docks via trailers or railcars.

Many refrigerated containers use hermetic or semi-hermetic compressors, as opposed to open-drive compressors and drive engines. Most container systems plug into the electrical power panel of the ship or dock facility. In addition, these systems may have integral generator units to provide electrical power when moving via truck or railcar. *Figure 14* shows nose-mount and under-mount generator sets.

Because many of these containers require external electrical power or generators, evaporator defrosting is often done with electric heating elements. Although the condenser is typically air cooled, most manufacturers offer an optional water-cooled condenser. Water-cooled condensers allow operation regardless of how the containers are stacked on the ship, saving space. Air-cooled condensers can struggle with high ambient temperatures and a

Figure 13 Refrigerated cargo container on a fitted chassis.
Source: Serjio74/iStock/Getty Images

(A) Under-Mount Generator Set **(B) Nose-Mount Generator Set**

Figure 14 Under-mount and nose-mount generator sets.
Source: Courtesy of Klinge Corporation

lack of airflow around the containers. However, water-cooled condensers must have a reliable water source and pumping system, and they are subject to fouling.

Like trailers, proper placement of the product in the container is vital. Air must be able to circulate around the product to absorb the heat and return to the evaporator. Unlike most trailers, airflow is routed through the container from the bottom, taking advantage of the natural tendency for warmer air to rise (*Figure 15*). The unit supplies the air through chutes, deep grooves, or T-shaped slots in the floor. The air cools the product as it rises to the top of the container where it is drawn back to the evaporator.

Advanced options for chilled containers include humidification as well as a feature marketed as *controlled atmosphere* (CA). CA is an agricultural storage technique through which the oxygen, nitrogen, and CO_2 levels inside the container are regulated and expelled as necessary to minimize ripening and extend the shelf life.

When approaching a CA-equipped container, you should see a sign similar to the one shown in *Figure 16*. The container doors should be fully opened for 30 to 45 minutes for the container's oxygen levels to return to normal prior to entry.

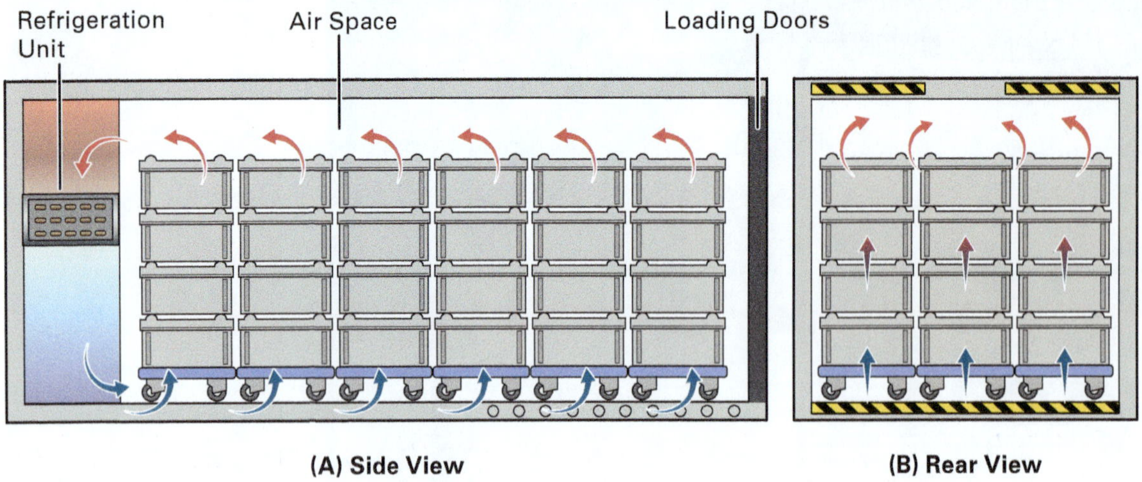

Figure 15 Airflow through a refrigerated cargo container.

Figure 16 CA container warning sign.

WARNING!

Never enter a CA container immediately after opening the door. The low levels of oxygen required to preserve produce can result in asphyxiation. Before entering the container, turn off the CA system using the control panel; open the doors fully, and wait 30 to 45 minutes before entering.

1.3.3 Refrigerated Railcars

Refrigerated railcars today typically use the same type of refrigeration unit used for trailers (*Figure 17*). Refrigerated railcars, also referred to as *reefers*, have heavier insulation and sturdier construction features than trailers. Railcars offer significantly more cargo space and can reach a distant destination much faster than a tractor/trailer combination. Refrigerated railcars can typically accommodate up to 130,000 lb of cargo and maintain temperatures well below 0°F (–18°C).

Figure 17 Refrigerated railcars.
Source: Thermo King LLC

In addition to refrigerated railcars, flatbed railcars can accommodate truck trailers (*Figure 18*) and cargo containers. The trailers have self-contained refrigeration systems, and refrigerated cargo containers that are not equipped with generators draw power from the train while in transit.

Figure 18 Refrigerated trailers on flatbed railcars.
Source: Thermo King LLC

Did You Know?

Help from Outer Space

Some refrigerated trailers, cargo containers, and railcars are equipped with global positioning system (GPS) units that provide instantaneous tracking and monitoring information. GPS tracking allows dispatchers to confirm the location and direction of cargo wherever it goes. The systems provide two-way communication, allowing a technician to monitor load temperature, refrigerant pressures, and drive-engine fuel level, as well as allowing changes to the unit setpoint, initiating a defrost, and downloading other operating data.

1.0.0 Section Review

1. Which of the following is a common cryogenic fluid?
   a. Water
   b. Helium
   c. Oxygen
   d. Hydrogen

2. True or False? Most fruits and vegetables can be successfully stored at the same temperature and humidity conditions to preserve quality.
   a. True
   b. False

3. Which defrost method is most commonly used for refrigerated trailers?
   a. Air defrost
   b. Hot-gas defrost
   c. Electric defrost
   d. Off-cycle defrost

2.0.0 Refrigeration System Components

Objective

Identify and describe various commercial and industrial refrigeration system components.

a. Identify and describe refrigeration compressor configurations.
b. Describe the application, control, and installation of air-cooled condensers.
c. Identify and describe various evaporator and display case configurations.
d. Identify and describe refrigeration system accessories, specialties, and support systems.

Performance Task

1. Install or troubleshoot a commercial refrigeration condensing unit and unit cooler system.

The components and operation of a mechanical refrigeration system are basically the same as those used in comfort cooling systems. Depending on the application though, some additional components are often needed.

Figure 19 shows a basic refrigeration circuit. Redundant, or *backup*, equipment is commonly available to prevent product and financial loss if the primary refrigeration system fails. The decision to add redundancy to a system is largely economic and depends on the nature of the product and the extent of the losses in the event of a failure.

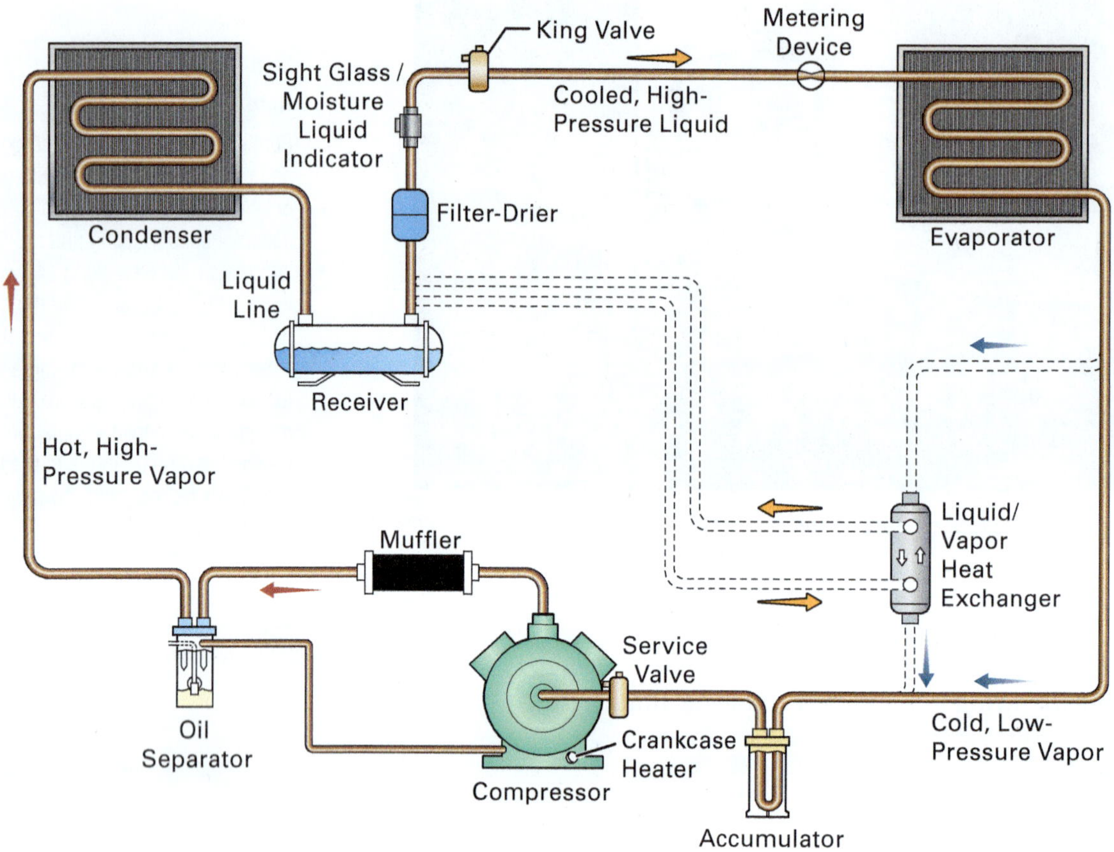

Figure 19 Basic refrigeration system circuit.

The primary refrigeration system components are the compressor, evaporator, condenser, and metering, or *expansion*, device. What differs from system to system is the refrigerant used, the operating temperatures, and the size and style of the components. The series of events that occurs as the refrigerant moves through the circuit is called the *refrigeration cycle*. For review, the operation of the basic refrigeration system and its components are described in detail in NCCER Module 03107, *Introduction to Cooling*.

Like comfort systems, the compressor and condenser for commercial refrigeration systems are often packaged into a condensing unit (*Figure 20A*). Condensing units can be air-cooled or water-cooled. Air-cooled units may contain two or more compressors and/or condenser fans.

Condensing units can be placed in mechanical rooms, mounted outdoors on the roof or at ground level, or placed directly on top of the refrigerated enclosure. Condensing units designed for indoor installation, like the one shown in *Figure 20A*, do not have weather-resistant enclosures. Of course, commercial condensing units and unit coolers grow much larger.

The evaporator, metering device, and indoor fans or blowers are also packaged into an assembly. This assembly is commonly referred to as a **unit cooler** (*Figure 20B*). Think of a unit cooler as the equivalent of a fan coil unit used in comfort cooling.

Unit cooler: A packaged refrigeration evaporator assembly containing the evaporator coil, expansion device, defrost heaters (freezers), condensate drain pan, and the indoor fans or blowers commonly used in chill rooms, walk-in coolers, and walk-in freezers.

Refrigeration equipment manufacturers classify equipment by *cooling class*, which is based on the evaporator operating temperature. *Table 1* shows common refrigeration applications and cooling classes assigned by one manufacturer. Not all manufacturers agree on the three classes, but they have similar temperature ranges.

(A) Refrigeration Condensing Unit

(B) Unit Cooler

Figure 20 Commercial refrigeration condensing unit and unit cooler.
Sources: Courtesy of Emerson (A); Hussmann Corporation (B)

TABLE 1 Refrigeration Applications and Temperature Classes

Application	Storage Temperature Range	Evaporator Class (Saturation Temperature) Low (–40°F to –10°F)	Medium (–10°F to 30°F)	High (30°F to 50°F)
Meat cutting and packaging; fresh fruit and vegetable storage; floral boxes	35°F to 60°F			X
Meat storage; extended storage of many fruits and vegetables	28°F to 34°F		X	X
Low-ceiling coolers for multiple products; walk-in coolers and warehouses for meat, dairy, and produce	35°F and up			X
Low-ceiling freezers for multiple products; walk-in freezers	–30°F to 30°F	X	X	

2.1.0 Refrigeration Compressors

Reciprocating, scroll, and rotary screw compressors are all used in refrigeration systems. The choice is largely determined by the size and scope of the refrigeration load. Hermetic and semi-hermetic reciprocating compressors have long been widely used for commercial refrigeration. However, improvements have increased the popularity of scroll compressors.

NOTE

The operating principles for all these compressors can be found in NCCER Module 03302, *Compressors*.

On Site

Energy Efficiency Ratio (EER)

The energy efficiency ratio (EER) for cooling systems was introduced earlier in this program, and you have likely become used to seeing and hearing the term at this point. The term is applied to comfort systems as the seasonal energy efficiency ratio (SEER), but the concept is similar. The EER is the number of Btus produced by a system per watt of energy consumed.

Open-drive reciprocating compressors are most often found in heavy commercial and industrial refrigeration applications. A soda bottling plant or dairy may have 50 or more such compressors in service. Rotary screw compressors offer far more capacity in a single compressor package than a reciprocating compressor for the largest applications. Centrifugal compressors are not suited for refrigeration applications, and rotary vane compressors are rarely used.

As you consider compressors used in refrigeration, keep in mind that refrigeration applications require more energy than comfort cooling applications. In refrigeration, lower evaporator temperatures (saturated suction temperatures) are required to produce colder results. This significantly increases the compression ratio and the energy consumption of the compressor. The energy efficiency ratio (EER) for systems operating at extremely low temperatures can be as low as 3. Thus, when engineers choose compressors, energy consumption to deliver the necessary results is a major factor.

2.1.1 Single Compressor Applications

Systems with one compressor are found in about half of the commercial refrigeration equipment in use. The compressor sizes commonly range between $\frac{1}{2}$ and 30 horsepower (hp). A single compressor is often used to serve multiple evaporators in display cases, reach-in units, and unit coolers. As shown in *Figure 21*, a single compressor can even serve loads at different temperatures.

Evaporators on a single system often experience different loads throughout the day. Late at night, the load imposed on the compressor may be very small. For this reason, the compressor should be equipped with unloaders to reduce capacity. This significantly reduces compressor cycling and improves energy efficiency.

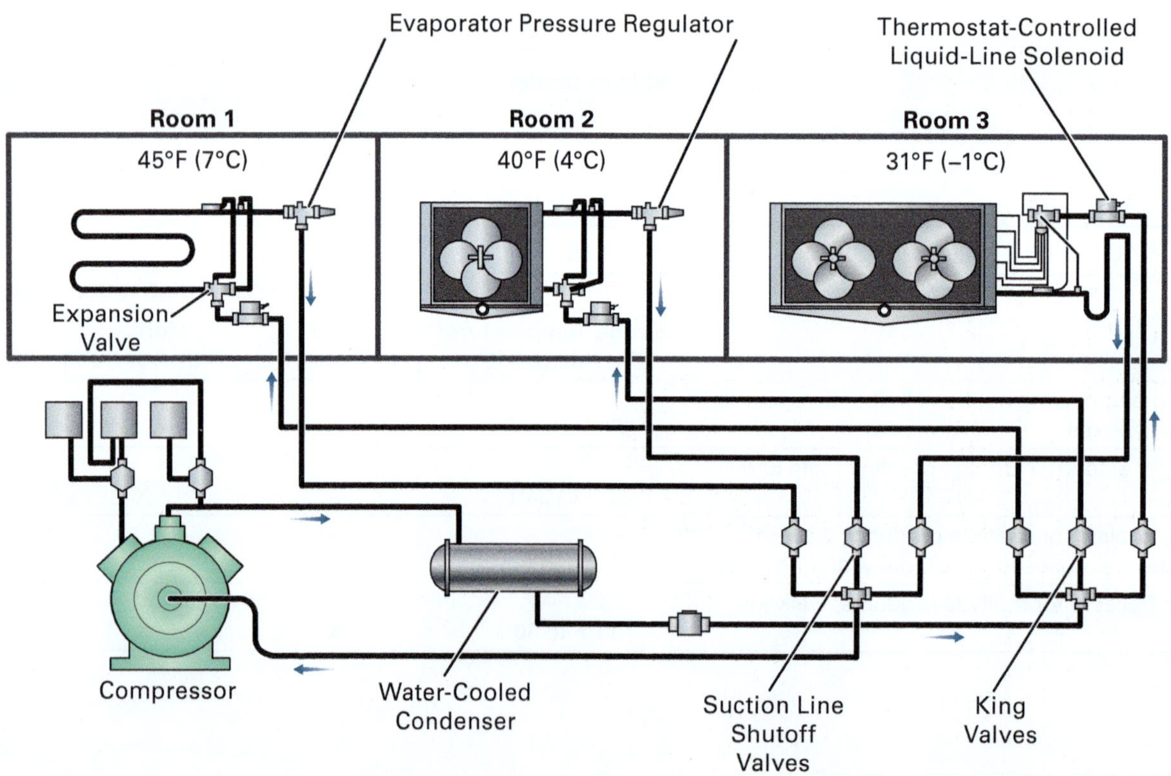

Figure 21 Single compressor system with multiple evaporators.

When a single compressor is used to serve loads operating at different temperatures, as shown in *Figure 21*, the compressor must be set up to operate at the evaporator temperature and corresponding suction pressure required by the coldest evaporator in the system. In *Figure 21*, this would be Room 3.

To prevent the other evaporators from operating at the same pressure, evaporator pressure regulators (EPRs) are installed on the suction lines (*Figure 22*) serving the warmer loads. The EPRs maintain a preset pressure in the evaporator, thereby controlling the saturation temperature. In more sophisticated applications, the compressor's suction pressure setpoint can be automatically reset and adjusted as colder loads are satisfied, preventing it from operating at a higher compression ratio than necessary to serve the current conditions.

NOTE

EPRs are discussed in detail in NCCER Module 03304, *Retail Refrigeration Systems*.

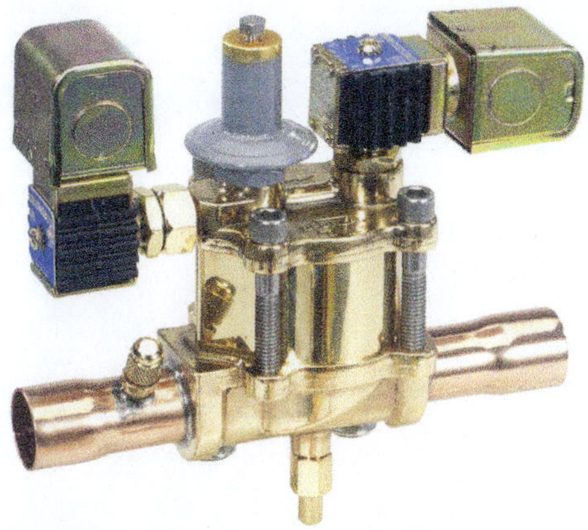

Figure 22 Evaporator pressure regulator (EPR).
Source: Courtesy of Sporlan Division - Parker Hannifin Corporation

2.1.2 Multistage Compressors

Conventional reciprocating compressors use a single-stage process to compress the refrigerant gas in one or more cylinders. *Single-stage compression* means the refrigerant gas is drawn into a compressor cylinder(s) on the piston downstroke, and then is compressed to the required pressure on the compression upstroke of the piston. The compressed gas is then discharged from the compressor cylinder(s) for routing to the condenser.

Two-stage compressors are used in ultra-low-temperature systems to compress vapor entering at a very low pressure, then compress the discharge vapor again to reach the necessary condensing pressure and temperature. Two-stage compressors are generally applied where the compression ratio of a single-stage compressor would exceed 10:1.

Two-stage compressors (*Figure 23*) compress the refrigerant gas in a two-step sequence. The system suction gas enters the first stage, where it is compressed to about the midpoint of the compression target. The gas is then discharged into the inlet of a second cylinder. The refrigerant gas is compressed again and raised to the final condensing pressure and temperature conditions. It is then discharged to the condenser.

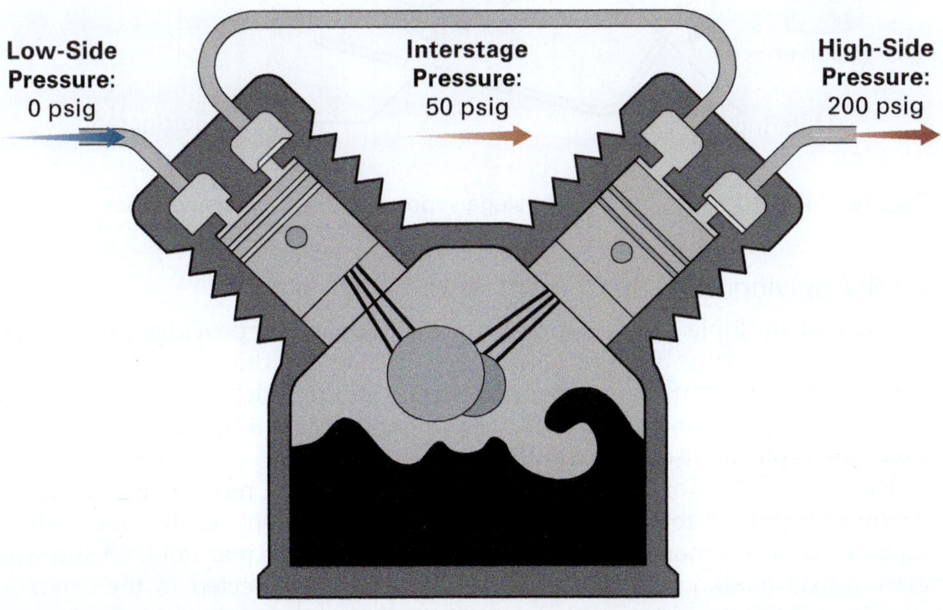

Figure 23 Two-stage reciprocating compressor operation.

The use of a two-stage compressor results in a lower compression ratio per stage than that with a single-stage compressor. In the example shown in *Figure 23*, the compression ratio of the first stage is 4.4:1, compared to 3.3:1 for the second stage. If a single-stage compressor were used with the same suction and discharge pressures, the compression ratio would be an impractical 14.6:1.

The compression ratio is calculated by dividing the absolute discharge pressure by the absolute suction pressure. To find the absolute pressure, add 14.7 psi to the gauged pressure. Here is the equation related to the first stage of compression shown in *Figure 23*:

$$\textit{Stage 1 compression ratio} = \frac{\textit{discharge pressure} + 14.7}{\textit{suction pressure} + 14.7}$$

$$Stage\ 1\ compression\ ratio = \frac{50 + 14.7}{0 + 14.7}$$

$$Stage\ 1\ compression\ ratio = \frac{64.7}{14.7}$$

$$Stage\ 1\ compression\ ratio = 4.4$$

One disadvantage of using two-stage compression is the increased temperature of the vapor entering the second stage, resulting from heat of compression in the first stage. This can overheat the high-stage compressor. One method used to eliminate this problem is to desuperheat the gas (*Figure 24*). This is done by drawing a small amount of liquid refrigerant from the receiver, passing it through a metering device, and injecting it into the compressor where it removes sensible heat from the vapor.

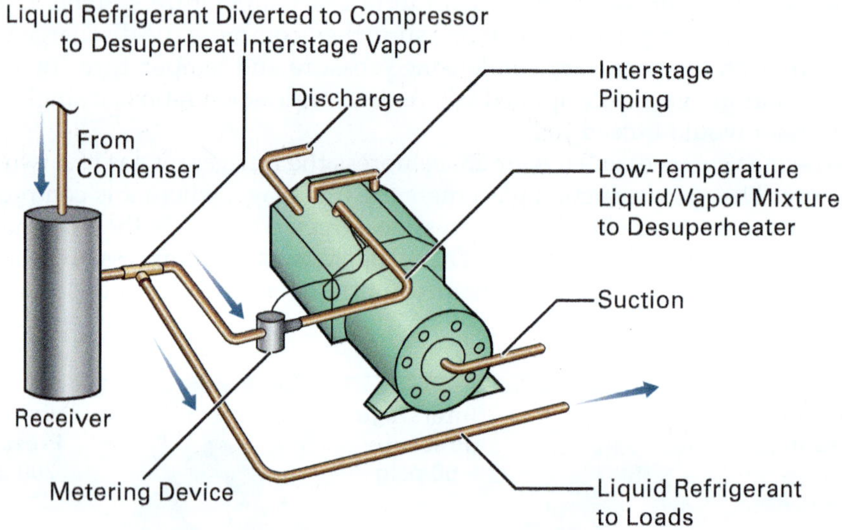

Figure 24 Desuperheating the interstage vapor in a two-stage compressor.

2.1.3 Applying Multiple Compressors

The use of multiple compressors connected in parallel provides greater system capacity and the ability to meet varying load conditions more effectively. Connecting two or three smaller compressors in parallel normally results in a higher Btu-per-horsepower ratio. Compressors in the 5 hp, 7.5 hp, and 10 hp range are typically used for parallel arrangements.

Parallel-compressor systems normally have a large receiver to ensure an adequate supply of refrigerant for all operating conditions. Each compressor is tapped into a common suction and discharge header or manifold. Each refrigeration load (walk-ins, open display cases, etc.) is connected to the common suction and liquid refrigerant header. *Figure 25* shows a system using two compressors connected in parallel.

The compressors can be the same size or different sizes. The compressors can be cycled on and off as needed for capacity control, staged on and off based on the system suction pressure. One or more of the compressors may also be equipped with unloaders for additional capacity control. Parallel compressors of different sizes can be staged to obtain more capacity steps than the same number of equally sized compressors. *Figure 26* shows an example of capacity control that can be obtained by using three parallel compressors—(1) 5 hp, (1) 7.5 hp, and (1) 10 hp.

In systems using multiple compressors, attention must be given to the return of discharged oil. Oil must return to each compressor under all operating conditions. The use of multiple compressors, condensers, and evaporators increases the volume of oil in the system. To the extent possible, piping layouts must be

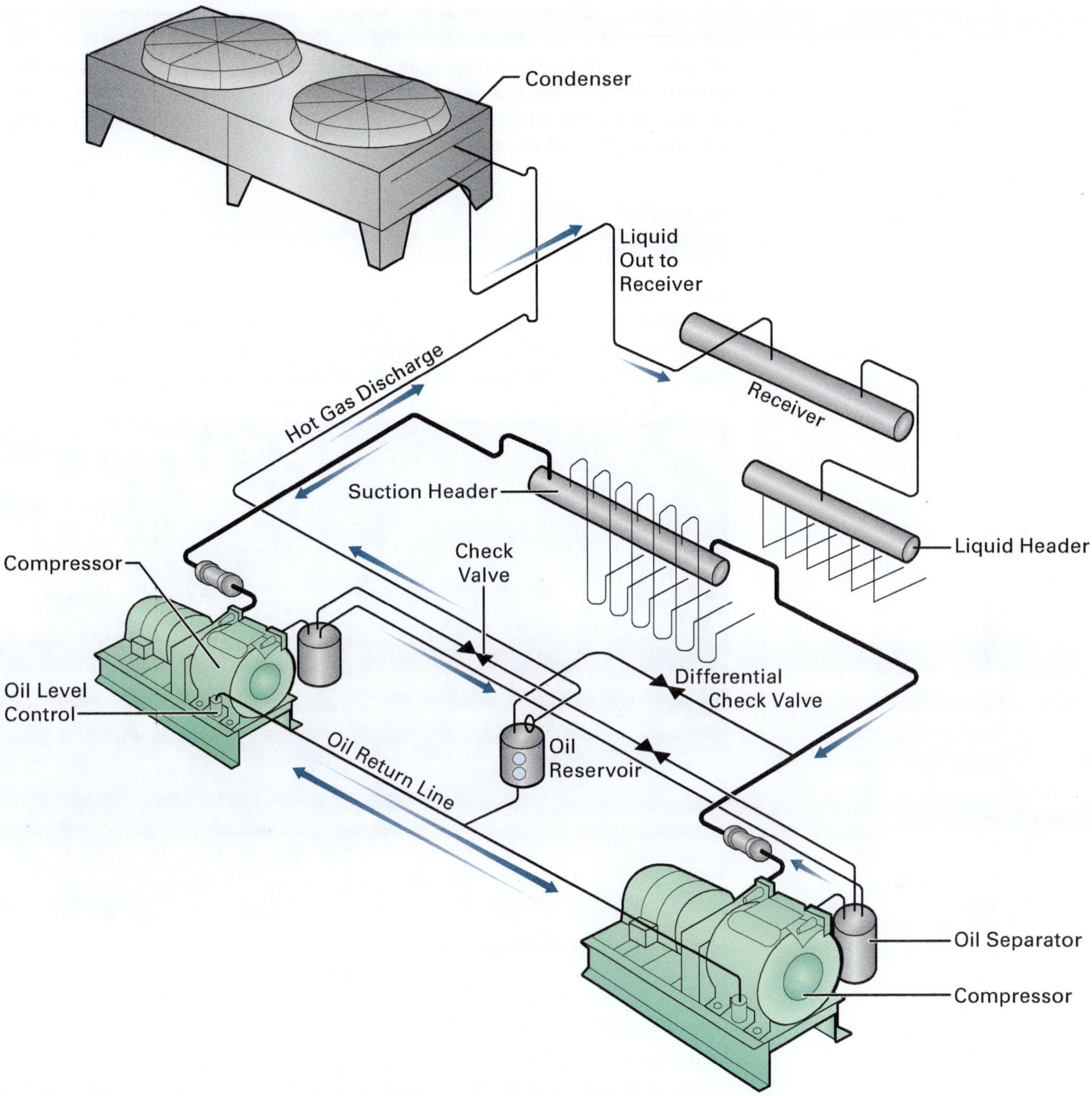

Figure 25 Parallel-connected compressor system with multiple evaporators.

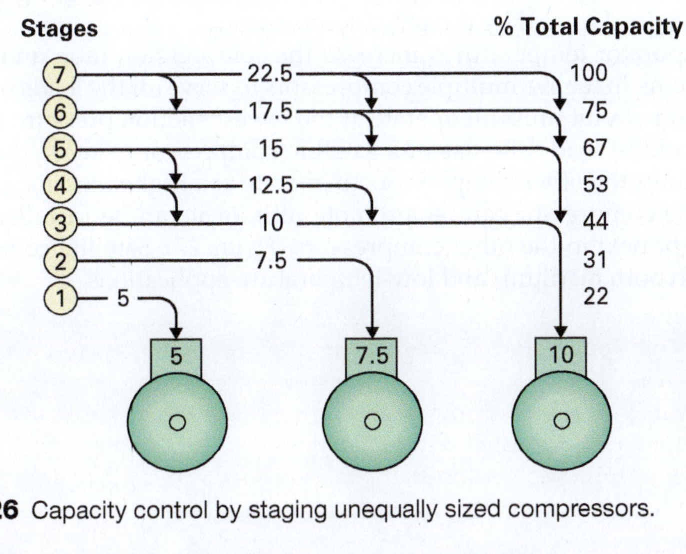

Figure 26 Capacity control by staging unequally sized compressors.

designed to avoid oil accumulation in inactive areas of the system. In *Figure 25*, oil is captured at the discharge of the compressors by the oil separator. The oil then moves to the oil reservoir where it is delivered to each compressor through the oil level control in the proper volume. Oil separators and reservoirs prevent the oil from ever leaving the mechanical room and ensure that each compressor reclaims its share of the oil.

On Site

Supermarket Parallel-Rack Systems

Packaged racks of compressors, complete with controls and safety devices, are readily available for installation. Features and available options include integral oil reservoirs with all required oil management devices, a wide selection of compressor types, and alarm systems that visually and audibly indicate a loss of refrigerant or oil as well as mechanical component failure.

Source: Courtesy of Hillphoenix

2.1.4 Satellite Compressors

Satellite compressor: A separate compressor that uses the same refrigerant source as that used by a related group of parallel compressors but operates independently and serves a refrigeration load that requires a lower temperature than that being maintained by the related parallel compressors.

A **satellite compressor** is a separate compressor tied to a refrigerant circuit shared with other compressors but does not operate at the same pressures and temperatures as the other compressors. The satellite compressor functions independently to serve a separate refrigeration load, which is usually significantly colder than the load served by the other compressors.

Lower evaporator temperatures increase the compression ratio and are less efficient. Systems that have multiple compressors to serve all the loads on a common refrigerant circuit must all operate at the lowest suction pressure required to serve the coldest load. The use of a satellite compressor to serve the coldest load alone allows the other compressors to operate at a higher suction pressure.

The satellite compressor can be a remote unit, or it can be installed on the same rack right next to the other compressors (*Figure 27*). Satellite compressors can be used in both medium- and low-temperature applications.

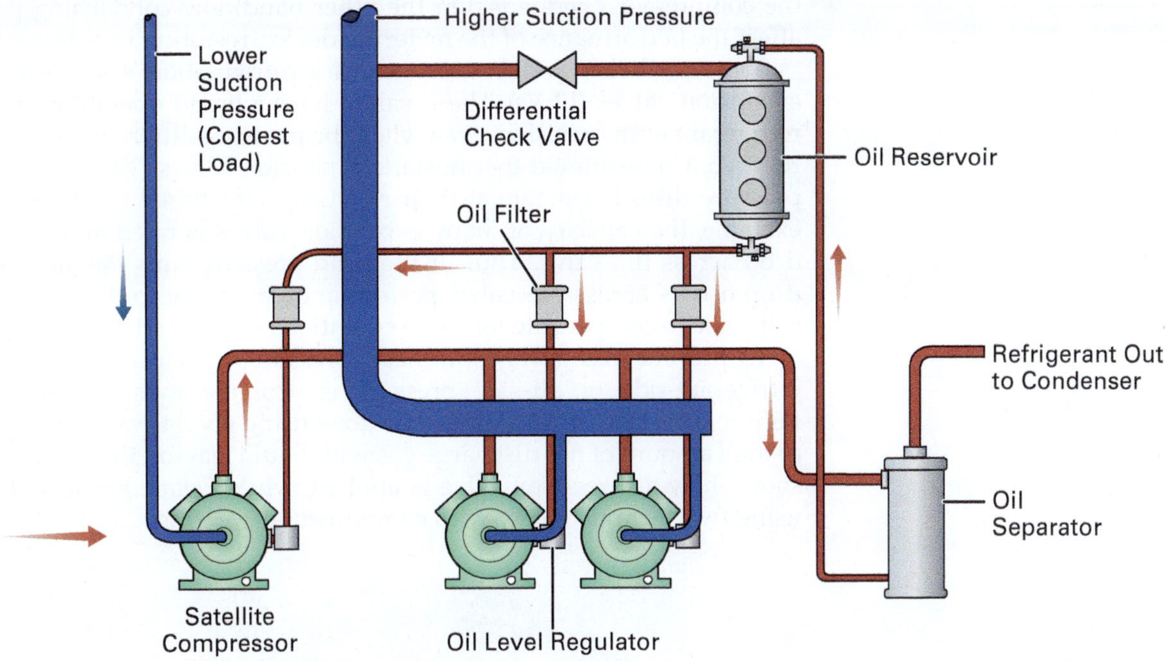

Figure 27 Satellite compressor serving the coldest load.

2.2.0 Air-Cooled Condensers

Some commercial and industrial refrigeration systems use evaporative condensers or water-cooled condensers with cooling towers. However, the most common type of condenser is the air-cooled condenser. An air-cooled condenser can be an integral part of a condensing unit, or it can be remote. Both single-circuit and multicircuit condensers are available.

2.2.1 Condenser Ratings

Condensers used in refrigeration systems are rated by their **total heat rejection (THR)** value. THR is the total heat removed in the desuperheating, condensing (latent heat), and subcooling of the refrigerant as it flows through the condenser.

Total heat rejection (THR): A value used to rate condensers. It represents the total heat transferred while desuperheating, condensing, and subcooling a refrigerant as it flows through the condenser.

The rating of an air-cooled condenser is based on the temperature difference (TD or ΔT) between the dry-bulb temperature of the air entering the condenser coil and the refrigerant's condensing temperature corresponding to the pressure. You may also hear this value referred to as the *condenser split.*

The typical air-cooled condenser splits for refrigeration systems are as follows:

- 10°F to 15°F (5.5°C to 8.3°C) for low-temperature systems
- 15°F to 20°F (8.3°C to 11.1°C) for medium-temperature systems
- 15°F to 30°F (8.3°C to 16.7°C) for high-temperature systems

A condenser is normally selected to match the capacity of the compressor(s). The THR capacity of a condenser is considered proportional to its TD. For example, the capacity of a condenser at a 30°F (16.7°C) split is about 50% greater than that of a condenser operating at a split of 15°F (8.3°C). Greater temperature difference between substances means faster heat transfer. To compensate, condensers selected for a lower split are typically larger with more surface area.

Condensers can also be rated in terms of *net refrigeration effect* (*NRE*), also known as *net heat rejection (NHR).* NRE is the total heat rejection minus the heat of compression added to the refrigerant by the compressor.

2.2.2 Air-Cooled Condenser Pressure Control

For a refrigeration system to operate properly, the condensing pressure and temperature must be maintained within certain limits. An increase in condensing temperature sacrifices capacity. If the condensing temperature rises too high, the compressor overloads. On the other hand, low condensing pressures can affect the performance of the metering device, resulting in a starved evaporator.

Some medium- and low-temperature refrigeration systems use electronic expansion valves (EEVs). These valves have a broad operating range and feed refrigerant with precision, even when the pressure difference across the valve is reduced. Conventional thermostatic expansion valves (TXVs) require a specific pressure drop to operate at their rated capacity, based on the refrigerant. For example, the capacity of many expansion valves is rated at a 100-psi pressure drop across the valve. Thus, if the head pressure runs low and only a 50-psi drop occurs across the valve, performance suffers accordingly. EEVs are typically able to compensate for these conditions without a loss of performance.

To prevent excessively low head pressures that can occur in the winter, refrigerant-side or air-side pressure control methods are commonly used. Refrigerant-side pressure control is done using special valves (*Figure 28*) to mix a small amount of hot discharge gas with liquid leaving the condenser. *Figure 29* shows how a three-way valve is applied, while *Figure 30* shows the approach using two separate valves with complimentary settings.

Figure 28 Three-way head-pressure regulating valve.
Source: Courtesy of Sporlan Division - Parker Hannifin Corporation

Adding a bit of hot vapor back to the liquid has a small effect on the pressure, but a far more significant affect is that the liquid refrigerant begins to back up in the condenser. This reduces the amount of surface available for condensing, increasing the discharge pressure. Once enough refrigerant is backed up in the condenser, hot gas flow ceases or slows dramatically. A large receiver is needed to ensure sufficient refrigerant is available to flood a portion of the condenser. This method results in a very stable head pressure in cold outdoor conditions.

Another common method of refrigerant-side pressure control is to use a condenser with two equal parallel sections, each handling 50% of the load during normal summer operation. During the winter, only half of the condenser remains active. Solenoid valves are used to isolate one of the condenser sections as well as any pump-down circuits and fans.

Due to its low initial cost, air-side control is a common method of head pressure control. It is accomplished by using one of three methods, or some combination of the three:

- Condenser fan on-off cycling
- Modulating condenser fan dampers
- Condenser fan speed control

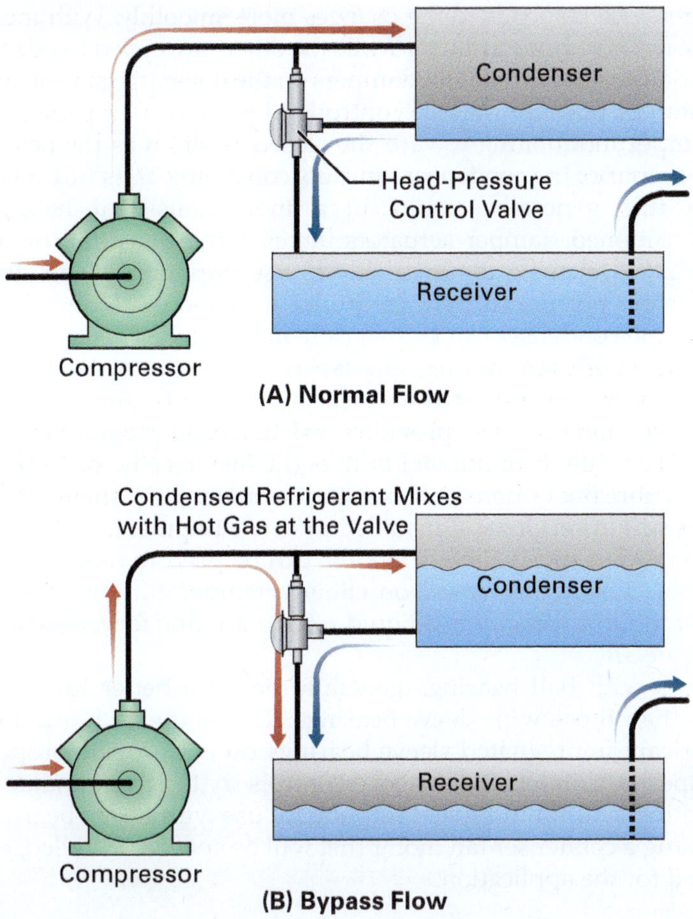

Figure 29 Head-pressure regulation with a three-way valve.

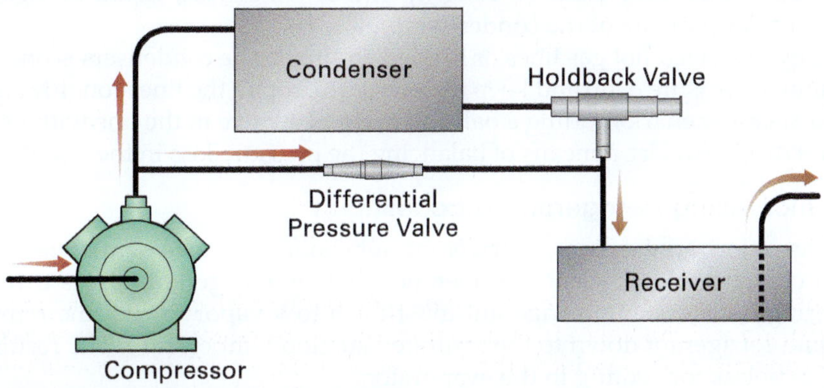

Figure 30 Head-pressure regulation with two valves.

Even on condensing units with a single fan, the fan can be cycled on and off to keep the head pressure up. As the pressure falls, a pressure switch opens and cycles the fan off. Once the head pressure rises to a predetermined setpoint, the switch closes and the condenser fan restarts.

Although simple and inexpensive, in cold weather this method often results in consistent changes in head pressure across the range of the switch. This, in turn, causes the metering device to constantly change its position to cope with the changing pressure differential and regulate refrigerant flow to the evaporator. In addition, wear and tear on the motor increases as the on and off cycles add up.

Condenser fan cycling provides better performance when used with condensers that have multiple fans. Although the head pressure will continue to change up and down between the on and off pressure settings, it is generally a slower process, and

Figure 31 Pressure-controlled condenser fan speed control.
Source: Courtesy of White-Rodgers by Copeland

the metering device responds to the changes more smoothly. With multiple fans available, one fan may run constantly while the others are cycled as needed.

Some condensers are fitted with dampers on the discharge side of one or more fans. The damper position may be controlled by refrigerant pressure directly, with the damper modulating toward the closed position as the head pressure falls. In this scenario, the condenser fan runs constantly at its full speed.

Although this generally results in a more consistent head pressure, refrigerant-positioned damper actuators increase the potential for refrigerant leakage. Today's systems are more likely to use pressure sensors and electric damper actuators positioned by a controller in response. Using dampers, air flow through the condenser can be precisely modulated. This method is sometimes combined with fan cycling; one fan is fitted with a damper while the other fans are cycled on and off as necessary.

Variable-speed fans can also provide consistent head pressure in all kinds of weather. Electronically commutated motors (ECMs) have helped this approach gain more traction. But options still exist for the many permanent split capacitor (PSC) motors still in service and installed in new equipment.

An electronic fan-speed control module can be used to modulate the speed of a compatible PSC motor based on either a temperature or pressure input. A model that mounts directly to a liquid refrigerant line for a pressure input is shown in *Figure 31*.

PSC motors with ball bearings generally perform better in speed control applications than those with sleeve bearings. This is due to lubrication issues, because lubricant-impregnated sleeve bearings do not provide proper lubrication at low speeds. Condenser fan-speed controls try to accommodate this issue by incorporating a minimum speed setpoint for use with sleeve-bearing motors. Before choosing a condenser fan motor that will be speed-controlled, ensure the motor is rated for the application.

2.2.3 Multiple Condensers

When more than one condenser is installed, the pressure drop in each hot gas line to the condensers must be roughly equal. Otherwise, liquid refrigerant tends to back up in one of the condensers.

Slightly oversized hot gas lines or headers to supply the condensers is one way to minimize pressure drop and keep the pressure drop in the lines roughly equal. In a worst-case scenario, adding a ball valve or globe valve in the line with lowest pressure drop provides a means of balancing the pressure loss in the two lines.

2.2.4 Increasing Refrigerant Subcooling

In refrigeration applications, refrigerant subcooling is an essential factor in system capacity and efficiency. Remember that, as refrigerant passes through a metering device, some refrigerant must flash to a vapor to cool the remaining liquid refrigerant down to the saturated suction temperature. The resulting vapor is useless for cooling in the evaporator.

To minimize flash gas, the refrigerant entering the metering device should be as close as possible to the saturated suction temperature of the system. Of course, it isn't practical in most applications to subcool the refrigerant to this point. For example, a freezer may operate at a saturated suction temperature of –15°F (–26°C). Subcooling the refrigerant to this temperature isn't practical unless another refrigeration system is used to do it. But it is important to gain as much subcooling as possible to maximize capacity and energy efficiency. A capacity increase of roughly 0.5% per 1°F of additional subcooling is achievable.

Did You Know?

Cascade Systems

Cascade refrigeration systems use multiple vapor-compression cycles, each using a different refrigerant, to achieve very low temperatures. Each cycle operates at progressively lower temperatures, transferring heat from one stage to the next. For example, the evaporator for a higher stage serves as the condenser for a lower stage.

One method used to increase subcooling is to add a suction-to-liquid heat exchanger (*Figure 32*). It is a simple double-walled heat exchanger, with suction vapor flowing through one side and liquid refrigerant flowing through the other, in opposite directions. The liquid refrigerant definitely does experience additional subcooling from the cool suction vapor, but it is argued that the heat energy never leaves the system since the heat is absorbed by the suction gas, increasing superheat.

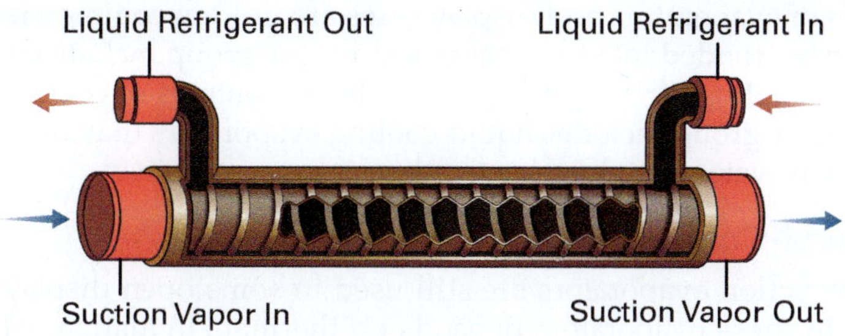

Figure 32 Suction-to-liquid heat exchanger.

However, in a refrigeration system that operates at a very low superheat, the suction-to-liquid heat exchanger can also help prevent liquid floodback. A variety of studies indicate that the overall effect is positive with some refrigerants and negative with others, so it is suggested that some research be done before applying a suction-to-liquid heat exchanger.

Additional subcooling can also be provided by adding a water-cooled condenser in the liquid line leaving the air-cooled condenser. The water becomes a **secondary coolant** that transfers heat much more quickly than air. In addition, the temperature of the water supply, even in the summer, can be significantly colder than the air temperature. Remember that if it is 100°F (38°C) outdoors, it is impossible to cool the refrigerant below that temperature with air alone.

Secondary coolant: A substance used as a heat transfer fluid that does not change its state as it transfers heat; also referred to as a *secondary refrigerant*. Examples include water/glycol mixtures and chilled water.

In many cases, process wastewater can be used with a water-cooled condenser. This method uses water that is otherwise wasted to reduce energy consumption and increase refrigeration capacity and efficiency with a limited investment.

2.2.5 Guidelines for Installing Air-Cooled Condensers

The most important factor to consider when installing air-cooled condensers or condensing units is the need for unimpeded airflow in and out of the condenser. It is also important to ensure that air leaving the condenser is not recirculated in the immediate area.

Always follow the manufacturer's recommendations concerning clearances from walls and other obstructions that must be maintained. It is also important to provide adequate space for service and replacement. Condensers must not be located in the vicinity of discharging steam, hot air, or exhaust fumes.

Condensers should be mounted where high levels of sound are not important. Roof-mounted units should be installed level on steel channels or I-beam frames capable of supporting the weight of the unit. For new installations, a structural engineer should determine whether the roof structure can support the equipment and how it should be modified. Vibration isolators may be necessary, especially for heavier units.

Additional guidelines that should be observed when installing air-cooled condensers and condensing units include the following:

- Never use the condenser coil manifolds, control panels, or return piping bends to lift or move the condenser unit.
- Route and support all piping so that vibration and the stress caused by thermal expansion and building settling are minimized.
- Ensure all roof penetrations are properly prepared and sealed.

- Route all liquid lines so the lines do not protrude more than a few inches from the unit before dropping to the roof level.
- Insulate hot discharge piping near the control enclosure to protect workers from burns.

2.3.0 Evaporators and Display Cases

Many kinds of evaporators and display cases are used in commercial refrigeration and can be divided into two main groups. One group includes the evaporators used for cooling air, which in turn cools the contents of coolers or display cases. The other group includes liquid-cooling evaporators that are used to cool liquids such as water, soft drinks, other beverages, and brines.

2.3.1 Natural-Convection Evaporators

Natural-convection evaporators are still used in some open display cases. Air circulation in these evaporators depends on thermal circulation, where warm air rises and dense, cold air descends. However, these evaporators are relatively inefficient, and it is easy for product placement to create some airflow problems, since the air is not forced.

2.3.2 Forced-Convection Evaporators

The unit coolers used in food preparation rooms, chill rooms, and walk-in coolers are good examples of forced-convection evaporators (*Figure 33*). It is a packaged assembly consisting of the evaporator coil, blower or fan, metering device, condensate collection pan, and defrost heaters (for electric defrost). The size and shape of the storage area determine the type and number of unit coolers used and where the coolers are best located.

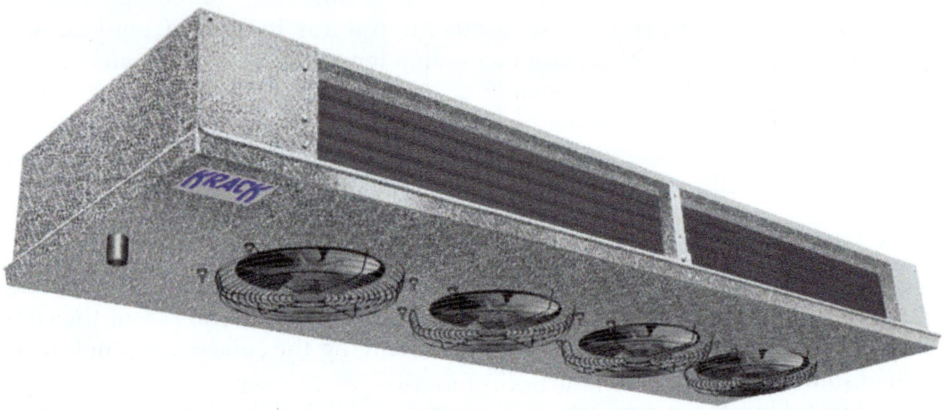

Figure 33 Typical low-profile unit cooler.
Source: Hussmann Corporation

The unit cooler in *Figure 33* has a low profile, making it a good choice for walk-in coolers or freezers with limited head room. This is a *blow-through* unit, meaning that the fans draw air from the enclosure and blow it through the evaporator coil. This also results in relatively gentle air movement in the enclosure.

Unit coolers can cool an enclosure quickly. However, significant air movement tends to cause dehydration of the product unless care is taken in the selection. Foods that are packaged in some way are not affected.

When storing most types of fresh produce, the humidity must be kept high to maintain product quality. Conversely, the humidity should be low in a room used to process and chill warm, red meat. This is necessary to avoid the formation of fog and to prevent condensed moisture from dripping onto the meat.

The selection of the proper evaporator coil is where humidity control begins. Unit coolers used in areas requiring high humidity levels have larger coils operating with a small air-to-refrigerant temperature difference, such as 4°F to

8°F (2.2°C to 4.4°C). In addition, the air volume for these coils is high for two reasons. Air moving rapidly across the coil does not have time to condense as much moisture on the coil, reducing dehumidification. Also, with a coil selected for a small temperature difference, more airflow is needed to provide the same amount of cooling.

When unit coolers are used in areas requiring low humidity, the evaporator coil can be smaller, operating with a higher air-to-refrigerant temperature difference and reduced airflow. For low humidity applications, a temperature difference of 20°F to 30°F (11.1°C to 16.7°C) is common. The rate of airflow across the evaporator is also reduced.

To summarize this topic, unit coolers are selected for capacity based on a specified air-to-refrigerant temperature difference. A given coil produces more refrigeration capacity at higher temperature differences, but the humidity in the enclosure falls as this temperature difference rises. For this reason, evaporator coils must be carefully selected for the application with consistent attention to the humidity level required to sustain the product.

Installing Unit Coolers

When installing unit coolers, always follow the manufacturer's recommendations concerning clearances from walls and obstructions around the top, bottom, and sides of the unit. Obstructing airflow through the coil will negatively affect performance. Guidelines that should be observed when installing unit coolers include the following:

- Make sure that the air pattern covers the entire enclosure. If the throw of the unit is too short, for example, products on the far side of the room may be warmer than desired.
- Avoid installing the unit directly above door openings in a walk-in. When the door opens, the coil is immediately hit with warm, moist air from outside the enclosure.
- Allow sufficient space between the rear of the unit cooler and the wall to permit free air movement, cleaning, and service.
- A sufficient gap should be left between the top of the unit cooler and the ceiling to allow for washdown in areas where open food is present.
- Always trap condensate drain lines to prevent vapor migration into the enclosure through the drain line. Traps on low-temperature units should be outside of the refrigerated enclosure. Traps and condensate lines that are exposed to freezing temperatures must be protected with heat tape and insulated.
- In coolers or freezers with glass display doors, be sure that discharge air from the coil blows above the doors, not directly at them.

2.3.3 Multiple Evaporators

Many different configurations can exist in a multiple evaporator system. Each coil must have its own metering device. Refrigerant flow to each coil is typically controlled by a liquid solenoid shutoff valve.

Each expansion valve sensing bulb must not be influenced by the refrigerant flowing through the common suction line. Out of many unit coolers on the same refrigerant circuit, perhaps one expansion valve fails to the open position. The resulting liquid entering the common suction line can influence all the other expansion valves and cause them to feed less refrigerant unless the suction line for each coil is properly routed. Placement of the sensing bulb for the valve is equally important.

For this reason, the suction line leaving each coil should drop into the top of the common suction line. This eliminates the possibility of liquid refrigerant in the common suction line influencing the other coils. This is especially true when the unit cooler or coil is below the common suction line. The suction riser must extend past the common suction line and turn back down into the top.

2.3.4 Display Cases

Open, self-service display cases (*Figure 34*) that operate at medium and low temperatures are used extensively in food marketing. Display cases are used to maintain the temperature of a product that has previously been cooled to the proper temperature.

Figure 34 Open, self-service display cases.

> **On Site**
>
> **Displaying Produce**
>
> Display cases must be loaded properly. The product on display should never be piled so high that it is out of the refrigerated zone. The load line recommended by the display case manufacturer should always be observed to maintain proper cooling. Also, the product should not be stacked so that circulation of the refrigerated air is blocked.

Display case operation is clearly and significantly affected by the temperature, humidity, and movement of the ambient air. The refrigeration capacity for supermarket display cases is normally rated by the manufacturer based on ambient summer design conditions of 75°F and 55% relative humidity.

Sources of radiant heat, product packaging, and density of display case loading also affect display case performance. If not controlled properly, heat sources in the store can raise the surface temperatures of the products being displayed, shortening the shelf life. It is important to note that these display cases contribute to store cooling. However, grocers know that one of the best ways they can improve fixture performance and energy efficiency is to keep the store temperature low, without being too cold for customers.

Most display cases used in markets have what is referred to as *endless construction*. Endless construction allows individual case modules to be joined together at the end to form a continuous line. Separate end sections are provided for the first and last units in the display bank. In *Figure 34*, a total of five display modules are visible.

Deck: A shelf, pan, or rack that supports refrigerated items stored in merchandising coolers and display cases.

A **deck** is a shelf, pan, or rack that supports the refrigerated items in a case. Display cases are often classified by the number of decks they contain (i.e., single-deck units versus multideck units). Open display cases keep the food cold using a blanket of cold air. The food is stored under the top of the air blanket. The evaporator in an open display case is normally located beneath the cabinet, with the decks above.

Compared to closed coolers, open display cases need additional refrigeration capacity because of the higher heat gain. The heat gain of an open display case is about twice that of a case with glass doors. Customers reaching into the case to remove product cause air movement, disturbing the blanket of air and temporarily increasing heat gain.

To minimize disturbances to the air blanket in open display cases, it is best to locate the cases away from all fans, entrance doors, HVAC terminals, and unit heaters.

Frozen food cases depend heavily on an effective blanket of cold air. Because the cases operate at temperatures of 0°F (–18°C) and below, a substantial temperature difference exists between the room air and the case. Moist air which becomes frost increases the load significantly, as it becomes an additional latent load on the evaporator.

As mentioned previously, the losses caused by the air outside the case do contribute to store cooling. As a result, supermarkets often require lesser cooling capacity per square foot than other buildings. In cooler geographic areas, the cooling effect of the open refrigeration cases can eliminate the need for comfort cooling in the summer. Conversely, open cases increase the winter heating load significantly. During the winter months, keeping the store warm works against the refrigerated cases, and energy costs rise as a result.

Open vertical display cases may have the evaporator coil at the bottom or in the back. *Figure 35* shows the airflow through a case with the evaporator at the bottom. *Figure 36* shows a case with the evaporator in the back. When the evaporator is in the back, the panel separating it from the food may not be perforated. In both *Figure 35* and *Figure 36*, note the line connecting the air discharge at the top to the return grille at the bottom. This represents the surface of the protective cold-air blanket.

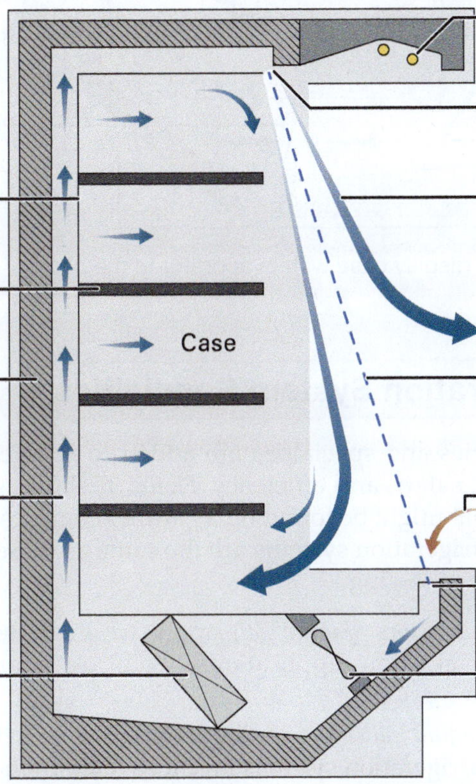

Figure 35 Open vertical display case with evaporator at the bottom.

Going Green

Supermarkets and the Environment

Supermarkets require a lot of refrigerant, and most of those refrigerants are currently hydrofluorocarbons (HFCs). Unfortunately, the HFCs in use have a high global warming potential (GWP). At this writing, refrigerant regulations continue to change, with mildly flammable and flammable substances entering the market to replace HFCs. These substances have much lower, and perhaps no, global warming potential. The world at large is working toward the elimination of refrigerants that harm the environment.

The Environmental Investigation Agency (EIA) is a 501(c)(3) nonprofit organization that is known for its use of innovative investigative techniques and research in environmental issues. The organization prepares a bi-annual report entitled, "Climate-Friendly Supermarket Scorecard," assessing major supermarket chains on efforts to reduce HFC use and transition to other refrigerants. To view the report, visit the EIA at *www.eia.org*.

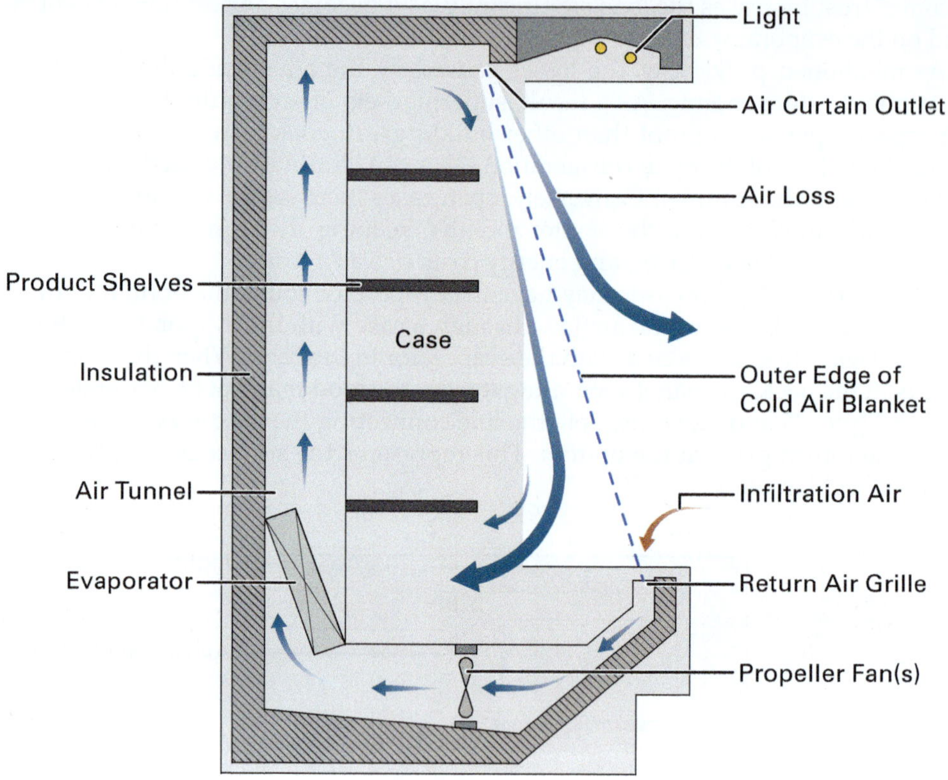

Figure 36 Open vertical display case with evaporator in the back.

On Site

Moisture Removal

For every pint of moisture that condenses on product in a self-service display case, about 1,000 Btuh of heat is added to the refrigeration load. The load increases dramatically if the product is frozen and the condensation freezes.

2.4.0 Refrigeration System Specialties

Refrigeration accessories and specialties are added to refrigeration systems to improve functionality, safety, and efficiency. *Figure 37* shows some of the many possible accessories that might be found on a parallel-compressor system. Many accessories used in refrigeration systems are the same as those used on comfort systems, including the following:

- *Filter-driers* — Filter-driers (*Figure 38*) apply to all systems. In commercial/industrial refrigeration applications, you are more likely to see replaceable-core filter-driers.
- *Sight glass/moisture-liquid indicators* — These windows into the refrigerant circuit are found on refrigeration systems far more often than comfort systems.
- *Suction line accumulators* — Accumulators are often found on heat pumps. They have the same function in refrigeration—protect the compressor from refrigerant floodback. On large industrial systems, the accumulator can be very large. It is not unusual to encounter accumulators that are 4' in diameter and 8' tall installed in large ammonia systems.
- *Crankcase heaters* — Crankcase heaters are essential to ensure refrigerant vapors do not condense in the crankcase when the compressor is off. However, many refrigeration systems also have additional solenoid valves in the piping that prevent refrigerant migration.
- *Pressure-relief valves and manifolds* — Large refrigeration systems often require more than one relief valve. Any pressure vessel with an internal volume of 10 ft^3 or more must have dual relief valves. The two valves are often mounted on a special manifold (*Figure 39*). This allows one relief valve to be isolated for service and calibration while the second valve remains active. The back-seating valve can isolate one valve or the other, or it can be placed in a mid-position to keep both valves active (normal position). However, no valve position exists that isolates both valves at the same time.

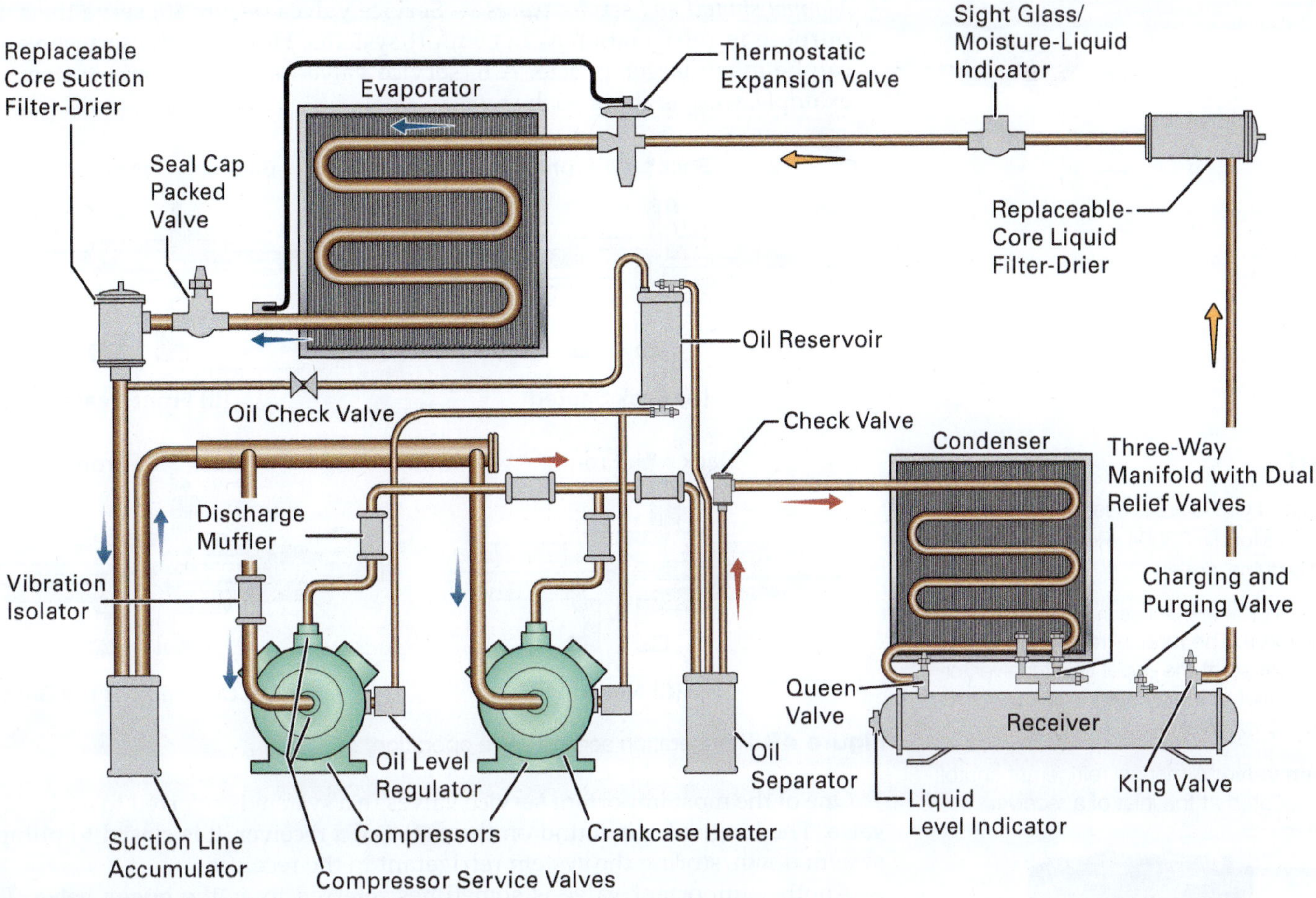

Figure 37 Parallel-compressor refrigeration system with accessories.

Figure 38 Filter-driers.
Source: Courtesy of Sporlan Division - Parker Hannifin Corporation

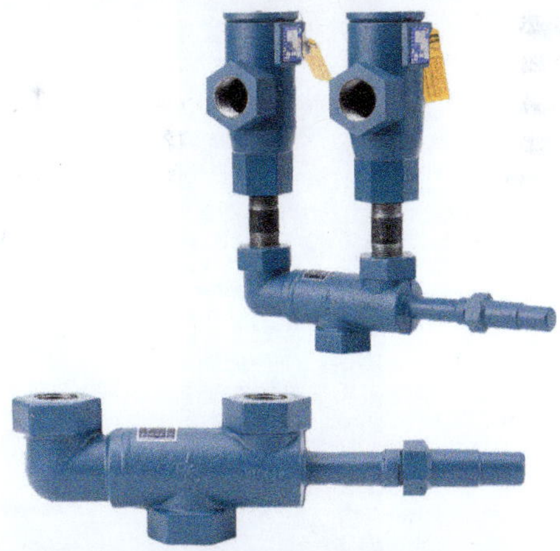

Figure 39 Pressure-relief valve manifold.
Source: Courtesy of Sporlan Division - Parker Hannifin Corporation

- *Compressor mufflers* — Mufflers are generally found on semi-hermetic reciprocating compressors.
- *Receivers* — Receivers are needed in air-cooled refrigeration systems that have large load variations. Additional refrigerant is also needed for systems equipped with refrigerant-side head pressure controls. The receiver can also be used to store the system charge during service procedures or prolonged shutdowns. Note that most water-cooled refrigeration systems do not use a receiver because it is an integral part of the water-cooled condenser. The receiver is installed in the liquid line between the condenser and the metering device.

- *Manual shutoff and service valves* — Service valves (*Figure 40*) serve the same purpose in refrigeration as in comfort systems. However, the service valves can be much larger in size. A 6" service valve for an ammonia system, for example, weighs about 90 lb.

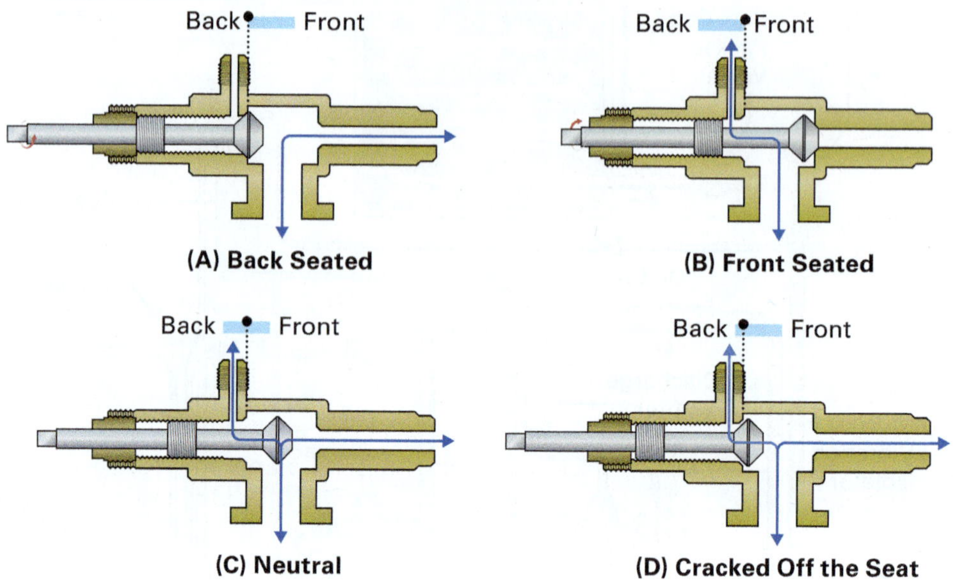

Figure 40 Refrigeration service valve operation.

NOTE

EPRs and CPRs are presented in detail in NCCER Module 03304, *Retail Refrigeration Systems*.

King valve: The manual refrigerant shutoff valve located at the outlet of a receiver, or at the outlet of the condenser for systems without a receiver.

Queen valve: A manual refrigerant shutoff valve located at the inlet of a receiver.

One of the most important service valves in a system is referred to as the **king valve**. The king valve is found on the outlet of a receiver. It is closed to pump a system down, storing the system refrigerant in the receiver.

Another important valve is sometimes referred to as the **queen valve**. The queen valve is located at the inlet to the receiver. Together, the king and queen valves can trap all the refrigerant charge in the receiver.

Large service valves generally have packing around the valve stems to prevent leaks. When working with these valves, it is important to test for leakage around the stem before reinstalling the cap. A gland nut on the valve stem is used to apply pressure to the packing to stop any leakage.

Evaporator pressure regulators (EPRs) and crankcase pressure regulators (CPRs) are almost exclusively used in refrigeration. Comfort cooling systems have no need for these components. EPRs are used on evaporator coils and unit coolers when multiple evaporators are connected to a single refrigeration system that must operate at different temperatures.

CPRs help regulate the suction pressure returning to a refrigeration compressor when a system completes a defrost cycle or is started up with the enclosure warm. Under these conditions, the compressor can easily be overloaded to the rise in compression ratio.

Two other systems commonly found in commercial/industrial refrigeration are oil collection and control systems and pump-down systems. These systems are presented in the sections that follow.

2.4.1 Oil Collection and Control Systems

Oil separators and control components are also used extensively due to the complexity of industrial refrigerant piping and the challenge of ensuring oil can return to the compressors once in the system at large.

Oil coats the inside of every refrigerant-circuit component it passes through. It reduces heat transfer in the evaporator and condenser. Oil separators (*Figure 41*) are installed in the hot gas discharge line as close to the compressor as practical. Separating the oil from the refrigerant vapor as it passes through, the oil separator collects the oil in a small sump. A float valve in the sump maintains a seal between the high-pressure and low-pressure sides of the system. Oil separators may be accessible like the one shown in *Figure 41*, or they may be hermetically sealed.

Figure 41 Accessible oil separator.
Source: Temprite

As the oil level in the sump rises, the float rises with it. When only one compressor is in the circuit, the rising float allows high-pressure refrigerant vapor to push the oil back into the low-pressure crankcase of the compressor. Once the oil is flushed out of the separator and back to the compressor, the float drops and seals off the oil return line. When used with a single compressor, the oil separator can function without additional system components except for an optional oil filter.

In circuits with more than one connected compressor, the correct oil level must be maintained in each of the compressors while all the compressors are discharging into a common hot gas line. The volume of oil leaving the individual compressors through the discharge line varies due to age, wear, and operating time. Thus, the oil captured by each oil separator must be collected into a common reservoir and returned to the compressors as needed.

To do this, the oil separators send the accumulated oil to a larger oil reservoir (*Figure 42*). To reduce the pressure in the reservoir to a level slightly higher than the pressure in the compressor crankcase, excess pressure is relieved through the oil differential check valve (*Figure 43*) back to the low-pressure side of the system. The valve shown is preset to an appropriate value to ensure that adequate pressure is maintained in the reservoir to push oil into the crankcase, but only enough pressure to accomplish the task.

Oil is delivered back to the individual compressors through the oil level control (*Figure 44*). One of the flanged ports on the control is used to mount it to the compressor crankcase, while the other is covered with a sight glass for viewing. As the oil level drops in the crankcase, the control allows oil to enter.

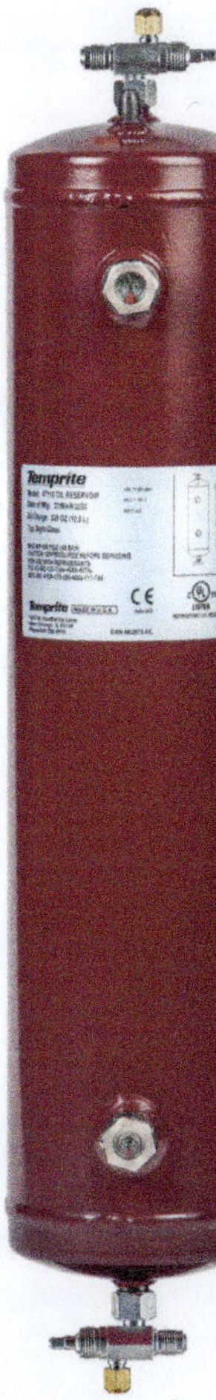

Figure 42 Oil reservoir.
Source: Temprite

On Site

Phantom Leak

When a system is diagnosed as being low on charge, the source of the leak must be found and corrected before additional refrigerant is added. If no leak is detected, examine the relief valve. This is especially true for systems that are repeatedly low on charge with no detectable leak. Once a relief valve has opened, it cannot reseat properly when it resets, or it can open at a lower pressure than that for which it was originally designed.

Figure 43 Oil differential check valve.
Source: Courtesy of Sporlan Division - Parker Hannifin Corporation

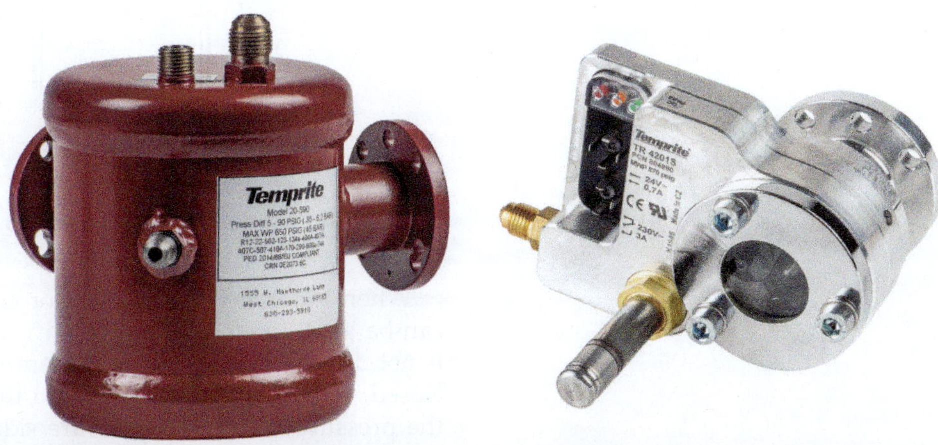

(A) Mechanical Oil Level Control

(B) Electronic Oil Level Control

Figure 44 Oil level control.
Source: Temprite

2.4.2 Pump-Down Control

When most comfort systems shut down, the system pressures typically equalize in a short period of time. Although the compressor stops, the refrigerant can still move around in the circuit. For example, when the compressor stops, the evaporator coil is still about 75% filled with liquid refrigerant. As it stagnates in the coil and warmer air circulates around the evaporator, the refrigerant vaporizes and may move to another part of the system that is cooler.

In a low-temperature refrigeration system, the temperature difference between the refrigerant in the evaporator and the air in the enclosure is often very low. What often happens here is that the liquid refrigerant remaining in the evaporator at shutdown doesn't vaporize. Instead, the vapor in the coil may condense, increasing the amount of liquid refrigerant in the coil.

In the low-temperature scenario, when the compressor restarts, the liquid in the evaporator rushes out quickly and floodback occurs. It is often enough liquid to mechanically destroy the compressor.

To alleviate this problem, a pump-down cycle is used at the end of every operating cycle. During the pump-down cycle, the compressor pumps the refrigerant from the low-pressure side of the system into the condenser and/or receiver. This clears most of the liquid refrigerant out of the low-pressure side of the system. Refer to *Figure 45*.

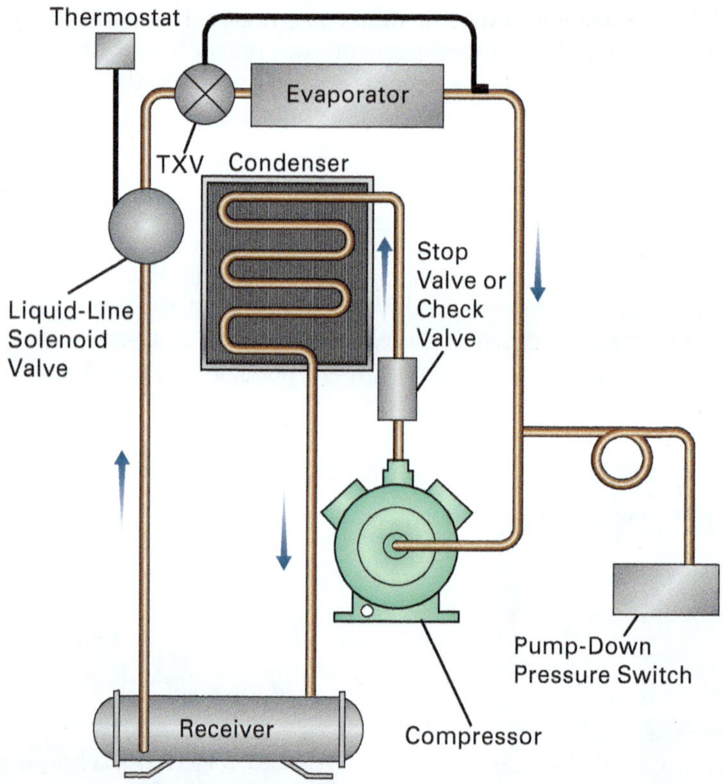

Figure 45 Pump-down system devices.

When the system thermostat or controller is satisfied and the compressor can be shut down, a solenoid valve in the liquid line closes. The compressor is not directly controlled by the temperature controls. With the solenoid valve closed, refrigerant is pumped into the condenser and/or receiver. As it does, the pressure on the low-pressure side falls until reaching the setpoint of the pump-down pressure switch. When the pressure switch opens, the compressor shuts down. The pressure switch setting varies from system to system, but it is usually near 0 psig.

The liquid-line solenoid valve is best located just upstream from the metering device and evaporator coil inlet. This prevents the liquid line from sweating due to vaporizing refrigerant. Note that the low-pressure switch for the system must be set at a lower value than the pump-down pressure control.

2.0.0 Section Review

1. When a low-temperature application requires a compression ratio of 14:1, what system characteristic or component will likely be needed?
   a. Two-stage compressor
   b. Oversized refrigerant piping
   c. Crankcase pressure regulator
   d. Evaporator pressure regulator

2. When using multiple air-cooled condensers on a system, what is the result of there being a significantly different pressure drop in the discharge lines feeding each condenser?
   a. The compressor will likely overload.
   b. Liquid refrigerant tends to back up in one of the condensers.
   c. The metering device feeds too much refrigerant to the evaporator.
   d. The superheat measured at the compressor becomes dangerously low.

3. Which of the following statements about unit cooler installation is *true*?
   a. Traps are not necessary in condensate lines serving low-temperature unit coolers.
   b. Unit coolers perform best when installed directly above the entrance door of a walk-in.
   c. Unit coolers in reach-ins with glass doors should push the air directly at the doors to keep the glass clear.
   d. A sufficient gap should be left between the top of the unit cooler and the ceiling to allow for washdown in areas where open food is present.

4. Any refrigerant pressure vessel with an internal volume of 10 ft^3 or more must have ______.
   a. a sight glass
   b. dual relief valves
   c. king and queen valves
   d. a pressure regulating valve

3.0.0 Defrost Systems

Objective

Identify and describe various types of defrost systems.

a. Describe an off-cycle defrost system.
b. Explain how electric defrost systems operate.
c. Identify and describe hot-gas defrost systems.

Performance Task

1. Install or troubleshoot a commercial refrigeration condensing unit and unit cooler system.

As long as a supply of water vapor is inside a refrigerated space, it will condense on the surface of an active evaporator when the coil temperature is below the dew point. However, when the coil surface is below freezing, the condensed moisture freezes and accumulates, impeding and eventually blocking airflow through the coil. The coil must be defrosted to restore proper operation.

Several common factors exist regarding frost accumulation on the evaporator coils. Unit coolers that were selected to operate at a high air-to-refrigerant temperature difference will generally build frost more rapidly. Fixtures that have a low air-to-refrigerant temperature difference will not accumulate frost as quickly.

Another factor in frost accumulation is related to the fixture type and the infiltration of outside air. Many refrigerated enclosures experience high consumer traffic. In hot, humid weather, a fixture that is opened many times each day for product access will be constantly exposed to hot, moist air. This significantly increases the volume of moisture in the enclosure. In addition, the infiltration of hot and humid air significantly increases the refrigeration system load. It requires a great deal of energy to condense water vapor.

Conditions like this can overwhelm the unit and interfere with performance. Consider the potential difference in performance of an identical frozen-food refrigerated fixture installed in two different locations—one in a small, remote store in Arizona with little consumer traffic, and the other in a consistently busy, 24-hour convenience store in coastal Florida.

It is important to note that all heat added to the enclosure during the defrost cycle must also be removed by the refrigeration unit. This has a significant impact on the refrigeration load, and frequent defrost cycles increase the refrigeration load substantially. In addition, the time a system spends in the defrost mode must be subtracted from the available time for refrigeration. For example, if the total refrigeration load for a freezer were to be calculated as 96,000 Btus over a 24-hour period (4,000 Btuh), but the unit is expected to operate in a defrost mode for 3 hours each day, then the refrigeration equipment must be able to generate 96,000 Btus in 21 hours instead of 24 hours. This requires 4,571 Btuh of refrigeration capacity.

In commercial/industrial refrigeration applications, there are three primary approaches to defrost:

- Off-cycle and/or timed defrost
- Electric defrost
- Hot-gas defrost

Going Green

Infiltration Load on Refrigerated Fixtures

Your responsibility to provide professional service on refrigeration systems does not end at the refrigeration equipment. You must also take proper care of the enclosure. Badly worn door gaskets and hinges can seriously impact the performance and energy efficiency of a refrigerated fixture. The constant infiltration of warm and moist outside air around poorly fitted doors and gaskets can prevent even the best fixture from reaching its setpoint, increase frost accumulation that impedes heat transfer, and increase equipment run time and operating costs. The maintenance and replacement of door gaskets and hinges that experience high wear and abuse are as important to the function of a fixture as the proper refrigerant charge.

Always be on the lookout for problems with the fixture that affect performance and waste energy. While the client may choose not to address the problem, it is your responsibility to report and document these problems.

3.1.0 Off-Cycle Defrost

Off-cycle defrost is a passive and simple process. It also avoids the consumption of additional energy to melt the ice, which must then be removed by the refrigeration unit. However, it is not an effective approach for many fixtures.

In fixtures maintained at or above 36°F (2°C), the coil can simply be allowed to defrost during the normal off cycle. When the thermostat is satisfied and the compressor shuts down, the air temperature in the box will allow accumulated frost to melt. This is referred to as *off-cycle defrost*.

Fixtures in this category will usually operate at a saturated suction temperature around 16°F to 31°F (–9°C to –1°C), depending on the desired humidity and the box design. At these coil temperatures, the coil will certainly accumulate frost, but the evaporator should defrost naturally once the compressor cycles off because the fixture temperature is several degrees above freezing.

Although it is important to ensure refrigerated fixtures are never set to maintain lower temperatures than necessary, this step is critical for a system that uses a natural defrost approach. Setting the box temperature even a few degrees colder than necessary can seriously impede and prevent a natural defrost process, resulting in frozen coils and rising box temperatures. Once this occurs, the fixture must be shut down for an extended period to rid the coil of accumulated ice.

Enclosures that operate at a slightly lower temperature, closer to 32°F (0°C) but not below freezing, require a longer defrost period to clear accumulated ice than a normal off cycle provides. The defrost cycles need to be better planned and defined to ensure the coil remains clear. A defrost timer (*Figure 46*) is the simplest way to accomplish this.

(A) Electronic Defrost Timer

(B) Mechanical Defrost Timer

Figure 46 Electronic and mechanical defrost timers.

A 24-hour defrost timer with normally closed contacts is placed in series with the fixture's temperature controls. The time clock simply breaks the circuit to turn off the compressor, leaving the evaporator fan running. The number of times per day a shutdown is required, as well as the length of each shutdown, depends a great deal on the environment and fixture use. The fixture temperature often rises a few degrees above normal during the defrost cycle, but it will quickly recover.

Mechanical defrost timers have a dial assembly allowing the timer to be configured for the current time and the duration, and number of defrost cycles each day (*Figure 47*). The inner dial sets the duration of the defrost period. A trip pin is screwed into each time that a defrost cycle is desired. On this clock, defrost cycles can be set as often as every two hours. The defrost cycle is terminated after the period has elapsed. The timer contacts then return to the normally closed position and refrigeration resumes.

As is true in all defrost scenarios, setting the defrost timer for periods that coincide with the fixture's lightest use (i.e., late at night) helps prevent the fixture temperature from rising too high. Always try to avoid scheduling defrost cycles at times of heavy use.

3.2.0 Electric Defrost

Electric defrost systems are frequently used in low- and medium-temperature commercial systems. Electric defrost components are relatively inexpensive, reliable, and simple to troubleshoot and repair.

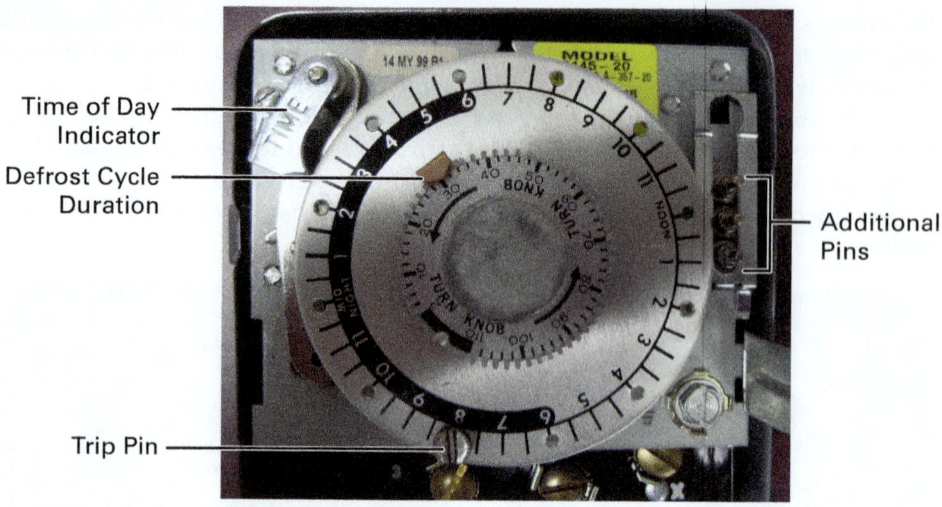

Figure 47 Mechanical defrost timer dial.

Figure 48 shows a typical arrangement for resistive defrost heaters installed on both the face and back of a unit cooler evaporator. To work well, the heater must be in good thermal contact with the coil fins to distribute heat across the coil face.

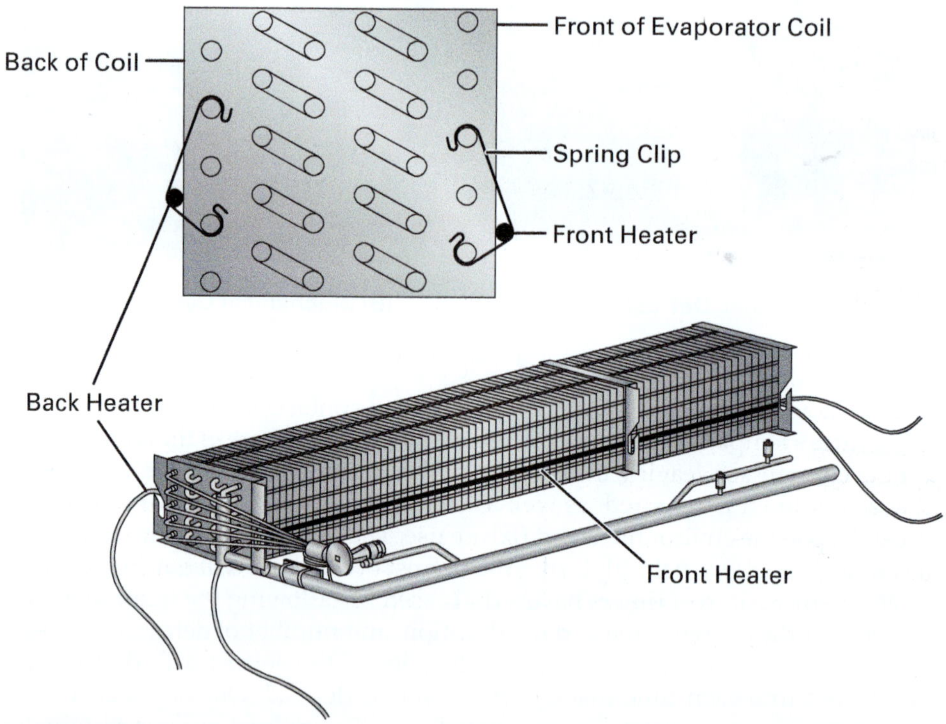

Figure 48 Electric defrost heater arrangement on a unit cooler.
Source: Hussmann Corporation

During the defrost period, a time clock stops the compressor and evaporator fans, then energizes the electric heat. In many cases, the size and shape of the defrost heater is unique to a given coil. This requires access to original parts when replacement is necessary. Some coil designs require that the heating element be slipped into a tube placed deep inside the coil assembly, making replacement more challenging. However, the tube is bonded to the coil fans, resulting in excellent heat transfer.

Some units are equipped with radiant electric heaters (*Figure 49*). The heater wire is wound like a spring and enclosed inside a quartz tube, with the wires

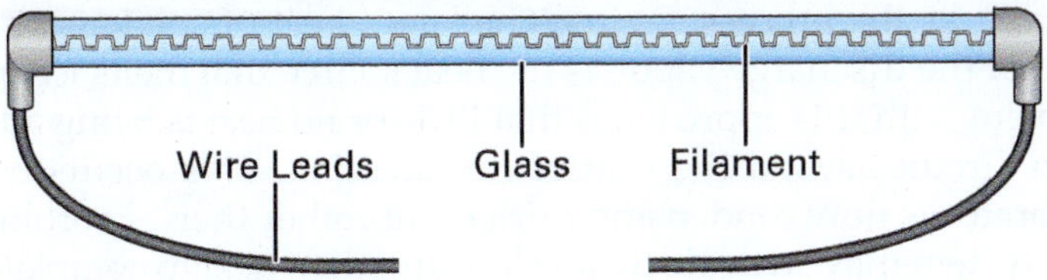

Figure 49 Infrared defrost heater.

passed through seals at the ends of the tube. Using radiant heat, the heater doesn't need to be in contact with the coil surface. These heaters reach a high temperature very quickly and are water- and corrosion-resistant. The quartz enclosure is fragile, as you might expect.

Melted frost must be kept above freezing until it has completely drained out of the refrigerated space. Condensate collection pans for freezers must also be equipped with electric heaters. These heaters are typically long, rigid elements that fit around the perimeter of the drain pan. Where the drain line must be routed through an area below freezing, the condensate drain line must be heated as well. Thin, braided drain line heating elements are available specifically for this purpose. The cable is placed inside the drain line.

Defrost termination can be time-only, temperature-only, or a combination of the two. When combined, either the time clock or a line-mounted thermostat can terminate the cycle. By monitoring the temperature of the coil, or the refrigerant suction line adjacent to the coil, the thermostat can indicate when the temperature is high enough to confirm a complete defrost. The defrost cycle can then be terminated before the time clock has reached the maximum time allowed. Defrost cycles that always continue until the time clock terminates the process are often an indication that the thermostat has failed closed. This is a condition that is easily overlooked by the user.

The proper setpoint for defrost termination by temperature will differ among products, as it is based on the manufacturer's chosen thermostat location and laboratory testing. Line-mounted thermostats used in this application are generally hermetically sealed and nonadjustable, so it is important that the correct replacement is chosen. Electronic time clocks may offer the option of a thermistor instead of a thermostat.

Defrost timers for freezers often have control over the evaporator fans. The evaporator fans are turned off during defrost so that airflow doesn't interfere with the process. When the defrost cycle is over, the evaporator fans do not restart immediately. The evaporator coil is wet, and starting the fans immediately causes water droplets to be blown on to the product and floor. The compressor starts first, allowing a short period of time for the remaining water to freeze on the coil before the evaporator fans start. The delay for the evaporator fans is usually provided by a separate thermostat that closes when the coil temperature indicates any remaining water should be frozen.

On Site

Door and Cabinet Heaters

In addition to defrost heaters, reach-in and walk-in freezers also have low-wattage heating elements routed through the door frame of the cabinet. Heavy condensation and frost developing around the doors is a sign that the heating elements or the power source has failed. These flexible heating elements that resemble common electrical wire are called *mullion heaters*.

The threshold of a walk-in freezer also requires heat, and it must also withstand a lot of abuse from footsteps and product movement.

3.3.0 Hot-Gas Defrost

Hot-gas defrost is the fastest and most efficient approach to defrost. Instead of having only the small contact area provided by electric heating elements on the coil, the entire surface of the coil tubing is actively engaged in the process. Hot-gas defrost is seven to eight times more efficient than using electricity to produce the same heating effect. However, possible issues in the refrigeration circuit must be considered and addressed. Due to the additional specialties required, and the greater risk to the compressor, care must be taken in its application.

Hot-gas defrost, in its simplest form, sends discharge vapor from the compressor directly to the evaporator inlet, bypassing the metering device. The coil pressure is allowed to rise to a predetermined level, but the coil isn't exposed to the full discharge pressure. The warm discharge vapor rapidly condenses in the cold evaporator as the process begins. The latent heat of condensation, not the temperature of the discharge vapor, is the heat source that melts ice on the coil.

One problem with this approach is that little or no heat is being added to the refrigeration circuit beyond the heat of compression. This occurs because the single evaporator is now condensing refrigerant rather than absorbing heat. As a result, the system may struggle to develop sufficient heat to complete a defrost cycle. This can be avoided to some degree by frequent but short defrost periods.

Another more serious problem is the potential for condensed liquid refrigerant to return to the compressor (floodback). For hot-gas defrost to work well, any refrigerant leaving the evaporator as a liquid must be fully vaporized before it reaches the compressor. This can be done in several ways.

Latent heat defrost: A method used in hot-gas defrost applications that relies on the heat absorbed by one or more active evaporators as a source of heat to defrost another evaporator.

One approach works well when two or more evaporators are connected to the same circuit. This approach is sometimes referred to as **latent heat defrost**. One or more evaporators remain in service, absorbing heat and transferring it to the refrigerant, while one coil is being defrosted.

As shown in *Figure 50*, Evaporator Coil 1 is in the defrost mode and is serving as the system condenser. The condensing refrigerant vapor melts the ice, and the condensed refrigerant then flows to the receiver. Liquid refrigerant continues to flow from the receiver to Evaporator Coil 2. The vaporized refrigerant from Evaporator Coil 2 takes its normal path back to the compressor. This approach works especially well on large supermarket systems where many evaporators are connected to a compressor rack system. Evaporator coils take turns defrosting throughout the course of the day.

An effective approach for larger systems with one evaporator coil relies on a tank of liquid, known as a *thermal bank*, that is warmed by the hot discharge gas during the refrigeration cycle (*Figure 51*). The liquid stores heat during the operating cycle and uses it to vaporize refrigerant as it returns from the evaporator coil during defrost.

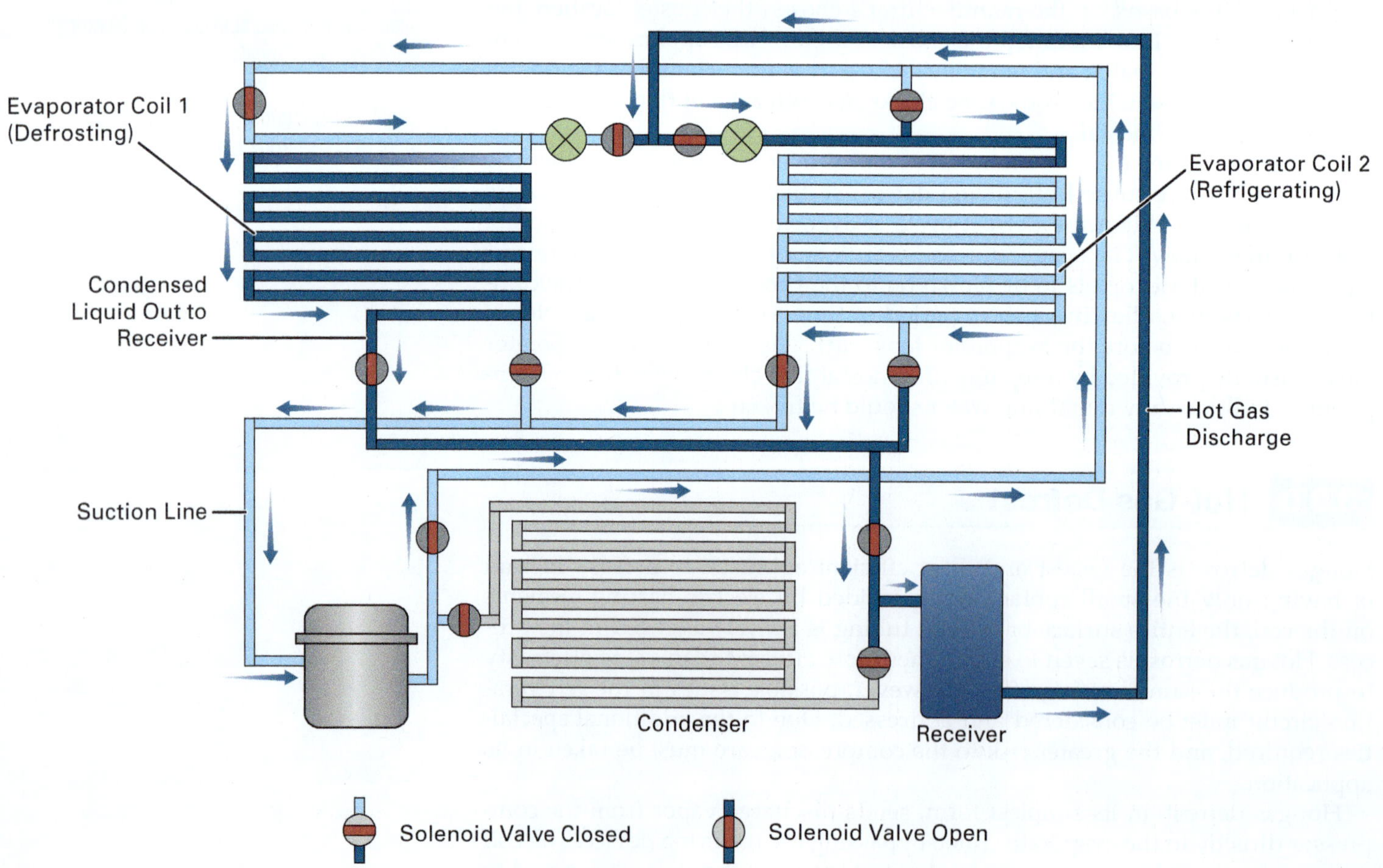

Figure 50 Dual evaporator system with active hot-gas defrost.

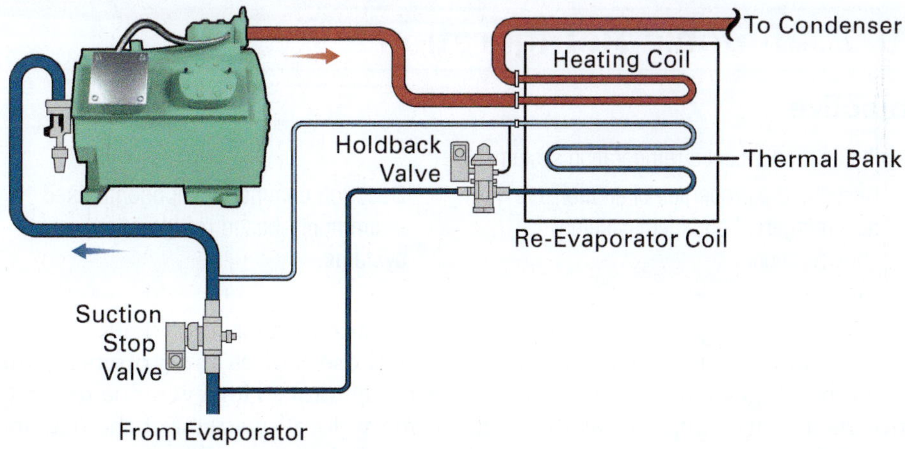

Figure 51 Hot-gas defrost with a thermal bank.

During the normal cycle, hot refrigerant vapor travels through a coil submerged in the thermal bank. The water is heated, and the refrigerant is pre-cooled before moving on to the condenser. When a defrost cycle begins, the hot gas is diverted from the condenser, passes through tubing on the evaporator drain pan, and then on to the evaporator coil. The ice melts and the hot vapor condenses to a liquid. At the same time, the suction stop valve closes and the holdback valve, which is a solenoid-operated pressure regulator, opens.

As this liquid/vapor refrigerant mixture exits the evaporator, it passes through the holdback valve, which limits the downstream pressure to avoid overloading the compressor. The condensed liquid vaporizes in the thermal bank, then moves on to the compressor in its vapor state. In some cases, the water in the thermal bank may even freeze during a defrost cycle. But it quickly thaws during the next refrigeration cycle.

3.0.0 Section Review

1. To ensure effective defrosting, a refrigeration fixture operating at a 34°F (1.1°C) setpoint will likely need to be equipped with _____.
   a. a defrost timer
   b. electric defrost heaters
   c. multiple evaporator coils
   d. hot-gas defrost components

2. True or False? As the defrost cycle for a freezer terminates, the evaporator fans start first and run for a short period of time before the compressor restarts.
   a. True
   b. False

3. To which type of system or characteristic does the term *latent heat defrost* apply?
   a. Systems with receivers
   b. Systems with multiple evaporators
   c. Systems with multiple compressors
   d. Systems with electric defrost heaters

4.0.0 Ammonia Refrigeration

Performance Tasks

There are no Performance Tasks in this section.

Objective

Describe ammonia-based refrigeration systems.

a. Describe the properties of ammonia as a refrigerant and identify safety considerations.

b. Describe common components used in ammonia-based refrigeration systems.

The choice of refrigerants used in low-temperature industrial refrigeration systems is somewhat limited. Ammonia is one of those choices and it offers a number of advantages when compared to other refrigerants. However, due to safety concerns and its pungent nature, codes in many locations restrict the use and storage of ammonia to applications that are publicly inaccessible.

Ammonia systems are usually custom designed for a particular application. Few, if any, packaged ammonia systems are ready for use. Ammonia refrigeration systems must be field constructed.

Ammonia systems offer a wide range of evaporating and condensing conditions. These systems can be used to maintain temperatures from –50°F to 50°F (–46°C to 10°C). Ammonia is used for blast freezing and to chill massive warehouses requiring diverse conditions to satisfy the needs of different commodities. Generally, the larger the application, the more likely that ammonia is the best option. Ammonia system construction and installation is more cost-effective as the size increases.

4.1.0 Properties of Ammonia

Anhydrous: Free of water.

Anhydrous ammonia, designated as R-717, offers excellent heat transfer qualities. Ammonia is a chemical compound of nitrogen and hydrogen (NH_3). Refrigerant-grade anhydrous ammonia is a clear, colorless liquid or vapor, free from visible impurities. It has a minimum purity of 99.95%. At atmospheric pressure, the boiling point of ammonia is –28°F (–33°C).

WARNING!

The term *anhydrous* refers to the lack of water in refrigerant-grade ammonia. Ammonia is extremely hydroscopic, seeking to react with water from any available source. A volume of vapor equal to 1,300 gallons of ammonia can be absorbed into a single gallon of water. Ammonia's affinity for water attracts it to all moist surfaces of the human body, such as the eyes, nasal passages, mouth, and lungs, causing these areas to absorb significant amounts of ammonia vapor.

Because of its low boiling point, ammonia can support refrigeration applications well below 0°F (–18°C) without the need to operate below atmospheric pressure (vacuum) in the evaporator. In the presence of moisture, ammonia chemically reacts with and attacks copper and bronze. As a result, these metals are avoided in ammonia systems. Steel piping and fittings are primarily used in ammonia systems. Aluminum tubing is often used in coil construction.

Immiscible: Unable to effectively mix with or dissolve in another liquid.

Ammonia is an **immiscible** refrigerant. This means that it does not dissolve in lubricating oil like many other refrigerants. This impedes oil return from the system. The oil separates from the ammonia and collects in the evaporator and other low-velocity areas of the system. The oil coats or fills the heat transfer surfaces, reducing system performance. Collected oil may need to be removed from these areas periodically. High-efficiency oil separation at the compressor discharge is a necessity for most systems, but a small amount of oil often passes by even the best oil separator.

Ammonia has a very pungent, recognizable odor. One small advantage to using ammonia as a refrigerant is that leaks are self-alarming; the odor brings immediate attention and alerts system operators to the presence of a leak. Concentrations in air as small as 3 to 5 parts per million (ppm) can be detected by humans.

Small ammonia leaks have traditionally been pinpointed by using *sulfur sticks*—small handheld wicks coated with sulfur (*Figure 52*). The wick is lit and smolders to produce a relatively colorless smoke. In the presence of leaking ammonia, the smoke turns white and very dense. Since ammonia is one of the oldest refrigerants in use today, the simple sulfur stick has been used to locate ammonia leaks for over 100 years.

Sulfur sticks are still used to pinpoint leaks since the sticks are simple and inexpensive. However, handheld electronic analyzers and detectors (*Figure 53*) are often used today as the sensitivity, availability, and reliability have improved over the years.

Figure 52 Sulfur sticks for leak detection.
Source: Prasanna Ravishankar, CC BY SA 4.0, via Wikimedia Commons

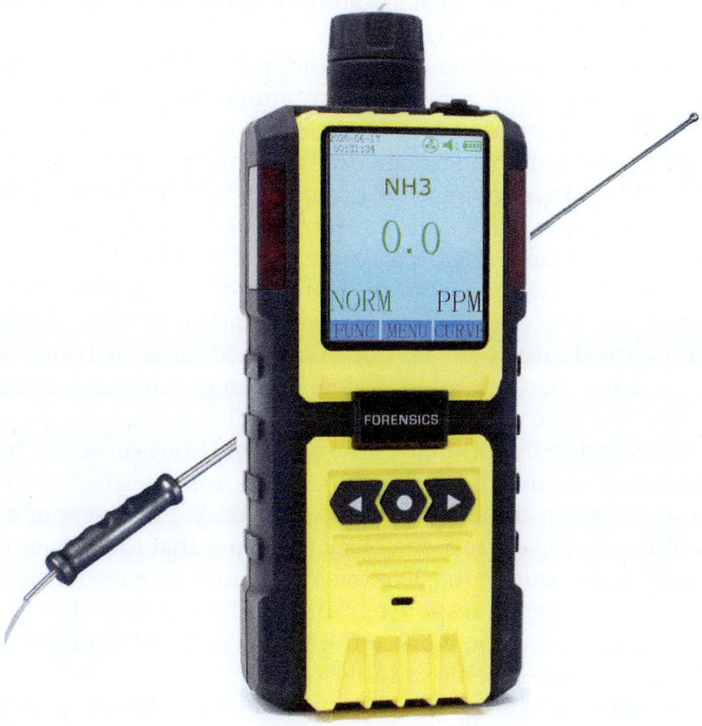

Figure 53 Ammonia analyzer and leak detector.
Source: Forensics Detectors

Case History

Ammonia Release

A cloud generated by 2,200 lb of ammonia was released from the relief valves of an industrial refrigeration system used to produce ice cream. The release occurred after runoff from melting snow leaked from an indoor drain line onto the electrical control panel for the refrigeration system's ammonia compressors. The panel shorted out, and the system shut down. System deficiencies resulted in excessive system pressure, causing the relief valves to open. The employees were evacuated from the building with no injuries being reported.

The operators of the facility neglected to inform the Environmental Protection Agency (EPA) about the incident, as required, until five days later. The company was fined $41,252 for failing to notify the EPA in a timely manner. The EPA requires immediate notification whenever there is a release of a reportable volume of a hazardous substance. The reportable release volume for ammonia is 100 lb.

4.1.1 Ammonia Safety Considerations

Ammonia has a harsh effect on the respiratory system. Only very small concentrations can be safely inhaled. The Occupational Safety and Health Administration (OSHA) sets the permissible exposure limit (PEL) for ammonia at 50 ppm averaged over an 8-hour workday. The National Institute for

Occupational Safety and Health (NIOSH) recommends an airborne exposure limit (REL) of 25 ppm averaged over a 10-hour workday, and a maximum of 35 ppm for any 15-minute period.

Ammonia is hazardous to life at 5,000 ppm and is flammable at a concentration of 150,000 to 270,000 ppm in air. Ammonia's pungent odor can be detected at just 3 ppm to 5 ppm. The odor becomes slightly irritating, making the eyes water, at 15 ppm. While high concentrations of ammonia are a hazard to human health, ammonia is considered environmentally safe.

Rusting pipes and vessels in older systems containing ammonia can create a safety hazard. Also, cold liquids should never be confined in a flooded pipe or vessel between closed valves where the liquid can warm up. The liquid expands as it warms, rapidly increasing the pressure when there is no space for expansion. This causes the pipe or vessel to burst with explosive force. While this does represent a significant hazard with ammonia, confining any cold liquid this way can lead to a similar result.

Figure 54 Ammonia concentration sensor.
Source: Forensics Detectors

The slight smell of ammonia is sometimes present due to minor leakage from valve packings and pump seals, or from ammonia vaporizing from recently drained oil. The scent of ammonia in the air is quite common near a large system.

Any significant ammonia leakage or release must be prevented to avoid personal injury. When an ammonia leak occurs in a facility, the odor provides some warning. In unoccupied areas, such as refrigerated storage rooms or equipment rooms, automatic sensors (*Figure 54*) are used to provide a warning of increasing concentrations. The sensitivity of each type of sensor varies with its design and intended use. The electrochemical sensor shown in *Figure 54* provides protection for occupied areas with a factory-calibrated range of 0 ppm to 1,000 ppm.

OSHA requires that ammonia system maintenance and service personnel who are expected to respond to an ammonia release receive appropriate training. Having experience working with ammonia is not sufficient to satisfy the OSHA requirement, and the training must be appropriate for the duties the individual is expected to perform in an emergency. End users of ammonia, as well as contractors, have a responsibility to ensure that the proper training has been received and documented. Contractors should take responsibility for general safety and emergency response training. The end user should ensure that all contractor personnel who work at the site, as well as their own maintenance employees, receive site-specific training.

Before working with ammonia, become familiar with the information contained on its Safety Data Sheet (SDS). You should also know the first-aid procedures required for ammonia exposure. The safety guidelines listed below should also be observed:

- Wear the proper protective equipment, such as a respirator mask, face shield, and gloves when working with ammonia to avoid inhalation and contact with your skin and eyes. A self-contained breathing apparatus (SCBA) is needed for extended work and rescue (*Figure 55*). A filtered breathing apparatus is sufficient to escape an enclosed space filled with ammonia vapors.
- If bodily contact with ammonia occurs, flush all affected areas with copious amounts of cool water.
- Remove clothing unless it is saturated with liquid ammonia. Once released to atmospheric pressure, liquid ammonia vaporizes quickly, freezing clothing to the skin.
- Do not wear contact lenses while working with ammonia. Even small concentrations of ammonia cause the eyes to water profusely.

4.2.0 Ammonia-Based Refrigeration Systems and Components

Ammonia systems fall into two classes: mechanical and non-mechanical. In the familiar mechanical system, a compressor provides the required pressure change to make the system work.

Figure 55 Self-contained breathing apparatus (SCBA).
Source: zhaubasar/iStock/Getty Images

Non-mechanical ammonia systems, called *absorption systems,* do not have a compressor. Absorption systems have many applications and are even used in recreational vehicle (RV) cooling and refrigeration systems. In an RV refrigerator, ammonia and water are heated in a boiler, then separated by distillation. The ammonia evaporates and cools the refrigerator, while the water returns to the boiler. The ammonia and water then recombine to start the cycle again. Ammonia absorption systems are also used in some comfort applications. However, large mechanical systems are the focus of this module.

4.2.1 Single-Stage and Compound Ammonia-Based Refrigeration Systems

Mechanical ammonia systems can be single-stage or multistage (compound) systems. Compound systems compress the gas from the evaporator to the condenser in several stages. Typically, compound systems are needed when maintaining coil temperatures around –15°F and colder. Single-stage ammonia systems consist of the same basic components used in a single-stage refrigeration system operating with other refrigerants.

Figure 56 shows a basic two-stage ammonia system. It has two compressors: a low stage and a high stage. It also has two cooling loads: a low-temperature evaporator and an intermediate-temperature evaporator.

Low-pressure vapor returned from the primary evaporator(s) is applied to the suction input of the low-stage compressor. There, the vapor is compressed and discharged at an intermediate pressure into an intercooler. The intercooler desuperheats the discharge gas from the low-stage compressor to prevent overheating of the high-stage compressor.

In *Figure 56*, note the metering device that feeds refrigerant to the intercooler to desuperheat the discharge gas passing through. The discharge gas from the low-stage compressor isn't condensed in the intercooler; the vapor is merely cooled to a reasonable temperature.

NOTE

The intercooler shown in *Figure 56* is an integral part of a two-stage refrigeration compressor. In multistage systems using multiple compressors, the intercooler must be a separate component.

4.2.2 Ammonia Liquid Recirculation Systems

Ammonia liquid recirculation systems (*Figure 57*) are also referred to as *liquid overfeed systems.* In this type of system, ammonia is supplied from a low-pressure receiver to the system evaporators. The low-pressure receiver is a vessel that stores liquid ammonia at a low pressure to supply the evaporators with liquid ammonia. A pump moves the liquid refrigerant through the evaporator coil and back to the low-pressure receiver.

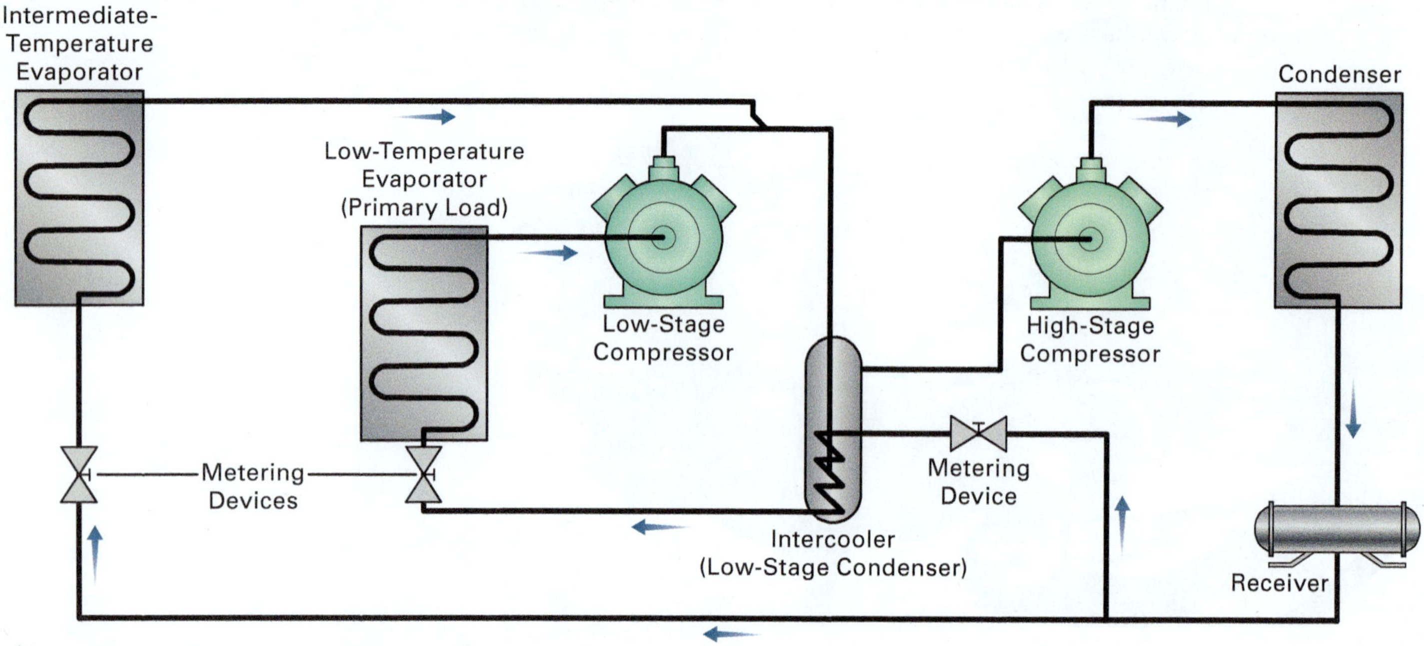

Figure 56 Basic two-stage ammonia refrigeration system with an intercooler.

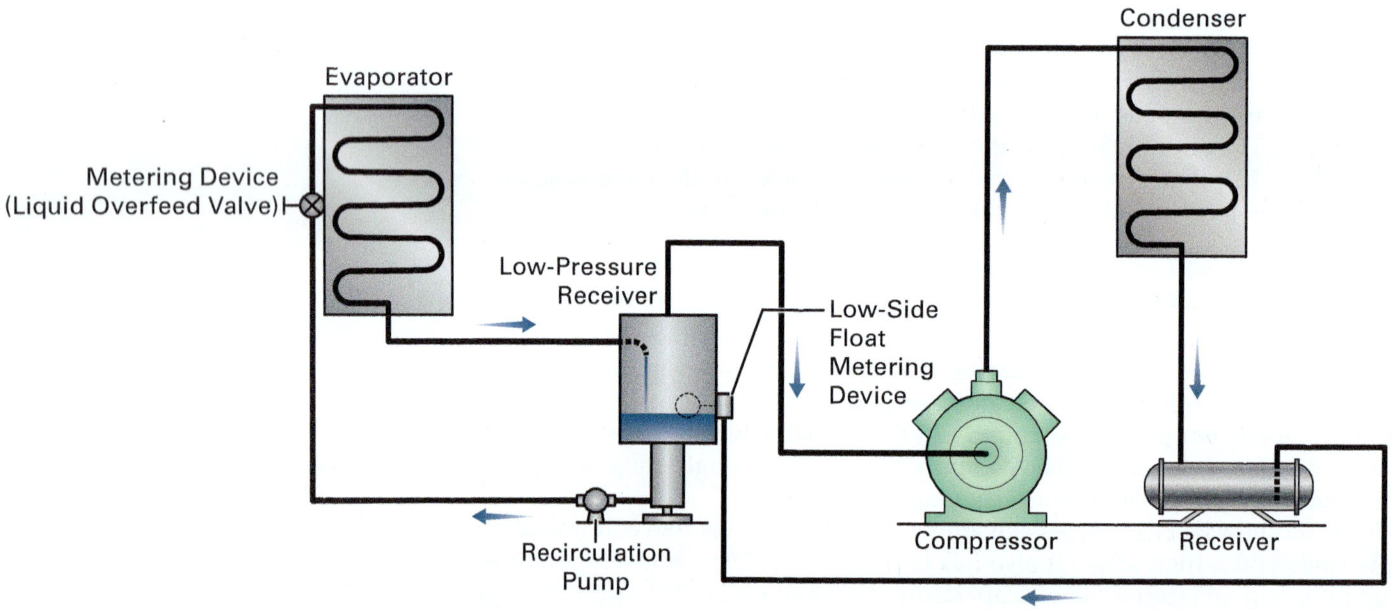

Figure 57 Simplified ammonia liquid-recirculation system.

The compressor suction line is connected to the low-pressure receiver, maintaining the pressure and, in turn, the temperature of the liquid refrigerant inside. The liquid overfeed valve is set to ensure that the evaporator coil remains largely flooded. It is set manually and does not respond to changing conditions.

Enough liquid refrigerant is fed to the coil that constant adjustment of the valve position isn't necessary. The evaporator coil is fed significantly more refrigerant than it can vaporize even when the load is significant. This is why the term *overfeed* is applied to these systems. As a result of the overfeed, a mixture of liquid and vapor exit the evaporator coil and return to the low-pressure receiver. The vapor returns to the compressor while the liquid remains in circulation.

Large suction accumulators are often incorporated to prevent compressor damage in the event the low-pressure receiver overfills with liquid. At startup, there is also some potential for liquid carryover into the suction line if the pressure is reduced too quickly, causing the ammonia liquid in the low-pressure receiver to boil violently. Therefore, the pressure must be reduced slowly in the low-pressure receiver when starting a warm system.

4.2.3 Compressors

Reciprocating, rotary vane, and rotary screw compressors are commonly used in ammonia systems. Reciprocating compressors (*Figure 58*) are most often used in smaller single-stage or multistage systems. However, reciprocating compressors are also used in large groups to provide as much or as little capacity as a large system needs. Centrifugal compressors are not suited for operation on ammonia.

Figure 58 Reciprocating compressors on a low-temperature ammonia system.
Source: Berg Chilling Systems Inc.

Rotary screw compressors (*Figure 59*) are used in singles for systems starting around 40 hp. Combinations of these compressors can provide lots of capacity. These compressors can also be used together for staged operation.

Figure 59 Rotary screw compressor serving an ammonia system.
Source: Berg Chilling Systems Inc.

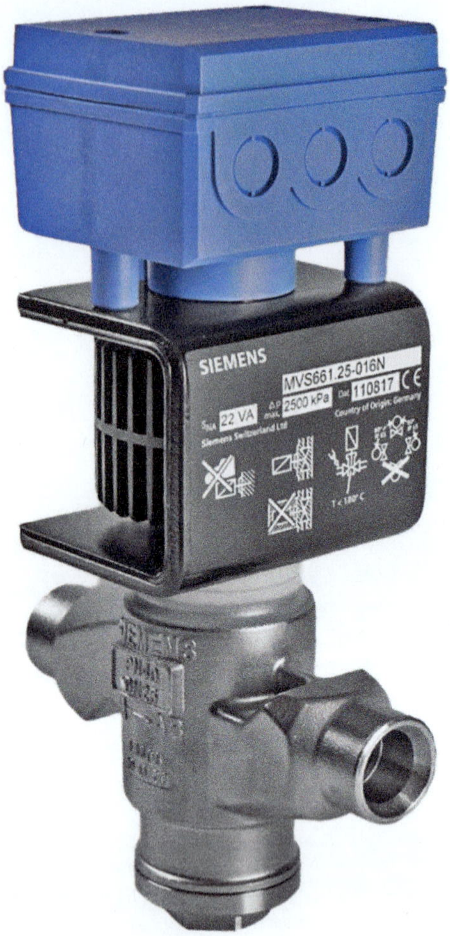

Figure 60 Electronic expansion valve suitable for ammonia.

Source: Courtesy of Sporlan Division - Parker Hannifin Corporation

A low-compression stage for a system might be provided by a rotary vane or rotary screw compressor while the high-compression stage might be a reciprocating or rotary screw compressor.

Due to the compression ratios, compressors often require oil cooling. When reciprocating compressors need an oil cooler, external air-cooled or water-cooled heat exchangers are most often used.

The oil for rotary screw compressors is usually cooled by the direct injection of liquid refrigerant into the compressor to cool the discharge gas, or with an external oil cooler. A *thermosyphon oil cooler* may also be used, which relies on system refrigerant to cool the oil in a plate-and-shell heat exchanger.

4.2.4 Condensers and Receivers

Water-cooled shell-and-tube condensers and evaporative condensers are commonly used with ammonia systems. Air-cooled condensers are rarely used, largely because of the heat rejection capacity needed. More than one condenser is often present, with multiple condenser water pumps available.

When using shell-and-tube condensers, the piping should be arranged so that the condenser tubes are always flooded with water. Air vents on the shell head, with hand valves for manual air purging, are important. Receivers should always be mounted below the condenser(s) so that the condensing surface can't be flooded with ammonia. Piping from the condensers to the receivers is generously sized to allow for liquid to flow down the pipe while displacing vapor in the receiver.

4.2.5 Evaporators and Metering Devices

Direct-expansion ammonia-rated unit coolers can be used, but their use is limited to systems with saturated suction temperatures of 0°F (–18°C) or higher. Flooded shell-and-tube evaporators are often used in ammonia systems where fluids such as water, brine, or glycol must be cooled.

Ammonia evaporator coils are made of either aluminum or steel tubing, because copper and brass are not compatible with ammonia. Manual, thermostatic, and electronic expansion valves (*Figure 60*) are all options as metering devices. Most ammonia systems are designed to take advantage of the efficiency of hot-gas defrost systems. On the largest of systems and applications, liquid overfeed and high- or low-side float valves are commonly used. These metering devices support flooded evaporator applications.

4.0.0 Section Review

1. What issue presents itself with ammonia due to it being immiscible?
   a. Watery eyes
   b. Problems with oil return
   c. Respiratory system damage
   d. Slow to vaporize in an evaporator

2. Which of the following compressor types are *not* suitable for ammonia?
   a. Centrifugal
   b. Rotary vane
   c. Rotary screw
   d. Reciprocating

Module 03408 Review Questions

1. Which commercial freezing method is accomplished by placing the product to be frozen between refrigerated metal plates?
   a. Freeze drying
   b. Contact freezing
   c. Air blast freezing
   d. Cryogenic freezing

2. In cryomechanical freezing applications, cryogenic freezing is combined with _____.
   a. freeze drying
   b. blast freezing
   c. submersion in a fluid
   d. product movement by conveyor

3. Lyophilization produces food products that _____.
   a. must be cooked before consumption
   b. require freezing and are consumed frozen
   c. are lightweight and require no refrigeration
   d. contain all the moisture originally found in the product

4. The temperature and humidity maintained in cold storage are determined by the type of food being stored and what other factor?
   a. Local climate
   b. Size of the storage room
   c. Weight of the product to be frozen
   d. Length of time it is to be held in storage

5. To maintain top quality, fruits and vegetables stored in chill rooms are normally stored _____.
   a. in airtight bins
   b. at a temperature above freezing
   c. between vertical plate evaporators
   d. at temperatures equal to their ripening temperature

6. At what temperature is ice cream typically stored for long-term storage?
   a. −20°F (−29°C)
   b. −5°F (−21°C)
   c. 15°F (9°C)
   d. 25°F (4°C)

7. What are controlled atmosphere (CA) refrigerated containers designed to do?
   a. Freeze unfrozen products
   b. Maintain a specific humidity value
   c. Manage gases that accelerate ripening
   d. Maintain multiple sets of conditions in a single container

8. Assuming all other factors are equal, which of the following applications would cause the refrigeration system to operate at the *lowest* EER?
   a. Chill rooms
   b. Ice cream freezers
   c. Meat cutting rooms
   d. Food preparation rooms

9. Refrigerant-side head pressure control relies on ______.
   a. modulating condenser fan speed
   b. cycling condenser fans on and off
   c. modulating dampers on the condensing unit
   d. reducing the active surface inside the condenser coil

10. Which of the following components is *only* needed for oil control when there is more than one compressor connected to a single circuit?
    a. Oil separator
    b. Oil reservoir
    c. Crankcase pressure regulator (CPR)
    d. Evaporator pressure regulator (EPR)

11. True or False? Defrost cycles ultimately affect the needed capacity of the equipment serving a refrigeration system.
    a. True
    b. False

12. Off-cycle defrost is only practical for fixtures that maintain temperatures at or above ______.
    a. 40°F (4.4°C)
    b. 36°F (2.2°C)
    c. 32°F (0°C)
    d. 28°F (–2.2°C)

13. Which term describes a hot-gas defrost method used with multiple evaporators where one evaporator is being defrosted while the others continue refrigerating?
    a. Off-cycle defrost
    b. Sequenced defrost
    c. Latent heat defrost
    d. Time-temperature defrost

14. What is the OSHA permissible exposure limit (PEL) for ammonia?
    a. 5 ppm averaged over 4 hours
    b. 35 ppm averaged over 4 hours
    c. 50 ppm averaged over an 8-hour workday
    d. 150 ppm averaged over an 8-hour workday

15. In an ammonia liquid recirculation system, to which component is the compressor suction line connected?
    a. Condenser
    b. Evaporator outlet
    c. Low-pressure receiver
    d. High-pressure receiver

Answers to odd-numbered questions are found in the Review Question Answer Key at the back of this book.

Answers to Section Review Questions

Answer	Section	Objective
Section One		
1. b	1.1.3	1a
2. b	1.2.0	1b
3. b	1.3.1	1c
Section Two		
1. a	2.1.2	2a
2. b	2.2.3	2b
3. d	2.3.2	2c
4. b	2.4.0	2d
Section Three		
1. a	3.1.0	3a
2. b	3.2.0	3b
3. b	3.3.0	3c
Section Four		
1. b	4.1.0	4a
2. a	4.2.3	4b

MODULE 03409

Alternative and Specialized Heating and Cooling Systems

Source: artas/iStock/Getty Images

Objectives

Successful completion of this module prepares you to do the following:

1. Identify and describe various alternative heating and cooling systems.
   a. Identify and describe common solid-fuel heating equipment.
   b. Describe waste-oil heating equipment.
   c. Recognize and compare passive and active solar heating systems.
   d. Describe evaporative coolers and the various applications.
2. Identify and describe various specialized heating and cooling systems and equipment.
   a. Recognize and describe the operation of direct-fired makeup air units.
   b. Describe computer room cooling systems and equipment.
   c. Identify and describe enclosure- and spot-cooling equipment.
   d. Describe the application of valance and chilled-beam cooling approaches.
   e. Identify and describe air-turnover and air-displacement systems.

Performance Tasks

This is a knowledge-based module. There are no Performance Tasks.

Overview

Alternative heating and cooling methods continue to evolve. As the demand for these systems increases, technicians can expect to encounter alternative systems and equipment in a variety of applications. This module explores alternative fuels and methods of providing comfort and also examines some of the unique applications of HVACR principles.

Digital Resources for HVACR

Scan this code using the camera on your phone or mobile device to view the digital resources related to this craft.

SCAN ME

1.0.0 Alternative Heating and Cooling Systems

Performance Tasks

There are no Performance Tasks in this section.

Objective

Identify and describe various alternative heating and cooling systems.

a. Identify and describe common solid-fuel heating equipment.
b. Describe waste-oil heating equipment.
c. Recognize and compare passive and active solar heating systems.
d. Describe evaporative coolers and the various applications.

Alternative heating and cooling methods and systems can take many forms. Alternative heating and cooling approaches presented in this section include the following:

- *Equipment that burns solid fuel* — This category includes stoves, furnaces, and boilers. Fuels commonly burned include wood, wood pellets, coal, and even shelled corn.
- *Waste-oil heaters* — These units burn waste motor oil, used transmission and hydraulic fluid, waste vegetable oil, and other petroleum-based fluids.
- *Solar heating systems* — These systems should not be confused with solar photovoltaic (PV) systems that generate electrical power. Solar heating systems can heat air or fluids.
- **Evaporative coolers** — This form of cooling relies solely on the cooling effect that occurs when water is evaporated. In some climates, it can provide adequate cooling for a structure without additional assistance from a mechanical refrigeration system.

Evaporative coolers: Cooling units that cool air through the evaporation of water, typically consisting of a blower, wetted media, and a water source.

1.1.0 Solid-Fuel Equipment

Solid fuels such as wood and coal have been used to provide heat since prehistoric times. In the United States, coal and wood were widely used for home heating until the middle of the twentieth century. At that time, solid fuels were replaced by cleaner and more convenient fuels, such as natural gas and fuel oil.

By the 1970s, rising energy costs forced people to take another look at solid fuels as an energy source. As a result, solid-fuel stoves, furnaces, and boilers regained some popularity, especially in rural areas where wood is plentiful and inexpensive.

1.1.1 Wood-Fired Stoves

Figure 1 Installed wood stove.

Wood-fired stoves (*Figure 1*) and fireplace inserts are used to provide area heating, often to supplement a conventional heating system while adding ambience to the environment. Wood-fired stoves and similar equipment may burn chunks of wood, wood pellets, coal, or shelled corn.

Stoves are typically constructed of welded steel plates or cast iron. The combustion chamber may be lined with firebrick for enhanced combustion and heat retention. Modern stoves are designed to burn the fuel through controlled combustion. Combustion can be controlled using a manual air damper or a thermally activated air damper. Thermostatically controlled, electrically positioned air dampers and/or induced draft blowers are also an option.

Wood stoves and fireplace inserts sold in the United States since 1988 must meet US Environmental Protection Agency (EPA) smoke emission standards. Today, these units are often equipped with a **catalytic combustor**. The catalytic combustor is a coated ceramic honeycomb that captures smoke and other particles and provides a substrate for them to be burned again.

Catalytic combustor: A ceramic structure resembling a honeycomb that provides a substrate for the high-temperature combustion of smoke and particulates, resulting in a cleaner discharge and increased heat production.

Catalytic combustors do degrade and require periodic replacement. Furnaces equipped with a catalytic combustor also require more effort to start and stabilize. Smoke emissions for these units cannot exceed 4.1 grams per hour.

Non-catalytic stoves cannot emit more than 7.5 grams of smoke per hour. These units rely on a variety of construction techniques such as internal baffles and insulated fireboxes to enhance combustion and reduce smoke emissions.

Wood stoves that meet emission standards, catalytic or non-catalytic, can become EPA-certified appliances (*Figure* 2). Note, however, that some states and localities have enacted more stringent standards than those of the EPA. It is important to be aware of any such standards in your area.

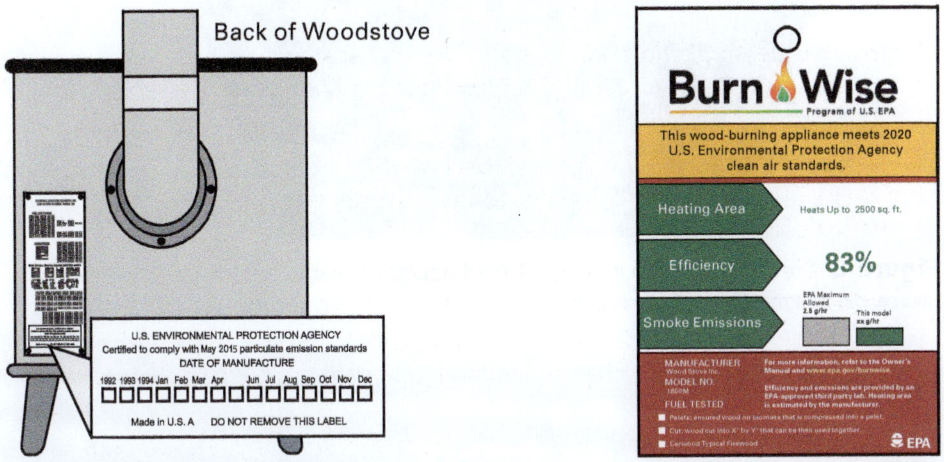

(A) Wood Stove Label **(B) Temporary Hang Tag**

Figure 2 EPA-certified wood stove label and temporary hang tag.
Source: United States Environmental Protection Agency (EPA)

On Site

EPA-Qualified Appliances

A significant difference exists between an *EPA-certified* appliance and one that is *EPA-qualified*. Wood stoves must meet specific emissions standards to be EPA-certified. These emission guidelines are not suggestions. To lawfully manufacture and sell the stove as an EPA-certified product, these guidelines must be met.

Fireplaces and outdoor, wood-fired boilers have no such EPA-required standards to meet. The EPA has established a set of voluntary emission standards for these appliances that manufacturers can choose to meet to enhance the manufacturer's status. If the voluntary standard is met, the appliance can become EPA-qualified.

Wood stoves provide heat through radiation. Some wood stoves can be fitted with an optional circulating fan that moves air over the surface of the stove to better distribute the heat, but they are not classified as furnaces.

Hardwoods and softwoods produce roughly the same heat per pound. However, hardwood is denser and provides more heat per volume of wood. Only properly seasoned wood should be burned in a wood-fired stove or furnace. When green or wet wood is burned, the released water vapor combines with other combustion products to form creosote in the chimney. The **creosote** buildup shown in *Figure 3* became a major fire hazard long before the photo was taken.

Creosote: A black, sticky combustible byproduct created when wood is burned, especially at lower temperatures, that collects on flue and chimney surfaces.

WARNING!

Creosote collects on the surfaces of a chimney and can ignite at high temperatures. Chimneys must be inspected and cleaned periodically, and users must be educated in the use of wood-fired appliances and how to avoid creosote accumulations.

Figure 3 Creosote buildup in a wood-fired appliance vent.
Source: digitalfarmer/iStock/Getty Images

Did You Know?

Seasoned Wood

Wood that is freshly cut is known as *green wood* and generally has a moisture content between 75% and 80%. Wood that has been allowed to air dry for a season is known as *seasoned wood* and generally has a moisture content of around 20%.

A number of factors affect how quickly wood dries naturally. These factors include the wood species, whether the logs are split or left whole, the ambient humidity, and how the wood is stacked.

On Site

Pellet Stoves

Special stoves are available that burn wood pellets or shelled corn. The pellets are made of compressed wood waste and sold in bags. The pellets are loaded into a hopper on the side of the stove and are fed automatically into the firebox as needed with a small auger. Today's wood pellet grilles use a similar approach.

Shelled corn can often be used instead of wood pellets. Both fuels are convenient to use and burn cleaner than wood chunks or logs.

1.1.2 Wood-Fired Furnaces

Wood-fired forced-air furnaces (*Figure 4*) are used as the primary source of heat in some structures. Some older wood-fired furnaces offered dual-fuel capability in a single unit, perhaps burning coal as well as wood. However, these furnaces no longer meet the EPA requirements.

Wood-fired furnaces can be installed as an add-on to an existing furnace or as a stand-alone source of heat. *Figure 5* shows how an add-on furnace is integrated with the ductwork of an existing furnace in a parallel arrangement. This arrangement requires dampers at the supply air discharge of each furnace. The dampers prevent air from flowing through the idle furnace when the other is in operation. Note that the two furnaces cannot be operated simultaneously.

Figure 4 Wood-fired forced-air furnace.
Source: Courtesy of HY-C Company

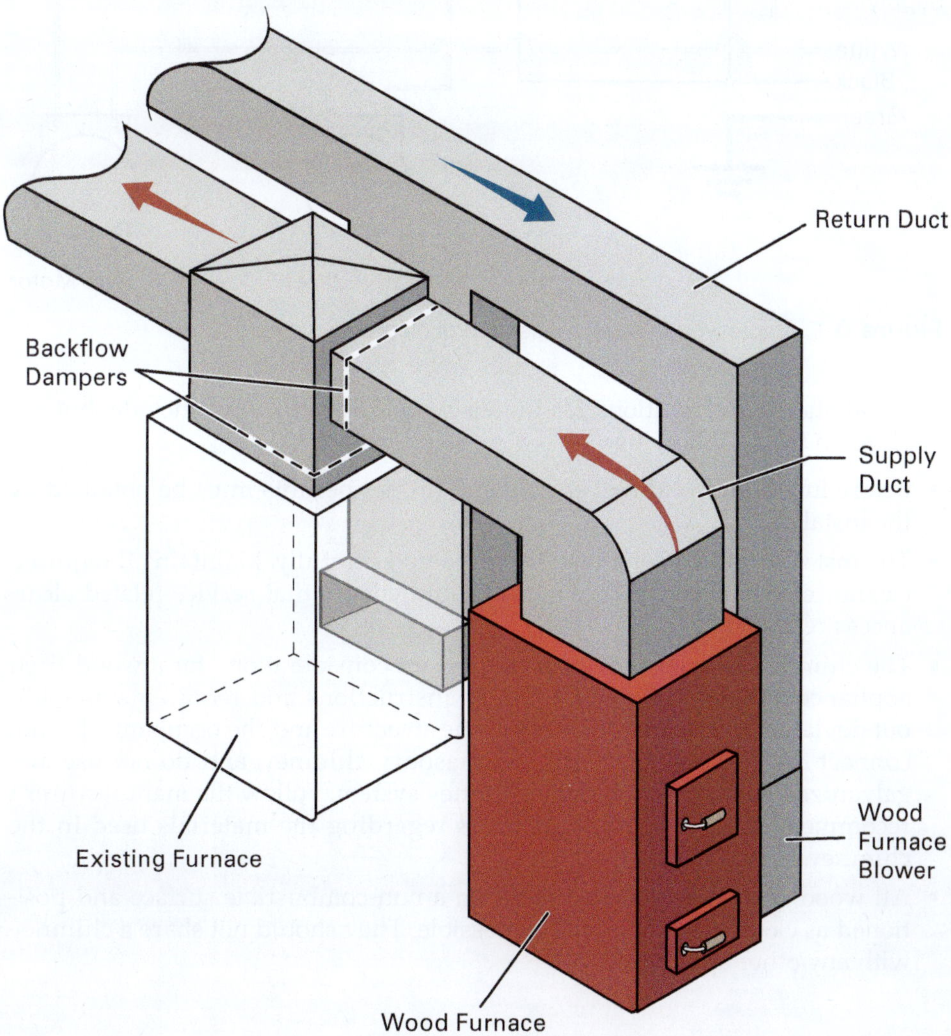

Figure 5 Add-on furnace in a parallel arrangement.

CAUTION

The ductwork configurations and control circuits used to parallel wood-fired appliances with other HVAC equipment can vary significantly. Before beginning an installation of this type, review the installation requirements and suggested control scheme carefully to ensure the system will operate safely and as expected. The nature of wood-fired equipment presents some unique challenges, including the fact that the combustion process cannot be stopped instantly.

Figure 6 shows an example control diagram for a wood-fired furnace. The draft inducer provides combustion air, but only when the thermostat calls for heat. Otherwise, combustion air is provided by natural draft to minimize the heat output and wood consumption. The indoor blower cycles on when the furnace temperature reaches the fan control setpoint.

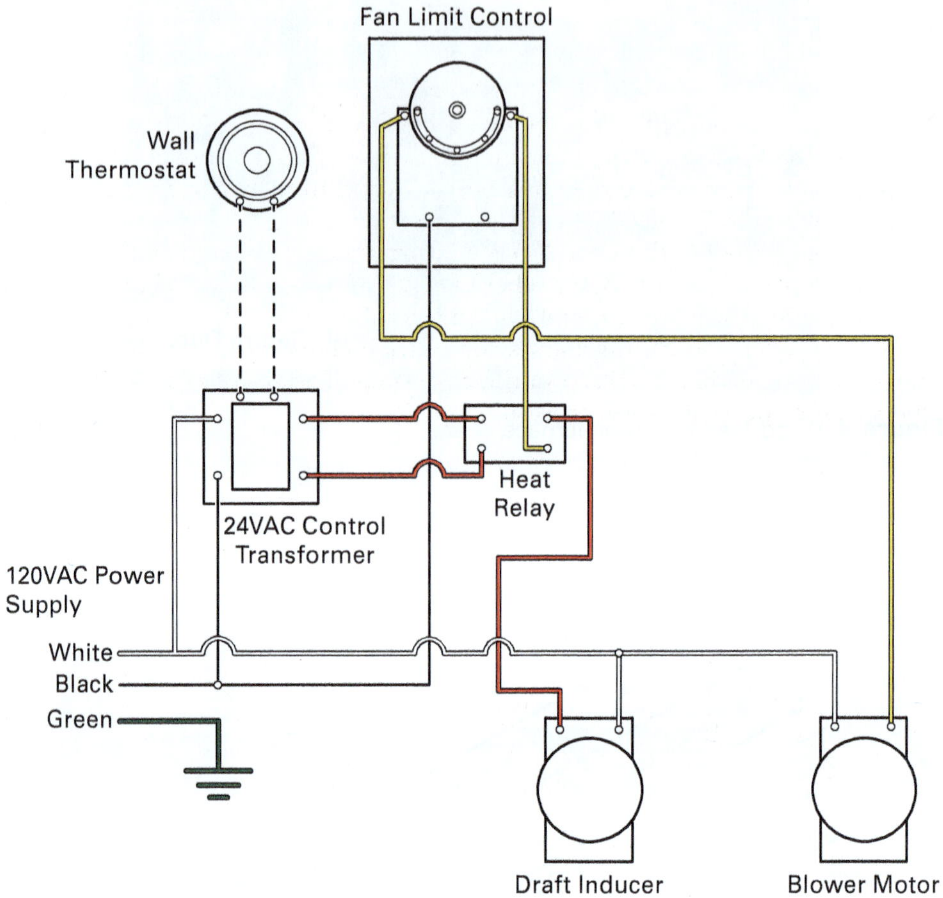

Figure 6 Example wood-fired furnace control diagram.

Installation considerations for indoor wood-fired furnaces include, but are not limited to, the following:

- Before installing solid-fuel equipment, proper permits must be obtained by the installer.
- The installation location must be considered carefully. Maintain all required clearances from combustible surfaces and ensure that service-related clearances are also met.
- The chimney is one of the most important considerations for a wood-fired appliance. Follow the manufacturer's instructions and requirements without deviation to ensure the safety of the structure and the occupants. Do not connect the furnace to an unlined masonry chimney, and do not use any galvanized steel material in the chimney system. Follow the manufacturer's recommendations and all local codes regarding the materials used in the chimney.
- All wood furnaces must be placed on a non-combustible surface and positioned as close to the chimney as possible. They should not share a chimney with any other appliance.

- Smoke and carbon monoxide (CO) detectors should be placed on all levels of a home.
- Combustion air is as important in wood-fired furnaces as it is with fossil-fuel furnaces. Ensure an ample supply of combustion air is available.
- Ensure that the user understands that only properly seasoned wood should be placed in the combustion chamber. Manufacturers often provide a long list of materials that must be avoided. Review the list carefully with the user.

Going Green

Renewable Energy Resource

Wood and corn used as fuel are considered renewable energy resources because these fuels can be replenished through natural processes. Oil and natural gas are not renewable fuels; once used, replacement requires centuries. The use of renewable energy resources reduces our dependence on oil and gas.

1.1.3 Wood-Fired Boilers

Wood-fired boilers (*Figure 7*) are often used as a primary source of heat and can also be used to supplement an existing fossil-fuel or electric boiler. An example piping schematic that incorporates both a wood-fired boiler and a gas-fired boiler is shown in *Figure 8*.

Note that the boiler shown in *Figure 7* incorporates **gasification** into its design. Gasification refers to a process where solid fuels such as wood and coal are heated to a high temperature but with minimal oxygen. This results in a flammable gas called *syngas*. A separate combustion chamber is used to ignite and burn the syngas mixed with additional oxygen. The result is a cleaner, more energy-efficient combustion process.

Gasification: A process that converts any carbon-based material into a fuel gas called synthesis gas, or syngas, that can also be burned, significantly reducing the emissions that would otherwise result from solid-fuel combustion.

Figure 7 Wood-fired boiler employing gasification.
Source: Tarm Biomass

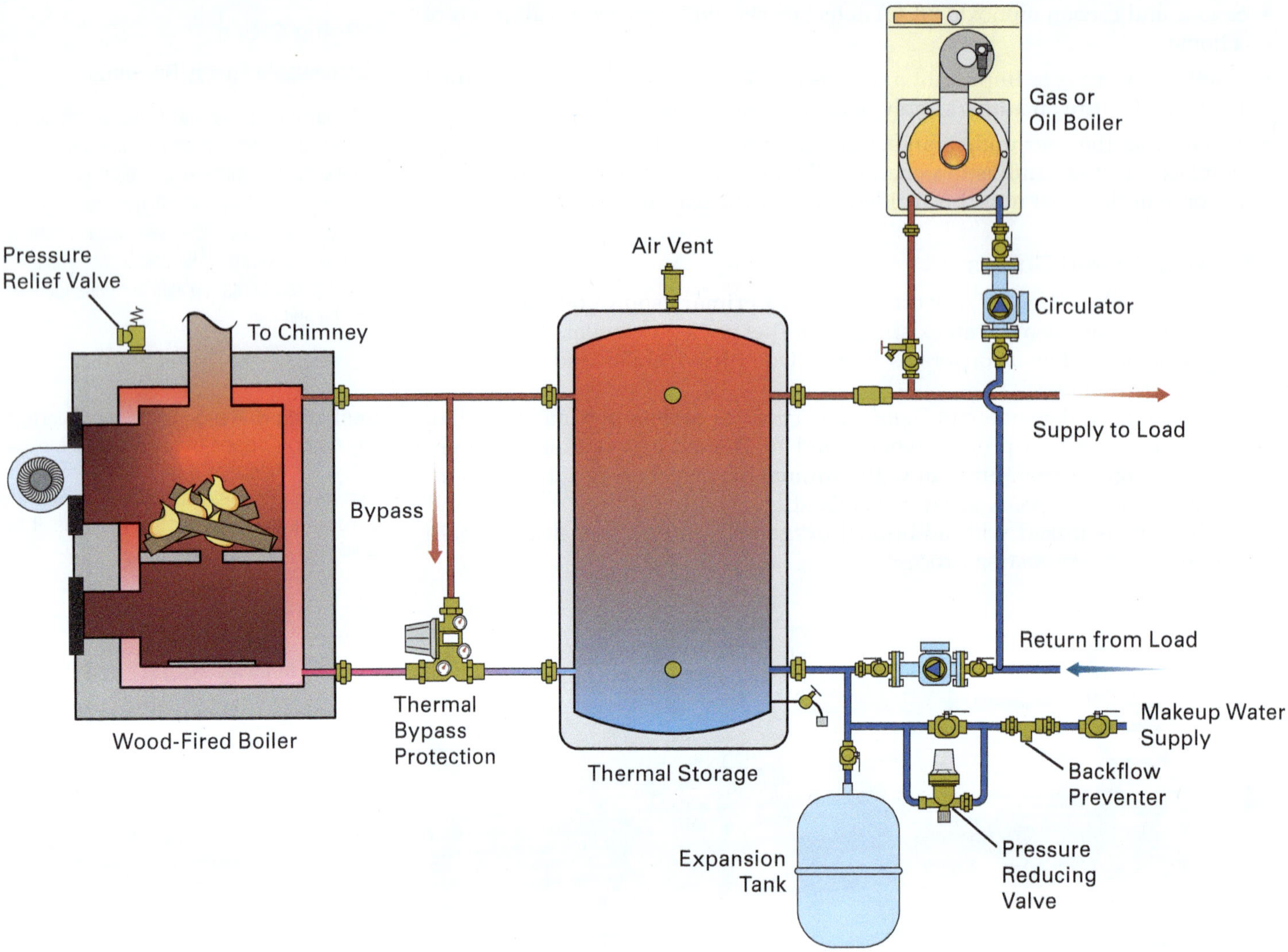

Figure 8 Wood-fired boiler piping with auxiliary fossil-fuel boiler.

The thermal storage tank shown in *Figure 8* increases the efficacy of wood boilers by storing hot water, thus storing heat for later use. Like wood-fired furnaces, these boilers can't be shut down quickly or cycled on and off like a gas appliance in a pattern that matches the heat loss. In other words, the consumption of fuel isn't easily controlled in sync with the load.

The thermal bypass valve shown in *Figure 8* acts as a protective temperature-control device, preventing the water in the boiler from reaching dangerously high temperatures by automatically regulating the water flow through it. The valve opens to allow more water to circulate when the water gets too hot, thus cooling it down, and closes when the temperature drops to a safe level.

On Site

Pellet Boilers

Like stoves, boilers that burn pellets and incorporate a self-feeding mechanism are available. The boiler shown here features automatic filling and automatic ignition. It also includes controls designed for easy integration into solar applications. Ash need only be removed every 300 to 500 hours of operation.

Source: Tarm Biomass

Wood-fired boilers can be installed outdoors in a self-contained weatherproof structure, complete with a chimney (*Figure 9*). The heated water is pumped through insulated pipes buried underground and connected to the existing hydronic heating system inside the structure. Advantages of this system are that it is cleaner (no ashes or wood debris inside the structure) and no new chimney or vent needs to be installed in the structure. However, it is important to remember that moving the boiler outdoors does not affect emission requirements. Many localities have stringent regulations regarding the placement and use of solid-fuel appliances of all types.

An outdoor dual-fuel, wood-fired boiler (*Figure 10*) combines a wood-fired boiler with a fossil-fuel burner in a single package. The fuel source can be chosen by the user as needed, and the boiler can be set up to automatically change to the auxiliary heat source when the wood is gone.

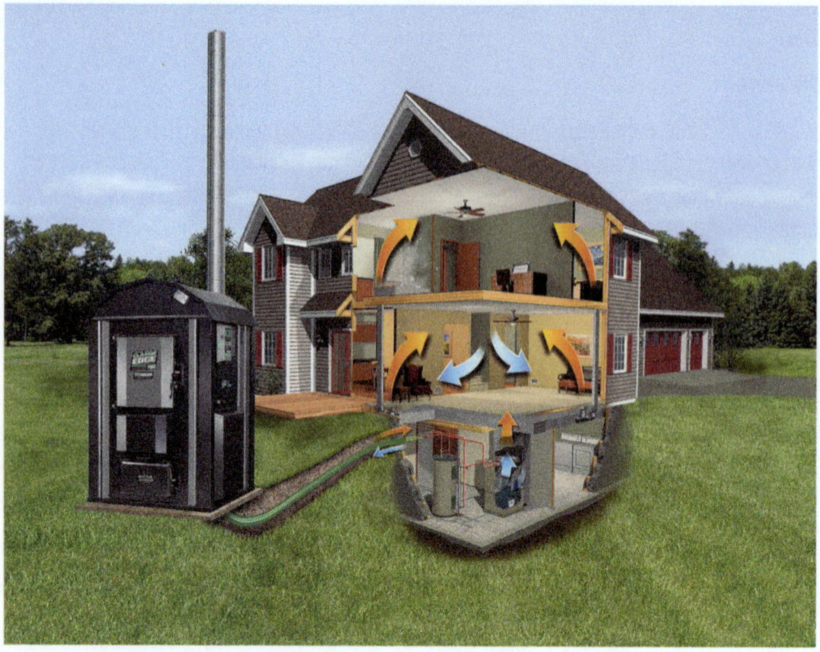

Figure 9 Outdoor wood-fired boiler.
Source: Central Boiler, Inc.

Figure 10 Dual-fuel boiler.
Source: Central Boiler, Inc.

On Site

Outdoor Boilers

Owners of wood-fired outdoor boilers can realize outstanding energy savings when they have a good source of inexpensive firewood. However, outdoor boilers are not without controversy. If unseasoned wood or different fuels, such as construction debris or old tires are carelessly burned, obnoxious smoke is produced. Since the chimneys on these boilers are relatively low to the ground, the smoke is released low to the ground. This has led to lawsuits that resulted in local ordinances to restrict or even ban the use of outdoor boilers. Restrictions include the size of the residential lot and minimal setback distances from lot lines for the boiler location.

1.1.4 Wood-Fired Appliance Installation and Maintenance

Heating appliances that burn wood or another solid fuel have some unique installation and maintenance requirements. Installation and maintenance considerations including the following:

- Clearance to combustibles
- Combustion air
- Venting
- Cleaning and maintenance

Safety and Service Clearances

Before installing any wood-fired appliance, always read the manufacturer's installation and service literature thoroughly. While many of the same considerations apply to fossil-fuel heating equipment, the requirements can differ significantly.

All fuel-burning appliances have specified installation clearances that must be maintained. For example, combustible materials must be kept away from certain portions of the equipment. Wood-fired appliances are no different. However, because wood-fired stoves and furnaces operate at higher temperatures than gas or oil furnaces, more clearance is generally required.

Wood-fired equipment must be placed on an approved, non-combustible base. When installed in a room, a wood stove must typically maintain 36" of clearance from the top, sides, back, and front to combustible surfaces. If the stove is equipped with an approved heat shield, these distances can potentially be reduced. This distance also allows room for service and maintenance.

The flue pipe of a wood stove, or *stovepipe*, must be at least three times the diameter of the pipe from a combustible surface. Thus, a 6" stovepipe must be at least 18" away from combustibles, per the example shown in *Figure 11*.

While these clearances are common, never rely on a textbook for that information. Check the manufacturer's literature so any specific clearance requirements also meet local codes. For the most up-to-date information on all aspects of installing wood-fired appliances at the national level, consult the National Fire Protection Association (NFPA) codes.

Combustion Air

Stoves and furnaces that burn solid fuel must be supplied with adequate air for combustion. In a loosely built structure, air entering through natural infiltration may be adequate to supply sufficient combustion air. If the structure is tightly built, outside air will likely have to be ducted into the structure to support combustion.

Outside air can be brought in through a duct into the area near the stove or furnace (*Figure 12*). The required duct size is typically greater than the size of the vent opening of the appliance. The combustion air supply should always be generous to ensure complete combustion and a good draft through the chimney. The appliance installation instructions will provide guidance in sizing. Note, however, that you must think carefully about where the combustion air duct terminates indoors, as it must be as far away from a system return air inlet as possible.

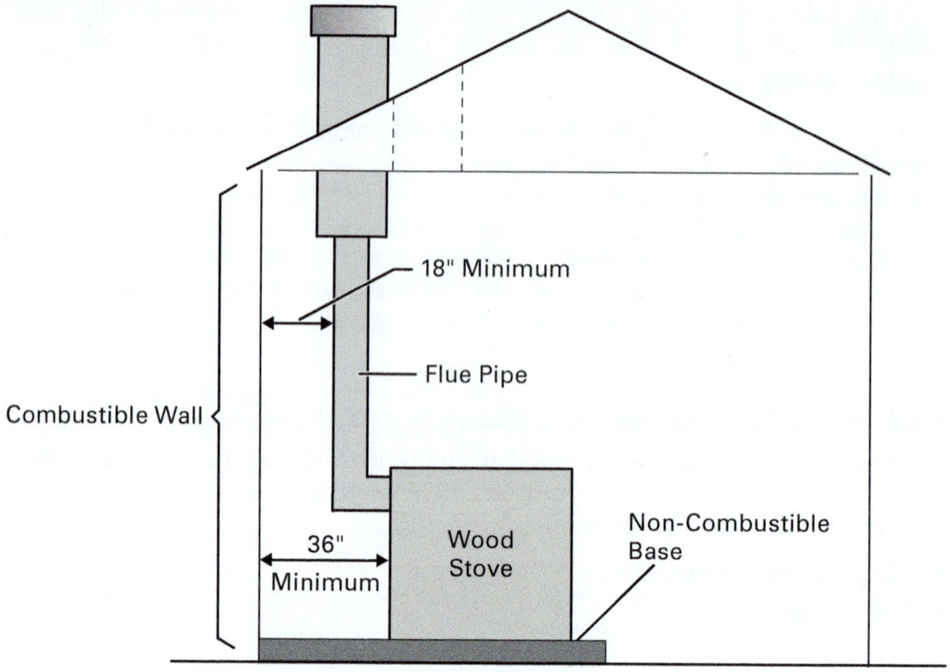

Figure 11 Common wood stove installation clearances.

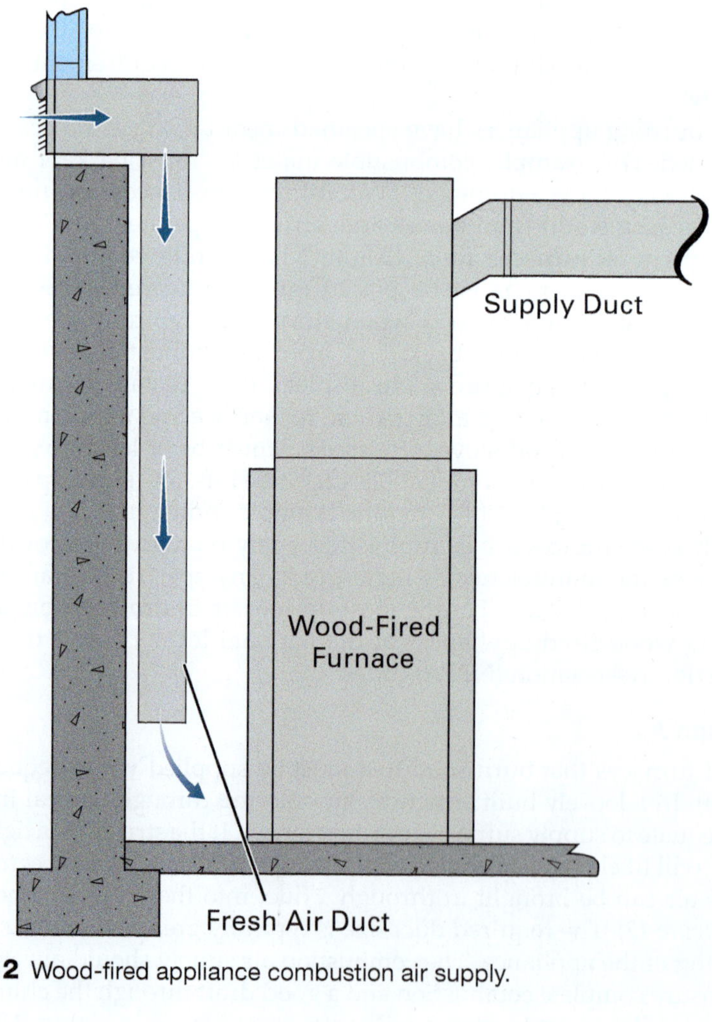

Figure 12 Wood-fired appliance combustion air supply.

If the appliance is installed in a confined space such as a utility room, combustion air intakes are generally required near both the floor and ceiling. Each grille must have a free air opening of 1 square inch (in^2) for each 1,000 Btuh of input capacity. For example, assume a combination wood/oil furnace has a maximum input capacity of 144,000 Btuh. Two grilles having at least 144 in^2 (1 ft^2) of free area each would be required. Be aware that free air area is not the same as the area of the face of the grille. Consult the grille manufacturer's specifications to see free area for any given grille.

Venting

Wood-fired stoves, furnaces, and boilers must be vented. The byproducts of combustion can be vented through a correctly sized, lined masonry chimney or through an *all-fuel* **Type HT vent** product (*Figure 13*). Type HT vents meet the requirements of UL 103, *Factory-Built Chimneys for Residential Type and Building Heating Appliances* and are rated to handle temperatures up to 1,000°F (538°C).

Type HT vent: A metal venting product capable of withstanding temperatures up to 1,000°F (538°C), commonly used to vent wood-fired stoves and furnaces.

WARNING!

Never vent a wood-fired stove, furnace, or boiler through a chimney or vent serving a fireplace or other fuel-burning appliance.

Figure 13 Type HT vent.

When installing a metal venting system, do not mix components from different manufacturers. Mixing components can create an unsafe condition. The rating of the assembled vent is only valid if rated components from the same manufacturer are used in a system.

Corn and wood pellet stoves generally do not have flue temperatures as high as those of wood-fired appliances. For that reason, stoves, boilers, and furnaces that burn only corn or wood pellets can sometimes be vented directly through a sidewall using the vent kit supplied by the equipment manufacturer. A special vent product, **Type PL vent**, is available for venting these specific stoves. A corn or pellet stove may be vented through an existing vent or chimney as long as it is not being used to vent another appliance and does not violate local codes.

Type PL vent: A metal venting product specifically designed for stoves that burn wood pellets or corn.

Cleaning and Maintenance

Wood-fired appliances should be cleaned and maintained in accordance with the manufacturer's instructions. Maintenance includes oiling motors, checking blower belts, and replacing air filters.

When wood burns, especially at lower temperatures, it creates creosote that builds up on the internal surfaces of the stove, furnace, and chimney. Creosote is combustible, and can ignite if left to accumulate. Once ignited, creosote can burn with extreme intensity, permanently damaging the stove and/or vent system and potentially causing a larger fire.

For that reason, chimneys and vents serving wood-fired appliances must be inspected and cleaned regularly (*Figure 14*). Since creosote buildup varies by intensity of use and the type of wood burned, there is no hard and fast rule for inspection intervals. It is not uncommon to clean a wood-fired furnace or boiler vent system every month or two if the equipment runs consistently.

Figure 14 Chimney sweep at work.
Source: Henfaes/iStock/Getty Images

On Site

Creosote Cleaning Products

Chemical cleaners are available that claim to clean the creosote from inside wood-fired stoves and chimneys. Often the cleaners take the form of powder that is sprinkled on the fire or sprayed up a chimney while a slight draft helps carry the powder to the top. A chemical agent neutralizes the acidic creosote, making it less corrosive. A dehydrator absorbs any oils contained in the creosote. A combustion inhibitor helps reduce the chance of chimney fires. Finally, a catalyst uses the heat to break down heavy creosote concentrations into ash.

However, some equipment manufacturers do not recommend the use of these products due to the potential for corrosion on the inner surfaces of the appliance. Also note that many of these products do not recommend their use in wood-fired equipment with a catalytic combustor. Note that the product shown here is not suitable for pellet stoves, although there are products designed for that specific application.

Source: Courtesy of Rutland

1.2.0 Waste-Oil Heaters

People today realize the importance and necessity of recycling and reusing resources to protect the environment. Heaters that burn waste oil are examples of products that use a resource that, in the past, was discarded.

Waste oil can come from a variety of sources. Automotive repair shops generate used motor oil, transmission fluid, and other petroleum-based lubricants and fluids. Restaurants and food processing facilities generate vegetable oils while cooking.

A waste-oil burner (*Figure 15*) is similar to a conventional gun-type oil burner. The burners can be used on furnaces and boilers. Because of the nature of the fuel, waste-oil burners are usually found in commercial and industrial applications where the waste oil is generated in the same location or by the same organization.

The quality and physical properties of waste oil vary significantly. Used motor oil is significantly different from used cooking oil. The fuel delivery system and the burner must be capable of handling these variables.

Although a variety of oil types can be burned, waste-oil burners need fuel at a consistent viscosity to fire properly. These burners have a built-in oil heater that preheats the fuel, keeping the viscosity more uniform. Compressed air, supplied by a built-in air compressor or by an external source, is also used to help atomize the fuel. The air volume requirement is relatively low, so the compressor is usually quite small. Otherwise, the installation and venting of waste-oil heaters is similar to that required for conventional oil burners.

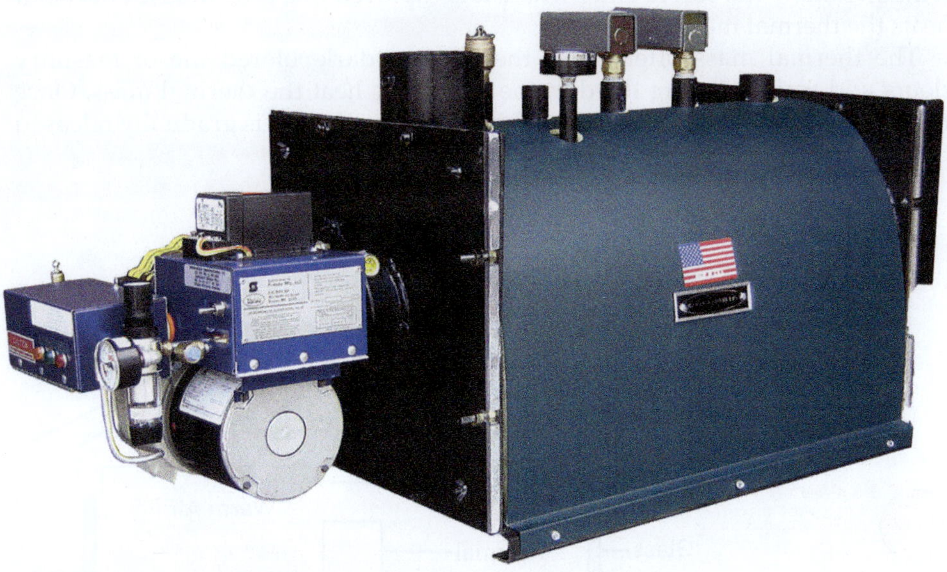

Figure 15 Waste-oil burner on a boiler.
Source: Columbia Boiler Company

1.2.1 Waste-Oil Heating Issues

Burning waste oil can provide significant cost savings but also presents challenges. The EPA has rules that cover the burning of waste oil. For example, regulations govern the conditions under which waste oil containing specific contaminants can be burned. Local codes may also regulate or restrict the burning of waste oil.

Waste oil can contain different contaminants that can affect its usefulness as a fuel. Once combusted, the contaminants become airborne. Used motor oil can contain water, gasoline, antifreeze, and traces of heavy metals. Used refrigeration oil can contain refrigerant and acids, while used cooking oil can contain water and food particles. The volume of a given contaminant in used oil can significantly affect its suitability as a fuel.

Burning a variety of fuels, along with the potential contaminants contained, usually results in decreased intervals between scheduled maintenance and more frequent equipment faults. The challenges as well as the potential cost savings should be carefully weighed before installing a waste-oil heater.

1.3.0 Solar Heating Systems

NOTE

This module focuses on solar hydronic heating. Solar photovoltaic (PV) systems that generate electrical power are not included in this discussion.

Harnessing the energy of the sun to provide heat is an important part of our energy future. It is estimated that each square mile of Earth receives enough solar energy to generate up to 5 billion kilowatt-hours (kWh) each year.

Solar systems fall into two categories: passive and active. Solar heating systems are popular in southwestern parts of the United States where winter sunshine is plentiful.

1.3.1 Passive Solar Heating Systems

The family cat is aware of passive solar heating. On winter days, the cat can often be found stretched out on the carpet in front of a south-facing window. The rays of the sun warm both the carpet and the cat.

Passive solar heating systems: Solar heating arrangements characterized by a lack of moving parts, pumps, and controls.

Passive solar heating systems (*Figure 16*) capture and store solar energy. Simple passive systems require two major components: large, south-facing windows, and a thermal mass to collect and store the heat. In the United States, the sun is low in the southern sky during the heating months. Large south-facing windows allow the sun to enter where it is captured in a greenhouse effect and heats the thermal mass.

The thermal mass often takes the form of dark-colored tile or masonry floors and walls. During the day, the sun's rays heat the thermal mass. Once the sun goes down, the heat retained in the thermal mass is gradually released into the space.

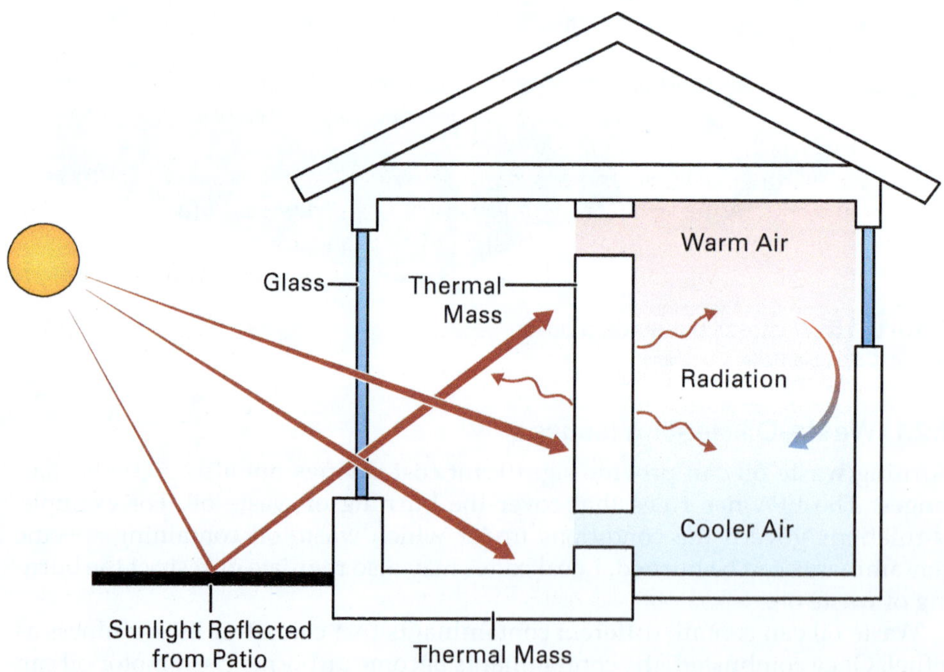

Figure 16 Passive solar heating system.

In some schemes, the thermal mass is a wall filled with water. The floor can also serve as the thermal mass if the sun can be directed to it. A common feature includes thermal drapes that automatically close when the sun goes down to prevent heat loss through the windows. Some passive designs are laid out so that the thermal mass induces gentle convection airflow throughout the room or structure. Too many architectural variations on the basic design exist to discuss all of them in this module. The one feature passive designs share is that no mechanical apparatus is required.

Passive solar hydronic heating systems are also available. In this scheme, fluid from a roof-mounted solar collector moves naturally to a storage tank, then back to the collector. A **thermosiphon system** (*Figure 17*) is typical of passive systems. It operates by the different densities between the water in the storage tank and the collector.

Thermosiphon system: A system in which a difference in fluid temperatures in a system causes fluid movement, with warmer fluids moving toward the top of the piping circuit.

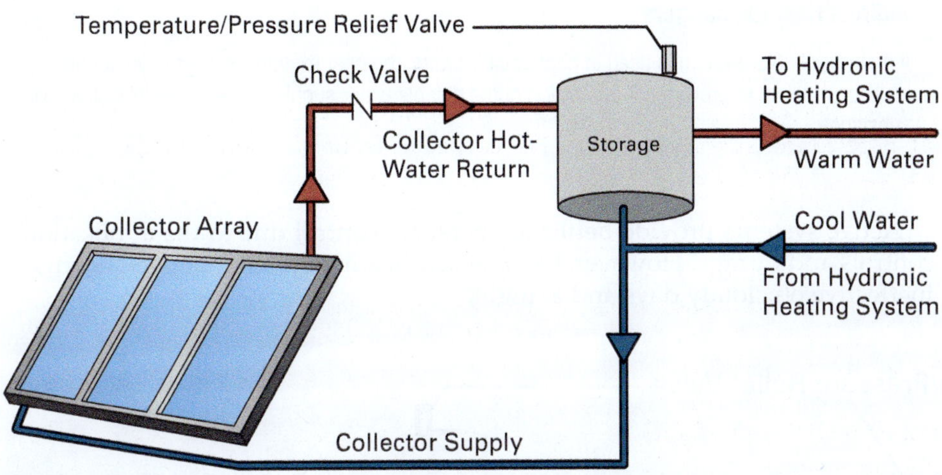

Figure 17 Passive solar thermosiphon hydronic heating system.

Cool water in the storage tank and hot water in the collector accelerates the flow of the fluid. Hot water in the storage tank and cool water in the collector results in little or no flow. This latter condition could possibly lead to *reverse thermosiphoning*. A check valve must be installed to prevent this backward flow of water through the system, as shown in *Figure 17*.

Thermosiphon systems are passive and do not have temperature controls, so the water temperature in the system will vary with changes in the weather. The warm water in the storage tank is best used to preheat water in a conventional hydronic heating system with a boiler, rather than act as a primary source of heat or domestic hot water.

Simplicity is the major advantage of passive systems. However, the lack of direct operational control is a significant disadvantage.

1.3.2 Active Solar Heating Systems

Active solar hydronic heating systems are more complex than passive systems. Active systems rely on pumps, valves, and other devices to initiate and control the flow of fluid through the solar collector and the rest of the system.

An **indirect solar hydronic heating system** uses an antifreeze solution of water and glycol as the circulating fluid in the system. The heated antifreeze solution is then passed through a heat exchanger where it transfers heat to water that is circulating through the hydronic heating system and/or domestic water piping and terminals. The water in the solar collector loop remains isolated from the primary hydronic system.

Active solar hydronic heating systems: Solar heating systems that pump fluids through exposed solar collectors to absorb heat and are equipped with controls.

Indirect solar hydronic heating system: An active solar hydronic heating system that heats fluid using a double-walled heat exchanger to prevent potentially toxic fluid in the solar collectors from contaminating a potable water system.

WARNING!

Ethylene glycol is a toxic chemical. Wear all required PPE and follow safe handling procedures when handling ethylene glycol.

Figure 18 shows a typical indirect system. The storage tank shown would commonly be piped in series with an existing hydronic system boiler or domestic water heater. The boiler may operate very little or not at all when the sun is shining. On cloudy days or at night, when little or no preheating occurs, the boiler must manage the load.

On Site

Indirect Heat Exchangers

If an indirect solar system is used to heat potable water, the heat exchanger must have double-wall construction to prevent contamination of the potable water supply with ethylene glycol, a toxic substance.

Active systems provide better temperature control due to the application of controls and pumps. However, like all solar heating systems, active systems are ineffective on cloudy days and at night.

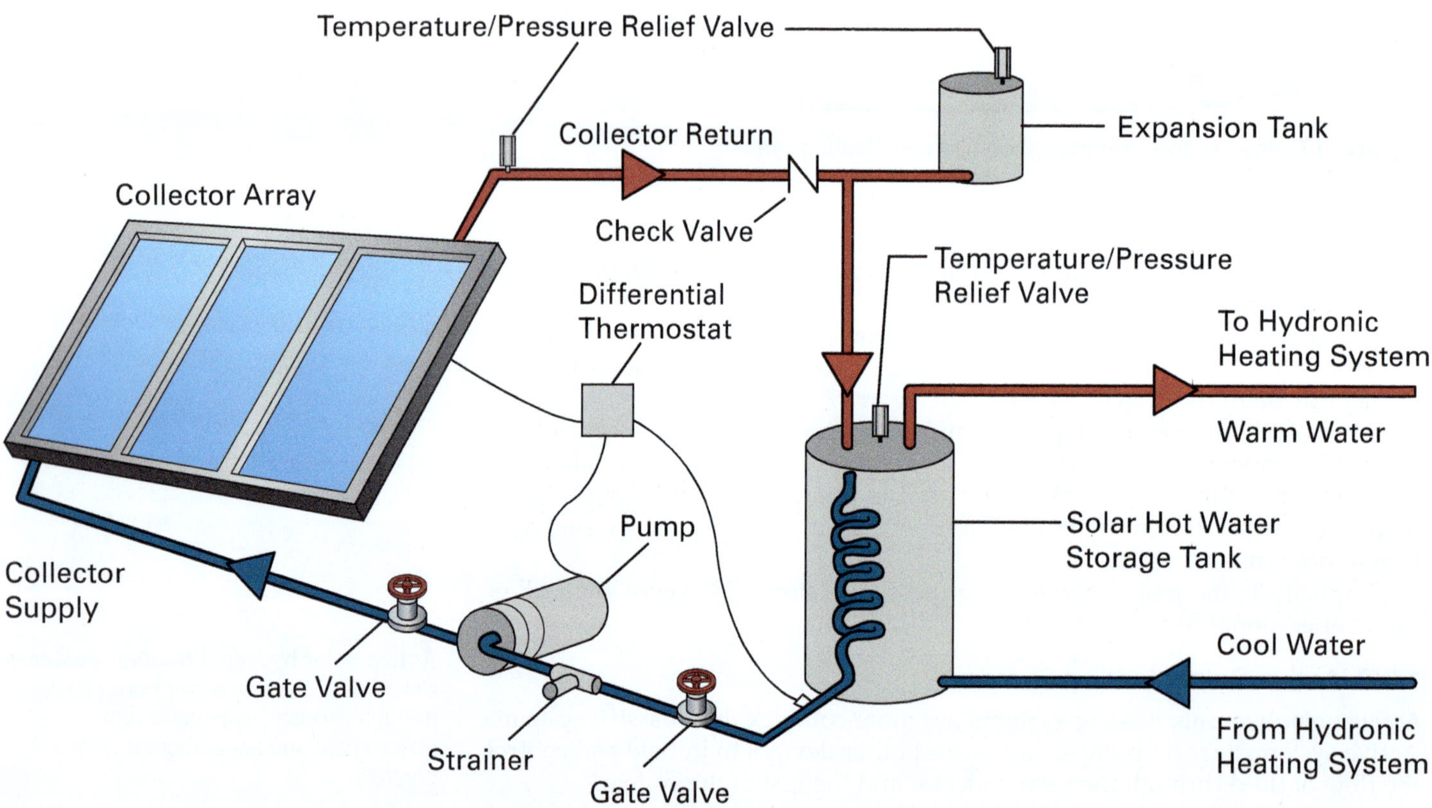

Figure 18 Active solar indirect hydronic heating system.

1.4.0 Evaporative Coolers

Evaporative coolers, often called *swamp coolers* (*Figure 19*), are used in some parts of the country for comfort cooling. Evaporative cooling offers lower operating costs than conventional air conditioners, because there are no significant electrical loads beyond the blower motor.

Evaporative coolers are very simple in their construction. These units operate on the principle that, as water is vaporized, heat is absorbed from the surrounding air. *Figure 19* shows both side-discharge and vertical-discharge units.

Figure 19 Evaporative coolers.
Source: Courtesy of Essick Air Products

Figure 20 Evaporative cooler media.
Source: iodrakon/iStock/Getty Images

A basic evaporative cooler consists of a louvered cabinet containing a blower assembly, water-absorbing pads, a water pump with a water distribution system, and a float valve to control the water level in the sump. The pump distributes water, keeping the pads wet. The blower assembly draws hot air across the wet pads, where the water evaporates and cools the passing airstream. Moisture is also added to the airstream, increasing the humidity in the conditioned space.

The pads used in evaporative coolers (*Figure 20*) were traditionally made of shredded aspen wood. Today, pads are available that are made of cellulose or man-made fibers that last longer.

Many evaporative coolers do not need a return duct because outdoor air is generally used. To prevent the structure from being pressurized, windows or doors must be opened, or a relief duct must be provided to relieve the pressure. When there is no means for the air to easily escape the building, the pressure indoors increases, and the airflow is significantly reduced.

Evaporative coolers are most effective when the outdoor air is very dry. That is why they are popular in the hot, dry climates of the southwestern United States. When the relative humidity outdoors is high, evaporative coolers become less effective because the air cannot absorb additional moisture. In those situations, a conventional air conditioner is needed for comfort cooling. Evaporative coolers are also used in greenhouses to maintain a high wet-bulb temperature while lowering the temperature inside, keeping the relative humidity high.

Portable models (*Figure 21*) are also used for spot cooling applications in hot work areas. A water source, garden hose, and power cord complete the setup. Other applications for evaporative cooling include commercial kitchen makeup air systems.

Figure 21 Portable evaporative cooler.
Source: Courtesy of Essick Air Products

Sometimes evaporative coolers are installed so the supply duct is shared with a conventional air conditioning system. Manual isolation panels or positive-closure dampers are used to prevent the conditioned air from one system being passed through to the other system.

The major advantage of evaporative coolers is the low installation and operating cost compared to a traditional air conditioner. The blower motor and water pump consume much less power than a compressor and condenser fan motor. However, evaporative coolers do consume water and increase the humidity in otherwise dry spaces.

Disadvantages include poor performance when the humidity is high, as well as the increased maintenance that is often associated with constantly wetted surfaces. Biological growth must be controlled, and the media requires periodic replacement.

1.0.0 Section Review

1. True or False? The catalytic combustor of an equipped wood stove captures and collects smoke particles in a special filter that is replaced following each burn cycle.
   a. True
   b. False

2. To maintain uniform viscosity of the fuels, a waste-oil heater may be equipped with a ______.
   a. catalytic combustor
   b. viscosity adapter
   c. fuel heater
   d. special filter

3. The thermal mass in a passive solar heating system is likely to be constructed of ______.
   a. tile or masonry
   b. glass
   c. wood
   d. sheet metal

4. Evaporative coolers perform well in greenhouses because they ______.
   a. maintain a low relative humidity
   b. maintain a high relative humidity
   c. reduce the wet-bulb temperature
   d. increase the dry-bulb temperature

2.0.0 Unique Heating and Cooling Systems

Performance Tasks

There are no Performance Tasks in this section.

Objective

Identify and describe various specialized heating and cooling systems and equipment.

a. Recognize and describe the operation of direct-fired makeup air units.
b. Describe computer room cooling systems and equipment.
c. Identify and describe enclosure- and spot-cooling equipment.
d. Describe the application of valance and chilled-beam cooling approaches.
e. Identify and describe air-turnover and air-displacement systems.

Many types of HVACR systems might be considered unique. It would require several volumes to cover all in depth. *Cascade systems*, for example, are special refrigeration systems that provide extremely low temperatures. Cascade systems may be used to cool product test chambers for electronic device testing, for example. These systems are comprised of two independent refrigeration systems operating on different refrigerants.

One system, the low-stage system, is used to provide the lowest evaporator temperature required: −50°F (−46°C), for example. The evaporator of the second-stage, or high-stage, acts as a refrigerant-to-refrigerant condenser for the low-stage system. The second-stage system has a more typical air- or water-cooled condenser. Different refrigerants are used in the two stages to create the right conditions. However, systems like this are not often encountered.

Some specialized systems will be encountered more often than others. Systems discussed in this section include the following:

- Heated makeup air units
- Computer room cooling systems
- Enclosure- and spot-cooling equipment
- Valance and chilled-beam systems
- Air-turnover systems

2.1.0 Direct-Fired Makeup Air Units

In many commercial and industrial applications, large quantities of air must be mechanically exhausted to purge harmful contaminants, smoke, and odors. For example, air near a loading dock can be contaminated by vehicle exhaust. A second example is an industrial paint booth, which requires a great deal of air to be exhausted.

Many possible industrial situations require high volumes of exhaust air, and the exhausted air must be replaced. Further, this makeup air must often be conditioned before entering the space to maintain reasonable comfort and process temperatures. Without mechanical equipment to replace exhaust air, significant negative pressures in the building result. Unconditioned outside air then enters without control through any available opening.

In many of these applications, cooling is not a requirement. In many locations, and in many different industrial applications, an indoor temperature equal to the outdoor temperature is acceptable. In those cases, unconditioned outdoor air is simply provided to replenish the exhausted air. However, in cold climates, the outside air needs to be heated.

When ensuring that the makeup air is filtered and warmed is a priority, a **direct-fired makeup air unit** (*Figure* 22) is used. The unit replaces the exhausted air, using 100% outside air, and heats the air as it passes through. These units are typically gas-fired, but they do not contain a heat exchanger or even a flue vent. The unique feature of the direct-fired unit is that the combustion process takes place directly in the primary airstream. All the byproducts of combustion remain in the airstream entering the building.

Direct-fired makeup air unit: A furnace that heats outdoor air and replaces indoor air that is exhausted through ventilation systems, helping maintain building pressure while warming the space in wintry conditions.

It is essential that the combustion byproducts be sufficiently diluted to ensure the safety of building occupants. Direct-fired systems are specifically designed with this key requirement in mind, and the level of combustion byproducts is sufficiently diluted by the high volume of air flowing through them. The velocity of the airstream across the burner must generally be in the range of 3,000 fpm to satisfy regulations. Precision static pressure switches are used to ensure that the airflow through the unit stays within the acceptable range for safety.

One of the principal advantages is that these units achieve 100% combustion efficiency since no heat is lost through a flue vent. However, direct-fired units are not 100% thermally efficient. On average, about 8% of the heat energy is lost to the moisture-related latent heat load of the outdoor air. When the outside air is very dry, their thermal efficiency rises.

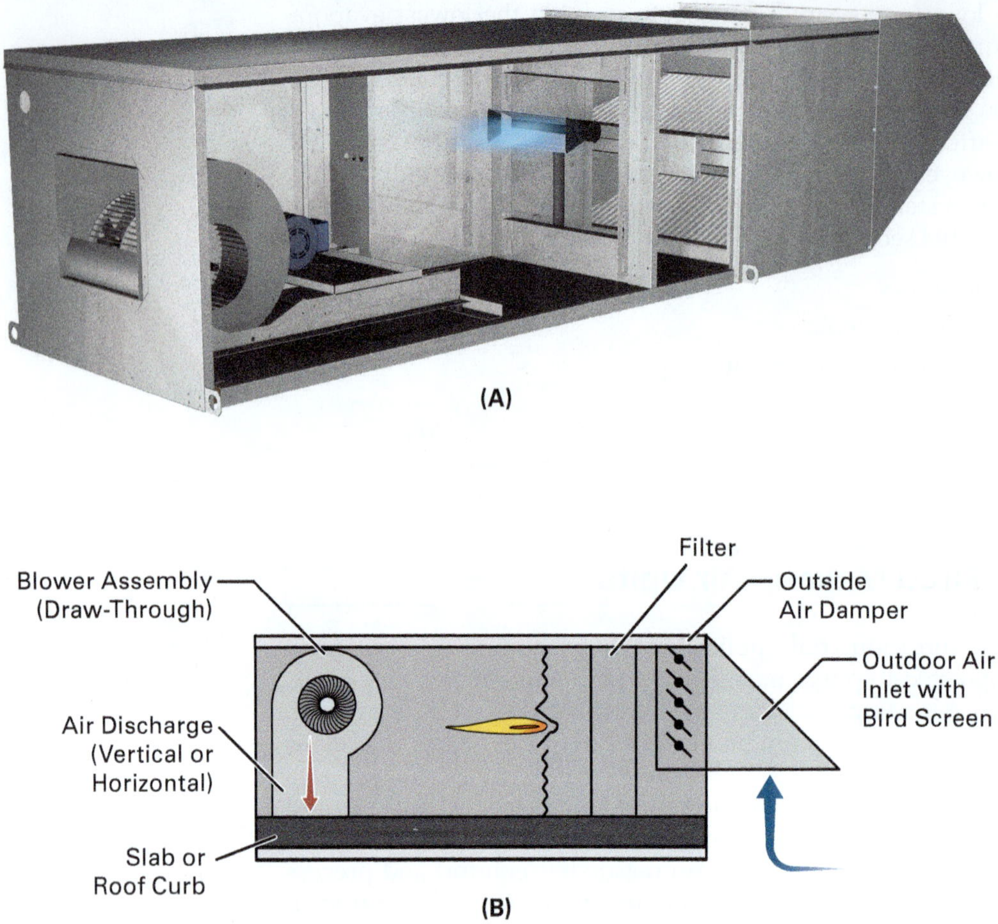

Figure 22 Direct-fired makeup air unit.
Source: Image courtesy of Greenheck (A)

> **On Site**
>
> **Indirect-Fired Makeup Air Units**
>
> Since the discussion here focuses on direct-fired systems, you may be asking yourself, "What about indirect-fired systems? Is there such a thing?"
>
> In fact, all common furnaces are indirectly fired. A heat exchanger separates the combustion process and its byproducts from the conditioned air supply.
>
> Some makeup air units do incorporate a heat exchanger, making them indirect fired. However, these units are simply large versions of common furnaces that are designed to operate on 100% outside air. This often requires a stainless-steel heat exchanger for durability.

Direct-fired units are generally controlled by a discharge-air temperature sensor and use a modulating gas burner. Unlike a typical residential furnace, which usually produces supply air temperatures of 110°F (43°C) or more, the discharge temperature of a direct-fired unit is often set at, or slightly above, the desired room temperature. However, direct-fired units are typically able to operate at a temperature rise ranging from 60°F to 120°F (16°C to 49°C).

Another common system feature is a control interlock between the makeup air unit and the related exhaust system. If the exhaust system is not in operation or fails, the makeup air unit is automatically shut down as well. Direct-fired units can be ducted or designed to simply blow freely into the space served.

Figure 23 shows a typical direct-fired burner assembly. Most utilize a cast-iron pipe as a manifold, drilled to incorporate the gas orifices, and stainless-steel

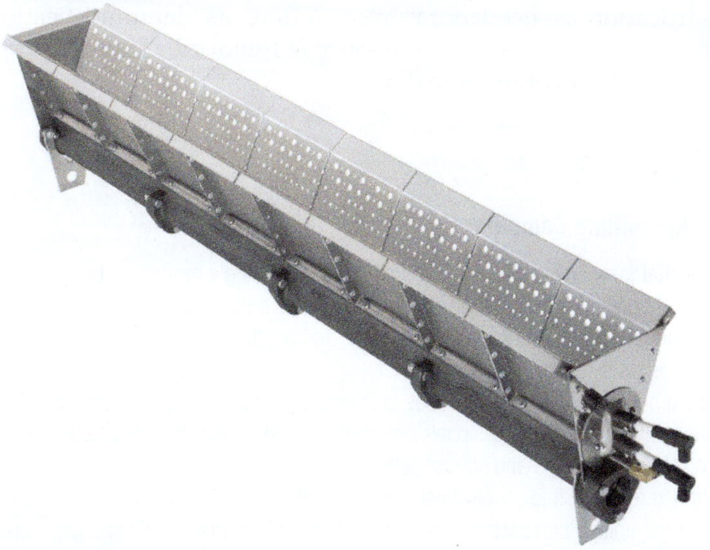

Figure 23 Direct-fired burner assembly.
Source: Midco International, Inc.

baffles. The baffles are also known as mixing plates. The burner must be designed to prevent the airstream from disturbing the combustion process, while simultaneously collecting sufficient combustion air to support the flame.

In applications that require the makeup air to be cooled and/or dehumidified as well, direct-expansion (DX) or chilled-water cooling systems can also be incorporated into the direct-fired makeup air unit. These are essentially rooftop units that operate on 100% outside air. These units can maintain reasonably precise discharge-air conditions regardless of the outdoor conditions.

Since high entering air temperatures are typically imposed on these systems, as well as high wet-bulb temperatures, the evaporator coil and refrigerant circuit design are somewhat different than the typical DX system. Standard cooling systems may easily be overloaded with entering air temperatures exceeding 90°F (32°C) for extended periods of time, especially when the air is also moisture laden.

2.2.0 Computer Room Cooling Systems

Computers and other electronic equipment generate a great deal of heat that must be addressed. If allowed to overheat, computers will malfunction or fail entirely. For security reasons, computers and servers are often installed in dedicated rooms and closets. The various components of the system are often stacked tightly together, allowing a great deal of heat to build up in a very small area. For this reason, dedicated equipment is used to cool computers and similar electronic devices. The cooling equipment must be able to reliably handle a significant and constant cooling load.

In most computer and electronic cooling applications, the sensible cooling load represents a larger portion of the total load than in common comfort applications. Since computers and other electronic devices produce only sensible heat, the only latent load is that which comes from infiltration or from moisture added by humidification systems.

Maintaining proper and consistent humidity levels is also very important, because electronic systems are sensitive to damage from static electricity. For this reason, humidification equipment and dehumidification cycles are generally incorporated into the cooling systems. However, dehumidification is not needed very often if the controls are calibrated and set up properly. When computer room systems are found operating in and out of this mode regularly, it is often caused by an unnecessarily tight setting of the humidity deadband.

Humidification is needed far more often, as dehumidification naturally occurs in the cooling mode. As moisture is removed, it must be replenished to keep the air from becoming too dry.

Going Green

Controlling Multiple Computer Room Units

It is not unusual for many computer room units to serve a single room or data center, each operating on its own internal controls. As a result, it is also not unusual to discover that one unit is cooling, one is heating, one is adding humidity, and another is dehumidifying. Obviously, energy costs skyrocket as a result.

This happens because of differences in control calibrations and inconsistent unit setpoints. Information technology (IT) professionals are primarily concerned with the results and rarely understand HVACR systems and control settings.

For these reasons, it is usually best to use a centralized control system to coordinate the operation of the units. Such systems can pay for themselves very quickly in energy savings.

Did You Know?

Dehumidifiers

For dedicated dehumidifiers using a DX evaporator coil, the coil is designed to maximize the latent heat exchange, which results in moisture removal. Unlike computer room coils, a dehumidifier coil is much thicker than a basic evaporator coil. This maximizes the time that the passing air spends in contact with the coil surface, resulting in a greater volume of condensate.

One important difference between standard DX cooling systems and those used specifically for computer rooms and data centers is found at the evaporator coil. The evaporator of computer room systems is typically designed to deliver as much sensible cooling as possible while minimizing moisture removal. This is done by increasing the surface area of the coil while decreasing the thickness. As air passes through the thinner coil, it has less time in contact with the coil surface, and therefore less moisture condenses. Although condensate is not eliminated, it is minimized this way.

The cooling load in a computer room or data center is typically consistent year-round, so computer room systems must be designed for operation at low-ambient conditions. This includes some form of head pressure control for winter, as well as crankcase heaters for compressors. Refrigerant migration outdoors during the off-cycle must also be prevented, generally by using a liquid-line solenoid valve that closes at the end of an operating cycle. Capacity control through compressor staging, compressor unloaders, and/or hot gas bypass is also a common feature.

Servers and other electronic equipment are often found in large racks. As a result, most of the heat generated in a computer room or data center is generated in small areas. Simply cooling the entire room from above is not as effective as delivering the air to these heat-generating locations. In addition, computer rooms and data centers are constantly growing or being rearranged. As a result, server racks grow in size and number and are often relocated. The air-distribution system must be flexible enough to accommodate the constant changes.

2.2.1 Computer Rack and High-Density Cooling

Small computer rack systems, isolated in small rooms, represent an interesting problem. The facility cooling system rarely has the capacity and ductwork to provide the cooling needed in these areas. In addition, the facility systems may be operating in a heating mode during the winter months, while the isolated racks continue to generate heat. In most cases, a rack system located in a small room needs a dedicated system to maintain reasonable operating conditions.

One way to do this is by using a dedicated rack-cooling system. The rack unit shown in *Figure 24* provides built-in chilled water cooling. Fans move the air across chilled-water heat exchangers inside the rack, transferring the heat to the water and out to a remote chiller. No compressor is in the enclosure; chilled water is provided by a separate chiller.

The fan speed modulates depending on the cooling load. The unique design isolates the heat exchanger from the electronics to avoid the significant damage that could occur in the event of a water leak.

Figure 24 Water-cooled server rack.
Source: nVent/Schroff

A liquid-coolant distribution unit (*Figure 25*) is often used to support direct liquid cooling of computer chips and processors. The liquid-coolant distribution unit provides a separate loop of cooling fluid, distributed to the individual processors through small tubing. An integral heat exchanger transfers heat to the facility chilled-water supply from the secondary loop.

Figure 25 Liquid-coolant distribution unit.
Source: nVent/Schroff

Refrigerant-based systems can also be used instead of chilled water. Indeed, many IT professionals prefer not to use the words "electronics" and "water" in the same sentence. One manufacturer makes special indoor modules that can be placed on top of a server rack. Liquid refrigerant at a low pressure and above the dewpoint temperature of the room is distributed to these units, eliminating the potential for condensation to develop inside the rack as the cooling process takes place. The liquid refrigerant absorbs heat and changes state to a vapor as usual, using this latent process to cool the air. No expansion device is needed at the cooling module to reduce the pressure and temperature of the refrigerant.

Because the refrigerant is above the dewpoint, it also results in a limited temperature difference between the air and the refrigerant. This approach works best by exposing the air to the refrigerant using a highly efficient microchannel evaporator coil, helping to compensate for the low temperature difference.

Small cooling systems similar to ductless units can also be used on server racks with a compatible indoor unit mounted on top of the rack. These systems may operate like any other refrigerant circuit, meaning that condensate may develop and must be drained away.

2.2.2 Open Data-Center Systems

Large, more expansive data centers (*Figure 26*) usually require a different strategy than systems designed for individual racks. Although one would think an air-distribution system for this type of facility would need to be complex, most are quite simple. The answer often lies in a raised floor.

Figure 26 Data center on a raised floor.
Source: NASA/Dominic Hart

Not only does a raised floor provide a means of delivering air throughout the room, but it also allows the air delivery to be very flexible. Looking again at *Figure 26*, note the tiled appearance of the floor. The floor is raised on legs connected by a metal grid. The grid supports 24" × 24" heavy-duty tiles (*Figure 27*). The floor provides an air-distribution plenum as well as an area for piping, power wiring, and network wiring. The height of the floor varies depending upon the needs and use of the room.

Looking at both *Figure 26* and *Figure 27* again, note the floor panels with perforations. These tiles act as the supply air diffusers for the system. By placing

Figure 27 Raised floor tile and removal tool.
Source: piyaphun/iStock/Getty Images

these tiles in the desired locations, cool air is provided to one side of the racks and moved around as needed to accommodate changing arrangements.

Hot/cold aisle configuration: A method used to install equipment cabinets in computer rooms to manage the flow of warm and cool air out of and into the cabinets.

Racks and cabinets are often arranged in what is known as a **hot/cold aisle configuration** (*Figure 28*). Most racks have small fans that draw air in through the front and discharge it through the back. Some electronic components also have their own fans built into the housing to pull air across the electronic devices.

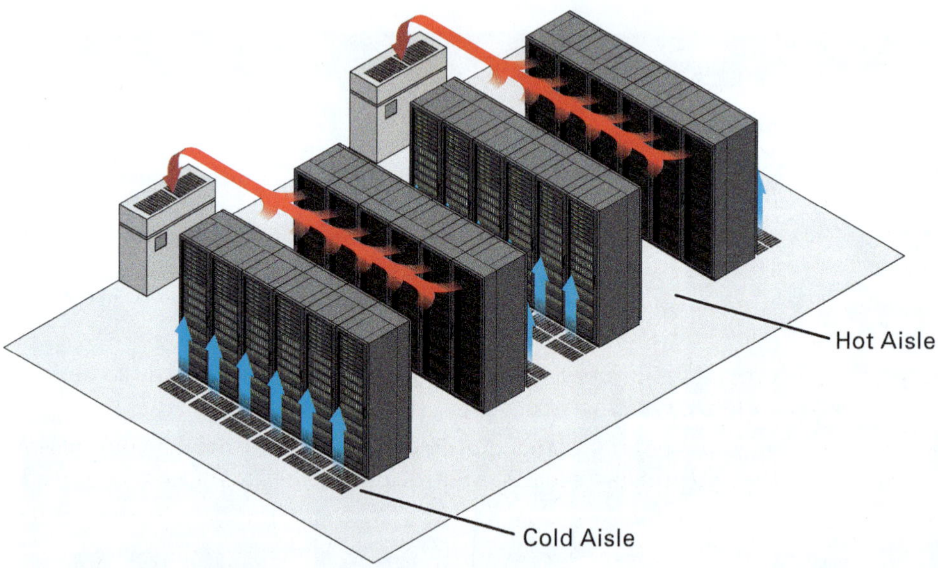

Figure 28 Hot-aisle/cold-aisle configuration.
Source: ZerLiew/iStock/Getty Images

The perforated floor tiles, acting as diffusers, are placed along the cold aisle to provide the conditioned air from the plenum below. In *Figure 29*, a ceiling has been installed and is being used as a return air plenum. In this example, the computer room unit (CRU) is placed outside the room.

In many installations, the unit is simply placed inside the room with the racks. Thus, return air can simply be drawn directly into the cabinet.

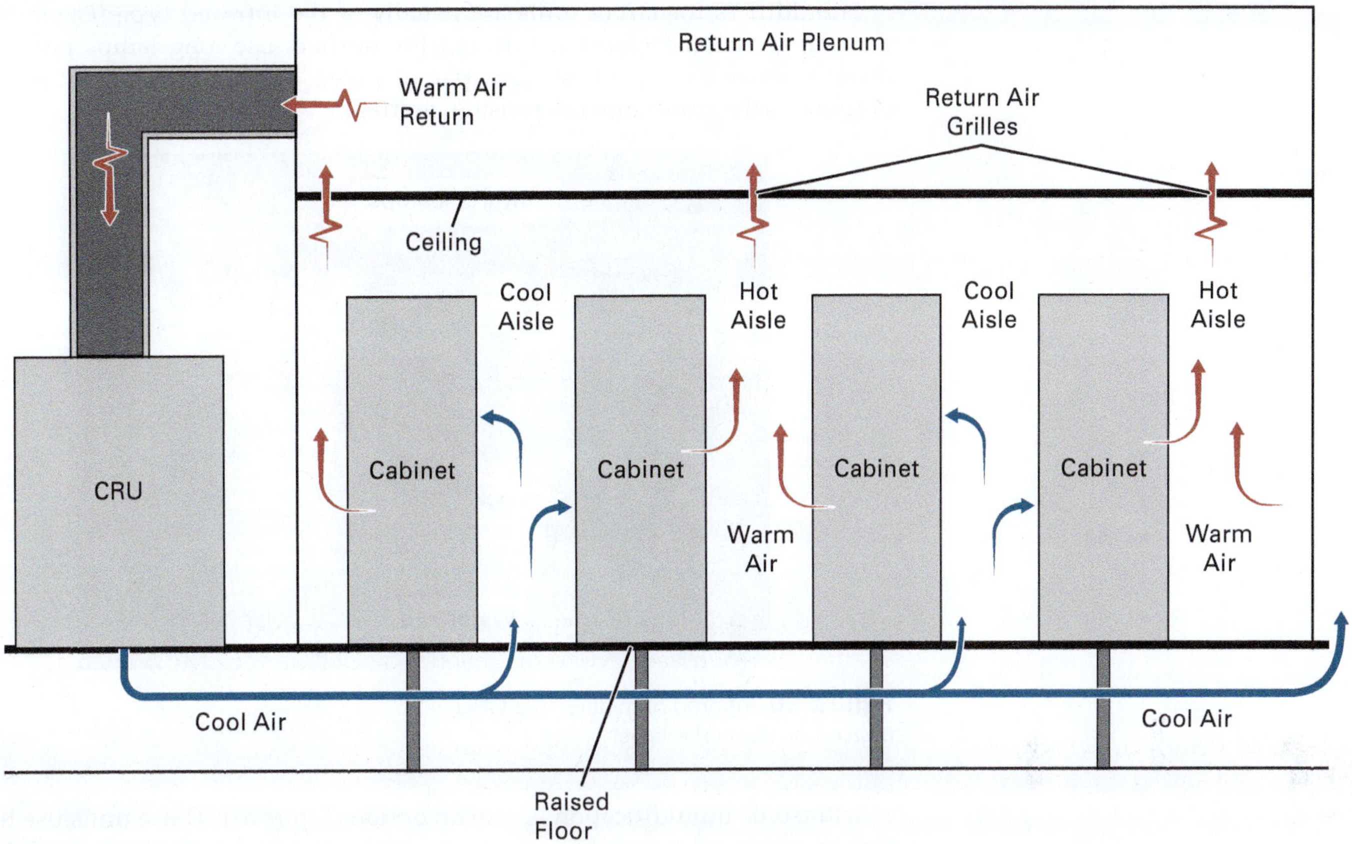

Figure 29 Hot-aisle/cold-aisle arrangement with the CRU located outside the room.

2.2.3 Computer Room Cooling Equipment

Computer room units are true environmental units. Sophisticated CRUs are typically equipped with the following:

- Precision temperature and humidity controls.
- Integral control panel with LCD/LED display or a graphical user interface (GUI).
- Belt- or electronically commutated motor-driven (ECM) blowers for flexibility in air volume adjustments.
- Electrode-type, steam-generating canisters or infrared humidifiers to be connected to an on-site water source.
- Multiple stages of cooling, using multiple compressors and/or compressor unloaders. For more precise capacity control, hot-gas bypass systems are used.
- Expansion valve metering devices. Newer models may be equipped with electronic expansion valves for even greater precision and energy efficiency.
- Electric reheat systems to maintain the room temperature setting when the unit operates extensively in the dehumidification mode.
- Heavy-duty condensate pumps that may also handle water that is periodically flushed from the humidifier.
- Integral air filter rack that can usually accommodate 2" or 4" filters.

Through the user control panel on the front of the unit, or through a building automation system (BAS), the operating parameters are easily programmed. Temperature and humidity setpoints are set, along with the range and sensitivity level of the setpoints. A variety of alarm setpoints can also be programmed. Of course, CRUs can also be networked and controlled as a group from a local or remote location.

Humidifiers for larger units are usually of the infrared type (*Figure 30*) or a steam-generating canister. Infrared humidifiers use long lamps positioned directly above a relatively shallow pan of water. The intense heat of the lamps evaporates the water into the passing airstream.

Figure 30 Infrared humidifier in a CRU.
Source: Courtesy of Humidifirst

Ultrasonic humidification is also an option (*Figure 31*). These units use high-frequency sound waves to excite water molecules and turn the water into an extremely fine mist. They typically use less energy and are very responsive, since the water does not need to become hot before moisture is vaporized.

Figure 31 Ultrasonic humidifier in a large air handling unit.
Source: Courtesy of Humidifirst

It is rare that CRUs are installed in a location that allows gravity drainage of the condensate. As a result, most are equipped with high-capacity condensate pumps. Due to the extended operating hours and the humidity levels that are maintained, these units can produce a lot of condensate. A pump is needed that is capable of pumping against a significant head while also moving a lot of water.

For obvious reasons, water is a great concern in the computer room and/or data center environment. A condensate pump switch is not always enough protection. The underfloor area under CRUs is often fitted with additional water protection. The use of water sensors under the floor (*Figure 32*), capable of sensing even a few drops of liquid, is a more comprehensive approach.

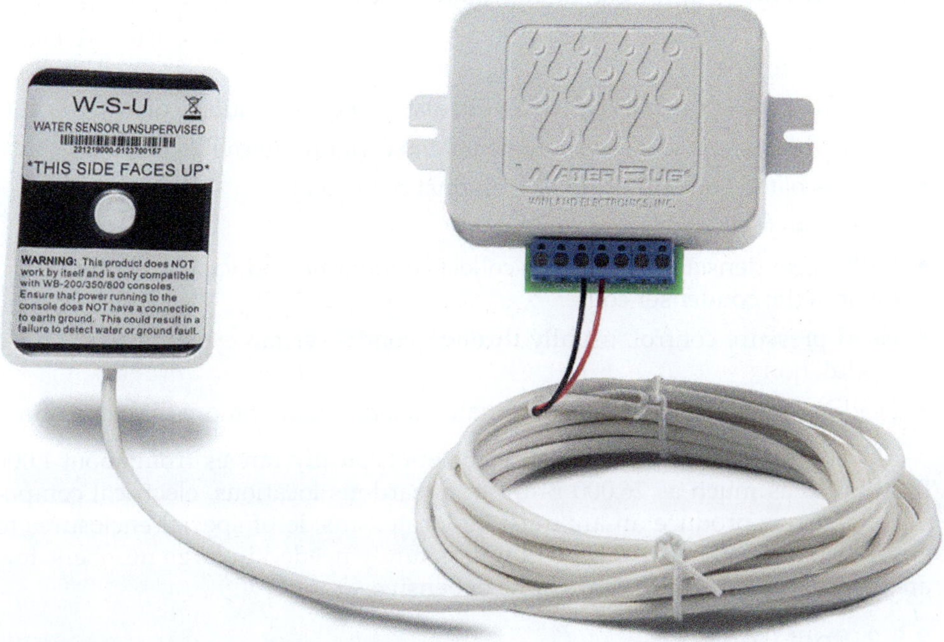

Figure 32 Underfloor water sensor.
Source: Courtesy of Winland Electronics, Inc.

This type of sensor is often placed in several locations, including under HVAC equipment and along the perimeter of the room. It can then provide protection against water seeping through the walls. Sensors can be networked and connected to a control system for instant alarm activation when water is present.

2.3.0 Enclosure and Spot Cooling Equipment

Many different heat-producing components are housed in enclosures, and for a variety of reasons. These components include contactors, relays, fuses, circuit breakers, and power disconnect switches. Transformers are also commonly found here, as well as any number of electronics devices. These cabinets, especially those that handle a considerable amount of power, can become extremely hot. To ensure reliability and longevity, the components inside the enclosures must be cooled.

Spot cooling equipment can be used in permanent applications for a problem area. However, spot coolers are more often used as an emergency back up to maintain reasonable conditions when an enclosure's primary cooling unit fails. Both enclosure units and spot coolers are introduced here.

2.3.1 Enclosure Cooling Equipment

Small self-contained air conditioning units are often used for enclosure cooling. Electronic cabinets related to cellular transmission sites are also candidates for enclosure cooling.

These small, compact units must operate in harsh environments that are dusty and hot, and are often needed where hazardous or explosive atmospheres are possible. To function in such environments, options such as the following are available:

- Electrical wiring that satisfies the requirements of National Electrical Manufacturers Association (NEMA) standards such as the following:
  - *NEMA 1* — indoor use only
  - *NEMA 12* — indoor but protected against light splashing of liquids and dust
  - *NEMA 3R* — rain-tight, indoor/outdoor use
  - *NEMA 4* — same as 3R, but also protected against dust and heavy wash-downs
  - *NEMA 4X* — same as NEMA 4, but also protected against corrosive agents
  - *NEMA 7* — explosion-proof and able to contain internal explosions
- Epoxy-coated coils for protection against corrosives
- Stainless-steel chassis and cabinet
- Built-in condensate vaporizer to collect condensate and vaporize it by exposing it to the condenser coil
- Head pressure control, usually through condenser fan cycling or fan speed modulation
- Lead-lag controllers, which coordinate the operation of more than one unit

The cooling capacity of enclosure coolers typically ranges from about 1,000 Btuh up to as much as 26,000 Btuh. In hazardous locations, electrical components that can produce an arc must be located inside of special enclosures to ensure isolation. Since this type of unit is rarely needed in large numbers, few are manufactured and they are quite expensive.

2.3.2 Spot Coolers

Spot coolers (*Figure 33*) are portable packaged air conditioners, usually on casters, that can be moved to an area to provide temporary or supplemental localized cooling. These units can be purchased or rented for emergency use.

(A) Portable Spot Cooler **(B) Condenser Air Duct Attched**

Figure 33 Portable spot cooler.
Source: MovinCool and Climate Pro are registered trademarks of DENSO Corporation.

Most of these units use air from inside the space as condenser air, which must then be discharged outside of the conditioned space. In *Figure 33B*, note the flexible discharge duct attached to the condenser air outlet. Discharging the condenser air back to the space works against the goal of cooling and often results in a warmer space due to the added heat of compression.

Larger, weatherproof spot coolers can be left outdoors and ducted into the space, avoiding this issue altogether. Portable models must be provided with flexible or rigid round duct routed outdoors or to another area where the added heat isn't an issue.

The flexible supply-air nozzles allow users to direct the cooled air onto a piece of critical equipment. Optional ducting kits allow more length and flexibility.

Condensate is collected in a tank at the base of the unit. The tank is fitted with a float switch to shut the unit down once the tank is full. The tank must then be emptied manually and reinstalled to restore operation. Another option is to hook a drain hose to the tank for consistent gravity drainage, perhaps to an available floor drain. A condensate pump kit is also an option.

2.4.0 Valance and Chilled-Beam Cooling

Valance and chilled-beam cooling approaches are rather unique answers to unusual situations. Valance systems, for example, are installed in applications where a conventional ducted air-distribution system is impractical. Chilled-beam systems offer another unique way to avoid an air-distribution system.

2.4.1 Valance Systems

Valance cooling system: A cooling system in which chilled water is circulated through finned-tube radiators located near the ceiling around the perimeter of a room. Convection currents move the cooled air instead of a blower assembly. A decorative valance conceals the system.

A **valance cooling system** is a chilled-water system that uses finned-tube radiation installed around the perimeter of a room, just below the ceiling and often concealed behind a decorative valance.

Each valance cooling unit consists of at least one section of finned-tube radiation (*Figure 34*). Each unit is connected to an insulated chilled-water supply

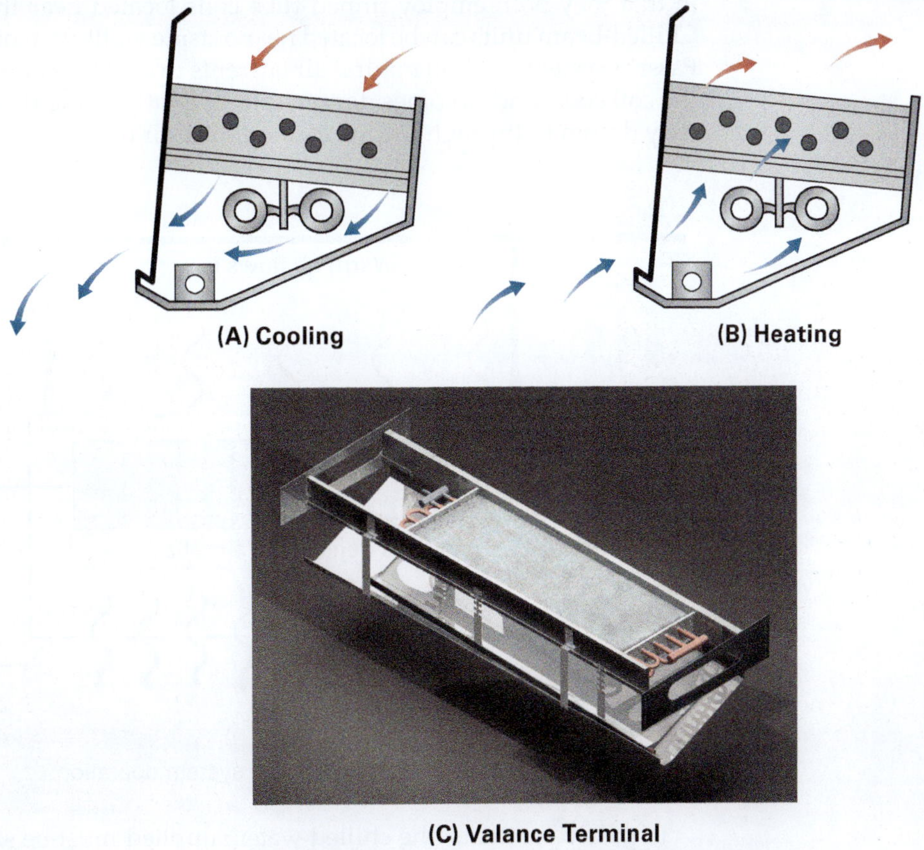

Figure 34 Valance heating/cooling terminal.

Source: Sigma Corporation (C)

and return line. A zone valve may be installed in the supply line. Beneath the finned-tube section is a pan to catch the condensate and drain it away for disposal. The condensate pan is often a part of the decorative valance. Valance units can also be connected to a hot-water system to provide heat, as shown in *Figure 34*, but performance is limited.

During cooling operation, chilled water flows through the coil on a call for cooling from the room thermostat. The first such signal received energizes the circulator pump and the zone valve, if equipped. As the air cools, the density increases and the air falls toward the floor. This sets up natural air currents in the room, with the dense falling air drawing warm air from the ceiling area through the coil. No circulating fans are needed or used.

Condensate must be collected and drained away. However, with proper engineering of the makeup air to the building and control of the chilled water temperature, condensate can be virtually eliminated.

Because valance systems do not use a circulating fan, the systems are very quiet, do not create uncomfortable drafts, and dust circulation is minimized. These systems provide energy-efficient air distribution because the circulating pump consumes less power than a blower motor. As with other types of hydronic systems, the installed cost of a valance cooling system is generally higher than a conventional forced-air cooling system.

2.4.2 Chilled-Beam Systems

Chilled-beam cooling systems: Cooling systems that employ coils mounted near the ceiling through which chilled water flows. Passive systems rely on convection currents for cooling, while active systems use mechanical ventilation to increase airflow over the beams.

Chilled-beam cooling systems have been used for years in Europe and Australia and are now used in the United States. Chilled-beam systems are similar in some respects to valance cooling systems. The name seems to imply that structural members of the building are cooled. In fact, the units used to cool the room are long, finned-tube radiators in an enclosure. They resemble beams suspended from the ceiling, thus the term *chilled beam*. Chilled-beam systems can be passive or active.

Passive chilled-beam systems (*Figure 35*) resemble valance cooling units in that they both employ finned-tube coils located near the ceiling of a room. Chilled-beam units can be located near outside walls or in other areas as needed. Passive systems rely on natural air currents to cool the room. Air in contact with the coil cools and becomes denser, falling gently to the floor. Warm air is naturally drawn in through the top as the cooler air drops.

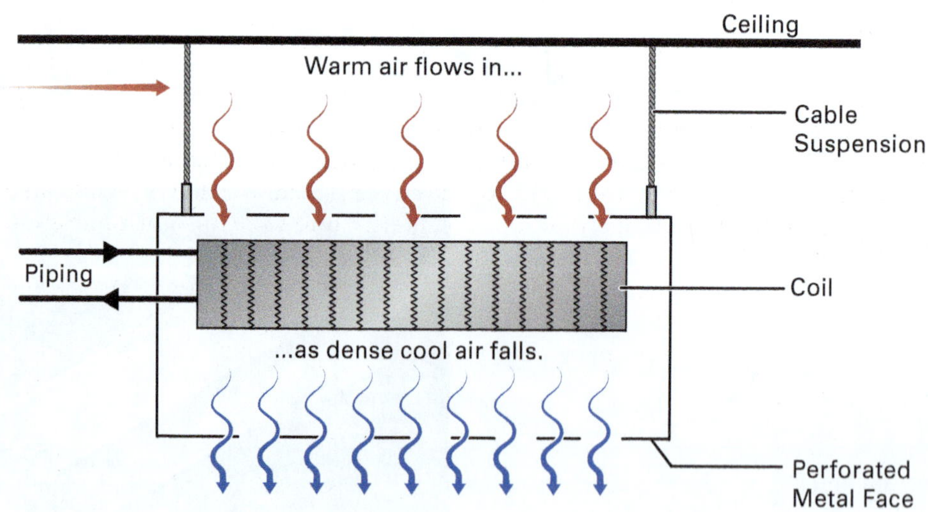

Figure 35 Passive chilled-beam cooling system operation.

In passive systems, the chilled water supplied must be slightly warmer than the dew point of the room air to prevent condensation from forming. This is the most significant difference between valance cooling and the chilled-beam approach. Since the apparatus can be suspended mid-room, there is no practical way to collect and dispose of condensate. Since the temperature difference between the chilled water and the air is relatively small, passive chilled-beam systems require more coil area to provide the same capacity as a valance system. However, as noted in the discussion of valance systems, they can be engineered to minimize condensate as well.

Active chilled-beam systems (*Figure 36*) use ducted, partially conditioned ventilation air to provide fresh air. The forced air, referred to as the *primary air*, is provided at a low pressure and discharged through jets or nozzles on the sides that increase the velocity. This creates a negative pressure in the center of the assembly, drawing in warm air through the center and across the chilled-water coils. The entering room air is referred to as the *secondary air*.

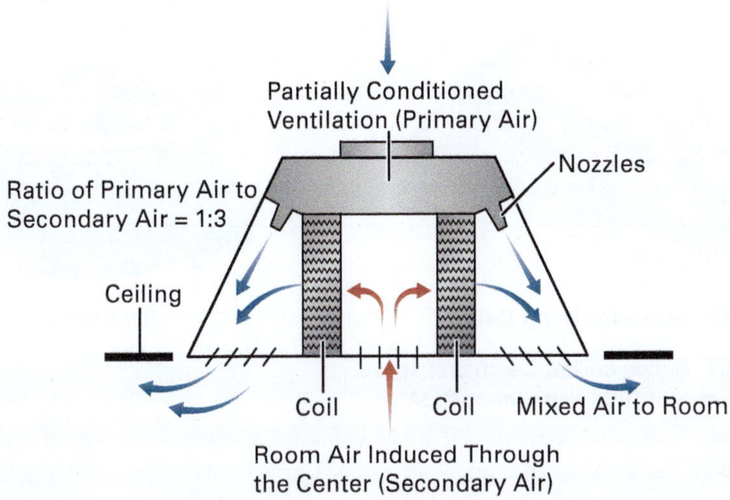

Figure 36 Active chilled-beam cooling system operation.

The primary air is partially cooled and dehumidified before it is sent to the chilled beams. This takes care of any latent load and prevents condensation on the chilled-beam coils. Thus, the function of the chilled beam is to satisfy only the sensible load in the room.

As shown in *Figure 36*, the ratio of ventilation air to induced air is around 1:3. The desired ratio determines the engineer's selection of coil fin spacing. Tightly spaced fins offer improved heat transfer but are more resistant to natural air movement.

Active chilled-beam fixtures can be suspended from the ceiling or tucked into the ceiling with the face flush. *Figure 37* shows various chilled-beam terminal styles. Note that heating can be provided through chilled beams as well. However, like valance systems, heating performance is less than ideal.

Chilled-beam systems are noted for energy efficiency. Active systems also offer quiet operation, excellent indoor air quality due to the use of ventilation air as a primary air source, and low maintenance.

Disadvantages include high installation costs compared to more conventional systems. A lot of insulated chilled-water piping is required. However, the advantages are attractive to building owners and developers seeking LEED certification.

2.5.0 Air-Turnover and Air-Displacement Systems

Stratification is possible in all indoor spaces, especially those with high ceilings and limited air movement. Warm, lighter air rises and tends to stagnate near the ceiling. Cool, heavier air moves toward the floor. These air masses develop naturally, creating different temperature zones and potentially result in discomfort. By preventing the air from stratifying, a more even temperature can be maintained throughout the space. Air distribution design strategies and forcing the air to move and mix using fans are the tools used to address stratification.

Stratification: The natural layering of undisturbed room air due to temperature differences. Warmer air is concentrated at the ceiling while cooler air is concentrated at the floor. Mechanical ventilation interrupts or eliminates natural stratification.

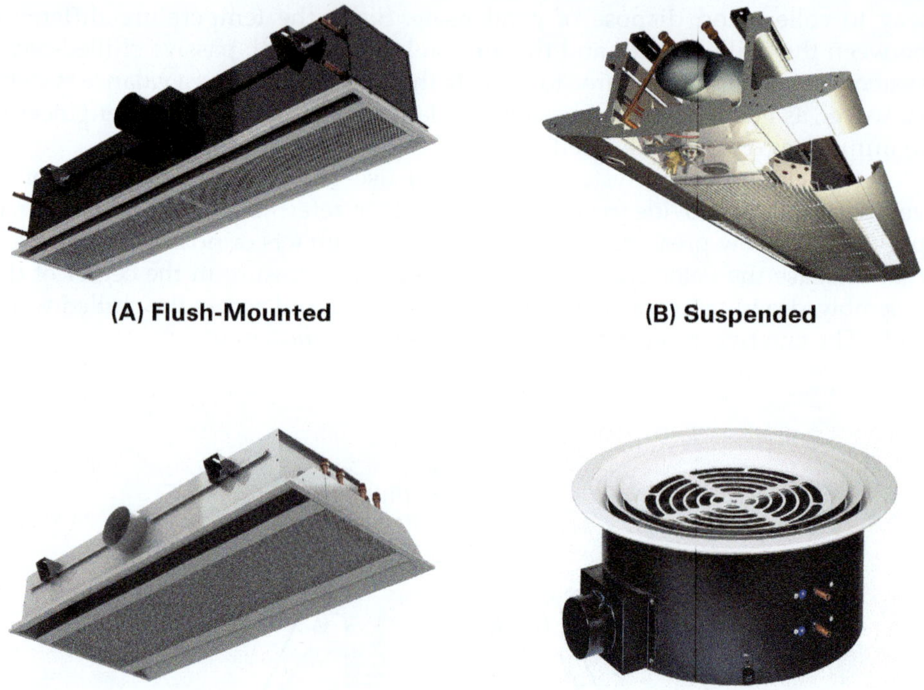

Figure 37 Active chilled-beam unit styles.
Source: Courtesy of TROX North America USA

On the other hand, stratification can also be an intentional strategy in some applications. Allowing a layer of hot air to collect and stagnate near the ceiling of a high manufacturing bay is of no consequence, and it reduces energy costs since that layer of air is not being conditioned.

Air-turnover systems are employed in large open spaces when stratification is undesirable. *Air-displacement systems* rely on stratification to reduce energy costs while still providing cooling where it is needed.

2.5.1 Air-Turnover Systems

One of the most efficient ways to prevent air stratification is to use ceiling-mounted air circulating fans, sometimes referred to as *paddle fans*. These fans have been used for over 100 years and are still used today in residential, commercial, and industrial buildings. While smaller paddle fans are sufficient for residential and commercial spaces, large-diameter high-volume, low-speed fans are preferable in high-bay industrial and commercial buildings because they efficiently move large volumes of air over expansive spaces. Fans like the one shown in *Figure 38* are available in sizes up to 24' in diameter.

Fans, though, can't produce conditioned air. To condition a large open space while also eliminating air stratification, an air-turnover unit is the answer.

Air-turnover units, also referred to as *air-rotation units*, are tall, vertical air handlers that can contain heating and/or cooling sections. The fans in air-turnover units gently circulate air, drawing air in at the floor level and discharging it at the top (*Figure 39*). Discharging the air at this height typically carries it over

Figure 38 High-volume, low-speed paddle fan.
Source: Courtesy of Big Ass Fans

Figure 39 Air-turnover concept for a warehouse.
Source: Courtesy of Johnson Air-Rotation® HVAC Systems

obstructions such as manufacturing operations and crowded warehouse storage racks. Propeller fans are generally used instead of centrifugal blowers since there is no duct to add resistance.

Air-turnover units (*Figure 40*) move large volumes of air continuously. This is the secret to their effective operation—moving all the air throughout a facility, from top to bottom, leaving no stagnant areas.

As the air passes through the unit, it is heated or cooled. Cooling can be achieved through DX coils or chilled-water coils. Heating is usually accomplished with gas burners, hot water, or steam. Gas-fired air-turnover systems are not direct-fired, since large volumes of outside air are rarely moved through the units. Thus, these systems use a traditional heat exchanger and flue piping. While most air-turnover units are located indoors, outdoor units are also available (*Figure 41*).

Figure 40 Air-turnover unit in a warehouse.
Source: Courtesy of Johnson Air-Rotation® HVAC Systems

Figure 41 Outdoor air-turnover units with close-coupled condensing units.
Source: Courtesy of Johnson Air-Rotation® HVAC Systems

Did You Know?

Taking Good Care of the Wine

The proper storage of wine as it ages is an interesting field of science. A certain amount of wine stored in barrels evaporates over time—as much as 12% of the contents may be lost to evaporation.

To limit the loss, wine in barrels needs to be stored in a very humid but cool environment. An area maintained at 55°F (13°C) and 80% relative humidity is considered desirable. However, mechanical cooling processes generally result in dehumidification, which works against the target conditions.

A famous California winery had literally hundreds of storage units used for aging. Rooftop-mounted cooling units were being used to maintain the proper conditions, but with limited success. Over $500,000 of product per year was being lost to evaporation losses alone.

Johnson Air-Rotation® proposed that air-turnover units be used and the organization guaranteed the results. How did that go, you ask? Wine evaporation was reduced by 75%, saving over $375,000 per year in evaporation losses at each of the company's storage units. Because the ductwork from the rooftop units was also eliminated, the horsepower requirement to move the needed volume of air was also cut in half.

Source: vanbeets/iStock/Getty Images

Applications vary for the use of an air-turnover unit. Warehouses, manufacturing areas, aircraft hangars, and gymnasiums are potential sites. A single, large air-turnover unit can replace a group of conventional rooftop units while also improving performance and energy efficiency. In addition, eliminating the physical load of many rooftop units from the roof structure can save substantially on construction costs.

2.5.2 Air-Displacement Systems

While air-turnover systems are used to prevent stratification, air stratification is sometimes employed as a load-reduction strategy. For example, a light manufacturing facility may have a roof height of 28'. If there is no concern for the conditions above 12', the air-distribution system—both supply and return ducts—can be dropped down to the 12' height.

This arrangement effectively allows hot air, which rises toward the roof, to stratify. This reduces the load on the cooling system, often reduces the required capacity of the equipment, and reduces operating costs. However, this particular strategy applies only to ducted systems. Unfortunately, lowering all the ductwork in this manner is usually a more costly approach.

In air-displacement systems, large quantities of supply air are discharged through special diffusers at low velocities near floor level (*Figure 42*). Thermal plumes drive the supply air up through the breathing zone, then returns or exhaust grilles at ceiling level remove heat and particulates.

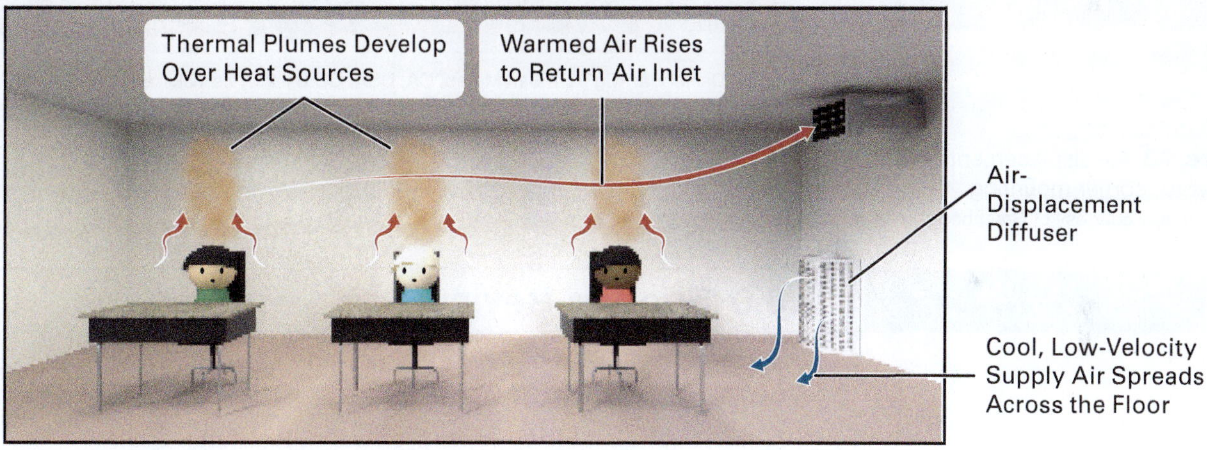

Figure 42 Airflow patterns in an air-displacement system.

The turbulence of conventional ventilation systems combines fresh incoming air with room air, resulting in a mixture of medium-quality air swirling throughout the room and toward the return air inlet. Displacement systems allow incoming air to gently displace the room air, creating two levels of *vertical stratification* in the room: fresh, cool air in the occupied zone, and warmer waste air in the unoccupied zone. The air only flows in one direction, from floor to ceiling, rather than moving about in a turbulent manner. The result is a higher-than-normal air turnover in the conditioned space, providing energy-efficient ventilation, improved air quality in the breathing zone, and thermal comfort.

Figure 43 shows an example of an air-displacement diffuser that would be placed in a corner at floor level. Models with a flat back are available for mounting along a wall. These diffusers are designed to deliver air at a low velocity that results in little or no turbulence.

Displacement ventilation works particularly well in large industrial applications with a significant heat load. It is also well-suited for theaters, concert halls, gyms, supermarkets, and educational facilities.

Figure 43 Air-displacement diffuser for corner mounting.
Source: Image courtesy of Greenheck

Around the World

Unique Fresh-Air Delivery Systems

Architects and HVACR engineers have become increasingly creative in their system design. For example, the Walt Disney Concert Hall in Los Angeles features supply air grilles under each seat that quietly and gently deliver fresh air using an air-displacement approach.

Source: Daniel Hartwig from San Mateo, CA, USA, CC BY 2.0, via Wikimedia Commons

2.0.0 Section Review

1. Which of the following statements about direct-fired makeup air units is *true*?
   a. Most direct-fired units use electric resistance heat.
   b. The byproducts of combustion remain in the airstream, rather than being exhausted through a flue pipe or chimney.
   c. Fuel oil is the most popular fuel for direct-fired units.
   d. The heat exchanger of a direct-fired unit is very complex and offers a high resistance to airflow.

2. Some rack-cooling systems in small computer rooms use pumped liquid refrigerant that is at a low pressure and _____.
   a. at a temperature above the dew point
   b. at a temperature below the dew point
   c. is superheated
   d. never changes its state

3. True or False? Equipment built to a NEMA 1 standard can be installed and operated in any location.
   a. True
   b. False

4. Valance cooling terminals _____.
   a. must be located above a drop ceiling
   b. are often installed in the center of the room
   c. are suspended around the perimeter of a room
   d. have no means of handling condensate collection or disposal

5. One common characteristic of an air-turnover unit is that it needs _____.
   a. both a supply and return duct
   b. only a supply duct
   c. only a return duct
   d. no ductwork at all

Module 03409 Review Questions

1. Wood stoves equipped with a catalytic combustor cannot have smoke emissions that exceed _____.
   a. 0.4 grams per hour
   b. 4.1 grams per hour
   c. 7.5 grams per hour
   d. 10 grams per hour

2. True or False? When a wood-fired furnace is added on to an existing fossil-fuel furnace, the two units cannot be operated simultaneously.
   a. True
   b. False

3. True or False? When installing a wood-fired furnace, the clearance to combustibles requirement is generally the same as that of a gas-fired furnace.
   a. True
   b. False

4. Which of the following vent types would be used to vent a combination gas/wood furnace?
   a. Type B
   b. Type PL
   c. PVC
   d. Type HT

5. For proper operation, waste-oil heaters must burn oil of a uniform _____.
   a. color
   b. temperature
   c. viscosity
   d. molecular weight

6. A thermal mass is an integral part of a(n) _____.
   a. passive solar heating system
   b. active solar heating system
   c. evaporative cooler
   d. waste-oil heater

7. Evaporative coolers are less effective when ______.
   a. it is hot outdoors
   b. it is humid outdoors
   c. the wet-bulb temperature is very low
   d. the water supply is warm

8. In a cascade system, the ______.
   a. low-stage system cools a falling stream of water
   b. high-stage system cools a falling stream of water
   c. high-stage evaporator acts as the condenser for the low-stage system
   d. low-stage evaporator acts as the condenser for the high-stage system

9. A direct-fired makeup air system typically uses ______.
   a. 35% outside air
   b. 50% outside air
   c. 75% outside air
   d. 100% outside air

10. True or False? Direct-fired makeup air units can be ducted or designed to free-blow into the space.
    a. True
    b. False

11. True or False? The evaporator coil of most computer room systems is designed to maximize the transfer of latent heat.
    a. True
    b. False

12. When computer room systems are found cycling in and out of dehumidification mode regularly, it is often caused by ______.
    a. moisture coming from operating electronics
    b. under-floor water leaks
    c. an unnecessarily tight setting of the humidity deadband
    d. a component failure in the refrigeration circuit

13. Dedicated computer room units are typically equipped with ______.
    a. direct-drive blower motors
    b. fixed-restrictor metering devices
    c. bypass humidifiers
    d. electric heat

14. True or False? Valance cooling system modules use a small fan to aid in air circulation.
    a. True
    b. False

15. A simple way to prevent air stratification in a building with a high ceiling is to use ______.
    a. window fans
    b. valance coolers
    c. paddle fans
    d. evaporative coolers

Answers to odd-numbered questions are found in the Module Review Answer Key at the back of this book.

Answers to Section Review Questions

Answer	Section	Objective
Section One		
1. b	1.1.1	1a
2. c	1.2.0	1b
3. a	1.3.1	1c
4. b	1.4.0	1d
Section Two		
1. b	2.1.0	2a
2. a	2.2.1	2b
3. b	2.3.1	2c
4. c	2.4.1	2d
5. d	2.5.1	2e

MODULE 03305

Commercial Hydronic Systems

Source: alacatr/iStock/Getty Images

Objectives

Successful completion of this module prepares you to do the following:

1. Identify basic water properties and characteristics and calculate various pressure values.
    a. Identify common water properties and calculate static pressure.
    b. Explain how pressure drop and head are calculated.
2. Describe various commercial hot-water heating system components and piping arrangements.
    a. Identify various types of hot-water boilers.
    b. Describe the characteristics and construction of various centrifugal pumps.
    c. Identify various types of valves and other commercial hydronic system components.
    d. Identify the characteristics of various commercial piping systems.
    e. Describe how a typical hydronic piping system is balanced.
3. Describe commercial/industrial chilled-water equipment and systems.
    a. Identify and describe various types of chillers.
    b. Identify and describe cooling towers and evaporative condensers.

Performance Tasks

Under supervision, you should be able to do the following:

1. Identify the major components in active commercial hot-water heating and chilled-water cooling hydronic systems.
2. Identify common piping configurations used in commercial hot-water and chilled-water hydronic systems.

Overview

In commercial and industrial hydronic systems, water is used as the heat-exchange medium for both heating and cooling. Chillers provide the cold water for cooling, while hot water for heating is provided by boilers. Commercial hydronic systems have unique servicing and maintenance requirements, including maintaining proper chemical balance of the water to prevent fouling, corrosion, and similar problems. The layout of the water distribution system varies with the size and type of building.

Digital Resources for HVACR

Scan this code using the camera on your phone or mobile device to view the digital resources related to this craft.

SCAN ME

1.0.0 Water and Movement

Performance Tasks

There are no Performance Tasks in this section.

Objective

Identify basic water properties and characteristics and calculate various pressure values.

a. Identify common water properties and calculate static pressure.
b. Explain how pressure drop and head are calculated.

Before addressing more complex hydronic topics, a review of basic terms and concepts related to water will prove helpful. Many problems with hydronic systems can be identified and resolved when the technician has a thorough understanding of basic concepts.

1.1.0 Water Properties

Water (H_2O) is a chemical compound consisting of hydrogen (H) and oxygen (O). It can exist as ice, water, or steam (solid, liquid, or vapor). The state, or *phase*, of water results from the temperature and pressure to which it is exposed. At the standard atmospheric pressure found at sea level, water begins to freeze (solidifies) at 32°F (0°C), and begins to boil (vaporizes) at 212°F (100°C).

In the liquid state, the density of water changes with the temperature. At a given pressure, the higher the temperature, the lower its density. This change in density (weight) per unit volume is due to molecules that move farther apart as the temperature rises. Fewer molecules in a specific space means less density.

For example, 1 cubic foot (ft^3) of water still in its liquid state weighs 62.41 lb at 32°F (0°C). At 200°F (93°C), 1 ft^3 of water weighs 59.45 lb—a difference of roughly 3.0 lb/ft^3.

Differing water densities cause natural circulation in a closed system, due to gravity. Cooler, heavier water will fall to the bottom while warmer, lighter water rises to the top. This natural circulation is referred to as *thermal circulation*.

1.1.1 Corrosion and Scale

Water is noncorrosive to most materials used in hydronic systems. The oxygen in the water molecule is basically inert. However, oxygen from air is also dissolved in water, and this oxygen is free to react with other elements. This free oxygen supports corrosion. Common rust, or iron oxide, is a compound formed from iron and oxygen. Oxygen also reacts with copper to form copper oxides, which you see rapidly develop when soldering and brazing.

For this reason, the amount of free oxygen in a closed system must be minimized. Once a system is filled with water, makeup water must be minimized to reduce the amount of free oxygen that enters with it. Only so much free oxygen is contained in a closed system (barring the addition of more), and once this oxygen has reacted to develop some form of corrosion, the corrosive process stops.

Open systems—systems that are open to the atmosphere—allow water to continuously interact with air and keep the water *oxygenated*. Open systems are not used for hydronic heating and cooling systems that circulate through a building. However, evaporative coolers and cooling towers are designed for exposure to, and interaction with, air to transfer heat.

Each time fresh makeup water enters a closed system, additional oxygen is admitted, and fresh corrosion begins until all the free oxygen is once again consumed. If fresh water is continually admitted, components in the system will corrode more quickly. In addition, more minerals enter the system, resulting in an increase in scale and sediment.

Because most systems use automatic valves to admit makeup water, leaks must be addressed quickly. Given time, a small leak can cause a lot of damage, in places you can't see. The problem of corrosion in a piping system is announced when strainers begin to fill with debris or operational problems begin. By then, a great deal of damage has been done, and it is not easy to fix. Larger systems often have water meters to monitor the volume of fresh water being added.

As water is heated in a boiler, dissolved air is liberated in the form of bubbles. Eventually, most of the air bubbles collect in high spots or pockets in the piping system. That air must be removed from the system for it to function properly. Air is not easily driven out of cold water, but the temperature affects corrosion too. Heat generally accelerates corrosive processes. This is not to say that corrosion doesn't occur in chilled water piping, but it does progress more slowly.

Certain substances called **oxygen scavengers** work to reduce dissolved oxygen by chemically combining the substance with the oxygen to form harmless compounds. These chemicals may be applied to systems by authorized craft professionals. Always check with the boiler manufacturer before adding any water treatment chemical to boiler water. The wrong chemical may damage the boiler and void any warranties.

Oxygen scavengers: Chemicals added to boiler feed water that bond with dissolved oxygen to inhibit corrosion within the system.

Because public and well water sources also contain varying amounts and types of minerals, scale can also form in the piping. Scale reduces heat transfer and inhibits flow, providing another reason to minimize the admittance of fresh water. Chemicals are also available to inhibit the ability of minerals to bond and to dissolve existing scale.

Clearly, it is important to minimize free oxygen and the admittance of makeup water to all closed systems to keep them in good health. Dirty, corroded piping and heat transfer surfaces are often the cause of a system's demise.

NOTE

The topic of water treatment is covered in NCCER Module 03308, *Water Treatment*.

1.1.2 Static Pressure

The change in water pressure as depth increases can be calculated using basic math. The math assumes the water is not moving—it is the change in *static* water pressure, meaning that no other forces are acting on it beyond gravity.

NOTE

Remember that the density of water, and thus its weight, rises and falls slightly as it is cooled or heated. Also note that values related to static water pressure herein have been rounded to a sufficiently accurate value for most HVACR-related applications.

To visualize static water pressure, refer to *Figure 1* and consider a cubic foot of water. Reference tables indicate that 1 cubic foot (ft^3) of water weighs 62.4 lb at standard atmospheric conditions (sea level). The bottom surface of the cubic foot covers 1 square foot (ft^2). Thus, the pressure exerted by the water at the bottom

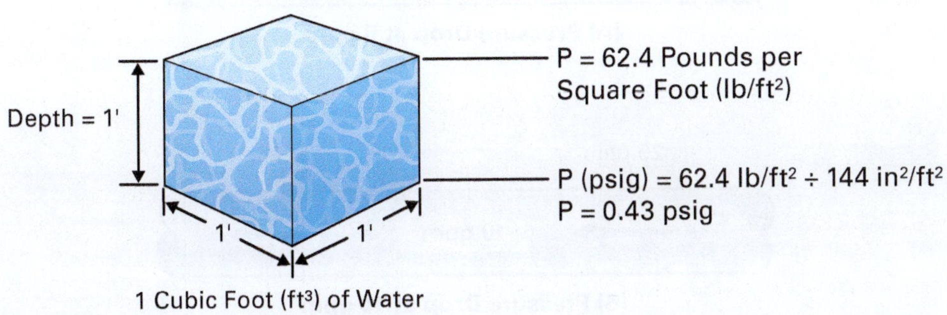

Figure 1 Static pressure of water.

is 62.4 pounds per square foot (lb/ft²). You can calculate this pressure using the following formula:

$$P = \text{pressure per foot of depth} \times D$$

Where:

P = pressure per foot of depth
D = depth in feet

As shown in *Figure 1*, at a 1' depth, the pressure is 62.4 lb/ft². If the depth is 3', multiply the weight per square foot by 3':

$$P = 62.4\ \text{lb/ft}^2 \times 3'$$
$$P = 187.2\ \text{lb/ft}^2$$

In HVACR work, technicians usually measure work with pressure in pounds per square inch (psi). To convert pressures in pounds per square foot to pounds per square inch, divide the result by 144, since there are 144 in² in a square foot. Thus, the static pressure of water per square inch at a depth of 1' is determined as follows:

$$P(\text{square inches}) = 62.4\ \text{lb/ft}^2 \div 144\ \text{in}^2/\text{ft}^2$$
$$P = 0.43\ \text{psi}$$

Since a column of water 1' high produces a pressure at its base of 0.43 psi, the static pressure increases by 0.43 psi for each additional foot of water depth. A column of water 100' deep produces a static pressure of 43 psi.

1.2.0 Pressure Drop and Head

Water flow in a piping system is affected by several factors. These factors include the resistance to water flow presented by the piping and the pressure available to overcome that resistance.

1.2.1 Pressure Drop

Pressure drop is the reduction in pressure between two points in a piping system due to pipe resistance and changes in elevation. Pressure drop is an important factor in hydronic system pipe sizing.

In the example shown in *Figure 2A*, there is a 5 psi drop in pressure between two distant points when the flow rate is 5 gallons per minute (gpm). However, as shown in *Figure 2B*, doubling the flow rate essentially quadruples the amount of pressure drop at the same conditions and pipe size. It's clear that inaccurate pipe sizing can result in an insufficient amount of pressure at the end of the line.

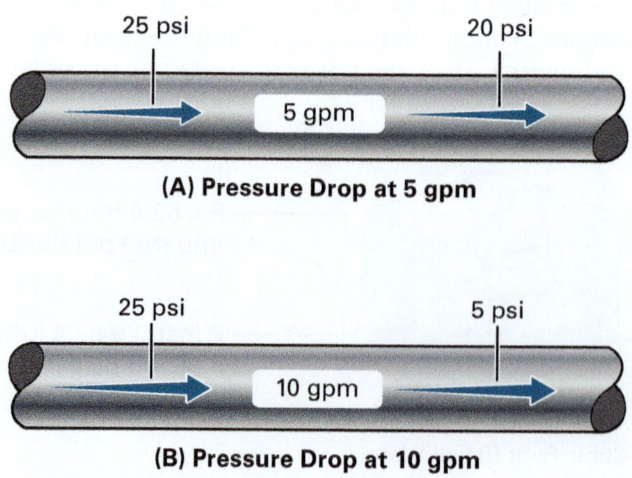

Figure 2 Pressure drop and the effect of increased flow.

The change in pressure drop resulting from a change in flow rate can be calculated as follows:

$$\text{Change in Pressure Drop} = \left(\frac{\text{New Flow Rate}}{\text{Original Flow Rate}}\right)^2 \times \text{Initial Pressure Drop}$$

Here is the math supporting *Figure 2B*:

$$\text{Change in Pressure Drop} = \left(\frac{\text{New Flow Rate}}{\text{Original Flow Rate}}\right)^2 \times \text{Initial Pressure Drop}$$

$$\text{Change in Pressure Drop} = \left(\frac{10\ gpm}{5\ gpm}\right)^2 \times 5\ psi$$

$$\text{Change in Pressure Drop} = (2\ gpm)^2 \times 5\ psi$$

$$\text{Change in Pressure Drop} = 4 \times 5\ psi$$

$$\text{Change in Pressure Drop} = 20\ psi$$

Hydronic system pumps must provide enough power to overcome the resistance encountered in the piping system. This is accomplished by creating a pressure difference on each side of the pump. The manufacturer's specifications for hydronic system components such as coils and heat exchangers include pressure drop data for various flow rates. The pressure drop of these components at the deign flow rate must be added to the pressure drop of the piping.

1.2.2 Head Pressure and Static Pressure

Head pressure is another important value, which is usually expressed in feet. Head pressure represents the pressure difference in a column of water due to the elevation. The term is often used in pump specifications, indicating the height of a water column that the pump can lift. You have probably noted this specification associated with condensate pumps, indicating the maximum distance these pumps can lift condensate to discharge it from the building.

Since a column of water 1' in height produces a static pressure of 0.43 psi, simple calculations prove that lifting a column of water 2.31' in height requires a pressure of 1 psi (1' ÷ 0.43 psi). The calculations are shown in *Figure 3*.

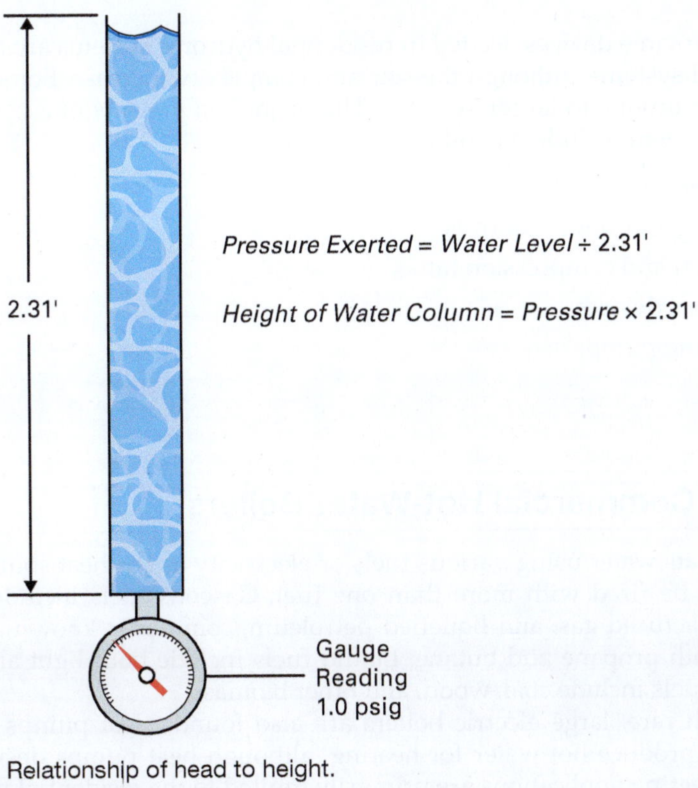

Figure 3 Relationship of head to height.

Head pressure is somewhat related to *static pressure*. Static pressure refers to the pressure exerted by a column of fluid when it is at rest.

NOTE

This formula is reasonably accurate for relatively low flow rates through large-diameter horizontal pipe with no substantial change in fluid temperature between the points. Significant changes in these factors affect the result.

On Site

Head Pressure

Practice your skills in determining static pressure and head:

- The water level in a water system reservoir is maintained at 150' above its point of discharge. What is the pressure reading of a gauge installed at the point of discharge?
- A gauge installed at a reservoir's point of discharge reads 64.5 psig. What is the head, in feet, at this point?

1.0.0 Section Review

1. What substance causes water to be corrosive to hydronic system components?
   a. Iron oxides
   b. Dissolved nitrogen
   c. Suspended particles
   d. Dissolved free oxygen

2. In a given pipe size, what effect does doubling the flow rate have on the pressure drop?
   a. The pressure drop is quadrupled.
   b. The pressure drop is also doubled.
   c. The pressure drop remains the same.
   d. The pressure drop is reduced by half.

2.0.0 Commercial Hydronic Heating System Components

Performance Tasks

1. Identify the major components in active commercial hot-water heating and chilled-water cooling hydronic systems.
2. Identify common piping configurations used in commercial hot-water and chilled-water hydronic systems.

Objective

Describe various commercial hot-water heating system components and piping arrangements.

a. Identify various types of hot-water boilers.
b. Describe the characteristics and construction of various centrifugal pumps.
c. Identify various types of valves and other commercial hydronic system components.
d. Identify the characteristics of various commercial piping systems.
e. Describe how a typical hydronic piping system is balanced.

NOTE

For an introduction to residential hydronic systems and common components, refer to NCCER Module 03203, *Introduction to Hydronic Systems*.

Many of the same devices needed in residential hydronic systems are needed in commercial systems, although the size and complexity increase. Some devices, though, are unique to larger systems. The major components of a commercial hot-water system include the following:

- Hot-water boilers
- Operating and safety controls
- Expansion and compression tanks
- Air control devices
- Circulating pumps
- Valves

2.1.0 Commercial Hot-Water Boilers

A boiler heats water using various fuels or electricity as the heat source. Some boilers can be fired with more than one fuel. Gaseous fuels include natural gas, manufactured gas, and liquefied petroleum, commonly known as LP. LP includes both propane and butane. Liquid fuels include both light and heavy oils. Solid fuels include coal, wood, and other biomass.

Although rare, large electric boilers are also found. Heat pumps can also be used to produce hot water for heating, although heat pumps designed for hydronic heating applications are primarily limited to the residential market.

Going Green

From Coal to Heat Pumps

For generations, the city of Esberg, Denmark has been warmed by a central heating plant that burns coal to heat hot water for wide distribution, releasing 120,000 tons of CO_2 into the atmosphere each year. Nearly 100,000 people depend on the facility for heat through the long winters. Who would think that a couple of heat pumps could replace a coal-fired central plant on that scale?

Two heat-pump systems supplied by MAN Energy Solutions Switzerland Ltd. can now harness the needed heat from the adjacent Wadden Sea. Shown here is one of the variable-speed, hermetically sealed compressors powering the heat transfer. Ironically, the system operates using CO_2 as a refrigerant. Denmark continues to be a world leader in decarbonization, with most of the country's electric power currently generated by wind.

A highly informative video about the new systems, presented by Watt Energy, can be found on YouTube at https://www.youtube.com/watch?v=-bIlAkTDw8Q.

Source: © Sebastian Vollmert/MAN Energy Solutions

The combustion process in commercial/industrial gas- and oil-fired boilers is basically the same as those serving the residential market. However, many larger gas-fired boilers use power burners (*Figure 4*) instead of induced-draft systems. Fuel ignition can be achieved by use of an electronic spark igniter, a large ignition transformer, or a hot-surface igniter (HSI). The presence of a pilot or burner flame is proven by a protective device before the main gas or oil supply valve is allowed to open.

Low-pressure boilers are the most widely used type of hot-water boiler. Low-pressure hot-water boilers may have working pressures of up to 160 psig. However, most are designed for a 30-psig maximum working pressure. The most common operating pressure is 12–15 psig.

Low-pressure hot-water boilers are generally limited to a maximum operating temperature of 250°F (121°C). Above this temperature, water at the common operating pressure will begin to boil, creating a dangerous condition. In most cases, low-pressure boilers operate at temperatures well below 200°F (93°C). This prevents the water from flashing to steam at the site of a leak or when the pressure relief valve opens.

Figure 4 Gas-fired power burner on an industrial boiler.
Source: Courtesy of Superior Boiler

Medium- and high-pressure hot-water boilers are built to operate at pressures above 160 psig and temperatures above 250°F (121°C). Keeping the pressure higher raises the boiling point of the water, allowing the higher temperatures. In large central plants, such as a military installation or college campus, these higher pressures and temperatures allow the water to be circulated great distances and arrive at the destination with plenty of remaining energy. Operating and working on these systems can also be extremely dangerous.

Superheated steam: Steam at a temperature above the saturation temperature corresponding to its pressure.

At extreme temperatures (over 450°F or 232°C in some cases) and under high pressure, water escaping through a small leak instantly flashes to steam. The **superheated steam** is absorbed so quickly that no steam plume is visible. This is due to its sudden exposure to atmospheric pressures at which water boils at 212°F (100°C). The escaping steam can cause serious burns and even cut flesh.

WARNING!

Walking near or moving a hand across the path of escaping steam from a high-pressure hydronic system can result in serious injury. Steam at any pressure can produce serious physical injury. Use extreme caution when looking for leaks or servicing high-pressure systems.

Most boilers in larger commercial applications are made of cast iron or steel. Commercial facilities often use boilers with a copper-finned tube heat exchanger for rapid heat transfer. This design continues to increase in popularity because of heat transfer efficiency, reduced installation cost, and application flexibility. Any number of these copper-finned tube boilers can be placed in parallel on a system to form a modular hot-water plant.

2.1.1 Copper-Finned Tube Boilers

Boilers using the copper-finned tube design offer some distinct advantages in commercial applications. Because traditional steel and cast-iron boilers generally store a significant volume of hot water, these boilers experience greater heat losses when heating needs are low. Traditional boilers also require more space for installation and service.

Copper-finned tube boilers do not store any water beyond what is flowing through the tubes—system flow must be established before the boiler can fire. These boilers are also known as *low-mass boilers* due to the light construction and lack of mass in the heat exchanger. When compared to traditional cast-iron and steel alternatives, copper-finned tube boilers are very lightweight.

With the copper-finned tube design, heat transfer is greatly enhanced, providing for rapid recovery and hot water on demand. Since all the heat is directed to the relatively small copper-finned tube heat exchanger, the required water velocity through the boiler is significantly higher than that required by a cast-iron or steel boiler with large heat-exchange surfaces.

Atmospheric gas-fired models, such as the one shown in *Figure 5*, offer modulating or staged-heating capacity to enhance operating efficiency and flexibility. One other unique feature these boilers can offer is outdoor installation. The unit pictured in *Figure 5* does not have a top connection for flue venting and can be installed outdoors without any additional protection.

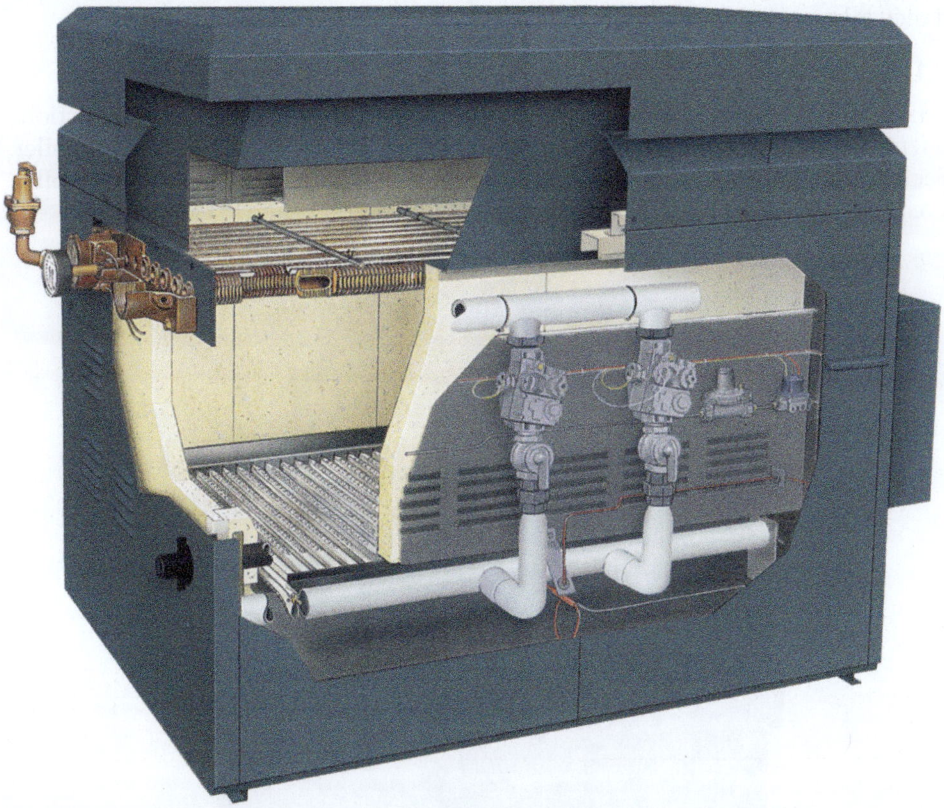

Figure 5 Commercial copper-finned tube boiler cutaway.
Source: Raypak, Inc.

However, boilers installed outdoors are subject to damage from freezing if the burner or pumping system fails. For outdoor installations, a glycol-based antifreeze solution is added to protect the boiler and water circuit from freeze damage. The addition of antifreeze does affect heat transfer and the viscosity of the water. Both issues must be considered in the pump selection process. Generally, system flow rates must be increased when antifreeze is used to transfer the same amount of heat.

Other copper-finned boiler models feature induced-draft combustion, allowing for additional installation flexibility and higher combustion efficiency. An example of an induced-draft model is shown in *Figure 6*. The induced-draft design allows the combustion chamber to be sealed; air for the combustion process can be ducted in from outdoors.

CAUTION

Proper water flow through a copper-finned tube boiler is critical to the life of the boiler. Consult the manufacturer's instructions and engineering guides to determine the correct pipe size and flow requirements.

Also note that galvanized piping is not recommended when a glycol-based antifreeze is used. The zinc in the pipe coating may react with the glycol to form sludge in the piping system.

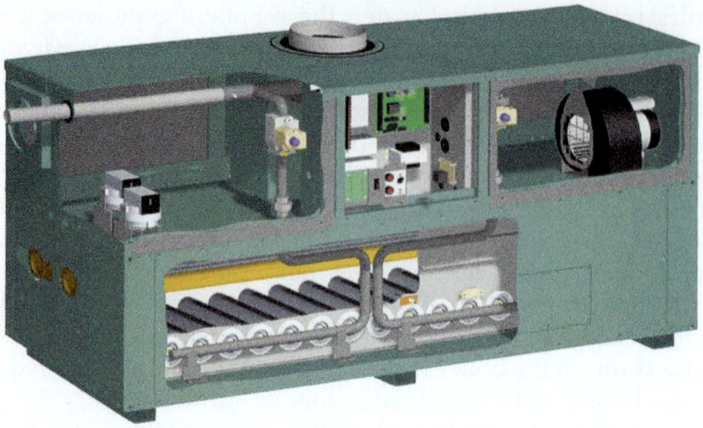

Figure 6 Induced-draft gas boiler with sealed combustion chamber.
Source: Raypak, Inc.

Although copper-finned tube units offer very high heating capacities, multiple units are often selected and piped in parallel to increase capacity and redundancy when one fails or is being serviced.

NOTE

The unit *MBtuh* is a combination of the Roman numeral M, which stands for the Latin word *mille*, meaning 1,000, and the more familiar British thermal units per hour, or Btuh. Therefore, 1 MBtuh = 1,000 Btuh.

2.1.2 Cast-Iron Sectional Boilers

Cast-iron boilers are assemblies of individual cast-iron heat exchanger sections (*Figure 7*). The number of sections used determines the capacity of the boiler. Sections can also be added if more capacity is needed from the same boiler. Cast-iron boiler capacity ranges as high as 13,000,000 Btuh (13,000 MBtuh).

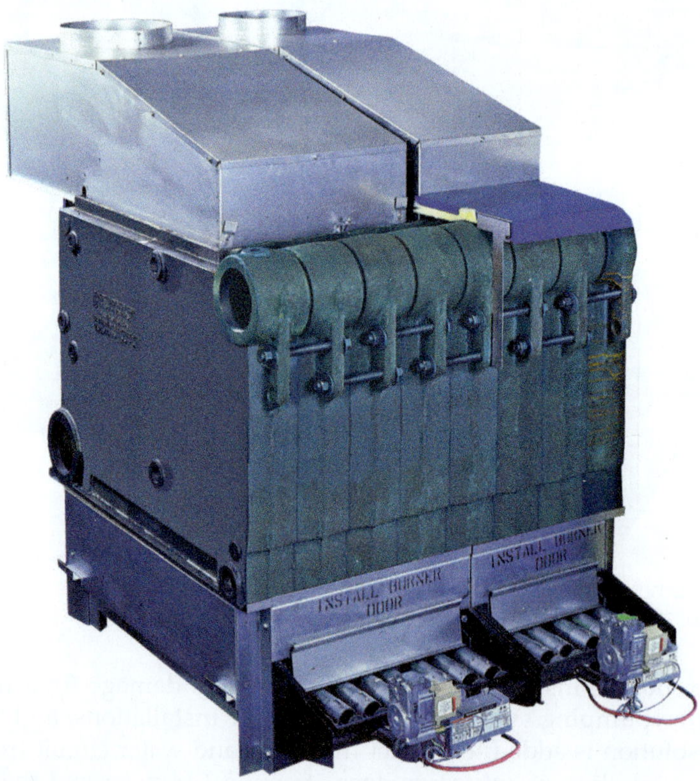

Figure 7 Cast-iron sections of hot-water boiler.
Source: ECR International, Inc.

In a cast-iron boiler, the water circulates inside the sections while the heat and flue gases flow around the outside of the sections. The cast-iron sections are usually mounted vertically, but they can be arranged horizontally as well. In either arrangement, the heat transfer surface is large relative to the volume of water. This allows the water to heat up quickly.

Smaller cast-iron boilers are supplied completely assembled and are referred to as *packaged boilers*. A packaged boiler typically includes the burner, boiler sections, and safety and operating controls. A circulating pump may be included. Larger boilers require on-site assembly.

The terms *dry-base*, *wet-leg*, and *wet-base* are used by manufacturers when describing cast-iron and other boilers. These terms refer to the location of the boiler's water-filled sections relative to the combustion chamber, or *firebox*:

- *Dry-base* — Firebox is located beneath the water vessels.
- *Wet-leg* — Firebox top and sides are enclosed by the water vessels.
- *Wet-base* — Firebox is surrounded by water vessels, except for necessary service openings.

2.1.3 Steel Firetube and Watertube Boilers

Steel boilers are generally manufactured and do not require assembly. However, the largest steel boilers may be built on site, rather than assembled. The heat exchanger surface is usually formed by vertical, horizontal, or slanted tubes.

Two common types of steel boilers are the **firetube boiler** (*Figure 8A*) and the **watertube boiler** (*Figure 8B*). In a firetube boiler, flue gases are contained inside the tubes, immersed in the water. In the watertube boiler, the water flows through the tubes, with the heat and flue gases passing around them. Smaller steel boilers are usually vertical firetube units. Medium- and large-capacity steel boilers normally use horizontal or slanted tubes. These larger boilers may be either firetube or watertube designs.

Firetube boiler: A boiler designed for the hot flue gases to flow through tubes that are immersed in the water to be heated.

Watertube boiler: A boiler designed for the water to be heated to flow through tubes while the hot flue gases and byproducts of combustion flow around the outside of the tubes.

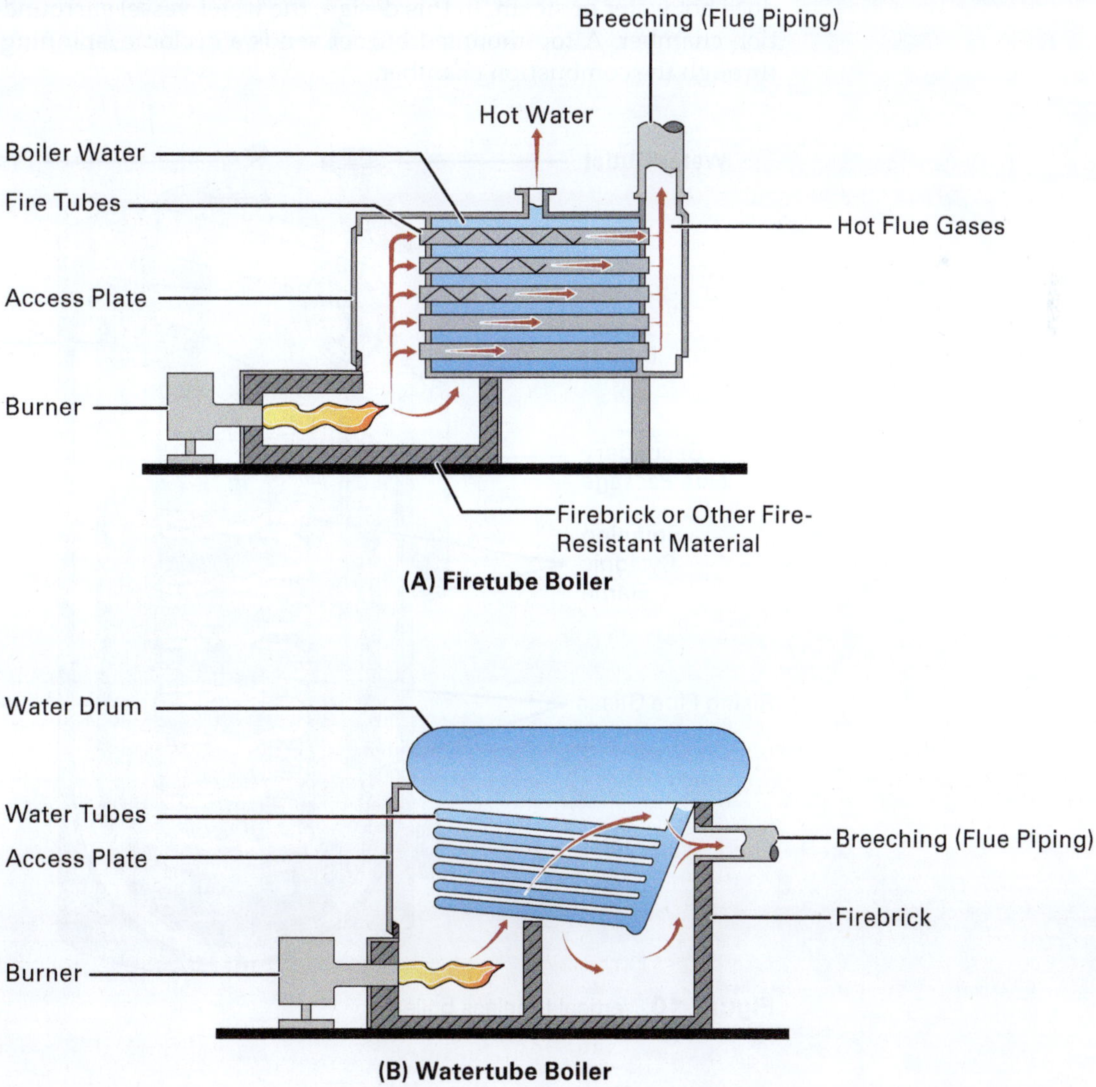

Figure 8 Firetube and watertube boilers.

A common steel firetube boiler design is the Scotch marine boiler (*Figure 9*). As with most boilers, Scotch marine boilers can produce either hot water or steam. These boilers have a low profile, allowing installation in areas where headroom is limited. The name originates from the common use aboard ships where headroom is often limited.

Figure 9 Scotch marine boiler.
Source: Burnham Commercial

2.1.4 Steel Vertical Tubeless Boilers

Steel vertical tubeless boilers (*Figure 10*) are another type of boiler used to produce hot water or steam. In this design, the water vessel surrounds the combustion chamber. A top-mounted burner sends a cyclonic (spinning) flame down through the combustion chamber.

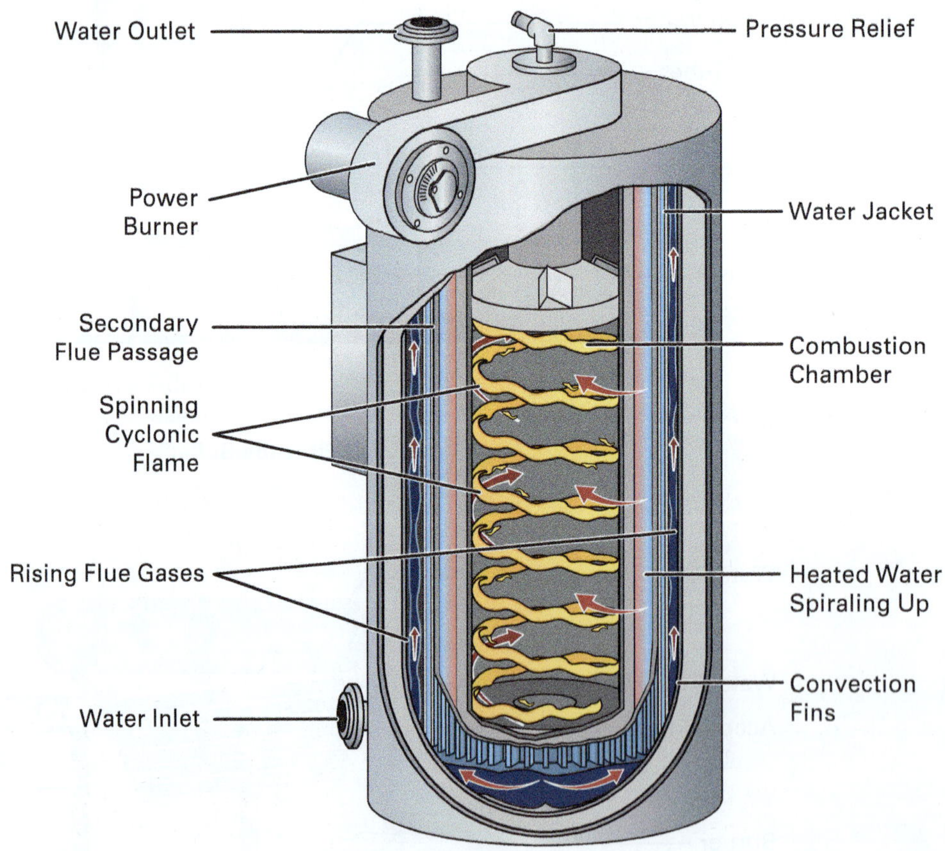

Figure 10 Vertical tubeless boiler.

As the flame swirls downward, the water on the jacket around the combustion chamber spirals upward. Hot gases from the burner are carried up the outside of the water vessel in a secondary flue passage convection area. Full-length convection fins welded on the outside of the water vessel help transfer heat to the water.

2.1.5 Electric Boilers

Electric boilers (*Figure 11*) use resistance heating elements to heat water. The elements are immersed in the water and are installed through flanges in the top of the boiler assembly. This feature makes the elements easier to remove for inspection and maintenance. Horizontal configurations are also common in larger sizes.

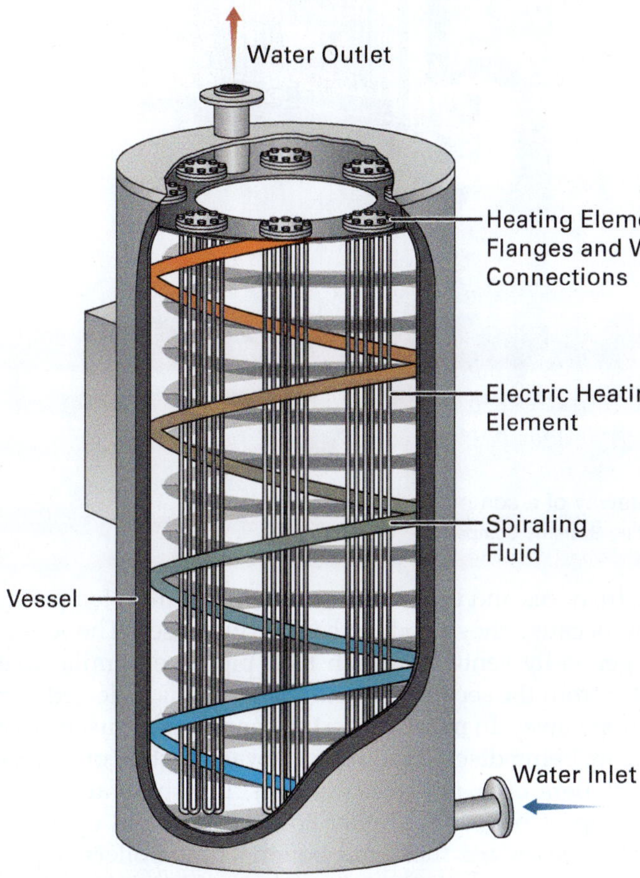

Figure 11 Electric hot-water boiler.

The capacity of an electric boiler may be as high as two megawatts, or 200 **boiler horsepower (BHP)**. An electronic controller controls the heating elements. The controller activates and deactivates individual heating elements in response to the load. The controller also loads the boiler incrementally to reduce power fluctuations and surges during start up.

Boiler horsepower (BHP): A capacity rating for boilers, representing the heat necessary to vaporize 34.5 lb of water per hour at 212°F (100°C) at atmospheric pressure. This is equivalent to 33,475 Btuh or 9.803 kW of energy.

2.1.6 Commercial Condensing Boilers

Condensing boilers operate under the same principles as residential condensing furnaces and operate just as efficiently. Condensing boilers (*Figure 12*) recover the latent heat in the flue gases by cooling the gas and condensing the water vapor. This increases the combustion efficiency to 98% and above. Multiple condensing boilers can be connected to create an efficient, expandable, modular hot-water system.

NOTE

Boiler horsepower (BHP) must not be confused with *brake horsepower* (bhp).

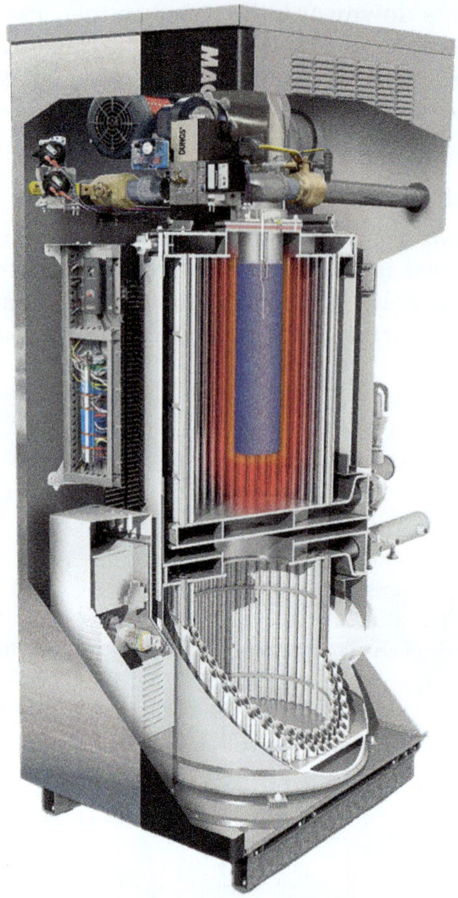

Figure 12 Cutaway of a commercial condensing boiler.
Source: Laars Heating Systems Company

Condensing furnaces and boilers do not require a high-temperature breeching or chimney because they produce cooler flue gases. The low-temperature flue gases are generally vented through PVC piping or similar materials. The condensed water from the secondary heat exchanger is collected in a sump and drained or pumped away. In many cases, the condensate must be treated to neutralize acids before being discharged into a sewage system or waterway.

Condensing boilers offer many advantages, but there are some disadvantages to the concept. Condensing boiler construction and operation are more complicated than a standard fuel-fired boiler. These boilers require a variety of safety interlocks to ensure the heat-recovery and exhaust system is working properly before igniting the main burners. In cold climates, the exhaust pipe is subject to collecting ice due to exiting water vapor. Proper design and installation of the exhaust system can eliminate this problem, but it can be a challenge.

2.1.7 Boiler Operating and Safety Controls

A variety of controls are needed to manage and maintain safe operating conditions on hot-water boilers and combustion systems. Most of these devices, such as the pressure/temperature gauge and pressure relief valve, are described in detail in NCCER Module 03203, *Introduction to Hydronic Systems*. Familiarize yourself with the common devices before proceeding.

A low-water cutoff control (*Figure 13*) is a device used to protect boilers from damage caused by operating with insufficient water. The device shown uses electronic probes to determine water level based on conductivity. Older mechanical controls are significantly larger and heavier, relying on a float that actuates an electrical switch to open the burner control circuit. Electronic models provide the same switching action but require a power supply to power the electronics.

Figure 13 Electronic low-water cutoff control.
Source: Courtesy of Xylem Inc.

The water level for the controls is factory-set and nonadjustable. The controls must be mounted to the boiler at ports provided by the manufacturer to ensure a safe water level is maintained. Some low-water fuel cutoff controls have an opening through which a tool can be inserted to simulate a low-water condition for testing. Electronic controls typically have a test button.

Copper-finned tube boilers, including condensing models, are sensitive to the water flow rate. The boilers will overheat very quickly if water is not flowing when fired. These units require a water flow switch (*Figure 14*) or a pressure differential switch to confirm water flow.

Figure 14 Flow switch.
Source: Courtesy of Xylem Inc.

Flow switches have a paddle placed into the water stream and can be adjusted for a specific minimum flow rate to close the switch and enable boiler operation. Flow switches are also used as a safety device for chillers. Sensitive pressure differential switches sense the pressure drop through the boiler between the inlet and outlet to prove water flow.

2.2.0 Centrifugal Pumps

A pump is a mechanical device designed to create a pressure difference that causes fluid flow. All pumps operate based on the principle that fluid naturally flows from a high-pressure region to a region of lower pressure. Like blowers, the highest pressure in a system is found at the pump discharge while the lowest pressure is found at its inlet. Pumps in hydronic systems are generally driven by electric motors.

Pumps are classified as either *positive-displacement* or *centrifugal*. Positive-displacement pumps are often used to pump high-viscosity liquids or liquids containing solids or particulate matter. Positive-displacement pumps generally develop relatively high pressures.

Centrifugal pumps are far more common in HVACR applications and are used to pump liquids with a low viscosity, such as water. Centrifugal pumps are generally used in applications that do not require extreme pressure differentials, though this is not necessarily a limitation of their design. Multistage pumps are more effective in developing high pressure differentials.

2.2.1 Centrifugal Pump Construction

Figure 15 shows a base-mounted centrifugal pump. Fluid entering a pump is rotated by an impeller. The spinning impeller creates a pressure difference between the inlet and outlet of the pump, resulting in fluid flow. During normal operation, the fluid velocity at the inlet and outlet should be the same.

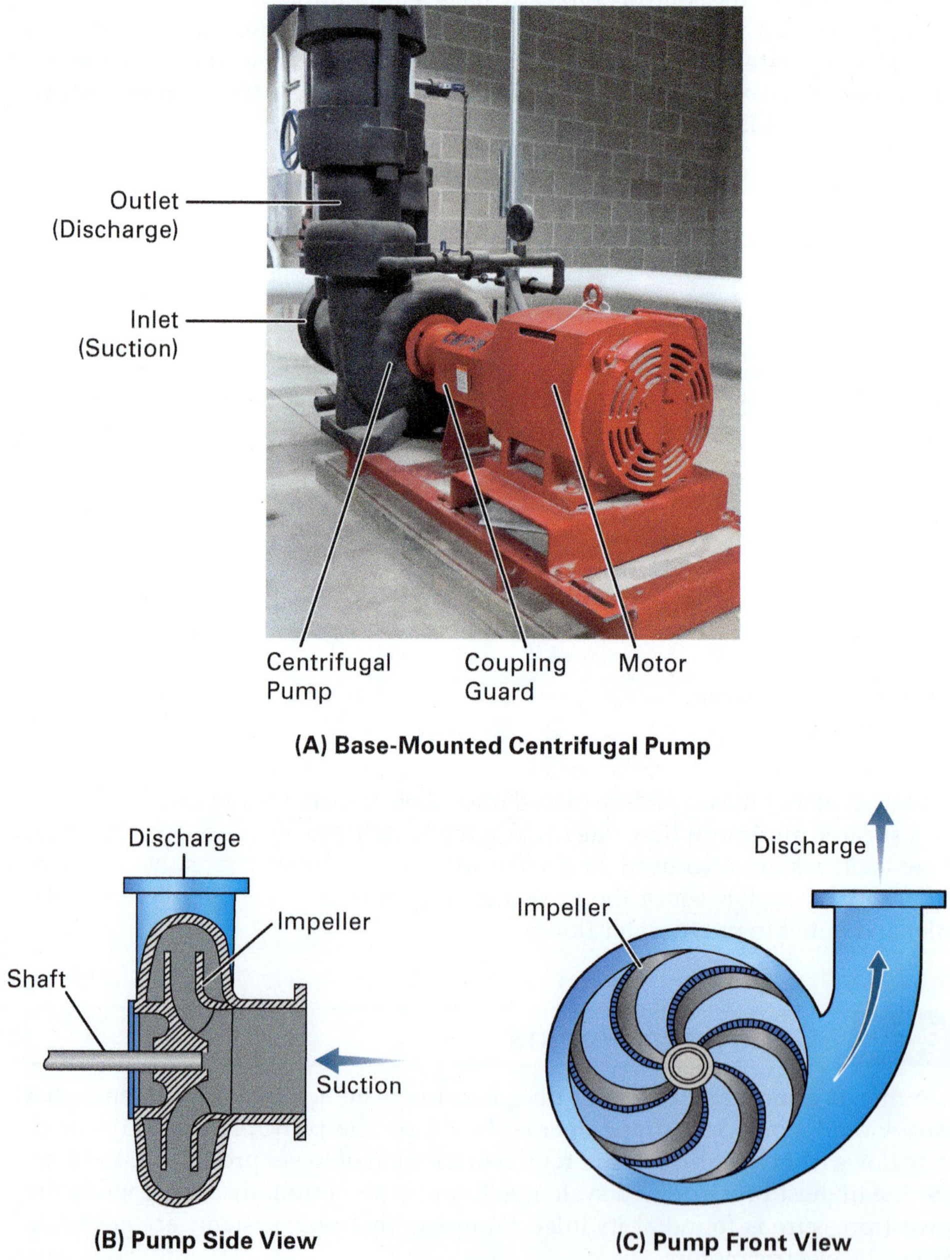

Figure 15 Typical centrifugal pump.

The parts of centrifugal pumps vary in size and shape depending on the manufacturer and application requirements, but the primary parts have the same functions. Centrifugal pumps include the following primary parts, shown in *Figure 16*:

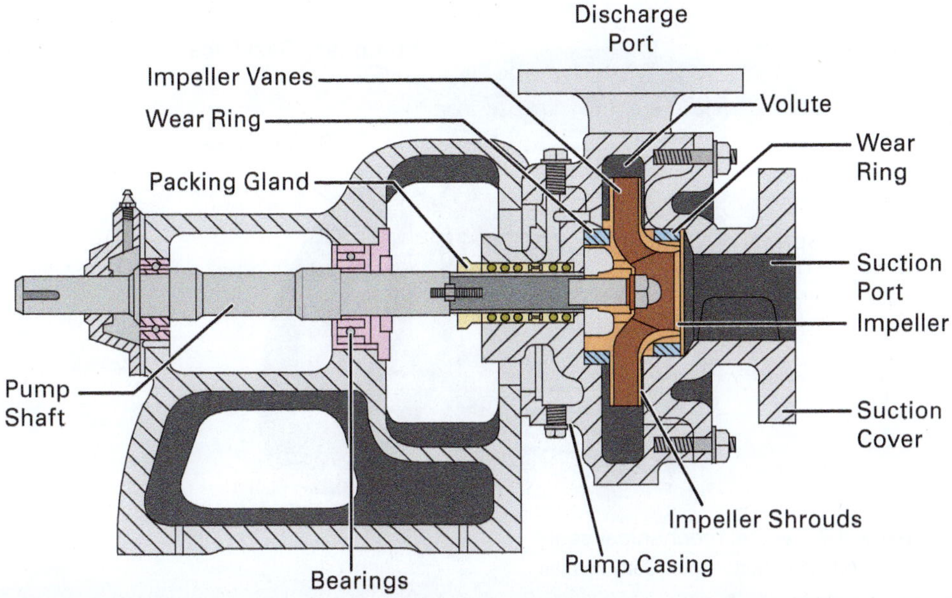

Figure 16 Common centrifugal pump components.

- *Bearings* — Parts that support the rotating shaft and impeller in the casing (*Figure 17*).
- *Discharge port* — Where fluid leaves the pump casing and enters the piping system.
- *Impeller shrouds* — Parts that enclose the impeller vanes and reduce recirculation inside the volute.
- *Impeller vanes or blades* — Parts of the impeller that accelerate the fluid.
- *Impeller* — The rotating part that accelerates the fluid and raises its pressure (*Figure 18*). There are many types of impellers used for different applications.

Figure 17 Typical ball bearing.
Source: urfingus/iStock/Getty Images

Figure 18 Pump impeller.
Source: Sergii Zhmurchak/iStock/Getty Images

- *Mechanical seal* — Creates a seal to prevent leakage between the rotating pump shaft and casing. Some pumps, like the one shown in *Figure 16*, rely on packing instead to eliminate leakage around the shaft. A mechanical seal is shown in *Figure 19*.

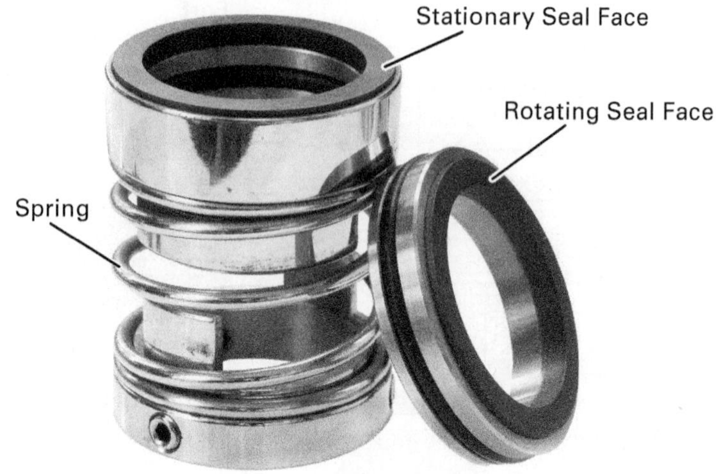

Figure 19 Typical mechanical seal.
Source: nayladen/Shutterstock/Getty Images

NOTE

Packing installation and removal is presented in NCCER Module 03215, *Basic Maintenance*.

Figure 20 Graphite-yarn braided packing.
Source: Image courtesy of SEPCO, www.sepco.com

- *Packing gland* — For pumps that use packing instead of a mechanical seal, the packing gland covers the packing (*Figure 20*) and compresses it against the shaft to create a water seal. A small amount of leakage past the packing is generally acceptable, as the water helps reduce friction and heat between the packing and the rotating shaft.
- *Pump casing* — The housing surrounding the shaft, bearings, packing gland, and impeller, enclosing the fluid inside the pump. Pump casings can be of split or single-piece castings.
- *Pump shaft* — The bearing-supported part that connects the motor to the impeller and rotates it.
- *Suction port* — Where fluid enters the pump casing.
- *Volute* — Sweeping shape within the pump casing that works together with open-vane impellers to help guide and accelerate the fluid toward the discharge port of the pump.
- *Wear rings* — Replaceable rings used in some pumps to maintain impeller alignment and make up for the lack of bearing support at the inlet of the impeller. Wear rings permit a small amount of fluid leakage between the impeller and the casing in the suction area. The leakage creates a hydraulic seal and helps the pump operate more efficiently.

A variety of impeller styles are available, selected based on the application and operating characteristics desired. *Figure 21* provides examples of impeller styles. The most common designs used in hydronic comfort systems are the single-suction closed impeller (*Figure 21C*) and the double-suction impeller (*Figure 21D*). Most impellers can be precisely machined to a specified diameter to provide the specified performance.

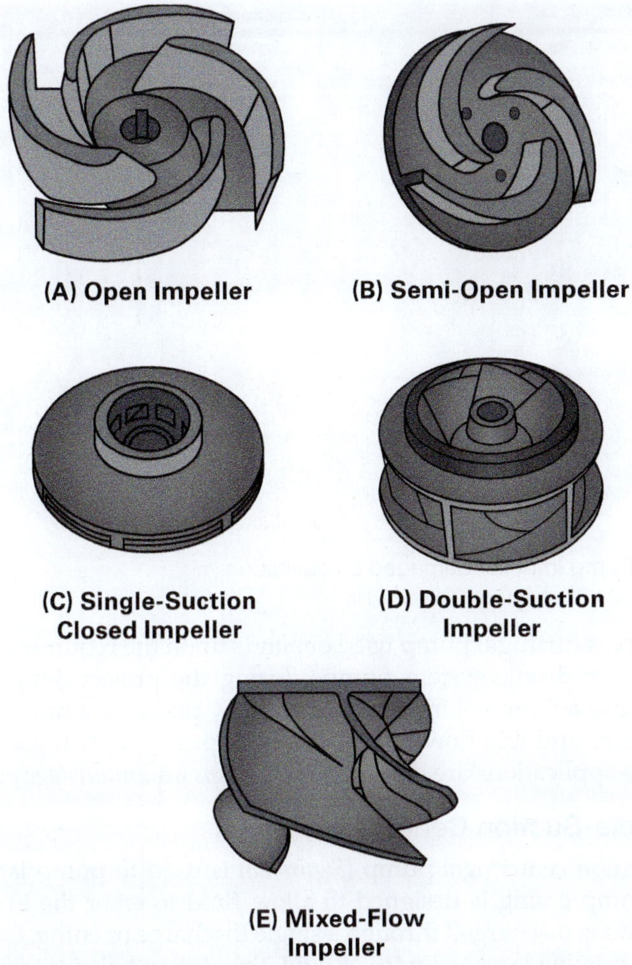

Figure 21 Impeller designs.

It is common for pump impellers to be removed and machined (trimmed) after a new system is started up to precisely match the specified flow for the system. Test data provides the information to determine how much of the impeller needs to be trimmed. Excessive water flow rates, especially in systems that run around the clock, are a significant waste of energy.

Cavitation can occur in a water-circulating pump when the pressure at the inlet of the impeller falls below the vapor pressure of the water, causing the water to vaporize and form bubbles. These bubbles are carried through the impeller to the area of higher pressure where the bubbles implode (collapse) with extreme force. The force is sufficient to pit and erode the impeller vanes and volute if the problem is not resolved (*Figure 22*).

Cavitation: A condition caused by liquids reaching their vapor pressure at the pump inlet, causing bubbles to form and then collapse violently as they pass through the pump.

The problem of cavitation can be eliminated by maintaining a sufficient suction pressure at the pump inlet to overcome the pump's internal losses. This minimum suction pressure is called the *net positive suction head required* (NPSHR). The net positive suction head (NPSH) is a combination of atmospheric pressure, water temperature, and the height of the column of water on the suction side of the pump.

Symptoms that indicate cavitation is occurring in a pump include the following:

- Snapping and crackling noises at the pump inlet
- Pump vibration
- Variations in pump sounds
- Reduced water flow
- Oscillations in discharge pressure

Figure 22 Pump impeller damaged by cavitation.
Source: Javaman24/Shutterstock/Getty Images

On Site

Pump Selection

The websites for most major pump manufacturers have an online app that can be used to help select a pump for a particular application. User inputs to this program typically include the required flow rate, total pump head, desired pump rotational speed, and the mounting requirements, such as inline or base mounted.

The type of centrifugal pump used depends upon the requirements. Engineers usually select hydronic system pumps during the project design phase. Some systems require a large volume of fluid at a low pressure while others require a higher pressure and less flow. Two common types of centrifugal pump designs used in large applications are *double-suction pumps* and *multistage pumps*.

2.2.2 Double-Suction Centrifugal Pumps

A double-suction centrifugal pump (*Figure 23*) is used to pump large volumes of fluid. The pump casing is designed to allow fluid to enter the impeller on both sides. The fluid is discharged through a single discharge opening. Due to the forces generated in moving large volumes of fluid, this design balances the axial force on

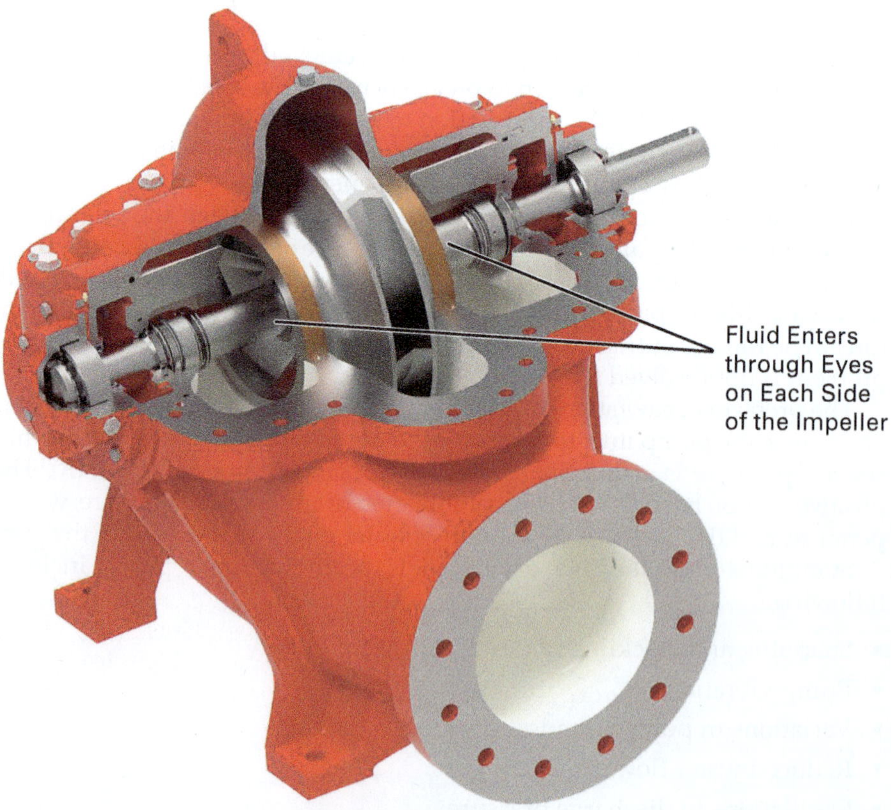

Figure 23 Double-suction centrifugal pump.
Source: Courtesy of Xylem Inc.

the shaft and motor, minimizing bearing wear. The pump shown in *Figure 23* is also referred to as a *horizontal split-case pump* due to the orientation of the pump cover.

2.2.3 Multistage Centrifugal Pumps

Multistage centrifugal pumps (*Figure 24*) contain two or more impellers and can be either single-suction or double-suction pumps. These pumps are used when higher pressure differences are required.

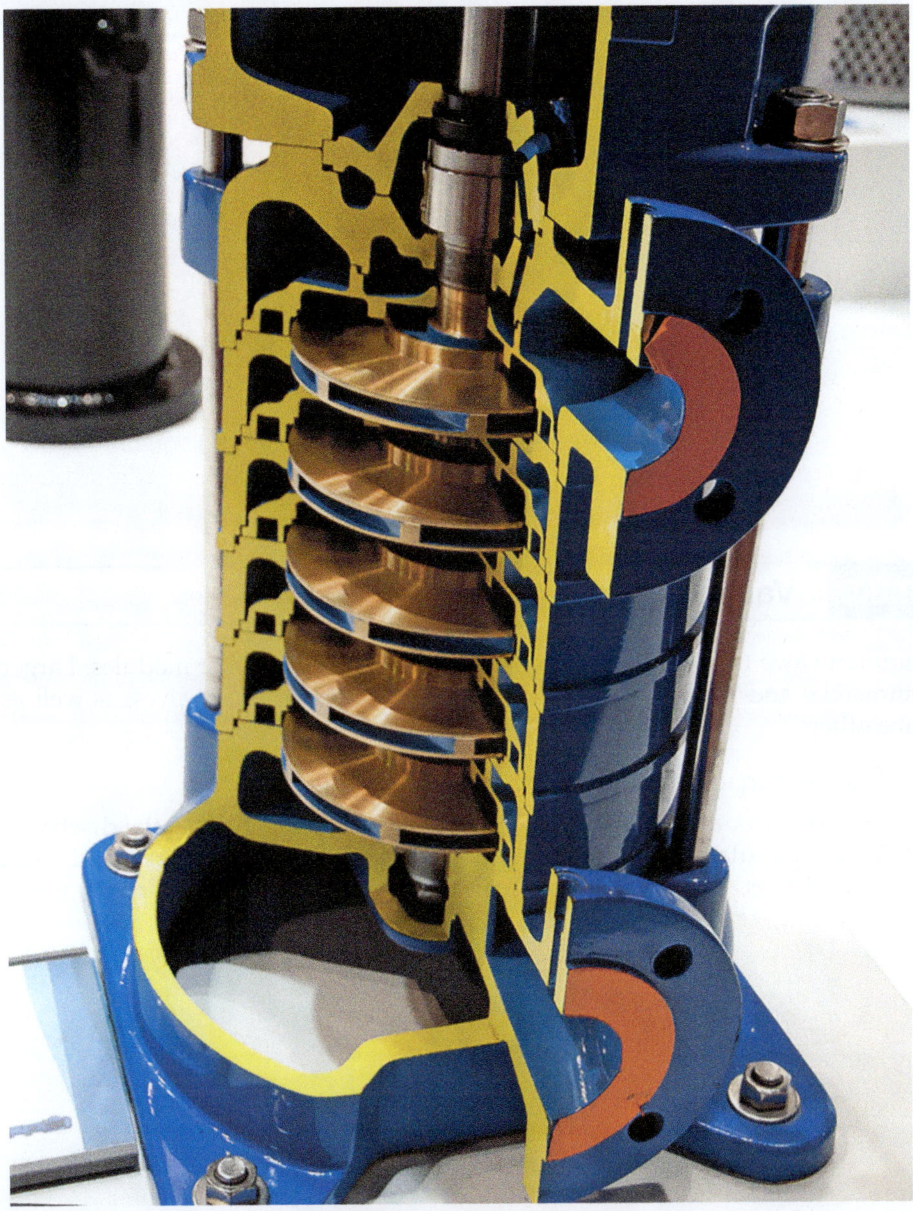

Figure 24 Multistage centrifugal feedwater pump.
Source: Adrian Wojcik/iStock/Getty Images

Each impeller stage increases the pressure of the fluid as it is received from the previous stage. Fluid is discharged from one stage to the next through passages inside the pump casing. Multistage feedwater pumps like the one shown in *Figure 24* are needed for large industrial boilers with high operating pressures. The feedwater pump must operate at a higher pressure than the boiler to force makeup water in.

Going Green

Variable-Speed Centrifugal Pumps

Energy-saving variable-speed centrifugal pumps are becoming more common in hydronic systems, especially in zoned systems where loads are constantly changing. More pumps are now being driven by electronically commutated motors (ECMs) that are already popular in residential and commercial HVACR systems. An electronic control senses temperature conditions in the hydronic system and its zones. The control then signals the pump to increase or decrease its speed. This, in turn, delivers the correct volume of water at the correct temperature to satisfy the load.

Source: Armstrong Fluid Company

2.3.0 Valves and Components

Common valve types and usage have been discussed in earlier modules. Larger commercial and industrial systems generally use these same valves, as well as some others.

2.3.1 Multipurpose Valves

In larger systems, a multipurpose valve is commonly installed in the discharge side of the circulating pump (*Figure 25*). These valves are also referred to as *triple-duty valves*. This single valve functions as a shutoff valve, a check valve,

Figure 25 Multipurpose, or triple-duty, valve.

and a balancing valve. It has a calibrated stem so the valve can be returned to the set balancing position after the valve has been used as a shutoff valve. When in the set position, it acts as a balancing valve.

The valve has tapped (threaded) ports to install two readout valves connected to a flow meter that reads the pressure drop through the valve and calculates a flow rate (*Figure 26*). A differential pressure gauge can also be connected. The measured pressure change can then be used with a conversion chart to find the corresponding flow rate through the valve.

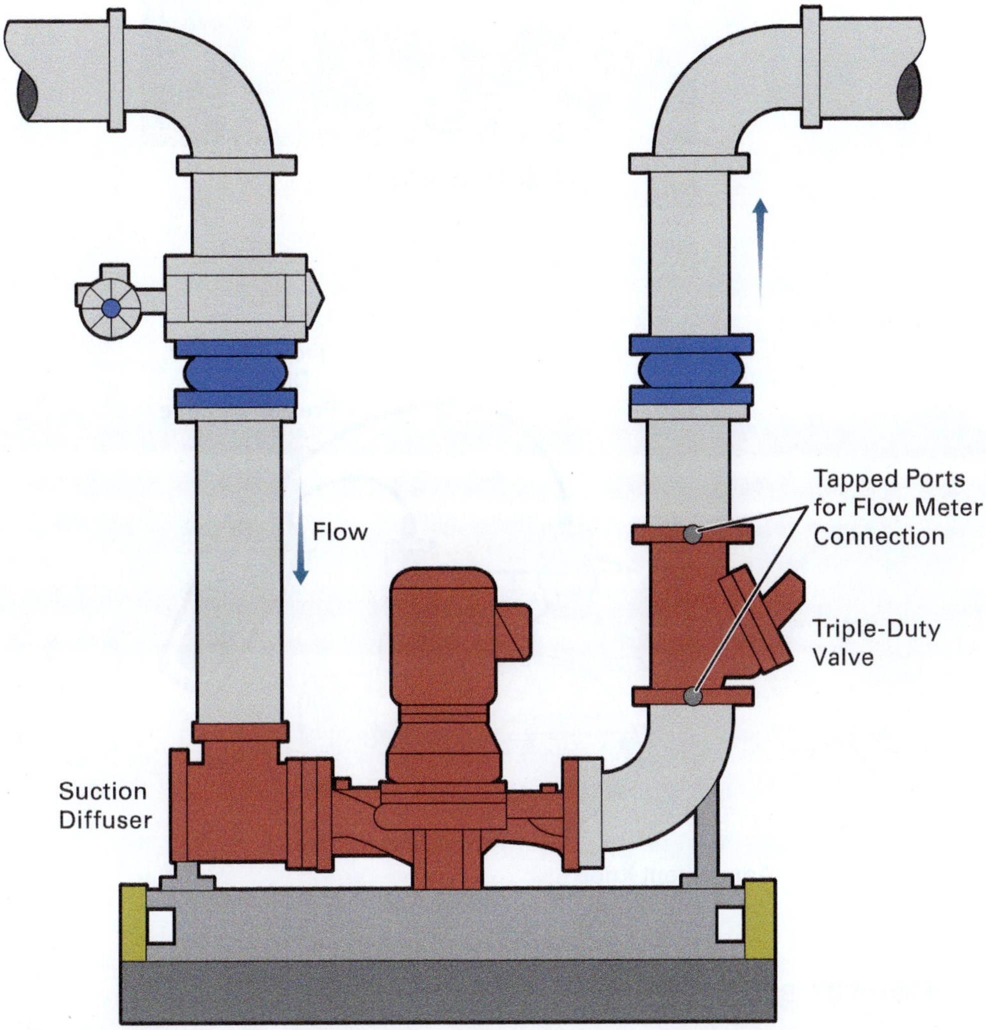

Figure 26 Multipurpose valve in pump discharge line.

2.3.2 Balancing Valves and Flow Meters

Balancing valves (*Figure 27*) and flow meters provide a means of balancing a water system. A balancing valve can function as both a precision balancing device as well as a shutoff valve.

As a balancing valve, it is adjusted to a setting that relates to the degree of resistance needed to achieve the desired water flow rate at a given pressure. The system designer usually determines the initial setting, which is based on a specific design pressure. The installer sets it accordingly.

When it is set to the closed position, the valve is fully closed and serves as a shutoff valve. Once the system is running, operating data such as the pressure and actual flow rate are measured and the balancing valve is adjusted as needed.

(A) Balancing Valve

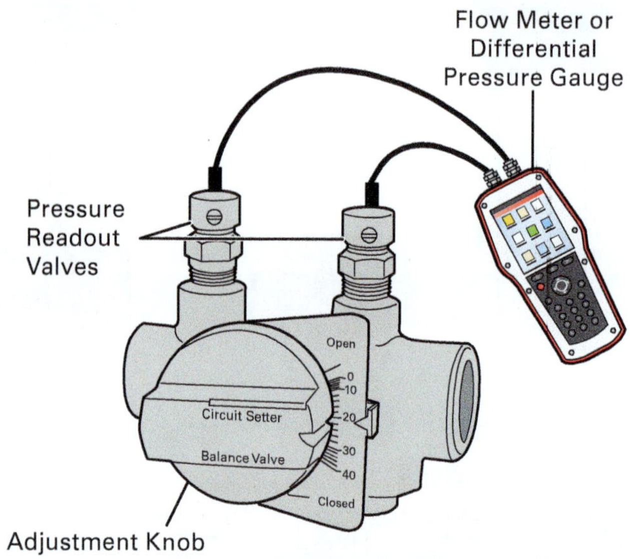

(B) Flow Meter Connection

Figure 27 Balancing valve.

Like multipurpose valves, most balancing valves are equipped with two ports that enable the valves to be connected to a flow meter. When a differential pressure gauge is used instead, the flow rate can be determined from charts that accompany the balancing valve.

2.3.3 Butterfly Valves

Butterfly valves, also known as *wafer valves*, are popular in large piping systems as a means of controlling flow. Butterfly valves generally weigh less than other types of valves because of the narrow body design.

A butterfly valve (*Figure 28*) has a round disc that fits tightly in its mating seat and rotates 90 degrees in one direction to open. A butterfly valve is operated by simply turning the hand wheel or lever one-quarter turn (90 degrees) to open or close it, similar to a ball valve. These valves can function in the open, closed, or partially open position for throttling operation. Although the valves can be used in a partially open position, butterfly valves are not designed for balancing. Remember that gate valves are also poorly designed for balancing service.

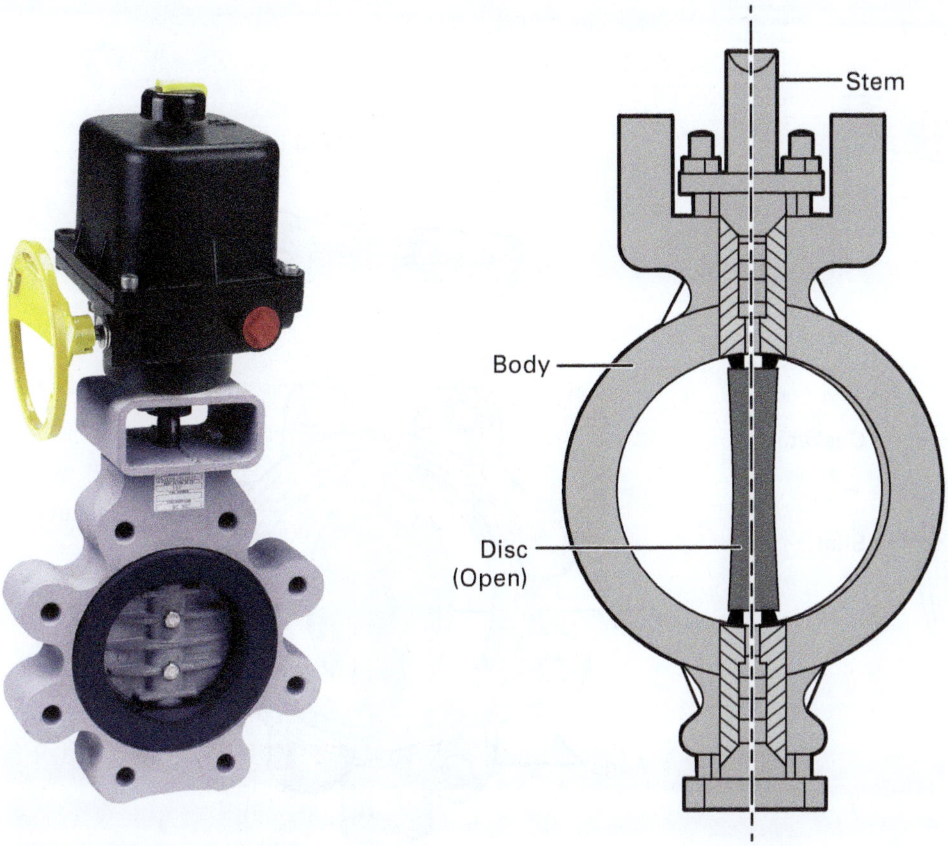

(A) (B)

Figure 28 Butterfly valve.
Source: Image courtesy Dwyer Instruments, LLC (A)

When the butterfly valve is equipped with a hand lever, the position of the lever indicates whether the valve is open, partially open, or closed. When the lever is perpendicular to the flow through the valve, the valve is closed; when the handle is in line with the flow, the valve is fully open.

Butterfly valves that are 12" diameter and larger are usually equipped with a handwheel or gear-operated actuator because of the large amount of fluid flowing through the valve and the great amount of pressure pushing against the seat when the valve is fully closed.

Butterfly valves have an arrow stamped on the side, indicating the proper direction of flow through the valve. The valves must be installed in the proper flow direction, or they will not seal as they should, allowing fluid to flow past a closed valve.

Three common types of butterfly valve bodies are in use: *wafer, wafer lug, and two-flange* styles. The wafer butterfly valve (*Figure 29*) is designed for quick installation between two flanges. No additional flange gasket is needed between the valve and the flanges because the valve seat is lapped over the edge of the body to make contact with the valve faces. Bolt holes are provided in larger wafer valves to help line up the valve with the flanges.

Wafer lug valves (*Figure 30*) are the same as wafer valves except that wafer lug valves have bolt lugs completely around the valve body. Like the wafer valve, no additional gasket is needed between the valve and the flanges because the valve seat is lapped over the edges of the body. The lugs are normally drilled to match standard 150-lb pipe flanges, unless the valve is rated for higher pressures. Higher pressures are rarely necessary in hydronic comfort systems.

The bolts securing the two flanges on each side of the valve normally pass through the lugs and are secured with a nut. However, the lugs on some valves are drilled and tapped so that the bolts from each flange thread directly into the valve body. When the valve is closed, downstream piping can be dismantled for cleaning or maintenance while the upstream piping remains intact.

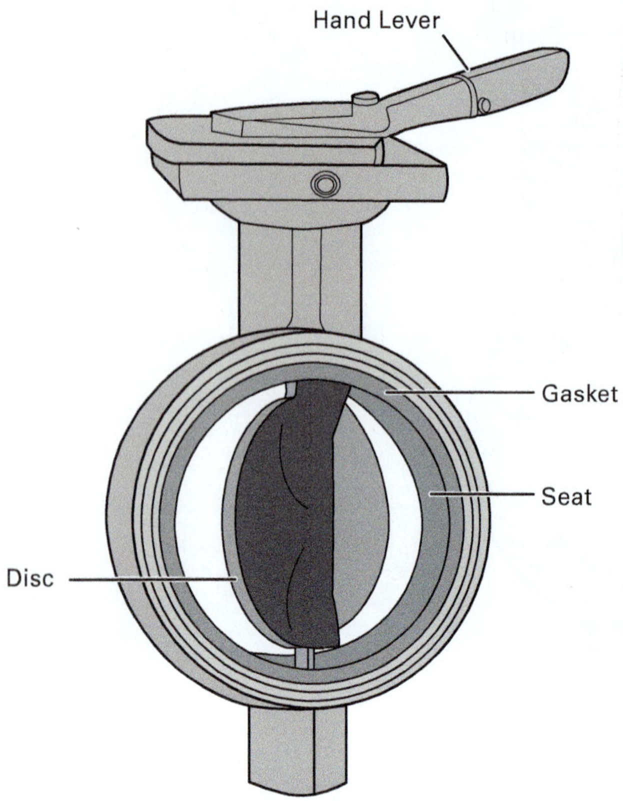

Figure 29 Wafer butterfly valve.

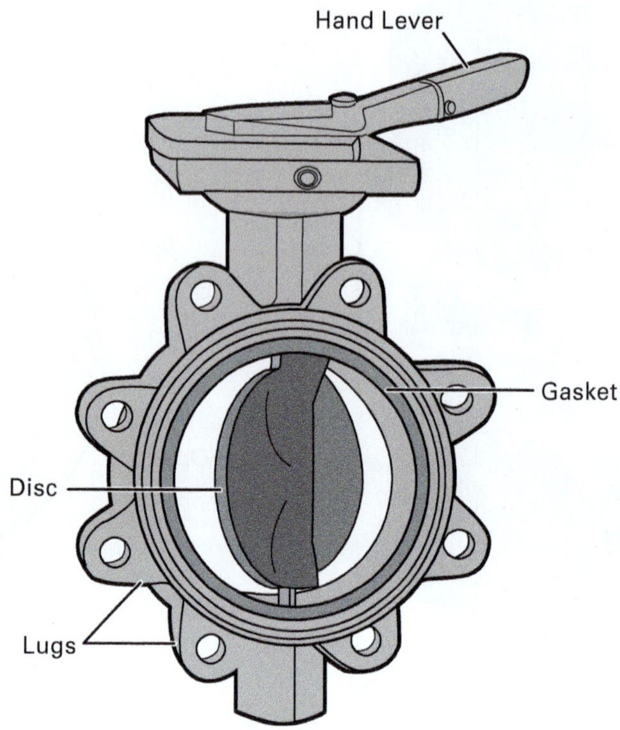

Figure 30 Wafer lug butterfly valve.

When installing both wafer and wafer lug valves, the valve must be in the open position prior to tightening the flange hardware to prevent damage to the valve seat.

Butterfly valves with full face flanges on each side are also available. Downstream piping on this style can also be dismantled when the valve is closed.

2.3.4 Expansion and Compression Tanks

Commercial hot-water systems incorporate compression or expansion tanks (*Figure 31*), as is the case with smaller systems. Expansion tanks are also referred to as *bladder tanks* or *diaphragm tanks*. Compression tanks are basically empty vessels.

Figure 31 Expansion tanks.
Source: Courtesy of Xylem Inc.

If an expansion tank fails to hold its air charge, suspect a leaking air valve or a leaking bladder. While small compression tanks are not repairable, the bladder of larger tanks is accessible through a flanged port on the top or shoulder (*Figure 32*).

> **On Site**
>
> **Waterlogged Compression Tanks**
>
> You should suspect a waterlogged compression tank if the boiler relief valve discharges whenever the boiler water warms up. With no room for the expanding water to go, the system pressure rises very quickly.

Figure 32 Expansion tank with accessible bladder.
Source: Courtesy of Xylem Inc.

2.3.5 Air Management

Air management in commercial hydronic systems is important to ensure proper water circulation and prevent noise. The same types of air-management devices used in residential applications are generally employed in commercial systems. However, because of the greater volume of water and the potential for much larger volumes of air to collect and interfere with circulation, large commercial systems usually have an air separator (*Figure 33*). This type of air separator is also used in chilled-water systems.

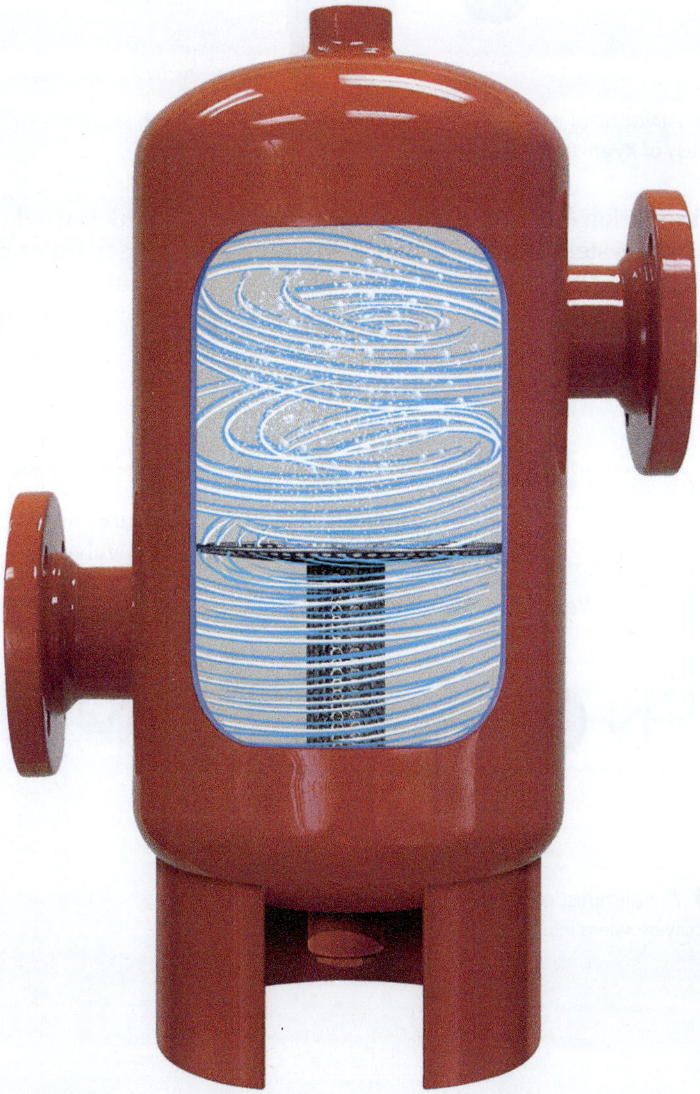

Figure 33 Hydronic system air separator.
Source: Courtesy of Xylem Inc.

Air separators efficiently separate and collect entrained air from water. The unit shown in *Figure 33* is available in sizes large enough to accommodate 36" pipe. Inside the separator, the velocity of the flowing water is reduced to a speed that allows the air bubbles to rise and burst. At the same time, the circulation pattern imparts a centrifugal spin in the water stream. This causes the heavier water to move outward and the lighter air to remain closer to the center. The air collects in the center and leaves through the top.

Air management in hydronic systems is accomplished through one of two strategies: *air control* or *air elimination*. In an air-control system (*Figure 34*), a compression tank is mounted on the air separator discharge. The air rises through any water in the tank and collects at the top. The air and water are in contact inside the tank, but the air is separated from the water in circulation.

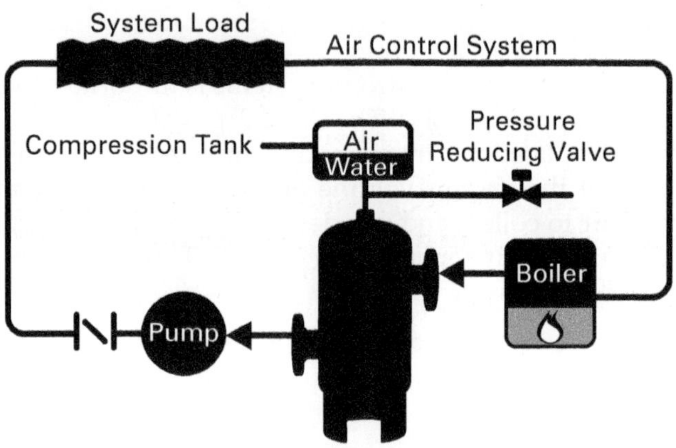

Figure 34 Air-control system component arrangement.
Source: Courtesy of Xylem Inc.

In air elimination (*Figure 35*), an expansion, or *bladder*, tank is used on the suction side of the system pump. The air collected by the air separator discharges through the top and out to the atmosphere through an automatic air vent. In *Figure 35*, note that the pressure reducing valve is providing makeup water to the system through the air separator.

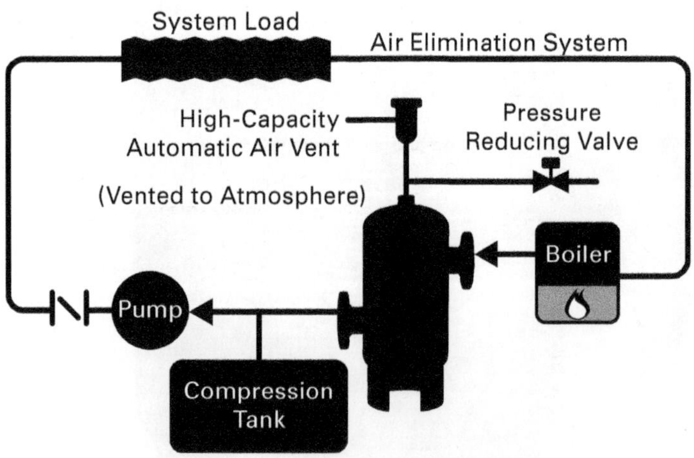

Figure 35 Air-elimination system component arrangement.
Source: Courtesy of Xylem Inc.

Did You Know?

The Science of Air Separation

The solubility of air in water at different temperatures and pressures is based on a law of physics known as Henry's law. This law describes how both the air pressure and water temperature affect the solubility.

As the system pressure increases, water can hold an increasing volume of air in solution. When the system pressure decreases, some of the air dissolved in the water forms bubbles and separates until the air and water again reach equilibrium at the new pressure.

An increase in water temperature decreases its ability to hold air in solution. As the temperature falls, the water can hold more air.

The result is that air separates more readily from water at higher temperatures and lower pressures. This also means that chilled-water systems that operate at higher pressures and lower temperatures represent a more challenging air-management application than hot-water systems.

2.4.0 Common Commercial Piping Arrangements

Piping arrangements used in residential and light commercial hydronic systems tend to be simple. Basic hydronic piping systems were presented in NCCER Module 03203, *Introduction to Hydronic Systems*.

Large commercial and industrial piping systems are more complex. Some areas of a large building might need cooling while another area requires heat. Providing both utilities simultaneously substantially increases the scope of the piping.

2.4.1 Four-Pipe Systems

A four-pipe system is used to create a dual-temperature piping system, providing both hot and chilled water. Because of its design simplicity, zone control, and operational economy, it is common to large commercial applications.

A four-pipe system (*Figure 36*) is constructed of two separate two-pipe circuits. One supply and return piping circuit carries the hot water, while the other supply and return circuit carries the chilled water. The chilled-water and hot-water streams never meet in a four-pipe system.

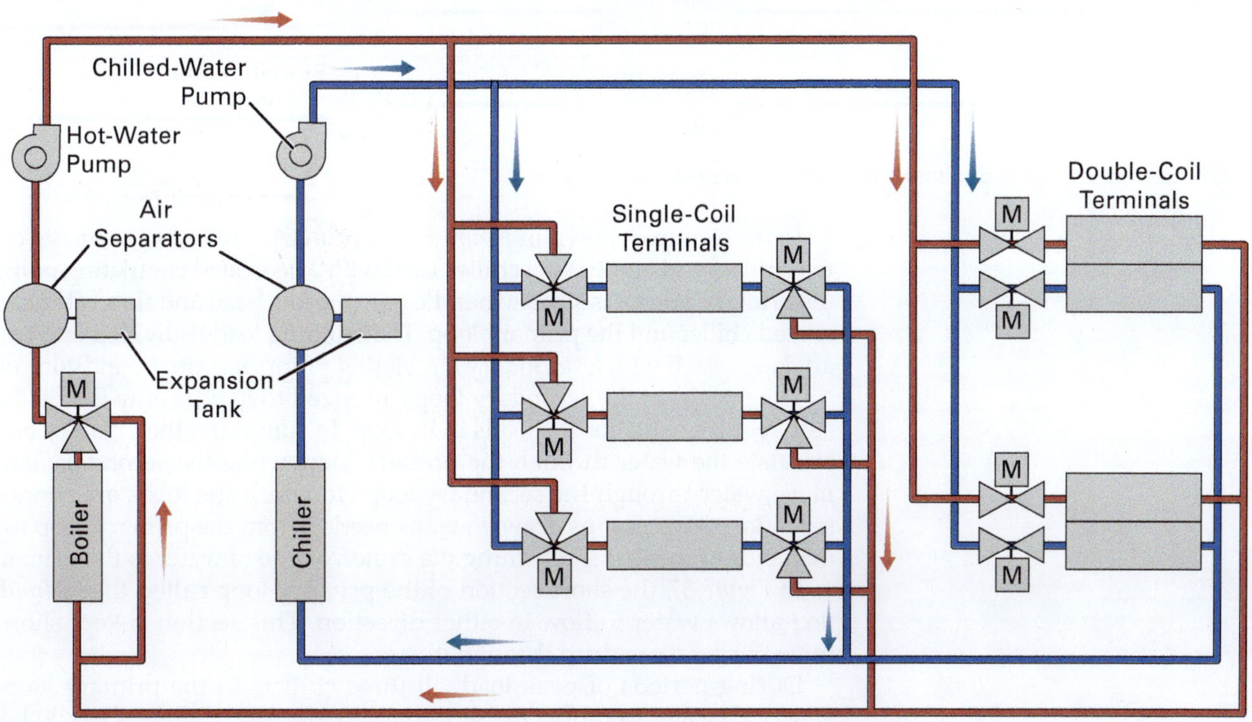

Figure 36 Example four-pipe water system.

As shown in *Figure 36*, three-way valves are often used at the input and output of a single-coil terminal. The system also uses terminals with separate chilled- and hot-water coils (double-coil terminals).

Two-way valves can be used with the heating and cooling coils of a double-coil terminal. However, when two-way valves close, all flow ceases in that portion of the circuit. When most or all the valves close, the pressure from the pump increases. To overcome this problem, the speed of the pump is modulated by a variable-frequency drive (VFD) or other means to maintain a consistent system pressure.

2.4.2 Primary-Secondary Piping Systems

Primary-secondary hot-water and chilled-water piping systems (*Figure 37*) are used in applications that have widely varying loads. These piping arrangements supply variable-flow, constant-temperature chilled and hot water while also conserving energy. *Figure 37* shows a chilled-water system arrangement. For a hot-water system, the chillers would be replaced by boilers.

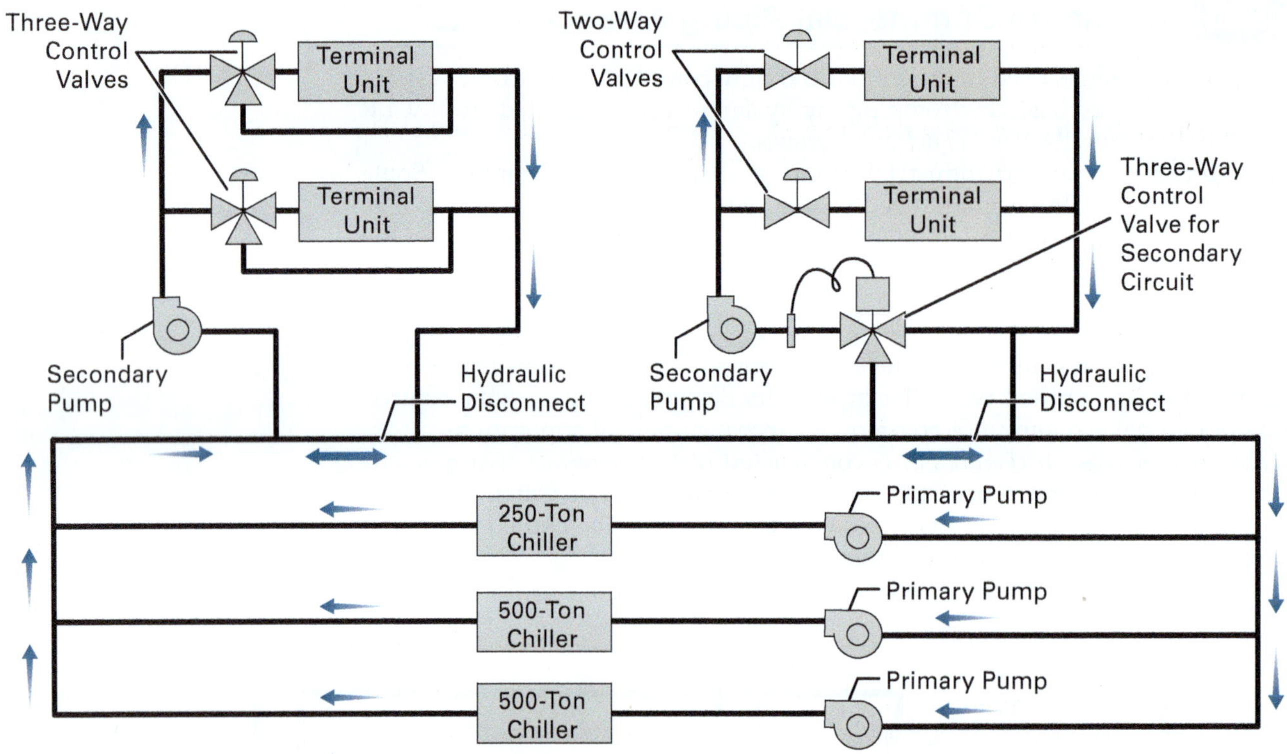

Figure 37 Example simplified primary-secondary chilled-water system.

In the example shown in *Figure 37*, the primary chiller loop consists of two 500-ton chillers and one 250-ton chiller, each with a dedicated circulating pump. Each of the primary pumps is sized to handle only the low head and flow rate needed for its related chiller and the primary loop. The building load is divided into two secondary loops, each with a secondary circulating pump, terminals, and control valves.

The pumps in the secondary loops are sized to handle only the flow and pressure requirements for the loads in its loop. In summary, the primary loop pumps circulate the water through the primary loop, while the secondary loop pumps move water through the secondary loops to which the loads are connected. The secondary loop pumps draw water as needed from the primary loop to meet the demands of the loads, returning the same volume of water to the primary loop.

In *Figure 37*, the short section of the primary loop called the *hydraulic disconnect* allows water to flow in either direction. This section is kept short to minimize the pressure drop through it.

During periods of peak load, all three chillers in the primary loop produce 45°F (7°C) water to supply the secondary loops. All the pumps would be running to meet peak demand. When the load is reduced, perhaps only the 250-ton chiller and its primary pump are needed to support the load of the secondary loops.

2.5.0 Water System Balancing

In a large building, the water balancing process can be a lengthy and complicated task. Hydronic system balancing is most often done by firms that specialize in the procedure, especially when a new building or system is first commissioned. Proper training beyond the scope of this program is required. In addition, especially when a new building is being commissioned, owners and mechanical engineering firms prefer to have the analysis of system performance provided by a company other than the installing contractor.

Before attempting to balance a water system, the boiler and/or chiller and related components should be operating in accordance with the manufacturer's instructions. The system startup procedures include making sure that the chiller is correctly charged with refrigerant and that the hot-water and chilled-water systems are filled with treated water. All air must be purged from the systems to ensure proper water flow throughout.

2.5.1 Pre-Balance Checks

Before beginning the balancing procedure, the following pre-balancing tasks are generally performed:

- Obtain all necessary technical data, including the original facility plans and specifications or the as-built plans pertaining to the hydronic equipment. This includes the following data:
  - Pump curves and performance data
  - Pressure, water flow and temperature characteristics, and other performance data for boilers, chillers, heat exchangers, and coils
  - Motor nameplate data and ratings
  - Operating instructions and performance characteristics of flow meters and differential pressure gauges
- Review the related piping diagrams very carefully. Obtain or prepare a schematic layout of each circuit. Identify connection points of primary and secondary piping circuits, manual valves, automatic valves, flow meters, air vents, coils, and expansion or compression tanks. Number all water-flow balance and measurement points for rapid identification and reporting.
- Locate and record the ratings and settings of system makeup water pressure-reducing valves, pressure relief valves, and thermal relief valves.
- Document the steps and sequence required for balancing and acquire the instruments needed for the task. Note current system characteristics that could contribute to an unbalanced condition, such as dirty coils on terminals or clogged air filters.
- Ensure that all necessary instruments are properly calibrated. If more than one instrument of a similar type is used, calibrate each one to the same level of precision.

2.5.2 Hydronic System Balancing Procedure

The following general procedure can be used for testing, balancing, and adjusting the flow in a hydronic system:

WARNING!

When working with hydronic systems, be aware of rotating, pressurized, and hot components. Follow all safety guidelines related to the equipment and provided by the equipment manufacturer. Use extreme caution when working near energized electrical loads and controls.

Step 1 Ensure all preliminary visual checks and maintenance tasks have been performed.

Step 2 Ensure all manual valves are open or properly preset.

Step 3 Verify that all automatic valves are as specified for balancing.

Step 4 Read and record the system static pressure at the pump with the pump off.

Step 5 Ensure the system has been commissioned and ready for operation. Ensure all the air has been bled from the system.

Step 6 Start the equipment and primary loop pumps.

Step 7 Measure and record the operating voltage and current of all electric motors.

Step 8 Slowly close the balancing cock (typically a multipurpose valve) in the pump discharge line with the pump running, then record the shutoff discharge and suction pressures at the pump gauge connections. It is best to use a single gauge to take the suction and discharge pressure readings. As soon as you have the readings, slowly reopen the valve.

WARNING!

Never operate a centrifugal pump with the discharge valve closed for more than a few moments. Without water moving through the pump, water in the pump volute becomes hot and can transition to steam over time. The excessive heat and pressure damage the pump seals and can lead to an explosion. Positive displacement pumps should never be deadheaded.

Step 9 Use the shutoff head to determine the pump operating curve for the impeller size and compare it with the design performance data. If there is a discrepancy in the pump output compared to the specifications, report the issue to the system engineer.

Step 10 With the multipurpose valve or shutoff valve fully open, record the pump discharge pressure, suction pressure, and total pump head in feet.

Step 11 Using the total head and the pump curve, read the flow rate from the curve. If the total head is greater than specified, the flow rate will be less than design. If the total head is less than specified, the flow rate will be higher than design. If the flow rate is higher than desired, the pump discharge head should be increased by adjusting the multipurpose balancing valve toward the closed position until the flow rate is about 110% of the job specification.

Step 12 Start the secondary loop pumps and readjust the multipurpose balancing valve in the primary circuit pump discharge if necessary. Record the new readings and multipurpose valve position.

Step 13 Take a complete set of pressure drop readings through all the primary equipment including chillers, boilers, hot-water coils, and chilled-water coils before adjusting any balancing valves associated with each piece of equipment. Compare the readings with the design specifications. Document which pieces of equipment are above or below design requirements, if any.

Step 14 Make a preliminary adjustment in the balancing valve on all units with excess water flow. Adjust them to reduce the flow rate to about 10% above the design flow rate.

Step 15 Perform another complete set of pressure and pump motor current readings. If the pump flow rates have fallen below design, adjust the balancing cock at the pump discharge to increase the flow as needed, bringing the flow rate within 105%–110% of the design flow rate.

Step 16 Decrease the flow a second time on all pieces of equipment that are still receiving more than the design flow rate. This will also increase water flow through units that are not presently receiving enough water.

Step 17 Repeat the procedure (adjusting unit balancing valves, followed by adjusting pump flow rates) until all units are within ±10% of the design water flow. This is the typical tolerance allowed in flow balancing, but job specifications may differ.

Step 18 Make a final reading of the flow rates for the pump and each piece of equipment, as well as the pump motor current readings. Document all final data.

Step 19 After balancing is complete, mark all balancing cocks, gauges, and thermometers at the final settings or range of operation for future reference.

Step 20 Check all water flow safety switches and differential pressure switches to ensure the related equipment shuts down when the flow rate falls below the minimum per the manufacturer's specifications.

Step 21 Finalize and submit all necessary reports.

These are the basic steps required to balance many commercial hydronic systems. However, some hydronic systems and subsystems require additional procedures. As noted earlier, a great deal of additional training and experience, along with certification, is typically needed to properly balance all types of hydronic systems.

2.0.0 Section Review

1. Although some are built to withstand higher pressures, the typical maximum working pressure of a low-pressure hot-water boiler is ______.
   a. 10 psig
   b. 30 psig
   c. 160 psig
   d. 250 psig

2. Which component of a centrifugal pump rotates to create the needed pressure difference for fluid to flow?
   a. Piston
   b. Volute
   c. Impeller
   d. Bearings

3. A butterfly valve can be fully opened or closed by rotating the handle ______.
   a. one full turn
   b. two full turns
   c. one-half turn
   d. one-quarter turn

4. When two-way control valves are used at heating or cooling terminals in a four-pipe system, how must the system pump respond to valve closure?
   a. The pump speed is modulated as needed to maintain a consistent system pressure.
   b. Large pumps are shut down and smaller pumps start to maintain a consistent system pressure.
   c. No change is necessary unless every valve closes, stopping all flow, and then the pump shuts down.
   d. Two-way valves have switches that close when the valve closes, signaling the pump controls to stop the pump and then restart at a new speed.

5. True or False? The typical target for balancing flow through hydronic system equipment is ±5% of the design flow rate.
   a. True
   b. False

3.0.0 Chilled-Water Systems

Objective

Describe commercial/industrial chilled-water equipment and systems.

a. Identify and describe various types of chillers.
b. Identify and describe cooling towers and evaporative condensers.

Performance Task

1. Identify the major components in active commercial hot-water heating and chilled-water cooling hydronic systems.

A chilled-water system eliminates the need to circulate refrigerant through a building to provide comfort cooling. Instead, a chiller (*Figure 38*) uses refrigerant to cool the water. The chilled water is circulated throughout the building and back to the chiller where the water is again cooled.

Figure 38 Packaged air-cooled chiller.
Source: Sakan Piriyapongsak/iStock/Getty Images

After chilled water is delivered to the building, it is usually circulated through finned coils in air handling units and fan coil units to deliver conditioned air.

3.1.0 Chillers

This section focuses on chilled-water systems that operate above 40°F (4°C) and are used for comfort cooling. Colder chilled-water systems may be used to support industrial processes and comfort systems that have an integrated ice-building system to provide cooling. These systems cool a brine or glycol/ water solution, operating at water temperatures below 32°F (0°C).

Figure 39 shows a simplified diagram of a simple chilled-water system. It consists of a chiller, coils, a chilled-water pump, and the related chilled-water piping. A system may have only a few coils or other terminals, or it may have hundreds.

For comfort cooling applications, the chilled-water temperature supplied to the building is generally around 45°F (7°C), with the returning water temperature expected to be around 10°F (6°C) warmer. Of course, the return water temperature varies as the load varies. The chilled water acts as a *secondary coolant*, which is any fluid used to transfer heat from the conditioned space without a change in its state (vaporizing).

Chillers fall into two classes: *mechanical* and *non-mechanical*. Mechanical chillers are by far the most common and use various types of compressors to provide cooling, relying on the familiar vapor compression cycle. Non-mechanical chillers approach cooling in a very different way. The primary example of a non-mechanical chiller is called an *absorption chiller*, which is presented later in Section 3.1.4, *Absorption Chillers*.

A packaged chiller is a piece of equipment that includes the compressor, condenser, evaporator, metering device, and basic controls. The system water pumps are rarely part of the package, as the pumps must be sized and selected specifically for each application.

The three common types of mechanical chillers presented in this section include reciprocating or scroll chillers, centrifugal chillers, and rotary screw chillers.

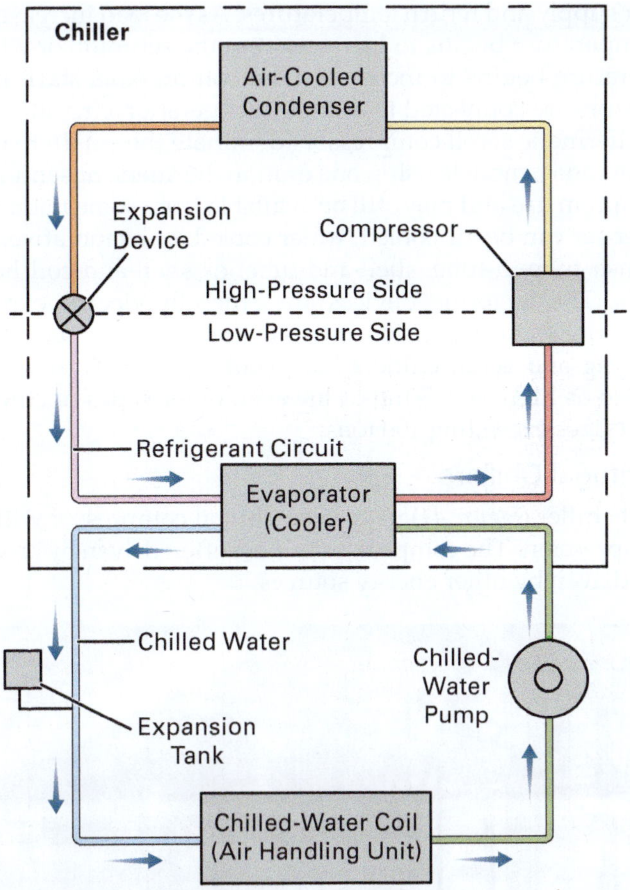

Figure 39 Basic chilled-water system.

3.1.1 Reciprocating and Scroll Chillers

Reciprocating and scroll chillers (*Figure 40*) use one or more compressors to provide the necessary cooling capacity. When multiple compressors are used, the compressors are generally staged by temperature controls monitoring the

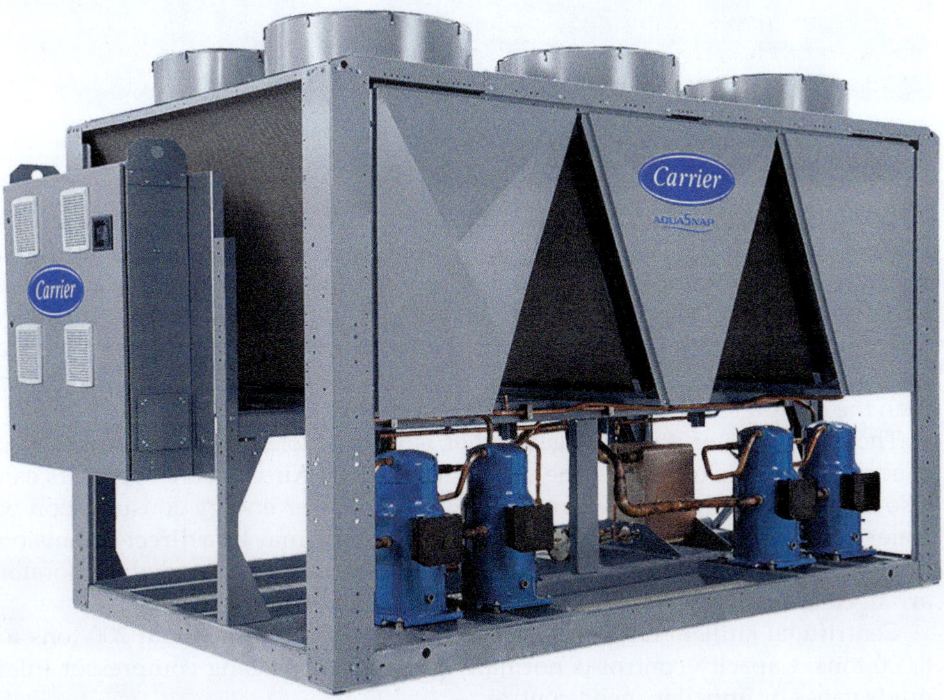

Figure 40 Packaged air-cooled scroll chiller.
Source: Carrier Corporation

chilled-water supply and return temperatures. As the load increases and the supply water temperature begins to increase from the setpoint, or when the return water temperature begins to increase, more compressors start. In some cases, two compressors are connected to a single refrigeration circuit.

Although hermetic scroll compressors dominate the small- to medium-sized chiller market today, models using one or more hermetic or semi-hermetic compressors remain in use and may still be available from some manufacturers. The chiller condenser can be air cooled, water cooled, or evaporative. Water-cooled condensers use tube-in-tube, shell-and-tube, or shell-and-coil heat exchanger arrangements. The metering devices are generally electric expansion valves (EEVs), but thermostatic expansion valves (TXVs) are also used.

Reciprocating and scroll chillers have cooling capacity sizes ranging from about 5 tons to as high as 225 tons. However, other types of chillers dominate the market in sizes exceeding 100 tons.

3.1.2 Centrifugal Chillers

A centrifugal chiller (*Figure 41*) uses a centrifugal compressor with one or more stages of compression. The compressor is most often driven by an electric motor, but it can be driven by other energy sources.

Figure 41 Centrifugal chiller.
Source: BrianBrownImages/iStock/Getty Images

Centrifugal compressors use a high-speed impeller with many blades that rotate in a spiral-shaped housing. Refrigerant vapor is fed into the housing at the center of the impeller. The impeller throws the incoming vapor outward in a circular path from between the blades and into the compressor housing. This centrifugal action adds energy to the refrigerant, increasing the pressure forcing it out the discharge port. However, the compression ratio is much lower than the compression ratio developed by other compressor types.

The condenser of a centrifugal chiller is water-cooled, with the refrigerant condensing on the outside of the water-filled tubes. Air-cooled condensers can also be used, but they cannot compete with the lower energy consumption of water-cooled models. The chiller evaporator, or *cooler*, may be a direct-expansion type with a familiar metering device, but most units have a flooded evaporator arrangement using a float valve to maintain the liquid level.

Centrifugal chillers have cooling capacities ranging from about 200 tons to 8,000 tons. Capacity control is normally accomplished using compressor inlet guide vanes or impeller speed control.

HCFC-123 and HFC-134a are common refrigerants used in centrifugal chillers. HCFC-123 is a low-pressure refrigerant, while HFC-134a is a high-pressure refrigerant. Today, chillers using more environmentally friendly refrigerants are being introduced. These refrigerants include HFO-514A and HFO-1233zd. Carbon dioxide (CO_2) is also an option.

Going Green

Central Plant Facilities

Central plant facilities that provide hot and chilled water to large service areas have long been used for college campuses, airports, high-technology manufacturing centers, and some municipalities. Central plant technology is experiencing renewed interest from energy-conscious system designers. Many new facilities combine cogeneration capability that uses the heat produced by stand-alone electricity generation to support hot-water generation or even to power chillers. Typical chiller tonnages for central plant facilities begin in the 3,000-ton range and may reach the 8,000- to 10,000-ton range.

Methods of increasing refrigerant subcooling without an energy penalty are often used to conserve energy. One arrangement introduces the warm, condensed liquid refrigerant into the condenser where it meets the coldest water tubes. This causes the condensed refrigerant to subcool further, increasing the refrigeration cycle efficiency and reducing power consumption. This arrangement is referred to as a **thermal economizer**.

A **flash economizer** is an option for multistage centrifugal chillers. *Intercooling* is a process in which the condensed refrigerant leaving the condenser is passed through a flash chamber before being sent to the evaporator (*Figure 42*). The pressure in this chamber falls between the condenser and evaporator pressures. Because its pressure is lower than the condensing pressure, some of the liquid refrigerant flashes to a vapor (flash gas). The flashing refrigerant serves to subcool the remaining liquid. The vapor is drawn from the flash chamber and routed to the inlet of the second-stage impeller, where it is mixed with discharge gas from the first-stage impeller. The flash gas passes through only one stage of compression to reach the condenser pressure, saving power.

Thermal economizer: An energy conservation arrangement used in chillers in which warm liquid refrigerant is brought into contact with the coldest water tubes in a condenser or heat exchanger. This causes additional subcooling of the condensed refrigerant.

Flash economizer: An energy conservation arrangement used in chillers that separates vapor (flash gas) from liquid leaving the condenser, then routes the vapor directly to the second stage of a multistage centrifugal chiller, bypassing the evaporator and the first stage of compression.

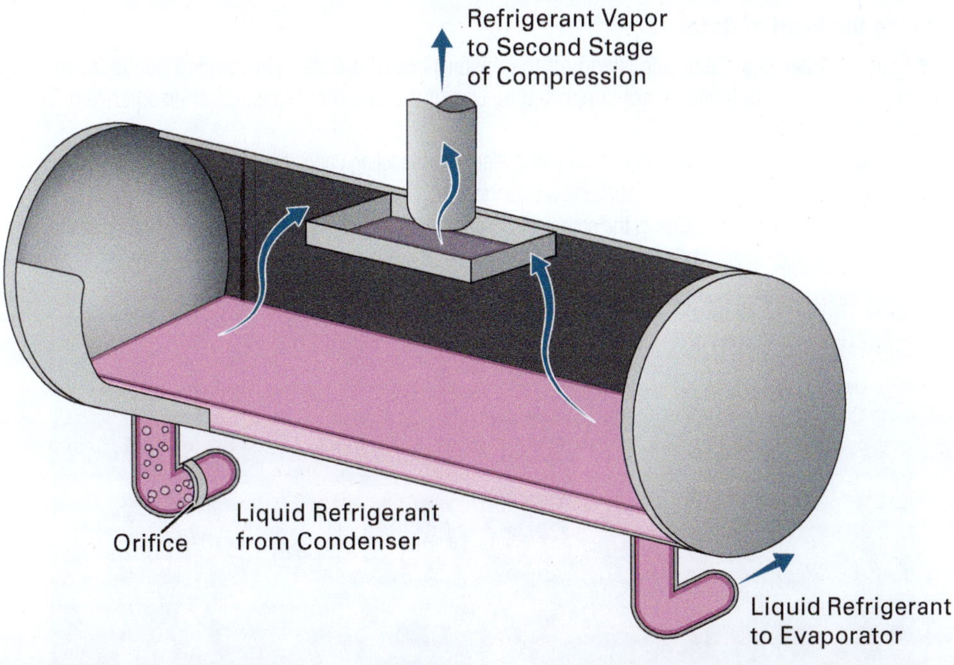

Figure 42 Flash economizer.

Purge units (*Figure 43*) are another device used on centrifugal chillers containing low-pressure refrigerants, such as HFO-1233zd, HFO-514A, and HCFC-123. Chillers using these refrigerants operate below atmospheric pressure on the low side of the refrigerant circuit. Because of this, air and moisture tend to leak into the system, rather than the refrigerant leaking out. A purge unit is used to separate and remove non-condensable gases and water vapor from the system.

Figure 43 Centrifugal-chiller purge unit.
Source: Trane International Inc.

Purge units are small, independent refrigeration units mounted on the chiller. These units draw in refrigerant vapor from the chiller, then chill the refrigerant vapor enough for it to condense. Any vapor that remains is considered a non-condensable gas and is released (purged) to the atmosphere. Any moisture in the refrigerant is captured by filter-driers. The condensed liquid refrigerant is then returned to the chiller.

Around the World

Cooling the Pearl of Qatar

The Pearl of Qatar is a manmade island off the eastern coast of Qatar. The cooling needs for the massive island and its facilities are supported by one of the world's largest centralized chilled-water plants.

The plant provides a nominal 130,000 tons of chilled-water cooling capacity using 46 chillers arranged in pairs. Each pair is part of a series counterflow arrangement. The primary pumping system moves 180,000 gpm of chilled water. Due to the scarcity of water in this part of the world and the demands of the water-cooled plant, a desalination plant is also part of the facility.

Source: Leonid Andronov/iStock/Getty Images

3.1.3 Rotary Screw Chillers

Rotary screw chillers (*Figure 44*) use one or more oil-injected hermetic or open-drive rotary screw compressors. Rotary screw compressors use a matched set of screw-shaped rotors, one male and one female, enclosed within a cylinder. The motor drives the male rotor, which in turn drives the female rotor.

As the two rotors turn in opposing directions, the gas is compressed between them. The vapor moves through a series of progressively smaller spaces between the lobes until the compressed vapor exits.

Rotary screw chillers can also be water-cooled (*Figure 45*), using a shell-and-tube condenser, with the refrigerant condensing on the outside of the water-filled tubes. However, air-cooled condensers are also very popular in this type of chiller, since air-cooled condensers are less expensive to install and require less maintenance. Water-cooled chillers are located indoors, in equipment rooms, while air-cooled chillers are located outside the building.

Figure 44 Air-cooled rotary screw chiller.
Source: Carrier Corporation

Figure 45 Water-cooled rotary screw chiller.
Source: Carrier Corporation

The evaporator may be either direct-expansion or flooded. When a direct-expansion evaporator is used, the refrigerant flows through the tubes, which are immersed in the water. With a flooded evaporator, the refrigerant fills the shell and the water flows through the tubes.

A flash economizer, similar to the one described for the centrifugal chiller, can be used to improve chiller efficiency. In the screw compressor, flash gas produced in the economizer is injected into the compressor at an intermediate point of the compression cycle, along the length of the screws instead of at the end.

Rotary screw chillers range from about 40 tons to 850 tons. Capacity control is normally accomplished with a hydraulically operated slide valve in each compressor, but units with multiple compressors use staging for capacity control as well.

3.1.4 Absorption Chillers

Absorption chillers (*Figure 46*) offer a completely different approach to chilled-water production. Although rarely encountered by most technicians, absorption chillers offer an energy-conscious solution in facilities where a large volume of waste heat or steam is available. To chill water, the absorption cycle requires the addition of heat.

Figure 46 Lithium-bromide absorption chiller.
Source: sunstock/Shutterstock

Lithium bromide (LiBr): A compound salt solution containing lithium and bromine that readily absorbs moisture.

Water and a solution of water containing **lithium bromide (LiBr)** are the working fluids of the absorption cycle. Lithium bromide is a caustic salt solution.

WARNING!

Lithium bromide is potentially harmful. The compound reacts rapidly with moisture on the skin to cause burns, damages the respiratory tract and lungs if inhaled, and is highly corrosive. Contact with the eyes can cause irritation, blurred vision, and permanent eye damage.

When handling lithium bromide, always use the appropriate PPE as recommended in the Safety Data Sheet (SDS). This includes safety goggles, thick rubber gloves, long-sleeve shirt, chemical-resistant suit, and an approved respirator where dust or fumes are present.

Absorption chillers are available in capacities ranging from 50 tons to 1,800 tons. As a rule, absorption chillers are used in applications where a significant amount of steam is readily available, often where refuse-burning facilities that generate vast amounts of steam are nearby, or onboard sea-going vessels with large steam boilers. Due to their rarity and nature of the absorption cycle, their operation is quite complex. For this reason, an in-depth study of their operation is beyond the scope of this program.

3.2.0 Cooling Towers and Evaporative Condensers

Water-cooled condensers are very common in commercial and industrial systems. For the condenser to operate properly, the condenser water must be at a temperature lower than the condensing temperature of the refrigerant in use. Water-cooled condensers reject the heat to circulating water, which is then circulated through a cooling tower or evaporative condenser.

On Site

Legionnaires' Disease

Regular maintenance is a must in any water system, but its importance in open systems cannot be stressed enough because of the health and safety issues related to Legionnaires' disease and similar illnesses. Poorly maintained cooling towers and evaporative condensers are known to support the growth of *Legionella* bacteria. The aerosol produced by the bacteria can potentially infect not only the people at the equipment site, but also those in the surrounding area.

Legionnaires' disease can be fatal, but our knowledge and understanding of the disease has largely eliminated fatalities today. However, it remains a concern, and outbreaks do still occur. If cooling towers are not properly maintained and treated, a building owner and/or the contractor responsible for maintenance may be held liable for outbreaks.

3.2.1 Cooling Towers

In a cooling tower, the heat absorbed by the water in the condenser is rejected into the atmosphere through evaporation. The cooled water is then recirculated back through the condenser.

Cooling towers are usually designed to reduce the water temperature by approximately 10°F (6°C) depending on the location of the system and local climate. A cooling tower is only able to cool water to about 3°F to 7°F (2°C to 4°C) above the ambient wet-bulb temperature. Because these towers depend upon the evaporation of water for a cooling effect, effectiveness improves and their size can be reduced in areas of the country where the humidity is low.

Cooling towers pass outdoor air over the warm water to remove the heat from the water. The cooler and drier the surrounding air, the better. Old cooling towers were constructed of wood and relied on a natural draft created by the difference in air density in the top and bottom of the tower. However, as the need for cooling capacity increased, cooling towers required more airflow than nature could provide. HVACR-related cooling towers must often be placed in limited spaces, so a larger size is not always a viable option. Thus, fans were added to towers to maximize performance in a smaller package.

On Site

Natural-Draft Cooling Towers

Although very few HVACR systems rely on natural-draft cooling towers today, these towers are still used in industrial applications and power plants. The tall hyperbolic towers associated with power plants rely on a natural draft. The immense height allows strong updrafts to be generated by the warm water sprayed into the air at the lower portion of the tower.

Source: TomasSereda/iStock/Getty Images

A *forced-draft tower* (*Figure 47*) is one in which fans force air up from the bottom of the tower and through the falling water droplets. In an *induced-draft tower* (*Figure 48*), the fans are mounted above the falling water, drawing (inducing) the air up from the bottom. Both types of towers are referred to as *counterflow towers* because the air and water are moving in opposing directions.

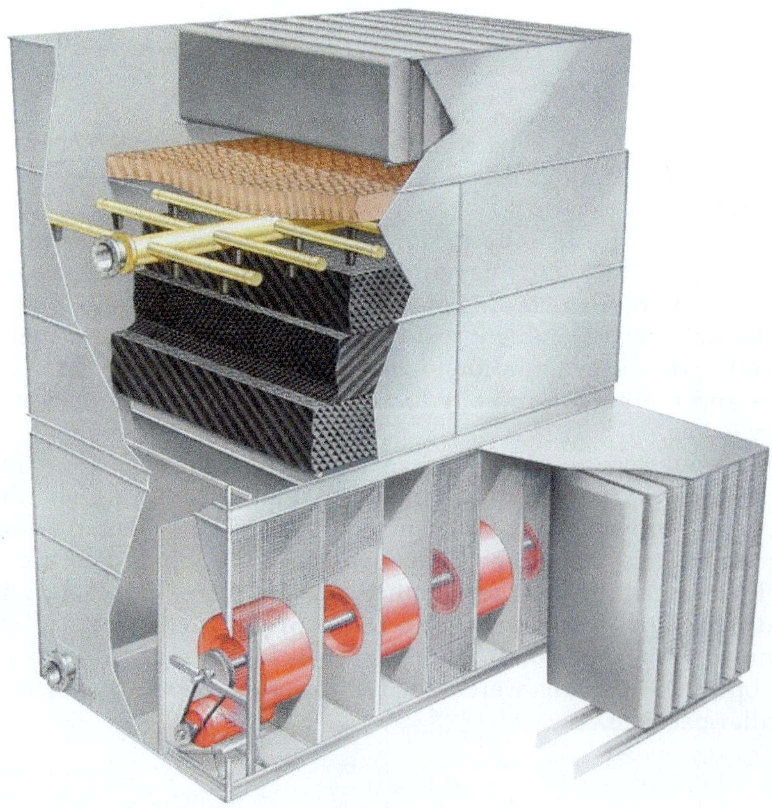

Figure 47 Forced-draft cooling tower.
Source: Photo provided courtesy of Marley/SPX Cooling Tech.

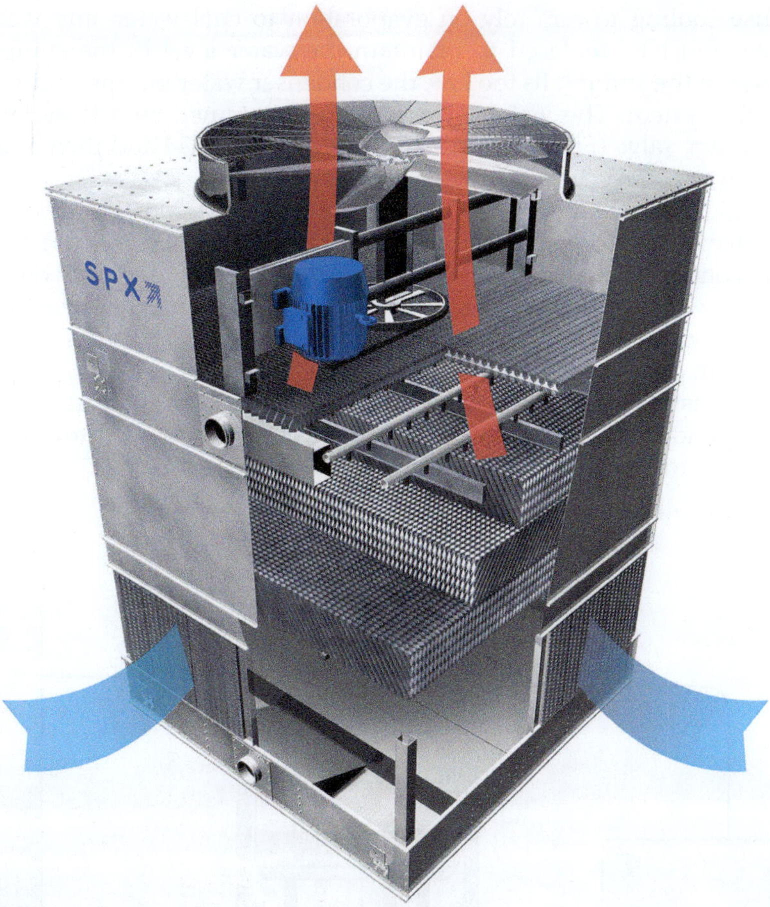

Figure 48 Induced-draft cooling tower.
Source: Photo provided courtesy of Marley/SPX Cooling Tech

A *crossflow tower* (*Figure 49*) distributes the hot water through fill placed vertically on the sides of the tower, perpendicular to the airflow. The water trickles down through the fill while the air is drawn horizontally through the fill by a fan on the top.

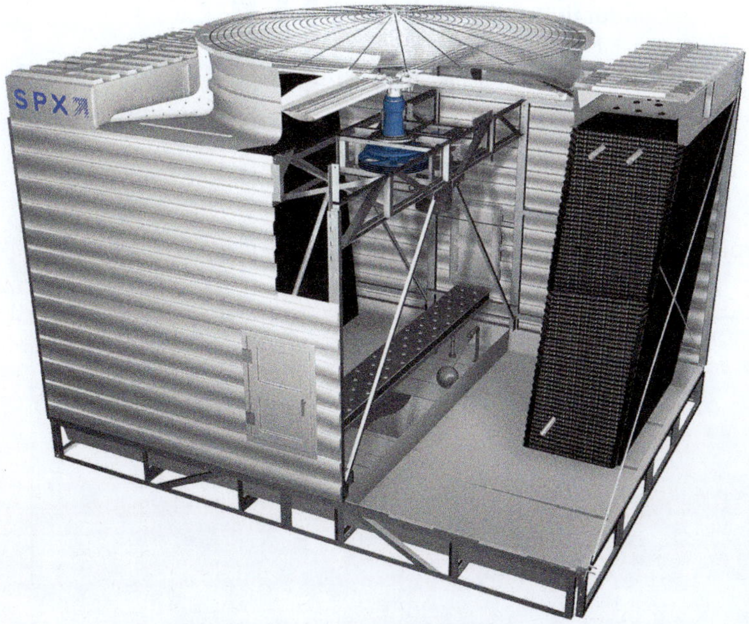

Figure 49 Crossflow cooling tower.
Source: Photo provided courtesy of Marley/SPX Cooling Tech.

Figure 50 Cooling tower float valve in a tower sump.
Source: Courtesy of EVAPCO, Inc.

Because cooling towers rely on evaporation to cool water, any water that evaporates must be replaced to maintain the water level in the sump. If the water level in the sump falls too low, the condenser water pumps begin to draw air into the system. The water level is often maintained by a float-controlled makeup water valve (*Figure 50*) that senses the water level and then opens and closes the water valve as the water level rises and falls.

Electronic water level sensing systems are also used. Most of these makeup water systems rely on a set of electronic probes (*Figure 51*) to sense the water level and control a solenoid valve in the water line. The probes rely on the conductivity of water to sense depth. The advantage to electronic sensing is the ability to provide better information regarding the tower status, as well as provide an alarm circuit.

In many installations, a bypass line is installed to bypass the sensor or float and allow rapid filling of the tower. A bypass line also allows the tower sump to be filled prior to startup before power is available.

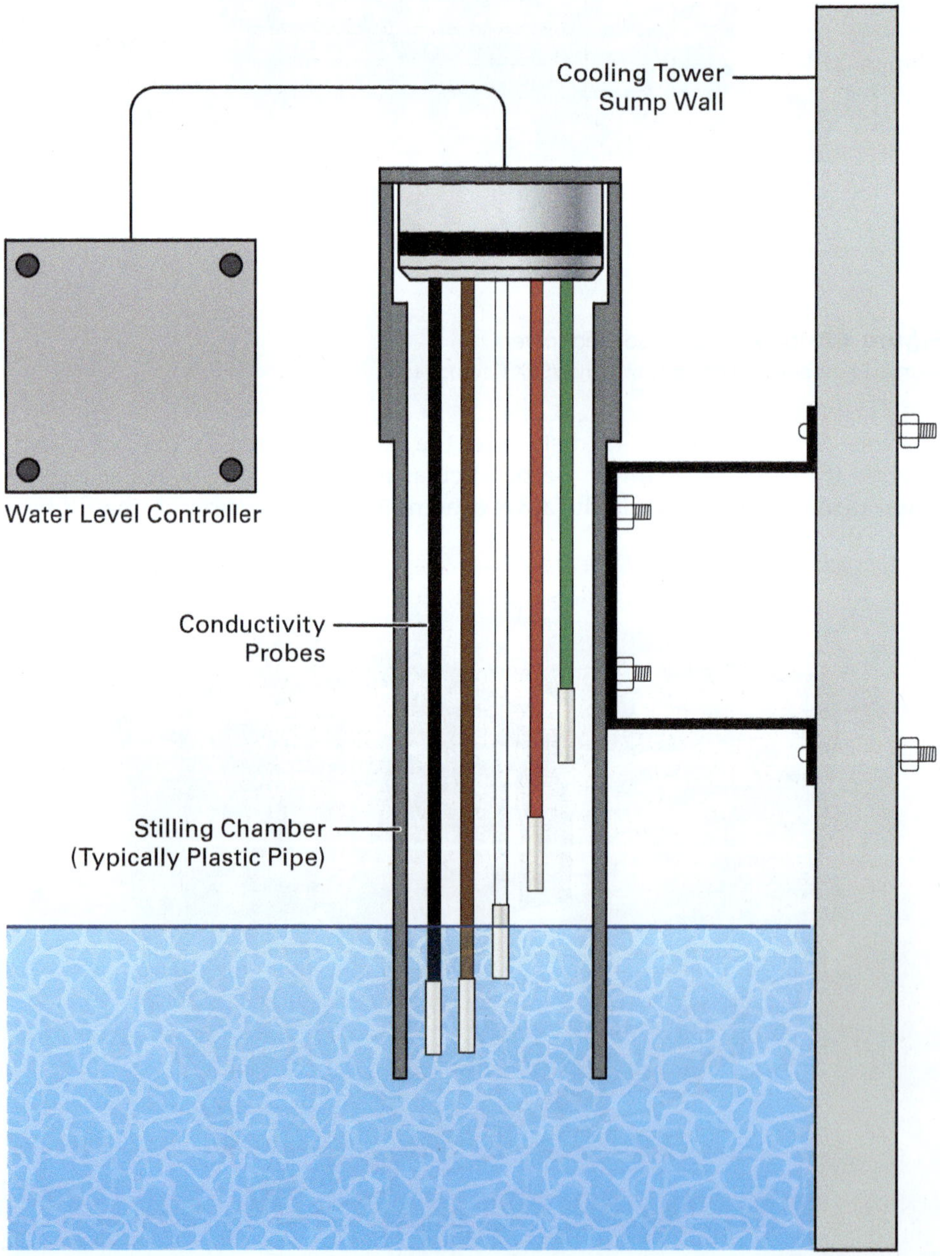

Figure 51 Cooling tower sump probe.

3.2.2 Evaporative Condensers

From the outside, evaporative condensers look like cooling towers. However, refrigerant lines are connected to the evaporative condenser as opposed to large water piping from a condenser water pump. Essentially, an evaporative condenser (*Figure 52*) is a cooling tower with the HVACR system condenser inside.

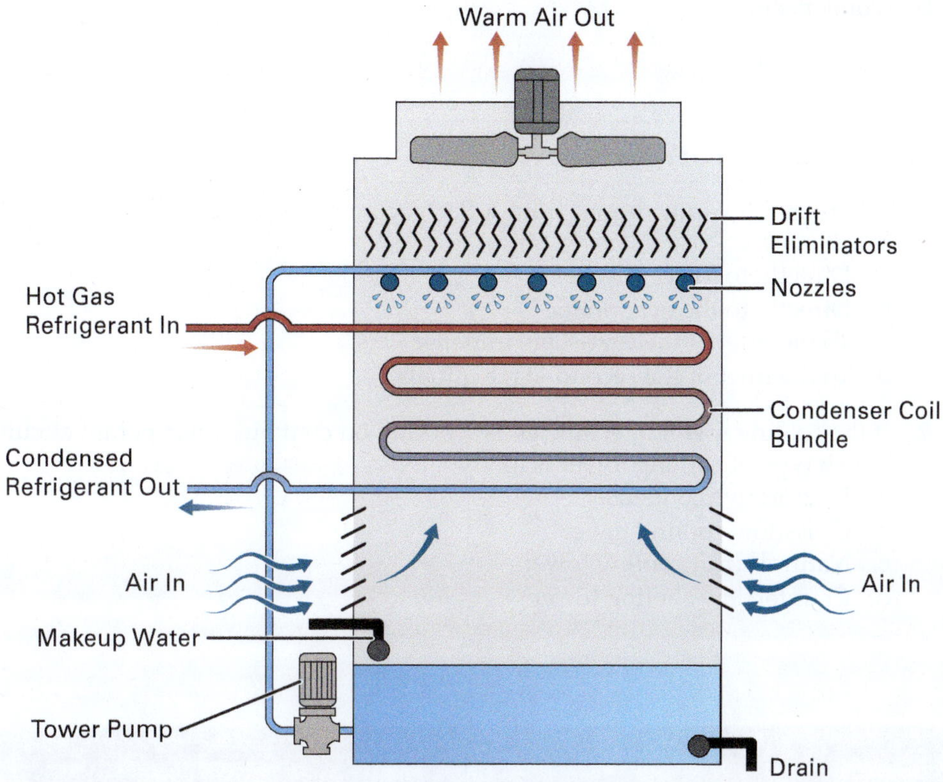

Figure 52 Evaporative condenser.

Air enters the bottom of the unit and flows upward over the refrigerant-filled condensing tube bundle (*Figure 53*). The tubes may be copper or steel, depending on the refrigerant used. Water from the sump is pumped to the top and sprayed over the tube bundle. Both the air and sprayed water absorb heat from the

Figure 53 Evaporative condenser tube bundle.
Source: Photo provided courtesy of Marley/SPX Cooling Tech

refrigerant flowing through the tube bundle. Water eliminators located beneath the fan keep water entrained in the rising air from escaping the condenser.

Placing the refrigerant condenser to the evaporative condenser requires additional refrigerant piping and a much larger refrigerant charge. Thus, there is a greater potential for damage and refrigerant leakage. Large ammonia systems for refrigerated storage warehouses and food processing often rely on evaporative condensers.

3.0.0 Section Review

1. Where is the vapor drawn from a flash economizer on a centrifugal chiller sent?
   a. Directly to the condenser
   b. Directly to the evaporator
   c. To the inlet of the first-stage impeller
   d. To the inlet of the second-stage impeller

2. If the cooling tower a technician is working on contains a refrigerant circuit, which type of cooling tower is it?
   a. Evaporative condenser
   b. Crossflow cooling tower
   c. Natural-draft cooling tower
   d. Counterflow cooling tower

Module 03305 Review Questions

1. Water in its liquid form changes its density when what other factor changes?
   a. Polarity
   b. Velocity
   c. Flow rate
   d. Temperature

2. How much static pressure does a column of water 100' high exert at its base? (*Round your answer to the nearest whole number.*)
   a. 22 psi
   b. 43 psi
   c. 62 psi
   d. 144 psi

3. A hydronic piping system was originally designed for a flow rate of 50 gpm with an initial pressure drop of 6 psi. If the flow rate is increased to 100 gpm, what would the pressure drop be after the change? (*Round your answer to the nearest whole number.*)
   a. 2 psi
   b. 12 psi
   c. 24 psi
   d. 72 psi

4. What is the typical maximum operating temperature for a low-pressure boiler?
   a. 180°F (82°C)
   b. 212°F (100°C)
   c. 250°F (121°C)
   d. 400°F (260°C)

5. Which of the following is a characteristic associated with a copper-finned tube boiler when compared to other boiler types?
   a. Lightweight
   b. Extended life
   c. Inefficient heat transfer
   d. Greater heating capacity

6. Which boiler type allows the heating capacity to be increased by adding more sections?
   a. Firetube
   b. Cast iron
   c. Watertube
   d. Vertical tubeless

7. True or False? In a steel firetube boiler, water flows through the tubes while the heat and flue gases flow around the outside of the tubes.
   a. True
   b. False

8. Which of the following statements related to condensing boilers is *correct*?
   a. Type HT flue venting materials are recommended for condensing boilers.
   b. The combustion efficiency range of a condensing boiler starts at about 80%.
   c. Condensing boilers do not perform well when placed in a modular arrangement.
   d. The condensate from condensing boilers often must be neutralized before it enters a drain.

9. In centrifugal pumps that have one, what does a packing gland do?
   a. Reduces pump shaft friction
   b. Stores lubricating oil for the packing
   c. Helps control fluid leakage at the shaft
   d. Attaches the impeller to the motor shaft

10. When a pump is *not* equipped with packing, what component provides the same service?
    a. Bearings
    b. Wear rings
    c. Impeller vanes
    d. Mechanical seal

11. To force feedwater into a high-pressure boiler, which type of pump is generally required?
    a. Multistage pump
    b. Double-suction pump
    c. Packed centrifugal pump
    d. Centrifugal pump with a mechanical seal

12. Commercial air separators enable air to separate from the water stream by significantly reducing the water _____.
    a. pressure
    b. velocity
    c. temperature
    d. flow volume

13. True or False? The cooling loads in a primary-secondary piping system are connected to the primary loop.
    a. True
    b. False

14. Which of the following is considered a non-mechanical chiller?
    a. Scroll chiller
    b. Centrifugal chiller
    c. Absorption chiller
    d. Rotary screw chiller

15. Where on the cooling tower is the fan or blower for a forced-draft tower located?
    a. Top
    b. Center of the side
    c. Bottom of the side
    d. Beneath the tower itself

Answers to odd-numbered questions are found in the Module Review Answer Key at the back of this book.

Answers to Section Review Questions

Answer	Section	Objective
Section One		
1. d	1.1.1	1a
2. a	1.2.1	1b
Section Two		
1. b	2.1.0	2a
2. c	2.2.1	2b
3. d	2.3.3	2c
4. a	2.4.1	2d
5. b	2.5.2	2e
Section Three		
1. d	3.1.2	3a
2. a	3.2.2	3b

MODULE 03201

Commercial Airside Systems

Source: Dmytro Kuznietsov/iStock/Getty Images

Objectives

Successful completion of this module prepares you to do the following:

1. Describe common approaches to commercial air distribution.
   a. Describe single-zone constant-volume systems.
   b. Describe multizone constant-volume systems.
   c. Describe variable volume and temperature (VVT) systems.
   d. Describe variable air volume (VAV) systems.
   e. Explain the need for ventilation and exhaust systems in commercial buildings.

Performance Tasks

This is a knowledge-based module. There are no Performance Tasks.

Overview

Commercial buildings have different needs than a residential environment. For example, some zones in a large building may require heat while others require cooling. The volume of conditioned air a large building needs far exceeds the requirements of a residence, and the air may also require more treatment to ensure it is clean and oxygenated. This module introduces commercial airside concepts and explains how the demanding requirements of a commercial space are met.

1.0.0 Commercial Air System Designs

Performance Tasks

There are no Performance Tasks in this section.

Objective

Describe common approaches to commercial air distribution.

a. Describe single-zone constant-volume systems.
b. Describe multizone constant-volume systems.
c. Describe variable volume and temperature (VVT) systems.
d. Describe variable air volume (VAV) systems.
e. Explain the need for ventilation and exhaust systems in commercial buildings.

Human comfort in a building is commonly associated with two air characteristics: temperature and relative humidity. However, the broader definition of comfort includes clean, oxygen-rich air that is free of odors and annoying drafts.

Figure 1 shows a typical office area. Heat flows into the office from outdoors during warm weather. Additional heat is generated by the people, lights, and equipment in the room. The office must be cooled to prevent an uncomfortable rise in room temperature. Because this office contains a thermostat, it is called a **control zone**.

Control zone: A building, group of rooms, single room, or part of a room controlled by a dedicated thermostat.

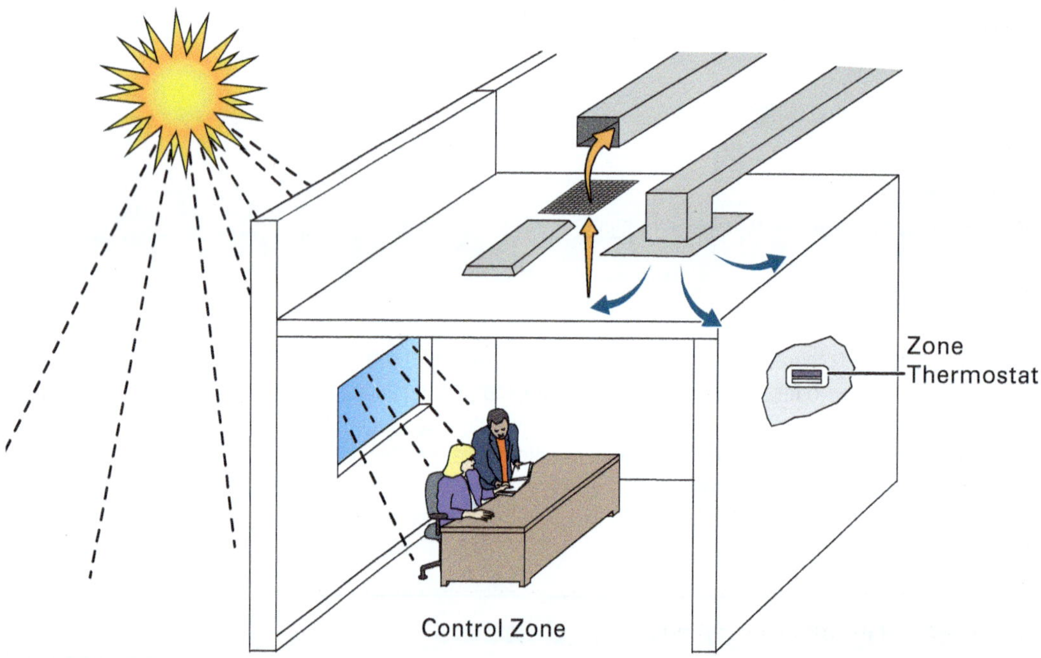

Figure 1 A typical control zone.

Each air conditioning system or terminal needs at least one control zone. A control zone may be a single room, several rooms, a floor, or an entire building. As the size of the building increases, the number of control zones also tends to increase. Some commercial buildings have a **zoned system** in which a single unit or system serves multiple control zones. These systems are also popular for residential applications.

Zoned system: An HVAC comfort system responsible for meeting the needs of two or more control zones.

Many air systems in commercial buildings are comprised of a split-system condensing unit or heat pump matched with an air handling unit (AHU) or fan coil unit. These are referred to as *unitary systems* by manufacturers. Larger buildings may have a chilled-water and/or hot-water system instead of a condensing unit or heat pump. These systems are supported with large AHUs located in equipment rooms. Such systems are referred to as *applied systems*.

Figure 2 shows a typical split system and identifies its major parts. These systems may also be packaged, like the average rooftop unit (RTU). Note that the

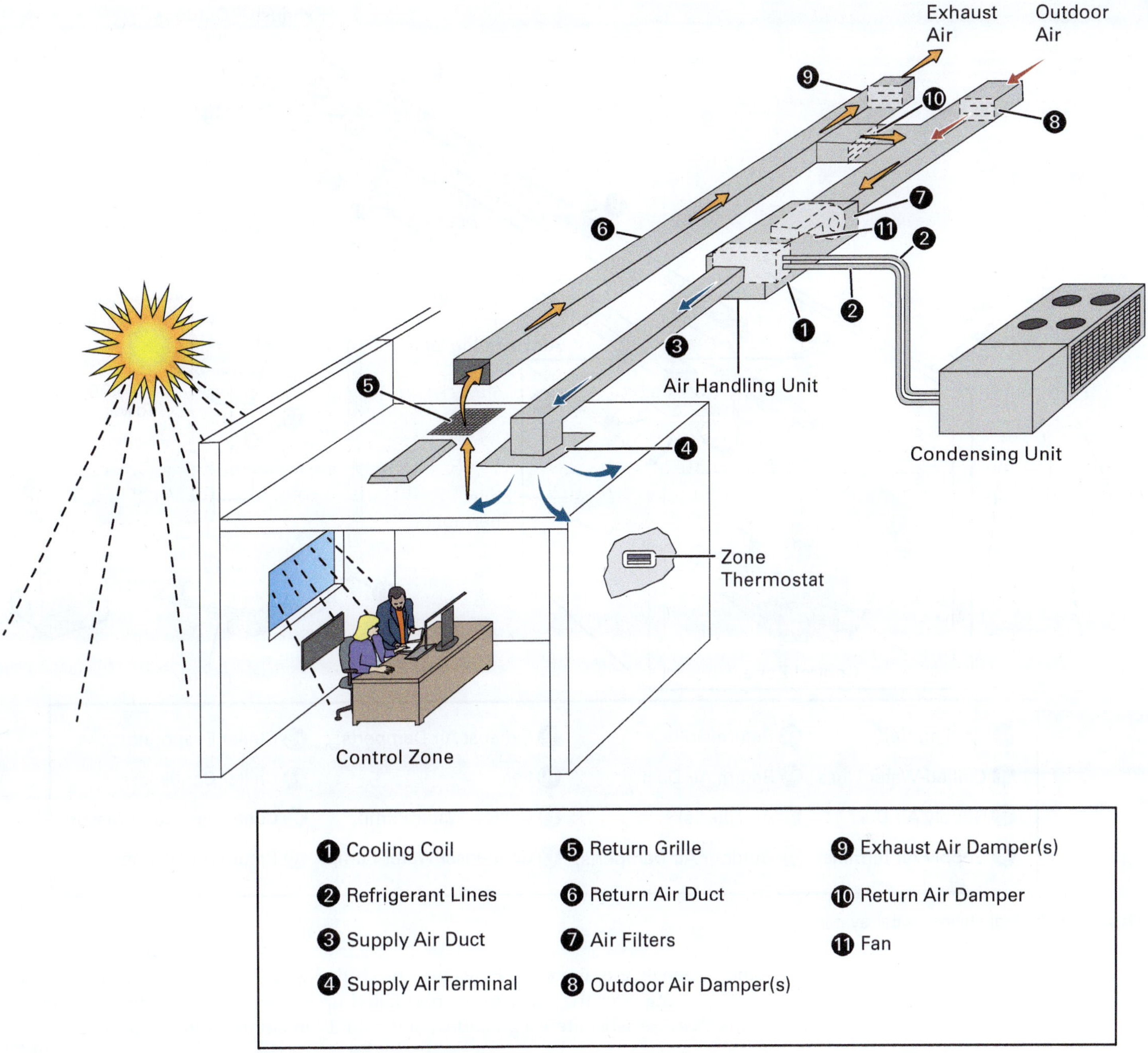

Figure 2 Typical split system in a commercial setting.

system has an outside air intake for ventilation. Bringing fresh outdoor air into the zone ensures that acceptable oxygen levels are maintained, and indoor pollutants are diluted. A set of dampers (8) is used to control the volume of outdoor air entering the system. Another set of dampers (9) is used to exhaust building air to maintain the proper building pressure and remove excess carbon dioxide (CO_2). The AHU mixes outdoor air with return air before conditioning the air and sending it downstream to the control zone.

Figure 3 shows a typical chilled-water system and identifies its components. A chilled-water coil (1) replaces the direct-expansion (DX) evaporator coil. Chilled-water systems are typically used in larger multistory buildings that require more than 100 tons of cooling capacity, but the economics can also be favorable in smaller applications.

Using chilled water as a cooling medium has some advantages. The temperature and volume of air flowing through a DX coil is critical to the reliable operation of the system. A reduction of airflow can lead to unstable operation, liquid floodback,

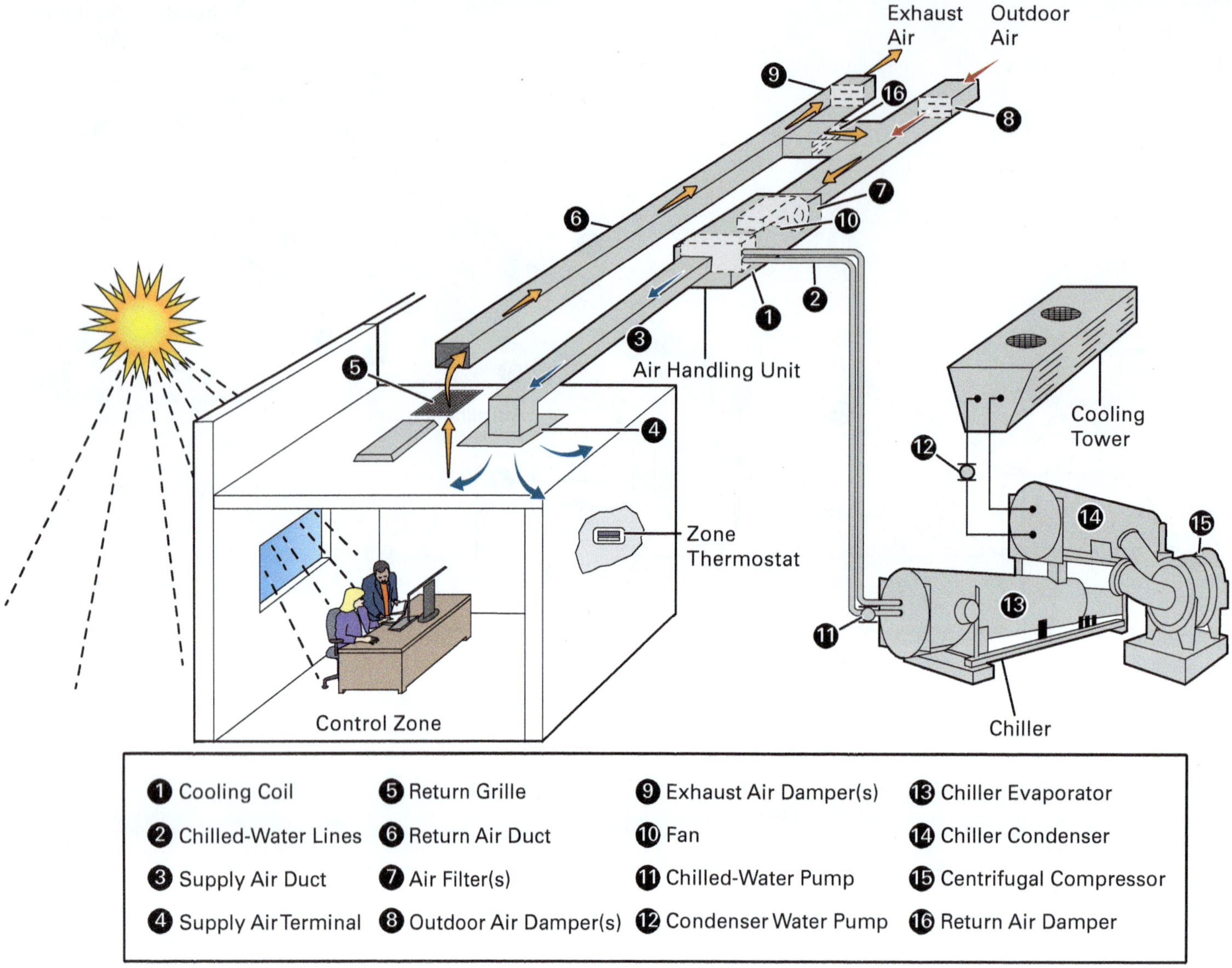

Figure 3 Typical chilled-water system.

and frozen evaporator coils. However, the chilled-water flow, as well as the air volume passing over the coil, can be modulated up or down without these concerns. This characteristic offers a great deal of flexibility in environmental control.

Did You Know?

Activated Carbon Filters and $KMnO_4$

Odor control is of particular concern today in commercial buildings because of tight buildings, new building materials, and outside air pollution. Filter devices used to control odors and other contaminants in all-air systems sometimes use activated carbon or potassium permanganate ($KMnO_4$) in the filtration section to adsorb odors. $KMnO_4$ is also used to disinfect drinking water and delay fruit ripening by adsorbing ethylene gas.

Four basic types of air systems are in use today. Each was developed to satisfy a building's zoning needs at a practical cost for the building type. The four types are as follows:

- Single-zone constant-volume
- Multizone constant-volume
- Variable volume and temperature (VVT)
- Variable air volume (VAV)

On Site

The Incredibly Simple Air Distribution System

Large commercial buildings without ceilings and walls don't need complex air distribution arrangements. The average home-improvement warehouse store is a perfect example. Packaged rooftop units can simply be equipped with concentric duct packages that drop straight down through the roof to a single supply/return grille. The supply diffuser can take various forms, with some fitted with heavy-duty louvers on each of the four sides to provide directional control of the airflow and a higher velocity airstream.

Concentric duct packages can be dropped down to a height a few feet above the material shelves, allowing hot air to stratify at the top of the building in the summer. Heat reclaim fans can be used in the winter to push the warmer air toward the floor.

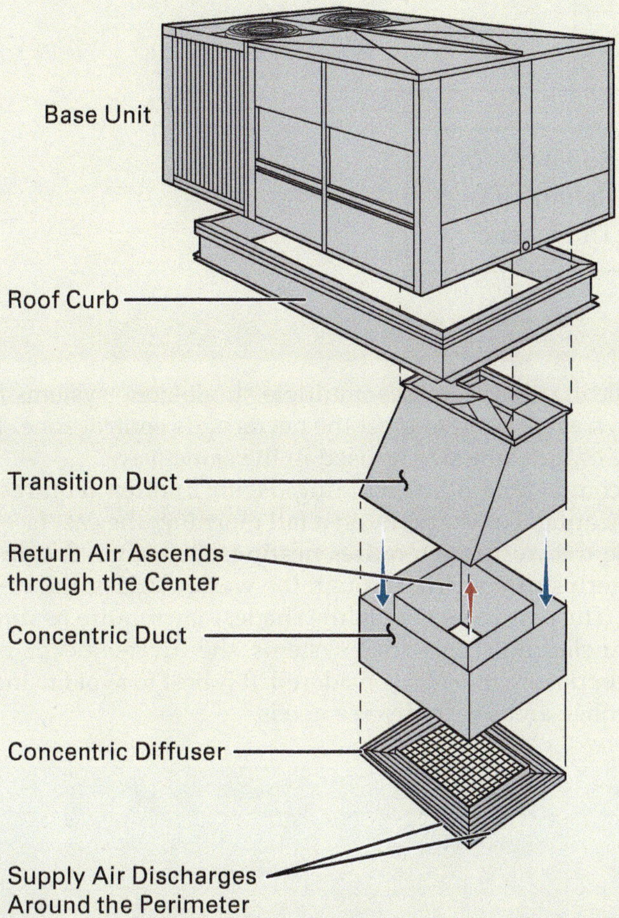

1.1.0 Single-Zone Constant-Volume Systems

Figure 4 shows a single-zone, constant-volume arrangement with three systems. Each system serves a single control zone. An area can be considered a control zone if it contains a thermostat, although the affected zone may extend well beyond a single room. Each RTU supplies a constant volume of air to its assigned control zone whenever the thermostat calls for heating or cooling. Three common equipment arrangements are typically used for single-zone, constant-volume systems: rooftop packaged units, vertical indoor packaged units, and split systems.

For small buildings where less than 10 tons of cooling capacity is needed, it is not unusual to see only one unit providing comfort for the entire facility. The building may have multiple rooms, but only one control zone.

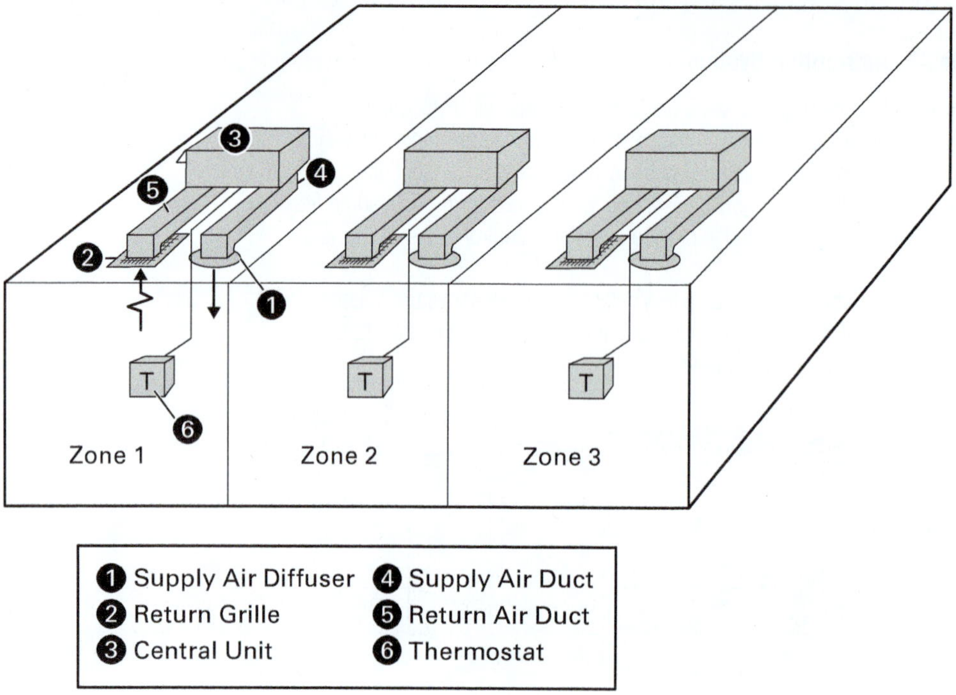

Figure 4 Single-zone, constant-volume systems.

As the building size increases, multiple single-zone systems are used. Each single-zone system is sized to meet the needs of its control zone. The zones may or may not be of the same size or used in the same way.

A typical commercial office building (*Figure 5*) often requires simultaneous cooling and heating. For example, on a fall morning, the east face might require cooling, while the west face requires heating. That same day, in the afternoon with the western sun pouring through the western glass, the west zone would need cooling. The east zone, now in the shade, may require heating. In all building designs, including single-zone systems, the diverse needs of the building and the solar exposure must be considered. It is best to avoid using a single-zone system to manage areas with diverse needs.

Figure 5 Low-rise commercial office building.
Source: qingwa/iStock/Getty Images

Despite the size of shopping malls, many use multiple single-zone units. Each retailer usually has its own single-zone system(s). Larger stores are provided with multiple units as needed, but the air distribution arrangement may mimic that of the smaller stores. If the larger store leaves, the mall can easily break the space back down into smaller spaces with a unit dedicated to each space. Likewise, the mall can easily combine smaller spaces into a single larger one. Mall tenants are often required to maintain the system(s) and pay the operating costs directly, but the mall owner retains responsibility for system replacement when needed.

1.1.1 System Control

Single-zone, constant-volume systems are controlled much the same way as a residential system. A single thermostat starts and stops the system as needed. Programmable thermostats are very common in commercial applications where no building automation system (BAS) is present. Lighting intensity or motion sensors (*Figure 6*) can also be used to determine occupied and unoccupied operating times.

Figure 6 Wall-mounted motion sensor.
Source: vchal/iStock/Getty Images

In the occupied cooling mode, the system responds to the setpoint. Some systems may be set to run the indoor fan continuously when the building is occupied. This may help balance temperatures across the control zone.

On Site

Humidity Control

Many commercial users operate the indoor blower throughout the occupied periods to help balance temperatures around the space, helping to eliminate hot and cold spots resulting from indoor loads, window glass, and similar sources. However, this strategy can also result in excess space humidity in the summer.

With the indoor blower running constantly, condensed water on the face of the coil and in the condensate pan are re-evaporated back to the space during each off cycle. If you encounter complaints about excess humidity and the problem is not related to the refrigerant charge or another deficiency, consider whether the indoor blower running continuously might be the source of the complaint.

The outside air dampers for these systems are often manually operated and set to a fixed position during system startup. These dampers remain in the set position year-round and in all operating modes. When the dampers are motorized, which also generally means the system is equipped with an economizer, the outside air dampers open to a predetermined position during the occupied mode.

Systems with an economizer use it as the first stage of cooling if the outdoor conditions are suitable. If the zone temperature continues to rise and/or an economizer is not used, capacity control is provided by cycling the compressor(s) on and off to maintain the zone setpoint. Larger systems in this category (above $7^1/_2$ tons) usually have multiple hermetic compressors, and often offer multiple stages of heating as well. Inverter-driven compressors modulate the compressor speed to match the system capacity for cooling and heat pump operation.

When the building is unoccupied, the setpoint changes to conserve energy. Motorized outside air dampers are driven closed. In a building with a BAS and economizer-equipped systems, a *nighttime free-cooling mode* may be activated in the summer. In this mode, the system indoor blower starts, and the outside air damper is driven 100% open. This purges the building with cool, fresh outdoor air. The nighttime free-cooling mode is usually scheduled for a time between 3 am and 7 am to pre-cool and refresh building air prior to occupancy.

1.2.0 Multizone Constant-Volume Systems

In commercial systems, the term *multizone* used to have a different meaning than it does today. *Figure 7* shows a traditional double-deck, multizone, constant-volume system. The system consists of a factory-assembled central AHU supported by remote condensing equipment or a compatible RTU (1). The

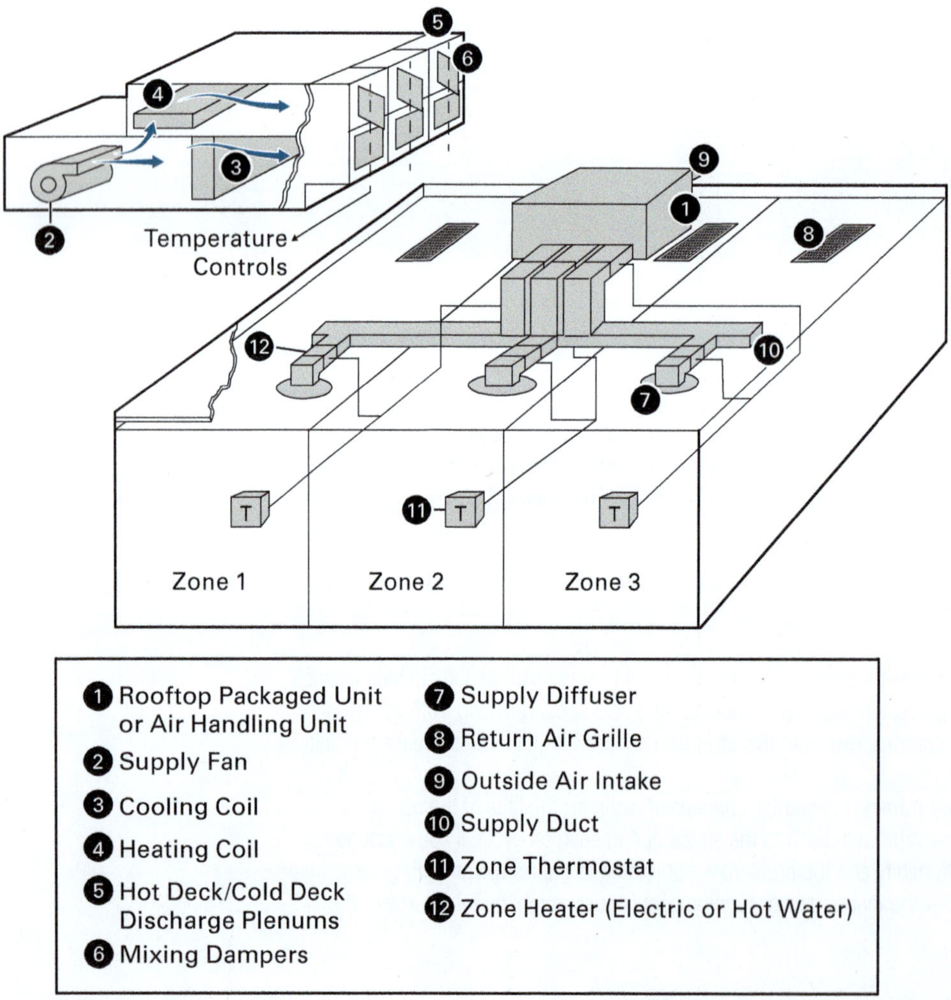

Figure 7 Multizone constant-volume system.

unit is equipped with a supply fan (2) that blows air through a cooling coil (3) or through a heating coil (4). The air then flows into a discharge damper section with multiple outlet plenums (5). This creates two separate airstreams (one cool, one warm) that flow to the control zones as needed. The area containing the cooling coil and associated dampers is referred to as the *cold deck*. The area housing the heating coil and associated dampers is called the *hot deck*.

Each supply air plenum is equipped with a set of mixing dampers (6). Each set of dampers, comprised of a hot-deck damper and a cold-deck damper, provides a constant air volume to an individual control zone. The air stream to a given zone can come from the cold deck, the hot deck, or be a mixture of the two airstreams. When the thermostat for a control zone is satisfied, the dampers move to provide a mixed airstream while the air volume remains constant.

Air is returned from each control zone through a return air grille (8). At the unit, an outside air intake (9) brings fresh air into the supply airstream. A thermostat (11) located in each control zone modulates its set of dampers (6) to provide cooling or heating as needed. When a zone has an unusual exposure and requires additional heat, an electric heater or hot-water coil can be installed in the ductwork (12).

On Site

Multizone Damper Controls

In a multizone AHU, the air supply to the zones is controlled by actuators. The actuators shown here are pneumatic, connected to a pneumatic control system for inputs and an air source to position the damper. The actuators may also be electric.

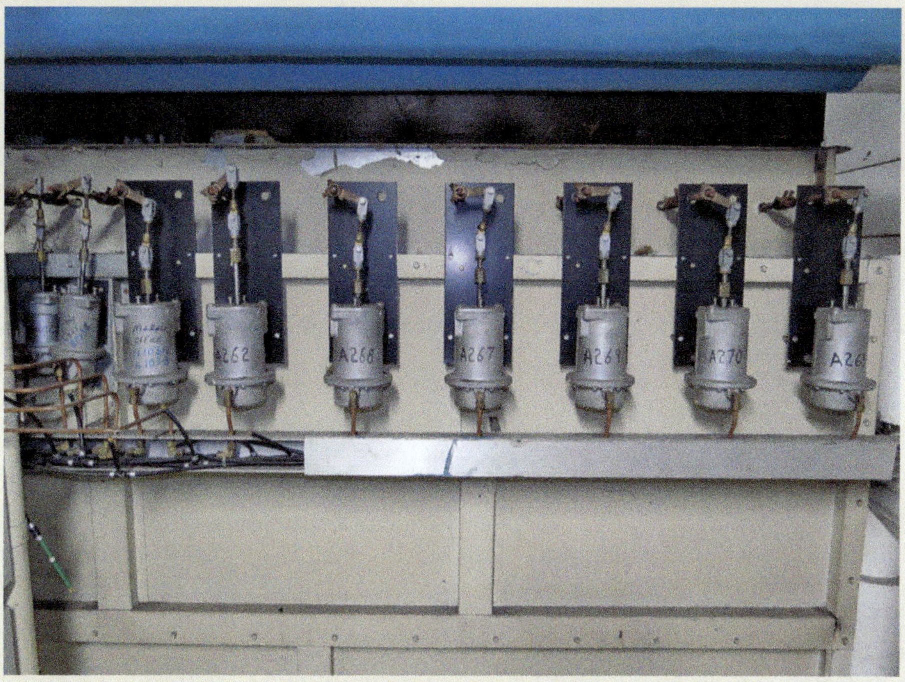

Typical applications for these systems included schools, libraries, museums, banks, laboratories, medical clinics, radio and television studios, department stores, and industrial facilities. Multizone systems like this typically provide between 30 and 100 tons of cooling capacity.

Few, if any, of these systems are installed today, although a few may remain in service. Better designs that offer more flexibility and energy efficiency have made multizone systems like these obsolete.

1.3.0 Variable Volume and Temperature (VVT) Systems

Variable volume and temperature (VVT) system: A VVT system is one that delivers a variable volume of air to each controlled zone, as the load dictates. The operating mode of the unit (heating or cooling) varies depending on the requests from the zones.

As the name implies, a **variable volume and temperature (VVT) system** can provide a varying volume of airflow to individual zones (*Figure 8*). However, this does not mean that one zone can be supplied with cool air while another receives warm air. A unit serving a VVT system can only operate in one mode or the other; it cannot provide simultaneous heating and cooling. The HVAC unit is the source of all conditioned air, and it must alternate between the two modes (heating and cooling) when the building has diverse needs.

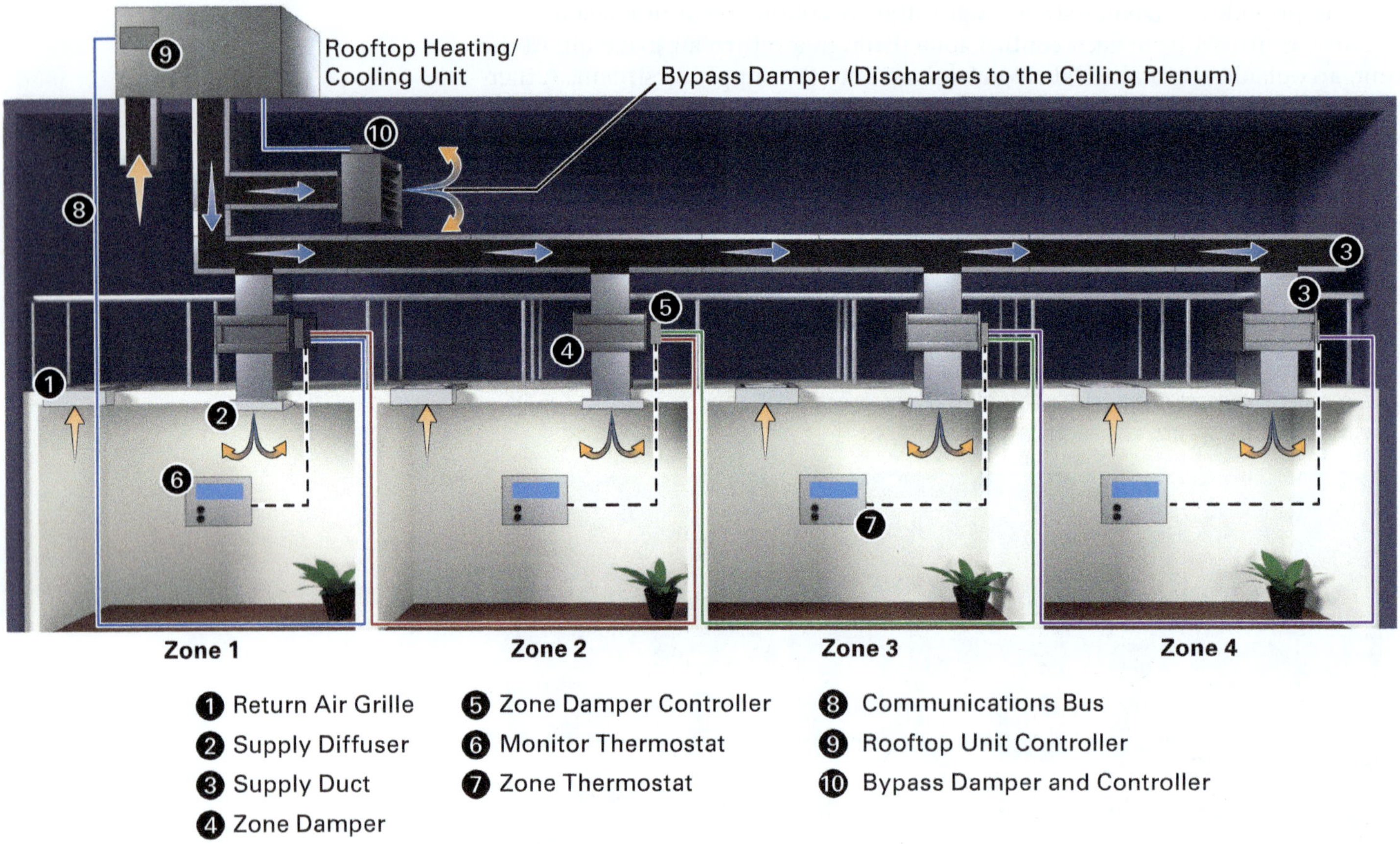

Figure 8 Variable volume and temperature (VVT) system.

NOTE

VVT systems are explored in detail in NCCER Module 03315, *Zoning, Ductless, and Variable Refrigerant Flow Systems*.

These systems generally operate democratically. Zones are polled and allowed to "vote" for what is needed, and the majority rules. If a zone wants heat but the system is busy satisfying a zones(s) in need of cooling, the damper controlling the zone needing heat simply closes until the other zones are satisfied and the unit can switch over to the heating mode.

In *Figure 8*, a packaged unit (9) provides conditioned air through a supply duct to VVT zone dampers (5). Each VVT terminal and zone is equipped with a thermostat. A *monitor thermostat* (6) is provided in one zone, while all the remaining zones have a zone thermostat (7). The zone controllers monitor zone temperature and modulate the VVT damper to maintain the setpoint. If the system is in the incorrect operating mode to satisfy the zone, the damper remains closed until the operating mode changes.

A bypass damper (10) maintains a constant air volume through the system indoor coil or furnace heat exchanger as VVT dampers close. The bypass damper opens further to ensure the unit has sufficient airflow moving through as VVT dampers begin to close. The bypass damper often discharges directly into a ceiling return air plenum. The bypass damper responds to the static pressure in the duct system, opening as the pressure rises.

The monitor thermostat (6) acts as the primary system interface module. It also contains a space sensor and maintains the conditions in its zone as do the other thermostats. The monitor thermostat provides the entire system with a time base and other programmed operating information. Information is exchanged between the monitor thermostat, zone thermostats, and the unit through a communication bus (8).

The three types of constant-volume equipment typically used with VVT systems include single-zone RTUs, vertical indoor packaged units, and traditional split systems. For small buildings requiring 5 to 20 tons of cooling, VVT systems are a popular way to achieve comfort control. Larger buildings can be broken down into 20-ton zones, each served by its own VVT system.

It is important to note that too much diversity in the needs of individual zones can create some comfort problems. For example, mixing common office zones with a radio station studio in its own zone that requires constant cooling may cause the other zones to wait long periods of time before the unit can change to the heating mode. VVT dampers are usually programmed to remain slightly open during occupied hours to ensure that adequate fresh air is flowing into each zone. Because of this, the other zones can easily be over-cooled while the studio continues to demand cooling.

1.3.1 VVT Terminals

Both round and rectangular VVT terminals (*Figure 9*) are available. VVT terminals are little more than motorized balancing dampers. The most economical installation cost is achieved when the damper size and shape match the dimensions of the branch duct in which it is installed. However, the terminal must be selected using the criteria of the manufacturer for the design air volume.

Figure 9 Rectangular VVT damper.
Source: Carrier Corporation

Round air terminals are commonly available in ratings from about 160 cfm to 1,200 cfm in capacity. The round dampers range from 6" to 16" in diameter to match typical branch duct sizes. Round dampers ship without insulation. Insulation and vapor retarders must be applied externally following damper installation.

Rectangular air terminals are typically made in sizes ranging from 8" × 10" to 8" × 24", with 8" representing the height of the terminal. Rectangular air terminals are available with a factory-insulated duct sleeve to simplify installation. The damper sleeve may also be fitted with duct liner on the inside. Opposed-blade dampers are used and are powered by a high-torque actuator.

1.4.0 Variable Air Volume (VAV) Systems

Variable air volume (VAV) system: An air distribution system that controls the temperature within a control zone by varying the volume of the conditioned supply air flowing into the zone.

Figure 10 shows a basic **variable air volume (VAV) system**. VAV systems are very popular in large commercial office buildings due to their performance and reasonable energy efficiency. VAV systems are more complex than other options, but this complexity provides flexibility and the ability to deliver warm or cool air to multiple areas simultaneously.

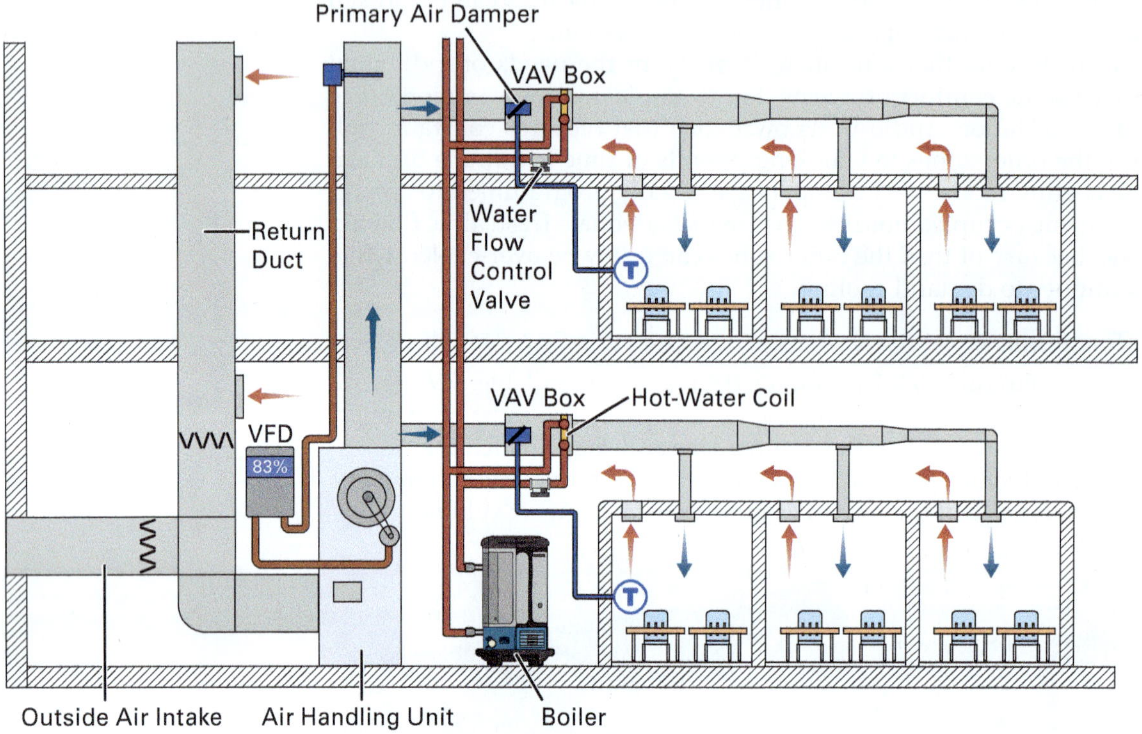

Figure 10 Variable air volume (VAV) system.

A central air source, such as an indoor packaged unit or RTU, provides cool air at a set temperature through a supply duct to the VAV air terminals. The terminals are referred to as *VAV boxes*. Each VAV box supplies air to a single control zone. The zone sensor modulates the position of the VAV box damper to maintain the zone setpoint.

The central unit is set to maintain a consistent supply-air temperature, typically around 55°F (13°C). The volume of airflow, however, changes dynamically as VAV box dampers downstream reposition the dampers. The blower motor speed is controlled by a variable frequency drive (VFD) that responds to the static pressure in the supply duct.

Cooling can be provided by chilled-water coils in the unit, or by packaged systems equipped with compressors and evaporator coils (*Figure 11*). Many of these units are water-cooled, but remote air-cooled condensers are also common. Refrigerant-based units must have versatile capacity-control capabilities to keep pace with the changes in air volume passing through the unit. Chilled-water coils offer a simpler option, the operation of which is virtually unaffected by the changing air volume. The chiller serving the system tracks the load precisely with its own capacity control.

The blower motor's VFD responds to the static pressure in the supply air duct, slowing or speeding up the blower as the duct pressure changes. The pressure in the primary supply duct is higher than you would find in a residential system. For example, a VAV system may be set to maintain a static pressure of 1.5" w.c. The pressure in the duct rises as VAV box dampers close and falls as the dampers open. The VFD responds as needed to compensate. A predetermined

Figure 11 Packaged, water-cooled VAV unit with multiple compressors.
Source: Carrier Corporation

volume of outdoor air is mixed with the return air stream to provide the required fresh air and dilute the CO_2 concentration.

However, the VAV box damper is not allowed to fully close during occupied hours. Since fresh air must be provided consistently, the damper must remain partially open. This means that all the zones receive some volume of cool air whether cooling is needed or not.

Heating in a VAV system is provided through the individual VAV boxes. The boxes are typically equipped with either electric heaters or hot-water coils. When a VAV zone needs heat, the VAV box heater cycles on to provide it.

As buildings become larger, the core of the building with no exposure to outside walls grows larger. The cooling load in the core area, assuming the space around it is conditioned, originates solely from people, lighting, and equipment. These cooling loads occur year-round and are not related to outside climate conditions. Thus, core areas usually require some level of cooling in all seasons, while the perimeter of the building requires heating in the winter to offset heat loss through the walls and glass. For this reason, it is common to find VAV boxes that are not equipped with any form of heat serving the core of a building.

1.4.1 Supply Air Temperature Reset

Although a supply air temperature of 55°F (13°C) may be needed when the heat load is at its peak, there are many times when a warmer temperature is sufficient to maintain comfort. From an energy conservation point of view, producing a consistent supply of air that is colder than necessary is wasteful. For that reason, many VAV systems incorporate *supply air temperature (SAT) reset* programming.

The simplest form of SAT reset is based on outdoor temperature. Using an outdoor temperature sensor as an input, the system adjusts the supply air temperature up as it gets colder outdoors. A high limit may also be programmed, providing a maximum SAT setpoint. If the setpoint is too high, the system is unable to dehumidify because the coil temperature rises above the dew point of the air.

A more complex SAT reset arrangement is based on what is referred to as *trim and response logic*. In this approach, the BAS consistently monitors the position of all the VAV box dampers. If most of the dampers are positioned at less than a predetermined position, less than 65% open for example, the system begins to increase the SAT setpoint. Setpoint changes are subtle; 1° at a time is common. The expectation is that some VAV box dampers will eventually open wider to satisfy the cooling load.

When the cooling load is high, signified by the majority of VAV box dampers being at or near fully open, the SAT setpoint is adjusted down to provide more cooling capacity. With colder air to work with, some VAV box dampers are expected to modulate slightly toward the closed position.

The BAS continues monitoring the damper positions and calculating an appropriate response. In some cases, if only a few dampers remain at or near fully open while most of the zones are performing satisfactorily, the system may increase the static pressure setpoint slightly, producing more air to increase cooling in the problem zones without reducing the SAT setpoint.

Some systems also prioritize specified zones. Perhaps there are zones where maintaining a specific temperature is more critical than others. When the system calculates current VAV damper position and performance, priority zones are given more weight. For example, a priority zone may be treated as two or even three zones with the same status.

It is important to note that not all VAV systems incorporate SAT reset. Most of today's VAV systems are capable of SAT reset by outdoor temperature if a temperature sensor is provided. Trim and response logic, though, requires a more sophisticated system that a user may decline if the cost doesn't seem worth the result. However, SAT reset is a feature that tends to pay for itself rather quickly in a large building.

1.4.2 VAV Box Types

Several types of VAV boxes are used, depending upon system design. More than one box type may even be used on the same system.

All VAV boxes are typically mounted above a suspended ceiling. Primary air from the system is routed into each VAV box at the design system pressure. The pressure is lower in the air distribution system connected to the outlet of each box.

The following VAV box types are commonly found in today's VAV systems:

- Single-duct VAV boxes
- Fan-powered parallel VAV boxes
- Fan-powered series VAV boxes.

Single-Duct VAV Boxes

A single-duct VAV box (*Figure 12*) consists of a primary air inlet connection, a round damper blade or set of opposed-blade dampers, a velocity sensor, and a damper actuator/controller mounted on the side of the box. A temperature sensor located in the control zone monitors space conditions. The damper position is based on the level of demand from the thermostat. As the room temperature gets closer to the setpoint, the damper modulates toward its minimum position.

The damper controller monitors the inlet air velocity using the velocity sensor. In *Figure 12*, note that the sensor receives its velocity inputs through the grid spanning the box inlet in two directions. The velocity sensor monitors the air speed

Figure 12 Single-duct VAV box with hot-water coil.
Source: Carrier Corporation

and mathematically converts the input to an air volume value. Monitoring the air volume allows the damper to adjust its position to compensate for changes in the primary duct static pressure. This feature is called *pressure-independent control.*

Zones that require heat and are served by a single-duct VAV box can also be fitted with an electric heating element or a hot-water coil mounted on the box outlet. However, bear in mind that the heater is warming the cool air provided by the central system that is around 55°F (13°C), rather than working with room air that is significantly warmer.

Fan-Powered Parallel VAV Boxes

Fan-powered VAV boxes are equipped with an individual small direct-drive blower. The integral blower allows the box to draw air from the ceiling plenum when the box switches to the heating mode. Unlike the single-duct VAV box, this warmer air is circulated across the heating elements or coil to heat the space in a fan-powered box.

A fan-powered parallel VAV box schematic is shown in *Figure 13*. In the cooling or neutral mode (*Figure 13A*), the fan is idle, and the backdraft damper remains closed. Conditioned air simply passes to the space through the inlet damper that modulates its position based on demand. These VAV boxes are also equipped with a velocity sensor to provide pressure-independent control.

When the zone temperature setpoint is satisfied, the controller positions the primary air damper to its minimum programmed position for ventilation. If the zone temperature begins to fall below the setpoint, the zone controller activates the integral fan. This can be considered the first stage of heat—simply adding air from the ceiling plenum to the incoming primary airstream. The fan draws warm ceiling-plenum air through an air filter and passes it through the backdraft damper into the primary airstream. The backdraft damper is not motorized; it remains closed by a lightly loaded spring. The pressure of the air behind it when the fan starts provides the opening force.

The warmer mixed airstream is directed to the space. The heating air volume delivered to the space is typically 50% to 60% of the design primary air volume. If the warmth of the air above the ceiling is not sufficient to offset the heating load, the electric heater or hot-water coil is activated (*Figure 13B*). This provides a significant source of heat to the zone.

The fan-powered parallel VAV box is ideal for perimeter zones in a building with a ceiling-plenum air return system. These boxes are certainly functional in core areas of a building, but the core of large commercial buildings rarely needs a significant amount of heat. Simpler single-duct VAV boxes are often a more cost-effective choice for core areas.

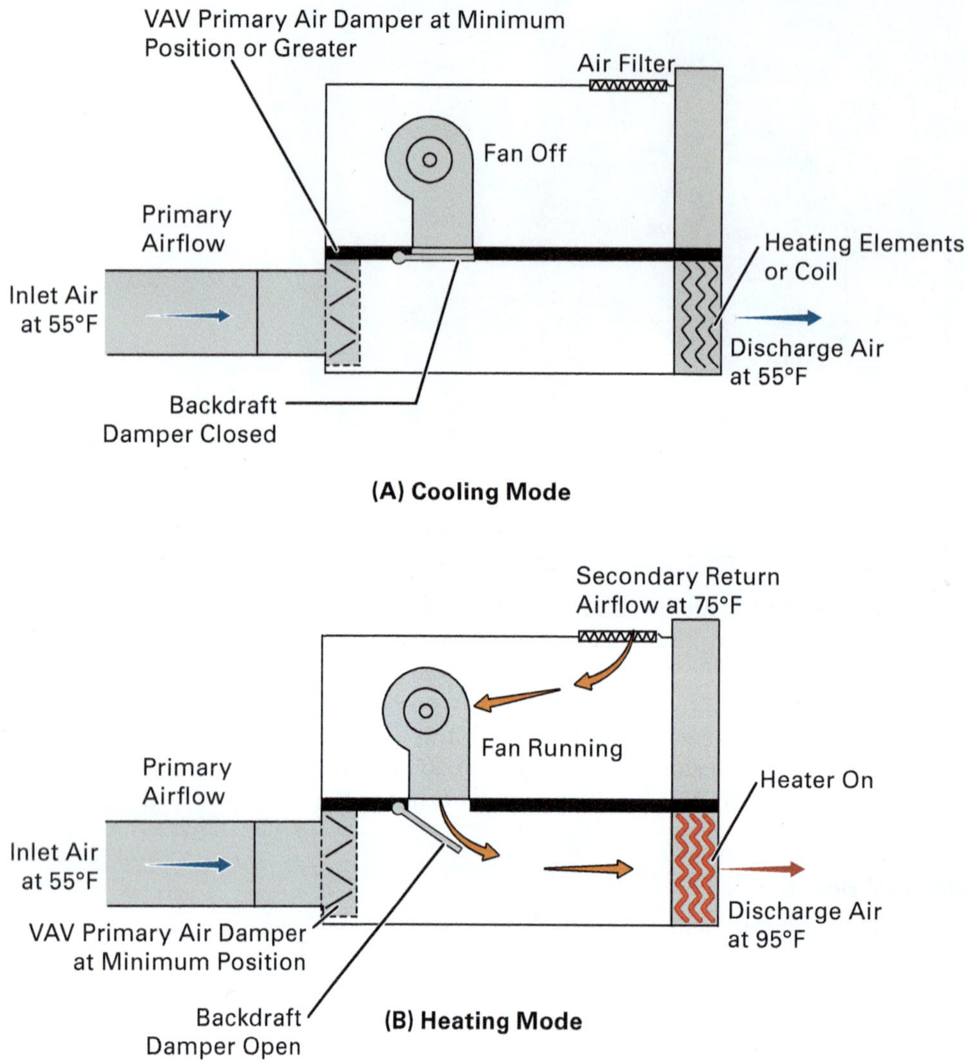

Figure 13 Fan-powered parallel VAV box operation.

Note that fan-powered VAV boxes from different manufacturers may have slightly different operating schemes.

Fan-Powered Series VAV Boxes

Another type of fan-powered VAV box positions the fan in series with the primary air damper and direction of airflow (*Figure 14*). Beyond this difference, the components comprising this VAV box are nearly identical to those of the parallel VAV box. However, with the fan now in the primary airstream, there is no need for a backdraft damper. In addition, the fan in the series box operates even when heating is not required.

The fan-powered series VAV box is characterized by the delivery of a constant volume of air to the control zone, regardless of the position of the primary air damper. Hotel atriums and public areas with high ceilings are good applications for the series VAV box. The consistent air volume makes the design of the downstream supply duct and the selection of grilles and registers easier.

Like the parallel unit, the fan-powered series VAV box uses the ceiling cavity as a return air plenum to supplement the primary airstream, or to provide warmer air for operation in the heating mode. The series box operates in a pressure-independent manner.

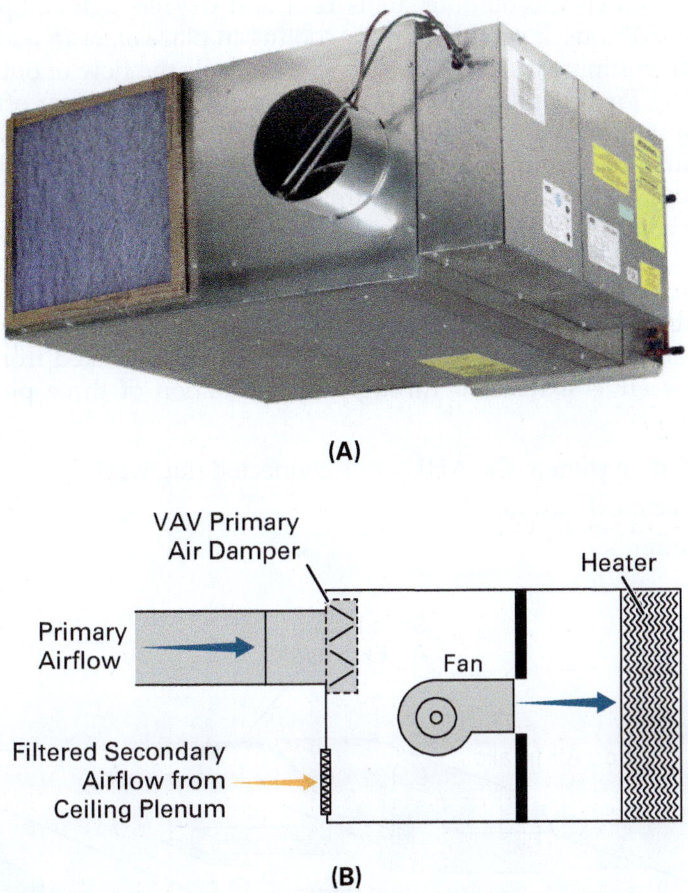

Figure 14 Fan-powered series VAV box.
Source: Carrier Corporation (A)

1.5.0 Ventilation and Exhaust Systems

Commercial air systems easily incorporate the introduction of outdoor air into the building. Outdoor air is introduced into commercial buildings for several reasons.

First, ventilation air is needed to dilute gaseous contaminants released within the conditioned space. Chemicals used in furniture, draperies, carpeting, wall coverings, and other furnishings may release gases for several years after installation. Without dilution, these gases accumulate and create a condition that can contribute to health problems and **sick building syndrome (SBS)**. In addition, people exhale CO_2 as they breathe. While not toxic, high concentrations of CO_2 can cause drowsiness and other unpleasant symptoms.

Sick building syndrome (SBS): A physical condition that exists when more than 20% of a building's occupants complain during a two-week period of common symptoms, including headaches, fatigue, nausea, eye irritation, and throat irritation, which are alleviated by leaving the building and are not confirmed to be caused by any specific contaminant.

Second, the ventilation air replenishes the oxygen used by people who occupy the conditioned space. The amount of air supplied varies with the building, level of activity, and location of the people. Common design practice is to provide a minimum of 20 cfm/person of ventilation air for office workers. In contrast, 30 cfm/person of ventilation air is recommended for people working in a dry-cleaning plant to ensure sufficient dilution of indoor pollutants. ASHRAE Standard 62.1, *Ventilation and Acceptable Indoor Air Quality* specifies minimum ventilation rates for commercial and public buildings. Although ASHRAE standards are not laws or codes, the standards are often incorporated into codes by reference and therefore become requirements.

The third reason outside air is often introduced is to save energy and operating costs. When the outdoor air is cool and dry enough compared to the building conditions, it can be used for cooling in place of, or in addition to, the mechanical equipment. The apparatus that controls the flow of outside air into the building for this purpose is the *economizer.* An economizer often requires only the system blower to cool an entire building, saving a great deal of energy.

The fourth use of outside air is to control the pressure within the building. Maintaining a slight positive pressure in the conditioned space prevents the infiltration of untreated outdoor air. Introducing slightly more outside air than is exhausted from a space to maintain a slightly positive indoor pressure prevents infiltration.

The outside air brought in must displace a nearly equal volume of air from the conditioned space. The stale indoor air must be exhausted from the building. Indoor air is exhausted through a combination of three possible paths (*Figure 15*):

- A relief air system in the AHU or its connected ductwork
- Mechanical exhaust fans
- Exfiltration

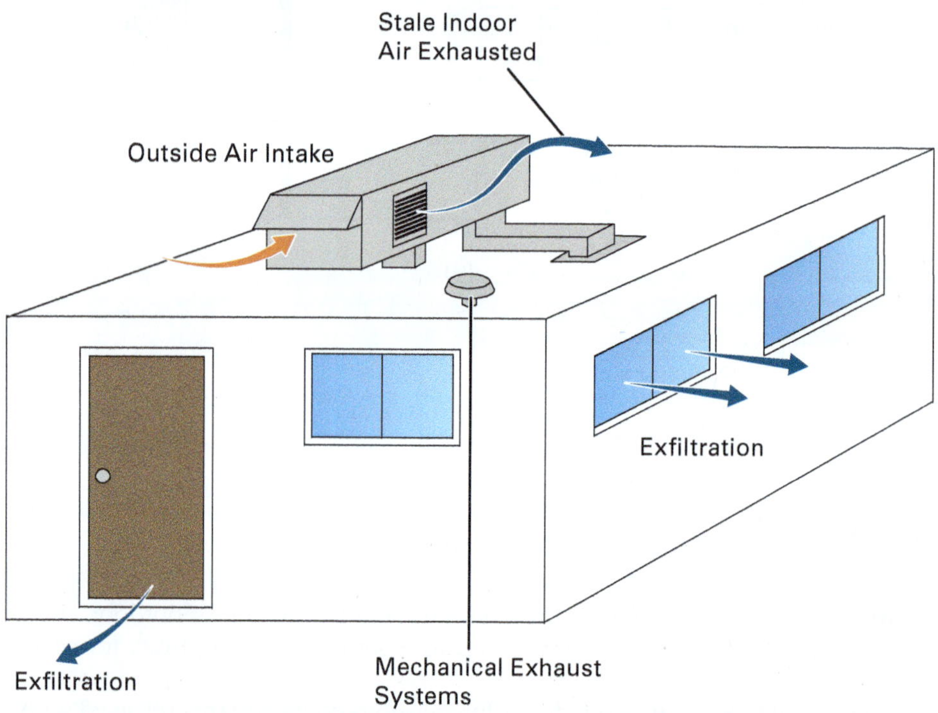

Figure 15 Building exhaust air paths.

Exhaust fans are common in all buildings. Restrooms and kitchens require exhaust fans, for example. When the building pressure is higher than outdoors, exfiltration occurs around window and door jambs and similar small cracks and crevices around the building. Exfiltration also occurs when the doors are opened.

The AHU and air distribution system usually provide the primary path for indoor air to escape, by design, although exhaust fans may be the primary path in various industrial settings.

On Site

Ceiling Return Plenums

In most residential systems and some commercial systems, the air delivered to the conditioned space is returned to the system through return air ducts. In commercial buildings with suspended ceilings, the space between the ceiling and the roof or deck above is often used as a return air plenum. This approach saves money by eliminating much of the cost of materials and labor associated with installing ductwork. As shown here, the return air from the conditioned zone enters the plenum above the suspended ceiling via the return air grilles mounted in the ceiling.

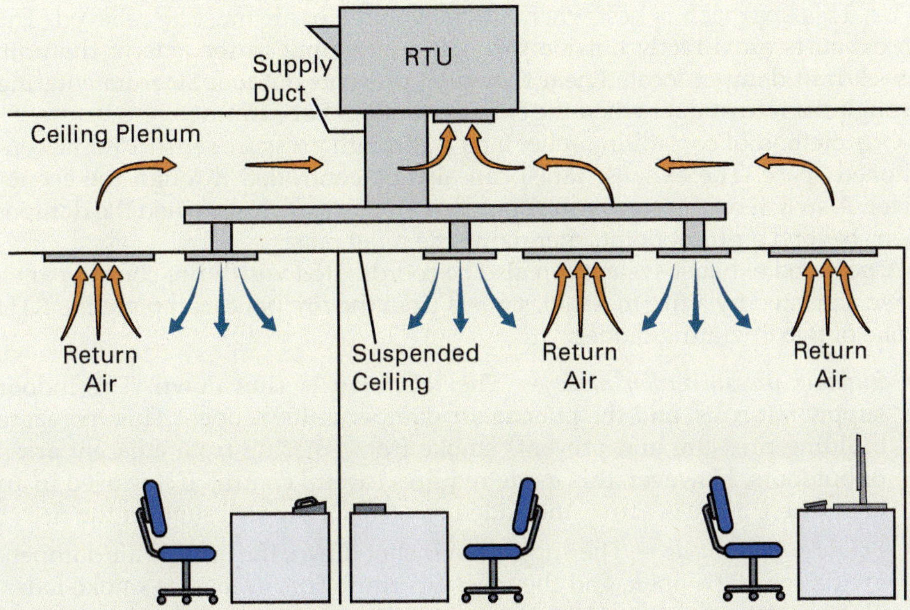

1.5.1 Relief Air Systems

Air pressure in a building is often reduced by simply providing the indoor air with a path outdoors through a *relief air damper* (*Figure 16*). A *barometric relief damper* opens automatically as the pressure in the building increases, without

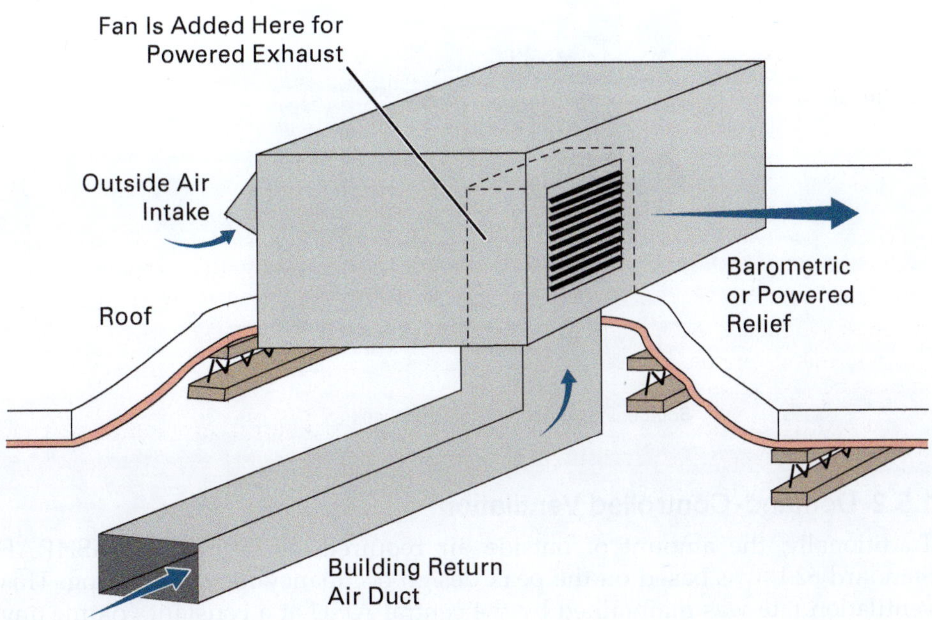

Figure 16 Exhausting air through the AHU.

Figure 17 Relief air fans in an RTU relief air hood.
Source: MicroMetl Corporation

mechanical assistance. When an economizer is in operation and the pressure in the building rises, a barometric relief damper will open fully in many cases. Otherwise, its position fluctuates as the indoor pressure changes.

A *motorized damper* can also be used to provide a path for pressure relief. The damper motor responds to inputs from building pressure sensors, opening and closing in response to pressure changes. These dampers are often sized to handle about 75% of the design outside air volume. The remainder of the air is expected to leave the building through exfiltration and common exhaust fans.

The relief air arrangement in the AHU or RTU can also be powered using a dedicated fan (*Figure 17*). This type of powered exhaust is an option for many RTUs. This approach is best when a large volume of air must be relieved. The fan exhausts air directly outside through an opening in the side of the unit. A backdraft damper located near the outlet prevents outdoor air from entering through the return duct when the exhaust fan(s) is not running.

One method of controlling relief fans is through pressure sensors in the conditioned space. The exhaust fan(s) can also be controlled through the economizer. A switch on the economizer outdoor air damper closes when the damper opens beyond a preset point, energizing the relief fans.

A powered exhaust system can also be coordinated with a fire control panel. When activated by a fire marshal, signals from the fire panel can place the RTU in one of two operating modes:

- *Building pressurization mode* — The relief fan is shut down. The indoor supply fan runs, and the outside air damper is 100% open. This increases building pressure and prevents smoke from entering from adjacent areas or outdoors. However, this mode requires careful control. If it is used in an area where a fire is active, the added oxygen supply will support the fire.
- *Smoke exhaust mode* — The supply fan is shut down, the outside air dampers are driven 100% open, and the relief fan runs. This evacuates smoke-laden air from the space.

On Site

Barometric Relief Dampers

Barometric relief air dampers can also be counterbalanced with weights to allow a specific positive pressure to develop in the building before the dampers begin to open.

Source: Image courtesy of Greenheck

1.5.2 Demand-Controlled Ventilation

Traditionally, the amount of outside air required, as defined by ASHRAE Standard 62.1, was based on the peak design occupancy for the building. This ventilation rate was maintained by the central AHU at a constant volume any time the building was occupied. Thanks to advancements in CO_2 sensing equipment and digital controls, this is no longer required.

Humans exhale CO_2 at a predictable rate, so the CO_2 level in a space can be used as a reliable indicator of the occupancy. Demand-controlled ventilation (DCV) is an approved ASHRAE Standard 62.1 control strategy that uses CO_2 sensors to match a building's ventilation rate to the current occupancy.

Each building control zone is provided with temperature and CO_2 sensors. Space CO_2 levels are monitored and the values are transmitted back to the central air source's controller. The central air source then compares outside air CO_2 levels to space CO_2 levels and adjusts the building outside air volume to meet the need.

On Site

Return Air Fans

Outside air ducts are generally very short and offer little resistance to airflow. A large central-station AHU, for example, is usually located in an equipment room with an adjacent outside wall. The outside air duct may only be a few feet long.

On the other hand, the return air ducts in a commercial building can be hundreds of feet long with many changes of direction, in addition to banks of high-efficiency air filters. This results in significant airflow resistance in the return duct. To compensate for the resistance and avoid equipping the main blower with a very large motor, a return air fan can be added. The return air fan is typically sized only to overcome the resistance of the return duct system.

1.0.0 Section Review

1. An area of a building can be classified as a control zone if it has which of the following?
   a. Occupants
   b. A thermostat
   c. A return air grille
   d. A supply air grille

2. Which of the following terms is associated exclusively with multizone constant-volume systems?
   a. Hot deck
   b. Motorized damper
   c. Primary air damper
   d. Mixed air temperature

3. In which type of system is a bypass damper required?
   a. VAV
   b. VVT
   c. Multizone constant-volume
   d. Single-zone constant-volume

4. Fan-powered terminals can be used with which type of system?
   a. VAV
   b. VVT
   c. Multizone constant-volume
   d. Single-zone constant-volume

5. A relief air arrangement in an RTU requires a backdraft damper to prevent outside air from entering if the RTU also has which of the following components?
   a. VFD
   b. Return air fan
   c. Powered exhaust
   d. Building automation system

Module 03201 Review Questions

1. True or False? A traditional heat pump split system with an indoor fan coil unit and an outdoor condensing unit is considered an applied system.
   a. True
   b. False

2. Which of the following is a characteristic that makes chilled water a good choice for cooling large commercial buildings?
   a. Chilled-water coils are unaffected by changes in system airflow.
   b. Outside air does not need to be introduced when chilled water is used.
   c. Chilled-water coils are substantially smaller than direct-expansion evaporator coils.
   d. Chilled-water systems require only half the energy of other systems to provide the same cooling capacity.

3. Single-zone constant-volume systems that have motorized outside air dampers are also typically equipped with ______.
   a. a VFD
   b. a gas furnace
   c. an economizer
   d. multiple thermostats

4. A single-zone constant-volume system with an economizer can provide a nighttime free-cooling mode if the facility also has ______.
   a. a BAS
   b. light sensors
   c. motion sensors
   d. a programmable thermostat

5. What do the hot-deck and cold-deck dampers do when a multizone constant-volume system is in use and a control zone thermostat is satisfied?
   a. Both the hot-deck and cold-deck dampers close.
   b. The cold-deck damper closes and the hot-deck damper opens.
   c. The hot-deck damper closes and the cold-deck damper opens.
   d. The hot-deck and cold-deck dampers move to a mid-position and a mixed airstream is provided.

6. In a VVT system, what is the primary source of heat for the control zones?
   a. Air terminal reheat
   b. The central HVAC unit
   c. Perimeter baseboard hydronic or electric heating
   d. Fan coil units above the ceiling for each control zone

7. VVT systems maintain a constant air volume across the indoor coil or furnace heat exchanger using which of these components as dampers begin to close?
   a. A VFD
   b. An economizer
   c. A bypass damper
   d. Belt-driven blower

8. Although the temperature setpoint is often reset to accommodate load conditions, what is the typical discharge-air setpoint for a VAV system during occupied hours?
   a. 45°F (7°C)
   b. 55°F (13°C)
   c. 65°F (18°C)
   d. 72°F (22°C)

9. Which of the following statements about VAV systems is *correct*?
   a. Heating in VAV systems is typically provided by natural gas or propane.
   b. The fan-powered parallel mixing box is a good choice for use in perimeter zones.
   c. The fan in a fan-powered series mixing box operates only during the heating mode.
   d. Multizone constant-volume systems are typically used in conjunction with VAV systems.

10. When does the integral blower in a fan-powered parallel VAV box operate?
    a. At all times
    b. When the zone needs cooling
    c. When the zone needs heating
    d. Only during unoccupied periods

Answers to odd-numbered questions are found in the Module Review Answer Key at the back of this book.

Answers to Section Review Questions

Answer	Section	Objective
Section One		
1. b	1.1.0	1a
2. a	1.2.0	1b
3. b	1.3.0	1c
4. a	1.4.2	1d
5. c	1.5.1	1e

MODULE 46100

Introduction to Leadership

Objectives

Successful completion of this module prepares you to do the following:

1. Explain business and organizational structures in the construction industry.
   a. Understand organizational charts.
   b. Describe organizational roles and relationships, including job descriptions and responsibility, authority, and accountability.
2. Describe frontline leadership and communication skills.
   a. Describe the qualities and role of a leader, including leadership styles and ethical leadership.
   b. Explain the communication process and distinguish between verbal, nonverbal, written, and visual communication types.
   c. Compare problem-solving and decision-making.
3. Identify key skills for team leadership and motivation.
   a. Choose strategies to motivate individuals and a team.
   b. Describe key approaches to building and leading teams.
   c. Recognize how to approach leadership challenges, such as transitioning from peer to leader, addressing performance issues, and resolving conflict.
4. Identify a frontline leader's safety leadership responsibilities.
   a. Assess the impact of incidents.
   b. Explain the purpose of OSHA and describe the role of OSHA in administering worker safety.
   c. Describe the importance of safety and health programs.
   d. Explain the frontline leader's role in incident preparedness and response.
   e. Recognize the importance of supporting your crew's total health.
5. Demonstrate an understanding of construction project phases, project planning and scheduling, field reporting, and cost and resource management.
   a. Identify construction project phases.
   b. Summarize crew planning and scheduling responsibilities, including creating lookahead schedules.
   c. Describe the frontline leader's role in field reporting.
   d. Explain approaches to managing costs and controlling project resources, such as materials, tools, and equipment.

Performance Tasks

Under supervision, you should be able to do the following:

1. Organize and deliver a safety-related toolbox talk, applying communication skills.
2. Complete a job safety analysis.
3. Develop and present a lookahead schedule.

Overview

As a frontline leader, you must be able to communicate effectively, provide direction to your crew, plan and schedule work, manage costs, and ensure the jobsite is safe for everyone. Whether you are already a frontline leader or want to become one, this module will help you learn more about the requirements and skills needed to succeed.

Digital Resources for HVACR

Scan this code using the camera on your phone or mobile device to view the digital resources related to this craft.

SCAN ME

NCCER Industry-Recognized Credentials

If you are training through an NCCER-accredited sponsor, you may be eligible for credentials from NCCER. The ID number for this module is 46100. Note that this module may have been used in other NCCER curricula and may apply to other level completions. Contact NCCER at 1.888.622.3720 or go to **www.nccer.org** for more information.

You can also show off your industry-recognized credentials online with NCCER's digital credentials. Transform your knowledge, skills, and achievements into credentials that you can share across social media platforms, send to your network, and add to your resume. For more information, visit **www.nccer.org**.

1.0.0 Business and Organizational Structures

Performance Tasks

There are no Performance Tasks in this section.

Objective

Explain business and organizational structures in the construction industry.

a. Understand organizational charts.

b. Describe organizational roles and relationships, including job descriptions and responsibility, authority, and accountability.

Crew leader: A person appointed leader of a team or crew, usually an experienced craft professional who has demonstrated leadership qualities. Also known as *frontline leader*, *frontline supervisor*, *lead person*, or *foreman*.

Superintendent: An on-site supervisor who is responsible for one or more crew leaders or frontline supervisors.

Project manager: Person responsible for managing one or more projects.

When a crew is assembled to complete a job, one person is appointed the leader. Different companies give this role different titles, such as **crew leader**, *frontline supervisor, lead person*, or *foreman*. In this module, this role is called the *crew leader* or *frontline leader*. A crew leader or frontline leader is often an experienced craft professional who has demonstrated leadership qualities.

A **superintendent** is essentially an on-site field supervisor who is responsible for one or more crew leaders or frontline supervisors. A **project manager** may be responsible for managing one or more projects. This training will concentrate primarily on the supervisory role of the frontline leader. These individuals work with others on projects as part of an organization.

1.1.0 Organizations and Organizational Structures

Organization: A group of people working together.

Organizational chart: A diagram that shows how the various management and operational responsibilities relate to each other within an organization. Named positions are ranked top to bottom, from those functional units with the most and broadest authority to those with the least. Lines connect positions and indicate chains of authority and other relationships.

An **organization** is a group of people working together. Businesses, clubs, and associations are examples of types of organizations. On a jobsite, the relationships among the people within the company or project create a type of organization.

1.1.1 Organizational Charts

A formal organization will have an **organizational chart** that shows all of the positions in the organization and how those positions are related. A detailed organizational chart, or org chart, may also show the name of the person in each position. *Figure 1* shows an example of an organizational chart for a construction company. Each position on that organizational chart represents an opportunity for career advancement for a crew leader.

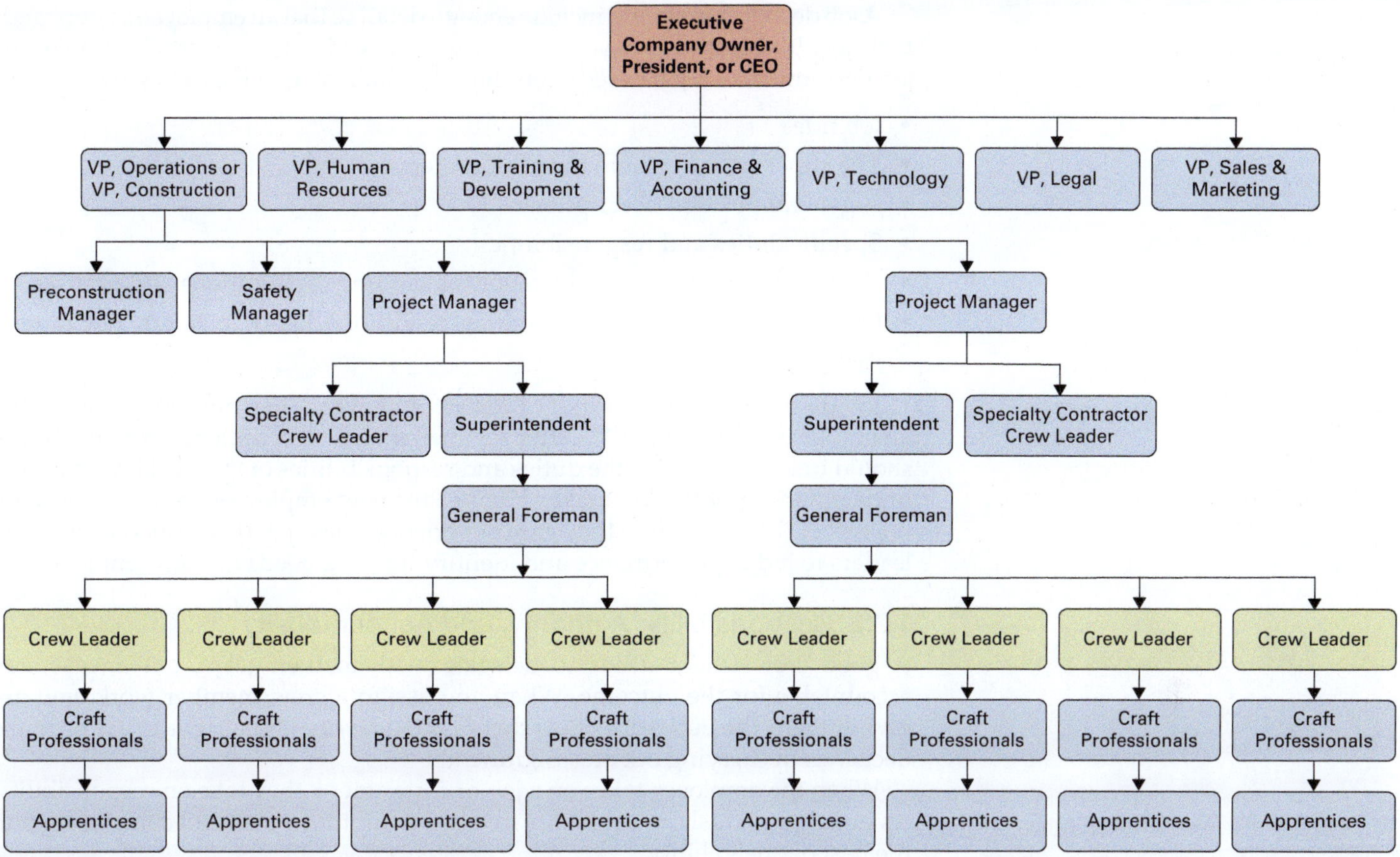

Figure 1 Sample organization chart for a construction company.

As shown in *Figure 1*, most companies have separate groups or departments that focus on running specific functions. Smaller companies may have one or two individuals who perform multiple functions. Larger companies will often create specific roles or departments to handle more complex projects and larger teams.

Companies may have additional roles or teams, such as architectural and engineering design functions. Depending on the size of these groups, they may become separate departments.

NOTE

Organizational charts vary widely based on the size of the company and the specific construction sector in which they operate. For example, an industrial company might have crew leaders who report to a plant manager. Some larger companies will have a superintendent per shift, who will report to a project superintendent, who reports to the project manager. If you are interested in learning more about your company's organizational chart, ask your supervisor or manager.

1.2.0 Organizational Roles and Relationships

As a crew leader, you will assign others the responsibility to perform specific jobs or tasks. Your crew and team members will perform best if the following is true:

- They have a clear job description that explains their job duties.
- They have the required skills and abilities to perform the task.
- They know who has **authority**, **responsibility**, and **accountability**.
- They have a clear understanding of expectations.
- They are trained on policies and procedures.
- They have all needed tools, equipment, and material.

Authority: The right to act or make decisions in carrying out an assignment.

Responsibility: The act of accepting a task and completing it to the best of one's ability.

Accountability: In relation to job completion, accepting responsibility and being personally answerable for the results.

1.2.1 Job Descriptions

As part of the hiring process, most companies give each employee a **job description**. A job description is a statement that describes the duties and responsibilities of a position with a company. A **duty** is a work task an employee is expected to perform. Responsibility involves accepting a task and completing it to the best of one's ability.

Job description: A description of the scope and responsibilities of a worker's job so that the individual and others understand what the job entails.

Duty: A work task an employee is expected to perform.

A job description should include enough detail so that an employee understands their job. It also should have any information needed to evaluate the employee's job performance. A job description should contain at least the following:

- Job title
- The supervisor to whom the position reports
- A general position description
- Specific duties and responsibilities
- Required and desired qualifications, including certifications and licenses
- Other information, such as working conditions, job type, hours, pay ranges, and more

Job descriptions set a standard for an employee. They outline the tasks that an employee should handle. After reviewing a job description, a new employee should understand all of the duties and responsibilities of the job. Having a clear understanding of the job makes it easier for new employees to work well with the rest of the crew while they gain experience. Clear job descriptions help crew leaders to judge performance and identify training needs for new employees.

1.2.2 Responsibility, Authority, and Accountability

A good crew leader needs to delegate work to their team, but always stay accountable for the outcome. When you assign a crew member work, you are giving them the authority to do that task. *Authority* is the right to act or make decisions in carrying out an assignment.

When the employee takes on a job or assignment, they take on responsibility. As noted previously, *responsibility* involves accepting a task and completing it to the best of one's ability.

In relation to job completion, *accountability* is accepting responsibility and being personally answerable for the results.

Each project or company has differences in the type and amount of authority, responsibility, and accountability a supervisor or worker has. As a crew leader, you must give sufficient responsibility and authority to allow employees to be successful at their job activities. You then must hold employees responsible for completing those assigned activities.

As shown in *Figure 2*, even though you may delegate authority and responsibility to crew members, the overall accountability for tasks assigned to a crew always rests with the frontline leader or crew leader.

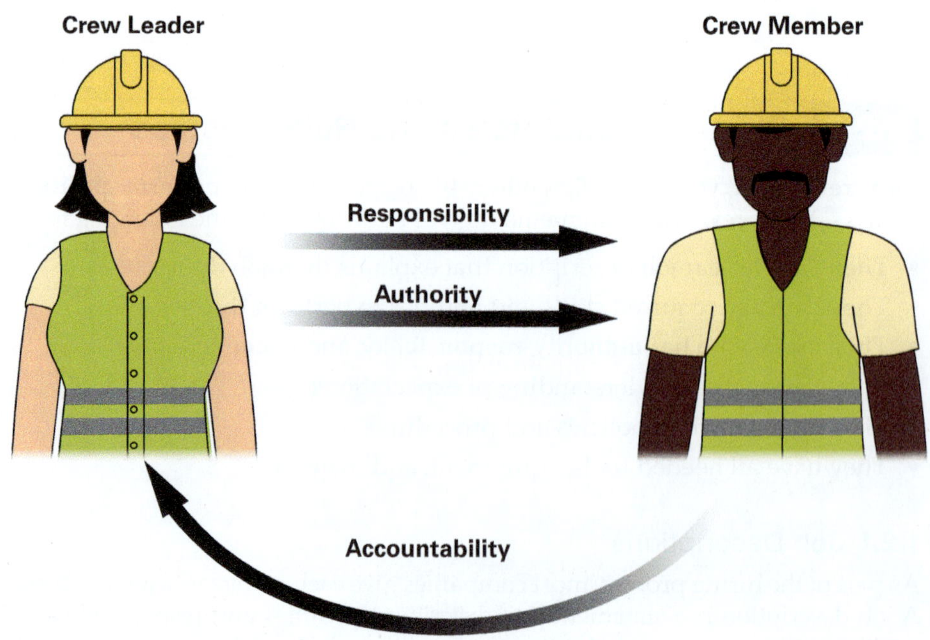

Figure 2 Accountability always rests with the frontline leader.

1.2.3 Policies and Procedures

Most companies have formal policies and procedures established to help crew leaders carry out their duties.

- A **policy** is a general statement establishing guidelines for a specific activity. Examples include policies on vacations, breaks, workplace safety, and checking out tools.
- **Procedures** are formal instructions to carry out and meet policies. For example, a procedure written to implement a policy on workplace safety would include guidelines for reporting incidents and following general safety practices.

Policy: A general statement establishing guidelines for a specific activity.

Procedures: Formal instructions to carry out and meet policies.

A crew leader must understand any company policies and procedures that relate to their work, especially safety practices. The **Occupational Safety and Health Administration (OSHA)** is a national public health agency that is part of the US Department of Labor. To promote a safe and healthy work environment, OSHA issues standards and rules for working conditions, facilities, equipment, tools, and work processes. When OSHA inspectors visit a jobsite, they often question crew leaders and employees about the company safety policies. During incident investigations, the inspector will verify if the responsible crew leader knew and followed the applicable company policy.

Occupational Safety and Health Administration (OSHA): A national public health agency that is part of the US Department of Labor. To promote a safe and healthy work environment, OSHA issues standards and rules for working conditions, facilities, equipment, tools, and work processes.

1.0.0 Section Review

1. Which of the following is shown on an organizational chart?
   a. How to organize materials on a jobsite
   b. How positions in an organization are related
   c. How prevailing wages are determined
   d. How to coordinate work with other crafts
2. Although you may delegate authority and responsibility for tasks to crew members, as a frontline leader, you have the overall _____.
   a. policy
   b. accountability
   c. job description
   d. authority

2.0.0 Leadership and Communication

Objective

Describe frontline leadership and communication skills.

a. Describe the qualities and role of a leader, including leadership styles and ethical leadership.
b. Explain the communication process and distinguish between verbal, nonverbal, written, and visual communication types.
c. Compare problem-solving and decision-making.

Performance Tasks

There are no Performance Tasks in this section.

In addition to being an experienced craft professional, a frontline leader also must demonstrate leadership qualities and strong morals. They must be able to communicate effectively, solve problems, and make good decisions. Frontline leaders are generally promoted up from a work crew. A worker's ability to accomplish tasks, get along with others, meet schedules, and stay within the budget have a significant influence on the selection process. The frontline leader must lead the team to work safely and provide a quality product.

This section explains the importance of developing and improving leadership skills as a frontline leader. It will cover the qualities and roles of a leader, as well as effective ways to communicate with coworkers and employees at all levels, solve problems, and make decisions.

2.1.0 The Qualities and Role of a Leader

There are many ways to define leadership. One simple definition of leadership is the ability of a person to influence others to achieve a common goal. A leader is someone who sees how to accomplish or improve things. A leader then rallies people to move toward that vision. Effective leadership requires communicating ideas effectively to engage others to act accordingly.

Leaders often have power based on a role or title. However, leadership is much more than power. An effective leader inspires others to act, while also directing the way that they act. A leader must have respect, so others follow their directions. A leader also must have the critical thinking skills to know the best way to use available resources.

Motivate: To make someone feel determined or enthusiastic or to make someone behave in a particular way.

Some people develop leadership qualities as part of their upbringing. Others may work hard to develop the qualities that **motivate** others to follow and perform. Leadership is a skill you can learn through experience and training, as well as hard work, focus, and practice.

2.1.1 Functions of a Leader

What is the purpose or function of a leader? The answer will vary based on the specific project, the work environment, the team they lead, and the tasks they must perform.

As shown in *Table 1*, a frontline leader has certain functions that apply to almost every work situation.

TABLE 1 Frontline Leadership Functions

Leadership Duties and Responsibilities	Leadership Functions
Planning the Work	Organize, plan, staff, direct, and control work
Safety and Compliance	Ensure that the team understands and follows company policies and procedures
Empathy	Be sensitive to the needs and qualities of a diverse workforce
Motivating and Coaching	Give team members the confidence to make decisions and take responsibility for their work
Conflict Resolution	Maintain a cohesive team by resolving tensions and differences
Professionalism	Represent the team or group and represent the company or organization
Accountability	Accept responsibility for the successes and failures of the team's performance

2.1.2 Leadership Traits and Behavior

Trait: A distinguishing quality of a person.

Behavior: The way a person responds to a specific situation.

As previously noted, leadership is a skill you can learn and develop through experience and training. As you continue to grow your leadership skills, remember that while each leader has some unique approaches, the most effective leaders share some common traits and behaviors. A **trait** is a distinguishing quality of a person. A **behavior** is the way a person responds to a specific situation.

Effective leaders share the following traits and behaviors:

- Follows company policies and procedures
- Has necessary technical knowledge

- Can plan, prioritize, and organize work
- Leads by example
- Is fair and ethical
- Solves problems, makes decisions, and assumes responsibility
- Communicates effectively
- Can promote an idea
- Is enthusiastic and shows strong initiative
- Is able to motivate individuals and teams
- Is loyal to their company and crew
- Extends respect and trust to all team members
- Can teach others and learn from others

First and above all, excellent leaders lead by example. They work and live by the standards that they set for their crew members. A leader will not ask someone to complete a task that they are not willing to do themselves.

Leaders tend to have a high level of drive and determination, as well as a persistent attitude. When faced with obstacles, effective leaders don't get discouraged. Leaders identify the potential problems and make plans to overcome them. They then work toward achieving the intended goal. In the event of failure, effective leaders learn from their mistakes and apply that knowledge to future situations. They also learn from their successes.

Respect is an important aspect of leadership. Leaders earn respect by being fair to employees and by listening to their complaints and suggestions. Using appropriate incentives and **rewards** to motivate a crew helps leaders earn respect. In addition, frontline leaders who have a positive attitude tend to gain the respect of their team and their peers. A **peer** is someone who is in a position or level in an organization equal or similar to yours.

Rewards: Tangible items given to an employee as recognition for good work.

Peer: Someone who is in a position or level in an organization equal or similar to yours.

Trust is an equally important part of being a great frontline leader. Trust is a group's belief that their leader is committed, capable, and will do what's best for all involved. To build trust, effective leaders create interactions and conversations with the team. They treat everyone on the team as an individual.

Leaders are good communicators who lay out the goals of a project to their crew members clearly. To do this, you may need to overcome language barriers, bias, or differences in personalities to ensure that each member of the crew understands the established goals of the project.

Effective leaders can motivate their team to work to their full potential. They will work to develop crew members' skills. They encourage them to improve and learn so they can contribute more to the team effort. Effective leaders strive for excellence from themselves and their team, so they work hard to provide the necessary skills and leadership.

It is important for leaders to have an expert knowledge of the activities they supervise. This is important because the crew members need to know that they have someone to turn to when they have a question or a problem, need some guidance, or must make modifications or changes on the job.

In addition, leaders must possess organizational skills. They know what needs to be accomplished and they use their resources to make it happen. Because they can't do it alone, leaders require the help of their team members to share in the workload. Effective leaders delegate work to their crew members. They also implement company policies and procedures to complete the work safely, effectively, and efficiently.

Finally, outstanding leaders have the experience, authority, and self-confidence that allow them to make decisions and solve problems. To accomplish their goals, leaders must be able to calculate risks, take in and understand information, set a course of action, make decisions, and assume the responsibility for those decisions.

2.1.3 Leadership Styles

Many terms are used to describe different leadership styles. As a frontline leader, you will never lead the exact same project twice. Having a leadership style that is also adaptable is important for you, the project, and your company's success.

Situational leadership: A leadership style that involves adjusting to the specific situation or task and the leadership needs of the group.

Situational leadership is a leadership approach that involves adjusting the way you lead to reflect the specific situation or task and the needs of the group. The main idea behind situational leadership is that there is no "one size fits all" or best style of leadership. Instead, you must adapt or adjust your leadership style to the situation.

As shown in *Figure 3*, situational leadership provides a model with four primary leadership styles:

- *Directing:* The leader tells a team what to do and how to do it.
- *Coaching:* The leader coaches individuals to improve skills. The leader encourages more back-and-forth between leaders and the team.
- *Supporting:* A leader supports the group, participating to help increase motivation. The leader offers less direction. Instead, the leader has team members take a more active role in coming up with ideas and making decisions.
- *Delegating:* A leader allows other people to own the task, and they facilitate it happening. Team members make most of the decisions and take most of the responsibility for what happens.

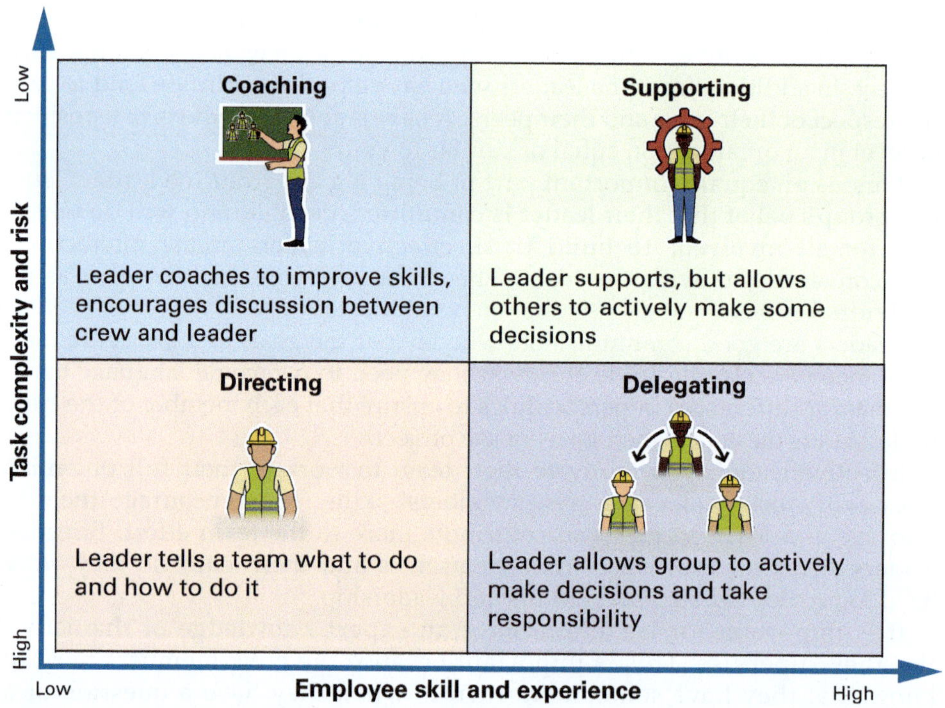

Figure 3 Situational leadership styles.

Knowing which leadership style to apply depends on several key factors:

- *Team relationships:* Consider your team, how well they work together, and how productive they are. A group that is not efficient or productive might benefit from a Directing style. That style emphasizes direction, organization, and clearly defined roles. On the other hand, a highly cohesive group might benefit from a Supporting style, which gives the team more inputs on decisions.
- *Team skills:* Consider the skills and abilities of the individuals completing the task. Giving a task to a team or team member who lacks the skills to complete the task is risky, is dangerous, and may end in failure.

- *Work tasks:* Consider the actual work or tasks at hand. You must have an expert knowledge of the tasks and activities to ensure they are done well, completely, and in the safest way possible. You may need to coach or train workers who do not have experience to help them complete the task.
- *Work situation:* Consider the situation you're facing. In some cases, the situation itself will override all other factors.

For example, if the crew does not have enough experience for the job ahead, you may want to use a *Directing* style. A Directing style works with people and teams who have limited experience or skill performing a task. It also works when team members are insecure or unmotivated to try a task. The Directing style requires your close supervision to spot signs of progress, which you can use to motivate ongoing development. If a job task or situation is very risky, a Directing style can help reduce the risk by providing clear and very specific direction.

If your team is working well and showing strong progress on work tasks, you might use a *Coaching* style. You still direct how and when specific tasks are completed, but you also discuss why the task matters and where it fits into the overall project or job. With this style, you recognize the enthusiasm, interest, and commitment of the team. You also reinforce the importance of learning and gaining task-related experience.

You may want to use a *Supporting* style if the team is reasonably skilled and work tasks are not complex, but commitment and attitude are an issue. The goal of the Supporting style is to create alignment. As a leader, you can discuss a person's commitment by asking open-ended questions. These questions should help you understand what is behind the performance challenges and create possible solutions. The Supporting style of leadership also works well when work is creative, as it is helpful to brainstorm and exchange ideas with crew members.

The *Delegating* style works best with an experienced, strong, committed crew on a well-defined project. The Delegating style of leadership is also effective when jobs have repetitive tasks that require few decisions. To use the Delegating style, however, your company must give you enough authority as the frontline leader to do the job.

Effective leadership takes many forms. The correct style for a particular situation depends on the team and the work you need to accomplish. Knowing the factors to consider can help you decide which type of leadership to apply in which situation. This flexibility allows you to meet each situation with the leadership style that brings out the best in your team and gives them the highest chance of success.

2.1.4 Ethics in Leadership

Ethics are the moral principles that guide a person's actions when dealing with others. *Figure 4* outlines three basic types of ethics. Frontline leaders must maintain strong ethics and the highest standards of honesty and legality. Every day, a leader makes decisions that have fairness and moral implications. When you make an unethical decision, it reflects poorly on you. It also impacts your team, other project team members, and the company as a whole.

Ethics: The moral principles that guide an individual's or organization's actions when dealing with others.

As a leader, you will find yourself in situations where you will need to weigh the ethical results of a choice or decision. For example, imagine a crew member shows symptoms of heat exhaustion. Do you keep that employee working just because the superintendent says the project is behind schedule? If you are aware that your crew did not properly erect the reinforcing steel, do you stop the pour and correct the situation? In both cases, you must act ethically as a leader and make the right choice. If supervision ever asks a frontline leader to carry out an unethical decision, that individual must inform the next level of authority of the unethical nature of the issue.

NOTE

Anyone who is aware of illegal activity can face legal concerns simply by not acting. This is true even if the activities were requested by one's supervisor.

Business or legal ethics	**Professional or balanced ethics**	**Situational ethics**
Adheres to all laws and published regulations related to business relationships or activities.	Carries out all activities in such a manner as to be honest and fair to everyone under one's authority.	Relates to specific activities or events that may initially appear to be a gray area. For example, you may ask yourself, "How will I feel about myself if my actions were going to be published in the news or if I need to justify my actions to my family, friends, and colleagues? Would I still do the same thing?"

Figure 4 Types of ethics.

2.2.0 Communication

Communication: The process of exchanging facts, information, and ideas. Communication comes in many forms, including verbal, written, visual, and nonverbal.

Successful crew leaders learn to communicate well with people at all levels of an organization. **Communication** is the act of accurately and effectively sharing facts, feelings, and opinions with another. Simply stated, communication is the process of exchanging information and ideas. Communication comes in many forms, including verbal, written, visual, and nonverbal.

2.2.1 Why Communication Matters

Effective communication on a jobsite or project is critical for many reasons (*Figure 5*). First and foremost, clear communication is a key part of safety. Everyone on a jobsite must receive and understand communications about safety policies and procedures. If this does not happen, it creates safety issues for everyone in the group.

Miscommunication is also expensive (*Figure 5*). One study indicated that over half of all rework is due to poor communication and bad information. If communication on a construction site fails, it can cause delays and quality issues for the

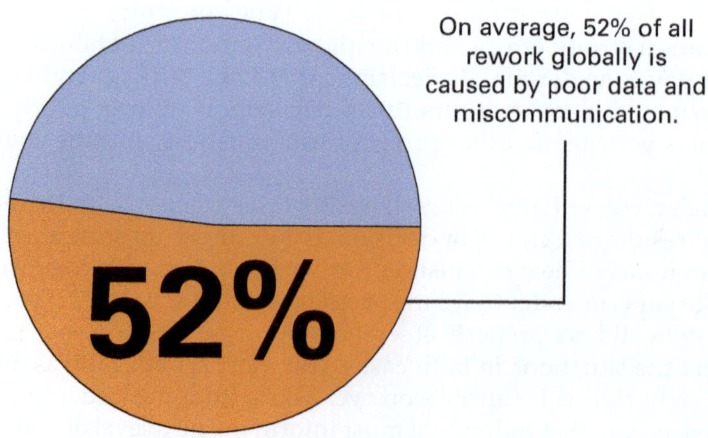

Figure 5 Industry impacts of miscommunication in the construction industry.

project. It can also result in damage to equipment or property. The good news is that communication is a skill that can be taught, practiced, and improved.

2.2.2 The Communication Process

Clear communication has three basic steps, as shown in *Figure 6*.

1. *Sender sends message to receiver.* A sender sends a message to a receiver, as a verbal, written, visual, or nonverbal communication. A message can be a set of directions, an opinion, or a personal message such as praise or a warning. Whatever it is, a message is an idea or fact that the sender wants the receiver to know.
2. *Receiver interprets message and repeats the message to provide feedback.* The receiver is the person who takes in the message. When the receiver gets the message, they figure out what it means by listening, viewing, or reading, and then repeat the message to provide feedback. *Feedback* is the receiver's communication back to the sender in response to the message. Feedback is a very important part of the communication process. It shows the sender how the receiver interpreted the message and if the receiver understood it as intended. In other words, feedback is a checkpoint to make sure the receiver and sender are on the same page.
3. *Sender validates that receiver correctly understood the message.* The sender confirms the receiver heard and understood the message.

This process is called **three-way communication**. Three-way communication is also sometimes called *repeat-back communication*.

Three-way communication is a simple way to make sure that communication is clear. If this process is easy, why is good communication so hard to achieve? When we try to communicate, many things can get in the way. These obstacles or **distractions** are anything that disrupts the communication flow between sender and receiver. Also called *noise*, distractions can include poorly worded messages, emotions, interpretations of a word, or the actual noise from a construction site. Distractions can happen at any point in the communication process.

Three-way communication: A form of communication with an initial message from a sender, which is then repeated by the receiver to confirm accuracy, and then validated by the sender. Also called *repeat-back communication*.

Distractions: Anything that disrupts the communication flow between sender and receiver. Also called *noise*.

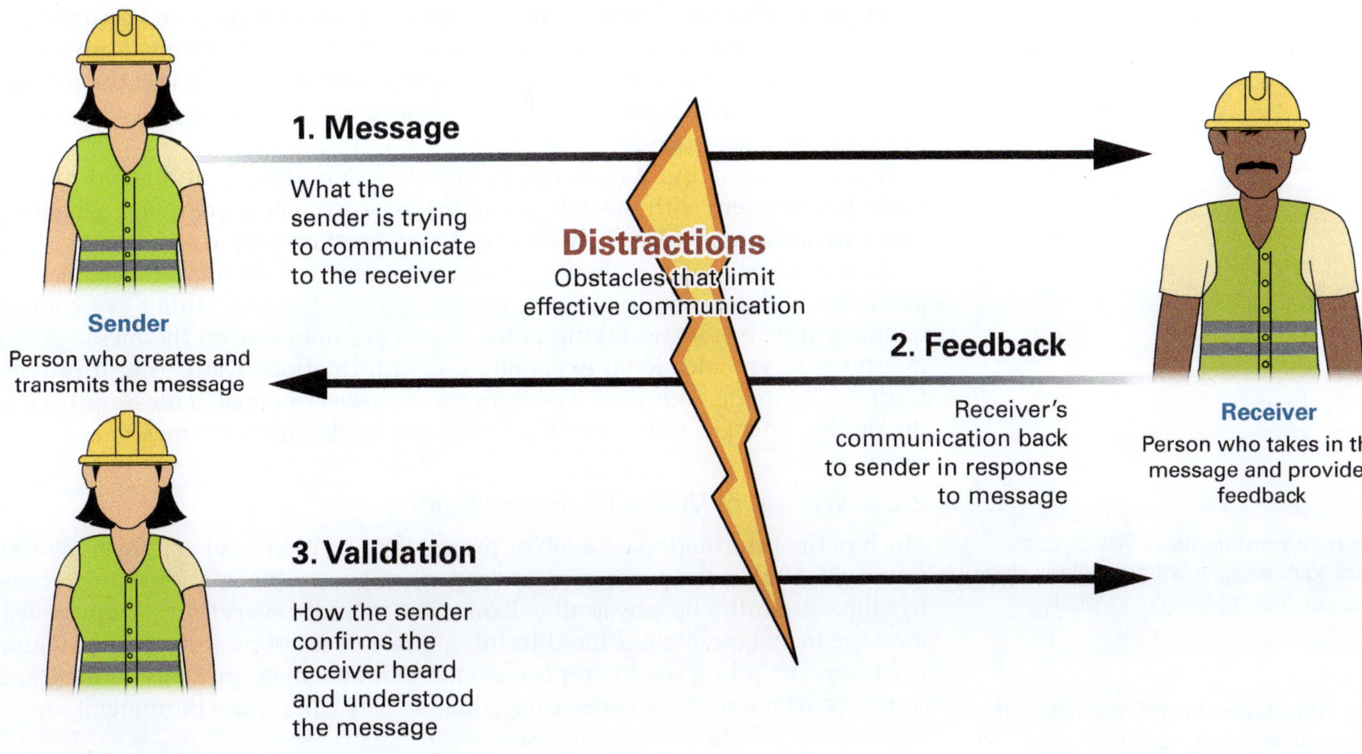

Figure 6 The three-way communication process.

Communication Types

Verbal
- Spoken or oral communication
- Examples: face-to-face conversations, telephone call, videoconference, voice chat

Written
- Messages or information sent using words
- Can be digital or physical
- Examples: text message, email, training guide, weekly report, change request, purchase order

Visual
- Using visual elements, such as photos, graphics, charts, colors, and fonts to convey messages
- Examples: sketches, diagrams, graphs, photos, construction drawings, signs

Nonverbal
- Things the speaker and the receiver see or hear while communicating
- Examples: appearance, personal environment, use of time, facial expressions, body movements, hand gestures, tone of voice, and eye contact

Figure 7 Types of communication.

As noted in *Figure 7*, there are four different types of three-way communication. The following sections review these different types of communication and discuss ways to increase the effectiveness of your communication, regardless of type.

2.2.3 Verbal Communication

Verbal communication: Messages that are sent using voice or other audible sounds. Also called *spoken communication* or *oral communication*.

Verbal communication includes messages that are sent face-to-face or by telephone, videoconferencing, voice chat, or other methods. Verbal communication is also called also *spoken* or *oral* communication.

With verbal communication, the way you speak is as important as the message itself. As a leader, you should assess your audience before you speak. Even in a meeting as standard as a daily huddle, using clear verbal communication sets your crew up for a successful day. Consider each person's knowledge of the topic and any language, gender, or cultural differences. Do not talk down to anyone in a group. Everyone, whether in a senior or junior position, deserves respect and courtesy.

You should also be sure to use words and phrases that the receiver can understand. For example, avoid technical language, acronyms, or slang if the receiver does not know these terms. In addition, speak at a reasonable pace, which gives the audience time to understand one point at a time.

If you are receiving a message, be sure to listen actively and avoid distractions that interfere with the delivery of the message. There are many barriers to effective listening, particularly on a busy construction jobsite.

As a receiver, you can show that you are actively listening. You can accomplish this without saying a word. Examples include maintaining eye contact, nodding your head, and taking notes. If you are not clear on the message, ask questions to get additional explanation or information. You can also provide feedback by using your own words, or *paraphrasing*, to repeat a message back to the sender. This lets the sender know that you understood the message.

2.2.4 Written or Visual Communication

Written communication: The process of sending messages or information using words, whether digitally or physically.

Visual communication: The process of using visual elements, such as photos, graphics, colors, and fonts, to convey messages.

Much of the information on a job or project is in written or visual form. **Written communication** is the process of sending messages or information using words, whether digitally or physically. Examples include everything from a text message to an email to a printed training guide. A lot of project documentation is written, including weekly reports, requests for changes, purchase orders, and contracts. Many of these items are written so they can form a permanent record for business and historical purposes.

Visual communication is the process of using visual elements to convey messages. Photos, diagrams, graphics, charts, and colors are all types of visual communication. Visual communication is widely used because it is an effective way to convey information quickly. Traffic signs, for example, use standard visual symbols because they are clear and well-understood. Many messages on a job also use visual communication. Examples include the project plans, drawings, or a video of a worksite incident. Visual communication can help workers who speak different languages or have limited language skills.

Written and visual communication used together are even more clear. For example, tags and signs on a project (*Figure 8*) use both written words and visual elements to help make them clear to all workers on a jobsite.

Figure 8 Communication tags and signs combine written and visual communication.
Source: Courtesy of AccuformNMC.com LLC

When writing or creating a written or visual message, you first should assess the receiver (the reader). The receiver must be able to read the message and understand the content. If they cannot, the communication process will be unsuccessful. Always consider the actual meaning of words or diagrams and how others might interpret them. *Table 2* and *Table 3* provide checklists to help make your written and visual communications effective.

TABLE 2 Checklist for Effective Written Communication

Did You . . .?	Yes/No
Assess your audience. Consider existing knowledge. Consider any language, gender, or cultural differences.	
State the purpose of the message clearly.	
Present the information in a logical manner.	
Provide an adequate level of detail.	
Avoid emotion-packed words or phrases.	
Avoid making judgments, unless asked to do so.	
Avoid using unfamiliar acronyms or abbreviations.	
Avoid using technical language unless you know your audience knows those terms.	
Make sure that the communication is clear and your handwriting is readable.	
Proofread your work. Check for typos, spelling errors, and grammatical errors. Do not assume spell-check or autocorrect got it right.	
Be prepared to provide a verbal or visual explanation if needed.	

TABLE 3 Checklist for Effective Visual Communication

Did You . . .?	Yes/No
Avoid complex graphics. (Simple is better.)	
Use colors and graphical elements to highlight key information.	
Ensure that the visual is large enough to be useful.	
Present the information in a logical order.	
Provide an adequate level of detail and accuracy.	
Provide a legend or written description if you include symbols.	
Be prepared to provide a written or verbal explanation of the diagram if needed.	

2.2.5 Nonverbal Communication

Nonverbal communication: Things the speaker and the receiver see or hear while communicating. Examples include facial expressions, body movements, hand gestures, tone of voice, and eye contact.

Unlike verbal or written communication, **nonverbal communication** does not involve spoken or written words or images. Nonverbal communication refers to things the speaker and the receiver see or hear while communicating. This includes appearance, personal environment, use of time, and body language. Examples include facial expressions, body movements, hand gestures, tone of voice, and eye contact.

Nonverbal communication can provide an external sign of someone's inner emotions. It often occurs at the same time as verbal communication. You can express feelings in many ways without realizing it. You may show you are happy by smiling and giving someone a warm handshake. You may show you are angry by raising your voice.

People observe your physical, nonverbal cues when you are communicating. In fact, nonverbal signals can have more impact than your verbal communication. Be aware of nonverbal cues to avoid miscommunication based on your posture, facial expression, or other signals. For example, if you have your arms crossed in a defensive posture while speaking, the person might assume you are angry, even if you are not.

On a jobsite, nonverbal communication is also a matter of safety. For example, a major aspect of rigging safety is having clear communication between a crane operator and the designated signal person on the ground. This communication relies on common signals, including nonverbal hand signals. The Emergency Stop signal used in rigging operations is a nonverbal communication. Anyone on the ground within sight of the crane operator can use this signal in the event of an emergency (*Figure 9*).

Figure 9 The Emergency Stop signal is an example of nonverbal communication.

Did You Know?

The Major Impact of Nonverbal Communication

Over the years, many studies have tried to measure how much of all communication is nonverbal. The specific numbers vary by study but indicate that between 55 to 93 percent of all communication is nonverbal. This means that your nonverbal signals send an even more powerful message than your verbal communication. Always remember that people might not remember what you say, but they do remember how you make them feel. Nonverbal cues have a major impact on how you make people feel when you communicate.

Did You Know?

Communication Is Number One

Research shows that the most common thing a supervisor does daily is communicate. They may do this as either a sender or receiver of messages. Eighty percent of a typical supervisor's day is spent communicating through writing, speaking, listening, or using body language. Of that time, studies suggest that around 20 percent is written communication, while 80 percent involves verbal communication.

2.2.6 Additional Communication Considerations

As a leader in the field, the way that you communicate has significant impact on the safety, quality, and productivity of your group. Regardless of your communication method, remember that everyone communicates differently. In some cases, you might need to communicate in multiple ways. You might also need to adjust your level of detail to ensure that everyone understands the intended meaning. For instance, a visual learner may need a map to comprehend directions. Another team member might just need a verbal explanation of which turns to take to a destination. *Figure 10* shows an example of ways to tailor the message to your audience.

Verbal Instructions (Experienced Crew) "Please drive to the supply shop to pick up our order."	**Verbal Instructions (Inexperienced Crew)** "Please drive to the supply shop. Turn right here and left at Route 1. It's at 75th Street and Route 1. Tell them the company name and that you're there to pick up our order."
Written Instructions 1. Turn right at exit. 2. Drive 2 miles to Route 1. Turn LEFT. 3. Drive 1 mile (past the tire shop) to 75th Street. 4. Look for supply store on right...	**Visual Instructions (Diagram/Map)** 7501 N. Highway 1 (800) 555-7567 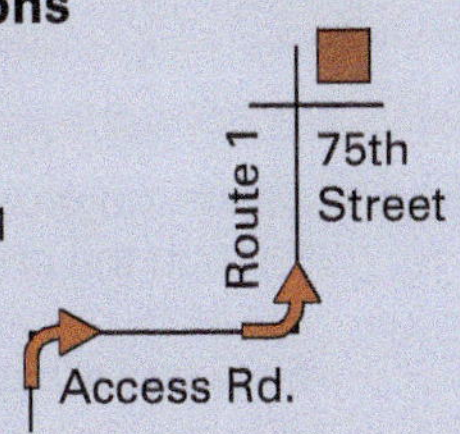

Figure 10 Assess your audience and then tailor your message.

The construction industry is a diverse and multicultural one. It is quite common to have a crew with people from different cultural backgrounds working together. Language and cultural differences can make communication even more challenging. Even if employees speak the same language, different understandings of words or symbols can create confusion or cause offense.

An important part of effective communication is to remember that people are different and to adjust your communication style to meet the needs of the person or people receiving your message. This means field leaders need to invest the time to learn how to communicate well with each individual. This requires you to know each crew member, recognize how each person learns, how each person delivers feedback, how each person works with others, and who they are as individuals, not just coworkers.

2.3.0 Problem-Solving and Decision-Making

Problem-solving and **decision-making** are a large part of every field leader's daily work. They are a part of life for all supervisors, especially in fast-paced, deadline-oriented industries like construction.

What is the difference between problem-solving and decision-making? Problem-solving is a process used to identify possible solutions to a situation. Decision-making refers to the action of making a choice. Decision-making involves starting or stopping a plan or choosing a different plan of action as

Problem-solving: A process used to identify possible solutions to a situation.

Decision-making: The action or process of making a choice. Decision-making involves starting an action, stopping one, or choosing a different plan or action.

appropriate for the situation. A decision may be complicated and involve many steps, but it is simply a choice.

The key difference between problem-solving and decision-making is that solving problems is a process, while making decisions is an action based on experience or information learned during the problem-solving process.

2.3.1 Problem-Solving

The ability to solve problems is an important skill in any workplace. It is especially important for craft professionals, whose workday is often not predictable or routine. Your group will look to you as a leader to solve problems regularly.

Figure 11 shows a five-step process for solving problems, which you can apply to any situation.

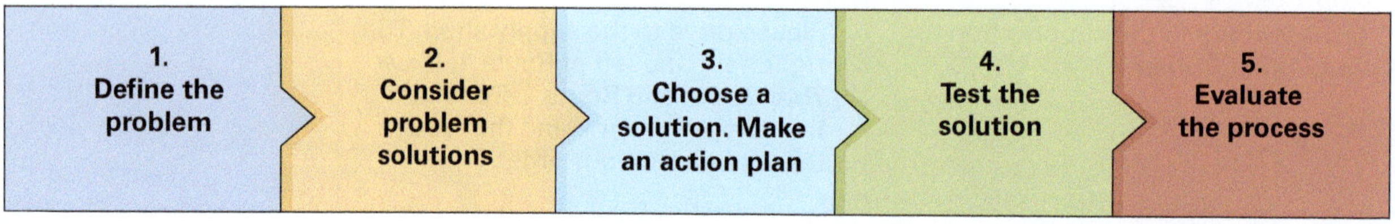

Figure 11 The problem-solving process.

Step 1 *Define the problem.* Defining a problem is not as easy as it sounds. Thinking through the problem often uncovers additional problems. Also, drilling down to the facts of the problem may mean setting aside your own biases or presumptions toward the situation or the individuals involved.

Step 2 *Consider possible solutions to the problem.* A problem may have more than one solution. Think through a number of solution options. As you do this, get inputs from people most affected by the problem, those who must correct the problem, and those who will be most affected by any potential solution.

Step 3 *Choose the solution that seems best and make an action plan.* After you consider several options to solve the problem, you must think through each possible solution and pick the best one. Again, you do not have to do this alone. Ask for feedback from others who have a stake in solving the problem well. Keep an open mind. The best solution might be taking pieces of two different solutions and combining them to create a new solution.

Step 4 *Test the solution to determine whether it actually works.* A solution might sound perfect before you try it but turn out not to be effective when you do. Alternatively, you might discover from trying to apply a solution that it is acceptable with a little change. If a solution doesn't work, think about how you could improve it, and then test your new plan.

Step 5 *Evaluate the process.* After your solution is in place, review the steps you took to discover and implement the solution. Could you have done anything better? The next time you are faced with a similar situation, you can use this experience to help solve the problem more effectively.

This problem-solving process can apply to almost any situation. *Figure 12* walks through a scenario to show how problem-solving works in practice. A key point to note is that it's important to receive input from anyone in your group affected by the problem or the solution. Each solution will have pros and

Problem Scenario

You are leading a team assigned to a new shopping mall project. The project will take about 18 months to complete. The only available parking is half a mile from the jobsite. The crew must carry heavy toolboxes and safety equipment from their cars and trucks to the work area at the start of the day, and then carry them back at the end of their shifts.

1. Define the problem	2. Consider problem solutions	3. Choose a solution Make an action plan	4. Test the solution	5. Evaluate the process
1. Problem: Workers are wasting time and energy hauling their tools and equipment to and from the worksite.	**2. Problem solution options:** Think about different ways to solve the problem. Several workers have proposed solutions: • Install lockers for tools and equipment closer to the worksite. • Have workers drive up to the worksite to drop off their tools and equipment before parking. • Bring in another construction trailer.	**3. Solution choice and action plan:** The worksite superintendent doesn't want an additional trailer in your crew's area. You and the team agree the shuttle service is the best option. It solves the time and energy issue *and* workers can keep their tools with them. The project supervisor plans to arrange a large van and driver to provide the shuttle service. **Note that this is a decision point in the problem-solving process.**	**4. Test the solution:** At first, the solution works, but a new problem arises. Everyone starts and leaves at the same time. The van can only hold half of the team and their equipment. To solve this problem, the supervisor schedules shuttle pickups every 15 minutes and adjusts work schedules so workers board the van at two different times. **Note that this is also a decision point in the problem-solving process.**	**5. Evaluate the process:** This process gave management and workers a chance to express an opinion and discuss the various solutions. Everyone feels pleased with the process and the solution.

Figure 12 Applying the problem-solving process, with decision points.

cons, so it's important to receive input from the workers affected by the problem. For example, workers will probably object to any plan (like the drop-off plan) that leaves their tools vulnerable to theft. Gaining the inputs of others will help you gain the trust and respect of your team.

2.3.2 Decision-Making

Figure 12 also shows where decisions are made as part of the problem-solving process. Decision-making is a choice made by using one's judgment. This judgment might be based on experience or on information learned during a problem-solving process. In general, you will have to make three types of decisions as a field leader, as shown in *Figure 13*.

No matter what the type, some decisions are simple or can be made based on past experiences. Other decisions are more difficult to make. These decisions require more careful thought about how to carry out an activity by using problem-solving techniques.

Note that decisions do have impacts, which you will need to consider. If you choose to use a shuttle to transport workers to a jobsite in shifts, as shown in *Figure 12*, using the shuttle may create a safety liability while workers are transported.

Decision Type	Description	Examples
Yes or no decisions 	Decisions in which the answer is yes or no. Also called *whether* decisions, because you have to decide whether to take an action or not.	• Whether to hire a van to use as a shuttle service • Whether or not to promote a team member • Whether to alert a supervisor to an unsafe condition • Whether to proceed with a concrete pour
Options decisions 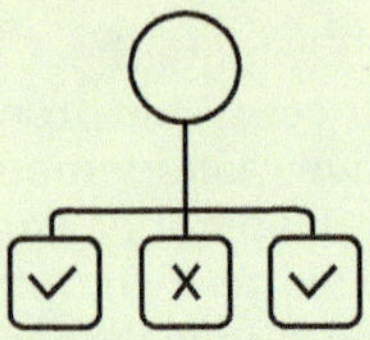	Decisions that require you to choose from a group of options. Also called *which* decisions, because you have to choose which option to take.	• Which of the three options will best address the problem of workers having to haul tools and equipment to and from the worksite? • Which toolbox talk would be most important to deliver to the team today? • Which team member should I have drive to the supply shop to pick up the order?
Contingent decisions 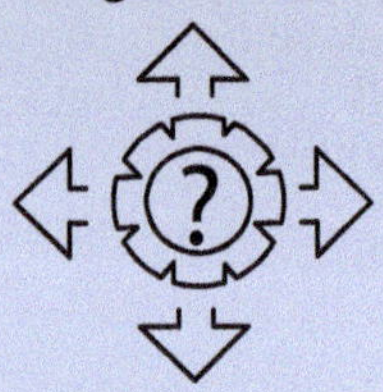	Decisions already made but on hold, until other specific events happen. These decisions can be very powerful because you can "pre-decide" what you will do in the event those events occur.	• If Carlo is late again two more times, I will set up a performance conversation. • If the materials arrive early, we will store them in the back of the jobsite. • If the team completes this task on time, I will find a clear way to recognize and reward the success.

Figure 13 Types of decisions.

2.0.0 Section Review

1. Which of the following is *not* a trait or behavior of effective frontline leaders?
   a. They lead by example.
   b. They are good communicators who lay out project goals to their crew.
   c. They motivate the team to work to their full potential.
   d. They have limited knowledge of the activities they supervise.
2. Which of the following is *not* a good practice to ensure effective communication?
   a. Speaking clearly at a reasonable pace
   b. Using highly technical language, acronyms, and slang
   c. Using visual cues, such as sketches or diagrams, to clarify instructions
   d. Using three-way communication to validate that a message was clear
3. While problem-solving is a process used to identify one or more possible solutions to a problem, the action of making a choice is called _____.
   a. decision-making
   b. a problem statement
   c. a decision statement
   d. action planning

3.0.0 Team Leadership and Motivation

Performance Tasks	Objective
There are no Performance Tasks in this section.	Identify key skills for team leadership and motivation. a. Choose strategies to motivate individuals and a team. b. Describe key approaches to building and leading teams. c. Recognize how to approach leadership challenges, such as transitioning from peer to leader, addressing performance issues, and resolving conflict.

In your role as a frontline leader, you will have incredible impact on the individuals on your team. Motivating individuals and building successful teams are important and challenging tasks for any leader. You will also face leadership challenges as a new or developing leader, including managing performance issues, handling conflict, and leading team members who were once your peers. In this section, you will learn strategies and approaches to succeed as a team leader.

3.1.0 Motivating Individuals

The ability to motivate others is a key skill that leaders must develop. To motivate is to make someone feel determined or enthusiastic or to make someone behave in a particular way. On a jobsite, a leader uses authority and influence to encourage individuals to put in their best effort to accomplish the project goals. Lack of motivation affects job quality and productivity, reducing profits and growth for companies, businesses, contractors, or organizations that rely on people to complete the work.

As a supervisor or crew leader, it is important to understand what motivates each individual on your team. Different things motivate different people in different ways. This means there is no single best approach to motivating all of your crew members, because what motivates one crew member may not motivate another. In addition, what works to motivate a crew member once may not motivate that same person again in the future.

How can you know what motivates the members of your project team? Effective communication skills, specifically active listening, are a key part of getting to know each person on your team. You can also learn more about what motivates an employee by looking at factors in the employee's performance. Examples of factors that may indicate an individual's motivation include the level or rate of unexcused absenteeism, the number of complaints, and the quality and quantity of work produced.

Generally, the factors that motivate individuals are the same as those that create job satisfaction. These include the following:

- Clear expectations and goals from leaders
- Training to help individuals be successful
- Consistent sense of **accomplishment** and growth
- Recognition and rewards
- A good work environment and culture

Accomplishment: An individual's need to set challenging goals and achieve them.

A crew leader's ability to provide strong support in these areas increases the likelihood of high **morale** within a crew. Morale refers to an individual's emotional outlook toward work and the level of satisfaction gained while performing the jobs assigned. High morale means that employees will be motivated to work hard, and they will have a positive attitude about coming to work and doing their jobs. A motivated team is also generally a more productive and safer team, which can have a positive effect on project success.

Morale: An individual's emotional outlook toward work and the level of satisfaction gained while performing the jobs assigned.

3.1.1 Clear Expectations and Goals

As a leader, it is important for you to clearly communicate your expectations. A lack of clear expectations can leave your team feeling lost and demotivated. To get the best performance from your team and to motivate them, you need

Did You Know?

Money Doesn't Always Buy Motivation

Many leaders consider money to be a good motivator. Money can drive motivation, but it often is only a temporary solution. If you give someone a raise, they tend to work harder for a period of time. However, research shows that this behavioral change does not last. Further, after someone has reached a level of financial security, factors other than money may come into play. Over time, too much reliance on money for motivation can negatively impact a worker's commitment. Remember, employees are individuals who are motivated by different things, including pride in their work, work environment, and more. For those people, a great workplace may mean more to them than better pay.

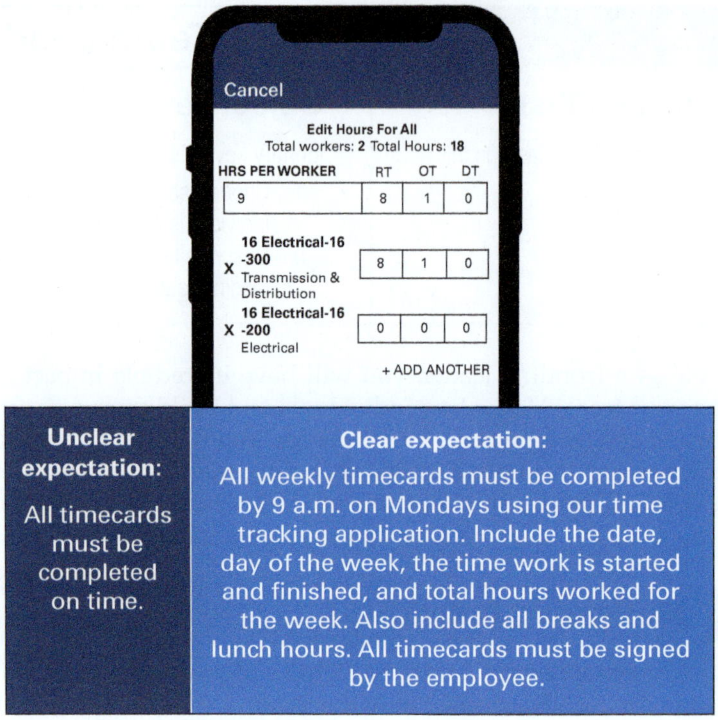

Figure 14 Examples of unclear and clear expectations.

to set clear goals for completing tasks. You also need to ensure that everyone on the team understands the deadlines for those tasks. *Figure 14* includes an example of clear expectations.

Set achievable short-term and long-term goals, including tasks to be done and expected timeframes. Having short-term goals gives you a chance to provide regular feedback, positive praise, and rewards. However, because short-term goals can change, also setting long-term goals gives the team additional opportunities for success. Do not set workers up for failure, as this leads to frustration, and frustration can lead to reduced productivity.

In addition to setting clear goals and expectations, you must also deliver timely feedback. People want and need feedback and positive reinforcement for their contributions and efforts. However, telling someone they did a good job last year or criticizing them for work from a month ago is meaningless. When sharing timely feedback, always include specifics to help the individual know what they did well or where they could improve.

You always should share negative feedback in private. When delivering negative feedback, approach the problem as a coaching moment, working with the individual to understand their challenges and working with them to correct the problem. Every failure or miss can be an opportunity for you to help that individual grow, if the conversation and feedback are constructive.

Positive feedback can be shared privately or publicly. When you praise a person in front of their team, it may also encourage other members of the team. Praise can be a valuable tool to reinforce behaviors that you want. (Not all people like to be publicly praised, so consider that before you offer public praise.)

3.1.2 Accomplishment and Growth

Accomplishment refers to an individual's need to set challenging goals and achieve them. A sense of accomplishment motivates most people. That is why setting and working toward recognizable goals tends to make employees more productive. A person with a sense of achievement is likely to be better motivated and help to motivate others.

A person who can align a task to their sense of meaning is much more likely to complete it. Frontline leaders can help their team feel a sense of accomplishment by keeping the work challenging and interesting. Workers who are completing the same task every day or performing work above or below their skill level can quickly become demotivated. For many individuals, overcoming the challenge of new or difficult assignments is much more satisfying than performing repetitive, boring tasks.

Remember, however, that too many or significant changes in work may have a negative impact on morale, because people often prefer some consistency and predictability in their work. Crew members react to change in many different ways and reasons. For some individuals, change creates anxiety and uncertainty, which can result in a negative attitude toward the work. To help manage change with your crew, remember to share why the change is occurring and explain what the employees and company will gain.

As a leader, you must ensure the individuals on your team feel that their skills and abilities are valued and make a difference. Employees who do not feel valued tend to have performance and attendance issues. Leaders should work to ensure the following is true for every crew member:

- They feel like an important part of the team with a skill set that contributes to the project success.
- They know that every position helps them build skills and experience that will last throughout a career.
- They have a chance to learn new things and develop their skill sets.
- They see opportunities to grow and progress within the company.

Professional and personal growth are other critical factors in motivation. **Professional growth** refers to gaining new skills and work experience that can help you reach a goal in your career. It is important that employees know that they aren't limited to their current jobs. Let them know that they have a chance to grow with the company and to be promoted as acknowledgment for excelling in their work when such opportunities occur.

Professional growth: Gaining new skills and work experience that can help one to reach a goal in one's career.

Whenever possible, show your team that you value their professional growth by giving them more advanced responsibilities when they are ready to perform them safely. People who are presented with challenging tasks often are motivated to work harder and improve. This approach is a win–win since the construction project benefits from a higher level of productivity, while the worker experiences professional growth.

Effective leaders encourage each of their crew members to work to their full potential. In addition, they share information and skills with their employees to help them advance within the organization.

Personal growth is about improving your talents and potential, both in and out of the workplace. For example, a team member may find it very rewarding to master a new competency on the job. Experiencing this type of personal growth at work helps to motivate, reduce boredom, and spark enthusiasm for additional growth.

Personal growth: Improving one's talents and potential, both in and out of the workplace.

Personal growth also takes place outside the workplace. Crew leaders should encourage their employees to identify goals for themselves both professionally and personally and then support their efforts to accomplish those goals. As an example, training can be a way to improve both professional and personal skills, such as time management, communication, financial management, and more.

3.1.3 Training and Upskilling

Another key factor in motivating individuals—and ensuring productive and safe projects—is providing sufficient training and opportunities for **upskilling**. Upskilling is a type of training that provides someone with more advanced skills through additional education. Upskilling provides employees with the tools and knowledge they need to perform their current job more effectively, more efficiently, or in a more modern way.

Upskilling: A type of training that provides someone with more advanced skills through additional education.

As people enter the construction workforce, many will have completed training programs but may not have the specific practical experience needed to complete a task on a project site. When a new employee joins your crew, you will need to take an active role in confirming their skills and helping with on-the-job training if there are gaps. A frontline supervisor on a construction jobsite only has a short time to evaluate their crew members on what tasks they can do or how much experience the crew member actually has.

In the first week, with any new employee, you should ensure that you train the new employee on any jobsite processes and allow time to do a task verification (*Figure 15*). After you have established the skills the individual has, you can work on a plan to leverage those skills while establishing new ones. If you identify gaps, you can create a training plan for that individual to address those areas.

3.1.4 Recognition and Rewards

Recognition: Acknowledgement of an employee's accomplishments through verbal and nonverbal communications.

Positive feedback is a form of **recognition**. Recognition acknowledges an employee's accomplishments through verbal and nonverbal communications. Individuals want to have good work appreciated, applauded, and acknowledged by others. You can accomplish this by thanking employees, offering an award for Employee of the Month, or giving tangible or monetary awards.

The following are tips to remember for giving recognition and praise:

- Be available on the jobsite so that you see good work as it is happening.
- Provide positive feedback publicly or in private.
- Give recognition and praise only when truly deserved. People quickly recognize insincere praise and will lose respect for the person delivering it.
- Encourage improvement by showing confidence in the ability of the crew members to do above-average work.
- Use goal setting as a tool to build confidence and create opportunities for praise.

Rewards are tangible recognition for good work. Rewards can be monetary, such as opportunities for raises, bonuses, or other incentive pay. Other types of rewards might include company gear, vouchers for new boots, gift cards, a team outing at the end of the week, or the best spot in the parking area for a month.

3.1.5 Good Work Environment and Culture

Having a good work environment and culture is critical for creating strong employee motivation and high morale. As a leader, you can create a good work environment and culture for your team in several ways.

Treat employees well. Be considerate, kind, and respectful. Treat people fairly. Mentor your employees. Coaching and supporting employees boosts their self-esteem, their self-confidence, and their motivation. Remember that small things matter. Having clean lunch areas, good housekeeping, and a comfortable workplace communicates that you respect the group.

Figure 15 When a new employee starts, be sure to evaluate the crew member's skills.
Source: kali9/E+/Getty Images

Communicate actively. As a leader, you should make yourself available for questions, opinions, and ideas. Actively engaging and communicating with workers plays a crucial role in motivating your team, as it shows that you appreciate and value their feedback. Involve employees in discussions so that they feel their opinions are valued. This, in turn, leads to a sense of pride and active participation. Making yourself available also prevents team members from feeling isolated from leadership.

When you communicate, be positive. Approaching your leadership responsibilities by threatening employees with negative consequences can result in higher employee turnover instead of motivation.

As much as possible, maintain a sense of humor, especially toward your own failings as a human being. No one is perfect. Your employees will appreciate that and be more motivated to encourage you.

Help individuals manage stress. In a job that can be demanding, fast-paced, and tiring at times, managers in the construction industry must take measures to ensure that their team is not exposed to undue stress. Work with your supervisors to ensure you have a safe and reasonable work environment. This means not only physical spaces, but also policies and procedures. For example, allowing workers to take time off to deal with personal matters or emergencies shows that the company prioritizes their well-being. Engage with your supervisors to help address resources and to organize shifts in such situations to ensure that project progress and construction costs are not negatively impacted.

Lead by example. Employees will be far more motivated to follow someone who is willing to do the same things they are required to do. Being a good leader and motivator means leading by example. Employees will not be motivated to work if upper management does not display the same level of commitment to the job that they are asked to have. For example, if you ask everyone to work late, but then leave early to go see a movie, this will be noticed and may create feelings of resentment and decrease team morale. As a leader, you can inspire your team by displaying the behaviors you would like to see from them, including working hard, showing personal and professional integrity, and having a can-do attitude.

3.2.0 Leading Teams

As a crew leader, you will need to lead a team of individuals from a range of backgrounds and experiences and create a high-performing team. To do so, you will need to delegate work to that team and implement policies and procedures that support the team. You will also need to address challenges like language barriers and **harassment**.

Harassment: Unwelcome conduct that is based on race, color, religion, sex, national origin, age, disability, or genetic information. Harassment includes verbal or physical conduct that impacts a person's work performance or creates an intimidating, hostile, or offensive work environment.

3.2.1 Delegating to Your Team

Once the various activities that make up the job have been determined, the crew leader must identify the person or persons who will be responsible for completing each activity. This requires that the crew leader be aware of the skills and abilities of the people on the crew. Then, you must put this knowledge to work by matching the crew's skills and abilities to specific tasks needed to complete the job.

After matching crew members to specific activities, you must then delegate the assignments to the responsible person(s). Generally, when delegating responsibilities, the crew leader verbally communicates directly with the person who will perform or complete the activity.

When delegating work, remember to do the following:

- *Delegate work to a crew member who can do the job properly.* If it becomes evident that the worker doesn't perform to the desired standard, either teach the crew member to do the work correctly or turn it over to someone else who can (without making a public spectacle of the transfer).
- *Make sure crew members understand what to do and the level of responsibility.* Be clear about the desired results, specify the boundaries and deadlines for accomplishing the results, and note the available resources.

- *Identify the standards and methods of measurement for progress and accomplishment.* Also identify the consequences of not achieving the desired results. Discuss the task with the crew member and check for understanding by asking questions. Allow the crew member to contribute feedback or make suggestions about how to perform the task in a safe and quality manner.
- *Give the crew member the time and freedom to get started.* Don't make them feel pressured by too much supervision. When making the work assignment, be sure to tell the employee how much time there is to complete it and confirm that this time is consistent with the job schedule.
- *Check in on progress.* Be sure to periodically check in on the crew member's progress. This will allow you to see how the task is going and allow the crew member to ask questions. Checking in along the way will also help you clear up any misunderstandings early and avoid rework.
- *Examine and evaluate the result.* After the task is complete, ask the crew member for their comments on how the work went. Provide your own feedback. Having this discussion will help you know what kind of work to assign that crew member in the future. It will also provide a means of measuring your effectiveness in delegating work.

There may be a time when someone else in the company issues written or verbal instructions to a crew member or your crew without going through you as the crew leader. If this happens, work with that individual to understand why they took that approach. Then, help that person understand why it is important for them to work through you for this type of instruction, whenever possible, so that you can manage the work of the entire crew safely and productively.

3.2.2 Implementing Policies and Procedures

Every company establishes policies and procedures that crew leaders are expected to implement and employees are expected to follow. Company policies and procedures are essentially guidelines for how the organization does business. They can also reflect organizational philosophies, such as putting safety first or making the customer the top priority. Examples of policies and procedures include safety guidelines, credit standards, and billing processes.

The following tips can help you effectively implement policies and procedures:

- Learn the purpose of each policy. This will help you follow the policy and apply it appropriately and fairly.
- If you're not sure how to apply a company policy or procedure, check the company manual or ask your supervisor.
- Always follow company policies and procedures. Remember that they combine what's best for the customer and the company. In addition, they provide direction on how to handle specific situations and answer questions.

NOTE

Try to obtain a supervisor's policy interpretation in writing or print out their email response so that you can append the decision to your copy of the company manual for future reference.

Crew leaders may need to issue directions to their crew members. A *direction* is a form of communication that initiates, changes, or stops an activity. Directions may be general or specific, written or verbal, formal or informal. It is up to you how you issue directions, but the policies and procedures of the company may govern your choice.

When issuing directions, do the following:

- Make them as specific as possible. Avoid being general or vague unless it is impossible to foresee all the circumstances that could occur in carrying out the directions.
- Recognize that you do not have to write directions for simple tasks unless your company requires this.
- Write directions for more complex tasks, tasks that will take considerable time to complete, or tasks that are permanent (standing) orders.

- Consider what is being said, the audience to whom it applies, and the situation under which it will be implemented to determine the appropriate level of formality for the direction.

3.2.3 Addressing Language Barriers

Leading a diverse team also means addressing language barriers. Millions of US workers speak languages other than English. Jobsites where two or more languages are used are common. In addition, even among English speakers, accents and dialects can present communication barriers.

As you have learned, any barrier in communication keeps people from being able to understand each other clearly. If a person does not fully understand the task a frontline supervisor gives them, the task may be completed incorrectly, which can cause delays, cause additional expenses, or create an unsafe situation. As a supervisor of a crew, you need to know that each crew member understands the work to be done. The following tips will help when communicating across language barriers:

- Be patient. Give workers time to process the information in a way they can comprehend.
- Avoid humor. Humor is easily misunderstood. The worker may misinterpret what you say as a joke at the worker's expense.
- Do not assume an individual is not intelligent simply because they don't understand what you are saying.
- Speak slowly and clearly. Avoid the tendency to raise your voice.
- Use face-to-face communication whenever possible. Other forms of communication can be more difficult when a language barrier is involved.
- Use the plans and drawings or make sketches to ensure people understand clearly.
- For critical written instructions, ensure the individual's literacy level matches the level of difficulty and language of the document. A person may speak clearly in a second language, but struggle with written understanding.
- Communicate in multiple ways or adjust your level of detail or terminology to ensure that everyone understands the meaning as intended.
- Confirm understanding. Ask for feedback with open-ended questions that require more than yes-or-no answers. Use three-way communication and have an individual repeat back what they heard to ensure mutual understanding and prevent miscommunication. If a worker is not fluent in English, ask the person to demonstrate their understanding through other means.

Many organizations offer language classes for supervisors and employees to help create more effective communication on jobsites. Whenever possible, take advantage of these training opportunities and encourage your crew to do so.

3.2.4 Preventing Harassment

As a frontline supervisor of a team, you must be aware of harassment as a potential challenge. As defined by the US Equal Employment Opportunity Commission (EEOC), harassment is unwelcome conduct that is based on race, color, religion, sex, national origin, age, disability, or genetic information. Harassment includes verbal or physical conduct that impacts a person's work performance or creates an intimidating, hostile, or offensive work environment.

The following are examples of activities that might qualify as harassment:

- Telling offensive jokes
- Calling someone offensive names
- Making verbal or physical advances or assaults

- Making threats or intimidating others
- Displaying offensive objects or pictures

Harassment becomes unlawful when a person must endure the offensive conduct to remain employed or the conduct is severe or common enough to create an intimidating, hostile, or abusive work environment. While not all acts of harassing behavior may be serious enough to be considered a violation of the law, all such behavior destroys teamwork and negatively affects jobsite safety and productivity.

Anyone can be a victim of harassment. Historically, many have thought of harassment as an act perpetrated by someone in power over another. However, a harasser can be a supervisor, a supervisor in another area, a coworker, or any other person in the work environment.

The EEOC enforces sexual harassment laws within industries. When investigating allegations of harassment, the EEOC looks at the whole record, including the circumstances and the context in which the alleged incidents occurred. Supervision may hold the crew leader responsible if they were aware of sexual harassment and did nothing to stop it. You should not only take action to stop harassment, but also should serve as a good example for the rest of the crew.

Prevention is the best way to eliminate harassment in the workplace. Your employer will have a harassment policy that clearly communicates zero tolerance for harassment and outlines a complaint or grievance process. This process should include taking immediate and appropriate action when an employee complains. This anti-harassment policy should apply to all employees (including managers, supervisors, and coworkers) and nonemployees (such as suppliers, owners, vendors, and others). Creating an open and inclusive work environment will discourage harassment.

As a leader, you must understand and uphold your employer's policies and procedures on harassment. If you believe that someone has violated the harassment policy, immediately notify your supervisor. If you are not comfortable reporting the incident to your immediate supervisor, you may report the complaint to any employee in a supervisory position or Human Resources. If you are not sure how to proceed in a situation, reach out to your Human Resources partners for guidance.

3.3.0 Leadership Challenges

As the leader of a team, you will need to manage leadership challenges and opportunities every day on the job. While the specific challenges will vary based on your team, the project, and other conditions, you must learn to address these when they arise. Such challenges can include navigating your initial transition from peer to leader, managing performance issues with your crew, or resolving conflicts that arise on the jobsite.

3.3.1 Transitioning from Peer to Leader

Making the transition from crew member to a crew leader can be difficult, especially when the new position involves overseeing a group of former peers. When you move from being a peer to a leader, your relationship with your former team members changes in important ways.

When you become a crew leader, you are no longer responsible for your work alone. You are now accountable for the work of a diverse team of individuals with varying skill levels, personalities, work styles, and cultural and educational backgrounds.

Crew leaders must learn to put personal relationships aside and work for the common goals of the entire crew. The crew leader can overcome these problems by working with the crew to set mutual performance goals and by freely communicating with them within permitted limits. Use their knowledge and strengths along with your own so that they feel like they are key players on the team. *Table 4* lists some additional suggestions on how to make the transition from peer to leader easier.

TABLE 4 Making the Transition from Peer to Leader

Strategy	Tactic
Reset boundaries with existing friends	Expectations and boundaries must be set with existing friends. Having a strong, professional working relationship is important, but you may need to reduce time spent with old peers outside of the project. Be clear and explain that, on the job, business comes before personal relationships.
Demonstrate that you have earned this leadership role	Quickly demonstrate good leadership for your team to earn employees' trust and respect. Be fair and consistent in everything you do. The perception of favoritism will tear a crew apart and negatively impact the outcome of any project.
Make the best decisions for the project and your company	Always make decisions that are best for the business. While it can be hard to make a decision that impacts someone who was a peer and is a friend, don't compromise to make a popular decision.

3.3.2 Managing Performance Issues

At the end of the day, if someone is simply not being accountable and completing their work, it might be time for a difficult conversation. In the workplace, a difficult conversation is one in which you have to manage emotions and information in a sensitive way to deal with a workplace issue. A difficult conversation may involve the following:

- Communicating business decisions
- Discussing poor performance or behavior
- Addressing conflict
- Handling complaints
- Ending employment or other changes in job status

As a leader and manager, you will need to have a difficult conversation at some point in your career. These conversations are never enjoyable, but you can handle them effectively by following some basic steps.

Identify the problem to resolve. Ensure that you have a clear understanding of the problem that you want to resolve. Before you start preparing for the conversation, ask: Why do you want to have the conversation? What do you want to achieve? Is the outcome you're looking for realistic?

Gather the information you need. Make sure you know the facts of the situation before approaching the employee. Bring any documents you will need to the meeting. For example, if you are discussing how an employee is late to work, you will want to have relevant policies and tardiness data at hand. If you are not sure about any internal workplace policies and procedures, you can always seek advice from a supervisor or Human Resources manager.

Think about the employee's point of view. Before you speak, you should think about the matter from your employee's point of view. Engage in the conversation with an open mind and truly seek to understand the situation from the employee's point of view.

Get the timing right. Try to schedule the conversation as soon as possible, but not if emotions are running high. Choose a time and place appropriate for the discussion, where you will not be rushed or interrupted. Generally, avoid having negative discussions at the end of the week and, if an employee needs time to process the conversation, offer to take a break and meet again later that day or the next.

Prepare what you are going to say. Write down the key points you need to cover, so that you can keep your conversation on track and stay in control. While it is important to prepare notes, avoid preparing a rehearsed script, as it may hamper your ability to listen effectively, react accordingly, or find alternative solutions.

Give the employee time to prepare. Make an appointment with the employee and provide some context to give them some time to prepare. Remember that knowing this conversation is coming may be stressful for the employee, so phrase the approach carefully and don't leave it too long. Examples of ways to phrase the appointment might be:

> "I'd like to talk about a series of quality issues I'm noting and get your point of view. Can you come see me tomorrow?"

"I think we have different perceptions about what it means to have a clean work area. I'd like to hear your thoughts on this. Are you free this afternoon?"

Seek support if you need it. If you're not comfortable having the conversation one-on-one with the employee, enlist help from your manager or Human Resources person.

After the conversation, be sure to take the following steps:

- *Document the discussion.* Write down any agreements that were reached, and indicate the date, time, and who was present. This will be helpful if you need to revisit the discussion at a future date. Your *job diary,* which you will learn about later, is a useful tool here.
- *Follow up with the employee.* Confirm what you had agreed in writing and send these to the employee. Be sure to give the employee time to actually act upon the steps they agreed to.
- *Take some time to reflect on the matter.* Learn from the experience and think about what you could do differently next time. Debrief with your supervisor or Human Resources if you feel you need to discuss the matter.

Also, remember that you should never discuss this matter with other staff members. You must treat the employee the same as you had prior to the conversation and remain completely professional.

If you do need to have this type of difficult conversation, do it as soon as you start to have concerns or see early signals of conflict. The worse a situation becomes, the harder it can be to manage and resolve. In the meantime, your project and the rest of the crew are being affected.

As previously noted, the best communication strategy is to have an open-door policy or informal one-on-one conversations so that issues can be dealt with as naturally as possible. Be approachable and keep in touch with your team, so they are more likely to come to you with problems before they escalate.

3.3.3 Conflict Resolution

Conflicts will happen during a project. When you have many groups, including owners, general contractors, and specialty contractors all working together to complete a project, disagreements are likely to happen. Conflicts can arise within your crew as well. Each person or group might have a different view or opinion on how something should be done. Those differing opinions often lead to conflicts.

As a leader of the team, you will have many opportunities to look for positive and proactive means to resolve conflict. Resolving conflicts can lead to a better understanding or a new way of doing things that improve the overall project. If you handle conflict well, it will further earn the respect and trust of your crew. The following are some strategies to use when addressing conflict:

Recognize that conflict is normal. Not all conflict is bad. Conflicts on the construction site can be good. If the individuals engaged in the conflict are mature, conflict can improve processes and approaches by creating conversation about new ways to do work.

Do not ignore the conflict. Take action early to avoid further complications down the line. Even a small dispute that may appear inconsequential can balloon into a larger problem if it goes unresolved. If you allow the parties involved to work it out on their own, you still need to follow up and make sure that there aren't any lingering issues.

Understand the real issue behind the conflict. You must understand the real issues causing the conflict to determine the best solution. Conflicts are often caused by different interpretations of plans and specifications, change orders, scheduling, and poor communication. Work stress can also be an underlying issue. Conflicts, for example, may happen more often later in the day or in a work shift, due to fatigue. Use your active-listening skills to listen to everyone involved. This shows everyone that their opinions matter and that you are committed to working with them to resolve the issue fairly.

Remain neutral and focus on business, not people. Conflicts can escalate quickly from a simple disagreement to a shouting match. If appropriate, call for a short break before starting the discussion again. Remind everyone that resolving conflicts is about making the best business decision for the project, not personal feelings. Both parties should sit down and stay focused on developing a solution that is in the best interest of the project.

Compromise and work with the team. Conflicts rarely have a clear-cut right and wrong. As you work to resolve the conflict, you will need to encourage others to be willing to compromise and work together to solve the problem. The effort here must stay focused again on the desired business result: "What is the best decision for this situation?" Construction is all about cooperation and collaboration. However, if the conflict involves safety rules and regulations or adhering to building codes, any solution must follow the proper policy, procedure, or regulation.

Avoid conflict through clear communications. Poor communication on the jobsite often leads to conflicts. Make sure that you clearly communicate expectations with everyone on the project. Keeping everyone up to date with any changes to the project or schedule can help avoid conflict.

Communicate the resolution and follow up. Resolving conflicts may not result in a win for everyone. It is important that once resolution is reached, everyone gets an explanation and is clear on what is expected moving forward. They might not always agree with the decision. Having a clear expectation will help others accept the decision. Whenever an agreement is reached, follow up to determine if both sides are working as discussed or if issues still exist. You also will remind everyone that, although the conflict might have been difficult, the resolution produced a good outcome for the project.

If you cannot resolve a conflict on your own, always reach out to your supervisor for guidance. Your supervisor or another senior leader can make a decision and communicate it to the appropriate individuals.

Remember, the goal of any construction project is to see the successful completion of the job on time and within budget. When conflicts arise, it can threaten the goal of on-time and on-budget job completion. As the leader of your crew, you must learn to address and resolve conflicts, so that the project work can proceed. Resolving conflicts can be challenging. It all comes down to making the best business decision to keep the project moving forward.

3.0.0 Section Review

1. Which of the following is *not* a way to motivate a team?
   a. Providing training and helping individuals feel a sense of accomplishment and growth
   b. Setting expectations without clear short-term and long-term goals
   c. Offering recognition and rewards when appropriate
   d. Having a good work environment and culture

2. One sign of a high-performing team is _____.
   a. the leader cannot delegate work to the team
   b. harassment is common on the worksite
   c. everyone on the team follows policies and procedures
   d. language barriers prevent effective communication

3. As a frontline leader, steps to successfully resolving conflict may involve any of the following *except* _____.
   a. remaining neutral and focusing on the business, not on people
   b. compromising and working with the team
   c. ignoring the conflict
   d. understanding the real issues causing the conflict

4.0.0 Safety and Safety Leadership

Performance Tasks

1. Organize and deliver a safety-related toolbox talk, applying communication skills.
2. Complete a job safety analysis.

Objective

Identify a frontline leader's safety leadership responsibilities.

a. Assess the impact of incidents.
b. Explain the purpose of OSHA and describe the role of OSHA in administering worker safety.
c. Describe the importance of safety and health programs.
d. Explain the frontline leader's role in incident preparedness and response.
e. Recognize the importance of supporting your crew's total health.

Hazards: Conditions or activities that, if left uncontrolled, have the potential for harm.

Incidents: Unplanned events that result in property damage, fatalities, injuries, illnesses, or near misses.

Craft professionals routinely face **hazards**. Examples of these hazards include falling from heights, working on scaffolds, using cranes in the presence of power lines, operating heavy machinery, and working on electrically powered or pressurized equipment. Despite these hazards, applying preventive safety measures can significantly reduce the number of **incidents**.

Safety matters because people matter. As a crew leader, some of your most important responsibilities are to enforce the company's safety program, make sure that all workers are performing their tasks safely, and ensure that everyone on the jobsite remains safe and goes home healthy (*Figure 16*). You must recognize safety risks and work to eliminate or reduce them.

To be successful, the crew leader should do the following:

- Be aware of the human and business impacts of incidents.
- Understand all federal, state, and local governmental safety regulations applicable to your work.
- Be the most visible example of the best safe work practices.
- Be involved in training workers in safe work methods.
- Conduct training sessions.
- Get involved in safety inspections and incident investigations.

Crew leaders are in an important position to ensure that their crew members perform all jobs safely. Providing employees with a safe working environment by preventing incidents and enforcing safety standards will go a long way toward maintaining the job schedule, ensuring a timely job completion within budget, and keeping your team healthy.

Figure 16 Committing to safety is a key part of a crew leader's role.
Source: RainStar/E+/Getty Images

4.1.0 The Impact of Incidents

An incident is an unplanned event that results in property damage, a fatality, injury, illness, or a near miss. A near miss is an unplanned event that does not result in personal injury but may result in property damage or is worthy of recording.

Incidents including work-related injuries, sickness, and fatalities cause immense suffering for workers and their families. Resulting project delays and budget overruns can cause huge losses for employers, and worksite incidents damage the overall morale of everyone on a project or jobsite.

Businesses lose billions of dollars every year because of on-the-job incidents. As just one measure of the business impact of incidents, construction companies in the United States pay over $1 billion per week in **workers' compensation** for nonfatal injuries. Workers' compensation, often called *workers' comp*, is insurance that companies have to cover medical expenses, lost wages, and other costs for employees who are injured or become ill while completing work for their job. It also pays death benefits to families of employees who are killed on the job.

Workers' compensation: Insurance that companies have to cover medical expenses, lost wages, and other costs for employees who are injured or become ill while completing work for their job.

Preventing injuries means protecting the lives of workers, while reducing significant costs for their employers. It is estimated that the full impact of workplace injuries and fatalities cost companies approximately $151 billion per year. These costs affect the individual employee, the company, and the construction industry as a whole.

Organizations encounter both direct and indirect costs associated with workplace incidents.

- Direct costs are the money companies must pay out to workers' compensation claims and sick pay.
- Indirect costs are all of the costs a company must account for as the result of a worker's injury or death, such as time off from work.

The direct and indirect costs of incidents are comparable to the visible and hidden portions of an iceberg, as shown in *Figure 17*. The visible part of the iceberg represents direct costs, some of which are covered by insurance. The hidden part of the iceberg represents the indirect costs, which are a greater financial burden than the direct costs.

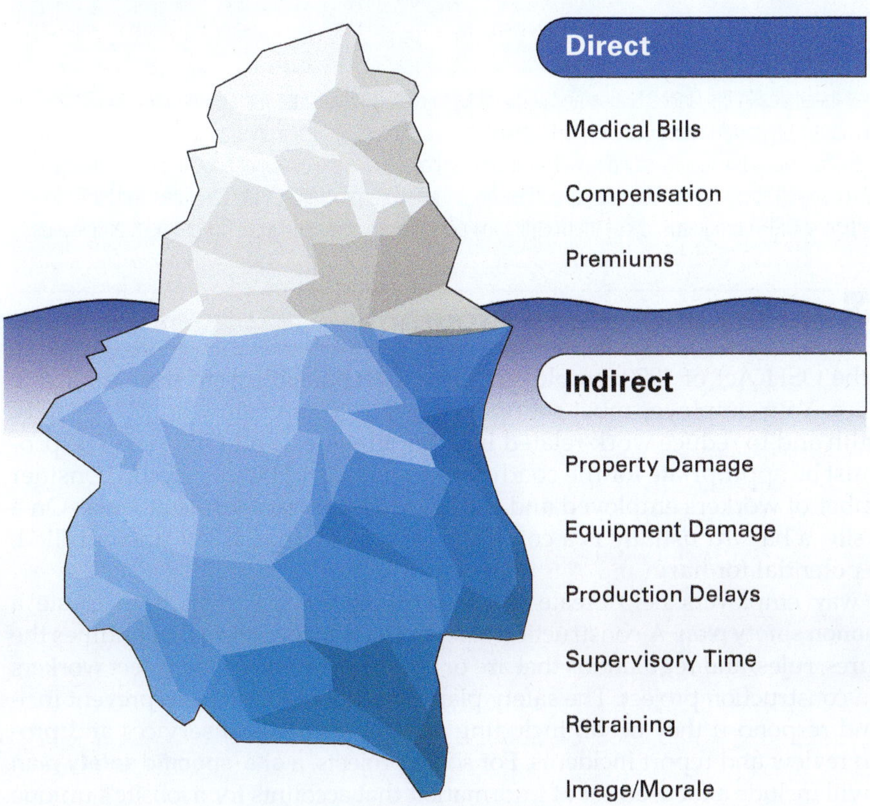

Figure 17 Incidents create both direct and indirect costs.

4.2.0 OSHA

Did You Know?

OSHA 10 and OSHA 30 Authorized Certification

OSHA helps to make safety training available through their Outreach Training Program's authorized 10-hour and 30-hour safety courses. The goal of these courses is to promote workplace safety and health and to make workers more knowledgeable about workplace hazards and their rights. As a crew leader, you may already have completed the OSHA-authorized 10-hour training program, which is designed for entry-level workers. The OSHA-authorized 30-hour training program provides more in-depth training and is intended for supervisors or others with some safety responsibility. If you want to take an OSHA-authorized 10-hour or 30-hour training program, work with your supervisor to obtain this training.

To reduce safety and health risks and the number of injuries and fatalities on the job, the federal government has enacted laws and regulations, including the Occupational Safety and Health Act of 1970 (OSH Act of 1970). This law created the Occupational Safety and Health Administration (OSHA), which is part of the US Department of Labor. OSHA also provides education and training for employers and workers. Through the administration of OSHA, the US Congress seeks to ensure safe and healthful working conditions for workers by setting and enforcing standards and by providing training, outreach, education, and assistance.

OSHA has many functions. OSHA issues standards and rules for working conditions, facilities, equipment, tools, and work processes. It does extensive research into occupational incidents, illnesses, injuries, and deaths to reduce the number of occurrences and adverse effects. In addition, OSHA regulatory agencies conduct workplace inspections to ensure that companies follow the standards and rules. The agencies also have the right to take legal action if companies are not in compliance with OSHA standards and rules.

OSHA has established significant monetary fines for the violation of its regulations. *Table 5* lists the maximum OSHA penalty amounts as of 2022. In addition to the fines, there are possible criminal charges for willful violations resulting in death or serious injury. The attitude of the employer and their safety history can have a significant effect on the outcome of a case.

TABLE 5 OSHA Maximum Penalty Amounts

Type of Violation	Maximum Penalty
Serious	$14,502 per violation
Other-Than-Serious	
Posting Requirements	
Failure to Abate	$14,502 per day
Willful or Repeated Violation	$145,027 per violation

Did You Know?

State OSHA Programs

Nearly half the states in the United States and some of its territories have state-run OSHA programs. These programs are set up under federal OSHA guidelines and must establish job health and safety standards that are at least as effective as the federal standards. The states/territories have the option of adopting more stringent standards or setting standards for hazards not addressed in the federal program. Of the 22 states and territories with nonfederal OSHA programs, six of them limit their coverage to state and local government employees.

4.3.0 Safety and Health Programs

Under the OSH Act of 1970, employers have a responsibility to provide a safe workplace. Your employer must set up a program to manage workplace safety and health and to reduce work-related injuries, illnesses, and fatalities. The program must be appropriate for the conditions of the workplace. It should consider the number of workers employed and the hazards they face while at work. On a project site, a hazard usually is a condition or activity that, if left uncontrolled, has the potential for harm.

Construction safety plan: A document that outlines the procedures, rules, and regulations that are or will be put in place to protect workers during a construction project.

Site-specific safety plan (SSSP): A safety plan that includes a focused set of information that accounts for a jobsite's unique hazards.

One way employers help create a safe working environment is to create a **construction safety plan**. A construction safety plan is a document that outlines the procedures, rules, and regulations that are or will be put in place to protect workers during a construction project. The safety plan should include steps to prevent incidents and respond if they occur, including emergency medical services and processes to review and report incidents. For some projects, a **site-specific safety plan (SSSP)** will include a focused set of information that accounts for a jobsite's unique hazards. Safety plans should be created before a construction project is underway, but also should be updated on a regular basis as conditions on the site change.

4.3.1 Education and Training

All employees should receive a safety policies and procedures manual and initial training when they are hired. A company's safety manual should cover the company's safety expectations and the procedures that must be followed to create a safe work environment and meet OSHA requirements. During orientation, appropriate company staff should guide new employees through the safety policies and procedures manual, focusing on the sections most important for their job. An employer must ensure that any barriers to worker understanding, such as language barriers, are removed or addressed.

In addition to formal training, safety is a topic you need to discuss with your crew every day on the jobsite. You need to help employees be aware of and avoid the hazards to which craft professionals are exposed. As a leader, you should plan to conduct frequent safety trainings and meetings with your crew, including holding toolbox talks (*Figure 18*). These meetings are important to building a strong safety culture and reinforcing your company's commitment to protecting your workers.

Toolbox talks are one way to effectively keep all workers aware and informed of safety issues and guidelines. Toolbox talks, sometimes referred to as *tailgate meetings* or *safety briefings*, are short, informal safety meetings held at the start of each day or shift. Toolbox talks are 5- to 10-minute meetings that review specific health and safety topics. *Table 6* provides a checklist for delivering effective safety briefings or toolbox talks.

Toolbox talks: Short, informal safety meetings held at the start of each day or shift. Sometimes referred to as *tailgate meetings* or *safety briefings*.

Always document your toolbox talks or other safety training. Even if certain OSHA standards do not require documentation of safety training, keep a record of each safety session signed by all attendees. In the record, include the date, a summary of the training session, the trainer, and attendees.

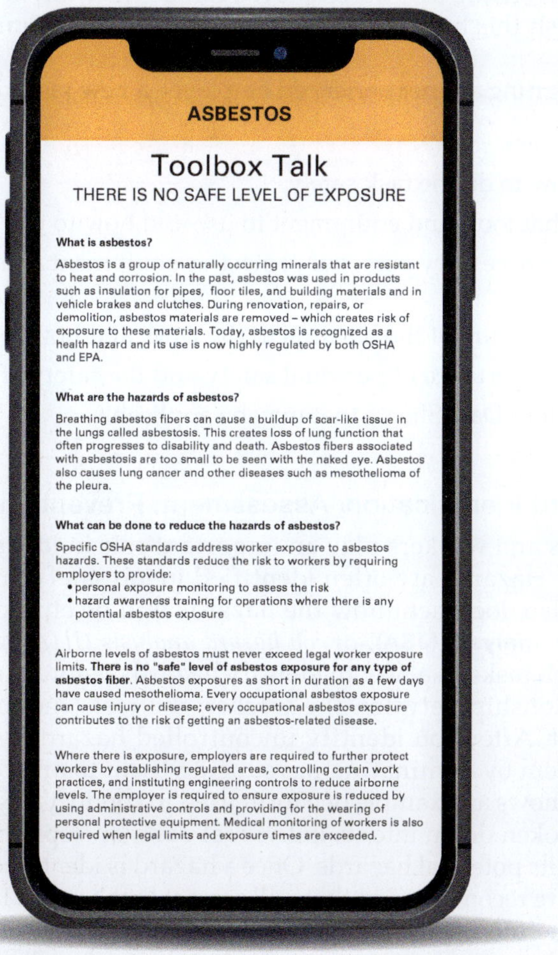

Figure 18 Safety management software can help companies deliver and track worker training. Some of the solutions help you schedule, track, and deliver toolbox talks.

TABLE 6 Delivering Effective Safety Briefings or Toolbox Talks

Did You . . .?	Yes/No
Schedule time in advance and create time to allow workers to participate.	
Speak clearly and directly at a natural pace.	
Try to find a spot free of noise and other distractions. If the workers cannot hear you talking, or are distracted by other activities in the area, they won't be focusing on your talk.	
Use props to help make your point and provide visual communication cues. For example, have a safety harness with you for your talk on fall protection.	
Reinforce safety basics, focus on high-risk scenarios, and inform workers about changes to the jobsite and working conditions that may have occurred since their last shift.	
Discuss any incidents or injuries that have occurred and how they could have been prevented.	
Focus the discussion on active work. For example, do not hold a toolbox talk on sitework when electrical work is the only thing being done that day.	
If appropriate, break out a toolbox talk into smaller groups to focus on individual crafts or work activities for certain individuals.	
Focus on the positive, too. Highlight safe working practices. Provide positive reinforcement and applaud workers following safety rules and regulations.	
Always give workers an opportunity to ask questions at the end of the toolbox talk.	

One-on-one or small group training is also a critical part of keeping everyone on the project safe. When you assign an inexperienced employee a new task, you must ensure that the employee can do the work in a safe manner. You can accomplish this by providing safety information or training for groups or individuals.

When assigning an inexperienced employee a new task, do the following:

- Define the task.
- Explain how to do the task safely.
- Explain what tools and equipment to use and how to use them safely.
- Identify the necessary personal protective equipment and train the employee in its use.
- Explain the nature of the hazards in the work and how to recognize them.
- Stress the importance of personal safety and the safety of others.
- Review Safety Data Sheets that may be applicable.

4.3.2 Hazard Identification, Assessment, Prevention, and Control

Crew leaders and workers play an important role in identifying and reporting hazards. Hazards are often identified in job safety or job hazard analyses, which also document how the hazards will be eliminated or mitigated. A **job safety analysis (JSA)**, or *job hazard analysis (JHA)*, is a technique that focuses on job tasks as a way to identify hazards before they occur. It focuses on the relationship between the worker, the task, the tools, and the work environment. After you identify uncontrolled hazards, you will take steps to control them by eliminating or reducing them to an acceptable risk level.

Job safety analysis (JSA): A technique that focuses on job tasks as a way to identify hazards before they occur. It focuses on the relationship between the worker, the task, the tools, and the work environment. Also known as *job hazard analysis (JHA)* or *activity hazard analysis (AHA)*.

Figure 19 shows an example of a form used to conduct a JSA. In a JSA, the task at hand is broken down into its individual parts or steps and then each step is analyzed for its potential hazards. Once a hazard is identified, certain actions or procedures are recommended that will correct that hazard. For example, during a JSA, it is determined that using a chain hoist to install a pump motor in a tight space would be safer than having a worker do it manually. By using the chain hoist, the chance that the worker's hand would get crushed during installation is reduced. Using the JSA process would help protect the worker from injury.

JOB SAFETY ANALYSIS

TITLE OF JOB OR TASK

TASK	START	END	HAZARDS	CONTROLS
1.				
2.				
3.				
4.				
5.				
6.				
7.				
8.				

Required Training: Required Personal Protective Equipment (PPE):

Job Name: ____________

Job Number: ____________

Supervisor: ____________

Date: ____________

Weekly Vehicle Checklist: ___ Tire Pressure ___ Transmission Fluid ___ Oil ___ Lights ___ Air Filter ___ Wkly Mileage

Name of Employees:

PRINT NAME	SIGN NAME	TOTAL HOURS

Figure 19 Job safety analysis form.

A job safety analysis can also be used during planning to ensure that safety is planned into the job. Taking this approach helps ensure that safety does not get compromised for the sake of schedule and productivity.

Crew leaders must protect workers from existing or potential hazards in their work areas. It is the crew leader's responsibility to determine what working conditions are unsafe and to inform employees of hazards and their locations. In addition, they should encourage their crew members to tell them about hazardous conditions. To accomplish this, crew leaders must be present and available on the jobsite.

4.3.3 Management Leadership and Employee Participation

A culture of commitment to safety and a strong safety and health program require management leadership and employee participation. As a frontline leader, the best way for you to encourage employee participation in safety and health is to lead by example. Your behavior sets standards for your crew members. If you cut corners on safety, you are telling your crew they may do so as well.

You should always be sure to do the following:

- Stay actively involved in your company's safety and health program.
- Exceed expectations when it comes to safety. Help others follow the rules and ask your crew if there are safer ways to do a task.
- Host safety training and meetings. Discuss safety at the start of each shift. Have regular toolbox talks. Include safety in all meetings, even if the primary purpose is schedule, cost, or production related.
- Be a visible participant on safety walks.
- Review and provide feedback on hazard analyses.
- Always follow up on safety concerns raised by the workers.

CAUTION

Team members often "follow the leader" when it comes to unsafe work practices. Sometimes, supervisory personnel engage in unsafe practices and take more risks because they are more experienced. However, inexperienced or careless workers who take the same risks won't be as successful in avoiding injury. As a leader, you must follow all safety practices to encourage your crew to do the same.

Your employer will use their policies and procedures for incident investigation and documentation. Be sure to follow any of these guidelines, when the situation arises.

4.4.0 Incident Preparedness and Response

While the goal is to have zero incidents on a worksite, you and your team must understand how to respond quickly to a safety incident. If an incident occurs, you should follow the safety policies and procedures in the construction safety plan. As a frontline leader, you are responsible for making sure everyone on your team understands the safety plan. You also may have reporting and investigation responsibilities after an incident.

4.4.1 Administering First Aid

First aid: Help given to a sick or injured worker on the worksite immediately after an injury occurs.

When someone is hurt or has a medical event, your first goal is ensuring the best outcome: minimizing the health impact to the person. **First aid** is help given to a sick or injured worker on the worksite immediately after an injury occurs. The primary purpose of first aid is to provide immediate and temporary medical care to employees involved in incidents, as well as employees experiencing non-work-related health emergencies, such as chest pain or breathing difficulty.

Many companies have very specific policies on administering first aid. If you are unsure of these policies, ask your supervisor and your safety team for guidance. You need to know how to handle these situations before an incident occurs. A quick response can be the difference between life and death. The victim of an injury or sudden illness at a jobsite may be at a remote location. The site may be far from a rescue squad, fire department, or hospital, presenting a problem in the rescue and transportation of the victim to a hospital.

AED (automated external defibrillator): A device used to help those experiencing sudden cardiac arrest.

Your employer must ensure prompt first aid treatment for any injured employees. To do this, they will provide access to readily available first aid kits. You and everyone on your team should be aware of the location and contents of first aid kits available on the jobsite. Many jobsites also have an **AED (automated external defibrillator)** that is used to help those experiencing sudden cardiac arrest (*Figure 20*). Many lives have been saved due to a prepared team who knew where the AED device was stored and how to use it.

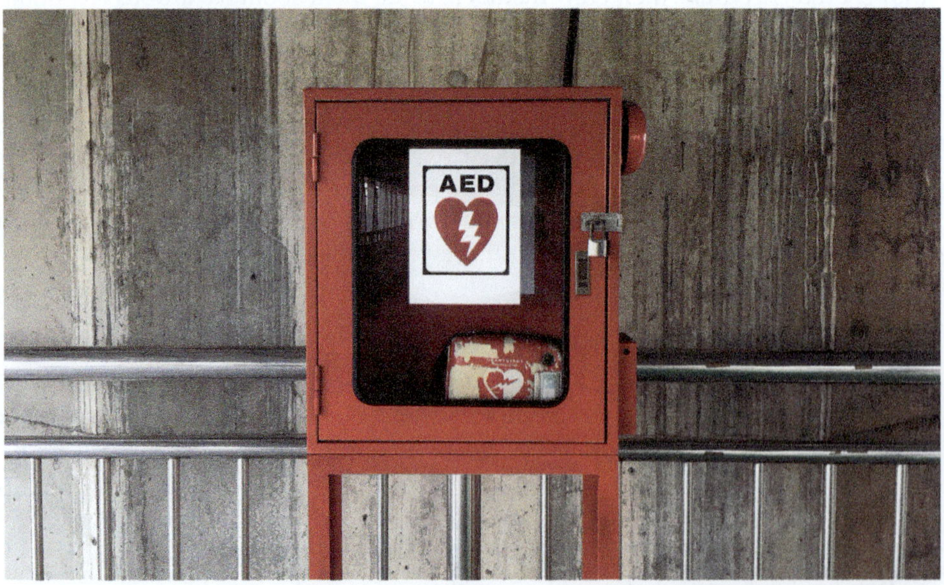

Figure 20 Many jobsites have an AED to help those experiencing sudden cardiac arrest. You should always know where AED devices are stored.
Source: PardiMer/Shutterstock

Your company will have an on-site medical professional or at least one member of the workforce who is trained and certified in first aid, including the use of an AED and CPR. **CPR (cardiopulmonary resuscitation)** is a lifesaving technique that is used when someone's heartbeat or breathing has stopped. The benefits of having personnel trained in first aid at jobsites include the following:

- Immediate and proper treatment of minor injuries. First aid may prevent minor injuries from developing into more serious conditions. OSHA does not require companies to record when simple first aid is used to address the injury or illness.
- Ability to determine if the injured person requires professional medical attention.
- Ability to help stabilize an injured person before professional medical care arrives.
- Help to eliminate or reduce medical expenses, lost work time, and sick pay.

If further medical treatment is needed, preparedness and a rapid communication process will mobilize the right people as soon as possible. Emergency numbers should be posted on the jobsite. As a leader, you should be very familiar with the construction safety plan and work with your Human Resources and safety leaders to ensure it stays updated. You need to locate your local off-site medical provider for emergency or nonemergency care prior to an incident occurring.

CPR (cardiopulmonary resuscitation): A lifesaving technique that is used when someone's heartbeat or breathing has stopped.

NOTE

The American Red Cross, Medic First Aid, and many other local organizations provide basic and advanced first aid courses at nominal costs. These courses include both first aid and CPR. The local area offices of these organizations can provide further details regarding the training available. CPR certifications must be renewed every two years.

4.4.2 Reporting and Investigating Incidents

As a frontline leader, you must know and follow your company's incident reporting processes. You also must ensure your team is following those policies. Your team should know from their training to immediately report any unusual event, incident, or injury. The policies and definitions around reporting should be included in safety manuals and construction safety plans.

If your employer is subject to the recordkeeping requirements of the OSH Act of 1970, it must maintain records of all recordable occupational injuries and illnesses using standard OSHA forms. In the event of an incident, the employer also is required to investigate the cause of the incident and determine how to avoid it in the future. According to OSHA regulations, the employer must investigate each work-related death, serious injury or illness, or incident having the potential to cause death or serious physical harm. The employer should document any findings from the investigation, as well as the action plan to prevent future occurrences. The company should complete these actions immediately, with photos or video if possible.

As a crew leader, you may be involved with an incident investigation. If required by company policy, you also will need to make a formal investigation and submit a report after an incident. An investigation looks for the causes of the incident by examining the circumstances under which it occurred and talking to the people involved. Investigations are perhaps the most useful tool in the prevention of future incidents. As shown in *Figure 21*, an incident investigation has four major parts.

NOTE

Three OSHA forms are used for injury and illness recordkeeping:

- OSHA Form 300, Log of Work-Related Injuries and Illnesses
- OSHA Form 300A, Summary of Work-Related Injuries and Illnesses
- OSHA Form 301, Injury and Illness Incident Report

All of these forms are available online at the OSHA website. OSHA has also launched their Injury Tracking Application (ITA) to allow for direct entry of OSHA Form 300A information. For more information, scan the QR code at the front of the module.

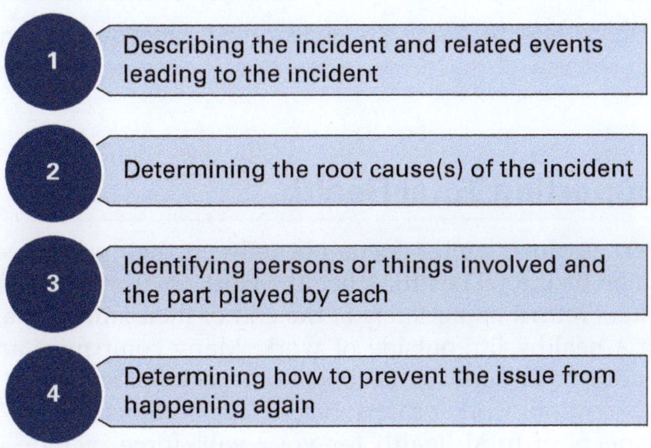

Figure 21 An incident investigation has four major parts.

Root cause: A core issue or fundamental reason for the occurrence of a problem.

A key step is determining the **root cause** of the incident. A root cause is a core issue or fundamental reason for the occurrence of a problem. Understanding the root cause is critical to avoiding similar incidents in the future. In many cases, the root cause was a flaw in the system that failed to recognize the unsafe condition or the potential for an unsafe act.

Root cause analysis (RCA): A method of problem-solving used for identifying the root causes of faults or problem.

Finding a root cause to a problem can be difficult. A **root cause analysis (RCA)** is a method of problem-solving used for identifying the root causes of faults or problems. A *Five Whys Analysis* is one type of root cause analysis that is helpful with incident investigations. The method is simple: When a problem occurs, you start with the broadest explanation for an incident and narrow it down to a more specific problem by asking "why" five times. The technique is even more effective when you include experienced team members who were actively part of the situation or have had similar experiences. It is especially useful when the root cause is not that obvious. *Figure 22* shows an example of how a Five Whys Analysis can apply to an incident investigation.

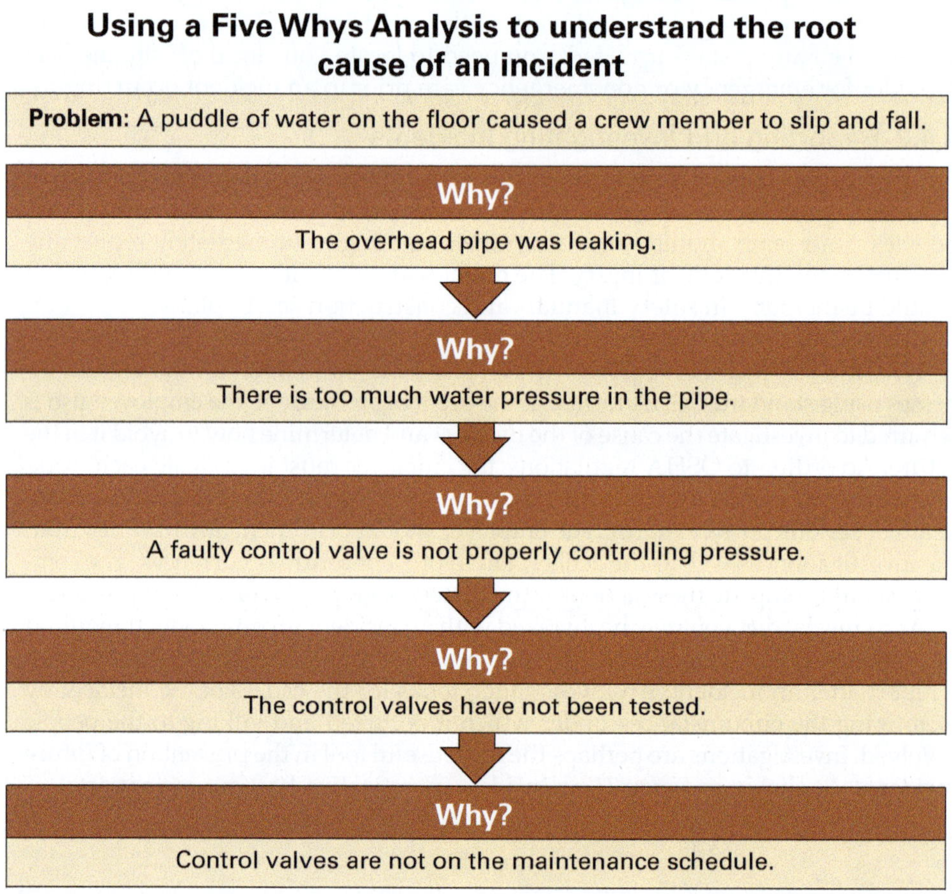

Figure 22 Using a Five Whys Analysis to understand the root cause of an incident.

4.5.0 Supporting Total Health

More than ever, employers must focus not only on their employees' physical well-being, but on their total health. A healthy work environment not only helps to ensure workers return home safely at the end of their shift, but also ensures they can have a healthy life outside of work. Many companies are adopting practices that address not only protection from work-related safety and health hazards, but also overall worker safety, health, and well-being.

Substance abuse: The inappropriate overuse of drugs and chemicals, whether they are legal or illegal.

Two key aspects of total health for your workforce include addressing **substance abuse** and mental health. As a frontline leader, you will not be able to address the root causes of these issues for your workers. You do, however, have the important responsibility to enforce company policies and procedures in these areas and work with your management or Human Resources Department to help employees in these situations.

4.5.1 Substance Abuse

Substance abuse is a serious safety issue in the workplace. Substance abuse is the inappropriate overuse of drugs and chemicals, whether they are legal or illegal. All substance abuse results in some form of mental, sensory, or physical **impairment**. Impairment is the point where someone's intake of substances affects their ability to perform appropriately. Alcohol can be abused by consuming to the point of intoxication. Legal prescription drugs can be abused by using the wrong dosage, using other people's medications, or self-medicating. Illegal drugs, such as cocaine, crystal meth, or marijuana in some states, can cause impairment.

Impairment: The point where someone's intake of substances affects their ability to perform appropriately.

Crew leaders must enforce company policies and procedures regarding substance abuse. There are legal consequences and safety implications associated with substance abuse. Suppose several crew members go out and smoke marijuana or drink during lunch. Then, they return to work to erect scaffolding for a concrete pour in the afternoon. If you can smell marijuana on the worker's clothing or alcohol on their breath, you must step in and take action. An impaired worker is more likely to cause an incident that could cause serious injury or death to themselves or others (*Figure 23*).

If you observe an employee showing impaired behavior for any reason, immediately contact your supervisor and/or Human Resources Department for assistance. You should work with management to deal with suspected drug and alcohol abuse and should not deal with these situations yourself. Reporting the concern helps you protect the business, and the employee's and other workers' safety.

It is often difficult to detect substance abuse because the effects can be subtle. The best way is to look for physical signs or sudden changes in behavior that are not typical of the employee. *Table 7* lists some examples of potential signs and behavioral changes that might indicate substance abuse.

Figure 23 An impaired worker is a dangerous one.
Source: vejaa/iStock/Getty Images

Did You Know?

Substance Abuse Does Not Pay

An employee who is involved in an incident while under the influence of drugs or alcohol may be denied workers' compensation insurance benefits.

TABLE 7 Potential Signs of Substance Abuse in a Worker

Physical Signs	Behavioral Changes
• Smelling of alcohol or drugs • Slurring of speech or an inability to communicate effectively • Wearing sunglasses indoors or on overcast days to hide dilated or constricted pupils, conditions which impair vision • Significant changes in personal appearance, cleanliness, or health	• Unscheduled absences • Failure to report to work on time • Significant changes in the quality of work • Unusual energy or tiredness • Sudden and irrational temper flare-ups • Sneaky behaviors, such as disappearing to wooded areas, storage areas, or other private locations • Attempting to borrow money from coworkers • Losing tools or company equipment

4.5.2 Mental Health

Mental health is a critically important, but often overlooked, part of overall construction safety and health. Construction workers have more risk factors than employees in other industries. These risk factors make workers more likely to experience mental health challenges, with a specific set of risks around suicide. Construction workers are six times more likely to die by suicide than they are to die from a fall.

With proper treatment, mental health conditions can be managed and overcome so those experiencing them can have full and productive lives. As with substance abuse, however, these conditions can lead to people being distracted, less productive than normal, and possibly unsafe and unable to perform their normal jobs. Because of this, mental health and suicide prevention need to be safety considerations.

As a leader, you cannot solve these challenges for your workers, but you can create a better environment. Use your toolbox talks to discuss mental health as part of overall total worker safety. In your safety huddles, check in on how your teammates are doing. Encourage your crew to talk to supervisors and colleagues when they feel stressed, anxious, or on the brink of burnout. Help ensure they are aware of any available counseling resources and that they are taking time off when they need it.

Even more so than substance abuse, it can be very difficult to identify a worker struggling with a mental health issue. *Table 8* lists some examples of potential signs and behavioral changes that might indicate someone is struggling with mental health issues.

If you observe an employee showing signs of mental health issues or have concerns for any reason, immediately contact your supervisor and/or Human Resources Department for assistance. You should work with management to deal with suspected mental health issues and should not deal with these situations yourself. Reporting the concern helps you protect the business, and the employee's and other workers' safety.

Did You Know?

Mental Health Is a Critical Part of Safety

Many organizations in the construction industry are focused on helping normalize conversations around mental health and make it less challenging to have these conversations. The National Institute for Occupational Safety and Health (NIOSH), for example, has a Total Worker Health® program focused on improving workforce well-being by protecting their safety and enhancing their health and productivity. Other focused nonprofits, like the Construction Industry Alliance for Suicide Prevention (CIASP), provide specific resources, like toolbox talks, that you can use to help with these conversations with your crew.

TABLE 8 Signs That May Indicate Someone Is Struggling with Mental Health Issues

Behavioral Changes	Performance Changes	Verbal Cues	Life Events
• Becoming withdrawn • Acting anxious or reckless • Misusing drugs and alcohol	• Missing work • Showing up late • Being less productive • Being unable to think clearly or solve problems • Having increased safety incidents	• Talking about feeling trapped • Talking about wanting to die • Saying they feel like a burden • Saying they feel like it would be better if they were gone	• Dealing with a divorce or breakup • Handling a death in the family • Working through a major illness or injury • Having financial issues

Source: Construction Industry Alliance for Suicide Prevention (ciasp.org)

4.0.0 Section Review

1. One of a crew leader's most important responsibilities to the employer is to ____.
   a. enforce company safety policies
   b. estimate material costs for a project
   c. make recommendations for setting up a crew
   d. provide input for fixed-price contracts
2. Which of these is *not* a function of OSHA?
   a. Issuing standards and rules for working conditions
   b. Researching occupational incidents and injuries
   c. Conducting workplace inspections
   d. Distributing company safety manuals
3. Which of the following should all employees receive upon being hired?
   a. A copy of the company's safety policies and procedures
   b. A new set of tools
   c. A leadership assessment to help map their career path
   d. First-aid and CPR training
4. A problem-solving method that is useful in investigating incidents and identifies the core issue of a problem is _____.
   a. problem-solving
   b. root cause analysis
   c. incident recordkeeping
   d. event playback
5. A focus on overall worker safety, health, and well-being, including mental health, is _____.
   a. home health
   b. complete worker safety
   c. occupational safety
   d. total health

5.0.0 Project Planning, Scheduling, and Cost and Resource Management

Objective

Demonstrate an understanding of construction project phases, project planning and scheduling, field reporting, and cost and resource management.

a. Identify construction project phases.
b. Summarize crew planning and scheduling responsibilities, including creating lookahead schedules.
c. Describe the frontline leader's role in field reporting.
d. Explain approaches to managing costs and controlling project resources, such as materials, tools, and equipment.

Performance Task

3. Develop and present a lookahead schedule.

This section describes the various phases of a construction project. It examines **estimating**, **planning**, **scheduling**, and **cost control** through managing material, people, tools, and other resources. All workers who participate in a job are responsible at some level for their assigned jobs. For example, the contractor's responsibility begins with obtaining the contract and it doesn't end until the client takes ownership of the project. The project manager is generally the person with overall responsibility for coordinating the project. Finally, the superintendent and frontline leadership, like a foreman, supervisor, or crew leader, are responsible for coordinating the work of one or more workers, one or more crews of workers within the company, or one or more crews of subcontractors.

Estimating: The process of calculating the cost of a project.

Planning: The process of laying out all details of a construction project before construction begins.

Scheduling: Involves defining which activities happen in what order and assigning dates and times to those tasks.

Cost control: The process that helps keep expenses under control by managing labor, material, and other costs to ensure that the project finishes on budget.

5.1.0 Construction Project Phases

A simple way to think about a construction project is in three phases: the *design and development phase*, the *preconstruction phase*, and the *construction phase*. *Figure 24* shows the flow of a typical project from beginning to end through these phases.

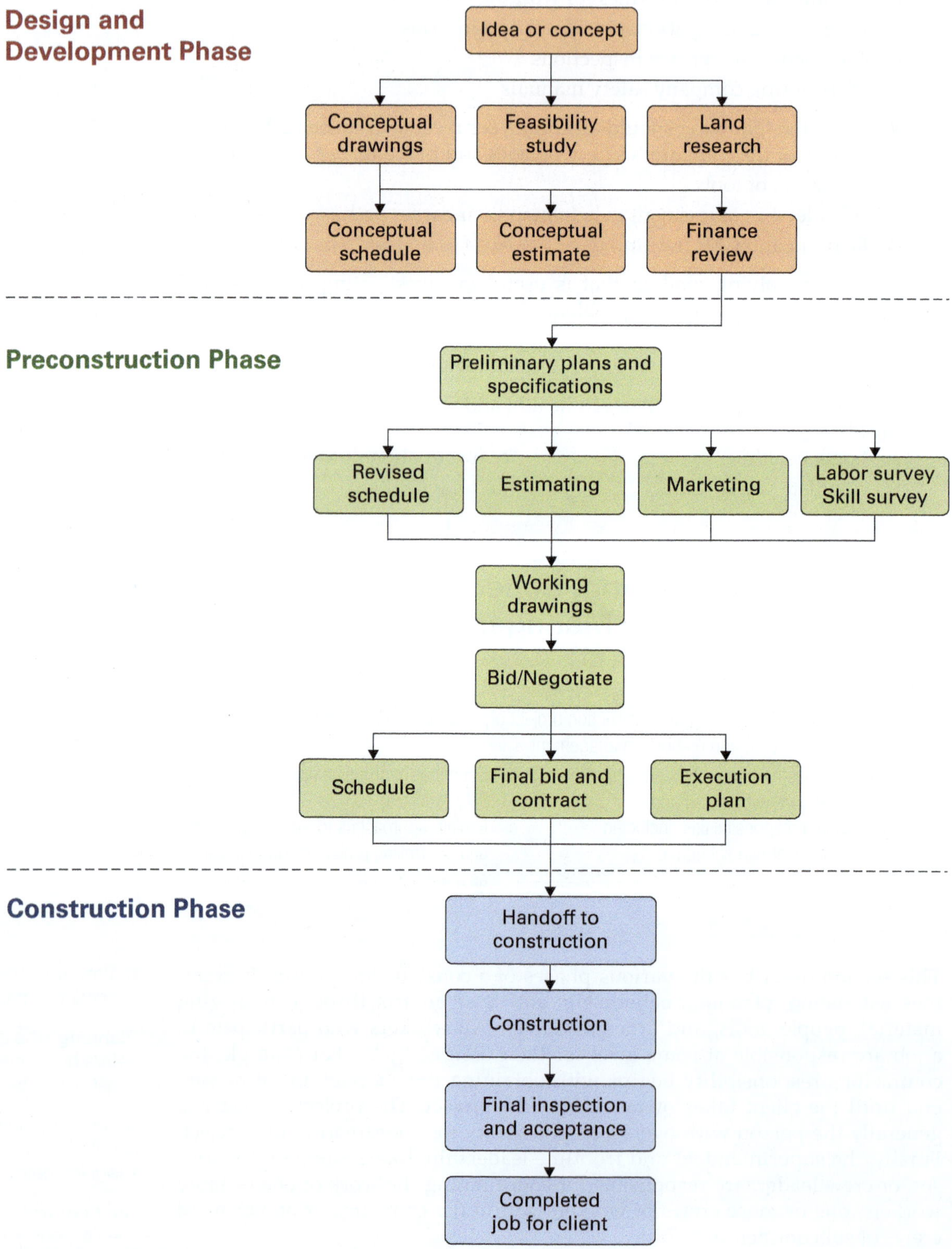

Figure 24 Construction project phases and flow.

5.1.1 Design and Development Phase

The design and development phase is the first stage of a project (*Figure 25*). It starts when an owner decides to build a new facility or add to an existing facility. This process involves land research and feasibility studies to ensure that the project has merit. A **feasibility study** is an analysis that considers all of a project's relevant factors—including budget, technical elements, legal aspects, and scheduling—to determine the likelihood that the project can be completed successfully.

Feasibility study: An analysis that considers all of a project's relevant factors—including budget, technical elements, legal aspects, and scheduling—to determine the likelihood that the project can be completed successfully.

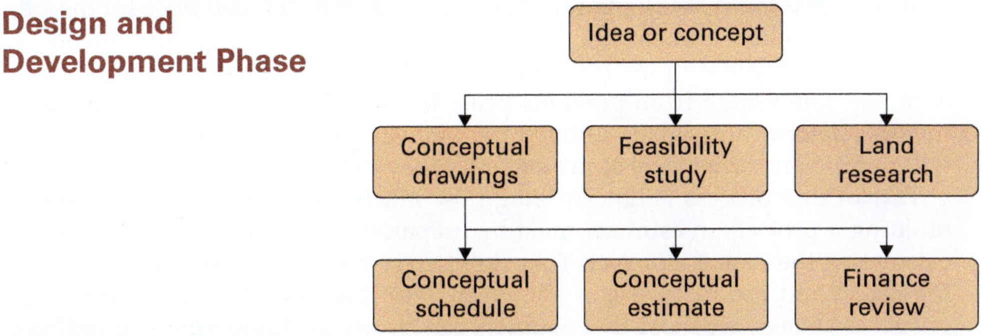

Figure 25 The design and development phase is the first stage of a project.

Architects or engineers develop the conceptual drawings that define the project graphically. They then provide the owner with sketches of room layouts and elevations and make suggestions about what construction materials to use.

During the design and development phase, architects, engineers, and/or the owner develop an estimate for the proposed project and establish a preliminary budget. Once that budget is established, the owner will determine how to pay for, or finance, the project. The owner will also complete a review to determine whether the project costs will exceed its market value and that the project provides a reasonable profit during its existence. If needed, the development team begins preliminary reviews for zoning, building restrictions, landscape requirements, and environmental impact studies.

5.1.2 Preconstruction Phase

If all steps in the design phase are completed successfully, the project moves to the preconstruction phase (*Figure 26*). The preconstruction stage includes planning that occurs before the start of construction. The preconstruction planning process does not always involve you, as a crew leader, but you should understand it to do your job well.

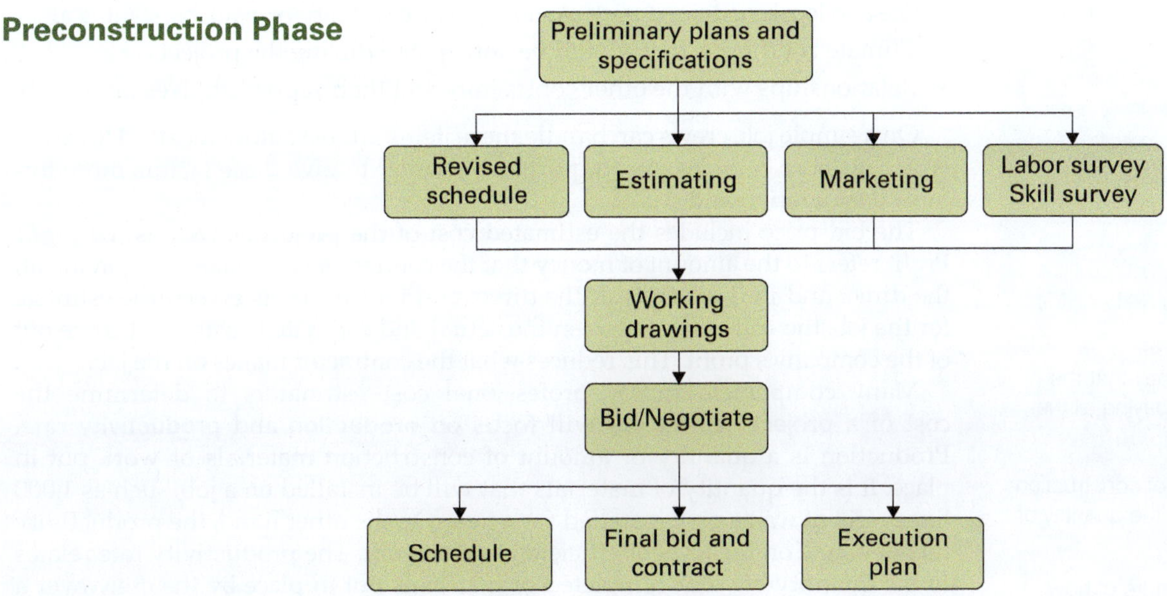

Figure 26 The preconstruction phase is the second stage of a project.

When the architects and engineers begin to develop the project drawings and specifications, they consult with other design professionals such as structural, mechanical, and electrical engineers. They perform the calculations, make a detailed technical analysis, and check details of the project for accuracy.

The design professionals create preliminary drawings and specifications. They use these drawings and specifications to communicate the necessary information to the contractors, subcontractors, suppliers, and workers who will contribute to the project.

A key next step is developing the proposal, bid, or negotiated price for the job. This is when the estimator, the project manager, and the field superintendent develop a preliminary plan for completing the work. They apply their experience and knowledge from previous projects to develop the plan. The process involves determining what methods, personnel, tools, and equipment the work will require and what level of productivity they can achieve.

Part of this process is determining how much the project will cost. Before building a project, an estimate must be prepared. Estimating is the process of calculating the cost of a project. There are two types of costs to consider, including direct and indirect costs as shown in *Table 9*. Direct costs, also known as *general conditions*, are those that planners can clearly assign to a budget. Indirect costs are overhead costs shared by all projects. Planners generally calculate these costs as an overhead percentage to labor and material costs.

TABLE 9 Examples of Direct and Indirect Costs

Direct Costs	Indirect Costs
• Materials	• Office rent
• Labor	• Utilities
• Tools	• Telecommunications
• Equipment	• Accounting
	• Office supplies and signage

Did You Know?

Profit Powers Businesses

Profit is the fuel that powers a business. It allows the business to invest in new equipment and facilities, to provide training, and to maintain a reserve fund for times when business is slow. In large companies, profitability attracts investors who provide the capital necessary for the business to grow. For these reasons, contractors can't afford to consistently lose money on projects. If they can't operate profitably, they are forced out of business. As a frontline leader, you can help your company remain profitable by managing budget, schedule, quality, and safety as outlined in the drawings, specifications, and project schedule.

Planners must also consider other factors when estimating project costs, including the following:

- Timing of all phases of work
- Types of materials to be installed and their availability
- Equipment and tools required and their availability
- Personnel requirements and availability
- Site and local conditions, such as soil types, accessibility, or available staging areas
- Climate conditions that should be anticipated during the project
- Relationships with the other contractors and their representatives on the job

On a simple job, crews can handle these items almost automatically. However, larger or more complex jobs require the planner to give these factors more formal consideration and study.

The bid price includes the estimated cost of the project as well as the **profit**. Profit refers to the amount of money that the contractor will make after paying all the direct and indirect costs. If the direct and indirect costs exceed the estimate for the job, the difference between the actual and estimated costs must come out of the company's profit. This reduces what the contractor makes on the job.

Many companies employ professional cost estimators to determine the cost of a project. Estimators will focus on **production** and **productivity rate**. Production is a quantity or amount of construction materials or work put in place. It is the quantity of materials that will be installed on a job, such as 1,000 linear feet of waste pipe installed for a task. On the other hand, the productivity rate depends on the level of efficiency of the work. The productivity rate relates to the quantity of work completed or materials put in place by the crew over a certain period of time. For example, productivity could be the amount of waste pipe installed each day by one worker or a crew.

Profit: The amount of money that the contractor will make after paying all the direct and indirect costs.

Production: The amount of construction materials put in place. It is the quantity of materials installed on a job.

Productivity rate: A measure of how much work gets done over a period of time, also called *output per labor hour*.

Production levels are set during the estimating stage. The estimator determines the total amount of materials to be placed based on the drawings, plans, and specifications. After the job is complete, supervision can assess the actual amount of materials installed and can compare the actual production to the estimated production.

During the estimating stage, the estimator uses company records of productivity to determine how much time and labor it will take to place a certain quantity of materials. From this information, the estimator calculates the productivity necessary to complete the job on time.

For example, say you need a crew of two people to paint 8,100 ft^2 in nine days. To calculate the required average productivity, divide 8,100 ft^2 by 9 days. The result is 900 ft^2/day. As the work progresses, you can compare the daily production of any crew of two painters doing similar work with this average (*Figure 27*).

NOTE

Many software programs and mobile apps are available to simplify the cost-estimating process. Estimating programs are typically set up to include a takeoff form and a form for estimating labor by category. Most of these programs include a database that contains current prices for labor and materials, so they automatically price the job and produce a bid. The programs can also generate purchase orders for materials on the job. Many of them are tailored to specific industries, such as construction or manufacturing, as well as specific trades within the industries, such as plumbing, electrical, HVACR, and insulating contractors.

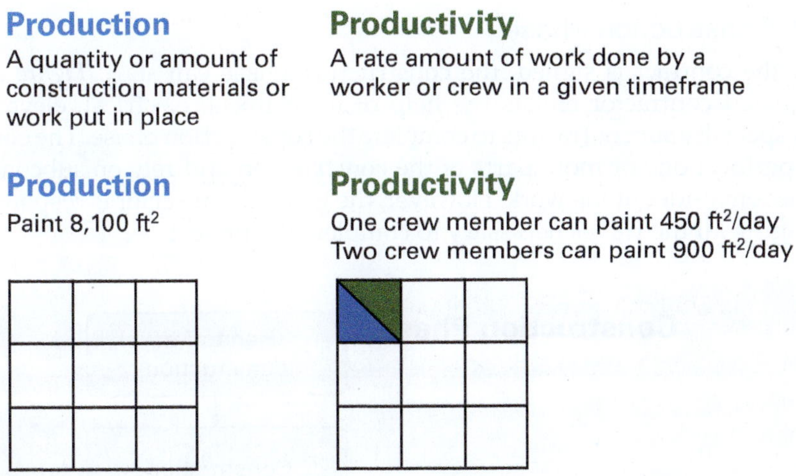

Figure 27 Production and productivity.

During the preconstruction phase, the owners hold many meetings to review and refine estimates, adjust plans to conform to regulations, and secure a construction loan. If the project is one that requires a marketing program, such as a condominium, office building, or shopping center, the marketing may start during preconstruction. In such cases, the selling of the building space often starts before actual construction begins.

Next, the design team produces a complete set of drawings, specifications, and bid documents. After this, the owner will select the method to obtain contractors. The owner may choose to negotiate with several contractors or select one through competitive bidding. Everyone concerned must also consider safety as part of the planning process. A safety crew leader may walk through the site as part of the pre-bid process.

Contracts for construction projects can take many forms. The most common types of contracts are *lump sum* and *time and materials* contracts.

In a *lump sum contract*, also called a *fixed price* or *firm-fixed price contract*, the owner generally provides detailed drawings and specifications, which the contractor uses to calculate the cost of materials and labor. The contractor also adds a percentage to these costs representing company overhead expenses, such as office rent, insurance, and accounting/payroll costs. At the end, the contractor adds a profit factor.

When submitting the bid, the contractor will state very specifically the conditions and assumptions on which the company based the bid. The **scope**, or *scope of work*, is the list of the commitments and activities that all contractors, subcontractors, and suppliers are expected to do during a project.

Scope: The list of the commitments and activities that all contractors, subcontractors, and suppliers are expected to do during a project. Also called *scope of work*.

In a *time and materials contract*, also known as a *T&M contract*, the owner pays back the contractor for labor, materials, and other costs encountered in the performance of the contract. Typically, the contractor and owner agree in advance on hourly or daily labor rates for different categories of worker. These rates include an amount representing the contractor's overhead expense. The owner also reimburses the contractor for the cost of materials and equipment used on the job. Since the buyer reimburses the contractor for the cost of materials and pays an hourly wage, any unexpected delays, roadblocks, and other changes to the scope of work are covered. T&M contracts thus work best for projects in which the scope of work is not well-defined.

NOTE

You may also hear about other types of contracts, such as cost-plus and guaranteed maximum price. Some types of contracts, such as unit-rate and hybrid, combine several aspects of these other contract types. Scan the QR code in the front of this module to access additional resources on construction contract types.

On a time and materials contract project, you play an important role by understanding the scope of the work and making sure that everything is tracked so that your company is reimbursed for the work. You must identify changes or problems that increase the scope or amount of planned labor or materials. You also must ensure you and your team are following your company policies and getting proper authorization before starting any work.

After the client awards the contract, planners create a final schedule and an execution plan that defines the actual work methods and resources needed to perform the work.

5.1.3 Construction Phase

After the contract is signed, the construction phase can start (*Figure 28*). The designated contractor enlists the help of mechanical, electrical, elevator, and other specialty subcontractors to complete the construction phase. The contractor may perform one or more parts of the construction and rely on subcontractors for the remainder of the work. However, the general contractor is responsible for managing all the trades necessary to complete the project.

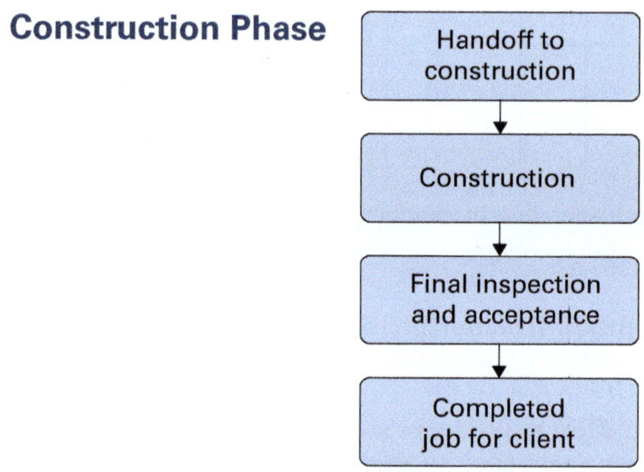

Figure 28 The construction phase is the third stage of a project.

The initial set of drawings for a construction project reflects the completed project as conceived by the architect and engineers. During construction, changes may be necessary because of factors unforeseen during the design phase. For example, when electricians must reroute cabling or conduit, or the installed equipment location is different than shown on the original drawing, such changes must be marked on the drawings. Without this record, technicians called to perform maintenance or modify the equipment later will have trouble locating all the cabling and equipment.

Project supervision must document any changes made during construction or installation on the drawings as the changes occur. Before digital drawings, architects would note changes on hard-copy drawings using a colored pen or pencil. These marked-up versions are commonly called **redline drawings**. With new digital drawings created using building information modeling tools, architects and engineers can revise and share the latest drawing versions almost instantly to mobile phones and tablets. After the drawings have been revised to reflect any redline changes, the final drawings are called **as-built drawings**. As-built drawings are a revised set of drawings submitted by a contractor upon completion of a project or job. These drawings, sometimes called *record drawings*, reflect all changes made during construction and show the exact specifications of the completed work.

Redline drawings: Marked-up versions of construction drawings documenting any changes made during construction or installation.

As-built drawings: A revised set of drawings that reflect all changes made during the construction process and show the exact specifications of the completed work.

The main part of your role as a frontline leader happens during the construction phase, including daily work planning and scheduling, cost control, and resource management, including material delivery and storage, tool control, and equipment control. All of these tasks drive production and productivity, as well as the overall profitability of the project.

As construction nears completion, the architect/engineer, owner, and government agencies start their final inspections and acceptance of the project. If the general contractor has managed the project, the subcontractors have performed their work, and the architects/engineers have regularly inspected the project to ensure it satisfies the local code, then the inspection process can finish up in a timely manner. This results in a satisfied client and a profitable project for all.

On the other hand, if the inspection reveals faulty workmanship, poor design, incorrect use of materials, or violation of codes, then the inspection and acceptance will take longer. It may result in a dissatisfied client or an unprofitable project.

Construction **project delivery methods** provide different ways to organize the services of development, planning, and construction in order to execute a project.

NOTE

Scan the QR code in the front of this module to access additional resources on construction project delivery methods.

Project delivery methods: Different ways to organize the services of development, planning, and construction in order to execute a project.

5.2.0 Planning and Scheduling

In the previous sections, you learned about the phases of a construction project, including the design and development phase, the preconstruction phase, and the construction phase. As noted, the main part of your role as a frontline leader happens during the construction phase.

During construction, you may be directly involved in daily work planning and scheduling. You also may be involved in cost control and resource management, including material delivery and storage, tool control, and equipment control.

5.2.1 Understanding Overall Schedules

Construction planning and scheduling refers to the process through which you map out what will take place during a project, what resources you need, when key activities will take place, which personnel will be involved, and anything else pertinent to the project.

Construction planning refers to the process of laying out all details of a construction project beforehand. Construction scheduling involves defining which activities happen in what order and assigning dates and times to those tasks. Construction schedules are linked to resources such as materials, labor, and equipment to help create the most efficiency in your project. Construction scheduling is deeply connected to planning.

Planning and scheduling are closely related and are both very important to a successful job. Planning identifies the required activities and how and in what order to complete them. Scheduling involves establishing start and finish times/dates for each activity.

A schedule for a project typically shows the following:

- Operations listed in sequential order
- Units of construction
- Duration, or length of time, of an activity
- Estimated dates to start and complete each activity
- Quantity of materials to be installed

Figure 29 shows an example of a high-level project schedule. As noted, the development of a schedule for a project starts as early as the design and development phase. As the project moves through each phase and the work is better defined, the schedule also becomes more clearly defined.

When a crew is close to completing specific work—say, within three weeks of working on Foundation 1—a **Level 4 schedule** is created. A Level 4 schedule offers additional detail to provide working schedules to the crews doing the

NOTE

Scan the QR code in the front of this module to access additional resources on schedule levels.

Level 4 schedule: A schedule that offers additional detail to provide working schedules to the crews doing the work, usually within three to six weeks of completing specific work.

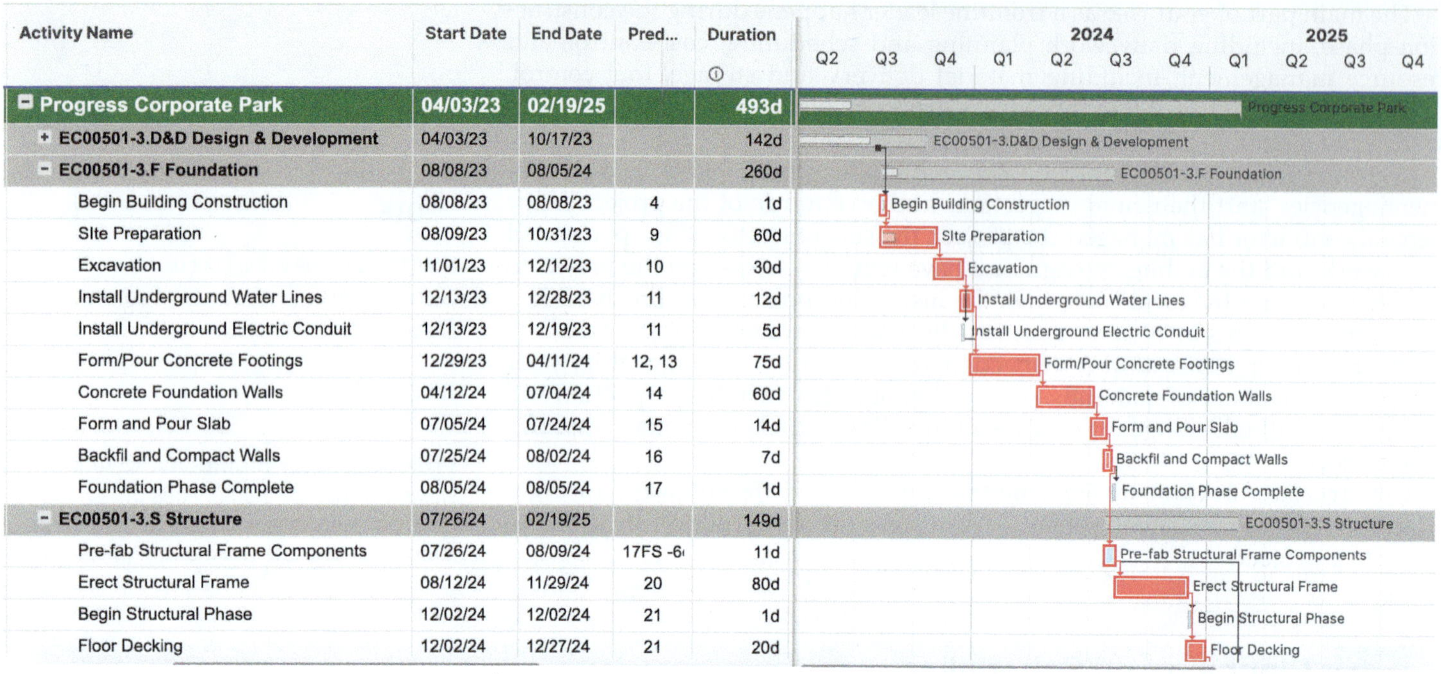

Activity Name	Start Date	End Date	Pred...	Duration
Progress Corporate Park	04/03/23	02/19/25		493d
EC00501-3.D&D Design & Development	04/03/23	10/17/23		142d
EC00501-3.F Foundation	08/08/23	08/05/24		260d
Begin Building Construction	08/08/23	08/08/23	4	1d
SIte Preparation	08/09/23	10/31/23	9	60d
Excavation	11/01/23	12/12/23	10	30d
Install Underground Water Lines	12/13/23	12/28/23	11	12d
Install Underground Electric Conduit	12/13/23	12/19/23	11	5d
Form/Pour Concrete Footings	12/29/23	04/11/24	12, 13	75d
Concrete Foundation Walls	04/12/24	07/04/24	14	60d
Form and Pour Slab	07/05/24	07/24/24	15	14d
Backfil and Compact Walls	07/25/24	08/02/24	16	7d
Foundation Phase Complete	08/05/24	08/05/24	17	1d
EC00501-3.S Structure	07/26/24	02/19/25		149d
Pre-fab Structural Frame Components	07/26/24	08/09/24	17FS -6	11d
Erect Structural Frame	08/12/24	11/29/24	20	80d
Begin Structural Phase	12/02/24	12/02/24	21	1d
Floor Decking	12/02/24	12/27/24	21	20d

Figure 29 Example of a construction project schedule.

work (*Figure 30*). A Level 4 schedule allows a supervisor to create a plan, schedule handoffs from crew to crew, and estimate all of the resources that must be in place for the work to happen.

In the schedule in *Figure 30*, ironworkers must fabricate and set the rebar and carpenters must build the forms before the handoff is made to place the concrete. Each of these activities is a critical path activity. A **critical path** is the sequence of dependent tasks that form the longest duration. In *Figure 29* and *Figure 30*, the critical path in the schedule is shown in red. In other words, until one activity is complete, workers cannot start the other tasks, and the project will likely be delayed by the amount of time the previous activity was delayed.

Critical path: In manufacturing, construction, and other types of creative processes, the required sequence of tasks that directly controls the ultimate completion date of the project.

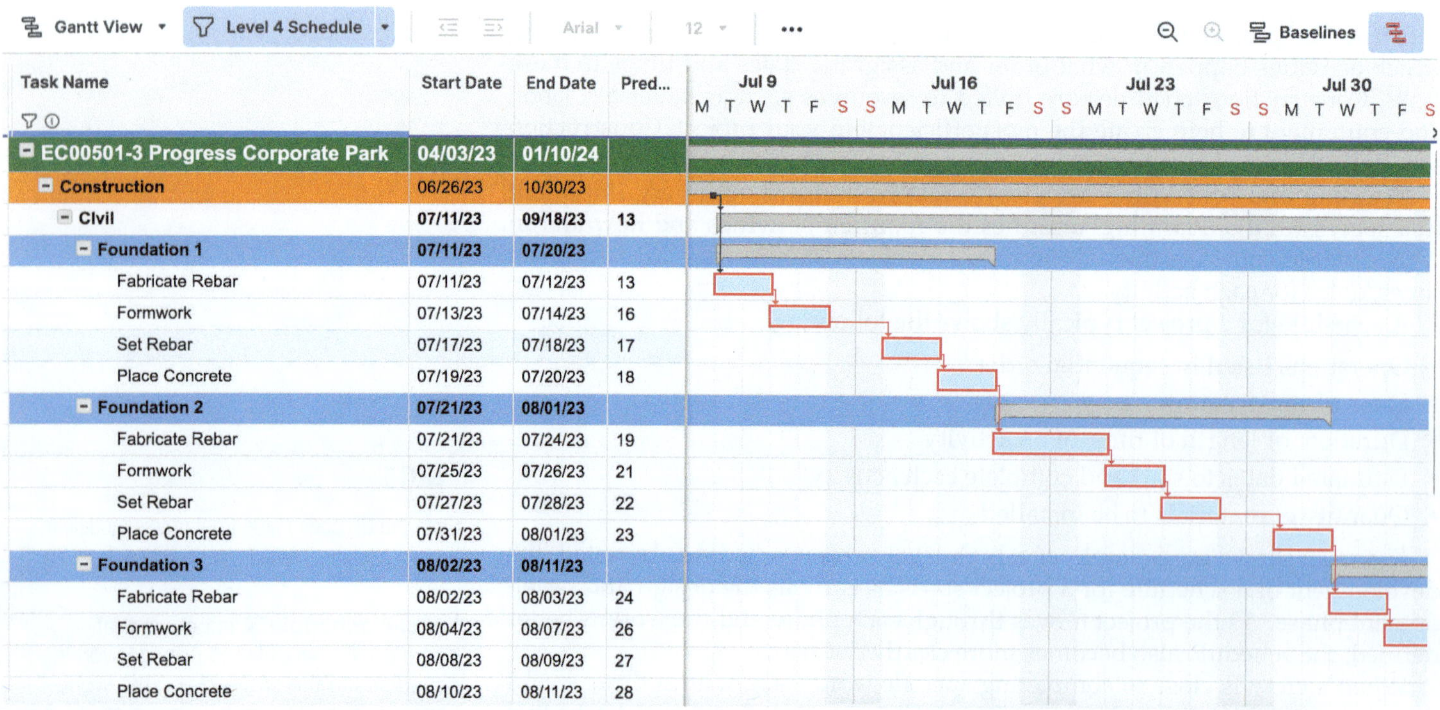

Task Name	Start Date	End Date	Pred...
EC00501-3 Progress Corporate Park	04/03/23	01/10/24	
Construction	06/26/23	10/30/23	
Civil	07/11/23	09/18/23	13
Foundation 1	07/11/23	07/20/23	
Fabricate Rebar	07/11/23	07/12/23	13
Formwork	07/13/23	07/14/23	16
Set Rebar	07/17/23	07/18/23	17
Place Concrete	07/19/23	07/20/23	18
Foundation 2	07/21/23	08/01/23	
Fabricate Rebar	07/21/23	07/24/23	19
Formwork	07/25/23	07/26/23	21
Set Rebar	07/27/23	07/28/23	22
Place Concrete	07/31/23	08/01/23	23
Foundation 3	08/02/23	08/11/23	
Fabricate Rebar	08/02/23	08/03/23	24
Formwork	08/04/23	08/07/23	26
Set Rebar	08/08/23	08/09/23	27
Place Concrete	08/10/23	08/11/23	28

Figure 30 A Level 4 schedule provides enough details for project foremen to forecast handoffs and needed resources.

Grid View | 3 Week Lookahead | Arial | 10

Task Name	Start Date	End Date	Materials Available	Assigned To	Notes	Status
Civil	**07/11/23**	**09/18/23**				
Foundation 1	**07/11/23**	**07/20/23**				
Fabricate Rebar	07/11/23	07/12/23	Yes	JKR		Not Started
Formwork	07/13/23	07/14/23	Yes	JKR		Not Started
Set Rebar	07/17/23	07/18/23	N/A	JKR	Quality signoff	Not Started
Place Concrete	07/19/23	07/20/23	Yes	JKR		Not Started
Foundation 2	**07/21/23**	**08/01/23**				
Fabricate Rebar	07/21/23	07/24/23	Yes	JKR		Not Started
Formwork	07/25/23	07/26/23	Yes	JKR		Not Started
Set Rebar	07/27/23	07/28/23	N/A	JKR	Quality signoff	Not Started
Place Concrete	07/31/23	08/01/23	Yes	JKR		Not Started
Foundation 3	**08/02/23**	**08/11/23**				
Fabricate Rebar	08/02/23	08/03/23	On Order	JKR		Not Started
Formwork	08/04/23	08/07/23	Yes	JKR		Not Started
Set Rebar	08/08/23	08/09/23	N/A	JKR	Quality signoff	Not Started
Place Concrete	08/10/23	08/11/23	Yes	JKR		Not Started

Figure 31 A step-by-step lookahead in a Level 4 schedule provide details for frontline leaders to plan handoffs.

For work to happen, you will need resources such as materials, available crew, available construction equipment, completed permits, and more. Developing a **lookahead schedule** helps ensure that all resources are available for the project when needed. This is often included in a Level 4 schedule. A lookahead schedule is useful for planning and scheduling work that is happening in the short term, usually within one to three weeks (*Figure 31*).

Lookahead schedule: For planning and scheduling work that is happening in the short term, usually within one to three weeks; used for anticipating material, labor, tool, and other resource requirements, as well as identifying potential schedule conflicts or other problems. Also called *short-term schedule*.

5.2.2 Creating a Short-Term or Lookahead Schedule

As a frontline leader of a crew, you must be able to create, read and interpret a lookahead schedule. You can use this information to plan work more effectively, set realistic goals, and compare the starts and completions of tasks to those on the schedule. As the leader of the team executing the work, you are a key expert to help produce a plan at this level of detail. *Table 10* provides a checklist for planning and scheduling.

TABLE 10 Crew Leader Planning and Scheduling Checklist

Did You…?	Yes/No
Determine the best method for performing the job.	
Identify the responsibilities of each person on the work crew.	
Determine the duration, or length in time, and sequence of each activity.	
Identify the tools and equipment needed to complete the job.	
Ensure that the required materials are at the worksite when needed.	
Make sure that heavy construction and other equipment is available when required.	
Work with other contractors in such a way as to avoid interruptions and delays.	

The following outlines the process you must complete to develop a plan and a schedule and to use the schedule to manage work.

Understand or Establish a Goal

A goal is a specific outcome that one works toward. The goals you must accomplish are often set in the higher-level schedules. For example, the project superintendent of a residential construction project could establish a goal to have a house dried-in or ready for the application of roofing and siding by a certain date. To meet that goal, you and the superintendent would need to agree to a goal to have the framing completed by a given date. You then establish subgoals for your crew to complete each element of the framing (floors, walls, roof) by a set time.

Identify the Required Work Activities and Divide Them into Tasks

To plan effectively, you must be able to break a work activity assignment down into smaller tasks. Large jobs include a greater number of tasks than small ones, but all jobs can be broken down into manageable components.

Make a list of all the activities that the plan requires to complete the job. This includes individual work tasks and special tasks, such as inspections or the delivery of materials. Next, start to divide work activities into tasks. As you do this, remember that almost every work activity has three general parts: preparing for the activity, performing the activity, and cleaning up after the activity.

Figure 32 shows a suitable breakdown for the work activity to install square vinyl floor tiles in a cafeteria. You could create even more detail by breaking down any one of the tasks into subtasks. In this case, however, that much detail is unnecessary and wastes your time and the project's money.

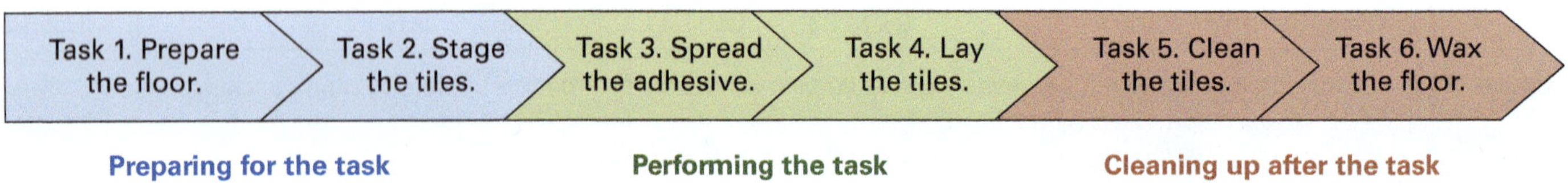

Figure 32 Example task breakdown of the work activity to install square vinyl floor tiles in a cafeteria.

When breaking down an activity into tasks, make each task *identifiable* and *definable*. A task is identifiable when you know exactly the types and amounts of resources it requires. A task is definable if it has a specific duration. The activity breakdown does not need to be overly detailed or complex unless the job has never been done before or must be performed with strictest efficiency.

At this point, you should just be concerned with generating a list, not with determining how to accomplish the activities, who will perform them, how long they will take, or the necessary sequence to complete them.

Plan Resources Needed for Each Task

Once a job has been broken down into activities and tasks, you must identify and account for all of the resources you need to complete them. Resource planning includes considerations for safety, materials, site-specific issues, equipment, tools, and labor (*Table 11*). Each of these is discussed in more detail in the sections that follow.

Organize the Work into a Logical Sequence

The next step in planning is to use the list of activities and resources needed to organize the work activities into a logical sequence. When doing this, keep in mind that certain steps can't happen before the completion of others. For example, concrete placement cannot happen until after rebar is set. Some construction work activities, such as installing 12" footing forms, are done so often that they require little planning. However, other jobs, such as placing a new type of mechanical equipment, require substantial planning.

Assign Responsibility for Each Task

After you have identified the tasks, you must determine who will complete each activity. This requires that you are aware of the experience, skills, and abilities of

TABLE 11 Key Resource Considerations During Planning

Resource Planning Area	Key Responsibilities
Safety Planning	Using the company safety manual as a guide, you must assess the safety issues associated with the job and take necessary measures to minimize any risk to the crew.
Materials Planning	You may be required to estimate quantities of materials during construction. You also have a key role in the receipt, storage, and control of the materials after they reach the jobsite.
Site Planning	Many planning elements are involved in sitework, including access roads, parking, site security, material and equipment staging and storage, sedimentation control, stormwater runoff, and more.
Equipment Planning	You may need to plan the types of equipment needed, the use of the equipment, and the length of time it will be on the site. You must work with the main office to make certain that the equipment reaches the jobsite on time. You must reserve time for equipment maintenance to prevent equipment failure.
Tool Planning	To plan tool usage, you must determine the tools required, tell your crew who will provide the tools, ensure that the crew is qualified to use the tools safely and effectively, and establish proper tool controls.
Labor Planning	In many companies, the project manager or job superintendent determines the size and makeup of the crew. Then, supervision expects you to accomplish the goals and objectives with the crew provided. If you are responsible for planning for labor, you must identify the skills needed to perform the work, determine the number of people having the required skills, and decide who will be on the crew.

everyone in your team. You must put this knowledge to work by matching the crew's skills and abilities to specific tasks. Your job is to draw from their expertise to get the job done well and in a safe and timely manner.

Assign a Duration to Each Task

Next, assign a duration or length of time that it will take to complete each activity and determine the start date and time for each. Then place each activity into a schedule format. This step is important because it helps you compare the task time estimates to the scheduled completion date or time.

Make note of any activities that cannot start until previous work has been completed and ensure those do not start before the other work is complete. You also will need to ensure you have not assigned too much work to any one person or team.

Some of the information to support short-term scheduling comes from the estimate or cost breakdown. *Figure 33* shows how you can use estimate or cost breakdown information to set short-term production goals for a team.

Scenario: A carpentry crew on a retaining wall project is about to form and pour catch basins and put up wall forms. The crew will work on both the basins and the wall forms at the same time. The crew has put in several catch basins, so the crew leader is sure that they can perform the work within the estimate. However, the crew leader is concerned about their production of the wall forms and wants to set a clear target for the amount of work to be done each day to hit the schedule.

The scheduling process in this scenario could be handled as follows:

1. Review the estimate or cost breakdown.	2. Consider available workers.	3. Determine the job goal.	4. Analyze how much work the existing workers can complete.	5. Communicate production goal with the crew.
• Production factor for wall forms = 16 worker-hours (w-h) per 100 ft^2 • Work to be done by measurement = 800 ft^2 • Total time: (800 ft^2 × 16 w-h/100 ft^2) = 128 w-h	The carpentry crew includes: • One carpentry crew leader • Four carpenters • One laborer	The crew leader determines the goal for the job should be set at 128 w-h (from the cost breakdown).	If the crew remains the same (six workers), the work should be completed in about 21 crew-hours (128 w-h ÷ 6 workers/crew = 21.3 crew-hours).	• Crew leader shares the production goal of completing 800 ft^2 in 21 crew-hours with the crew. • Crew leader encourages them to work together to meet the goal of getting the forms erected within the estimated time.

Figure 33 Crew leaders can use lookaheads or short-term scheduling to set production goals.

In this example, the crew leader used the short-term schedule to translate production into work hours and to schedule work so that the crew can accomplish it within the estimate. A crew can improve production when they know the amount of work to accomplish and the time available to complete the work, and can provide input when setting goals. In addition, setting production targets provides the motivation to exceed schedule targets.

Communicate the Schedule and Assigned Responsibilities

After the schedule is complete, you must then assign work responsibilities to the crew. Each person in the crew should know what they are responsible for accomplishing on the job.

Use the Schedule to Monitor and Control Work

NOTE

Most scheduling apps and programs include this feature that automatically helps you focus on the lookahead period for a job.

Schedules only have value if they are actively used as tools to monitor and control work. A lookahead or short-term schedule provides you with the information you need to plan for work and track progress. Before starting a job, actively review the schedule to:

- Determine the materials, tools, equipment, and labor needed to complete the job.
- Determine when the various resources are needed.
- Follow up to ensure that the resources are available on the jobsite when needed.

You should also verify the availability of needed resources three to four working days before the start of the job. This should occur even earlier for larger jobs. Advance preparation will help avoid situations that could potentially delay starting the job or cause it to fall behind schedule. As work continues, use your schedule to monitor whether actual production begins to slip behind estimated production and develop plans to recover the schedule as needed.

Follow Up and Confirm Work Is Completed Well

After you delegate tasks to the appropriate crew members, you must follow up to confirm that the crew completed the tasks correctly and efficiently. To do this effectively, you must be present on the jobsite to make sure all the resources are available to complete the work, to confirm that team members are working on their assigned activities, to answer any questions, and to help to resolve any problems that occur while the work is being done.

5.3.0 Crew Leader Field Reporting Responsibilities

A field-reporting system consists of a series of forms that are completed by you and other leaders on the project to share updates on labor, materials, tools, equipment, and more.

As a frontline leader, you may not be directly involved in updating overall schedules. However, the daily field or progress reports you submit will be used by the company to keep the schedule up to date.

The following sections provide some examples of field reporting. Each company has its own forms and methods for obtaining information. You should always follow your company's policies and procedures for field reporting.

5.3.1 Completing Daily Reports

Job diary: A written record that a supervisor maintains periodically (usually daily) of the events, communications, observations, and decisions made during the course of a project.

You will be involved with many activities on a day-to-day basis. As you work each day, you will want to keep a **job diary** to track events such as job changes, interruptions, and visits. A job diary, also called a *daily report, site diary, construction log,* or *site journal,* is a notebook or digital application in which you record activities or events that take place on the jobsite that may be important later (*Figure 34*).

July 8, 2023

Weather: Hot and Humid

Project: Opal Garage Annex

- The paving contractor crew arrived late (10 am).
- The owner representative inspected the footing foundation at approximately 1 pm.
- The concrete slump test did pass. Two trucks had to be ordered to return to plant, causing a delay.
- David Lopez had an incident on the second floor. I sent him to the on-site medical personnel for medical first aid treatment. The cause of the incident is being investigated.

Figure 34 Sample page from a job diary.

For this comparison process to be of use, the information you and others submit from the field must be accurate. In the event of a legal or contractual conflict with the client over costs, job diaries can be helpful as evidence in legal proceedings or in reaching a settlement. *Table 12* outlines an example scenario where your action of accurate field reporting will help benefit your company.

TABLE 12 Accurate Field Reporting Benefits Your Company

Scenario	Field Reporting Action
You are running a crew of five concrete finishers for a specialty contractor. When you and your crew show up to finish a slab, the general contractor (GC) informs you, "We're a day behind on setting the forms, so I need you and your crew to stand down until tomorrow."	Follow your company's field reporting policy to report the change. For example, you might first call your office to let them know about the delay. Then, you would record it in your job diary. A six-man crew for one day represents 48 worker-hours. If your company charges $30 an hour, that's a potential loss of $1,440, which the company would want to recover from the general contractor. If there is a dispute, your entry in the job diary could result in a favorable decision for your employer.

Today, many companies use mobile applications that allow you to complete daily reporting on your mobile phone or tablet while you are on-site. You can quickly enter information throughout the day and attach photos and other documents to add more information. They generally also make it easy to send to your supervisor via email or text message (*Figure 35*). Your supervisor will be notified when the report is complete.

Whether print or digital, when making entries in a job diary, make sure that the information is accurate, factual, complete, consistent, organized, and up to date. Do not include any emotional details; keep the report focused on facts. This is especially true if documenting personnel problems. If the company

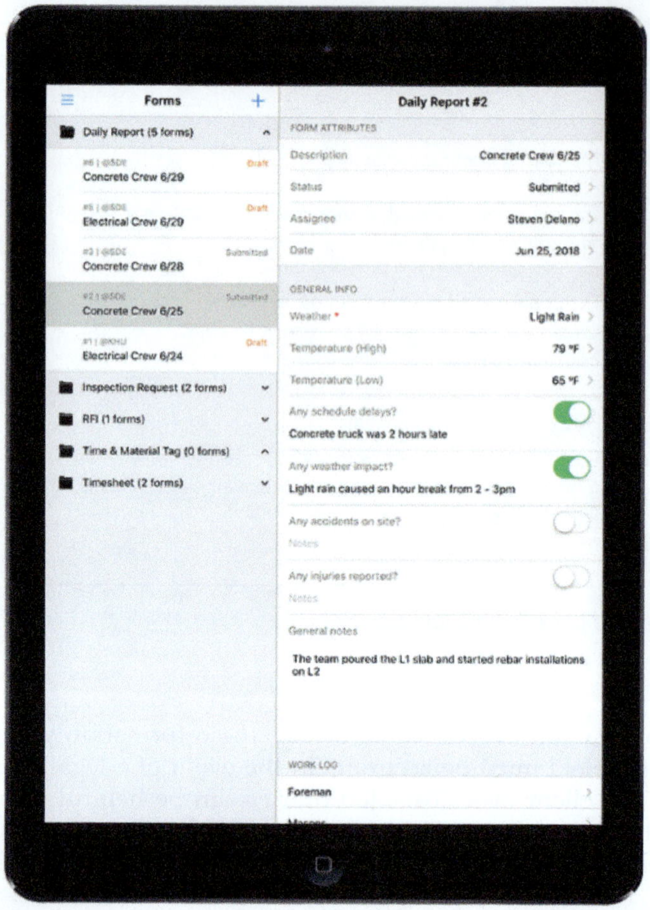

Figure 35 Mobile apps make it easier to complete daily reporting on-site.
Source: Fieldwire, https://www.fieldwire.com/

requires maintaining a job diary, follow company policy in determining which events and what details you should record. The information needs to be detailed enough so that, if needed, you could determine what happened on any day years later. If you are unsure what to include, it is better to have more information than too little. Check in with your supervisor to confirm if your daily reporting is including the right information.

5.3.2 Tracking Employee Costs Through Timesheets

Work breakdown structure (WBS): A method of organizing the entire scope of work in a logical manner based on how the work will be executed.

When the project budget is set up, many companies create a work breakdown for each project. A **work breakdown structure (WBS)** organizes the entire scope of work in a logical manner based on how the work will be executed. Within the WBS, planners assign each major task a unique charge number or code (*Figure 36*). Anyone working on that task charges that number or code on their timesheet, so that project managers can readily track cost performance.

As a frontline leader, you not only need to ensure everyone is completing their timesheets, but also that they are coding those timesheets to the proper codes in the WBS. Doing so ensures you and your company are tracking and managing labor costs against specific work.

For example, as shown in *Figure 36*, if all timesheets are correctly coded, you and the company can determine the labor cost to install each of the three types of pipe: carbon steel piping, stainless steel piping, and the high-density polyethylene (HDPE) piping. You can also calculate how much it costs to install the piping for the entire project.

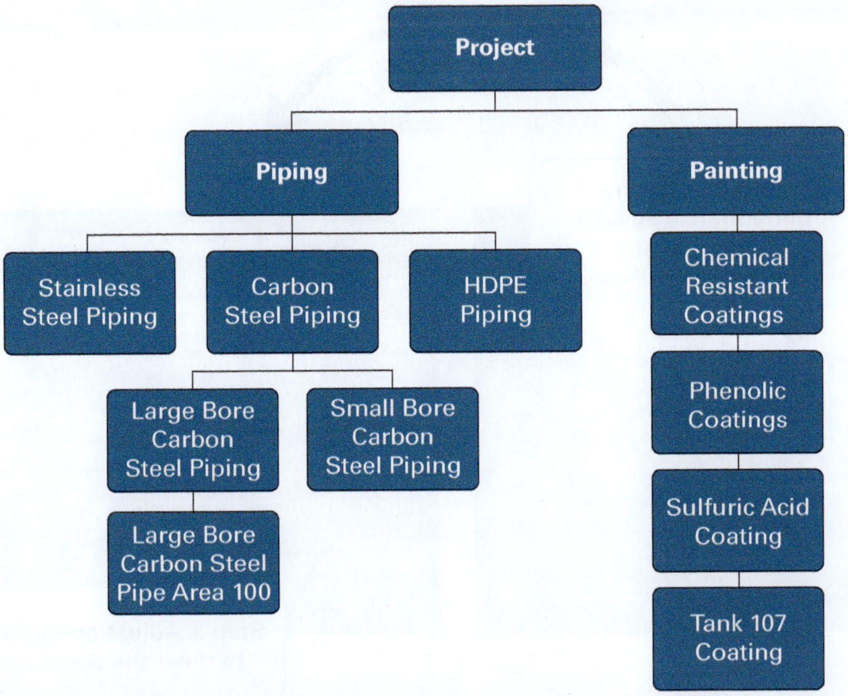

Level	Work	Account Code
Level 0	Piping	B
Level 1	Stainless Steel Piping	B-SS
Level 1	Carbon Steel Piping	B-CS
Level 1	HDPE Piping	B-HP
Level 2	Large Bore Carbon Steel Piping	B-CS-L
Level 2	Small Bore Carbon Steel Piping	B-CS-S
Level 3	Large Bore Carbon Steel Pipe Area 100	B-CS-L-100
Level 0	Painting	L
Level 1	Chemical Resistant Coatings	L-CR
Level 2	Phenolic Coating	L-CR-P
Level 3	Sulfuric Acid Tanks Coating	L-CR-P-ST
Level 4	Tank 107 Coating	L-CR-P-ST-107

Figure 36 An example of a work breakdown structure (WBS) for piping and painting on a project.

5.4.0 Frontline Leader Cost and Resource Management Responsibilities

Being aware of costs and controlling them is the responsibility of every employee on the job. As a frontline leader, it is your job to ensure that employees uphold this responsibility. Cost control is the process by which you keep expenses under control by managing labor, material, and other costs to ensure that the project finishes on budget.

5.4.1 Managing Costs

As you have learned, profit refers to the amount of money that the contractor will make after paying all the direct and indirect costs. If the costs exceed the estimate for the job, the difference between the actual and estimated costs must come out of the company's profit. This reduces what the contractor makes on the job.

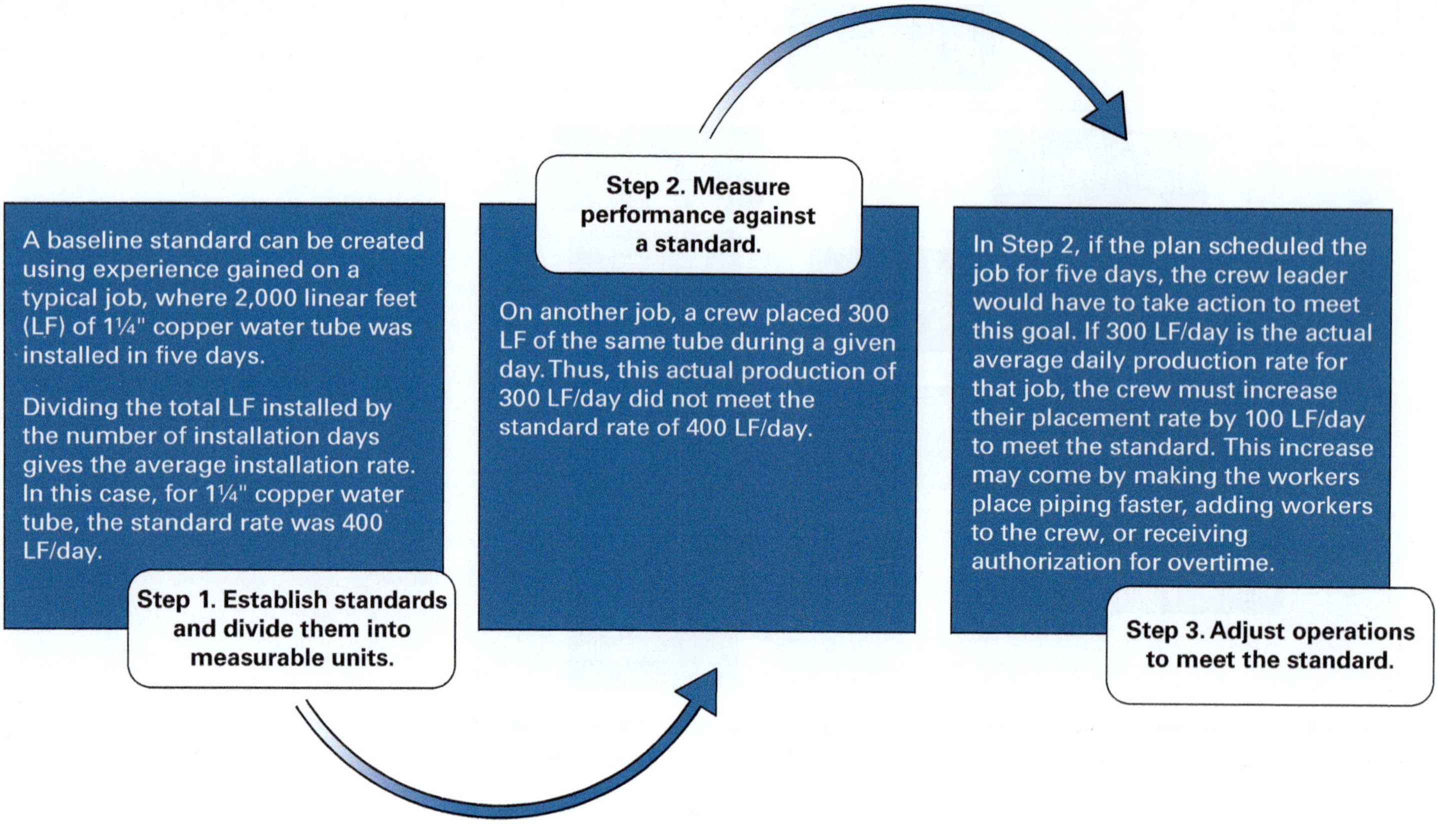

Figure 37 An effective cost control process monitors production and productivity to help manage costs.

NOTE

Before starting any action, you should check with company supervision to see that the action proposed is acceptable and within the company's policies and procedures.

Early identification of cost changes is very important and a key part of your role. As shown in *Figure 37*, an effective cost control process to monitor production and productivity includes several key steps. Identifying changes provides enough notice to correct the causes of overruns and manage those costs. You must learn to anticipate changes from plans and specifications based on experience and take measures to prevent them from occurring. If a situation cannot be fully corrected, early notification allows time for communicating and for managing expectations. No one likes surprises when it comes to increased project costs.

Table 13 lists some of the factors that can make actual costs exceed estimated costs, along with some potential actions you can take to avoid these situations.

A key factor in managing costs is controlling waste. Waste in construction can add up to loss of critical and costly materials and may result in job delays. You need to ensure that every crew member knows how to complete the work assigned and use materials efficiently. One example of waste is a carpenter who saws off a piece of lumber from a full-sized piece, when he could have found the length needed in the lumber scrap pile.

Doing rework is another form of waste (*Figure 38*). For example, less-experienced or less-trained crew members may make mistakes that cause work to have to be done over. You can help avoid these errors by providing on-the-spot training and by checking on newer, less experienced workers more often. Rework also can result from a conflict in craft scheduling that requires removal and reinstallation of finished work. Rework also increases safety risks on a job. Rework often involves demolition, time pressure, and frustration from the team, all of which create a higher likelihood of an incident.

Some issues that cause rework, such as mistakes on drawings, may be outside of your control. However, you can manage this risk by ensuring that you always have the most current set of drawings and examining the drawings carefully before you start work. Finding errors before the work begins increases safety and saves time, labor, materials, and more, which saves costs overall.

TABLE 13 Factors That Drive Cost Changes

Factor	Proactive Action
Client-related changes	Clients may want to change the scope of a project after it has started. You must be able to assess the potential impact of such changes and, if necessary, confer with the employer to determine the course of action. If contractor-related losses are occurring, you and the superintendent will need to work together to get the costs back in line.
Late delivery of materials, tools, and/or equipment	Review and update your lookahead schedules on a regular basis. Plan ahead to ensure that job resources will be available when needed.
Inclement weather	Work with the superintendent and other field leaders to have alternate plans ready.
Higher-than-expected wages or unexpected overtime due to low manpower	Plan ahead to ensure that labor resources will be available when needed. Ensure you are accurately recording daily work using your field reporting systems and job diary. Work with your supervisor or the superintendent to identify other staffing approaches.
Contractor or owner delays due to work availability	Ensure you are accurately recording daily work using your field reporting systems and job diary. Work with the superintendent and other field leaders to have alternate plans ready.
Unmotivated workers	Apply the leadership and team motivation strategies discussed earlier in this course. Remember, different things motivate different people in different ways. This means there is no single best approach to motivating all of your crew members.
Incidents	Continue to lead by example, reinforce the critical nature of safety, and follow the guidelines discussed earlier in the course. Work with the superintendent and other field leaders to help evaluate and improve the safety and health program, if needed.
Owner-related changes	Owners may want to change the scope of a project after it has started. You must be able to assess the potential impact of such changes and, if necessary, confer with the employer to determine the course of action. If contractor-related losses are occurring, you and the superintendent will need to work together to get the costs back in line.
Waste due to rework, material usage, and more	Train less-experienced workers on how to complete work. Review and update your lookahead schedules on a regular basis. Ensure crew members know how to use materials efficiently. Ensure you have the latest drawings and reference them daily.

Figure 38 Rework is another form of waste on a construction site.
Source: Vladimir Cetinski/iStock/Getty Images

5.4.2 Ensuring Crew Productivity

People typically represent more than half the cost of a project and thus have an enormous impact on profitability. For that reason, it is essential for you to manage your crew and their work environment in a way that maximizes their productivity.

One way to manage costs and increase productivity is to make sure you have the right skills on your crew to execute the level of tasks that you have to complete. *Crew mix* is a concept that describes the number of experienced or journey-level workers relative to the number of apprentices or less-experienced workers in a crew. For safety, quality, and productivity reasons, you do not want a worker trying to complete a task for which they are not skilled. However, you also do not want to have higher-wage crew members doing work that can be done by lower-wage crew members. For example, paying a journeyman electrician to pull wire costs more than paying an electrician helper who is skilled enough to do that work under the supervision of a journeyman electrician. As a frontline supervisor, you may not have control of your specific crew mix, but you can work to align their skills levels most appropriate to the specific work. Doing this well can reduce the overall project costs.

Other practices can help to ensure productivity:

- Keeping the crew adequately staffed at all times to avoid job delays
- Ensuring that all workers have the required resources when needed
- Ensuring that everyone knows where to go and what to do after each task is completed
- Making reassignments as needed
- Ensuring that all workers have completed their work properly

The time on the job should be for business, not for taking care of personal problems. Anything not work related should be handled after hours, away from the jobsite, if possible. Even the planning of after-work activities, arranging social functions, or running personal errands should occur after work or during breaks. Limited and necessary exceptions to these rules should be permitted (for example, making or meeting medical appointments), and these should be spelled out in company or crew policies.

Given the nature of construction work, frequent breaks can be needed to ensure worker safety. However, if crew members are often taking unscheduled breaks or goofing off, you will need to confront the behavior and coach these workers. Document any discussions in your job diary. If needed, you should redirect repeated violations to the attention of higher management as guided by company policy.

You can minimize or avoid delays caused by other crews or contractors by carefully tracking the project schedule. In doing so, you can anticipate delays that will affect your work and either take action to prevent the delay or redirect the crew to another task.

The availability and location of materials, equipment, and tools can have a serious effect on a crew's productivity. In the sections that follow, you will learn more about how to manage these resources to minimize such problems.

5.4.3 Handling Material Delivery and Storage

Your specific responsibilities for material delivery and storage will depend on the policies and procedures of the company. In general, you will be responsible for ensuring on-time delivery of materials, verifying the materials were delivered, and storing the materials (*Figure 39*). You also must prevent the waste and theft of materials.

NOTE

Scan the QR code in the front of this module to access additional resources on the steps involved in the development of an estimate, including creating a quantity takeoff sheet.

You will also be involved in planning materials for tasks such as job-built formwork and scaffolding or acquiring materials if you run out. If you run out of a specific material, be sure to consult the appropriate supervisor on how to proceed, because most companies have specific purchasing policies and procedures.

It is essential that the materials required for each day's work be on the jobsite when needed. You should confirm in advance the placement of orders for all

Figure 39 The receipt, storage, and control of materials after they reach the jobsite is an important part of a frontline leader's work.
Source: July Alcantara/E+/Getty Images

Source: metamorworks/Shutterstock

Did You Know?

Just-in-Time (JIT)

Just-in-time (JIT) delivery is a strategy in which materials arrive at the jobsite when needed. This means that a crew may install the materials right off the truck. This method reduces the need for on-site storage and staging areas. It also reduces the risk of loss or damage while moving products about the site. Other modern material management methods include the use of radiofrequency identification (RFID) tags that make it easy to inventory and locate material in crowded staging areas. Using an RFID inventory control system requires special scanners.

materials and that they will arrive on schedule. A week or so before the delivery date, follow up to make sure there will be no delayed deliveries or items on backorder. If other people are responsible for providing the materials for a job, you must follow up to make sure that the materials are available when needed. Otherwise, delays can occur as crew members wait for the delivery of materials.

You may be responsible for the receipt of materials delivered to the worksite. When this happens, you should require a copy of the shipping invoice or similar document and check each item on the invoice against the actual materials to verify delivery of the correct amounts.

You should also check the condition of the materials to verify that nothing is defective before signing the invoice. This can be difficult and time-consuming because it requires the crew to open cartons and examine their contents. However, this step cannot be overlooked, because a signed invoice indicates that the recipient accepted all listed materials and in an undamaged state. If the crew leader signs for the materials without checking them, and then finds damage, no one will be able to prove that the materials came to the site in that condition (*Figure 40*).

Figure 40 When receiving materials, check the condition of the materials to verify that nothing is defective before signing the invoice.
Source: kali9/E+/Getty Images

After checking and signing the shipping invoice, you should give the original or a copy to the superintendent or project manager. The company then will file the invoice for future reference because it serves as the only record the company has to check bills received from the supplier company.

Another essential element of materials control is selecting where to store materials on a jobsite. *Table 14* outlines three factors to consider when selecting an appropriate material storage location.

TABLE 14 Considerations When Selecting a Material Storage Location

Factor	Description
Access	If possible, store materials near where the crew will use them, so the crew can access them easily. Not having to carry the materials long distances will greatly reduce time, effort, and installation costs.
Damage Protection	Store materials in a secure area to avoid damage. Be sure that the storage area suits the materials. For instance, temperature-sensitive materials, such as chemicals or paints, should be stored in climate-controlled areas to prevent waste.
Security	Theft or vandalism of construction materials increases direct and indirect costs. These include replacement materials, lost production time while replacing needed materials, and the costs of additional security precautions. In addition, a contractor's insurance costs may increase due to theft and vandalism.

The best way to avoid theft and vandalism is to have a secure jobsite. At the end of each workday, store unused materials and tools in a secure location, such as a locked construction trailer. If the jobsite is fenced or the building has lockable access, the materials can be stored within. Many sites have security cameras and/or intrusion alarms to help minimize theft and vandalism.

5.4.4 Planning and Controlling Tool Usage

Among companies, various policies govern who provides hand and power tools to employees. Some companies provide all the tools, while others furnish only the larger power tools. You should be familiar with and enforce any company policies related to tools.

As a crew leader, you are responsible for planning tool usage for a job. This task includes the following:

- Determining the tools required
- Informing the workers who will provide the tools (company or worker)
- Making sure the workers are qualified to use the tools safely and effectively
- Determining what controls to establish for tools

Tool control involves controlling the issue, use, and maintenance of all tools provided by the company. It also involves storing the tools in a location that provides easy access to your crew (*Figure 41*).

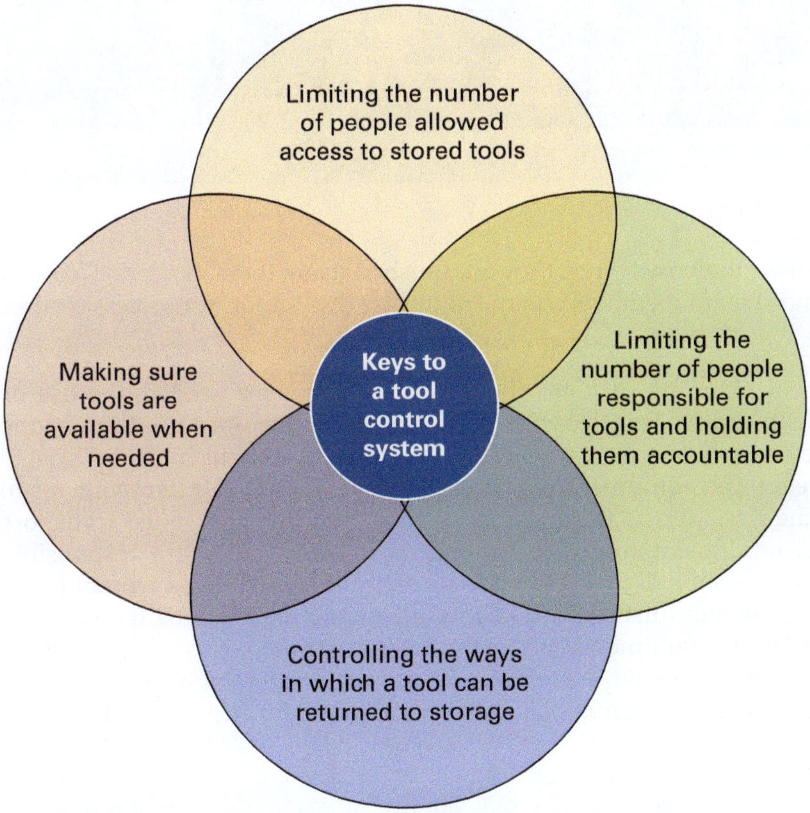

Figure 41 Studies show there are four keys to an effective tool control system.

Using the proper tools correctly saves time and energy. In addition, proper tool use reduces the chance of damage to the tool during use, as well as injury to the user and to nearby workers. This is especially true of edge tools and those with any form of fixed or rotating blade. A dull tool is far more dangerous than a sharp one.

Tools must be adequately maintained and properly stored. Making sure that tools are cleaned, dried, and lubricated prevents rust and ensures that the tools are in the proper working order. This applies to tools supplied by the company as well as tools that belong to the workers.

If a tool is damaged, it is essential to repair or replace it promptly. Otherwise, an incident or injury could result. Users should take care of company-issued tools as if they were their own. Workers should not abuse tools simply because they are not their property.

One of the major sources of low productivity on a job is the time spent searching for a tool. To prevent this from occurring, supervision should establish a storage location for company-issued tools and equipment. You should make sure that crew members return all company-issued tools and equipment to this designated location after use. Similarly, workers should organize their personal toolboxes so that they can readily find the appropriate tools and return their

NOTE

Regardless of whether a worker or the company owns a tool, OSHA holds companies responsible for the consequences of using a tool on a jobsite. The company is accountable if an employee receives an injury from a defective tool. Therefore, the crew leader needs to be aware of any defects in the tools the crew members are using.

Did You Know?

Connected Tools Are Controlled Tools

Tool companies are using technology to help you track tools and equipment on a jobsite. These smart tool systems have a software application that allows you to add tools enabled with Bluetooth® technology to the tracking system. You can also attach powered tags to large equipment or materials and add those to the tracking system. As crew members sign out tools, the assignment is tracked in the system, so you can track which tools are in use, by whom, and where they are on the jobsite at any point. You can also track materials and equipment in the same way. By helping to ensure all of your equipment, tools, and materials are accounted for and easy to find, these systems help companies control costs by saving time and replacement costs.

Source: DEWALT Industrial Tool Co.

tools to their toolboxes when they are finished using them. If needed, coach less-experienced team members on how to master this important professional skill.

5.4.5 Ensuring Equipment Control

Equipment planning includes the types of equipment needed, the use of the equipment, and the length of time it will be on the site. As a frontline leader of a crew, you may not be responsible for long-term equipment control.

However, the equipment required for a specific job is often your responsibility. The first step is to schedule when a crew member must transport the required equipment from the shop or rental yard. You are responsible for informing the shop of the location of its use and confirming a worker returns it to the shop when finished with it. You must work with the main office to make certain that the equipment reaches the jobsite on time.

Coordinating the use of the equipment is also very important. Some equipment operates in combination with other equipment (*Figure 42*). For example,

Figure 42 Coordinating equipment requires knowing which equipment operates in combination with other equipment.

Source: RyanOverman/iStock/Getty Images

dump trucks are generally required when loaders and excavators are used. You should also coordinate equipment with other contractors on the job. Sharing equipment can save time and money and avoid duplication of effort.

It is common for equipment to lie idle at a jobsite because of a lack of proper planning for the job and the equipment arriving early. For example, if wire-pulling equipment arrives at a jobsite before the conduit is in place, this equipment will be out of service while awaiting the conduit installation. In addition to the wasted rental cost, the equipment could also be damaged, lost, or stolen while awaiting use.

You need to control equipment use, ensure that the crew operates the equipment according to its instructions, and make sure they are adhering to time and cost guidelines. You must also assign or coordinate responsibility for maintaining and repairing equipment as indicated by the applicable preventive maintenance schedule. Delaying maintenance and repairs can lead to costly equipment failures.

You must also ensure that crew equipment operators are properly trained. This includes ensuring that the equipment operators have the necessary credentials to operate the equipment, including applicable licenses.

You are also responsible for the proper operation of all other equipment resources, including cars and trucks. Reckless or unsafe operation of vehicles will likely result in damaged equipment, leading to disciplinary action for the workers, fines and repair costs levied by the vehicle vendors, and a loss of confidence in your ability to control the actions of your crew.

You should also arrange to secure all equipment at the close of each day's work to prevent theft. If continued use of the equipment is necessary for the job, you should make sure to lock it in a safe place. If it is not, return it promptly.

5.0.0 Section Review

1. Which of these activities occurs during the design and development phase of a project?
   a. Architect/engineer sketches are prepared, and a preliminary budget is developed.
   b. Final inspections are completed.
   c. Detailed project drawings and specifications are prepared.
   d. Contracts for the project are awarded.

2. Which of these activities is *not* part of creating a lookahead schedule?
   a. Identifying the required work activities and dividing them into tasks
   b. Planning resources needed for each task
   c. Organizing the work into a logical sequence
   d. Assigning cost estimates to each task

3. A job diary should typically record any of the following *except* _____.
   a. job interruptions and visits
   b. personal details and emotions
   c. changes needed to project drawings
   d. the actual time each job task took to complete

4. Which of the following is a *correct* statement regarding project cost?
   a. Cost is handled by the Accounting Department and isn't a concern of the crew leader.
   b. The difference between the estimated cost and the actual cost affects a company's profit.
   c. Wasted material is factored into the estimate and is never a concern.
   d. The contractor's overhead costs aren't included in the cost estimate.

Module 46100 Review Questions

1. Who is generally the on-site field supervisor who is responsible for one or more crew leaders?
   a. Project Manager
   b. Lead Engineer
   c. Superintendent
   d. Pre-Construction Manager

2. Accepting responsibility for job completion and being personally answerable for the results is known as _____.
   a. responsibility
   b. delegation
   c. authority
   d. accountability

3. Which of the following is *not* true of situational leadership?
   a. It involves adjusting the way you lead to the specific situation and needs of the group.
   b. It assumes all situations do not require the same leadership.
   c. It allows you to adapt your leadership style to the situation.
   d. It has four styles, including directing, coaching, supporting, and delegating.

4. When you communicate with your team verbally, be sure to _____.
   a. assess your audience before you speak
   b. use technical language as much as possible
   c. speak quickly to save time
   d. assume everyone has the same basic understanding of the topic

5. The *first* step in problem solving should be _____.
   a. choosing a solution to the problem
   b. evaluating the solution to the problem
   c. defining the problem
   d. testing the solution to the problem

6. Which of these is a way to lower the motivation and morale of your employees and team?
   a. Focusing on task completion only, not professional growth
   b. Giving recognition and rewards
   c. Setting clear expectations and goals
   d. Providing opportunities for training and upskilling

7. During an employee's first week of work on your crew, you should be sure to set aside time to _____.
   a. eat lunch together
   b. do a task verification
   c. quiz the employee on company policies
   d. evaluate social engagement

8. Which of these steps is important when delegating work to an individual or a team?
   a. Letting individuals set their own standards of measurement
   b. Checking in only after the work is complete
   c. Ensuring crew members understand what to do and have the skill to do it
   d. Allowing the team to set their own deadlines and get their own resources

9. Which common jobsite documents can help a crew leader overcome a language barrier with a worker?
   a. OSHA incident records
   b. Contracts and schedules
   c. Plans and drawings
   d. Permits

10. As a leader, you can help to manage performance issues by _____.
    a. ignoring the problem that needs resolving
    b. gathering information and preparing well for the employee conversation
    c. discussing the performance issues publicly with the rest of the team
    d. having the conversation when the employee is unprepared

11. The negative impact of incidents may include all of the following *except* _____.
    a. lower workers' compensation costs
    b. damage to the overall morale of everyone on a project or jobsite
    c. increased project costs to the company
    d. project delays

12. OSHA was established in 1970 to _____.
    a. collect dues for membership
    b. ensure safe and healthful working conditions for workers
    c. manage workers' compensation claims
    d. conduct medical research

13. Which of the following statements regarding safety training sessions is *correct*?
    a. The project manager usually holds them.
    b. They are held only for new employees.
    c. They should be conducted frequently by a frontline leader.
    d. They are required only after an accident has occurred.

14. A key step in an incident investigation is to determine the incident's _____.
    a. root cause
    b. secondary driver
    c. cost impacts
    d. design flaws

15. If you observe an employee showing impaired behavior or signs of mental health issues, you should work with _____.
    a. OSHA
    b. the safety inspector
    c. the owner or client
    d. your supervisor and/or Human Resources

16. What is completed during the design phase to ensure a project can be delivered successfully?
    a. Feasibility study
    b. As-built drawings
    c. Installation work packages
    d. Project schedule

17. A contract in which the client pays the contractor for their labor, materials, equipment, and other costs they incur during the performance of the contract is known as a _____.
    a. fixed-price contract
    b. time-spent contract
    c. time and materials contract
    d. cost-reimbursable contract

18. Which of the following would be tracked in a job diary?
    a. Job changes, interruptions, and visits
    b. The crew's time off
    c. Training needs and career development
    d. Personal information provided by an employee

19. Which of the following is a source of data in the field-reporting system?
    a. The cost-estimate for the project
    b. The daily job diary
    c. The Level 1 schedule
    d. The lookahead schedule

20. To prevent job delays due to late delivery of materials, the crew leader should _____.
    a. demand a discount from the supplier to compensate for the delay
    b. tell a crew member to go look for the delivery truck
    c. refuse the late delivery and re-order the materials from another supplier
    d. check with the supplier in advance of the scheduled delivery

Answers to odd-numbered review questions are found in the Module Review Answer Key at the back of this book.

Answers to Section Review Questions

Answer	Section	Objective
Section One		
1. b	1.1.2	1a
2. b	1.2.2	1b
Section Two		
1. d	2.1.2	2a
2. b	2.2.3	2b
3. a	2.3.0	2c
Section Three		
1. b	3.1.1	3a
2. c	3.2.0	3b
3. c	3.3.3	3c
Section Four		
1. a	4.0.0	4a
2. d	4.2.0	4b
3. a	4.3.1	4c
4. b	4.4.2	4d
5. d	4.5.0	4e
Section Five		
1. a	5.1.1	5a
2. d	5.2.2	5b
3. b	5.3.1	5c
4. b	5.4.1	5d

APPENDIX 03402A

Examples of Completed Air Balance Forms

Mailing Address
P.O. Box 5782
Greenville, SC 29606-5782

165 South Hammett Road
Greer, SC 29651

Phone (864)877-6832

Fax (864)877-5490

email palmettoab@aol.com

Palmetto Air&Water BALANCE

Raising the level of efficiency and comfort

AIR APPARATUS TEST REPORT

PROJECT: CHARTER COMMUNICATIONS

LOCATION: 1ST FLOOR MECHANICAL ROOM

SYSTEM/UNIT: SCU-1

UNIT DATA

Make	TRANE
Model No.	SCWFN5242DOAo11
Serial No.	T02C18736
Arr./Class	DRAW THRU/1
Discharge	TOP VERITCAL
Sheave Diameter	2B5V124
Sheave Bore	B 2.44
No. Belts/make/size	2/GOODYEAR/5VX630
No. Filters/type/size	2/TA/20X20X2 4/TA/16X20X2 4/TA/20X25X2 8/TA/16X25X2

MOTOR DATA

Make/Frame	BALDOR/254T
H.P.	15
RPM	1760
Voltage	460
Phase/Hertz	3 / 60
F.L. Amps	20.30
S.F.	1.25
Sheave Diameter	2B5V52
Sheave Bore	B 1.63
Sheave Center Distance	7.25

TEST DATA	DESIGN	ACTUAL
Total CFM	17645	18235
Fan RPM		648
Motor Volts		486
Motor Volts		486
Motor Volts		486
Motor Amps T-1		21.7
Motor Amps T-2		21.7
Motor Amps T-3		21.7
C.F.L.A.		19.2
B.H.P.		16.9

TEST DATA	DESIGN	ACTUAL
Make/Frame		
H.P.		2.26
RPM		–0.60
Voltage		2.83
Phase/Hertz		
F.L. Amps		
S.F.		
		2.2
Sheave Diameter		
Sheave Bore		
Sheave Center Distance		

Remarks:

Filename: CCG047

Test Date: Jan–03

Readings By: GF

Source: Palmetto Air & Water Balance, Inc.

Mailing Address
P.O. Box 5782
Greenville, SC 29606-5782

165 South Hammett Road
Greer, SC 29651

Phone (864)877-6832

Fax (864)877-5490

email palmettoab@aol.com

Palmetto Air&Water BALANCE

Raising the level of efficiency and comfort

RECTANGULAR DUCT TRAVERSE TEST REPORT

PROJECT: CHARTER COMMUNICATIONS

Zone: DT-1 SUPPLY

SYSTEM/ UNIT: SCU-1

Correction Factor: 1.00

Width: 30" Height: 22" — Duct Area: 4.583 Sq Ft

Actual — Vel. 1640 CFM 7516

Position	1	2	3	4	5	6	7	8	9	10	11	12	13	14
1 >>>	1487	1689	1893	1818	1716	1933								
2 >>>	1690	1656	1505	1614	1730	1833								
3 >>>	1619	1589	1539	1416	1472	1573								
4 >>>	1713	1381	1467	1740	1673	1610								
5 >>>														
6 >>>														
7 >>>														
8 >>>														
9 >>>														
10 >>>														
Subtotal	6509	6315	6404	6588	6591	6949	0	0	0	0	0	0	0	0

Total FPM	/ Number of Readings	= Average Velocity	X Duct Area	= CFM	Final S.P.
39356	24	1640	4.583	7516	1.31

Remarks: *SUPPLY CFM TOTAL = DT-1 + DT-2
Filename: CCG049
Test Date: Jan–03
Readings By: GF

Source: Palmetto Air & Water Balance, Inc.

Mailing Address
P.O. Box 5782
Greenville, SC 29606-5782

165 South Hammett Road
Greer, SC 29651

Phone (864)877-6832

Fax (864)877-5490

email palmettoab@aol.com

Palmetto Air&Water
BALANCE
Raising the level of efficiency and comfort

RECTANGULAR DUCT TRAVERSE TEST REPORT

PROJECT: CHARTER COMMUNICATIONS
Zone: DT-2 SUPPLY

SYSTEM/ UNIT: SCU-1
Correction Factor: 1.00

Width: 34"
Height: 24"
Duct Area: 5.667 Sq Ft
Actual Vel. 1892 CFM 10719

Position	1	2	3	4	5	6	7	8	9	10	11	12	13	14
1 >>>	1314	2351	2436	2231	1348	–412								
2 >>>	2745	2551	2260	2308	1619	335								
3 >>>	2585	2337	2368	2419	1572	461								
4 >>>	2537	2423	2509	2479	1825	796								
5 >>>														
6 >>>														
7 >>>														
8 >>>														
9 >>>														
10 >>>														
Subtotal	9181	9662	9573	9437	6364	1180	0	0	0	0	0	0	0	0

Total FPM	/ Number of Readings	= Average Velocity	X Duct Area	= CFM	Final S.P.
45397	24	1892	5.667	10719	1.62

Remarks:
Filename: CCG050
Test Date: Jan–03
Readings By: GF

Source: Palmetto Air & Water Balance, Inc.

Mailing Address
P.O. Box 5782
Greenville, SC 29606-5782

165 South Hammett Road
Greer, SC 29651

Phone (864)877-6832

Fax (864)877-5490

email palmettoab@aol.com

Palmetto Air&Water
BALANCE
Raising the level of efficiency and comfort

VAV TEST REPORT

PROJECT: CHARTER COMMUNICATIONS
SYSTEM/ UNIT: SCU-1

Box Number	Type	Design CFM	All Boxes Maxed	Damper Position	Flow Corr
FPB-1-1	SUPPLY	1015	1006	67%	0.729
FPB-1-2	SUPPLY	360	*	0%	0.579
FPB-1-3	SUPPLY	1980	1960	66%	0.647
FPB-1-5	SUPPLY	2400	2424	65%	0.753
FPB-1-7	SUPPLY	2400	2156	100%	0.630
FPB-1-8	SUPPLY	1465	1478	70%	0.587
FPB-1-9	SUPPLY	1550	1414	100%	0.597
FPB-1-10	SUPPLY	1550	1630	67%	0.604
VAV-1-2	SUPPLY	1005	1020	55%	0.737
VAV-1-3	SUPPLY	1600	1670	100%	0.755
VAV-1-4	SUPPLY	360	352	54%	0.732
VAV-1-5	SUPPLY	1615	1540	100%	0.462
VAV-1-6	SUPPLY	750	*	0%	0.728
VAV-1-7	SUPPLY	730	710	65%	0.799
TOTAL					

Remarks: *CLOSED FOR DIVERSITY
Filename: CCG048
Test Date: Jan–03
Readings By: GF

Source: Palmetto Air & Water Balance, Inc.

Mailing Address
P.O. Box 5782
Greenville, SC 29606-5782

165 South Hammett Road
Greer, SC 29651

Phone (864)877-6832

Fax (864)877-5490

email palmettoab@aol.com

Palmetto Air&Water
BALANCE
Raising the level of efficiency and comfort

AIR OUTLET TEST REPORT

PROJECT: CHARTER COMMUNICATIONS
SYSTEM/ UNIT: SCU-1

Area Served	Number	Type	Design CFM	Preliminary CFM	Final CFM	Percent Design
128	1	SUPPLY	800	885	868	109%
131	2	SUPPLY	800	1124	852	107%
134	3	SUPPLY	800	363	720	90%
FPB-1-7 TOTAL			2400	2372	2440	102%
134	1	SUPPLY	265	156	292	110%
133	2	SUPPLY	400	416	379	95%
133	3	SUPPLY	400	411	361	90%
132	4	SUPPLY	200	266	220	110%
132	5	SUPPLY	200	309	195	98%
FPB-1-8 TOTAL			1465	1558	1447	99%
135	1	SUPPLY	280	250	270	96%
135	2	SUPPLY	280	222	252	90%
135	3	SUPPLY	280	272	253	90%
135	4	SUPPLY	355	524	361	102%
135	5	SUPPLY	355	445	356	110%
FPB-1-9 TOTAL			1550	1713	1492	96%
135	1	SUPPLY	280	318	279	100%
135	2	SUPPLY	280	285	272	97%
135	3	SUPPLY	280	268	255	91%
135	4	SUPPLY	100	192	105	105%
135	5	SUPPLY	255	336	237	93%
135	6	SUPPLY	355	601	363	102%
FPB-1-10 TOTAL			1550	2000	1511	97%
TOTAL						

Remarks:
Filename: CCG002
Test Date: Jan–03
Readings By: GF

Source: Palmetto Air & Water Balance, Inc.

APPENDIX 03407A

Sample Residential Load Calculation

wrightsoft

Load Short Form
Entire House
Bertie Heating and Air Conditioning

Job:
Date: [redacted]
By: Ryan Johnston

1730 NE 23rd Ave, Gainesville, FL 32609 Phone: 352-331-2005 Email: ryanj@bertieair.com Web: www.bertieair.com License: CAC058522

Project Information

For: [redacted]
Gainesville, FL

Design Information

	Htg	Clg
Outside db (°F)	33	92
Inside db (°F)	70	75
Design TD (°F)	37	17
Daily range	-	M
Inside humidity (%)	50	50
Moisture difference (gr/lb)	32	47

Infiltration

Method	Simplified
Construction quality	Semi-tight
Fireplaces	0

HEATING EQUIPMENT

Make	Carrier
Trade	15 SEER2 HP
Model	GH5SAN43600AA0
AHRI ref	210998688

Efficiency	7.5 HSPF2	
Heating input		
Heating output	34000	Btuh @ 47°F
Temperature rise	27	°F
Actual air flow	1140	cfm
Air flow factor	0.033	cfm/Btuh
Static pressure	0.50	in H2O
Space thermostat		

Capacity balance point = 35 °F

Backup:
Input = 10 kW, Output = 34601 Btuh, 100 AFUE

COOLING EQUIPMENT

Make	Carrier
Trade	15 SEER2 HP
Cond	GH5SAN43600AA0
Coil	FJ4DNXB36L
AHRI ref	210998688

Efficiency	12.0 EER2,14.3 SEER2	
Sensible cooling	25650	Btuh
Latent cooling	8550	Btuh
Total cooling	34200	Btuh
Actual air flow	1140	cfm
Air flow factor	0.045	cfm/Btuh
Static pressure	0.50	in H2O
Load sensible heat ratio	0.87	

ROOM NAME	Area (ft²)	Htg load (Btuh)	Clg load (Btuh)	Htg AVF (cfm)	Clg AVF (cfm)
Bed 1	223	4485	3926	147	175
Family Room	378	5960	5003	195	224
Bath 1	112	2042	844	67	38
Toilet	23	59	48	2	2
WIC	81	218	174	7	8
Laundry	43	123	786	4	35
Dining	89	3111	1941	102	87
Kitchen	198	1968	2547	65	114
Mech	15	0	0	0	0
Bath 3	54	606	337	20	15
Flex Room	156	3358	1773	110	79
Hall	99	0	0	0	0
Bed 3	188	2437	1698	80	76
WIC 3	33	728	264	24	12

Calculations approved by ACCA to meet all requirements of Manual J 8th Ed.

Source: Bertie Heating and Air Conditioning, LLC

WIC 2	30	662	240	22	11
Bath 2	58	364	205	12	9
Hall 2	35	0	0	0	0
Bedroom 2	204	6028	4398	198	197
Foyer	63	2611	1327	86	59
Entire House	2079	34759	25513	1140	1140
Other equip loads		0	0		
Equip. @ 0.97 RSM			24722		
Latent cooling			3680		
TOTALS	2079	34759	28402	1140	1140

Calculations approved by ACCA to meet all requirements of Manual J 8th Ed.

wrightsoft Right-Suite® Universal 2024 24.0.01 RSU59482
...ureate Village Lot 64 - Derby - Manual JD&S.rup Calc = MJ8 Front Door faces: W

2024-Jul-02 15:07:48
Page 2

Source: Bertie Heating and Air Conditioning, LLC

wrightsoft

Project Summary
Entire House
Bertie Heating and Air Conditioning

Job:
Date: [redacted]
By: Ryan Johnston

1730 NE 23rd Ave, Gainesville, FL 32609 Phone: 352-331-2005 Email: ryanj@bertieair.com Web: www.bertieair.com License: CAC058522

Project Information

For: [redacted]
Gainesville, FL

Notes:

Design Information

Weather: Gainesville Rgnl, FL, US

Winter Design Conditions

Outside db	33	°F
Inside db	70	°F
Design TD	37	°F
Ventilation Method	MJ8	

Heating Summary

Structure	27548	Btuh
Ducts (R-6.0)	7212	Btuh
Central vent (0 cfm)	0	Btuh
Humidification	0	Btuh
Piping	0	Btuh
Equipment load	34759	Btuh

Infiltration

Method	Simplified
Construction quality	Semi-tight
Fireplaces	0

	Heating	Cooling
Area (ft²)	2079	2079
Volume (ft³)	20790	20790
Air changes/hour	0.22	0.11
Equiv. AVF (cfm)	76	38

Heating Equipment Summary

Make	Carrier
Trade	15 SEER2 HP
Model	GH5SAN43600AA0
AHRI ref	210998688

Efficiency	7.5 HSPF2	
Heating input		
Heating output	34000	Btuh @ 47°F
Temperature rise	27	°F
Actual air flow	1140	cfm
Air flow factor	0.033	cfm/Btuh
Static pressure	0.50	in H2O
Space thermostat		

Capacity balance point = 35 °F

Backup:
Input = 10 kW, Output = 34601 Btuh, 100 AFUE

Summer Design Conditions

Outside db	92	°F
Inside db	75	°F
Design TD	17	°F
Daily range	M	
Relative humidity	50	%
Moisture difference	47	gr/lb

Sensible Cooling Equipment Load Sizing

Structure	17701	Btuh
Ducts (R-6.0)	7812	Btuh
Central vent (0 cfm)	0	Btuh
Blower	0	Btuh
Use manufacturer's data	n	
Rate/swing multiplier	0.97	
Equipment sensible load	24722	Btuh

Latent Cooling Equipment Load Sizing

Structure	2209	Btuh
Ducts	1471	Btuh
Central vent (0 cfm)	0	Btuh
Equipment latent load	3680	Btuh
Equipment Total Load (Sen+Lat)	28402	Btuh
Req. total capacity at 0.75 SHR	2.7	ton

Cooling Equipment Summary

Make	Carrier
Trade	15 SEER2 HP
Cond	GH5SAN43600AA0
Coil	FJ4DNXB36L
AHRI ref	210998688

Efficiency	12.0 EER2,14.3 SEER2	
Sensible cooling	25650	Btuh
Latent cooling	8550	Btuh
Total cooling	34200	Btuh
Actual air flow	1140	cfm
Air flow factor	0.045	cfm/Btuh
Static pressure	0.50	in H2O
Load sensible heat ratio	0.87	

Calculations approved by ACCA to meet all requirements of Manual J 8th Ed.

Source: Bertie Heating and Air Conditioning, LLC

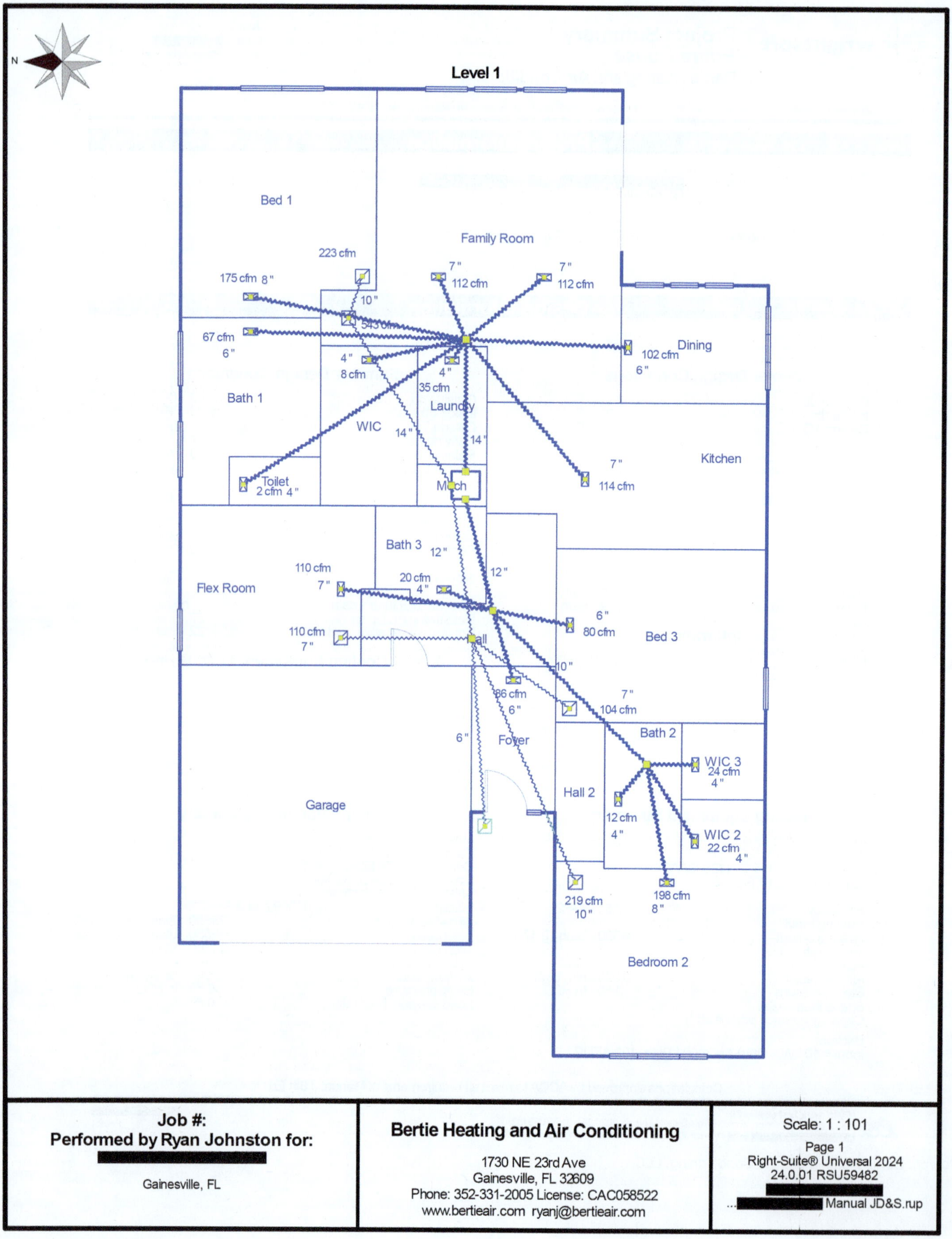

Job #:
Performed by Ryan Johnston for:
[redacted]
Gainesville, FL

Bertie Heating and Air Conditioning
1730 NE 23rd Ave
Gainesville, FL 32609
Phone: 352-331-2005 License: CAC058522
www.bertieair.com ryanj@bertieair.com

Scale: 1 : 101
Page 1
Right-Suite® Universal 2024
24.0.01 RSU59482
[redacted]
...[redacted] Manual JD&S.rup

Source: Bertie Heating and Air Conditioning, LLC

wrightsoft

Manual S Compliance Report

Entire House

Bertie Heating and Air Conditioning

Job:
Date: [redacted]
By: **Ryan Johnston**

1730 NE 23rd Ave, Gainesville, FL 32609 Phone: 352-331-2005 Email: ryanj@bertieair.com Web: www.bertieair.com License: CAC058522

Project Information

For: [redacted]
Gainesville, FL

Cooling Equipment

Design Conditions

Outdoor design DB:	91.9°F	Sensible gain:	25513	Btuh		Entering coil DB:	77.1°F
Outdoor design WB:	76.2°F	Latent gain:	3680	Btuh		Entering coil WB:	63.5°F
Indoor design DB:	75.0°F	Total gain:	29192	Btuh			
Indoor RH:	50%	Estimated airflow:	1140	cfm			

Manufacturer's Performance Data at Actual Design Conditions

Equipment type: Split ASHP
Manufacturer: Carrier Model: GH5SAN43600AA0+FJ4DNXB36L
Actual airflow: 1140 cfm
Sensible capacity: 25650 Btuh 101% of load
Latent capacity: 8550 Btuh 232% of load
Total capacity: 34200 Btuh 117% of load SHR: 75%

Heating Equipment

Design Conditions

Outdoor design DB:	33.4°F	Heat loss:	34759	Btuh	Entering coil DB:	68.6°F
Indoor design DB:	70.0°F					

Manufacturer's Performance Data at Actual Design Conditions

Equipment type: Split ASHP
Manufacturer: Carrier Model: GH5SAN43600AA0+FJ4DNXB36L
Actual airflow: 1140 cfm
Output capacity: 34000 Btuh 98% of load Capacity balance: 35 °F
Supplemental heat required: 759 Btuh Economic balance: -99 °F

Backup equipment type: Elec strip
Manufacturer: Model:
Actual airflow: 1140 cfm
Output capacity: 10.1 kW 100% of load Temp. rise: 50 °F

Meets all requirements of ACCA Manual S.

Source: Bertie Heating and Air Conditioning, LLC

wrightsoft

Duct System Summary
Entire House
Bertie Heating and Air Conditioning

Job:
Date: [redacted]
By: Ryan Johnston

1730 NE 23rd Ave, Gainesville, FL 32609 Phone: 352-331-2005 Email: ryanj@bertieair.com Web: www.bertieair.com License: CAC058522

Project Information

For: [redacted]
Gainesville, FL

	Heating		Cooling	
External static pressure	0.50	in H2O	0.50	in H2O
Pressure losses	0.22	in H2O	0.22	in H2O
Available static pressure	0.28	in H2O	0.28	in H2O
Supply / return available pressure	0.163 / 0.117	in H2O	0.163 / 0.117	in H2O
Lowest friction rate	0.097	in/100ft	0.097	in/100ft
Actual air flow	1140	cfm	1140	cfm
Total effective length (TEL)		288 ft		

Supply Branch Detail Table

Name		Design (Btuh)	Htg (cfm)	Clg (cfm)	Design FR	Diam (in)	H x W (in)	Duct Matl	Actual Ln (ft)	Ftg.Eqv Ln (ft)	Trunk
Bath 1	h	2042	67	38	0.116	6.0	0x 0	VIFx	25.0	115.0	st1
Bath 2	h	364	12	9	0.101	4.0	0x 0	VIFx	27.0	135.0	st3
Bath 3	h	606	20	15	0.128	4.0	0x 0	VIFx	12.1	115.0	st2
Bed 1	c	3926	147	175	0.116	8.0	0x 0	VIFx	25.3	115.0	st1
Bed 3	h	2437	80	76	0.127	6.0	0x 0	VIFx	13.8	115.0	st2
Bedroom 2-A	h	6028	198	197	0.097	8.0	0x 0	VIFx	32.4	135.0	st3
Dining	h	3111	102	87	0.120	6.0	0x 0	VIFx	21.0	115.0	st1
Family Room	c	2502	98	112	0.126	7.0	0x 0	VIFx	14.4	115.0	st1
Family Room-A	c	2502	98	112	0.124	7.0	0x 0	VIFx	16.6	115.0	st1
Flex Room	h	3358	110	79	0.121	7.0	0x 0	VIFx	19.3	115.0	st2
Foyer	h	2611	86	59	0.127	6.0	0x 0	VIFx	13.5	115.0	st2
Kitchen	c	2547	65	114	0.118	7.0	0x 0	VIFx	22.6	115.0	st1
Laundry	c	786	4	35	0.129	4.0	0x 0	VIFx	11.3	115.0	st1
Toilet	c	48	2	2	0.113	4.0	0x 0	VIFx	28.6	115.0	st1
WIC	c	174	7	8	0.124	4.0	0x 0	VIFx	16.7	115.0	st1
WIC 2	h	662	22	11	0.099	4.0	0x 0	VIFx	30.3	135.0	st3
WIC 3	h	728	24	12	0.100	4.0	0x 0	VIFx	27.3	135.0	st3

Supply Trunk Detail Table

Name	Trunk Type	Htg (cfm)	Clg (cfm)	Design FR	Veloc (fpm)	Diam (in)	H x W (in)	Duct Material	Trunk
st3	Peak AVF	255	228	0.097	468	10.0	0 x 0	VinlFlx	st2
st2	Peak AVF	551	458	0.097	701	12.0	0 x 0	VinlFlx	
st1	Peak AVF	589	682	0.113	638	14.0	0 x 0	VinlFlx	

Source: Bertie Heating and Air Conditioning, LLC

Return Branch Detail Table

Name	Grille Size (in)	Htg (cfm)	Clg (cfm)	TEL (ft)	Design FR	Veloc (fpm)	Diam (in)	H x W (in)	Stud/Joist Opening (in)	Duct Matl	Trunk
rb6	0x 0	223	223	77.3	0.151	409	10.0	0x 0		VIFx	rt1
rb4	0x 0	219	207	120.1	0.097	402	10.0	0x 0		VIFx	rt2
rb3	0x 0	104	88	109.7	0.107	388	7.0	0x 0		VIFx	rt2
rb5	0x 0	110	79	110.6	0.106	412	7.0	0x 0		VIFx	rt2
rb2	0x 0	484	543	84.2	0.139	691	12.0	0x 0		VIFx	rt1

Return Trunk Detail Table

Name	Trunk Type	Htg (cfm)	Clg (cfm)	Design FR	Veloc (fpm)	Diam (in)	H x W (in)	Duct Material	Trunk
rt1	Peak AVF	707	766	0.139	716	14.0	0 x 0	VinlFlx	
rt2	Peak AVF	433	374	0.097	552	12.0	0 x 0	VinlFlx	

Source: Bertie Heating and Air Conditioning, LLC

Module Review Answer Key

MODULE 01 (03308)

Answer	Section
1. c	1.1.0; *Figure 1*
3. d	1.2.1
5. c	1.3.0
7. b	2.1.2
9. d	2.2.2
11. b	3.1.3
13. a	3.2.2
15. d	3.3.5

MODULE 02 (03404)

Answer	Section
1. d	1.1.0; *Figure 2*
3. b	1.2.2
5. b	1.3.0; *Figure 12*
7. c	1.5.0
9. c	1.7.1

MODULE 03 (03405)

Answer	Section
1. d	1.0.0; *Figure 1*
3. a	1.1.0
5. b	1.2.1
7. d	1.3.3
9. b	2.3.4
11. b	3.1.0
13. b	3.2.1
15. a	4.1.2
17. c	4.2.2
19. b	5.1.3

MODULE 04 (03402)

Answer	Section
1. a	1.1.0; *Figure 2*
3. b	1.2.2
5. b	2.1.3
7. d	2.3.1
9. b	3.1.0
11. b	3.2.4
13. c	3.4.1; *Figure 41*
15. a	3.4.3

MODULE 05 (03406)

Answer	Section
1. a	1.1.0
3. d	1.2.0
5. d	1.2.0
7. d	2.2.3
9. c	2.3.1

MODULE 06 (03401)

Answer	Section
1. b	1.1.0
3. d	1.3.1
5. c	1.3.2
7. b	1.4.1
9. d	1.4.3
11. d	1.5.0
13. b	1.7.0
15. c	2.2.0

MODULE 07 (03407)

Answer	Section
1. c	1.1.1
3. b	2.0.0
5. b	2.1.4
7. a	2.2.2
9. c	3.0.0
11. d	3.2.0
13. b	4.2.2
15. b	4.2.4

MODULE 08 (03408)

Answer	Section
1. b	1.1.2
3. c	1.1.4
5. b	1.2.1
7. c	1.3.2
9. d	2.2.2
11. a	3.0.0
13. c	3.3.0
15. c	4.2.2

MODULE 09 (03409)

Answer	Section
1. b	1.1.1
3. b	1.1.4
5. c	1.2.0
7. b	1.4.0
9. d	2.1.0
11. b	2.2.0
13. d	2.2.3
15. c	2.5.1

MODULE 10 (03305)

Answer	Section
1. d	1.1.0
3. c	1.2.1
5. a	2.1.1
7. b	2.1.3
9. c	2.2.1
11. a	2.2.3
13. b	2.4.2
15. c	3.2.1; *Figure 47*

MODULE 11 (03201)

Answer	Section
1. b	1.0.0
3. c	1.1.1
5. d	1.2.0
7. c	1.3.0
9. b	1.4.2

MODULE 12 (46100)

Answer	Section
1. b	1.1.1
3. d	2.1.3
5. c	2.3.1
7. b	3.1.2
9. c	3.2.5
11. a	4.1.0
13. c	4.3.1
15. d	4.5.2
17. c	5.1.2
19. b	5.3.1

GLOSSARY

Accomplishment: An individual's need to set challenging goals and achieve them.

Accountability: In relation to job completion, accepting responsibility and being personally answerable for the results.

Active solar hydronic heating systems: Solar heating systems that pump fluids through exposed solar collectors to absorb heat and are equipped with controls.

AED (automated external defibrillator): A device used to help those experiencing sudden cardiac arrest.

A_k factor: The net free area of an air distribution outlet or an inlet stated in square feet. The terminal device manufacturer specifies this value to aid in airflow calculations.

Algorithm: A set of mathematical rules and logic statements used in a computer or microprocessor to solve a specific problem. In HVAC applications, algorithms are typically used to control a wide range of system functions based on varying conditions.

Alkalinity: A measure of water's ability to neutralize or resist a change to increased acidity.

Anhydrous: Free of water.

As-built drawings: A revised set of drawings that reflect all changes made during the construction process and show the exact specifications of the completed work.

Authority: The right to act or make decisions in carrying out an assignment.

Backwashing: A procedure used periodically to reverse the normal direction of waterflow through a strainer or filter to flush accumulated debris.

Baud rate: The rate at which information is transmitted across communication lines.

Behavior: The way a person responds to a specific situation.

Bits: Abbreviation for binary digits; the smallest element of data that a computer typically handles. A bit represents an off or on state (zero or one) in a binary system.

Bleed-off: Consistently draining (bleeding off) a controlled amount of water from a water circuit to carry away minerals and other impurities that remain as water evaporates from a cooling tower or evaporative condenser, reducing the concentration of the minerals in the remaining water.

Block load calculation: A load calculation that considers the entire structure as one single area, essentially considering the structure to be a single room.

Boiler horsepower (BHP): A capacity rating for boilers, representing the heat necessary to vaporize 34.5 lb of water per hour at 212°F (100°C) at atmospheric pressure. This is equivalent to 33,475 Btuh or 9.803 kW of energy.

Boyle's law: States that, given a constant temperature, the pressure on a confined gas varies inversely with the volume the gas occupies. Similarly, at a constant temperature, the volume a confined gas occupies varies inversely with the pressure applied.

Buffering capacity: A measure of water's ability to resist reductions in its pH, which lead to acidity.

Building management system (BMS): A centralized, computer-controlled system for managing the various systems in a building, such as HVAC, lighting, and security. Also known as a *building automation system (BAS)*.

Bus: A data communication path, often relying on a multiwire cable, that links all the components in a computer or between computers.

Carryover: Refers to any contaminant that is transferred along with steam, typically from a boiler to a system, but contaminants can also be transferred into a boiler that is at a lower pressure than the steam header. Carryover can be comprised of contaminants in solid, liquid, or vapor form.

Catalytic combustor: A ceramic structure resembling a honeycomb that provides a substrate for the high-temperature combustion of smoke and particulates, resulting in a cleaner discharge and increased heat production.

Cavitation: A condition caused by liquids reaching their vapor pressure at the pump inlet, causing bubbles to form and then collapse violently as they pass through the pump.

Change order: The documentation of a change in the originally agreed-upon and contracted scope of work, usually related to additions or deletions from the scope with corresponding alterations to the completion date and/or the contract value.

Charles' law: States that, given a constant pressure, the volume of a confined gas varies directly with a change in temperature. Similarly, given a constant volume of gas, the pressure varies directly with a change in temperature.

Chilled-beam cooling systems: Cooling systems that employ coils mounted near the ceiling through which chilled water flows. Passive systems rely on convection currents for cooling, while active systems use mechanical ventilation to increase airflow over the beams.

Colloidal substance: A jelly-like material comprised of very small, insoluble, non-diffusible particles that are larger than molecules, but small enough to remain suspended in a fluid without settling.

Commodities: Raw materials or agricultural products, such as iron and grain, that are commercially bought, sold, and traded.

Communication: The process of exchanging facts, information, and ideas. Communication comes in many forms, including verbal, written, visual, and nonverbal.

Concentration: An indication of the strength or relative amount of a substance present in a volume of water.

Construction safety plan: A document that outlines the procedures, rules, and regulations that are or will be put in place to protect workers during a construction project.

Control point: The term for each input and output device wired to a digital controller.

Control zone: A building, group of rooms, single room, or part of a room controlled by a dedicated thermostat.

Coolers: Enclosed refrigerated storage systems maintained at temperatures above freezing. Coolers range in size from reach-in cabinets to drive-in warehouses.

Coordination drawings: Elevation, location, and other drawings produced for a project by the individual contractors for each craft to prevent a conflict regarding the installation of each craft's materials and equipment. Coordination drawing development evolves through a series of meetings between the various contractors.

Cost control: The process that helps keep expenses under control by managing labor, material, and other costs to ensure that the project finishes on budget.

CPR (cardiopulmonary resuscitation): A lifesaving technique that is used when someone's heartbeat or breathing has stopped.

Creosote: A black, sticky combustible byproduct created when wood is burned, especially at lower temperatures, that collects on flue and chimney surfaces.

Crew leader: A person appointed leader of a team or crew, usually an experienced craft professional who has demonstrated leadership qualities. Also known as frontline leader, frontline supervisor, lead person, or foreman.

Critical path: In manufacturing, construction, and other types of creative processes, the required sequence of tasks that directly controls the ultimate completion date of the project.

Cryogenic fluids: Fluids that exist as a liquid or gas at temperatures below –250°F (–157°C).

Cryogenics: Field of engineering and refrigeration related to extremely low temperature applications, typically below –250°F (–157°C).

Cut lists: Information sheets that are derived from shop drawings. They are the shop's guide for fabricating items such as duct sections and pipe subassemblies.

Cycle of concentration: A measurement of the ratio of dissolved solids contained in water to the volume of dissolved solids contained in the makeup water source. The term *cycle* is used to express the ratio.

Dalton's law: States that the pressure of a mixture of confined gases is equal to the sum of the partial pressures of the individual component gases. The partial pressure is the pressure that each gas would exert if it alone occupied the same volume of the mixture at the same temperature.

Deadband: For a chiller, the tolerance around the chilled-water temperature setpoint. For example, a 1°F deadband controls the water temperature to within ±0.5°F of the control point temperature (0.5°F + 0.5°F = 1°F deadband).

Decision-making: The action or process of making a choice. Decision-making involves starting an action, stopping one, or choosing a different plan or action.

Deck: A shelf, pan, or rack that supports refrigerated items stored in merchandising coolers and display cases.

Desiccants: Substances such as silica gel or calcium chloride ($CaCl_2$) that readily attract and hold water and water vapor, limited by the water-holding capacity of the material.

Detail drawing: An enlarged and detailed view of a specific feature or area of work that provides more elaborate information than is available in a large format.

Dew point: The temperature at which air becomes saturated and water vapor begins to condense and fall freely from the air.

Digital controller: An electronic device that uses an input module, a microprocessor to determine appropriate responses to sensed conditions, and an output module to perform control functions.

Direct digital control (DDC): A system that uses digital controllers to maintain precise control over system components; an alternate term for *building management system* and *building automation system*.

Direct-fired makeup air unit: A furnace that heats outdoor air and replaces indoor air that is exhausted through ventilation systems, helping maintain building pressure while warming the space in wintry conditions.

Dissolved gases: Gases such as oxygen and carbon dioxide that are dissolved in a water sample.

Dissolved solids: Solid substances such as calcium, magnesium, chloride, and sulfates dissolved as tiny particles in a water sample. Dissolved solids contribute to corrosion and scale formation in a water system.

Distillation: The process of purifying a liquid by successive operations to heat and vaporize the liquid, then cool it to allow the vapor to condense back to its liquid form. Solid contaminants remain behind as the liquid is initially vaporized.

Distractions: Anything that disrupts the communication flow between sender and receiver. Also called *noise*.

Duty: A work task an employee is expected to perform.

Electrolysis: The process of changing the chemical composition of a metal by passing an electrical current through it that initiates an exchange of molecules with a second metal.

Electrolyte: A substance that conducts an electric current between positive and negative ions when dissolved in water.

Elevation: A view that depicts a vertical side of a building or room, usually designated by the direction that side is facing; for example, right, left, east, or west.

Enthalpy: The total heat content (sensible and latent) of a refrigerant or other substance.

Equal-friction method: A method of sizing air distribution systems by designing for a consistent amount of pressure loss per unit length of duct (usually 100').

Estimating: The process of calculating the cost of a project.

Ethernet: A system of wired connections used to create a local area network, including the communication protocol to be used and the method of avoiding simultaneous information transfer by two or more network devices.

Ethics: The moral principles that guide an individual's or organization's actions when dealing with others.

Evaporative coolers: Cooling units that cool air through the evaporation of water, typically consisting of a blower, wetted media, and a water source.

Feasibility study: An analysis that considers all of a project's relevant factors—including budget, technical elements, legal aspects, and scheduling—to determine the likelihood that the project can be completed successfully.

Firetube boiler: A boiler designed for the hot flue gases to flow through tubes that are immersed in the water to be heated.

Firmware: Software permanently stored on a device's read-only memory (ROM) to guide basic operation.

First aid: Help given to a sick or injured worker on the worksite immediately after an injury occurs.

Flash economizer: An energy conservation arrangement used in chillers that separates vapor (flash gas) from liquid leaving the condenser, then routes the vapor directly to the second stage of a multistage centrifugal chiller, bypassing the evaporator and the first stage of compression.

Floor plan: A plan view showing a horizontal section of the floor arrangement, usually conceived as if the viewer is positioned just beneath the ceiling of the same floor.

Freezer: Refrigerated enclosure or apparatus that operates below freezing and either freezes products or maintains frozen products in storage.

Gasification: A process that converts any carbon-based material into a fuel gas called synthesis gas, or syngas, that can also be burned, significantly reducing the emissions that would otherwise result from solid-fuel combustion.

Gateway: A network device that allows two or more applications or networks operating under different protocols to share information.

Harassment: Unwelcome conduct that is based on race, color, religion, sex, national origin, age, disability, or genetic information. Harassment includes verbal or physical conduct that impacts a person's work performance or creates an intimidating, hostile, or offensive work environment.

Hardness: A measure of the amount of calcium and magnesium contained in water, which affects the formation of scale in a water system. As hardness increases, the potential for scale development increases.

Hazards: Conditions or activities that, if left uncontrolled, have the potential for harm.

Homogeneous: Describes a mixture, substance, or material that is uniform throughout its composition.

Hot/cold aisle configuration: A method used to install equipment cabinets in computer rooms to manage the flow of warm and cool air out of and into the cabinets.

Hypertext Transfer Protocol (HTTP): The base protocol used for transferring data between clients and servers on the internet.

Immiscible: Unable to effectively mix with or dissolve in another liquid.

Impairment: The point where someone's intake of substances affects their ability to perform appropriately.

Incidents: Unplanned events that result in property damage, fatalities, injuries, illnesses, or near misses.

Indirect solar hydronic heating system: An active solar hydronic heating system that heats fluid using a double-walled heat exchanger to prevent potentially toxic fluid in the solar collectors from contaminating a potable water system.

Internet Protocol address (IP address): A unique address that electronic devices use to identify themselves and communicate with each other on the internet.

Job description: A description of the scope and responsibilities of a worker's job so that the individual and others understand what the job entails.

Job diary: A written record that a supervisor maintains periodically (usually daily) of the events, communications, observations, and decisions made during the course of a project.

Job safety analysis (JSA): A technique that focuses on job tasks as a way to identify hazards before they occur. It focuses on the relationship between the worker, the task, the tools, and the work environment. Also known as *job hazard analysis (JHA)* or *activity hazard analysis (AHA)*.

King valve: The manual refrigerant shutoff valve located at the outlet of a receiver, or at the outlet of the condenser for systems without a receiver.

Latent heat defrost: A method used in hot-gas defrost applications that relies on the heat absorbed by one or more active evaporators as a source of heat to defrost another evaporator.

Layup: Refers to the status of a boiler when it is shut down for a significant period and not considered immediately operable.

Legionnaires' disease: A serious type of pneumonia developing 2 to 14 days after exposure to the airborne bacteria, although many healthy people exposed do not get sick. The *Legionella* bacteria was identified and named following a 1976 outbreak at an American Legion convention in Philadelphia.

Level 4 schedules: Schedules that offer additional detail to provide working schedules to the crews doing the work, usually within three to six weeks of completing specific work.

Lithium bromide (LiBr): A compound salt solution containing lithium and bromine that readily absorbs moisture.

Local area network (LAN): A server/client network connecting multiple computers and devices within a building or building complex.

Local interface device: A portable device with a keypad and display for data entry, used to connect directly to a digital controller via cable.

Longitudinal section: A section drawing on which the cut is made parallel to the longest dimension of a structure.

Lookahead schedule: For planning and scheduling work that is happening in the short-term, usually within one to three weeks; used for anticipating material, labor, tool, and other resource requirements, as well as identifying potential schedule conflicts or other problems. Also called a *short-term schedule*.

Lyophilization: A dehydration process that begins with freezing a substance, then reducing the pressure while adding heat to drive out moisture by sublimation; also known as *freeze drying*.

Monel®: An alloy that is very resistant to corrosion, made from nickel, copper, iron, manganese, silicon, and carbon.

Morale: An individual's emotional outlook toward work and the level of satisfaction gained while performing the jobs assigned.

Motivate: To make someone feel determined or enthusiastic or to make someone behave in a particular way.

Network: A structured linking of computers and other devices to allow communication and data sharing, which may rely on either wired or wireless connections.

Nonverbal communication: Things the speaker and the receiver see or hear while communicating. Examples include facial expressions, body movements, hand gestures, tone of voice, and eye contact.

Occupational Safety and Health Administration (OSHA): A national public health agency that is part of the US Department of Labor. To promote a safe and healthy work environment, OSHA issues standards and rules for working conditions, facilities, equipment, tools, and work processes.

Open protocol: A communication standard that is not owned by a particular entity and is not limited to specific products, allowing interoperability between different devices and manufacturers.

Organization: A group of people working together.

Organizational chart: A diagram that shows how the various management and operational responsibilities relate to each other within an organization. Named positions are ranked top to bottom, from those functional units with the most and broadest authority to those with the least. Lines connect positions and indicate chains of authority and other relationships.

Oxygen scavengers: Chemicals added to boiler feed water that bond with dissolved oxygen to inhibit corrosion within the system.

Passive solar heating systems: Solar heating arrangements characterized by a lack of moving parts, pumps, and controls.

Peer: Someone who is in a position or level in an organization equal or similar to yours.

Personal growth: Improving one's talents and potential, both in and out of the workplace.

pH: Abbreviation for *potential of hydrogen*, which is a measure of the alkalinity or acidity of water. The pH scale ranges from 0 (extremely acidic) to 14 (extremely alkaline), with a pH of 7 considered neutral. The pH defines the relative concentration of hydrogen and hydroxide ions.

Plan views: Views of an object or structure from overhead. A floor plan is a common example of a plan view.

Planning: The process of laying out all details of a construction project before construction begins.

Policy: A general statement establishing guidelines for a specific activity.

Problem-solving: A process used to identify possible solutions to a situation.

Procedures: Formal instructions to carry out and meet policies.

Product-integrated controllers (PICs): Digital controllers designed to control specific equipment, typically installed by the equipment manufacturer at the factory level; also known as a *product-specific controller*.

Production: The amount of construction materials put in place. It is the quantity of materials installed on a job.

Productivity rate: A measure of how much work gets done over a period of time, also called *output per labor hour*.

Professional growth: Gaining new skills and work experience that can help one to reach a goal in one's career.

Profit: The amount of money that the contractor will make after paying all the direct and indirect costs.

Project delivery methods: Different ways to organize the services of development, planning, and construction in order to execute a project.

Project manager: Person responsible for managing one or more projects.

Protocol: An established set of rules that specify how to format and exchange data transmitted between devices in a network, allowing connected devices to communicate with each other, regardless of differences in the internal structure and design.

Psychrometrics: The study of the physical and thermodynamic properties of air-vapor mixtures.

Pulse-width modulation (PWM): A common control technique based on the conversion of an analog signal into discrete parts. The signal produced appears as a chain of pulses that form square waves.

Queen valve: A manual refrigerant shutoff valve located at the inlet of a receiver.

Reagents: Substances that facilitate or initiate a predictable chemical reaction, used to test for the presence of other substances.

Recognition: Acknowledgement of an employee's accomplishments through verbal and nonverbal communications.

Recycle shutdown mode: A chiller mode of operation in which automatic chiller shutdown occurs when the compressor is operating at minimum capacity, but the chilled-water temperature drops below the chilled-water temperature setpoint. The chilled-water pump(s) continues running.

Redline drawings: Marked-up versions of construction drawings documenting any changes made during construction or installation.

Responsibility: The act of accepting a task and completing it to the best of one's ability.

Rewards: Tangible items given to an employee as recognition for good work.

Riser diagram: A one-line schematic typically depicting system connections of a piping, plumbing, or electrical system, shown as an elevation and often incorporating more than one floor of the building.

Root cause: A core issue or fundamental reason for the occurrence of a problem.

Root cause analysis (RCA): A method of problem-solving used for identifying the root causes of faults or problems.

Runaround loop: A closed-loop energy recovery system in which finned-tube water coils in supply and exhaust airstreams, connected by piping, exchange heat.

R-value: A number identifying the thermal resistance of a given thickness of insulating material; higher numbers reflect greater resistance to heat flow.

Satellite compressor: A separate compressor that uses the same refrigerant source as that used by a related group of parallel compressors but operates independently and serves a refrigeration load that requires a lower temperature than that being maintained by the related parallel compressors.

Scale: A layered coating of mineral matter that separates from water, settling on and adhering to internal water-system surfaces. Scale development is often accelerated by rising or falling water temperatures.

Schedules: Tables that identify and specify the types and sizes of components or equipment required for the construction of a building.

Scheduling: Involves defining which activities happen in what order and assigning dates and times to those tasks.

Scope: The list of the commitments and activities that all contractors, subcontractors, and suppliers are expected to do during a project. Also called *scope of work*.

Secondary coolant: A substance used as a heat transfer fluid that does not change its state as it transfers heat; also referred to as a *secondary refrigerant*. Examples include water/glycol mixtures and chilled water.

Section drawing: A drawing that depicts a feature of a building as if it were cut through the middle, exposing otherwise unseen details.

Shock treatments: Water system treatments with a single substantial dose of chemical(s), as opposed to consistent dosing over time.

Shop drawings: Drawings that indicate how to fabricate and/or assemble specific components and subassemblies of a construction project. Shop drawings may be drafted from the construction drawings of a project by one or more of the subcontractors or be provided by a manufacturer.

Sick building syndrome (SBS): A physical condition that exists when more than 20% of a building's occupants complain during a two-week period of common symptoms, including headaches, fatigue, nausea, eye irritation, and throat irritation, which are alleviated by leaving the building and are not confirmed to be caused by any specific contaminant.

Site plan: A civil drawing that indicates the location and position of a structure on its construction site.

Site-specific safety plan (SSSP): A safety plan that includes a focused set of information that accounts for a job site's unique hazards.

Situational leadership: A leadership style that involves adjusting to the specific situation or task and the leadership needs of the group.

Specific density: The weight (mass) of air per cubic foot. Per ASHRAE standards, standard air weighs 0.075 lb/ft^3.

Specific heat: The amount of heat required to change the temperature of one mass unit of a substance by one degree.

Specific volume: The space occupied by one pound of standard air. Per ASHRAE standards, one pound of standard air occupies a volume of 13.33 ft^3.

Static pressure: The pressure exerted uniformly in all directions by air within a duct system. Total pressure minus velocity pressure equals static pressure.

Stratification: The natural layering of undisturbed room air due to temperature differences. Warmer air is concentrated at the ceiling while cooler air is concentrated at the floor. Mechanical ventilation interrupts or eliminates natural stratification.

Sublimation: A change of state that progresses directly from a solid to a gas, such as the changing of ice to water vapor, without passing through the liquid state at any point.

Submittals: Written and illustrated documents related to project equipment and materials provided to the general contractor for review by the subcontractor responsible for the item to ensure the submitted components meet the specifications.

Substance abuse: The inappropriate overuse of drugs and chemicals, whether they are legal or illegal.

Superheated steam: Steam at a temperature above the saturation temperature corresponding to its pressure.

Superintendent: An on-site supervisor who is responsible for one or more crew leaders or frontline supervisors.

Suspended solids: Individual particles suspended in water that give the water a cloudy appearance.

Takeoff: The result of reviewing, measuring, and quantifying all the materials and equipment needed for a project, as indicated by the drawings. The information from a takeoff is then used to initiate purchases.

Tenant billing: The process used to track resource usage and charge a building tenant for specified uses of the building's HVAC system, as well as other systems.

Thermal conductivity: A number reflecting the heat flow per hour, in Btuh, through one square foot of 1" thick homogeneous material per 1°F of temperature difference between the two surfaces. The actual conductance of the material is changed by the actual temperature difference and material thickness.

Thermal economizer: An energy conservation arrangement used in chillers in which warm liquid refrigerant is brought into contact with the coldest water tubes in a condenser or heat exchanger. This causes additional subcooling of the condensed refrigerant.

Thermal shock: A potentially powerful and destructive response to major and rapid changes in temperature due to expansion or contraction, resulting in metal cracking or distorting.

Thermosiphon system: A system in which a difference in fluid temperatures in a system causes fluid movement, with warmer fluids moving toward the top of the piping circuit.

Three-way communication: A form of communication with an initial message from a sender, which is then repeated by the receiver to confirm accuracy, and then validated by the sender. Also called *repeat-back communication*.

Toolbox talks: Short, informal safety meetings held at the start of each day or shift. Sometimes referred to as *tailgate meetings* or *safety briefings*.

Total heat rejection (THR): A value used to rate condensers. It represents the total heat transferred while desuperheating, condensing, and subcooling a refrigerant as it flows through the condenser.

Total pressure: The sum of the static pressure and the velocity pressure in an air duct; the pressure produced by the supply air fan or blower.

Total solids: A measure of the volume of both dissolved and suspended solids contained in a water sample.

Trait: A distinguishing quality of a person.

Transverse section: A section drawing on which the cut is made along the short dimension of the structure.

Traverse readings: A series of velocity readings taken at several points over the cross-sectional area of a duct or grille.

Type HT vent: A metal venting product capable of withstanding temperatures up to 1,000°F (538°C), commonly used to vent wood-fired stoves and furnaces.

Type PL vent: A metal venting product specifically designed for stoves that burn wood pellets or corn.

Unit cooler: A packaged refrigeration evaporator assembly containing the evaporator coil, expansion device, defrost heaters (freezers), condensate drain pan, and the indoor fans or blowers commonly used in chill rooms, walk-in coolers, and walk-in freezers.

Upskilling: A type of training that provides someone with more advanced skills through additional education.

U-value: A heat-transfer coefficient rating of how well a building assembly, such as a wall consisting of multiple materials, conducts heat. The U-value is used as a multiplier in load calculations.

Valance cooling system: A cooling system in which chilled water is circulated through finned-tube radiators located near the ceiling around the perimeter of a room. Convection currents move the cooled air instead of a blower assembly. A decorative valance conceals the system.

Variable air volume (VAV) system: An air distribution system that controls the temperature within a control zone by varying the volume of the conditioned supply air flowing into the zone.

Variable volume and temperature (VVT) system: A VVT system is one that delivers a variable volume of air to each controlled zone, as the load dictates. The operating mode of the unit (heating or cooling) varies depending on the requests from the zones.

Velocity pressure: The pressure in a duct that results from the movement of the air; total pressure minus the static pressure equals velocity pressure.

Verbal communication: Messages that are sent using voice or other audible sounds. Also called *spoken communication* or *oral communication*.

Visual communication: The process of using visual elements, such as photos, graphics, colors, and fonts to convey messages.

Watertube boiler: A boiler designed for the water to be heated to flow through tubes while the hot flue gases and byproducts of combustion flow around the outside of the tubes.

Wide area network (WAN): A server/client network spread over a larger geographical area, generally with a radius of half a mile or more.

Work breakdown structure (WBS): A method of organizing the entire scope of work in a logical manner based on how the work will be executed.

Workers' compensation: Insurance that companies have to cover medical expenses, lost wages, and other costs for employees who are injured or become ill while completing work for their job.

Written communication: The process of sending messages or information using words, whether digitally or physically.

Zeolite: A natural volcanic mineral that possesses a rare natural negative charge. Its honeycomb structure and negative charge allows zeolite to absorb liquids and adsorb elements.

Zoned system: An HVAC comfort system responsible for meeting the needs of two or more control zones.

REFERENCES

Module 1 03308 Water Treatment

ASHRAE Handbook – HVAC Applications. 2023. Peachtree Corners, GA: The American Society of Heating, Refrigerating and Air-Conditioning Engineers.

ASHRAE Handbook – HVAC Systems and Equipment. 2024. Peachtree Corners, GA: The American Society of Heating, Refrigerating and Air-Conditioning Engineers.

HVAC Water Chillers and Cooling Towers: Fundamentals, Application, and Operation, 2nd edition. Herbert W. Stanford III. Boca Raton, FL: CRC Press, Taylor & Francis Group.

Lessons Learned: Commercial & Industrial Steam Boiler Operation. Ray Wohlfarth. Pittsburgh, PA: Independently Published.

Low Pressure Boilers, 5th edition. Fredrick M. Steingress, Daryl R. Walker. Orland Park, IL: American Technical Publishers.

Module 2 03404 Energy Conservation Equipment

ASHRAE Handbook—HVAC Systems and Equipment. 2024. Peachtree Corners, GA: The American Society of Heating, Refrigerating and Air Conditioning Engineers.

ASHRAE Handbook—HVAC Applications. 2023. Peachtree Corners, GA: The American Society of Heating, Refrigerating and Air Conditioning Engineers.

Module 3 03405 Building Management Systems

Building Management Systems Explained: Understanding Controllers and Field Devices. 2021. Robert O'Connor. Independently Published.

Building Automation Systems A to Z: How to Survive in a World Filled With BAS. 2016. Phillip G. Zito. Independently Published.

BACnet: The Global Standard for Building Automation and Control Networks. 2013. H. Michael Newman. New York, NY: Momentum Press-IG Publishing.

Module 4 03402 System Air Balancing and Psychrometrics

ACCA Manual P®, *Psychrometrics: Theory and Applications*. 2005. Hank Rutkowski. Alexandria, VA: Air Conditioning Contractors of Americ (ACCA) Association, Inc.

ASHRAE Handbook—*HVAC Applications*. 2023. Peachtree Corners, GA: The American Society of Heating, Refrigerating and Air Conditioning Engineers.

ASHRAE Handbook—*Fundamentals*. 2021. Peachtree Corners, GA: The American Society of Heating, Refrigerating and Air Conditioning Engineers.

ASHRAE Standard 111-2008 (RA 2017), *Measurement, Testing, Adjusting, and Balancing of Building HVAC Systems*. 2017. Peachtree Corners, GA: The American Society of Heating, Refrigerating and Air Conditioning Engineers.

Testing and Balancing HVAC Air and Water Systems, 6th edition. Samuel C. Sugarman. Gistrup, Denmark: River Publishers.

Understanding Psychrometrics, 3rd edition. Donald P. Gatley. Peachtree Corners, GA: American Society of Heating, Refrigerating and Air Conditioning Engineers.

Module 5 03406 System Startup and Shutdown

ASHRAE Handbook—HVAC Applications. 2023. Peachtree Corners, GA: The American Society of Heating, Refrigerating and Air Conditioning Engineers.

ASHRAE Handbook—HVAC Systems and Equipment. 2024. Peachtree Corners, GA: The American Society of Heating, Refrigerating and Air Conditioning Engineers.

Process Steam Systems: A Practical Guide for Operators, Maintainers, Designers, and Educators, 2nd edition. Carey Merritt. Hoboken, NJ: Wiley.

Module 6 03401 Construction Drawings and Specifications

Builder's Essentials: Plan Reading and Material Takeoff. 2015. Wayne J. DelPico. Greenville, SC: RSMeans.

Construction Specifications Writing: Principles and Procedures, 6th edition. Harold J. Rosen, Mark Kalin, Robert S. Weygant, and John R. Regener, Jr. Hoboken, NJ: Wiley.

Reading Architectural Work Drawings, 6th edition. Edward J. Muller and Phillip A. Grau III. Upper Saddle River, NJ: Prentice Hall.

Module 7 03407 Comfort System Design and Planning

ACCA Manual D®, *Residential Duct Systems*. Latest edition. Alexandria, VA: Air Conditioning Contractors of America (ACCA) Association, Inc.

ACCA Manual J®, *Residential Load Calculation*. Latest edition. Alexandria, VA: Air Conditioning Contractors of America (ACCA) Association, Inc.

ACCA Manual N®, *Commercial Load Calculation*. Latest edition. Alexandria, VA: Air Conditioning Contractors of America (ACCA) Association, Inc.

ACCA Manual S®, *Residential Equipment Selection*. Latest edition. Alexandria, VA: Air Conditioning Contractors of America (ACCA) Association, Inc.

HVAC Duct Construction Standards—Metal and Flexible. Latest edition. Chantilly, VA: Sheet Metal and Air Conditioning Contractors' National Association (SMACNA).

Module 8 03408 Commercial/Industrial Refrigeration

ASHRAE Handbook—*Refrigeration*. 2022. Peachtree Corners, GA: The American Society of Heating, Refrigerating and Air-Conditioning Engineers.

ASHRAE Handbook—*HVAC Systems and Equipment*. 2024. Peachtree Corners, GA: The American Society of Heating, Refrigerating and Air-Conditioning Engineers.

NCCER Module 03107, *Introduction to Cooling*.

NCCER Module 03302, *Compressors*.

NCCER Module 03304, *Retail Refrigeration Systems*.

Module 9 03409 Alternative and Specialized Heating and Cooling Systems

ASHRAE Handbook—HVAC Applications. Latest edition. Peachtree Corners, GA: The American Society of Heating, Refrigerating and Air-Conditioning Engineers.

ASHRAE Handbook—HVAC Systems and Equipment. Latest edition. Peachtree Corners, GA: The American Society of Heating, Refrigerating and Air-Conditioning Engineers.

Design Considerations for Active Chilled Beams. September 2008. Darren Alexander, P.Eng., and Mike O'Rourke. Peachtree Corners, GA: The American Society of Heating, Refrigerating and Air-Conditioning Engineers.

Wood Pellet Heating Systems, 2nd edition. 2021. Dilwyn Jenkins. Oxfordshire, England: Routledge.

Solar Water Heating—A Comprehensive Guide to Solar Water and Space Heating Systems, 2nd edition. July 2010. Bob Ramlow and Benjamin Nusz. Gabriola Island, BC, Canada: New Society Publishers.

Module 10 03305 Commercial Hydronic Systems

Cooling Towers: Principles and Practice, 3rd edition. G. B. Hill, E. J. Pring, and Peter D. Osborn. Oxford, UK: Butterworth–Heinemann.

HVAC Water Chillers and Cooling Towers: Fundamentals, Application, and Operation, 2nd edition. Herbert W. Standford III. Boca Raton, FL: CRC Press.

NCCER Module 03203, *Introduction to Hydronic Systems*.

NCCER Module 03215, *Basic Maintenance*.

NCCER Module 03308, *Water Treatment*.

Refrigeration and Air Conditioning: An Introduction to HVAC/R. Latest edition. Larry F. Jeffus. New York, NY: Pearson College Division.

Module 11 03201 Commercial Airside Systems

ASHRAE Standard 62.1, *Ventilation and Acceptable Indoor Air Quality*.

HVAC Systems and Equipment Handbook. 2024. Peachtree Corners, GA: American Society of Heating, Refrigerating and Air Conditioning Engineers (ASHRAE).

NCCER Module 03315, *Zoning, Ductless, and Variable Refrigerant Flow Systems*.

Variable Air Volume (VAV) Systems: Optimizing HVAC Efficiency. Charles Nehme. 2023. Independently published.

Variable Air Volume Systems, 2nd edition. Leo A. Meyer. Hayward, CA: Lama Books

Module 12 46100 Introduction to Leadership

Architecture, Engineering, and Construction Industry (AEC), *www.aecinfo.com*.

The Center for Leadership Studies, *www.situational.com*.

Conflict Resolution Playbook: Practical Communication Skills for Preventing, Managing, and Resolving Conflict. 2020. Jeremy Pollack. Rockridge Press.

Construction Industry Alliance for Suicide Prevention, *www.ciasp.org*.

The Constructor Construction Encyclopedia, *theconstructor.org*.

Crucial Conversations: Tools for Talking When Stakes Are High. 2002. Kerry Patterson, Joseph Grenny, Ron McMillan, and Al Switzler. Maidenhead, England: McGraw-Hill Contemporary.

Dare to Lead: Brave Work. Tough Conversations. Whole Hearts. 2018. Brene Brown. New York, NY: Random House Publishing. *www.penguinrandomhouse.com*.

Engineering News-Record, *www.enr.com*.

Equal Employment Opportunity Commission (EEOC), *www.eeoc.gov*.

The Five Dysfunctions of a Team. 2002. Patrick M. Lencioni. London, England: Jossey-Bass.

Mentoring for Craft Professionals. 2016. NCCER. New York, NY: Pearson Education, Inc.

National Association of Women in Construction (NAWIC), *www.nawic.org*.

National Census of Fatal Occupational Injuries (NCFOI), *www.bls.gov*.

National Institute of Occupational Safety and Health (NIOSH), *www.cdc.gov/niosh*.

National Safety Council, *www.nsc.org*.

Occupational Safety and Health Administration (OSHA), *www.osha.gov*.

Smart Choices: A Practical Guide to Making Better Decisions. 2015. John S. Hammond, Ralph L. Keeney, Howard Raiffa. Boston, MA: Harvard Business Publishing Education, *hbsp.harvard.edu*.

Society for Human Resources Management (SHRM), *www.shrm.org*.

Total Worker Health, *www.cdc.gov/niosh*.

United States Census Bureau, *www.census.gov*.

United States Department of Labor, *www.dol.gov*.